Partial Differential Equations

A First Course

Pure and Applied
UNDERGRADUATE TEXTS · 54

Partial Differential Equations

A First Course

Rustum Choksi

The cover image is courtesy of Harald Schrader.

2020 *Mathematics Subject Classification.* Primary 35Axx, 35Cxx, 35Dxx, 35Exx, 35Fxx, 35Jxx, 35Kxx, 35Lxx, 35Pxx, 35Qxx.

For additional information and updates on this book, visit
www.ams.org/bookpages/amstext-54

Library of Congress Cataloging-in-Publication Data

Names: Choksi, Rustum, 1965– author.
Title: Partial differential equations : a first course / Rustum Choksi.
Description: Providence, Rhode Island : American Mathematical Society, [2022] | Series: Pure and applied undergraduate texts, 1943-9334 ; volume 54 | Includes bibliographical references and index.
Identifiers: LCCN 2021040894 | ISBN 9781470464912 (v. 54; paperback) | ISBN 9781470468675 (v. 54; ebook)
Subjects: LCSH: Differential equations, Partial. | AMS: Partial differential equations – General topics. | Partial differential equations – Representations of solutions. | Partial differential equations – Generalized solutions. | Partial differential equations – Equations and systems with constant coefficients. | Partial differential equations – General first-order equations and systems. | Partial differential equations – Elliptic equations and systems. | Partial differential equations – Parabolic equations and systems. | Partial differential equations – Hyperbolic equations and systems. | Partial differential equations – Spectral theory and eigenvalue problems. | Partial differential equations – Equations of mathematical physics and other areas of application.
Classification: LCC QA374 .C475 2022 | DDC 515/.353–dc23
LC record available at https://lccn.loc.gov/2021040894

Visit the AMS home page at `https://www.ams.org/`

10 9 8 7 6 5 4 3 2 1 27 26 25 24 23 22

For Ron

"The profound study of nature is the most fruitful source of mathematical discovery. Not only does this study, by proposing a fixed goal for our research, have the advantage of excluding vague questions and calculations without result. It is also a sure way of shaping Analysis itself, and discovering which parts it is most important to know and must always remain part of the subject. These fundamental principles are those which are found in all natural phenomena. ... mathematical analysis is as extensive as nature herself. It defines all observable relationships, measures time, space, force, temperature. This difficult science grows slowly but once ground has been gained it is never relinquished. It grows, without ceasing, in size and strength, amid the constant error and confusion of the human spirit."

–Jean-Baptiste Joseph Fourier[a]

[a] Théorie Analytique de la Chaleur, 1827. English translation taken from Fourier Analysis by T. W. Körner.

"It seems to be one of the fundamental features of nature that fundamental physical laws are described in terms of a mathematical theory of great beauty and power, needing quite a high standard of mathematics for one to understand it. You may wonder: Why is nature constructed along these lines? One can only answer that our present knowledge seems to show that nature is so constructed. We simply have to accept it. One could perhaps describe the situation by saying that God is a mathematician of a very high order, and [God] used very advanced mathematics in constructing the universe. Our feeble attempts at mathematics enable us to understand a bit of the universe, and as we proceed to develop higher and higher mathematics we can hope to understand the universe better."

– Paul Dirac[a]

[a] The Evolution of the Physicist's Picture of Nature, *Scientific America*, May 1963. Available Online.

Contents

Preface

Partial Differential Equations
Open Windows into Our Complex World

Partial differential equations (PDEs) are a fundamental component of modern mathematics and science. In any system wherein the independent variables (e.g., space and time) take on a continuum of values, the laws governing the system usually result in a partial differential equation (PDE) for the quantity, or quantities, of interest. Since models used to approximate many physical, biological, and economic systems are PDE-based, having some proficiency in PDEs can provide a means of addressing these complex systems. In pure mathematics, PDEs are intrinsically related to fundamental structures in analysis, geometry, and probability.

This text presents a one- to two-semester course on PDEs for **senior undergraduates** in mathematics, physics, and the other sciences.

0.1. Who Should Take a First Course in PDEs?

There are primarily three groups. Students might find themselves identifying with more than one group. Listed below are these groups that we identify, each having different reasons, and therefore focuses, for pursuing studies in PDEs.

[1] **Students of "Pure" Mathematics**

Reason for an interest in PDEs: PDEs are fundamental mathematical objects in the fields of mathematical analysis, geometry, probability, and dynamical systems.

Their focus: Acquiring a foundation in PDEs and the appropriate mathematical tools in order to pursue future studies (perhaps at the graduate level) in mathematics. The primary focus of students of pure mathematics lies in finding and proving **theorems** related to existence, uniqueness, stability, and properties of the solutions (e.g., regularity, integrability, asymptotic behavior). They also benefit from studying theoretical structures in PDEs in order to gain insight into other areas of mathematics and the connections between them.

[2] **Students of "Applied" Mathematics**

Reason for an interest in PDEs: PDEs are fundamental mathematical objects in the continuum modeling of any complex system.

Their focus: Acquiring a foundation in PDEs and the appropriate mathematical tools in order to pursue further courses in scientific computing and mathematical modeling. In particular, applied mathematics students are interested in the **development** and application of analytical and computational tools in order to gain insight into the behavior of many complex systems.

[3] **Students of the Physical Sciences (Physics and Chemistry), Computer Science, Statistics, Biology, Engineering, and Economics**

Reason for an interest in PDEs: As for group [2], PDEs are fundamental mathematical objects in the continuum modeling of any complex system. They are ubiquitous throughout the physical sciences and engineering and are increasingly appearing in such fields as mathematical biology, mathematical medicine, economics, data science and machine learning, image and signal processing, and finance.

Their focus: Increasingly, students in this group want direct information about a relevant PDE, for example, an exact, approximate, or numerical form of the solution. They are less interested in methodology (tools) and issues related to their correctness or preciseness, but rather they want **direct answers** to their problems. Today, these "direct answers" do not usually come from hand-done calculations, but rather from the computer. **But here is the catch:** Fruitful interactions with the computer (i.e., what you input and how you interpret the output) require some foundation in PDEs and the associated mathematical objects and tools.

0.2. A Text for All Three Groups: Grounding in Core Concepts and Topics

It is still quite common that undergraduate PDE courses are based upon one method: **separation of variables** (also known as Fourier's method) for boundary value problems. For all cohorts of students, an undergraduate PDE text or course which focuses mainly on separation of variables has only limited appeal. Indeed, our experience has shown that an **entire course** devoted to the extensive details behind this technique can be summarized as follows:

- It is of limited interest to budding pure mathematicians in their preparation for further studies in the analysis and geometry of PDEs. With regard to the basic premise behind separation of variables, they would be better served with a "complete" and proper mathematical theory for eigenfunction expansions associated with certain differential operators (something which is usually only presented at the graduate level).
- It is of limited use to budding applied mathematicians, who require many analytical and computational techniques as well as a broad exposure to the character and behavior of different classes of PDEs and their solutions.

- It is of limited use to budding scientists (the future practitioners of PDEs), who are increasingly more interested in computational aspects of the theory. Moreover, much of the relevant phenomena that modern science aims to tackle is inherently nonlinear and these classical techniques have a rather limited scope.
- It gives the undergraduate student from all three camps the false impression that PDEs are an old-fashioned subject rooted in long and tedious calculations with infinite series and special functions. Nothing could be further from the truth, as PDEs form a major component of modern mathematics and science.

In our view, a clear need for a solid **grounding** in objects, concepts, tools, and structures which are ubiquitous in the theory of PDEs is common to **all three groups**. Separation of variables and Fourier series are certainly included in these core concepts/tools, but our previous point is that they should encompass chapters, **not** full books or courses. By grounding, we mean a foundation (or basis) to move forward in future courses, research, and computational excursions with PDEs. This foundation can only be achieved by developing a solid understanding of, and appreciation for, certain **core material**.

It is true that we have chosen **not** to address, in any detail, computational aspects of PDEs, an increasingly large subject of cardinal importance to the second and third groups. However, it is our feeling that all successful excursions into the computation of PDEs, from the development of novel methods to the direct application of well-known methods, require a firm grasp of the core material introduced in this text.

The need for a grounding in objects (such as the delta function), concepts, tools, and structures is particularly important in our present day when information is both ubiquitous and readily (in fact, instantaneously) accessible. Indeed, more and more, Wikipedia provides perfectly suitable definitions of objects and results relevant to any science course. **We live in an age where information is plentiful and cheap, but understanding anything well is priceless.**

This text is by no means encyclopedic nor comprehensive, but it contains core material on which to construct a one- or two-semester course. Our vision of the core material is based upon the following:

(i) **First-order equations**, the notion of a **characteristic**, and the method of characteristics. Here, one appreciates the fundamental difference between linear and nonlinear PDEs (Chapter 2).

(ii) The nature of **wave propagation** as dictated by the second-order **wave equation**, its predictions of causality in space dimensions one (Chapter 3) and three (Chapter 4).

(iii) The **Fourier transform**, its properties and uses (Chapter 6).

(iv) The nature of **diffusion** as described by the **diffusion equation**, its consequences and relationship to basic objects and concepts in probability (Chapter 7).

(v) The nature of **harmonic functions** and PDEs involving the **Laplacian** (Chapter 8).

(vi) The **fundamental solution** of the Laplacian and why it is so "fundamental". The notion of a **Green's function** for boundary value problems involving the Laplacian (Chapter 10).

(vii) **Fourier series** and the **Separation of Variables Algorithm** (Chapters 11 and 12).

Fundamental to most of these topics is a clear grasp of the **delta function** in one and several space variables, in particular:

- why this object is not a function in the usual sense, yet encapsulates concentration at a point;
- how it appears in differentiating (in a generalized sense) functions with discontinuities in both one and several independent variables;
- how it appears as the limit of different sequences of functions which concentrate;
- its crucial role in Fourier series and the Fourier transform;
- its crucial role in fundamental solutions (Green's functions) for the diffusion, Laplace, and Poisson equations.

To this end, we have not been shy about tackling **distributions** and, in particular, what it means to differentiate a function in the sense of distributions, as well as to take a limit of functions in the sense of distributions. These ideas form an important component of this text and are presented in Chapters 5 and 9.

Also fundamental to **all** these topics is a solid understanding of basic objects and techniques in **advanced calculus**. It is our hope that in learning these fundamentals, the student will, at the very least, acquire a grounding and proficiency in the geometric and physical meaning of the gradient, the divergence, and the power of integration and differentiation for functions of several variables.

Throughout this text, we have tried hard to clearly explain all steps, often with pre-motivations and post-reflections. In many ways, we have aimed towards a text for **self-learning**. We can at times be wordy in this exposition and have no scruples against repeating a key point or argument several times. Information (facts, methods, examples) is everywhere and cheap(!); again, our sole reason for writing this book is to facilitate understanding.

0.3. Basic Structure of the Text

0.3.1. Presentation and Modularization of Material. The first thing many will notice when faced with this text is its **length**. To this end, several comments are in order:

- We have strived to **modularize** and streamline material — be it concepts, results, or ideas. Each chapter is based upon a particular equation, class of equations, or a particular idea or tool. Chapters begin with an introduction, end with a summary, and are comprised of **sections** which, in turn, are sometimes divided into **subsections**. Each section (or subsection) is the basic **unit/module** of the text and addresses a particular issue/topic in a short and concise fashion. The vast majority of these modules (sections or subsections) are no more than a few pages, hence, the rather long table of contents!

- It is true that many sections do not contribute the same weight towards what we call the core. Hence, we have chosen to highlight with a **bullet** • all sections which we deem to be the "basics" (fundamentals) for a first course in PDEs. Of course, one will not be able to cover all these bulleted sections in one semester! These bulleted sections also provide the necessary prerequisite material for further chapters. The additional (nonbulleted) sections are obviously there for a reason and, in our view, they are important; but they can be browsed, or read carefully, at the discretion of the reader/instructor.
- **The long table of contents** thus provides an essential and invaluable guide to navigating this text, and the reader/instructor is encouraged to refer to it often.

0.3.2. Choice for the Orderings of Chapters. Whereas the reader/instructor need not follow our sequence of chapters (see the next subsection), let us briefly address our philosophy behind this order. The most contentious choice is to leave Fourier series and separation of variables for boundary value problems (Chapters 11 and 12) to the end. While the majority of these two chapters can be covered at any time and, indeed, are often the first topics one encounters in a PDE course, we feel that the inherent character of the **wave, diffusion, and Laplace equations (the so-called "big three")** are obscured by simply focusing on infinite series solution representations for the respective boundary value problems. It is far more enlightening to first consider the big three divorced from any boundary condition and focus, respectively, on the central character of wave propagation, diffusion, and harmonic functions.

After a short introductory chapter (Chapter 1) on general definitions for PDEs, the book begins not with one of the big three equations, but rather with a long chapter on general first-order equations and the method of characteristics (Chapter 2). There are many reasons for this: (i) It is natural to start with PDEs involving only first-order derivatives; (ii) there is a general notion (a characteristic) central to **all** first-order PDEs coupled with a general method (the method of characteristics) to solve them; (iii) characteristics are directly connected with topics and concepts from previous courses, namely the gradient and directional derivative from Calculus III and ordinary differential equations (ODEs); (iv) the subject allows us to address not just linear PDEs, but nonlinear PDEs as well. Whereas the remaining chapters will all focus on linear PDEs, nonlinear PDEs are fundamental in mathematics and the sciences, and we feel it is beneficial to introduce them early on, even if the study is confined to first-order equations.

After first-order equations, we address the second-order wave equation, the member of the big three whose structure is closest to the characteristics of the previous chapter. We study the wave equation first in one space dimension (Chapter 3) and then in three and two space dimensions (Chapter 4).

After the wave equation, we take a two-chapter **detour** before tackling the diffusion and Laplace equations. **Why?** Our methods for solving and analyzing first-order equations and the second-order wave equation are all based on a property which we can loosely sum up in one phrase: "**finite propagation speed**". To handle (i.e., solve) PDEs with finite propagation speed, the basic tools and objects from advanced calculus, e.g., differentiation and integration, suffice. These tools are all based upon **local** (in space) calculations. To address the diffusion, Laplace, and Poisson equations we must embrace a **new paradigm** wherein this principle of "finite propagation speed" is **false**. Here the solution to these problems at any point in the domain will involve a certain weighted average of **all** the data (initial values or boundary values). There are several ideas/concepts that are fundamental to address these classes of PDEs: **concentration** and the effect singularities have on differentiation and **nonlocal operations**. They are most likely **new** to the reader and will be analyzed, respectively, with the following mathematical objects, tools, and machineries: **the Dirac delta function, distributions in one space dimension** (Chapter 5), and **convolution** and **the Fourier transform** (Chapter 6). However, as outlined in the next subsection, it is only the very basics of Chapters 5 and 6 which are required for the remainder of the text; in fact, very little of Chapter 6 is actually required. That said, we feel strongly that having a solid grounding in the delta function and the Fourier transform will prove tremendously useful to all students in future studies/applications.

With these concepts in hand, we address the ubiquitous diffusion equation (Chapter 7). Next, we present Chapter 8 (with almost no prerequisites) on the Laplacian and properties of harmonic functions (solutions to Laplace's equation). This opens the way to solving boundary value problems involving the Laplacian (Chapter 10). Here, the crucial tools are the fundamental solution and Green's functions, functions whose Laplacian is concentrated at a multidimensional delta function. Hence, we first pause to address partial differentiation in the sense of distributions (Chapter 9).

0.3.3. Codependence and Different Orderings of Chapters.

The reader/instructor need not follow our precise ordering of the chapters. Let us first document the required prerequisites from previous chapters. Here we do not include any reference to Chapter 1 (basic PDE definitions and terminology) which contains definitions and notions used throughout the text.

Chapter 2. First-Order PDEs and the Method of Characteristics.
Prerequisites: None.

Chapter 3. The Wave Equation in One Space Dimension.
Prerequisites: Basic notion of a characteristic and the transport equation in 1D, which can be found in the Prelude to Chapter 2 (Section 2.1).

Chapter 4. The Wave Equation in Three and Two Space Dimensions.
Prerequisites: The wave equation in 1D — Chapter 3 (bulleted sections).

Chapter 5. The Delta "Function" and Distributions in One Space Dimension.
Prerequisites: None.

Chapter 6. The Fourier Transform.
Prerequisites: Definition of the delta function δ_0 and convergence in the sense of distributions to δ_0 as in Sections 5.2 to 5.5.

Chapter 7. The Diffusion Equation.
Prerequisites: The solution formula derived using the Fourier transform (Section 6.8), **or** this can be derived directly using similarity solutions (cf. Exercise **7.5**). Definition of the delta function δ_0 and convergence in the sense of distributions to δ_0 found in Sections 5.2 to 5.5. The basic notion of convolution discussed in Sections 6.3.1 and 6.3.2.

Chapter 8. The Laplacian, Laplace's Equation, and Harmonic Functions.
Prerequisites: None.

Chapter 9. Distributions in Higher Dimensions and Partial Differentiation in the Sense of Distributions.
Prerequisites: Chapter 5 (bulleted sections).

Chapter 10. The Fundamental Solution and Green's Functions for the Laplacian.
Prerequisites: Sections 8.1 and 8.5 and Sections 9.1, 9.3, and 9.5.

Chapter 11. Fourier Series.
Prerequisites: Only a very basic notion of the delta function (cf. Section 5.2).

Chapter 12. The Separation of Variables Algorithm for Boundary Value Problems.
Prerequisites: Chapter 11 (bulleted sections).

This relatively weak codependence presents many avenues for the reader/instructor. For example:

(i) Fourier series and the basics of separation of variables can be covered at any stage. Chapter 8 on Laplace's equation and harmonic functions can also be covered at any stage.

(ii) With a very short introduction to the delta function, convergence in the sense of distributions (Sections 5.2 to 5.5), and convolution (Sections 6.3.1 and 6.3.2), one can go directly to Chapter 7 on the diffusion equation.

While it is true that, except for the short Sections 6.3.1 and 6.3.2 introducing convolution, the long Chapter 6 on the Fourier transform is not required for the vast majority of this text; this material is vital for further studies in PDE, analysis, applied math, and science in general.

0.4. Prerequisites

Given that the scope of this text is quite vast, it is remarkable that from a purely technical point of view, the prerequisites, essentially **proficiency in advanced calculus**, are rather minimal.

0.4.1. Advanced Calculus and the Appendix.

One of the main difficulties students encounter in a PDE course (even at the graduate level) is "indigestion" of basic multivariable calculus. In particular, the student should be, or become, as the course progresses, comfortable and proficient with the geometric and physical meaning of the following:

- bulk (volume) and surface integrals,
- the gradient of a function of several variables and directional derivatives,

- the divergence of a vector field and the notion of flux,
- the Divergence Theorem.

In the Appendix to this book we detail the necessary concepts and tools from advanced calculus that we will need. We strongly advise the reader to read the appropriate parts as needed. Some readers may benefit by reading the first few sections of the Appendix before embarking on the subsequent chapters. To further this point, we often start a chapter or a section by reminding the reader of the relevant section(s) of the Appendix.

Basic exposure to ordinary differential equations (ODEs) is also important, in particular, what exactly an ODE is and why, in general, they are so difficult to solve. We will occasionally need some very basic techniques to solve simple ODEs.

There are some proofs in this text, and on occasion, exposure to a first undergraduate-level course in real analysis could prove helpful (for example, epsilon delta proofs and uniform convergence of functions). However, this is not necessary for the vast majority of the material, and when we do use language/approaches from real analysis, we attempt to be as self-contained and gentle as possible; see our comments below on breadth and nonrigidity!

0.4.2. Breadth and Nonrigidity.

"*Those who know nothing of foreign languages know nothing of their own.*"[a] - **Johann Wolfgang von Goethe**

[a]The beauty of this translated quote by the great German poet Goethe (1749–1832) is that the sentence resonates equally well with "languages" replaced by ... academic disciplines, doctrines, religions, ... The original reads, "*Wer fremde Sprachen nicht kennt, weiß nichts von seiner eigenen*", and is taken from *Maximen und Reflexionen* (1833).

PDEs are an intrinsically multidisciplinary subject and a *first course* should embrace this wonderful trait. This does require a certain amount of breadth and nonrigidity from the student. Some mathematics students may initially be put off by "too much physics" while some nonmath students may initially complain about too much mathematical rigor and the occasional dreaded "proof". Math majors must keep in mind that PDEs are intimately connected to physics, and physical intuition will go a long way in guiding us through the analysis. To the nonmath majors, there is sometimes a need for precision, and being precise can mean being rigorous, i.e., proving things. In particular, on occasion there is a fundamental need for mathematical precision to provide meaning to otherwise ill-defined and confusing objects (such as the delta function), where informal intuition and calculations may not be sufficient to gain the necessary proficiency. Moreover, science has taken on a huge computational component and **interactions with the computer require a degree of precision**.

For all cohorts of students, it is our opinion that what future academics, scientists, engineers, and quantitative analysts will need is breadth, flexibility, and diversity; being rigid at an early stage of one's education can prove rather toxic in the future. A first course in PDEs presents an ideal way to *foster* this diverse perspective. It is worth noting that this message is hardly novel; look no further than the **opening quotes at**

the beginning of the book wherein two founding giants of the past make an eloquent case about the following:

- Mathematics (as a discipline) **needs** nature (physics and the other sciences) to direct, guide, and illuminate it.
- Mathematics is the basis for the underlying structure of nature and all the sciences **need** mathematics to make quantitative and qualitative assertions and conclusions.

0.5. Acknowledgments

First, we would like to thank the many undergraduate students from PDE classes over the last 24 years. They have helped shape and improve this text.

We also express sincere thanks to the following:

- **Nathaniel Leitao**, who in his final undergraduate year at McGill spent a considerable amount of time on the text with editorial suggestions, content enhancements, and creating additional figures. Moreover, Nathaniel had a substantial role in the writing of Sections 6.12, 8.8, and 10.6.
- McGill student **David Knapik**, who spent considerable time with detailed proofreading, making several editorial suggestions, and improving many of the figures.
- McGill student **Marc Mekhanik** for providing detailed feedback, editorial suggestions, and improving many of the figures.
- The following McGill students for their comments, both editorial and content oriented, which helped with the preparation and development of this text: Miguel Ayala, Gabriel Martine La Boissonière, Robert Gibson, Ivan Gonzalez, Elias Hess-Childs, Yucong Huang, Hwi Lee, Mikhail Mamaev, Mihai Marian, Tudor Manole, Geoffrey McGregor, Mark Perlman, Gabriel Rioux, and Alistair Russell.
- Many colleagues for comments and suggestions on previous drafts: Almut Burchard, Linan Chen, Albert Cohen, Razvan Fetecau, Gerald Folland, David Muraki, Jean-Christophe Nave, Nilima Nigam, Jessica Lin, Govind Menon, Adam Oberman, Keith Promislow, Dejan Slepčev, Peter Sternberg, Ihsan Topaloglu, Christopher Towse, Konstantina Trivisa, Yen-Hsi Richard Tsai, Gantumur Tsogtgerel, Raghav Venkatraman, Andy Wan, and Benedikt Wirth.
- From the American Mathematical Society: **Ina Mette** for her constant encouragement and support; **Arlene O'Sean** for her excellent work as production and copy editor; and **Brian Bartling** for much help overcoming many LaTeX issues.
- **Ron Tuckey** for many editorial suggestions and corrections and for his patience and support during the writing of this text.

Finally it is important to acknowledge that over the years we have learned much about the subject from two excellent modern texts:

- At the undergraduate level, **Walter A. Strauss's** *Partial Differential Equations: An Introduction*, Wiley.
- At the mathematics graduate level, **Lawrence Craig Evans's** *Partial Differential Equations*, American Mathematical Society.

While the novelties in the exposition, content, organization, and style of this text will, hopefully, speak for themselves, it would be impossible for the text not to, on occasion, share certain similarities in approaches and style with these two works. Indeed, we owe much to these two authors.

Chapter 1

Basic Definitions

Definitions and notation for sets, boundaries, and domains
are presented in Section A.1 of the Appendix.

In this short first chapter you will find some basic definitions concerning partial differential equations and their solutions. In particular, we will define notions of:

- linear vs. nonlinear (semilinear, quasilinear, and fully nonlinear)
- order
- scalar PDEs vs. systems
- homogeneous vs. inhomogeneous
- what exactly we mean by a solution to a PDE
- general solutions, arbitrary functions, and auxiliary conditions
- initial value problems (IVP) and boundary value problems (BVP)
- well-posedness: existence, uniqueness, and stability.

We will also discuss a common strategy in solving PDEs. Some readers may choose to proceed directly to further chapters, referring back to these definitions as needed.

1.1. • Notation

We use standard notation for partial derivatives; for example, if $u(x, y)$ is a function of two variables, then

$$u_x = \frac{\partial u}{\partial x} = \partial_x u, \qquad u_{xx} = \frac{\partial^2 u}{\partial x^2}, \qquad u_{xy} = \frac{\partial^2 u}{\partial x \partial y},$$

with the understanding that these partial derivatives are also functions of (x, y). To begin with, the fundamental object associated with a PDE is an unknown **function of more than one independent variable**. Spatial independent variables will typically be denoted as x, y, z or $x_1, \ldots, x_n$. Each of these variables takes on real values (i.e., real numbers). We often denote an n-tuple (or vector) of independent spatial variables as **x**. Time will be denoted by t. For PDEs (as opposed to ODEs) there will always be **more than one** independent variable. We will most often use u to denote the *unknown* function, i.e., the dependent variable. So, for example, in purely spatial variables we would be dealing with $u(x, y), u(x, y, z)$, or more generally $u(\mathbf{x})$. When time is also relevant, we will deal with $u(x, t), u(x, y, t), u(x, y, z, t)$, or more generally $u(\mathbf{x}, t)$.

1.2. • What Are Partial Differential Equations and Why Are They Ubiquitous?

"The application of calculus to modern science is largely an exercise in the formulation, solution, and interpretation of partial differential equations. ...Even at the cutting edge of modern physics, partial differential equations still provide the mathematical infrastructure."

- **Steven Strogatz**[a]

[a] Strogatz is an applied mathematician well known for his outstanding undergraduate book on the qualitative theory of ODEs: *Nonlinear Dynamics and Chaos*, Westview Press. This quote is taken from his popular science book *Infinite Powers: How Calculus Reveals the Secrets of the Universe*, Houghton Mifflin Harcourt Publishing Company, 2019.

We begin with a precise definition of a partial differential equation.

Definition of a Partial Differential Equation (PDE)

Definition 1.2.1. A partial differential equation (PDE) is an equation which relates an unknown function u and its partial derivatives together with independent variables. In general, it can be written as

$$F(\text{independent variables}, u, \text{partial derivatives of } u) = 0,$$

for some function F which captures the structure of the PDE.

For example, in two independent variables a PDE involving only first-order partial derivatives is described by

$$F(x, y, u, u_x, u_y) = 0, \tag{1.1}$$

where $F : \mathbb{R}^5 \to \mathbb{R}$. Laplace's equation is a particular PDE which in two independent variables is associated with $F(u_{xx}, u_{yy}) = u_{xx} + u_{yy} = 0$.

Note that often there is no explicit appearance of the independent variables, and the function F need not depend on all possible derivatives. If we are only dealing with one independent variable, the PDE becomes an **ordinary differential equation**.

Partial differential equations are ubiquitous in the physical, natural, and social sciences. In modeling any complex system where the independent variables (for example, space and time) take on a continuum of values, PDEs will undoubtedly pop up. Why? In a complex system we wish to determine how a quantity (labeled as u) depends on the independent variables which are usually space and time. There is usually some physical or natural law dictating how u should change. We can often measure the values of u at some fixed time or some subset of space, hoping this information, together with a physical or natural law, will completely determine u everywhere. What do laws of nature tell us? They do **not** tell us explicitly the dependence of u on the space and time; rather, mother nature provides laws which pertain to how u should **change** with space and time. Note that, in our continuum context, we measure changes in u by its **partial derivatives**. But now, we give the fundamental complication responsible for why PDEs are so difficult to solve; the laws do not relate the partial derivatives of u directly to the independent variables, rather they give relations (equations) between these partial derivatives and, in certain cases, the quantity u itself. To illustrate this, we turn to a simple example which will eventually (cf. Section 7.1) lead us to the diffusion equation.

Modeling the Variation of Temperature in a Room: Consider the temperature in a large room. The quantity of interest, the temperature u, can depend on the position in the room (x, y, z) and time t. It is thus a function of four variables. We could measure the temperature in the room at time $t = 0$ or perhaps at the boundary of the room. From this information, we would seek to determine the temperature at all times and places in the room. At any fixed time $t \geq 0$, the temperature can vary from place to place, and this change is (instantaneously) measured by the partial derivatives $u_x(x, y, z, t)$, $u_y(x, y, z, t)$, and $u_z(x, y, z, t)$. On the other hand, at any fixed place the temperature can change with time, and this change is measured by the partial derivative $u_t(x, y, z, t)$. Next, we turn to the "law of nature" for heat flow. Why should temperature change over space and time? One reason could be sources and sinks; for instance people in the room giving off heat or open windows where heat escapes. Let us assume these additional complications are absent and focus solely on the essential law which guides temperature variations at any position in the room: This is where the PDE emerges. Note that if there were no sinks and sources and the temperature was initially constant throughout the room, i.e., $u_x = u_y = u_z = 0$ at $t = 0$, then there would be no temporal change in temperature; that is, $u_t = 0$ for all places in the room and all later times. Moreover, we would also have no spatial variations, i.e., $u_x = u_y = u_z = 0$, at all later times $t > 0$. However, if there was somewhere, at some time, a **nonzero spatial temperature gradient**, heat would flow from hot places to cold places and, hence, temperature at a given point would change with time. Thus, this simple observation, dictating changes in temperature from hot to cold, provides **some relation/equation** between u_t at a given place and time **and** the spatial gradient $\nabla u = \langle u_x, u_y, u_z \rangle$ at the same place and time. Hence **the punchline** here is that the value of $u_t(x, y, z, t)$ **depends on** $\nabla u(x, y, z, t)$. As we shall see in Section 7.1, a little modeling with temperature flux, combined with the Divergence Theorem, will yield the diffusion equation $u_t = \alpha \operatorname{div} \nabla u = \alpha (u_{xx} + u_{yy} + u_{zz})$, for some constant $\alpha > 0$.

A far more complex interplay between the partial derivatives of an unknown function (and the function itself) is at the heart of the continuum modeling of any **material** (a fluid or a solid). In this scenario, Newton's Second Law is applied to the *internal forces* (stress and pressure) of the material, resulting in a system of PDEs (cf. Section 2.8). These systems are extremely difficult to solve, yet their study is fundamentally important.

One could perceive the PDEs encountered in science and mathematics as rather *convoluted* statements about a, as yet unknown, function; however, they simply represent the essence of our complex, yet beautifully constructed, universe!

1.3. • What Exactly Do We Mean by a Solution to a PDE?

Definition of a Solution to a PDE

Definition 1.3.1. A solution (more precisely, a **classical solution**) to a PDE in a domain $\Omega \subset \mathbb{R}^N$ (where N is the number of independent variables) is a sufficiently smooth[a] function $u(\mathbf{x})$ which satisfies the defining equation F for all values of the independent variables in Ω.

[a] If the highest derivatives occurring in the PDE are of order k, then by *sufficiently smooth* we mean C^k in all the variables.

Thus a solution to $F(x, y, u, u_x, u_y) = 0$ on some domain $\Omega \subset \mathbb{R}^2$ is a C^1 function $u(x, y)$ such that for every $(x, y) \in \Omega$, $F\big(x, y, u(x, y), u_x(x, y), u_y(x, y)\big) \equiv$ [1] 0. For example, let us check that the function $u(x, y) = \sin(3x - 2y)$ is a solution to the PDE $2u_x + 3u_y = 0$ on $(x, y) \in \mathbb{R}^2$. For any point $(x, y) \in \mathbb{R}^2$, we compute using the chain rule

$$u_x(x, y) = \cos(3x - 2y)\,3, \qquad u_y(x, y) = \cos(3x - 2y)\,(-2),$$

and, hence,

$$2u_x + 3u_y = 2(3\cos(3x - 2y)) - 3(2\cos(3x - 2y)) = 0.$$

On the other hand, let us check that the function $u(x, t) = \frac{x}{t}$ is a solution to the PDE $u_t + uu_x = 0$, for $x \in \mathbb{R}$ and $t > 0$. For any (x, t) with $t > 0$, we compute

$$u_x(x, t) = \frac{1}{t} \quad \text{and} \quad u_t(x, t) = -\frac{x}{t^2}; \qquad \text{hence,} \quad u_t + uu_x = -\frac{x}{t^2} + \frac{x}{t}\frac{1}{t} = 0.$$

For most PDEs, it is impossible to guess a function which satisfies the equation. Indeed, for all but trivial examples, PDEs cannot be simply "solved" by direct integration. You should recall that this was already the case with ordinary differential equations.

We will see later that for a large class of PDEs there is a **weaker notion of a solution** called a "solution in the sense of distributions". It will always be very clear when we mean this weaker notion; we will either say "weak solution" or "solution in the sense of distributions" to differentiate it from a classical solution. Classical solutions will always be weak solutions but the converse is not true. In fact, we will be able to speak of weak solutions which are not smooth, or even continuous functions. In fact, we will be able to speak of weak solutions which are not even functions!

[1] Throughout this text, we often use the equivalence sign $\equiv$ (three parallel lines) to emphasize that equality holds for all $\mathbf{x}$ in the domain.

1.4. • Order, Linear vs. Nonlinear, Scalar vs. Systems

1.4.1. • Order.

Definition of the Order of a PDE

Definition 1.4.1. We define the **order** of a PDE to be the order of the highest derivative which appears in the equation.

In this text, we will deal primarily with first- and second-order PDEs. Note that this definition is irrespective of the number of dependent variables.

1.4.2. • Linear vs. Nonlinear. A central dichotomy for PDEs is whether they are linear or not (nonlinear). To this end, let us write the PDE in the following form: *All terms containing u and its derivatives = all terms involving **only** the independent variables.*

We write the left-hand side as $\mathcal{L}(u)$, thinking of it as some (differential) operator $\mathcal{L}$ operating on the function u. Some examples of specific operators are $\mathcal{L}(u) = u_x + u_y$ and $\mathcal{L}(u) = u_{xx} + xu_{yy} + uu_x$. Thus any PDE for $u(\mathbf{x})$ can be written as

$$\mathcal{L}(u) = f(\mathbf{x}), \tag{1.2}$$

for some operator $\mathcal{L}$ and some function f.

Definition of a Linear and Nonlinear PDE

Definition 1.4.2. We say the PDE is **linear** if $\mathcal{L}$ is linear in u. That is,

$$\mathcal{L}(u_1 + u_2) = \mathcal{L}(u_1) + \mathcal{L}(u_2)$$

and

$$\mathcal{L}(cu_1) = c\mathcal{L}(u_1).$$

Otherwise, we say the PDE is **nonlinear**.

There are different types of nonlinearities in a PDE and these have a profound effect on their complexity. We divide them into three groups: semilinear, quasilinear, and fully nonlinear.

Definition of a Semilinear, Quasilinear, and Fully Nonlinear PDE

Definition 1.4.3. A PDE of order k is called:

- **semilinear** if all occurrences of derivatives of order k appear with a coefficient which only depends on the independent variables,
- **quasilinear** if all occurrences of derivatives of order k appear with a coefficient which only depends on the independent variables, u, and its derivatives of order strictly less than k,
- **fully nonlinear** if it is not quasilinear.

The reader should note the focus on the highest derivative in the PDE; for example, a fully nonlinear PDE is one which is nonlinear in its highest derivative. By definition, we have the strict inclusions:

$$\text{linear PDEs} \subset \text{semilinear PDEs} \subset \text{quasilinear PDEs}.$$

For first-order PDEs in two independent variables x and y, linear means the PDE can be written in the form

$$a(x,y)\,u_x(x,y) + b(x,y)\,u_y(x,y) = c_1(x,y)\,u + c_2(x,y),$$

for some functions a, b, c_1, c_2 of x and y. Semilinear means the PDE can be written in the form

$$a(x,y)\,u_x(x,y) + b(x,y)\,u_y(x,y) = c(x,y,u),$$

for some functions a and b of x and y, and a function c of x, y, and u. Quasilinear means that the PDE can be written in the form

$$a(x,y,u)\,u_x(x,y) + b(x,y,u)\,u_y(x,y) = c(x,y,u),$$

for some functions a, b, and c of x, y, and u. Note that in all cases, the coefficient functions a, b, and c need **not** be linear in their arguments.

So for example,

$$\begin{aligned}
(xy)\,u_x + e^y\,u_y + (\sin x)\,u &= x^3y^4 && \text{is linear;}\\
(xy)\,u_x + e^y\,u_y + (\sin x)\,u &= u^2 && \text{is semilinear;}\\
u\,u_x + u_y &= 0 && \text{is quasilinear;}\\
(u_x)^2 + (u_y)^2 &= 1 && \text{is fully nonlinear.}
\end{aligned}$$

Note that we are not concerned with nonlinearities in the independent variables. For example, the first example which is linear has nonlinear terms in the independent variables but is linear with respect to the dependent variables (u and its partial derivatives). This is what counts.

The general setup (1.2) allows for another definition.

Definition of a Homogeneous and Inhomogeneous PDE

Definition 1.4.4. If $f \equiv 0$ in (1.2), then we say the PDE is **homogeneous**. Otherwise, we say it is **inhomogeneous**.

For example, a homogeneous linear first-order PDE in two independent variables has the general form $a(x,y)\,u_x(x,y) + b(x,y)\,u_y(x,y) = 0$, for some functions a, b of x and y.

1.4.3. • The Principle of Superposition for Linear PDEs.

Linear PDEs share a property, known as *the principle of superposition*, which as we shall see is very useful. For homogeneous linear PDEs we can formulate the property as follows: If u_1 and u_2 are two solutions to $\mathcal{L}(u) = 0$ and $a, b \in \mathbb{R}$, then $au_1 + bu_2$ is also a solution to $\mathcal{L}(u) = 0$.

We can also apply it to an inhomogeneous linear PDE in the form $\mathcal{L}(u) = f$: If u_1 is a solution to the inhomogeneous linear PDE and u_2 is any solution to the associated homogeneous PDE $\mathcal{L}(u) = 0$, then $u_1 + u_2$ is also a solution to the inhomogeneous linear PDE. More generally, if u_1 is a solution to $\mathcal{L}(u) = f_1$ and u_2 a solution to $\mathcal{L}(u) = f_2$, then $au_1 + bu_2$ is a solution to $\mathcal{L}(u) = af_1 + bf_2$.

1.4.4. • Scalar vs. Systems. Thus far we have dealt with PDEs for a *scalar* unknown u; these equations will be the focus of this text. However, analogous equations for vector-valued functions (more than one function) are also ubiquitous, important, and, in general, very difficult to solve. We call such equations **systems** of partial differential equations, as opposed to *scalar PDEs*. As with systems of linear algebraic equations, the difficulty here lies in that the partial differential equations are *coupled*, and one cannot simply solve **separately** the scalar equations for each unknown function. For example,

$$\begin{cases} u_x + v\,u_y = 0, \\ u\,v_x + v_y = v \end{cases} \tag{1.3}$$

is an example of a system of two equations for unknown functions $u(x, y), v(x, y)$ in two independent variables.

Famous examples of systems of PDEs are the linear **Maxwell equations** (cf. Section 4.1.1) and the quasilinear **Euler and Navier Stokes equations** (cf. Section 2.8).

1.5. • General Solutions, Arbitrary Functions, Auxiliary Conditions, and the Notion of a Well-Posed Problem

1.5.1. • General Solutions and Arbitrary Functions. You will recall from studying ordinary differential equations the notion of a **general solution** vs. a **particular solution**. An ODE has an infinite number of solutions and this infinite class of solutions (called the general solution) is **parametrized** (labeled) via **arbitrary constants**. For example when we say to find the general solution to $y'(t) = y(t)$, we are asking to describe **all possible solutions**. We do this by saying that the general solution is $y(t) = Ce^t$, for any constant C. Given an initial condition, i.e., the value of the solution at a single point, we determine which constant we want for our particular solution. A second-order ODE would require two constants to parametrize (label) all solutions.

PDEs will also have an infinite number of solutions but they will now be **parametrized (labeled)** via **arbitrary functions**. This type of parametrization or labeling can easily become the source of great confusion, especially if one becomes guided by notation as opposed to concept. It is important to digest this now as we illustrate with a few examples where we "find" the general solution by describing, or more precisely characterizing, all solutions.

Example 1.5.1. Find the general solution on the full domain $\mathbb{R}^2$ for $u(x, y)$ solving

$$u_x = 0.$$

The only restriction that the PDE places on the solution is that the derivative with respect to one of the variables x is 0. Thus, the solution cannot vary with x. On the other hand, the PDE says nothing about how u varies with y. Hence:

- If f is any function of one variable, then $u(x, y) = f(y)$ solves the PDE.
- On the other hand, any solution of the PDE must have the form $u(x, y) = f(y)$, for some function f of one variable.

These two statements together mean that $u(x, y) = f(y)$ for any function f of one variable is the general solution to the PDE $u_x = 0$. If we want to solve the same PDE but in three independent variables (i.e., solve for $u(x, y, z)$), then the general solution would be $u(x, y, z) = f(y, z)$, for any function f of two variables.

Example 1.5.2. Find the general solution on the full domain $\mathbb{R}^2$ for $u(x, y)$ solving

$$u_{xx} = 0.$$

The PDE tells us that the x derivative of $u_x(x, y)$ must be 0. Hence, following the logic of the previous example, we find $u_x = f(y)$. Integrating in x gives $u(x, y) = f(y)x + g(y)$, since any "constant" in the integration needs only to be constant in x (not necessarily in y). Thus the general solution for $u(x, y)$ to the $u_{xx} = 0$ is $u(x, y) = f(y)x + g(y)$, for any two functions f and g of one variable.

Example 1.5.3. Find the general solution on the full domain $\mathbb{R}^2$ for $u(x, y)$ solving

$$u_{xy} = 0.$$

In this case, the y derivative of $u_x(x, y)$ must be zero. Hence, $u_x = f(x)$ for any function f. Integrating with respect to x gives $u(x, y) = F(x) + g(y)$ where $F'(x) = f(x)$. Since f can be any function of one variable, so can its primitive F. Thus we can say that the general solution for $u(x, y)$ to the $u_{xy} = 0$ is $u(x, y) = f(x)+g(y)$, for any two functions f and g of one variable.

1.5.2. • Auxiliary Conditions: Boundary and Initial Conditions. PDEs will often be supplemented by an **auxiliary condition** wherein we specify, in some subset of the domain, the value of the solution u and/or its partial derivatives. Our hope is that enforcing this **auxiliary condition** on the general solution will yield a unique solution. Generally speaking, if there are N independent variables, then the auxiliary condition is a set of specified values of u (and/or derivatives of u) on an $(N-1)$-dimensional subset Γ of the domain Ω. For example, if $N = 2$, the auxiliary conditions are specified on a curve.

If the set Γ is chosen appropriately, we are able to determine the arbitrary function(s) in the general solution and obtain a single (unique) solution. For instance, if $u(x, y)$ solves $u_x = 0$, we could specify the values of u on any line which is not parallel to the x-axis and thereby determine exactly the function f. For $u(x, y, z)$ solving $u_x = 0$, the same holds true for specifying u on any plane which does not contain lines parallel to the x-axis. This will become much clearer in the next chapter on first-order PDEs and characteristics.

There are two natural classes of auxiliary conditions. They lead, respectively, to **initial value problems (IVPs)** and **boundary value problems (BVPs)**. For the former, we consider problems in which one of the independent variables represents time t. Hence, our PDE is for a solution $u(\mathbf{x}, t)$ and a natural condition is to specify the solution (and/or its time derivatives) at $t = 0$. Two famous examples are, respectively, the initial value problems for the wave and the diffusion (heat) equation in one space dimension:

$$\textbf{IVP Wave} \qquad \begin{cases} u_{tt} = c^2 u_{xx} & \text{for } -\infty < x < \infty, t > 0, \\ u(x,0) = \phi(x), \quad u_t(x,0) = \psi(x) & \text{for } -\infty < x < \infty; \end{cases} \tag{1.4}$$

$$\textbf{IVP Diffusion} \qquad \begin{cases} u_t = c^2 u_{xx} & \text{for } -\infty < x < \infty, t > 0, \\ u(x,0) = f(x) & \text{for } -\infty < x < \infty. \end{cases} \tag{1.5}$$

Here we have stated these problems for one-dimensional space $x \in \mathbb{R}$. The analogous problems can also be stated for two-, three-, or even N-dimensional space.

For problems in which all the independent variables are spatial, we often look for a solution in some bounded (or unbounded) region Ω of space where we specify the solution on the boundary $\partial\Omega$. This gives rise to a boundary value problem. A famous example is the so-called Dirichlet[2] problem for the Laplacian[3] where in 2D, we look for a solution in the unit ball $\Omega = \{(x,y) \mid x^2 + y^2 < 1\}$ where we specify the solution on the boundary circle $\partial\Omega = \{(x,y) \mid x^2 + y^2 = 1\}$:

$$\textbf{BVP Laplace} \qquad \begin{cases} u_{xx} + u_{yy} = 0 & \text{for } (x,y) \in \Omega, \\ u = f & \text{on } \partial\Omega. \end{cases} \tag{1.6}$$

Initial value problems involving a physical spatial domain which is not the entire real line (or space) will also have boundary value specifications. Hence they are both initial and boundary value problems.

1.5.3. • General Auxiliary Conditions and the Cauchy Problem.

A more general framework for combining an auxiliary condition with a PDE is to provide data for a PDE in N variables on some $(N-1)$-dimensional subset Γ of the domain. For example, in 2D we provide data on some (possibly bounded) curve; in 3D, we provide data on some (possibly bounded) surface. Finding a solution to the PDE and this auxiliary condition is known as **the Cauchy problem**[4]. The (or perhaps better said "a") Cauchy problem may be neither an initial value problem nor a boundary value problem.

---

[2] Named after the German mathematician **Johann Peter Gustav Lejeune Dirichlet** (1805–1859).
[3] Named after the great French mathematician and scientist **Pierre-Simon Laplace** (1749–1827).
[4] Named after the great French mathematician **Augustin-Louis Cauchy** (1789–1857).

1.5.4. • A Well-Posed Problem. Following **Hadamard**[5], we make the following definition.

Definition of a Well-Posed Problem

Definition 1.5.1. We say a PDE with one or more auxiliary conditions constitutes a well-posed problem if the following three conditions hold:

- **Existence** — for a given choice of auxiliary condition(s) wherein the data is chosen from some function class, there exists a solution to the PDE which satisfies the auxiliary condition(s).
- **Uniqueness** — there is, in fact, only one such solution.
- **Stability** — if we perturb slightly[a] the auxiliary condition, then the resulting unique solution does not change much. That is, small changes in the auxiliary condition(s) lead only to small changes in the solution.

[a]Note the ambiguity in the use of "perturb slightly" and "small changes". For any given problem, this terminology will need to be made precise.

The first two conditions ensure that, together with the PDE (which is a law for how a function should change), the auxiliary condition is exactly the right amount of information (no more, no less) that we require to unambiguously determine the solution. The third condition is also important. In the context of any physical problem, the auxiliary condition entails a measurement of some physical quantity u at a certain place (or place and time). There will always be small errors in these measurements, and indeed, one can only record a certain number of significant figures in any numerical value. This stability condition is also vital to any numerical technique for the simulation of a solution. Indeed, numerically computing a solution requires a certain discretization of data, capturing only an approximation of the data. Hence, it is vital that the problem we are trying to simulate is stable with respect to such approximations.

One may recall from ODEs that the number of conditions needed in order to have a unique solution was exactly the order of the ODE. Naively, one might expect the analogous situation for PDEs. However, the situation is far more subtle. As we shall see, the two IVPs and the BVP referenced in the previous subsection all constitute well-posed problems within appropriate classes for the solution. They are all second-order PDEs; however, the first requires two auxiliary conditions to create a well-posed problem while the other two require only one.

1.6. • Common Approaches and Themes in Solving PDEs

1.6.1. • Assuming We Have a Solution and Going Forward. Given a well-posed (PDE) problem, one would ideally like a **closed form solution** u, i.e., an **explicit formula** for u in terms of its independent variables. While, in general, it is very rare to have explicit solutions, we will be able to find explicit formulas solving many of the PDEs encountered in this book. There is a common approach to finding these explicit solutions which, at first, might seem a bit *like cheating*.

[5]**Jacques Hadamard** (1865–1963) was a French mathematician who made major contributions in PDEs as well as number theory, complex analysis, and differential geometry.

A Common Approach to Solving PDEs

- We start by assuming that a solution (a function) u to the PDE and auxiliary conditions exists and is sufficiently smooth (cf. Section A.2).
- Using only the assumption that u solves the PDE and auxiliary conditions, we *work with and analyze* this, as yet unknown, solution u. In doing so, we derive some basic results for its structure.
- In certain cases, this analysis of its structure will, in fact, yield an explicit formula for u. In other words, we know everything about u.
- This formula was derived on the basis (assumption) of the existence of a smooth solution. However, with the formula in hand one can now check that this indeed gives a solution. This last step is necessary and implies that there is no cheating in the end!

Hence, throughout this text we will often speak of a solution u at the very start of our analysis. Many of our results/theorems will be of the following form:

"**If** u solves an IVP or BVP, **then** $u = \ldots$."

Of course, with the formula in hand, one can then check (verify) the converse:

"**If** $u = \ldots,$ **then** u solves the IVP or BVP."

Often it is straightforward to check this converse; after all, one just has to check that the formula for u satisfies the PDE and the auxiliary condition(s). The fact that it satisfies the auxiliary conditions is often immediate (e.g., the solution to the IVP of the 1D wave equation). In other cases (e.g., the diffusion equation and the BVPs for Laplace's equation), it will not be so obvious (or immediate) that our derived formula satisfies the initial or boundary conditions, and a deeper investigation is required.

1.6.2. • Explicit vs. Nonexplicit Solutions, No Solutions, and Approximate Solutions. We will spend much time on the "so-called" **big three: the wave, diffusion, and Laplace equations**. Within certain domains, these equations can be explicitly solved; that is, we can find explicit formulas for their solutions in terms of either known functions or, possibly, infinite sums of known functions (power series). However, it is very important to note that **most** PDEs of any scientific interest **cannot** be explicitly solved in terms of known functions, or even in terms of possibly infinite sums of known functions (power series). For these PDEs, mathematicians often spend their entire careers addressing the following issues: (i) proving existence, uniqueness, and well-posedness without deriving explicit formulas (we refer to these as nonexplicit solutions) and (ii) proving qualitative properties of the solution, for example, regularity (how smooth the solution is) and the nature of its discontinuities (e.g., blow-up). As with finding explicit formulas, mathematicians often approach these tasks by first assuming the existence of a solution and then deriving certain estimates (bounds) known as *a priori estimates*.

Unlike with ODEs, there is **no** robust **general existence theorem for PDEs**, that is, a theorem which asserts, under certain assumptions on a PDE and auxiliary condition, the existence of a solution. The one exception to this rule (the only *general* existence theorem in PDE) is the **Cauchy-Kovalevskaya Theorem**[6] which asserts the local existence of a solution to the Cauchy problem under very strong assumptions on the PDE and data. To provide a few details, we consider the Cauchy problem for a PDE defined on an N-dimensional domain Ω with data provided on an $(N-1)$-dimensional subset $\Gamma \subset \Omega$. We look for a **local solution**, which means a solution in some subdomain Ω' of Ω which still contains Γ, in other words a solution in some "neighborhood" of the data set Γ. The assumption on the PDE and the data are based upon the notion of **analyticity**. A function is analytic on its domain if it can be expressed as a Taylor (power) series about any point in its domain. Such functions are extremely special and indeed very "regular"; in fact there are C^∞ functions which are not analytic. A very informal way of stating the Cauchy-Kovalevskaya Theorem is to say, "*If everything in sight is (or can be described by) an analytic function, then we are good to go with the existence of a local solution.*" Not surprising the approach to the proof is to expand *everything* in a power series; see Section 4.6 of [**10**] for a precise statement and proof.

Do all PDEs have a solution? The reader will soon see in the next chapter that it is easy to construct a PDE and an auxiliary condition on a subset of the domain which are incompatible; that is, the resulting problem (PDE + data) has no solution. However, a natural question to ask is: Does a PDE (with no additional constraint) involving smooth functions of u, it's derivatives, and the independent variables always have a solution in a neighborhood of a point in its domain? The answer is **no** but this is hardly obvious. In fact, in 1958 **Hans Lewy**[7] provided for the first time a *startling* example of a simple **linear** PDE with no solution. Lewy's example is best presented for a complex-valued (cf. Section 6.1) function $u(x, y, t)$:

$$u_x + iu_y - 2i(x + iy)u_t = f(t),$$

where $f(t)$ is a continuous (or even smooth) function on $\mathbb{R}$ which is not *analytic* at $t = 0$.[8] This PDE has no solution which is continuously differentiable. In Chapter 9 we will see how to interpret solutions to PDE in "the sense of distributions". This much weaker notion of a solution allows for solutions which are discontinuous or, moreover, solutions which are not even functions. However, Lewy was able to show that similar examples of linear PDEs exist which even have no solution in the sense of distributions. **Punchline:** There are linear PDEs with no solution to speak of!

---

[6]**Cauchy** proved a special case of the result while the full theorem was later extended by Russian mathematician **Sofia Vasilyevna Kovalevskaya** (1850–1891) who made important contributions to analysis, partial differential equations, and mechanics. She was one of the pioneers for women in science.

[7]**Hans Lewy** (1904–1988) was a German-born American mathematician, known for his work on PDEs and on complex variables theory. The relevant paper here is "An example of a smooth linear partial differential equation without solution", Annals of Mathematics **66** (1957), no. 1.

[8]For example,

$$f(t) = \begin{cases} e^{-\frac{1}{t}} & \text{if } t > 0, \\ 0 & \text{if } t \le 0. \end{cases}$$

Approximate Solutions via the Computer. Physicists, engineers, computer and data scientists, chemists, biologists, economists, etc., often encounter or derive PDEs for which no explicit solution formula exists. They are usually not interested in the above-mentioned "abstract" mathematical questions. Rather, they seek approximations to the solution. In many cases, this is now possible due to the computational power of modern computers. There are classes of general methods for numerically solving PDEs, and many numerical software packages are well developed and widely used by these practitioners. However, for many nonlinear PDEs, one often needs to develop tailor-made numerical methods designed for the specific PDE. This, undoubtedly, requires a mathematical understanding of the PDE's structure, and this is the focus of many applied mathematicians working in the fields of **numerical analysis and scientific computation**.

Exercises

1.1 For each PDE:
(i) State the order and decide if it is linear, semilinear, quasilinear, or fully nonlinear.
(ii) By looking online, describe a physical or geometric context in which the PDE is relevant. In particular, what do the dependent variable u and the independent variables model, and what do the parameters in the PDE (either the constants or specified functions) mean. Unless specified otherwise, $x \in \mathbb{R}$ and $\mathbf{x} \in \mathbb{R}^3$.
(iii) Based on your research, first decide what is a natural range for the independent variables (i.e., the domain of the solution u). For example, should it be all of the line/space or the particular domain Ω in $\mathbb{R}^N$? Should it be for all t or just $t > 0$? Then suggest appropriate auxiliary conditions in order to create a well-posed problem. These could be initial values, boundary values, or both.
(iv) If the PDE bears the name of someone — look them up online.
(a) The **Schrödinger Equation**: $u(\mathbf{x}, t)$ solves $i\hbar\, u_t = -\frac{\hbar^2}{2m}\Delta u + V(\mathbf{x}, t)u$ where $V : \mathbb{R}^3 \times \mathbb{R} \to \mathbb{R}$ is given and $\hbar$ is the reduced Planck constant.
(b) The **(Inviscid) Burgers Equation**: $u(x, t)$ solves $u_t + uu_x = 0$.
(c) The **Full Burgers Equation**: $u(x, t)$ solves $u_t + uu_x = \epsilon u_{xx}$.
(d) The **Hamilton-Jacobi Equation**: $u(\mathbf{x}, t)$ solves $u_t + H(\nabla u, \mathbf{x}) = 0$ where $H : \mathbb{R}^N \times \mathbb{R}^N \to \mathbb{R}$ is given.
(e) The **KdV Equation**: $u(x, t)$ solves $u_t + u_{xxx} - 6uu_x = 0$.
(f) The **Eikonal Equation**: $u(\mathbf{x})$ solves $|\nabla u| = f(\mathbf{x})$.
(g) The **Porous Medium Equation**: $u(\mathbf{x}, t)$ solves $u_t = \Delta(u^m)$ for some $m > 1$.
(h) The **Beam Equation**: $u(x, t)$ solves $u_{tt} + k^2 u_{xxxx} = 0$.
(i) The **Black-Scholes Equation**: $u(S, t)$ solves $u_t + \frac{1}{2}\sigma^2 S^2 u_{SS} + rSu_S - ru = 0$ where r and σ are constants.
(j) The **Monge-Ampère Equation**: $u(\mathbf{x})$ solves $\det(D^2 u) = f(\mathbf{x})$ where f is given and $D^2 u$ denotes the Hessian matrix (also denoted by $\mathbf{H}[u]$).

(k) The **Minimal Surface Equation**: Here let $\mathbf{x} \in \mathbb{R}^2$ and $u(\mathbf{x})$ solve

$$\operatorname{div}\left(\frac{\nabla u}{\sqrt{1+|\nabla u|^2}}\right) = 0.$$

(l) The **Klein-Gordon Equation**: $\hbar^2 u_{tt} - \hbar^2 c^2 \Delta u + m^2 c^4 u = 0$, where $\hbar$ is the reduced Planck constant.
(m) The **Sine-Gordon Equation**: $u_{tt} - u_{xx} + \sin u = 0$.
(n) The **Sivashinsky-Kuramoto Equation**: $u_t + u_{xxxx} + u_{xx} + uu_x = 0$, $x \in [-L, L]$.
(o) The **Fisher-KPP Equation**: $u_t - au_{xx} = bu(1-u)$, $a, b > 0$.
(p) The **Chaplygin Equation**: $u_{xx} + \frac{y^2}{1-y^2/c^2} u_{yy} + yu_y = 0$.

1.2 Find the general solution $u(x, y, z)$ solving each of the following: (a) $u_x = 0$, (b) $u_{xy} = 0$, and (c) $u_{xyz} = 0$.

1.3 Let $u(x, t) = f'\left(\frac{x}{t}\right)$, where f is a C^1 function with $f(0) = 0$. Show that u solves $u_t + f(u)_x = 0$ on $\Omega = \{(x, t) \mid t > 0\}$.

1.4 **(a)** Show that for any $\epsilon > 0$, $u(x, y) = \log \sqrt{x^2 + y^2}$ solves $\Delta u = 0$ on $\Omega = \{(x, y) \mid x^2 + y^2 > \epsilon\}$.
(b) Show that for any $\epsilon > 0$, $u(x_1, x_2, x_3, x_4) = \frac{1}{x_1^2+x_2^2+x_3^2+x_4^2}$ solves $\Delta u = 0$ on $\Omega = \{\mathbf{x} \in \mathbb{R}^4 \mid |\mathbf{x}| > \epsilon\}$.

1.5 Show that for any choice of constants A, B, C, and D,

$$u(x, y) = A(3y^2 + x^3) + B(y^3 + x^3 y) + C(6xy^2 + x^4) + D(2xy^3 + x^4 y)$$

is a polynomial solution to the **Euler-Tricomi equation** $u_{xx} + xu_{yy} = 0$.

1.6 We address the three issues associated with a well-posed problem in the context of a standard problem in linear algebra: Given an $N \times N$ matrix $\mathbf{A}$ and a vector $\mathbf{b} \in \mathbb{R}^N$, **solve for x where A x = b.**
(a) What properties of $\mathbf{A}$ imply existence and uniqueness of this problem for any choice of $\mathbf{b}$?
(b) Write down the property of stability. What condition on $\mathbf{A}$ insures stability and what does stability have to do with the size (absolute value or complex modulus) of the smallest eigenvalue of $\mathbf{A}$?

1.7 Consider the **Fisher-KPP equation** for $u(x, t)$ $u_t = u_{xx} + u(1-u)$.
(a) Show that if a solution exists of the form $u(x, t) = \phi(x - ct)$ for some constant $c > 0$, then ϕ must solve the ODE $\phi'' + c\phi' + \phi(1-\phi) = 0$. Such a solution is called a *traveling wave* solution. Why? What is the interpretation of c?
(b) If, in your previous ODE class, you have had some exposure to dynamical systems and phase plane analysis: How would one qualitatively address (i.e., say something about) the solutions of this ODE?

1.8 Directly verify that for any real constants a and b, $u(x, y) = f(ax+by)$ is a solution to the fully nonlinear **Monge-Ampère equation** $u_{xx}u_{yy} - u_{xy}^2 = 0$. Interpret this PDE in the context of the Hessian and graph of the surface $z = u(x, y)$.

1.9 (**1-D Transition Layer / Diffuse Interface**) This relatively simple exercise has amazingly rich consequences in nonlinear PDEs, physics, and materials science. It is a simple model for a phase transition between two "states" $u = -1$ and $u = +1$. Consider the nonlinear ODE for $u(x)$:

$$u'' + u(1 - u^2) = 0. \tag{1.7}$$

(a) Show that this ODE reduces to $(u')^2 = \frac{1}{2}(1 - u^2)^2 + C$, for some constant C. **Hint:** Multiply (1.7) by u' and show that the left-hand side is a perfect derivative.
(b) Find a solution to (1.7) such that:
(i) for all x, $-1 < u(x) < 1$;
(ii) $\lim_{x\to-\infty} u(x) = -1$; and
(iii) $\lim_{x\to\infty} u(x) = 1$.
Plot this solution and interpret it as a transition layer from -1 to 1.
(c) Now let $\epsilon > 0$ be a small number and find the analogous solution to $\epsilon u'' + u(1 - u^2) = 0$. Plot this solution for $\epsilon = 1, 0.5, 0.1$, and 0.01. What happens to the solution as $\epsilon \to 0^+$?

1.10 Consider the IVP on $\mathbb{R}$, $u_t = u_{xxxx}$ with $u(x, 0) = g(x)$. Show that this problem is not well-posed. Hint: Look at solutions of the form $u_n(x) = \frac{1}{n} e^{n^4 t} \sin nx$ for any integer n.

1.11 (**Soliton Solution to the KdV Equation**)
(a) Show that, for any constant $c > 0$, the function

$$u(x, t) = \frac{c}{2} \operatorname{sech}^2\left[\frac{\sqrt{c}}{2}(x - ct)\right]$$

is a solution to the KdV equation $u_t + u_{xxx} - 6uu_x = 0$. Here sech denotes the hyperbolic secant function; that is, $\operatorname{sech} y = \frac{2}{e^y + e^{-y}}$.
(b) Using any software, plot this solution for $c = 2$ and $t = 0, 1, 2, 5$.
(c) At any fixed time t, $u(\cdot, t)$ denotes the shape of the impulse ("wave"). Interpret the role of c both in terms of speed and shape of the impulse.
(d) This solution is called a **soliton**. Look up online the notion of a soliton and what it means in the context of the KdV equation and water waves.

Chapter 2

First-Order PDEs and the Method of Characteristics

The notion of a directional derivative is central to this chapter. The reader may benefit by first reading Section A.3 of the Appendix.

First-order PDEs can be viewed as, potentially nonlinear and complicated, equations involving the components of the **gradient vector field**; hence they are statements about **directional derivatives** of an unknown function u. There exists a general method for integrating along these directional derivatives which in certain cases (i.e., the cases presented in this chapter) result in an explicit solution to the PDE. This method is known as *the method of characteristics*. What exactly is a characteristic? Suppose we are given a PDE in two independent variables x and y and we wish to find a solution u defined on the upper half-plane $\Omega = \{(x, y) \mid y \geq 0\}$, with data (the auxiliary condition) given on the x-axis (i.e., when $y = 0$). A characteristic is a very special curve which lies in the upper half of the xy-plane (the domain of the, as yet, unknown solution) with the following property: Knowing only the value of u at the point of intersection of the curve with the x-axis, the PDE allows us to determine the values of u at all other points on the curve. In some sense, one could say "the PDE degenerates into an ODE" along this curve. Alternatively, viewing the PDE as a statement about directional derivatives of u, they are precisely the curves which follow these directions. They are not just **any** curves which lie in the domain Ω, but rather curves which are **characteristic of the PDE**.

With these characteristic curves in hand, the PDE can be viewed as a family of ODEs for u along these characteristic curves, with the auxiliary condition on the x-axis providing the "initial values" for these ODEs. So one will need to solve ODEs to find the characteristics themselves lying in Ω **and** ODEs corresponding to the values of the solution u **along** the characteristics. Further, in the case of nonlinear PDEs, these families of ODEs will be **coupled**.

Now suppose the following:

(i) We can "find" all the characteristic curves which are nonintersecting and *fill up* the domain Ω.

(ii) All the characteristic curves "start" at a point on the curve where the data is provided (say for simplicity this data curve is the x-axis). Moreover, given any point in the domain Ω, the unique characteristic curve which passes through the point only intersects the x-axis once.

(iii) We can solve the ODEs corresponding to the solution u along these characteristics using the data provided on the x-axis.

Then, in principle, we know u everywhere in Ω and hence have "solved" the PDE.

At first glance this should all seem a bit suspicious. Indeed, it is based upon a rather presumptuous assumption: At any point in the domain Ω, the solution to the PDE only depends on the data at one single point on the x-axis. This is not the case with all PDEs (for example, the diffusion and Laplace equations), but a consequence of the PDE being first order, and hence a collective statement about directional derivatives. These directional derivatives give rise to ODEs and, taken with a grain of salt, one could say *a first-order PDE behaves like (or is equivalent to) a tapestry of ODEs.*

The method of characteristics can be applied for any number of independent variables and on any general domain $\Omega \subset \mathbb{R}^N$. The data can in principle be specified on any part of its closure $\overline{\Omega}$ but usually Γ is the boundary or a part of the boundary of Ω. In fact, much of the time, we specify data on a coordinate axis or coordinate hyperplane.

One of the main difficulties in mastering the method of characteristics is **staying in control of the notation and not letting the notation take control of you**. You are the driver, not the notation. By this we mean clearly understanding the role that the variables (letters!) play in the description of the characteristics and how they relate to the notation associated with the solution and the independent variables. If at any time you feel the notation guiding you and not the other way around, stop and return to the first few examples of the next section.

The chapter will follow a path of increasing generality:

- Section 2.1: We start with a prelude involving some simple examples to illustrate what exactly a characteristic is and its relationship to an ODE. If you are still puzzled by what exactly a characteristic is, you will find clarification in this section. Advanced readers may skip this section, proceeding directly to the next.
- Section 2.2: With this intuition in hand, we present the method of characteristics for **linear** first-order equations in 2 and 3 independent variables. We give several examples and also take the opportunity to discuss (or derive) general transport equations and the continuity equation in 3D.
- Section 2.3: Before tackling nonlinear equations, we address a famous example (the inviscid Burgers equation) to demonstrate a fundamental difference in the structure of characteristics for nonlinear vs. linear first-order PDEs.
- Section 2.4: We then present the method of characteristics for **quasilinear** first-order equations in 2 and 3 independent variables.

- Section 2.5: Finally, we present the method of characteristics in **full generality** for an arbitrary number of independent variables. The method not only encapsulates the previous cases of linear and quasilinear equations but also applies to **fully nonlinear** first-order equations. We give a few important examples which include Hamilton-Jacobi equations, the eikonal equation, and the level set equation. This section can be skipped on first reading.

A Note on Solving ODES: The method of characteristics rests upon our ability to solve ODEs. The reader has most likely already taken a course devoted to solving ODEs, learning a variety of analytical and visual techniques, some of which were somewhat labor intensive. However, for almost all the examples and exercises you will encounter in this book, the resulting ODEs are either trivial or fairly straightforward to solve.

2.1. • Prelude: A Few Simple Examples Illustrating the Notion and Geometry of Characteristics

In this section we present a few preliminary, yet insightful, examples.

Example 2.1.1. Consider the PDE

$$au_x + bu_y = 0,$$

where $a, b \in \mathbb{R}$. We wish to "find" all functions $u(x, y)$ which satisfy this equation, i.e., to find the general solution. The key observation is that the PDE can be written as

$$\langle a, b\rangle \cdot \nabla u = 0, \quad \text{or equivalently} \quad D_{\mathbf{d}}u = 0,$$

where $D_{\mathbf{d}}u$ denotes *the directional derivative* of u in the direction of unit vector

$$\mathbf{d} := \frac{1}{\sqrt{a^2 + b^2}}\langle a, b\rangle.$$

Note that since the directional derivative is 0, the length of the vector $\langle a, b\rangle$ plays no role (as long as it is nonzero). Hence the rate of change of the solution u in the direction parallel to $\langle a, b\rangle$ is zero (see Figure 2.1). Therefore the PDE tells us that on any line parallel to vector $\langle a, b\rangle$, the solution u must be constant. That constant, however, can change from line to line. Here lies our degree of freedom of the general solution. Since these lines are "characteristic" of the PDE, we call them **characteristics**.

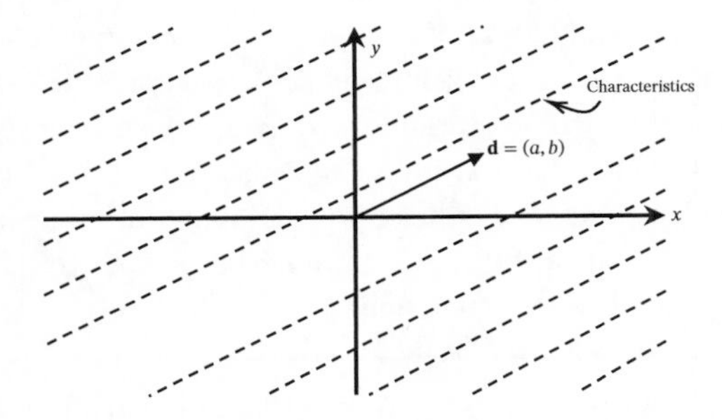

Figure 2.1. Characteristic lines with direction **d**.

How can we write this down mathematically? Lines which are parallel to $\langle a, b\rangle$ can be written in the form $bx - ay = c$, for some c. The choice of c determines the line. In other words, two points (x_1, y_1) and (x_2, y_2) lie on the **same** line if and only if $bx_1 - ay_1 = bx_2 - ay_2$. The value the solution takes on such a line is not dictated by the PDE and can change from line to line (i.e., with the value of c). Herein lies our degree of freedom. Thus we can write the general solution as

$$u(x, y) = f(bx - ay)$$

for any function f. This is precisely the statement, albeit in purely mathematical terms, that the only requirement for u to be a solution is that it is constant on lines parallel to $\langle a, b\rangle$. Indeed, we can summarize our reasoning quite succinctly: Since the PDE dictates that the values of u do not change on any line parallel to $\langle a, b\rangle$, a solution is simply a general function of such lines. Each such line is characterized by a *constant* c where $c = bx - ay$. Therefore,

$$u(x, y) = \text{"}f(\text{line})\text{"} = f(bx - ay).$$

Convince yourself of this before proceeding further.

Solution to Example 2.1.1 via changing variables. It is instructive to resolve the PDE using a **change of variables** (change of coordinates). In changing variables, we will henceforth often use the following standard abuse of notation.

Standard abuse of notation when changing variables

Consider a dependent variable u which is a function of x and y. We use $u(x, y)$ to denote this function, i.e., this input-output machine, which takes an input of two real numbers and outputs another real number. We often want to "change variables"; that is, we want to "describe" (more precisely analyze) the dependent variable u in terms of different independent variables ζ and η which are in turn functions of x and y. This amounts to considering a **new function** different from $u(x, y)$, i.e., a new input-output machine. For example, the function

$$u(x, y) = (4x + y)^2 + (x - y)^2, \quad \text{i.e.,} \quad (x, y) \longrightarrow (4x + y)^2 + (x - y)^2,$$

in the new variables

$$\zeta = 4x + y \quad \text{and} \quad \eta = x - y$$

becomes the new function (the input-output machine):

$$(\zeta, \eta) \longrightarrow \zeta^2 + \eta^2.$$

Hence, technically, we should not write the above function as $u(\zeta, \eta)$ which, strictly speaking, would result in an output of $(4\zeta + \eta)^2 + (\zeta - \eta)^2$; rather, we should adopt a new dependent variable name, for example $w(\zeta, \eta)$, to describe this different function. However, for most of our purposes, we simply want to perform change of variable calculus via the chain rule; hence, it is harmless and standard to "abuse notation" by simply writing "$u(\zeta, \eta)$". One should read this, not as the u input-output machine applied to ζ and η, but rather as "the dependent variable u described in terms of ζ and η". From here on we adopt this standard abuse of notation.

Returning to our example, let us introduce two new variables (ζ, η) where

$$\zeta = ax + by \qquad \text{and} \qquad \eta = bx - ay. \tag{2.1}$$

Based upon the previous geometric argument, one might expect that in the new variables (ζ, η), the PDE $au_x + bu_y = 0$ simply reduces to the PDE $u_\zeta = 0$. Indeed, let us verify this using the chain rule. One finds that

$$u_x = \frac{\partial u}{\partial x} = \frac{\partial u}{\partial \zeta}\frac{\partial \zeta}{\partial x} + \frac{\partial u}{\partial \eta}\frac{\partial \eta}{\partial x} = au_\zeta + bu_\eta$$

and similarly

$$u_y = bu_\zeta - au_\eta.$$

Substituting back into the PDE we find

$$au_x + bu_y = (a^2 + b^2)u_\zeta = 0$$

which, since $(a^2 + b^2) > 0$, implies $u_\zeta = 0$. The general solution $u(\zeta, \eta)$ to $u_\zeta = 0$ is $u(\zeta, \eta) = f(\eta)$. In terms of original variables x and y, this implies

$$u(x, y) = f(bx - ay).$$

Example 2.1.2 (The Previous Example with an Auxiliary Condition). We now consider a case in which data is prescribed (an auxiliary condition) to illustrate how to choose the arbitrary function f. Consider the PDE with auxiliary condition (data):

$$3u_x + 2u_y = 0 \qquad u(x, 0) = x^3. \tag{2.2}$$

We know from the previous example that the general solution to the PDE is $u(x, y) = f(2x - 3y)$. Therefore our goal is now to use the auxiliary condition to determine exactly what the function f should be and then to apply this function to $(2x - 3y)$. By the auxiliary condition, we have $u(x, 0) = f(2x) = x^3$. If we let $\zeta = 2x$ (hence $x = \frac{\zeta}{2}$), then we can describe the function f by $f(\zeta) = \frac{\zeta^3}{8}$. This formula tells us what the unknown function f must be. It is instructive to describe in words what the function is doing: It takes an input, cubes it, and then divides the result by 8. Hence, the solution to (2.2) is

$$u(x, y) = f(2x - 3y) = \frac{(2x - 3y)^3}{8}.$$

You should now verify that this function solves both the PDE and the auxiliary condition.

Here we specified data on the x-axis. We could have specified data on the y-axis instead and in a similar way determined a unique solution. In fact, to obtain a unique solution, we simply need to specify data on any curve with the property that it intersects each line parallel to the vector $\langle 3, 2 \rangle$ exactly once. So for example, specifying data on the curve $y = x^3$ would also give rise to a unique solution.

What **would present a problem** is if we specified data on a line which was parallel to $\langle 3, 2 \rangle$, i.e., specified data on a characteristic curve. These lines would be "characteristic" of the PDE in that the PDE tells us how the solution evolves along the line. For this PDE, the solution must be constant on the line. Hence, if we specified nonconstant data on the line $2x - 3y = 3$, there would exist **no** solution to the problem (PDE + data). On the other hand, if the data was constant along the line, then it would not contradict the PDE; however, there would be infinitely many solutions to the problem (PDE + data). Either way, prescribing data on a characteristic curve results in a problem which is *not well-posed*.

Our next example is again essentially the same as our first; however, it is phrased from a temporal (time) perspective and we cannot overemphasize this importance.

Example 2.1.3 (Temporal Perspective and the Transport Equation). It will prove very useful and instructive to consider the y independent variable as time t. To this end, consider the PDE with initial data:

$$u_t + cu_x = 0, \qquad u(x,0) = g(x).$$

We think of the PDE as an evolution equation (law) for a "signal" $g(x)$. Before continuing let us do a dimensional accounting. Regardless of the dimensions of u, we are adding a derivative with respect to time t with a derivative with respect to spatial x. In order for u_t and cu_x to have the same dimensions, the constant c must carry the dimensions of length $\times$ time^{-1}. In other words, it must have the dimensions of a velocity (speed).

To solve the PDE, note again that the PDE is equivalent to

$$\langle c, 1\rangle \cdot \langle u_x, u_t\rangle = 0.$$

As in Example 2.1.1, this has general solution $f(x - ct)$, for any function f. But if we apply the initial condition, this function f must be exactly g. So the solution to the initial value problem is

$$u(x,t) = g(x - ct).$$

In particular,

$$u(x,0) = g(x), \quad u(x,0.5) = g(x-0.5c), \quad u(x,1) = g(x-c), \quad u(x,2) = g(x-2c), \text{ etc.}$$

We say the "signal" $g(x)$ is *transported* with speed c and the PDE is often referred to as a **transport equation**. For any evolution equation, it is often illustrative to plot **profiles** for fixed times; i.e., for fixed t_* we plot $u(x,t_*)$ as a function of x. Here the profiles are simply translations of the signal $g(x)$. Note that c is exactly the *speed* of propagation of the signal (see Figure 2.2).

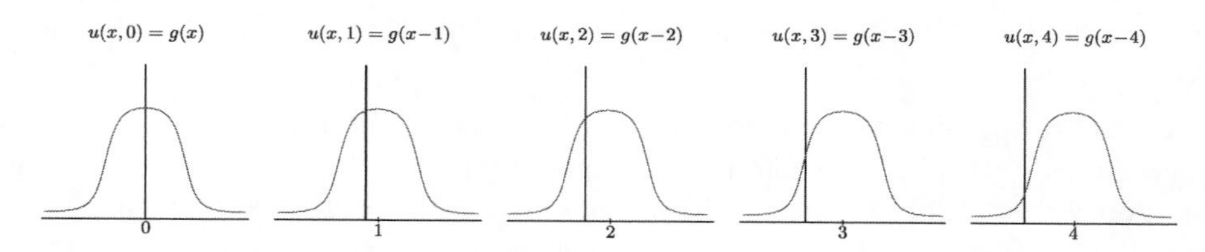

Figure 2.2. Transport in 1D space with speed $c = 1$.

Example 2.1.4. We now consider an example with variable coefficients:

$$u_x + yu_y = 0. \tag{2.3}$$

We can interpret this as

$$\langle 1, y\rangle \cdot \nabla u = 0,$$

or the directional derivative in the direction $\langle 1, y\rangle$ is equal to zero. As before, the PDE (2.3) dictates that any solution must not change in certain directions; however, **now these directions change over the xy-plane**. If we can find curves $y(x)$ which "move" in these directions, then the solution should be constant on these curves. How do we find these curves? At any point $(x, y(x))$ on such a curve we want its tangent vector $\left\langle 1, \frac{dy}{dx}\right\rangle$ to be parallel to the direction $\langle 1, y\rangle$. Consequently, we should look for curves which solve the ODE

$$\frac{dy}{dx} = \frac{y}{1} = y. \tag{2.4}$$

Along any curve $(x, y(x))$ which satisfies (2.4), the solution u will be constant. If you remain skeptical, let us do a reality check.

Take a curve $y(x)$ which solves (2.4), and consider the solution u to (2.3) on the graph of the curve. In other words, consider $u(x, y(x))$. We want to make sure that u is the same for all x; i.e., $u_x(x, y(x)) = 0$. To this end, we use the chain rule to compute

$$\frac{d}{dx}\big[u(x, y(x))\big] = u_x + u_y \frac{dy(x)}{dx} = u_x + u_y\, y = 0.$$

Here we utilized two facts: that $y(x)$ solves (2.4) and that u solves (2.3). The general solution to (2.4) is $y = Ce^x$, for any constant C, and these are our characteristic curves. Now we must find the general solution to (2.3) using only the information that any solution u must be constant along any curve of the form $y = Ce^x$. Two points (x_1, y_1) and (x_2, y_2) lie on the same curve iff they correspond to the same value of "C"; i.e., $y_1 e^{-x_1} = y_2 e^{-x_2}$. What value the solution u takes on this curve is arbitrary ((2.3) tells us nothing about this). Thus, we can write the general solution as

$$u(x, y) = f(e^{-x} y),$$

for any function $f(\cdot)$.

We next provide an example where the solution is not constant along the characteristics.

Example 2.1.5. Consider

$$au_x + bu_y + u = 0.$$

We make a similar change of variables as we did in Example 2.1.1:

$$\zeta = ax + by, \qquad \eta = bx - ay$$

which leads to

$$(a^2 + b^2) u_\zeta(\zeta, \eta) = -u(\zeta, \eta) \qquad \text{or} \qquad u_\zeta = -\frac{u}{a^2 + b^2}.$$

Solving the ODE we find

$$u(\zeta, \eta) = f(\eta) e^{-\frac{\zeta}{a^2+b^2}},$$

which in terms of the original variables gives

$$u(x, y) = f(bx - ay) e^{-\frac{ax+by}{a^2+b^2}}.$$

So far we have encountered some prototypical examples with two independent variables in which the following hold:

- The characteristics were lines upon which the solution was constant.
- The characteristics were not lines but curves upon the solution was constant.

- The characteristics were lines upon which the solution satisfied a simple ODE and grew exponentially.

For the latter case, we changed coordinate systems (i.e., changed variables) and then solved the ODE in one of the new variables.

Our next example should be one where the characteristics are **not** lines and where the solution is **not** constant on the characteristics; for example,

$$u_x + yu_y + u = 0,$$

together with some auxiliary condition, for example, $u(0, y) = y^2$. What do we do now? This may seem like a daunting task but this equation is readily solved via a simple and systematic approach. This approach, called *The Method of Characteristics*, is presented next for linear (and semilinear) equations. Whereas in the previous examples we proceeded by first finding the general solution, the method of characteristics is based upon a direct link to the auxiliary condition ("initial" condition).

2.2. • The Method of Characteristics, Part I: Linear Equations

We first illustrate this method for two independent variables. Suppose we are given a linear PDE to be solved in some domain $\Omega \subset \mathbb{R}^2$ with data given on some curve $\Gamma \subset \overline{\Omega}$. Usually Γ is a subset of the boundary of the set Ω, and often it will just be one of the coordinate axes. The most general linear PDE in the independent variables x and y takes the form

$$a(x, y)u_x + b(x, y)u_y = c_1(x, y)u + c_2(x, y),$$

for some **given** functions a, b, c_1, and c_2 of (x, y). Thus we consider the problem

$$\begin{cases} a(x, y)u_x + b(x, y)u_y = c_1(x, y)u + c_2(x, y), \\ u(x, y) \ \text{ given for } (x, y) \in \Gamma. \end{cases} \tag{2.5}$$

For any point $(x, y) \in \Omega$, we look for a curve (a characteristic), which connects this point to a point on the boundary Γ, **upon which** the PDE degenerates into an ODE (cf. Figure 2.3).

Parametrizing the characteristic curves: In our previous examples we described the characteristic curves in the form y as a function of x. While this was fine for a few simple examples, it is

(a) too restrictive as many curves cannot be written in this form and

(b) does not generalize for curves in N-space where $N \geq 3$.

So, henceforth, let us describe curves parametrically with parameter $s \in \mathbb{R}$. That is, we describe a curve in the xy-plane via $(x(s), y(s))$.

Adopting this notation for the characteristics, we note that the left-hand side of the PDE is a statement about the directional derivative of u in the direction of $(a(x, y), b(x, y))$. We find the characteristics, i.e., the curves which follow these directions, by solving

$$\frac{dx}{ds} = a\big(x(s), y(s)\big) \qquad \frac{dy}{ds} = b(x(s), y(s)). \tag{2.6}$$

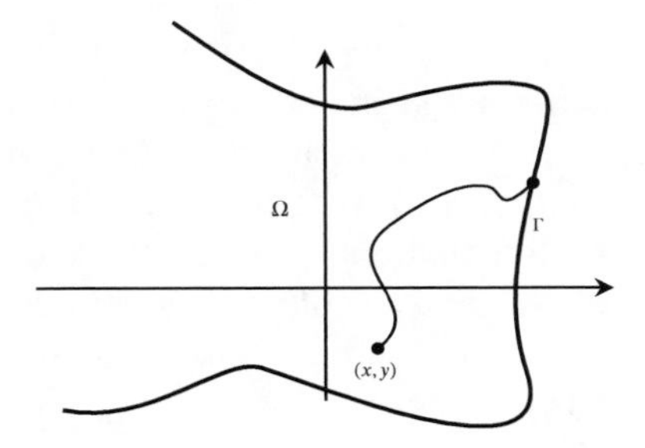

Figure 2.3. A characteristic curve lying in the domain Ω of the, as yet unknown, solution u. The curve connects a point $(x, y) \in \Omega$ to a point on Γ where data is specified.

Now suppose u is a solution to the PDE. Let $z(s)$ denote the values of the solution u along a characteristic; i.e.,

$$z(s) \;:=\; u(x(s), y(s)).$$

Then by the chain rule, we have

$$\frac{dz}{ds} \;=\; u_x(x(s), y(s))\,\frac{dx}{ds}(x(s), y(s)) \;+\; u_y(x(s), y(s))\,\frac{dy}{ds}(x(s), y(s)).$$

Using the defining ODEs (2.6) on the right-hand side gives

$$\frac{dz}{ds} \;=\; u_x(x(s), y(s))\,a\big(x(s), y(s)\big) \;+\; u_y(x(s), y(s))\,b\big(x(s), y(s)\big).$$

All right, so what? Well look at the right-hand side of the above; it looks like the left-hand side of the PDE (2.5) evaluated at particular points (x, y) which lie on the curve $(x(s), y(s))$. But the PDE holds at all $(x, y) \in \Omega$ and hence along the curve $(x(s), y(s))$; consequently the PDE implies that

$$\frac{dz}{ds} = c_1\big(x(s), y(s)\big)z(s) + c_2\big(x(s), y(s)\big).$$

This analysis yields the following three ODEs for the characteristic curves $x(s), y(s)$ and the solution u defined on them, $z(s)$:

$$\begin{cases} \dot{x}(s) \;=\; a\big(x(s), y(s)\big), \\ \dot{y}(s) \;=\; b\big(x(s), y(s)\big), \\ \dot{z}(s) \;=\; c_1\big(x(s), y(s)\big)z(s) + c_2\big(x(s), y(s)\big). \end{cases} \tag{2.7}$$

Here (and henceforth) we use the dot above the dependent variable to denote differentiation with respect to the parameter s; i.e.,

$$\dot{x}(s) \;=\; \frac{dx}{ds}(s).$$

We refer to the **closed system** of ODEs (2.7) as **the characteristic equations**. There are three dependent variables x, y, and z and one independent variable s. We say the system is closed since the right-hand sides of (2.7) only depend on these three dependent variables and s. In fact because the PDE is linear, the first two equations form their own closed system; that is, one can solve the first two without solving for $z(s)$. One can also think of $z(s)$ as a characteristic curve[1], and hence one often refers to the characteristic curves $(x(s), y(s))$ lying in the domain of the solution Ω as the **projected characteristics**. Note that $z(s)$ is simply the solution u along (or "above") the projected characteristics. We give a geometric depiction of these curves and their relationship with the solution u in Figure 2.4. You should become very comfortable with this picture before proceeding on.

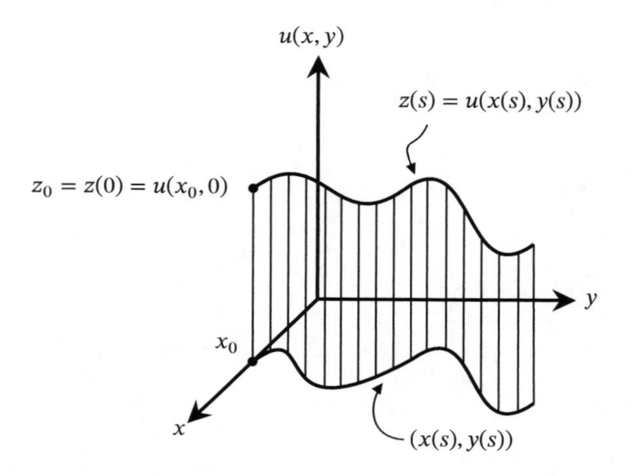

Figure 2.4. Geometric depiction for the characteristics associated with solving a first-order PDE for $u(x, y)$ where data is given on the x-axis; that is, at any point x_0 on the x-axis, $u(x_0, 0) = z_0$ is given. The solution u is described by the surface defined by $z = u(x, y)$. The method of characteristics asserts that, emanating from any point x_0 on the x-axis, there is a curve $(x(s), y(s))$ in the xy-plane, upon which we can calculate the solution $z = u(x(s), y(s))$. In other words, knowing only the structure of the PDE, x_0, and z_0, we can solve ODEs to find the part of the solution surface which lies above the curve.

To solve the characteristic equations, we can **first** solve the equations for $x(s)$ and $y(s)$ and then solve the equation for $z(s)$. To solve the system which constitutes the first two equations, we will have two constants to choose; equivalently, we will need to choose appropriate initial conditions $x(0)$ and $y(0)$ corresponding to $s = 0$ for the ODEs. The word "constants" is misleading; indeed, they will not all be **constant** throughout the problem. Rather, they will be used as *new parameters* (or rather continuous indices) which we will use to find a formula for the solution u at any point (x, y). We will choose these new parameters in such a way that $(x(0), y(0))$ lies on the data curve Γ. As we will see shortly in the examples, this will leave one degree of freedom (one parameter) which will capture the intersection point of a characteristic with the data curve.

Now, assume we can solve for $x(s)$ and $y(s)$ with "initial" conditions $x(0)$ and $y(0)$. The ODE for $z(s)$ is a first-order linear ODE which is readily solved via an integrating

[1] In fact, one can simply view the full triple $(x(s), y(s), z(s))$ as a curve in three space.

factor. Since $(x(0), y(0)) \in \Gamma$, we know the value of

$$z(0) = u(x(0), y(0)).$$

This provides the "initial" condition for the z ODE.

Finally, we *piece everything together* to actually solve (2.5); that is, we derive a formula for $u(x, y)$ involving only x and y. *Piecing everything together* should hopefully be easy since we have labeled/parametrized our characteristics in a constructive way. This may seem a bit abstract at the moment but as soon as you have done a few examples, it will be clear.

2.2.1. • A Few Examples.

Example 2.2.1. Recall the example

$$\begin{cases} au_x + bu_y = 0, \\ u(x, 0) = x^3, \end{cases}$$

where a, b are fixed real numbers. We have done this one before but let us redo it using the method of characteristics. The characteristic system of ODEs is

$$\dot{x}(s) = a, \qquad \dot{y}(s) = b, \qquad \dot{z}(s) = 0.$$

Solving the x and y equations first gives

$$x(s) = as + c_1, \qquad y(s) = bs + c_2.$$

What do we take for the "constants" c_1 and c_2? In other words, how do we choose the initial conditions for the ODEs? We choose them so that at $s = 0$, $(x(0), y(0))$ corresponds to a point on the data curve, here the x-axis. Thus we take $c_2 = 0$. The value of c_1 will then correspond to the intersection point on the x-axis. It will index the different lines (characteristics). Let us also use more descriptive notation for c_1 and denote it by x_0 since it corresponds to the point of intersection between the characteristic and the x-axis. Equivalently, we are solving the ODEs with initial values $(x(0), y(0)) = (x_0, 0)$.

Now the solution to the z equation is simply

$$z = c_3,$$

but since $z(0)$ corresponds to $u(x(0), y(0)) = u(x_0, 0) = x_0^3$, we must have $c_3 = x_0^3$. Thus

$$x(s) = as + x_0, \qquad y(s) = bs, \qquad z = z_0 = x_0^3.$$

OK, what do we have so far? We have equations for the characteristic curves $(x(s), y(s))$ which lie in the domain of the unknown solution. These curves are indexed (labeled) by x_0, the x coordinate of their intersection with the x-axis. We also know the value of the solution u along any one of these curves. This value is different on each curve and hence the appearance of x_0 in the equation for z. Surely, we can piece this all together to find a formula for $u(x, y)$ in terms of x and y (no x_0 or s in the formula!). Indeed, given (x, y), we need to know

- what characteristic this point lies on (i.e., which x_0) and
- where along this characteristic the point lies (i.e., the value of s).

This means we need to find x_0 and s such that

$$\begin{cases} x = as + x_0, \\ y = bs. \end{cases}$$

Solving for s and x_0 gives

$$s = \frac{y}{b} \qquad \text{and} \qquad x_0 = x - \frac{ay}{b}.$$

The value of the solution $u(x, y)$ is simply x_0^3 for the x_0 *corresponding to* the point (x, y). That is,

$$u(x, y) = x_0^3 = \left(x - \frac{ay}{b}\right)^3.$$

Example 2.2.2. Now we consider an example where the characteristics are not straight lines and the solution is not constant along them:

$$\begin{cases} -yu_x + xu_y = u, \\ u(x, 0) = g(x) \end{cases}$$

for $x > 0$. In this case Ω is the right half-plane and Γ is the positive x-axis. The characteristic system of equations is

$$\dot{x} = -y, \qquad \dot{y} = x, \qquad \dot{z} = z.$$

The general solution to the first two equations is[2]

$$x(s) = c_2 \cos s - c_1 \sin s, \qquad y(s) = c_1 \cos s + c_2 \sin s.$$

We, again, choose one of these "constants" to ensure that at $s = 0$ the characteristic lies on the x-axis. That is, let x_0 denote a point on the x-axis and set $x(0) = x_0$ and $y(0) = 0$. Thus

$$x(s) = x_0 \cos(s), \qquad y(s) = x_0 \sin(s).$$

Next, we solve the z equation to find

$$z(s) = c_3 e^s.$$

Since $z(0) = g(x_0)$, we have

$$z(s) = g(x_0) e^s.$$

Finally, we pick a point (x, y) in the right half-plane (dictated by $x > 0$). We need to know which characteristic (which in this case is arcs of circles) passes through (x, y) and the corresponding value of s. That is, given (x, y), we need to solve for x_0 and s such that

$$x = x_0 \cos(s), \qquad y = x_0 \sin(s).$$

We find

$$x_0^2 = x^2 + y^2, \qquad s = \arctan \frac{y}{x}.$$

Plugging into $z(s)$, we obtain

$$u(x, y) = g\left(\sqrt{x^2 + y^2}\right) e^{\arctan \frac{y}{x}}.$$

[2]Recall this fact from your previous ODE class.

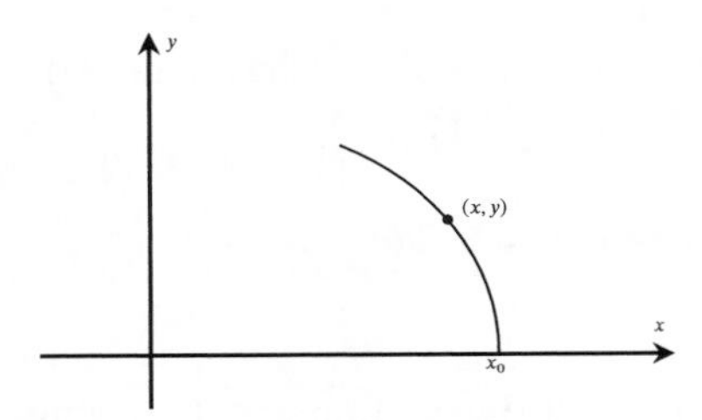

Figure 2.5. Arbitrary point (x, y) on the parametrized curve starting at $(x_0, 0)$, or $s = 0$.

Note that the solution is not defined at the origin $(0, 0)$. This is where the coefficients of u_x and u_y in the PDE (the components of the direction of the directional derivative of u) simultaneously vanish.

In the previous example we did not need the equations for x and y to determine z. As shown in the next example, in general this is not always true.

Example 2.2.3. Find the solution of

$$u_x + u_y + u = e^{x+2y}, \qquad u(x,0) = 0.$$

The characteristic ODEs for $x(s), y(s)$ are trivial and with $x(0) = x_0$ and $y(0) = 0$, they are solved as

$$x(s) = s + x_0, \qquad y(s) = s.$$

The characteristic equation for $z(s) = u(x(s), y(s))$ becomes

$$\dot{z}(s) = -z(s) + e^{3s+x_0}.$$

This ODE is solved with the integrating factor e^s; that is, multiply both sides by e^s and integrating gives

$$z(s) = \frac{1}{4}e^{3s+x_0} + c_1e^{-s}.$$

Using the fact that $z(0) = 0$, we find $c_1 = -e^{x_0}/4$ which leads us to

$$z(s) = \frac{1}{4}\left(e^{3s+x_0} - e^{x_0-s}\right).$$

Finally, given (x, y) we determine the corresponding x_0 and s:

$$s = y \qquad \text{and} \qquad x_0 = x - y.$$

Thus our solution is

$$u(x,y) = \frac{1}{4}\left(e^{x+2y} - e^{x-2y}\right).$$

2.2.2. • Temporal Equations: Using Time to Parametrize the Characteristics. In many cases, time is one of the independent variables and we are given the values of the solution at $t = 0$. In these cases, the linear PDE for $u(x,t)$ can always be written in a form where the coefficient of u_t is 1; otherwise, we can divide through the original equation by the coefficient of u_t. In fact, most of the time the PDE will be presented in the form

$$u_t + \cdots = 0.$$

Since the data will be given on the $t = 0$ axis, the characteristic ODE for t,

$$\dot{t}(s) = 1,$$

will be solved with condition $t(0) = 0$. Thus we have

$$t(s) = s \qquad \text{or} \qquad t = s,$$

and we may as well use t (time) as the parameter of our characteristics, eliminating the variable s. In this case, the characteristics $(x(s), t(s)) = (x(t), t)$ in the x- vs. t-plane can be written with the scalar equation for $x = x(t)$. We give an example.

Example 2.2.4 (The Transport Equation with Variable Speed). Consider the IVP for $u(x,t), t \geq 0$,

$$u_t - xu_x = 0, \qquad u(x,0) = \frac{1}{1+x^2}.$$

The graph of the initial disturbance $u(x,0)$ is a simple (noncompactly supported) **blip** which is centered at $x = 0$ and has height 1. Using t to parametrize the characteristics, we arrive at

$$\frac{dx}{dt} = -x, \qquad \frac{dz}{dt} = 0.$$

These are solved as

$$x = x_0 e^{-t}, \qquad z = \frac{1}{1+x_0^2}.$$

Hence, given (x,t), we have $x_0 = xe^t$ and our solution derives to

$$u(x,t) = \frac{1}{1+x^2e^{2t}} = \frac{e^{-2t}}{x^2+e^{-2t}}.$$

Plots of the profiles of u for $t = 0, 0.5, 1, 2,$ and 3 are displayed in Figure 2.6. The reader should note the effect on the profiles of the variable propagation speed $-x$.

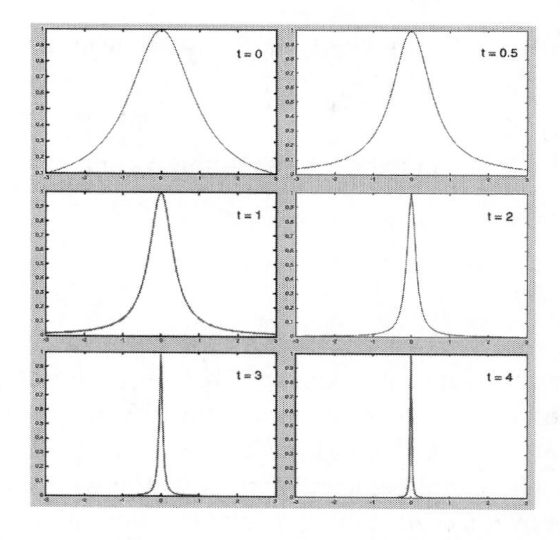

Figure 2.6. Profiles for the transport equation with variables speed.

2.2.3. • More Than Two Independent Variables. Our method can readily be performed when the unknown function depends on more than two variables. A few things change. First, since the domain of our solution u is a subset of $\mathbb{R}^N$ with $N \geq 3$, the number of equations for the characteristics will increase. Note that the characteristic will still be curves in $\mathbb{R}^N$ — one-dimensional objects. The "one" comes from the order of the PDE. There will, of course, be more than one parameter needed to label these characteristics. In fact, there will be $N-1$ such parameters, and data will now be given on an $(N-1)$-dimensional hypersurface (e.g., a two-dimensional surface for $N=3$). These $N-1$ parameters will label all characteristic curves via their intersection with the data hypersurface. The following provides an example with $N=3$.

Example 2.2.5. Find the solution for $u(x,y,z)$ to

$$xu_x + 2yu_y + u_z = 3u, \qquad u(x,y,0) = \sin(x+y).$$

Note here the danger of using z to denote one of the spatial variables. Since we are currently using z to denote the values of the solution u along a characteristic, let us rewrite the equation in (x_1, x_2, x_3) notation. That is,

$$x_1 u_{x_1} + 2x_2 u_{x_2} + u_{x_3} = 3u, \qquad u(x_1, x_2, 0) = \sin(x_1 + x_2).$$

The characteristic equations are

$$\dot{x}_1(s) = x_1, \quad \dot{x}_2(s) = 2x_2, \quad \dot{x}_3(s) = 1.$$

At $s=0$, we want the characteristics to lie on the x_1- vs. x_2-plane where the data is given; hence, let

$$x_1(0) = x_1^0, \quad x_2(0) = x_2^0, \quad x_3(0) = 0$$

and take note that we have two parameters to denote a point on the x_1- vs. x_2-plane. Also notice that since we are using the x_i notation for the independent variables, we denote these parameter values with superscripts of 0 rather than subscripts.

We now have

$$x_1(s) = x_1^0 e^s, \quad x_2(s) = x_2^0 e^{2s}, \quad x_3(s) = s.$$

The z equation is given by $\dot{z}(s) = 3z$, and we solve this equation with $z(0) = \sin(x_1^0 + x_2^0)$. Thus

$$z(s) = \sin(x_1^0 + x_2^0) e^{3s}.$$

Finally, given a point (x_1, x_2, x_3), we easily find that $s = x_3$ and $x_1^0 = x_1 e^{-x_3}$, $x_2^0 = x_2 e^{-2x_3}$ which leads us to

$$u(x_1, x_2, x_3) = \sin(x_1 e^{-x_3} + x_2 e^{-2x_3})\, e^{3x_3}.$$

2.2.4. • Transport Equations in Three Space Dimensions with Constant Velocity. In this subsection, we consider an important class of PDEs analogous to the transport equation (Example 2.1.3) but in three space dimensions. Recall that in one space dimension, we viewed the PDE as an equation for the evolution of a signal $g(x)$. We now consider the evolution of a signal in 3D $g(\mathbf{x})$ which, unfortunately, we cannot as easily visualize.

Let us begin by fixing an initial signal $g(\mathbf{x}) = g(x_1, x_2, x_3)$ and a constant vector $\mathbf{a} = \langle a_1, a_2, a_3 \rangle$. We wish to find a solution to the PDE

$$\begin{cases} u_t + a_1 u_{x_1} + a_2 u_{x_2} + a_3 u_{x_3} = 0, \\ u(x_1, x_2, x_3, 0) = g(x_1, x_2, x_3) \end{cases} \tag{2.8}$$

in the region of space-time

$$\Omega = \{(x_1, x_2, x_3, t) \mid t \geq 0\}.$$

If we denote ∇u to be the spatial gradient, i.e., $\nabla u = \langle u_{x_1}, u_{x_2}, u_{x_3} \rangle$, then the PDE can be written as

$$u_t + \mathbf{a} \cdot \nabla u = 0, \qquad u(\mathbf{x}, 0) = g(\mathbf{x}).$$

As we have previously seen, it is natural to use t instead of s to parametrize our characteristics. Then the characteristic equations can be written for $x_1(t)$, $x_2(t)$, and $x_3(t)$ as

$$\dot{x}_1(t) = a_1, \quad \dot{x}_2(t) = a_2, \quad \dot{x}_3(t) = a_3.$$

We solve these equations with initial conditions

$$x_1(0) = x_1^0, \quad x_2(0) = x_2^0, \quad x_3(0) = x_3^0$$

to find

$$x_1(t) = a_1 t + x_1^0, \quad x_2(t) = a_2 t + x_2^0, \quad x_3(t) = a_3 t + x_3^0.$$

The equation for z is simply

$$\dot{z}(t) = 0,$$

which we solve with initial condition

$$z(0) = g(x_1^0, x_2^0, x_3^0),$$

to give

$$z(t) = g(x_1^0, x_2^0, x_3^0).$$

Finally, given x_1, x_2, x_3, and $t > 0$, we have

$$x_1^0 = x_1 - a_1 t, \qquad x_2^0 = x_2 - a_2 t, \qquad \text{and} \qquad x_3^0 = x_3 - a_3 t,$$

and hence

$$u(x_1, x_2, x_3, t) = g(x_1 - a_1 t, x_2 - a_2 t, x_3 - a_3 t).$$

Succinctly, in vector notation, we have

$$u(\mathbf{x}, t) = g(\mathbf{x} - \mathbf{a}t).$$

The interpretation is again easy to grasp; the signal g is simply transported along with constant velocity vector $\mathbf{a}$. Hence, we refer to the PDE (2.8) as a **transport equation** associated with a constant velocity.

Before leaving this simple transport equation, we provide a simple physical interpretation. Consider a body of air whose motion, at any point in space, is given by a **constant** velocity vector $\mathbf{a}$ (the components of which have physical dimensions of length per time). Now let's introduce some particles in the air (think of a "cloud" of particles). The particles have an initial density given by $g(\mathbf{x})$ with physical dimensions of mass per volume. Since the particles move with the air, at any time t, their density will be given by $u(\cdot, t)$ where $u(\mathbf{x}, t)$ solves (2.8). It is simple to interpret the solution (i.e., the flow): The particle "cloud" is transported in space with velocity $\mathbf{a}$.

Next, let's consider the case of a general inhomogeneous right-hand side which we call a "forcing term". For a given function $f(x_1, x_2, x_3)$, we wish to solve

$$\begin{cases} u_t + a_1 u_{x_1} + a_2 u_{x_2} + a_3 u_{x_2} = f, \\ u(x_1, x_2, x_3, 0) = g(x_1, x_2, x_3), \end{cases} \tag{2.9}$$

or in vector notation

$$u_t + \mathbf{a} \cdot \nabla u = f, \qquad u(\mathbf{x}, 0) = g(\mathbf{x}).$$

The characteristic equations are the same for $x_i(t)$ yielding

$$x_1(t) = a_1 t + x_1^0, \quad x_2(t) = a_2 t + x_2^0, \quad x_3(t) = a_3 t + x_3^0.$$

The z equation is now

$$\dot{z}(t) = f(x_1(t), x_2(t), x_3(t)) = f(a_1 t + x_1^0, a_2 t + x_2^0, a_3 t + x_3^0)$$

and is solved with initial condition

$$z(0) = g(x_1^0, x_2^0, x_3^0).$$

This is an elementary ODE to solve since the right-hand side involves only the independent variable t. Thus, direct integration yields

$$z(t) = \int_0^t f\big(a_1\theta + x_1^0, a_2\theta + x_2^0, a_3\theta + x_3^0\big)\, d\theta + g(x_1^0, x_2^0, x_3^0).$$

Finally, given x_1, x_2, x_3, and $t > 0$, we have

$$x_1^0 = x_1 - a_1 t, \qquad x_2^0 = x_2 - a_2 t, \qquad \text{and} \qquad x_3^0 = x_3 - a_3 t,$$

and, hence, the solution to (2.9) is

$$\begin{aligned} u(x_1, x_2, x_3, t) &= \int_0^t f\big(x_1 + a_1(\theta - t), x_2 + a_2(\theta - t), x_3 + a_3(\theta - t)\big)\, d\theta \\ &\quad + g(x_1 - a_1 t, x_2 - a_2 t, x_3 - a_3 t). \end{aligned}$$

Succinctly, in vector notation we have

$$u(\mathbf{x}, t) = \int_0^t f\left(\mathbf{x} - \mathbf{a}(t - \theta)\right) d\theta + g(\mathbf{x} - \mathbf{a}t).$$

This time, in addition to the transport of the signal g with constant velocity $\mathbf{a}$, we have the accumulation (the integral) of the inhomogeneous f along the characteristic.

2.2.5. Transport Equations in Three Space Dimensions with Space Varying Velocity.

For the next two subsections, the reader is encouraged to first review Sections A.6 and A.9 from the Appendix.

Building on the previous subsection, let us consider transport equations where the velocity **depends on the spatial position**. In the context of dust in air, this means that the air velocity at a point $\mathbf{x}$ in space is given by $\mathbf{a}(\mathbf{x})$. The transport equation can then be written as

$$u_t + \mathbf{a}(\mathbf{x}) \cdot \nabla u = 0. \tag{2.10}$$

Solving this via the method of characteristics depends on the specified vector field $\mathbf{a}(\mathbf{x})$. In this short section, we will not actually solve this general equation; rather we will address a fundamental constraint (**divergence-free**) on the vector field $\mathbf{a}(\mathbf{x})$ which is directly related to a conservation principle. It can be viewed either as **conservation of mass** or, as we will now explain, conservation of **volume**.

Consider the IVP

$$u_t + \mathbf{a}(\mathbf{x}) \cdot \nabla u = 0, \qquad u(\mathbf{x}, 0) = g(\mathbf{x}), \tag{2.11}$$

where g is a nonnegative, continuous function with compact support; i.e., there exists an $r_0 > 0$ such that g vanishes for $|\mathbf{x}| \geq r_0$.

We make the following two assumptions on the vector-valued function $\mathbf{a}(\mathbf{x})$:

- $\mathbf{a}(\mathbf{x}) \in C^1(\mathbb{R}^3, \mathbb{R}^3)$ and is bounded (i.e., there exists a constant C such that $|\mathbf{a}(\mathbf{x})| \leq C$ for all $\mathbf{x} \in \mathbb{R}^3$);
- $\mathbf{a}(\mathbf{x})$ is divergence-free; i.e., $\operatorname{div} \mathbf{a}(\mathbf{x}) = 0$ for all $\mathbf{x}$.

The first assumption is for convenience only and can be weakened; the second assumption has an important consequence and as we shall now see, the divergence-free constraint is directly tied to a conservation principle. Indeed, we show that for any $t > 0$, we have

$$\iiint_{\mathbb{R}^3} u(\mathbf{x}, t)\, d\mathbf{x} = \iiint_{\mathbb{R}^3} g(\mathbf{x})\, d\mathbf{x}. \tag{2.12}$$

To prove (2.12) we apply two very common methods in PDE: differentiation under the integral sign and the application of calculus (integration by parts). While we will often see this procedure in subsequent chapters, this is our first implementation. Our goal is to show that for any time $t > 0$, we have

$$\frac{d}{dt} \iiint_{\mathbb{R}^3} u(\mathbf{x}, t)\, d\mathbf{x} = 0, \tag{2.13}$$

and, hence, the spatial integral of u is constant in time[3]. Since, assuming smoothness, this spatial integral is a continuous function on $t \in [0, \infty)$, it must equal its value at $t = 0$, yielding (2.12).

Step (i). Differentiation under the integral sign and using the PDE: By differentiation under the integral sign (cf. Section A.9), we have

$$\frac{d}{dt} \iiint_{\mathbb{R}^3} u(\mathbf{x}, t)\, d\mathbf{x} = \iiint_{\mathbb{R}^3} u_t(\mathbf{x}, t)\, d\mathbf{x} = -\iiint_{\mathbb{R}^3} \mathbf{a}(\mathbf{x}) \cdot \nabla u(\mathbf{x}, t)\, d\mathbf{x}, \tag{2.14}$$

where the second equality followed by using the PDE (2.11).

Step (ii). Localization of the information: Let us **fix a** $t > 0$. The initial data vanishes for $|\mathbf{x}| \geq r_0$ and information travels along characteristics which have speed $|\mathbf{a}|$

[3]One similarly show that $\iiint_{\mathbb{R}^3} |u(\mathbf{x}, t)|^2\, d\mathbf{x}$ is constant in time (cf. Exercise **2.29**).

no larger than C. Thus at any time t, $u(\mathbf{x}, t)$, as a function of $\mathbf{x}$, has compact support; in particular, we must have

$$u(\mathbf{x}, t) \equiv 0 \quad \text{for all } |\mathbf{x}| \geq Ct + r_0.$$

Step (iii). Integration by parts: We choose some $R > Ct + r_0$ and apply the vector field integration by parts formula (A.15) (from Section A.6.4) on the domain $B(\mathbf{0}, R)$ with "$\mathbf{u} = \mathbf{a}$ and $v = u$" to find

$$\iiint_{B(\mathbf{0},R)} \mathbf{a}(\mathbf{x})\cdot\nabla u(\mathbf{x}, t)\, d\mathbf{x} = -\iiint_{B(\mathbf{0},R)} (\operatorname{div}\mathbf{a}(\mathbf{x}))\, u(\mathbf{x}, t)\, d\mathbf{x} + \iint_{\partial B(\mathbf{0},R)} (u(\mathbf{x}, t)\mathbf{a}(\mathbf{x}))\cdot\mathbf{n}\, dS.$$

Since $u(\mathbf{x}, t) = 0$ on $\partial B(\mathbf{0}, R)$ and $\mathbf{a}$ is divergence-free, the right-hand side is 0. Moreover, since $u(\mathbf{x}, t) \equiv 0$ outside $B(\mathbf{0}, R)$, we have

$$0 = \iiint_{B(\mathbf{0},R)} \mathbf{a}(\mathbf{x}) \cdot \nabla u(\mathbf{x}, t)\, d\mathbf{x} = \iiint_{\mathbb{R}^3} \mathbf{a}(\mathbf{x}) \cdot \nabla u(\mathbf{x}, t)\, d\mathbf{x}.$$

Combining this with (2.14) yields (2.13).

Equation (2.12) can be interpreted as a statement of "conservation of mass". Alternatively, it can be interpreted as a statement of "conservation of volume". To this end, we can view the PDE (2.11) as modeling the evolution of media of constant density (e.g., a homogeneous liquid). Let V_0 be a bounded region of space wherein the liquid initially lives. For $\mathbf{x}_0 \in V_0$, let $\mathbf{x}(t; \mathbf{x}_0)$ be the solution to the characteristic equations for the projected characteristics:

$$\begin{cases} \frac{d\mathbf{x}}{dt} = \mathbf{a}(\mathbf{x}), \\ \mathbf{x}(0) = \mathbf{x}_0. \end{cases}$$

Note that we use the second argument in $\mathbf{x}(t; \mathbf{x}_0)$ as a parameter indexing the initial data. For any $t > 0$, let $V(t) := \{\mathbf{x}(t; \mathbf{x}_0) \mid \mathbf{x}_0 \in V_0\}$. Convince yourself that $V(t)$ (a subset of $\mathbb{R}^3$) is exactly where the liquid has moved to at time t. We claim that div $\mathbf{a}(\mathbf{x}) \equiv 0$ implies that at any later time t the following hold:

- Trajectories starting from different initial positions $\mathbf{x}_0$ terminate in distinct places (in other words, the trajectory paths do not intersect).
- We have

$$\text{Volume of } V(t) = \text{Volume of } V_0. \tag{2.15}$$

After a little thought, we notice that (2.15) is simply (2.12) in the specific case where the initial data is the characteristic (or indicator) function of V_0

$$g(\mathbf{x}) = \chi_{V_0}(\mathbf{x}).$$

Unfortunately, our path to (2.12) required the smoothness of g and, hence, does not immediately transfer to the discontinuous characteristic functions. Nevertheless, we will soon learn a wonderful tool in dealing with the convolution of functions which will allow us to "approximate", to the desired accuracy, such discontinuous functions with smooth functions. With this tool in hand, one can readily prove (2.15).

2.2.6. The Continuity Equation in Three Space Dimensions: A Derivation. A ubiquitous PDE, related to the transport equation (2.10), is known as the continuity equation. It is usually written with the dependent variable denoted by $\rho(\mathbf{x}, t)$ (instead of $u(\mathbf{x}, t)$) and takes the form

$$\boxed{\rho_t + \operatorname{div}(\rho \mathbf{v}) = 0} \tag{2.16}$$

where $\mathbf{v}(\mathbf{x}, t)$ is a vector field. The equation models the concentration ρ of a quantity which is being transported by a velocity field $\mathbf{v}$.

In this section, we assume that $\mathbf{v}$ is a **known** (specified) vector field, and hence the equation is linear. While we will not describe or solve the equation via characteristics, let us take the opportunity to derive the equation. We work in the context of the mass density $\rho(\mathbf{x}, t)$ of some fluid (a gas or liquid) which is being transported via a velocity field $\mathbf{v}(\mathbf{x}, t)$. The dependent variable ρ gives the density of fluid particles at position $\mathbf{x}$ and time t and has dimensions of mass per time. The velocity field $\mathbf{v}$ gives the velocity vector of the fluid particle which is at position $\mathbf{x}$ at time t; hence, $|\mathbf{v}|$ has dimensions of length per time. The continuity equation will simply be a direct consequence of **conservation of mass.**

Now consider W to be an arbitrary region of three space where the fluid lives. The total mass of the fluid in W at time t is precisely

$$m_W(t) := \iiint_W \rho(\mathbf{x}, t)\, d\mathbf{x}.$$

Note that $m_W(t)$ has dimensions of mass. Differentiating $m_W(t)$ with respect to t gives

$$\frac{d}{dt} m_W(t) = \iiint_W \rho_t(\mathbf{x}, t)\, d\mathbf{x}.$$

How can $m_W(t)$, the total mass of fluid in W, change with time? Since mass cannot be created or destroyed, the only way $m_W(t)$ can change is if fluid enters or leaves W across the boundary ∂W. The movement of fluid is described by the velocity field $\mathbf{v}(\mathbf{x}, t)$. Hence, at any point $\mathbf{x}$ on the boundary and time t, the flow of fluid across the boundary is instantaneously $\mathbf{v}(\mathbf{x}, t) \cdot \mathbf{n}$, where $\mathbf{n}$ denotes the outer unit normal to W. This means that

$$\iiint_W \rho_t(\mathbf{x}, t)\, d\mathbf{x} = -\iint_{\partial W} \rho(\mathbf{x}, t)\, \mathbf{v}(\mathbf{x}, t) \cdot \mathbf{n}\, dS.$$

Note that the minus sign on the right accounts for the fact that we are using the outer normal; hence, places on the boundary where $\mathbf{v}(\mathbf{x}, t) \cdot \mathbf{n} > 0$ contribute to a loss of mass. Note further that both sides of the equation have dimensions of mass per time. Using the Divergence Theorem (cf. Section A.6.1) on the right-hand side, we obtain

$$\iiint_W \rho_t\, d\mathbf{x} = -\iiint_W \operatorname{div}(\rho \mathbf{v})\, d\mathbf{x} \qquad \text{or} \qquad \iiint_W (\rho_t + \operatorname{div}(\rho \mathbf{v}))\, d\mathbf{x} = 0.$$

Since the above is true for **any** region W of space, we may invoke the IPW Theorem, Theorem A.6, to conclude that that ρ must satisfy the continuity equation (2.16).

Two important remarks:

- **Conservation of Mass vs. Conversation of Volume and the Relationship with the Transport Equation.** In the previous subsection, we considered the transport equation (2.10) under a divergence-free constraint and referred to both the conservation of mass and volume. The continuity equation is a general equation for mass conservation whereas the particular transport equation (2.10) is a reduction of the continuity equation in the case where the prescribed velocity field $\mathbf{v}$ is divergence-free; i.e.,

$$\operatorname{div} \mathbf{v}(\mathbf{x}, t) \;=\; 0 \qquad \text{for all } \mathbf{x} \in \mathbb{R}^3,\; t \geq 0.$$

  To see this we note that the vector identity (cf. Exercise **A.13** in the Appendix)

$$\operatorname{div}(\rho \mathbf{v}) \;=\; (\operatorname{div} \mathbf{v})\,\rho \;+\; \mathbf{v} \cdot \nabla \rho$$

  reduces the continuity equation to the transport equation

$$\rho_t \;+\; \mathbf{v} \cdot \nabla \rho \;=\; 0. \tag{2.17}$$

  This is simply (2.10) for the case where the velocity field, previously labeled with $\mathbf{a}$, is time dependent. Thus the divergence-free constraint reduces the continuity equation to the transport equation. As we discussed in Section 2.2.5, this equation entails **conservation of volume**, that is, incompressible flow as one would expect for a liquid like water.

- In an actual fluid (gas or liquid) one would expect that the velocity field $\mathbf{v}$ is not prescribed, but found via certain laws resulting in PDEs for $\mathbf{v}$. Indeed, by invoking conservation of linear momentum as described by Newton's Second Law, one derives a system of PDEs (for which the continuity equation is just one equation) called the **Euler equations** (see Section 2.8 for more details). A subclass of these equations, called the incompressible Euler equations, pertains to the case where $\operatorname{div} \mathbf{v} = 0$. These equations present a model for the dynamics of a liquid, like water. Here, one often imposes the **additional** assumption of **homogeneity**; the density ρ is constant in space and time. For a homogeneous liquid, the continuity equation reduces to the constraint $\operatorname{div} \mathbf{v} = 0$.

2.2.7. Semilinear Equations.

Consider a PDE of the form

$$a(x, y)u_x + b(x, y)u_y \;=\; c(x, y, u).$$

Strictly speaking, this is a nonlinear PDE since c can be a nonlinear function of u. However, the fact that there are no nonlinear terms involving derivatives of u implies that the PDE behaves more like a linear PDE. For this reason, we call such equations **semilinear**. Our previous method of characteristics will apply with one difference: Once we have solved for $x(s)$ and $y(s)$, we will have to solve a **nonlinear** ODE

$$\dot{z}(s) \;=\; c(x(s), y(s), z(s)).$$

Note that these observations generalize to semilinear PDEs in any number of independent variables. We provide one example in two variables.

Example 2.2.6. Find the solution to

$$u_x + 2u_y = u^2, \qquad u(x,0) = h(x).$$

Solving the characteristic equations for $x(s)$ and $y(s)$ with $x(0) = x_0$ and $y(0) = 0$, we find

$$x(s) = x_0 + s, \qquad y(s) = 2s.$$

The equation for $z(s) = u(x(s), y(s))$ is

$$\dot{z}(s) = z^2(s)$$

which one can check has solution $z(s) = \frac{1}{c-s}$, for any constant c. Using the fact that $z(0) = h(x_0)$, the "constant" c must be $1/h(x_0)$. So we have

$$z(s) = \frac{h(x_0)}{1 - h(x_0)s}.$$

Finally, given (x, y) we find that $s = y/2$ and $x_0 = x - y/2$ and hence a solution

$$u(x,y) = \frac{h(x - y/2)}{1 - h(x - y/2)y/2}.$$

2.2.8. Noncharacteristic Data and the Transversality Condition.

In our first example

$$au_x + bu_y = 0,$$

we noted that data given on any line which was not parallel to $\langle a, b\rangle$ would yield a well-posed problem. On the other hand, data given on a line parallel to $\langle a, b\rangle$ would either yield no solution or infinitely many solutions, the latter in the case of constant data. More generally, we could specify data on any curve Γ which was never tangent to $\langle a, b\rangle$. We call this **noncharacteristic data**, and the condition it must satisfy is known as **the transversality condition**. Let us now provide a precise statement of the transversality condition for the general first-order linear PDE in two variables

$$a(x,y)u_x + b(x,y)u_y = c_1(x,y)u + c_2(x,y),$$

with data supplied on some curve Γ. To describe a general curve Γ we need a parameter which, in order to not confuse it with the parameter s used for the characteristics, we call τ (cf. Figure 2.7). Thus, the data curve Γ is described in parametric form: $(x_0(\tau), y_0(\tau))$. We use the subscript of 0 since the data curve is usually associated with the $s = 0$ starting point of the characteristics.

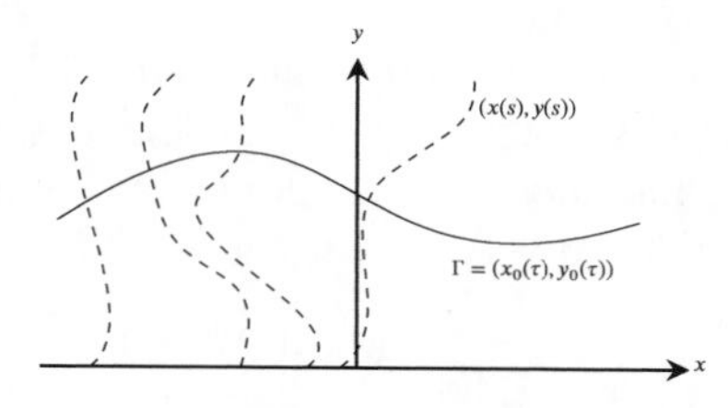

Figure 2.7. The data curve Γ, parametrized with the new variable τ, intersecting with the characteristics $(x(s), y(s))$.

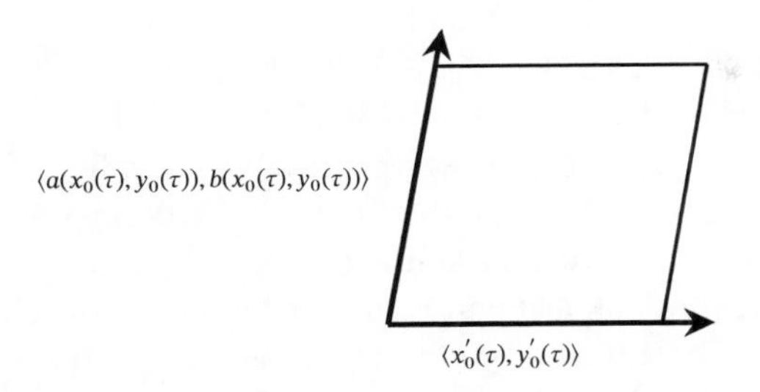

Figure 2.8. A geometric interpretation of the transversality condition in two independent variables with the horizontal vector tangent to the data curve Γ and the upward vector tangent to the characteristic. The determinant of the matrix corresponds to the area of a parallelogram created by the two vectors. If the vectors are parallel, the parallelogram will have no area.

Then we require that for all τ, the tangent vector to Γ, $\langle x_0'(\tau), y_0'(\tau)\rangle$, is **not parallel** to the tangent vector of the characteristic which passes through the point $(x_0(\tau), y_0(\tau))$. By definition of the characteristics, this tangent vector is given by

$$\langle a(x_0(\tau), y_0(\tau)),\ b(x_0(\tau), y_0(\tau))\rangle.$$

Hence this condition, called the transversality condition, can be rephrased in terms of the determinant of a matrix: for all τ,

$$\det \begin{bmatrix} a(x_0(\tau), y_0(\tau)) & x_0'(\tau) \\ b(x_0(\tau), y_0(\tau)) & y_0'(\tau) \end{bmatrix} \neq 0.$$

The analogous transversality condition can be formulated for any number of independent variables, for example, for a noncharacteristic surface in the case of three independent variables.

For general first-order linear PDEs, there can be issues with the solution to the characteristic ODEs, and it could be the case that Γ, while noncharacteristic, may not intersect **all** the (projected) characteristics. However, as long as the coefficient functions ($a(x, y)$, $b(x, y)$, and $c_i(x, y)$ for two variables) are smooth (for example, C^1), we can at least **prove the existence of a local solution**. By a local solution, we mean a solution to the PDE in a **neighborhood** of Γ, a smaller region of the xy-plane containing Γ, which satisfies the auxiliary condition on Γ. See Section 2.4.3 for a few more details on proving the existence of local solutions.

2.3. • An Important Quasilinear Example: The Inviscid Burgers Equation

Before we present the method of characteristics for general quasilinear and fully nonlinear PDEs, we pause to discuss an important example. To introduce this equation, first recall the linear transport equation of Example 2.1.3:

$$u_t + cu_x = 0, \qquad u(x, 0) = g(x), \tag{2.18}$$

for some $c > 0$. The characteristics in x- vs. t-plane are lines with slope $\frac{dx}{dt} = c$, and, hence, the solution at any later time t is simply $u(x, t) = g(x - ct)$. This means that at time t, the initial signal (shape) $g(x)$ is transported to the right with speed c. Note that the shape of the signal (the wave) does not change with time.

Now consider the shape of water waves traveling towards a beach. Do they propagate ("transport") without changing their shape? Certainly not. On the other hand, recall the transport equation with variable speed (Example 2.2.4) wherein the shape did change. However, if you look at Figure 2.6, it again does not reflect what one would observe on a beach. There, one observes a form of *the breaking of waves* wherein the top part overtakes the bottom (cf. the cartoon of Figure 2.9).

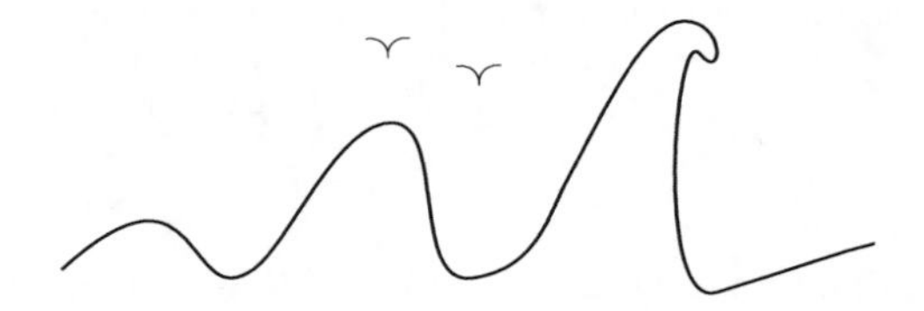

Figure 2.9. A cartoon depicting the breaking of water waves.

As we shall see, this phenomenon is a byproduct of **nonlinearity**, as captured by the **Burgers equation**[4]:

$$\boxed{u_t + uu_x = 0.} \tag{2.19}$$

Note that the nonlinearity stems from the product of u and u_x, and as such the PDE is quasilinear. Based upon our intuition from the transport equation, we can thus think of the Burgers equation as modeling transport in which **speed is proportional to the value of the solution.**

To give more detail, consider the initial value problem for Burgers's equation:

$$\begin{cases} u_t + uu_x = 0, \\ u(x,0) = g(x). \end{cases} \tag{2.20}$$

As with Example 2.1.3, we use t to parametrize the characteristics in this IVP. Following our previous logic, one would be tempted to write the characteristic equation for x as

$$\frac{dx}{dt} = u$$

or, in terms of the notation in the method of characteristics where $z(t) = u(x(t), t)$,

$$\frac{dx}{dt} = z(t). \tag{2.21}$$

Since we do not know what the solution is, we cannot start by solving (2.21). Rather, we proceed as follows. First, let us first assume that there is an (as yet unknown) solution u and it is smooth. Then the PDE (alternatively the ODE for z) tells us that

$$\frac{dz}{dt} = \frac{d}{dt}u(x(t), t) = 0. \tag{2.22}$$

[4]Following the historically inaccurate tradition in the PDE community, we will call this evolution equation Burgers's equation. However, we should really think of (2.19) as either Burgers's equation with vanishing viscosity or the Inviscid Burgers equation.

Hence, **along a characteristic** $x(t)$ **solving** (2.21)**, the solution must be constant.** This is our first observation. Before continuing, let us just check that the logic behind (2.22) is indeed correct. Suppose $u(x, t)$ is any function which solves (2.20). Then with u in hand, we can solve

$$\frac{dx}{dt} = u(x(t), t).$$

But then according to the PDE,

$$\frac{d(u(x(t)), t)}{dt} = u_t + u_x \frac{dx}{dt} = u_t + uu_x = 0.$$

In other words, the solution must be constant along a characteristic. Good!

Now comes the crucial punchline: Since the solution is constant along a characteristic, the right-hand side of (2.21) is constant. This leads us to the revelation that **the characteristics must be straight lines!** This was the case with the simple linear transport equation (2.18) **but now there is a catch**:

The slope of the characteristic line $=$ the value the solution takes on the line.

Since the solution is known at the point $(x_0, 0)$ to be $g(x_0)$, the solution must be $g(x_0)$ along the entire characteristic line. Putting this all together: The characteristic through point $(x_0, 0)$ is a line with slope $g(x_0)$, and this slope is the same number as the value of u along this line (cf. Figure 2.10, left). Rather convoluted you say? Welcome to our nonlinear world!

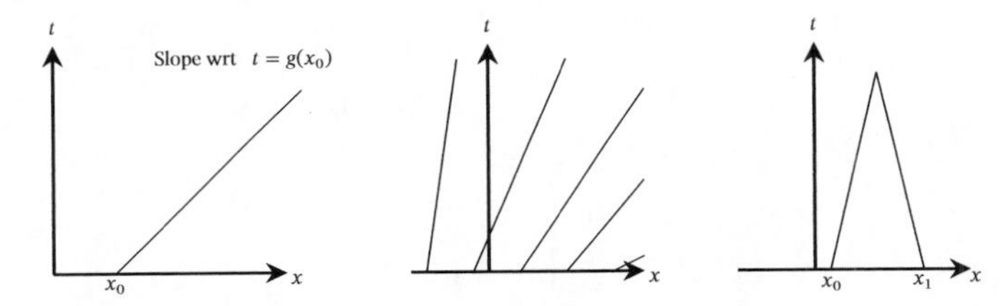

Figure 2.10. Left: A characteristic through point (x_0, t) has slope $u(x_0, t) = g(x_0)$. Note here that slope is with respect to t (i.e., $\frac{dx}{dt}$). **Middle:** For an increasing initial data $g(x)$, the characteristics do not intersect and fill out the upper half t-plane. **Right:** When $g(x)$ is decreasing on some interval, the characteristics will collide and consequently, the solution becomes overdetermined (or multivalued).

Now, if $g(x)$ is an increasing function of x, then we encounter no issue as the characteristics will span out space-time, and a unique solution can be determined (cf. Figure 2.10, middle). However, if $g(x)$ is not an increasing function of x, then characteristics can collide (cf. Figure 2.10, right). Since the solution equals the slope of the characteristic, the solution becomes **overdetermined**.

Let us now re-examine what is happening from the perspective of the solution profiles. We may view Burgers's equation as governing the propagation of an initial signal $g(x)$ which, for the sake of illustration, looks like a blip $g(x) = \frac{1}{1+x^2}$. In the linear transport problem (2.18), "information" (i.e., the height of the signal $g(x)$) will move through 1D space (the line) with a constant speed c, thus preserving the shape of the signal at any later time. For Burgers's equation, the height of the signal, $g(x_0)$ at a point x_0, will move through space at a speed equal to $g(x_0)$.

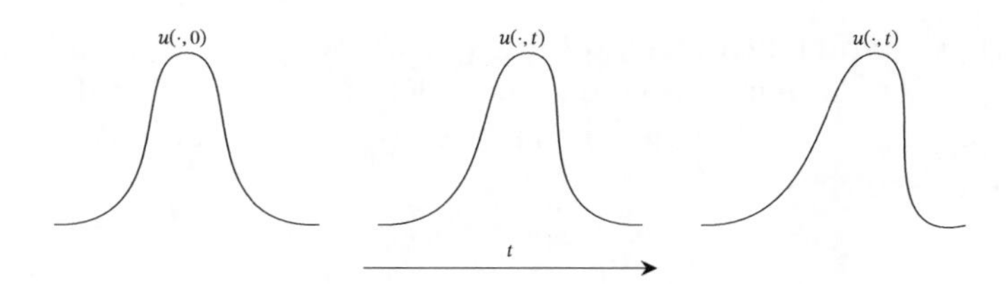

Figure 2.11. A signal transported by Burgers's equation.

Thus the higher the signal (wave), the faster it travels through 1D space; as a result the solution profile changes its shape with time (see Figure 2.11). Eventually, the "top" will overtake the "bottom" and this turns out to be exactly where the characteristic will begin to collide. The reader should agree that this does mirror the **initial stages** observed just before water waves break when approaching the beach: Here the top of the water wave overtakes the bottom and eventually causes the wave to break (the water "splashes over"); cf. Figure 2.9.

Let us consider the following explicit example to further illustrate this "breaking of waves".

Example 2.3.1. We solve

$$\begin{cases} u_t + uu_x = 0, \\ u(x,0) = g(x), \end{cases} \tag{2.23}$$

where

$$g(x) = \begin{cases} 1 & \text{if } x \le 0, \\ 1-x & \text{if } 0 \le x \le 1, \\ 0 & \text{if } x \ge 1. \end{cases}$$

A graph of g is shown on the left in Figure 2.13. Figure 2.12 shows the characteristic lines emanating from the $t = 0$ axis. We see that there is no issue for $t \le 1$; that is, through each point $(x,t), t \le 1$, there is a unique characteristic connecting the point with the $t = 0$ axis. For $t > 1$, characteristics collide and the solution is overdetermined. For $t \le 1$, we can determine exactly the solution formula from the characteristic picture (Figure 2.12). On the $t = 0$ axis, we see that characteristics emanating from $x \le 0$ have slope (with respect to t) equal to 1, while for $x \ge 1$, they have slope (with respect to t) equal to 0. For $0 \le x \le 1$, the characteristic slopes decrease (with x) in a linear fashion from 1 to 0 and all meet at the point $(1,1)$. If we pick a point (x,t) in this intermediate triangular region of the xt-plane, then to find the solution, we need only compute the slope of the line connecting the point to the point $(1,1)$, i.e., $\frac{1-x}{1-t}$.

Hence for $t \le 1$, we can explicitly write down the solution:

$$u(x,t) = \begin{cases} 1 & \text{if } x \le t, \\ \frac{1-x}{1-t} & \text{if } t < x < 1, \\ 0 & \text{if } x \ge 1. \end{cases}$$

We plot profiles of u for times $t = 0, 0.5, 0.75$, and 1 in Figure 2.13. As expected, information in the wave profile which is higher travels faster. Eventually at $t = 1$, this forces the wave profile to break and become discontinuous. Before continuing, the

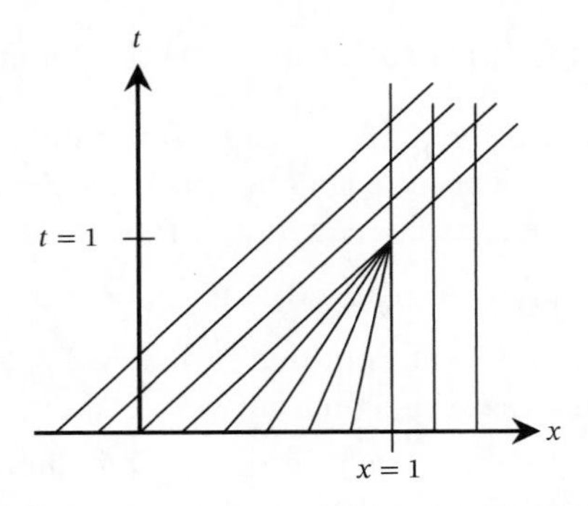

Figure 2.12. Characteristics for (2.23) collide after time $t = 1$.

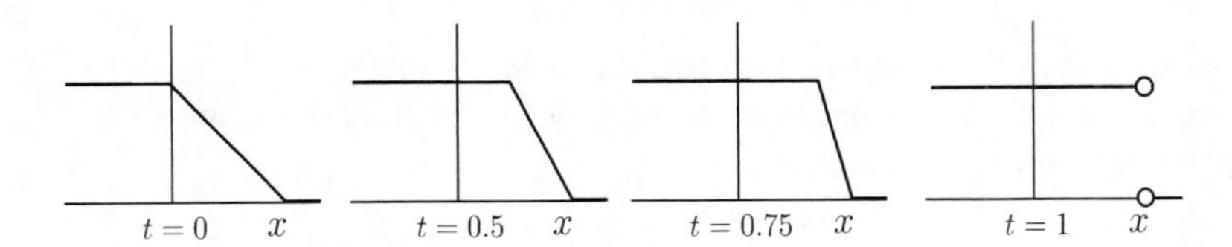

Figure 2.13. Solution profiles for (2.23).

reader must **see** the **direct** connection between the plot of the characteristics from $t = 0$ to $t = 1$ in Figure 2.12 and the profiles of Figure 2.13. Do **not** continue until this is clear.

In Section 9.7.2, we will see that the collision of characteristics is associated with jump discontinuities in the full solution. These discontinuities, which propagate through space-time, are called **shock waves**. In the context of the previous example, the discontinuity created at $t = 1$ will propagate to the right with a speed determined by the PDE.

At this stage, the punch line to take from (the inviscid) Burgers equation is the following:

> For nonlinear equations (which includes quasilinear equations) there is **an interplay** between the structure of the characteristics **and** the solution itself.

This was not the case for the previous examples, as they were all linear PDEs (or semilinear). For linear and semilinear PDEs, the structure of the characteristics is **entirely** determined by the structure of the PDE. In particular, one can determine the characteristics without knowledge of any auxiliary condition. For nonlinear equations, the structure of the characteristics depends on the auxiliary condition through the PDE.

2.4. • The Method of Characteristics, Part II: Quasilinear Equations

We start with two independent variables x and y (or x and t). Quasilinear first order means that the equation is linear in u_x and u_y, but their coefficients can depend on u in addition to x and y. So we consider equations of the form

$$a\big(x, y, u(x, y)\big)\, u_x \;+\; b\big(x, y, u(x, y)\big)\, u_y \;=\; c\big(x, y, u(x, y)\big), \tag{2.24}$$

for some known functions a, b, and c of three variables. We are interested in the PDE in some domain Ω with data specified on a curve Γ.

As before, we assume we have a solution $u(x, y)$ and view the PDE as a statement about the directional derivative of u along certain directions. Following these directions amounts to finding the characteristic curves $(x(s), y(s))$ which solve

$$\begin{aligned} \dot{x}(s) &= a\big(x(s), y(s), u(x(s), y(s))\big), \\ \dot{y}(s) &= b\big(x(s), y(s), u(x(s), y(s))\big). \end{aligned} \tag{2.25}$$

But now we cannot solve these two equations by themselves, as the left-hand sides involve u along the characteristics. After denoting $z(s)$ as the value of the solution along the characteristic $(x(s), y(s))$, i.e.,

$$z(s) := u(x(s), y(s)),$$

we turn to the PDE for insight on how $z(s)$ should evolve with respect to s (i.e., how u evolves along the characteristics). Note that

$$\dot{z}(s) = \frac{d}{ds} u(x(s), y(s)) = \dot{x}(s)\, u_x(x(s), y(s)) + \dot{y}(s)\, u_y(x(s), y(s)).$$

Since the characteristics satisfy (2.25), we have

$$\begin{aligned} \dot{z}(s) = \;& a\big(x(s), y(s), u(x(s), y(s))\big)\, u_x(x(s), y(s)) \\ & + b\big(x(s), y(s), u(x(s), y(s))\big)\, u_y(x(s), y(s)). \end{aligned}$$

Since the PDE holds at all points in Ω, it holds along a characteristic. This means

$$\begin{aligned} a\big(x(s), y(s), u(x(s), y(s))\big)\, u_x(x(s), y(s)) + \; & b\big(x(s), y(s), u(x(s), y(s))\big)\, u_y(x(s), y(s)) \\ = \; & c\big(x(s), y(s), u(x(s), y(s))\big). \end{aligned}$$

We arrive at our ODE for $z(s)$:

$$\dot{z}(s) = c\big(x(s), y(s), z(s)\big).$$

This, combined with the two equations in (2.25), gives a **closed system of three equations in three unknown functions**:

$$\begin{cases} \dot{x}(s) = a\big(x(s), y(s), z(s)\big), \\ \dot{y}(s) = b\big(x(s), y(s), z(s)\big), \\ \dot{z}(s) = c\big(x(s), y(s), z(s)\big). \end{cases} \tag{2.26}$$

There is now an interplay between the characteristics $(x(s), y(s))$ and the value the solution u takes on them (i.e., $z(s)$). In particular, to solve (2.26) we need to include the initial values for $x(0) = x_0$ and $y(0) = y_0$ where $(x_0, y_0) \in \Gamma$ with the associated initial values for $z(0) = u(x_0, y_0)$. The latter comes from the auxiliary condition giving the value of u on Γ. With these initial values in hand, one can attempt to solve the system (2.26).

2.4.1. • A Few Examples.

We begin by revisiting the example of the last section.

Example 2.4.1. Recall Burgers's equation from Section 2.3:

$$u_t + uu_x = 0, \quad u(x,0) = g(x).$$

We want to find a solution in Ω, the upper half-plane $x \in \mathbb{R}$, $t \geq 0$. Our data is given on Γ, the $t = 0$ axis. This is certainly of the form (2.24) with

$$a(x,t,u) = u, \qquad b(x,t,u) = 1, \qquad c(x,t,u) = 0,$$

and, hence, the characteristic equations (2.26) become

$$\dot{x}(s) = z, \qquad \dot{t}(s) = 1, \qquad \dot{z}(s) = 0.$$

As usual, we will take $t = s$ and parametrize the characteristics with time t. Thus we have the two equations

$$\frac{dx}{dt} = z(t) \qquad \text{and} \qquad \frac{dz}{dt} = 0,$$

which we solve with initial conditions $x(0) = x_0$ and $z(0) = z_0 = g(x_0)$.

The solutions are

$$z(t) = g(x_0) \quad \text{and} \quad x(t) = g(x_0)t + x_0.$$

Finally, given (x,t) with $t > 0$, we have

$$u(x,t) = g(x_0) \quad \text{where} \quad \frac{x - x_0}{t} = g(x_0). \tag{2.27}$$

The difference now is that we have an **implicit** description involving the data g for the dependence of x_0 on (x,t). This is the price of dealing with a nonlinear PDE. But if we can solve (2.27) uniquely for x_0 in terms of (x,t), then we are done! Here is the catch as we cannot always do so. In fact, we can only do this if $g(\cdot)$ is a nondecreasing function. For example, if $g(x) = x$, we have

$$u(x,t) = x_0 \qquad \text{where} \qquad \frac{x - x_0}{t} = x_0.$$

We can solve uniquely for x_0 as a function of x and t,

$$x_0 = \frac{x}{t+1}, \quad \text{and hence the solution is } u(x,t) = \frac{x}{t+1}.$$

You can explicitly check that this is indeed the solution to the PDE and initial condition on the entire upper half-plane $t \geq 0$. If, on the other hand, we set $g(x) = -x$, we would find

$$u(x,t) = -\frac{x}{1-t}.$$

This solution is only defined in the strip $\{(x,t) \,|\, 0 \leq t < 1\}$. Indeed at $t = 1$ characteristics collide and a shock (discontinuity) forms. We will briefly address such shocks in Section 9.7.2 and study them in more detail in a future volume planned for this subject.

Example 2.4.2. Consider

$$(x+u)\,u_x \,+\, yu_y \;=\; u \,+\, y^2 \quad \text{with} \quad u(x,1)=x.$$

We seek the solution on $\Omega = \{(x,y)\,|\,y \geq 1\}$.

The PDE is in the form (2.24) with

$$a(x,y,u) = x+u, \qquad b(x,y,u) = y, \qquad c(x,y,u) = u + y^2,$$

which means the characteristic equations (2.26) are

$$\dot{x}(s) = x(s) + z(s), \qquad \dot{y}(s) = y(s), \qquad \dot{z}(s) = z(s) + y^2(s).$$

We want to solve these ordinary differential equations with "initial" conditions

$$y(0) = 1, \quad x(0) = x_0, \quad \text{and} \quad z(0) = x_0.$$

We start with the easiest, the y equation, to find $y(s) = e^s$, which satisfies $y(0) = 1$. Next, we turn to the z equation (why?) which is now

$$\dot{z}(s) = z(s) + e^{2s}.$$

This ODE is readily solved by introducing the integrating factor[5] e^{-s}; multiplying each side by the integrating factor we find

$$\big(\dot{z}(s) - z(s)\big)e^{-s} = e^s, \quad \text{or} \quad \frac{d}{ds}\big(\dot{z}(s)e^{-s}\big) \;=\; e^s.$$

Hence,

$$z(s)e^{-s} \;=\; e^s + c \qquad \text{or} \qquad z(s) \;=\; e^{2s} + ce^s,$$

for a constant c which is chosen so that $z(0) = x_0$. This gives

$$z(s) \;=\; e^{2s} + (x_0 - 1)e^s.$$

The x equation becomes

$$\dot{x}(s) = x(s) + e^{2s} + (x_0 - 1)e^s.$$

Multiplying by the same integrating factor as before and using the initial condition $x(0) = x_0$, we determine

$$x(s) \;=\; e^{2s} + (x_0 - 1)e^s(s+1).$$

Finally, given (x,y) $(y \geq 1)$, we attempt to solve for s and x_0 such that $(x(s), y(s)) = (x,y)$. This is easy:

$$s = \log y \qquad \text{and} \qquad x_0 = 1 + y\frac{x - y^2}{\log y + 1}$$

yielding the solution

$$u(x,y) \;=\; y^2 + \frac{x - y^2}{\log y + 1}.$$

Note that this solution is defined on the entire half-plane $\{(x,y)\,|\,y \geq 1\}$.

This example was **carefully chosen**. Indeed, not only were we able to solve the ODEs in a simple sequence, but we were able to uniquely solve for x_0 and s in terms of any x and y. As a good exercise, consider separately each of the following changes to see the complications which could arise if one: (i) changes the initial data to $u(x,1) = x^2$;

[5]Quickly review first-order linear ODEs and integrating factors by looking them up online.

(ii) changes the coefficient of u_x to $x^2 + u$; (iii) changes the coefficient of u_y to x; (iv) change the right-hand side to $u + x^2$.

2.4.2. • More Than Two Independent Variables. The method readily extends to quasilinear equations in any number of independent variables. The following is an example with three independent variables.

Example 2.4.3. We find a solution $u(x, y, t)$ to the initial value problem

$$u_t + u u_x + t u_y = y, \quad u(x, y, 0) = x$$

on $\Omega = \{(x, y, t) \mid t \geq 0\}$, the upper half-space.

Using t to parametrize the characteristics, the characteristic equations become

$$\dot{x}(t) = z(t), \qquad \dot{y}(t) = t, \qquad \dot{z}(t) = y(t),$$

with initial conditions $x(0) = x_0, y(0) = y_0$, and $z(0) = x_0$. We solve first the y equation followed by the z equation to find

$$y(t) = \frac{t^2}{2} + y_0 \qquad \text{and} \qquad z(t) = \frac{t^3}{6} + y_0 t + x_0.$$

Solving the x equation yields

$$x(t) = \frac{t^4}{24} + y_0 \frac{t^2}{2} + x_0 t + x_0.$$

Finally, given x, y, and $t > 0$, we have

$$x_0 = \frac{x - \frac{t^4}{24} - y_0 \frac{t^2}{2}}{t + 1} \qquad \text{and} \qquad y_0 = y - \frac{t^2}{2},$$

and hence (after a little algebra),

$$u(x, y, t) = \frac{24x - 12yt^2 + 5t^4}{24(1 + t)} + yt - \frac{t^3}{3}.$$

2.4.3. Pausing to Reflect on the Inherent Logic Behind the Method of Characteristics and Local Solutions of Quasilinear Equations. Before continuing to general first-order equations, let us draw reference to the general flow of logic we followed in the method of characteristics for linear and quasilinear equations. This logic and approach are very common in PDEs, as discussed in Section 1.6.

- We started by assuming that a solution u exists to the PDE and auxiliary condition. This was required in order to write down the z equation for the method of characteristics.
- Noting that the PDE is a statement about directional derivatives of u in certain directions, we wrote down the ODEs for
  (i) the characteristics in the domain of u which follow these directions and
  (ii) the value of u itself along these characteristics.

The only result that this analysis actually **proves** is the following statement: **If** u solves the PDE and auxiliary solution **and** $(x(s), y(s))$ solve the appropriate characteristic ODEs, **then** $z(s)$, defined to be $u(x(s), y(s))$, must solve the characteristic ODE for z. However, these steps give us a way of *synthesizing* the solution via the following procedure. We attempt to uniquely solve the characteristic ODEs with the hope that the projected characteristics both fill up the domain in question and do not intersect. This will be the case if we can **invert** the solutions for the projected characteristics in order to solve uniquely for the variables used to parametrize the characteristics (e.g., x_0 and s) in terms of the independent variables (e.g., x and y). Our canonical example of Burgers's equation illustrates that this inversion step can fail. However, in cases where one can invert these equations and find an explicit formula for u, we can (of course) check that it does indeed solve the PDE (though we usually skip this last step).

A natural general question to ask is: Assuming smoothness of all the involved ingredients — the coefficient functions of the PDE, the data curve (where the data is specified), and the data itself — does there at least **exist a local solution** (a solution defined in a neighborhood of the data curve)? The answer is yes assuming the data is noncharacteristic in the sense that the **transversality condition** holds. For the general quasilinear equation (2.24) in two independent variables with data g on the curve Γ described in parametric form by

$$\big(x_0(\tau), y_0(\tau)\big),$$

the transversality condition is

$$\det \begin{bmatrix} a\big(x_0(\tau), y_0(\tau), g(x_0(\tau), y_0(\tau))\big) & x_0'(\tau) \\ b\big(x_0(\tau), y_0(\tau), g(x_0(\tau), y_0(\tau))\big) & y_0'(\tau) \end{bmatrix} \neq 0.$$

The key tool in the proof is the **Inverse Function Theorem**.

2.5. The Method of Characteristics, Part III: General First-Order Equations

We now describe the method of characteristics in full generality. This section is **optional** so do **not** attempt to read it until you are very comfortable with the method of characteristics for linear and quasilinear equations (and have successfully completed several exercises). For general fully nonlinear first-order equations, interpreting the PDE as an expression for the directional derivative of u is far less clear then it was for the previous cases of linear and quasilinear equations. Here, we take a more **implicit** path to finding the characteristic equations, with an exposition which closely follows Evans [**10**], who in turn follows Carathéodory [**8**].

2.5.1. The Notation.

Your biggest foe in following this section is notation. If you can be in control of the notation and not let it get the better hand (i.e., confuse you to no end), you will be fine. The independent variables will be denoted by $x_1, \ldots, x_N$. We use $\mathbf{x} = (x_1, \ldots, x_N)$ to denote the full vector of independent variables which will lie in some potentially unbounded domain of $\mathbb{R}^N$. Perhaps, in an application, we would want to treat one of the independent variables as a temporal variable t; however, here for ease of notation, let us here simply denote **all** independent variables by x_i. We will look for a solution $u(x_1, \ldots, x_N)$ defined on some domain $\Omega \subset \mathbb{R}^N$ with data $g(\mathbf{x})$ supplied on some subset Γ of the boundary of Ω. Since Ω is N-dimensional, Γ will be $(N-1)$-dimensional.

We must write down a generic form of the most general first-order PDE. This might seem like a mess but it is actually very easy. We write the equation involving the independent variables, the dependent variable u, and its first-order partial derivatives simply as

$$\boxed{F(\nabla u(\mathbf{x}), u(\mathbf{x}), \mathbf{x}) = 0} \tag{2.28}$$

for some choice of C^1 function $F : \mathbb{R}^N \times \mathbb{R} \times \mathbb{R}^N \to \mathbb{R}$. The unknown solution $u(x_1, \ldots, x_N)$ will solve (2.28) at all $\mathbf{x}$ in its domain and satisfy $u(\mathbf{x}) = g(\mathbf{x})$ on Γ.

It is convenient to denote the first two arguments of the function F with variables $\mathbf{p}$ and z, respectively, where $\mathbf{p}$ denotes a vector in $\mathbb{R}^N$. For the PDE, we input $\nabla u(\mathbf{x})$ for $\mathbf{p}$ and $u(\mathbf{x})$ for z. Thus, we may write the PDE simply as

$$F(\mathbf{p}, z, \mathbf{x}) = 0. \tag{2.29}$$

Note that a particular PDE is equivalent to specifying the function F. Hence, for a particular PDE, F and its partial derivatives with respect to p_i, z and x_i are either given or can be readily computed. For example, the quasilinear PDE for $u(x_1, x_2, x_3)$

$$uu_{x_1} + u_{x_1}u_{x_2} + x_1 u_{x_3} + u^3 = x_2 x_3$$

is equivalent to (2.29) with

$$F(\mathbf{p}, z, \mathbf{x}) = F(p_1, p_2, p_3, z, x_1, x_2, x_3) = zp_1 + p_1 p_2 + x_1 p_3 + z^3 - x_2 x_3.$$

2.5.2. The Characteristic Equations. We proceed as before by parametrizing curves in the domain Ω of the solution by s. These characteristic curves (the projective characteristics) are denoted by $\mathbf{x}(s)$ and have components

$$\mathbf{x}(s) = (x_1(s), \ldots, x_N(s)), \quad s \in \mathbb{R}.$$

We will also denote derivatives with respect to s as a dot above the variable. Assume that we have a C^2 solution u to the PDE (2.28) on a domain Ω. Consistent with our previous notation, define $z(s)$ and $\mathbf{p}(s)$ to denote, respectively, the values of the solution and the gradient of the solution along a characteristic $\mathbf{x}(s)$; i.e.,

$$z(s) := u(\mathbf{x}(s)) \qquad \text{and} \qquad \mathbf{p}(s) := \nabla u(\mathbf{x}(s)).$$

Note again that $z(s)$ is scalar-valued whereas $\mathbf{p}(s)$ is vector-valued. We denote the component functions as

$$\mathbf{p}(s) = (p_1(s), \ldots, p_N(s)).$$

The reader might wonder why we have introduced the new characteristic curves p_i which keep track of partial derivatives along the projected characteristics $\mathbf{x}(s)$. In the previous cases of linear and quasilinear equations, we only needed $z(s)$ and $\mathbf{x}(s)$. However, for nonlinear equations the direction in which the projected characteristics should move can be tied to components of ∇u. Hence it is imperative that we keep them in the picture.

From our experience in the last few sections, we know that if the method of characteristics is going to work, then we should be able to compute the evolutions of u and its first-order derivatives along a characteristic $\mathbf{x}(s)$ using **only** information about their values along $\mathbf{x}(s)$, i.e., the values $z(s)$ and $\mathbf{p}(s)$. Let us first focus on the $\mathbf{p}(s)$, the gradient of u along $\mathbf{x}(s)$. Suppose we wish to compute how it should evolve along the characteristic. That is, we want to compute

$$\dot{p}_i(s) := \frac{dp_i}{ds}(s), \qquad i = 1, \ldots, N.$$

By the chain rule, we have

$$\dot{p}_i(s) = \frac{d}{ds}(u_{x_i}(\mathbf{x}(s))) = \sum_{j=1}^{N} u_{x_i x_j}(\mathbf{x}(s))\,\dot{x}_j(s), \qquad i = 1, \ldots, N. \tag{2.30}$$

Before continuing, let us pause for reflection since the crucial step will descend upon us quite quickly. Thus far we have assumed that u is a solution to the PDE and considered some, as yet to be specified, curve $\mathbf{x}(s)$ lying in the domain Ω of u. We defined the value and gradient of the solution along $\mathbf{x}(s)$ by $z(s)$ and $\mathbf{p}(s)$, respectively, and found that $\dot{p}_i(s)$ is given by (2.30). However, now we encounter an issue: The right-hand side contains second derivatives, which we do not keep track of along a characteristic. The only remedy to this issue can come from the mother PDE. The idea is as follows:

$$\begin{cases} \text{Choose } \mathbf{x}(s) \text{ in such a way that the PDE allows us to compute} \\ \dot{\mathbf{p}}(s) \text{ only from information about } \mathbf{x}(s),\ z(s), \text{ and } \mathbf{p}(s). \end{cases}$$

What information does the PDE give us? Note that the PDE (2.28) holds for all $\mathbf{x} \in \Omega$. In particular, using the chain rule we can differentiate (2.28) with respect to x_i to find

$$\sum_{j=1}^{N} \left(\frac{\partial F}{\partial p_j}(\nabla u, u, \mathbf{x})\, u_{x_j x_i} \right) + \frac{\partial F}{\partial z}(\nabla u, u, \mathbf{x})\, u_{x_i} + \frac{\partial F}{\partial x_i}(\nabla u, u, \mathbf{x}) = 0.$$

This equation must hold for all $\mathbf{x} \in \Omega$. In particular, it must hold along $\mathbf{x}(s)$ and hence

$$\sum_{j=1}^{N} \left(\frac{\partial F}{\partial p_j}(\mathbf{p}(s), z(s), \mathbf{x}(s)) u_{x_j x_i}(\mathbf{x}(s)) \right) + \frac{\partial F}{\partial z}(\mathbf{p}(s), z(s), \mathbf{x}(s)) p_i(s)$$
$$+ \frac{\partial F}{\partial x_i}(\mathbf{p}(s), z(s), \mathbf{x}(s)) = 0. \tag{2.31}$$

Note that since u is a C^2 function, we always have $u_{x_j x_i} = u_{x_i x_j}$.

Now comes the **key observation**: If we want to be able to compute the right-hand side of (2.30) using only the information from (2.31), we will need to choose $\mathbf{x}(s)$ such that

$$\boxed{\dot{x}_j(s) = \frac{\partial F}{\partial p_j}(\mathbf{p}(s), z(s), \mathbf{x}(s)), \qquad j = 1, \ldots, N.} \tag{2.32}$$

In doing so, we can use (2.31) to find that

$$\dot{p}_i(s) = -\frac{\partial F}{\partial x_i}(\mathbf{p}(s), z(s), \mathbf{x}(s)) - \frac{\partial F}{\partial z}(\mathbf{p}(s), z(s), \mathbf{x}(s)) p_i(s), \qquad i = 1, \ldots, N. \tag{2.33}$$

In fact, any other choice of $\dot{x}_j(s)$ would result in a curve upon which we will **not** be able to compute $\dot{p}_i(s)$, for a given value of s, from $\mathbf{x}(s)$, $z(s)$, and $\mathbf{p}(s)$ using the PDE alone.

Turning to $\dot{z}(s)$, we have

$$\dot{z}(s) = \frac{d}{ds}u(\mathbf{x}(s)) = \sum_{j=1}^{N} \frac{\partial u}{\partial x_j}(\mathbf{x}(s))\dot{x}_j(s) = \sum_{j=1}^{N} p_j(s)\frac{\partial F}{\partial p_j}(\mathbf{p}(s), z(s), \mathbf{x}(s)). \tag{2.34}$$

So, what do we have so far? Equations (2.32), (2.33), and (2.34) constitute a system of $2N+1$ ordinary differential equations in $2N+1$ unknown functions. In other words, by choosing the characteristics to move in the direction dictated by (2.32), we obtain a **closed** system for all the relevant quantities. We summarize these below.

The Characteristic Equations

The system of equations

$$\begin{cases} \dot{x}_i(s) = \frac{\partial F}{\partial p_i}(\mathbf{p}(s), z(s), \mathbf{x}(s)), \\ \\ \dot{z}(s) = \sum_{j=1}^{N} p_j(s)\frac{\partial F}{\partial p_j}(\mathbf{p}(s), z(s), \mathbf{x}(s)), \\ \\ \dot{p}_i(s) = -\frac{\partial F}{\partial x_i}(\mathbf{p}(s), z(s), \mathbf{x}(s)) - \frac{\partial F}{\partial z}(\mathbf{p}(s), z(s), \mathbf{x}(s))p_i(s), \end{cases} \tag{2.35}$$

where for $i = 1, \ldots, N$, are known as **the characteristic equations**. They are coupled ordinary differential equations for: (i) the characteristics which are curves lying in the domain space of the solution, (ii) the value of the solution along the characteristics, and (iii) the first derivatives of the solution along the characteristics. We can write the characteristic equations in vector form as

$$\begin{cases} \dot{\mathbf{x}}(s) = \nabla_{\mathbf{p}}F(\mathbf{p}(s), z(s), \mathbf{x}(s)), \\ \\ \dot{z}(s) = \nabla_{\mathbf{p}}F(\mathbf{p}(s), z(s), \mathbf{x}(s)) \cdot \mathbf{p}(s), \\ \\ \dot{\mathbf{p}}(s) = -\nabla_{\mathbf{x}}F(\mathbf{p}(s), z(s), \mathbf{x}(s)) - \frac{\partial F}{\partial z}(\mathbf{p}(s), z(s), \mathbf{x}(s))\, \mathbf{p}(s). \end{cases} \tag{2.36}$$

Here $\nabla_{\mathbf{p}}F$ denotes the gradient of F with respect to the $\mathbf{p}$ components. The curves $\mathbf{x}(s)$ which lie in the domain Ω of the solution u are often called the **projected characteristics** in order to differentiate them from the curves $z(s)$ and $\mathbf{p}(s)$, which are also often referred to as characteristics.

We make two comments:

1. It is true that once we know z and $\mathbf{x}$ we are in principle done; i.e., we can attempt to solve for u. However, for fully nonlinear PDEs, solving for $\mathbf{p}$ is imperative to finding z and $\mathbf{x}$. As we see next, if the PDE is linear or quasilinear, we do not need the $\mathbf{p}$ equation to find z and $\mathbf{x}$; the z and $\mathbf{x}$ equations decouple from the $\mathbf{p}$ equation, making the latter redundant.
2. After deriving the characteristic equations (2.36), it is insightful to reflect on what exactly we have shown (proven). Precisely, we have proven that **if** u is a C^2 solution to the PDE (2.28) on a domain Ω and if $\mathbf{x}(s)$ solves the system of ODEs given

by (2.32) **with**

$$\mathbf{p}(s) = \nabla u(\mathbf{x}(s)) \quad \text{and} \quad z(s) = u(\mathbf{x}(s)),$$

then $\mathbf{p}(s)$ solves (2.33) and $z(s)$ solves (2.34).

Of course, in the method of characteristics we attempt to do the reverse, i.e., solve the characteristic ODEs and use their solutions to **construct** u, the solution to the PDE (2.28).

2.5.3. Linear and Quasilinear Equations in N independent variables.

The previous cases of linear, semilinear, and quasilinear equations all fall into our general framework. In fact, now we can consider these equations in N independent variables, denoted here by the vector $\mathbf{x}$.

A general **linear** equation takes the form

$$F(\nabla u, u, \mathbf{x}) = \mathbf{a}(\mathbf{x}) \cdot \nabla u(\mathbf{x}) + b(\mathbf{x})u(\mathbf{x}) + c(\mathbf{x}) = 0,$$

for some vector-valued function $\mathbf{a}$ and scalar-valued functions b and c of $\mathbf{x}$. In our notation, we have

$$F(\mathbf{p}, z, \mathbf{x}) = \mathbf{a}(\mathbf{x}) \cdot \mathbf{p} + b(\mathbf{x})z + c(\mathbf{x}),$$

which means $\nabla_{\mathbf{p}} F = \mathbf{a}(\mathbf{x})$. Hence the characteristic equations (2.36) for $\mathbf{x}$ and z are

$$\dot{\mathbf{x}}(s) = \mathbf{a}(\mathbf{x}(s)), \qquad \dot{z}(s) = \mathbf{a}(\mathbf{x}(s)) \cdot \mathbf{p}(s).$$

However, the PDE tells us that $\mathbf{a}(\mathbf{x}(s)) \cdot \mathbf{p}(s) = -b(\mathbf{x}(s))z(s) - c(\mathbf{x}(s))$. Hence, the z and $\mathbf{x}$ equations form a closed system

$$\begin{cases} \dot{\mathbf{x}}(s) = \mathbf{a}(\mathbf{x}(s)), \\ \dot{z}(s) = -b(\mathbf{x}(s))z(s) - c(\mathbf{x}(s)), \end{cases} \tag{2.37}$$

and we do not need $\mathbf{p}$ or its equations to find u. Of course, one could write down the $\mathbf{p}$ equations but they would be redundant and will be automatically satisfied by the solution u along the characteristics $\mathbf{x}$. A very important observation about the ODEs for z and $\mathbf{x}$ in (2.37) is that they are **decoupled**; we can first solve for $\mathbf{x}(s)$ and then solve for $z(s)$. Another way of saying this is that the structure of the PDE alone (i.e., the function F) tells us exactly what the characteristics $\mathbf{x}(s)$ are.

A general **quasilinear** equation takes the form

$$F(\nabla u, u, \mathbf{x}) = \mathbf{a}(\mathbf{x}, u) \cdot \nabla u(\mathbf{x}) + b(\mathbf{x}, u) = 0$$

which in our notation can be written as

$$F(\mathbf{p}, z, \mathbf{x}) = \mathbf{a}(\mathbf{x}, z) \cdot \mathbf{p} + b(\mathbf{x}, z)$$

for some vector-valued function $\mathbf{a}$ and scalar-valued function b of $(\mathbf{x}, z)$. The characteristic equations (2.36) for $\mathbf{x}$ and z are

$$\begin{cases} \dot{\mathbf{x}}(s) = \mathbf{a}(\mathbf{x}(s), z(s)), \\ \dot{z}(s) = -b(\mathbf{x}(s), z(s)), \end{cases}$$

and we still do not need the $\mathbf{p}$ equations to find u. However, this time the $\mathbf{x}$ and z equations are coupled. Thus, we need information about the solution z to find the characteristics $\mathbf{x}(s)$. In particular, we cannot say anything about $\mathbf{x}(s)$ unless we are given some auxiliary condition (data) for the solution. In the next subsections we will see the role of the **p** equations for several fully nonlinear examples.

2.5.4. Two Fully Nonlinear Examples.

We present two fully nonlinear examples wherein the **p** equation becomes vital.

Example 2.5.1 (from Evans [**10**]). Solve for $u(x_1, x_2)$ where

$$u_{x_1} u_{x_2} = u \quad \text{on} \quad \Omega = \{(x_1, x_2) \mid x_1 > 0\}, \qquad \text{with} \quad u(0, x_2) = x_2^2.$$

In our notation, $N = 2$ and $F(\mathbf{p}, z, \mathbf{x}) = F(p_1, p_2, z, x_1, x_2) = p_1 p_2 - z$. Hence, the characteristic equations (2.35) become

$$\begin{cases} \dot{x}_1(s) = p_2(s), \qquad \dot{x}_2(s) = p_1(s), \\ \dot{z}(s) = 2p_1(s)p_2(s), \\ \dot{p}_1 = p_1, \qquad \dot{p}_2 = p_2. \end{cases}$$

We solve these with the conditions

$$x_1(0) = 0, \qquad x_2(0) = x_2^0,$$

and

$$z(0) = z_0 = \left(x_2^0\right)^2, \qquad p_1(0) = p_1^0 = u_{x_1}(0, x_2^0), \qquad p_2(0) = p_2^0 = u_{x_2}(0, x_2^0).$$

Let us assume for the moment that we will be able to find p_1^0 and p_2^0. It is easy to solve the **p** equations first and obtain $p_1(s) = p_1^0 e^s$ and $p_2(s) = p_2^0 e^s$. With these in hand, we can solve the **x** and z equations to find

$$x_1(s) = p_2^0(e^s - 1), \qquad x_2(s) = x_2^0 + p_1^0(e^s - 1),$$

$$z(s) = z_0 + p_1^0\, p_2^0(e^{2s} - 1).$$

What do we take for p_1^0 and p_2^0? The only information we have is either from the data or from the PDE. From the data, we note that $u = x_2^2$ on Γ, and, therefore $p_2^0 = u_{x_2}(0, x_2^0) = 2x_2^0$. On the other hand, the PDE $u_{x_1} u_{x_2} = u$ implies that $p_1^0\, p_2^0 = z_0 = \left(x_2^0\right)^2$ so that $p_1^0 = x_2^0/2$.

Hence, we have

$$x_1(s) = 2x_2^0(e^s - 1), \qquad x_2(s) = \frac{x_2^0}{2}(e^s + 1), \qquad z(s) = \left(x_2^0\right)^2 e^{2s},$$

and we no longer need p_1 and p_2.

Finally, given a point in the plane (x_1, x_2) with $x_1 > 0$, we solve for the *corresponding* x_2^0 and s. Don't forget that this means that we find the characteristic curve indexed by x_2^0 which contains the point (x_1, x_2) and the value of s where it passes through (x_1, x_2). Hence, we need to solve

$$\begin{cases} x_1 = 2x_2^0(e^s - 1), \\ x_2 = \frac{x_2^0}{2}(e^s + 1) \end{cases}$$

for x_2^0 and s. Note, however, e^s also appears in z so we can save ourselves some work by solving instead for x_2^0 and e^s. To this end, we find

$$x_2^0 = \frac{4x_2 - x_1}{4} \qquad \text{and} \qquad e^s = \frac{x_1 + 4x_2}{4x_2 - x_1},$$

and the resulting solution is

$$u(x_1, x_2) = z(s) = \frac{(x_1 + 4x_2)^2}{16}.$$

Example 2.5.2. We solve for $u(x,t)$ on $x \in \mathbb{R}$, $t \geq 0$,

$$u_t + (u_x)^2 = 0, \qquad u(x,0) = x^2.$$

This is a simple example of a Hamilton-Jacobi equation (cf. Section 2.5.6). Here $\mathbf{x}$ is the 2D vector (x,t) and p_2 will denote the derivative of the solution u with respect to t. The PDE becomes $F(\mathbf{p}, z, \mathbf{x}) = F(p_1, p_2, z, x_1, t) = (p_1)^2 + p_2$.

As before, we use t to parametrize the characteristics. Denoting the t-derivative with the overhead dot, the characteristic equations (2.35) are

$$\dot{x} = 2p_1 \text{ with } x(0) = x_0, \qquad \dot{z} = 2p_1^2 + p_2 \text{ with } z(0) = (x_0)^2,$$

$$\dot{p}_1 = 0 \text{ with } p_1(0) = 2x_0, \qquad \dot{p}_2 = 0 \text{ with } p_2(0) = -4(x_0)^2.$$

Here we obtained the initial values of p_1 and p_2 by the exact same reasoning as in the previous example. Thus, $p_1(t) = 2x_0$, $p_2(t) = -4(x_0)^2$, and consequently

$$x(t) = 4x_0 t + x_0.$$

For the z equation we have $\dot{z}(t) = 4(x_0)^2$ and so $z(s) = 4(x_0)^2 t + (x_0)^2$. Finally, given x and $t > 0$ we have

$$x_0 = \frac{x}{4t+1}$$

and consequently

$$u(x,t) = (4t+1)(x_0)^2 = \frac{x^2}{4t+1}.$$

This solves the problem on $x \in \mathbb{R}$, $t \geq 0$.

2.5.5. The Eikonal Equation. For illustration purposes, we focus on dimension 2. Let Ω be a domain in $\mathbb{R}^2$ with boundary $\Gamma = \partial\Omega$. For example, Ω could be a disk with $\partial\Omega$ as its boundary circle. Given any function $f(x,y)$ defined in Ω, consider the PDE

$$|\nabla u| = f \quad \left(\text{or equivalently } \sqrt{(u_x)^2 + (u_y)^2} = f\right) \quad \text{in } \Omega,$$

with

$$u(x,y) = 0 \text{ for } (x,y) \in \partial\Omega.$$

This fully nonlinear first-order PDE is known as an **eikonal equation**. The German word *eikonal* is derived from the Greek word for *image*. The equation is important in the study of wave propagation, particularly in geometric optics (a description of the propagation of light by "rays"). Note that this is our first example in this chapter where data is supplied on the entire boundary of a bounded set — a true boundary value problem (BVP). We will consider the special case $f(x,y) \equiv 1$, which gives the BVP

$$\begin{cases} |\nabla u| = 1 & \text{in } \Omega, \\ u = 0 & \text{on } \partial\Omega. \end{cases} \tag{2.38}$$

We will keep here the x, y notation instead of x_1, x_2 but will still denote the respective gradients by p_1 and p_2. In our convention, the PDE is

$$F(p_1, p_2, z, x, y) = \sqrt{(p_1)^2 + (p_2)^2} - 1 = 0,$$

and hence, the characteristic equations (2.35) become

$$\dot{x}(s) = \frac{p_1(s)}{\sqrt{p_1^2(s) + p_2^2(s)}}, \qquad \dot{y}(s) = \frac{p_2(s)}{\sqrt{p_1^2(s) + p_2^2(s)}},$$

$$\dot{z}(s) = 1, \qquad \dot{p}_1(s) = 0, \qquad \dot{p}_2(s) = 0.$$

Since the PDE tells us that $\sqrt{p_1^2(s) + p_2^2(s)} = 1$, the first two equations become

$$\dot{x}(s) = p_1(s), \qquad \dot{y} = p_2(s).$$

The "initial" points $x(0) = x_0, y(0) = y_0$ are chosen such that $(x_0, y_0) \in \partial\Omega$. The auxiliary condition $u = 0$ on $\partial\Omega$ tells us two things about $z(0)$ and $p_1(0) = p_1^0$, $p_2(0) = p_2^0$. The first is that $z(0) = 0$ and the second is that the gradient of u along the boundary must be perpendicular to $\partial\Omega$ (a consequence of u vanishing on $\partial\Omega$). Thus, (p_1^0, p_2^0) is orthogonal to $\partial\Omega$ at the point (x_0, y_0), and the solutions to these ODEs are

$$p_1(s) = p_1^0, \qquad p_2(s) = p_2^0, \qquad z(s) = s, \qquad x(s) = p_1^0 s, \qquad y(s) = p_2^0 s.$$

In particular, the characteristics $(x(s), y(s))$ are lines which are orthogonal to $\partial\Omega$ at their intersection point (x_0, y_0).

What can we infer from these equations? As usual, given a point $(x, y) \in \Omega$, we need to find the characteristic (i.e., the corresponding $(x_0, y_0) \in \partial\Omega$) which passes through the point and the corresponding value of parameter s. Since (p_1^0, p_2^0) is a unit vector, s is exactly the distance along the line segment starting at $(x_0, y_0) \in \partial\Omega$ and ending at (x, y). Hence, the solution $u(x, y)$ is simply the **minimum distance** from (x, y) to the boundary $\partial\Omega$. However, we have a problem when this minimum distance to the boundary is attained at two or more points on the boundary. In this instance, characteristics intersect and a discontinuity emerges in derivatives of the solution. Therefore the **distance function** provides us with a continuous but possibly nonsmooth solution to the eikonal equation, i.e., a solution except at the set of points in Ω whose closest point to $\partial\Omega$ is attained at more than one point. Figure 2.14 illustrates the distance function solution for the case where Ω is the inside of the unit square.

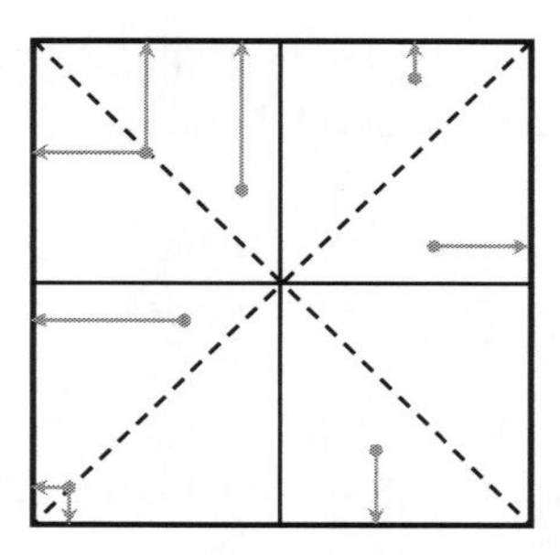

Figure 2.14. Some characteristics for the eikonal equation (2.38) where Ω is the unit square. Note that the distance function to the boundary is singular along the dashed lines.

One can also extend this distance function solution to all of $\mathbb{R}^2$ by considering the **signed distance function**: At any point $(x, y) \in \mathbb{R}^2$, the function is simply the closest distance to $\partial\Omega$ with the convention that it is positive if $(x, y) \in \Omega$ and negative if $(x, y) \in \Omega^c$.

In the context of optics, the characteristics are *light rays* — lines (or curves) perpendicular to the wave front of light. For the simple eikonal equation (2.38), the wave fronts are level sets (curves) of the distance function (to the boundary). The eikonal equation can be derived as an approximation for the wave equation in a certain asymptotic regime appropriate for visible light. This approximation and derivation are known as *geometric optics* and will be addressed in Section 4.4.4.

One can also see the appearance of the distance function and its characteristics in many, many other contexts, for example in **agates**[6]

2.5.6. Hamilton-Jacobi Equations. We consider Hamilton-Jacobi equations[7] in one space variable. To this end, let H be a function (called the **Hamiltonian**) of two scalar variables and consider the initial value problem for $u(x, t)$ with $t \geq 0$

$$\begin{cases} u_t + H(u_x, x) = 0, \\ u(x, 0) = g(x). \end{cases} \tag{2.39}$$

Our previous Example 2.5.2 found an explicit solution for the case $H(p, x) = p^2$ and $g(x) = x^2$. Finding solutions for general H and g is far more complicated. In general, solutions can develop discontinuities, and one needs a weaker notion of a solution together with new techniques beyond the method of characteristics. We will discuss this notion in a future volume planned for this subject. At this point, let us just

[6]**Agate** is banded *chalcedony*, a microcrystalline form of the ubiquitous mineral *quartz* (silica or silicon dioxide SiO_2). Microcrystalline means that the crystals are so small that they can only be seen microscopically and not by the naked eye. Agates were formed millions of years ago in cavities of volcanic rock where silica-rich water seeped through the boundary of the host rock and deposited minerals on the inside of the cavities. Over time, these layers of silica-rich material built up and filled the cavity. These layers often have subtle differences in mineral content giving the agate the banded structure. In many specimens (often referred to as geodes), the cavity is not completely filled and the inner layer consists of a (macro) crystalline structure of quartz, for example, amethyst (a purple variety of quartz). Agates are found all over the world. The ones on the book cover come from the Sonoran Desert of Southern California.

[7]**William Rowan Hamilton** (1805–1865) was an Irish mathematician who made many important contributions to classical mechanics, algebra, and optics. **Carl Gustav Jacob Jacobi** (1804–1851) was a German mathematician who made many important contributions to differential equations and dynamics, algebra, and number theory.

write down the characteristic ODEs (2.35), two of which will constitute a fundamental system of equations called a **Hamiltonian system**.

The only issue here is to overcome notational difficulties associated with the fact that our independent variables $\mathbf{x}$ are denoted in component form by (x, t). Hence p_1 refers to u_x and p_2 to u_t with the vector $\mathbf{p}$ denoting (p_1, p_2). The Hamilton-Jacobi equation in algebraic form is simply

$$F(p_1, p_2, z, x, t) = p_2 + H(p_1, x) = 0.$$

Hence, the characteristic equations (2.35) become

$$\dot{x}(s) = \frac{\partial H(p_1(s), x(s))}{\partial p_1}, \qquad \dot{t}(s) = 1,$$

$$\dot{z}(s) = p_1(s)\frac{\partial H(p_1(s), x(s))}{\partial p_1} + p_2(s), \qquad \dot{p}_1(s) = -\frac{\partial H(p_1(s), x(s))}{\partial x}, \qquad \dot{p}_2(s) = 0.$$

As usual $t = s$ and we use t to parametrize the characteristics. Of central importance are the two key coupled equations

$$\begin{cases} \frac{dx(t)}{dt} = \frac{\partial H(p_1(t), x(t))}{\partial p_1}, \\ \\ \frac{dp_1(t)}{dt} = -\frac{\partial H(p_1(t), x(t))}{\partial x}. \end{cases} \tag{2.40}$$

These equations are known as a **Hamiltonian system**. Using the PDE, we can write the z equation as

$$\frac{dz(t)}{dt} = p_1(t)\frac{\partial H(p_1(t), x(t))}{\partial p_1} - H(p_1(t), x(t)).$$

Hence, once the Hamiltonian system (2.40) is solved, we can find $z(t)$ by direct integration with respect to t. Take note that we did not need the equation $\dot{p}_2(t) = 0$ which tells us that p_2 is constant along a characteristic. However, this information highlights an important fact about the characteristics; since the PDE states that $p_2(t) + H(p_1(t), x(t)) = 0$, it follows that

$$H(p_1(t), x(t)) \text{ is independent of } t.$$

Hence, **the Hamiltonian is conserved along a characteristic**.

2.5.7. The Level Set Equation and Interface Motion. The following derivation and discussion can be done in any space dimension; for simplicity we choose two dimensions. We wish to describe the evolution of curves in $\mathbb{R}^2$. It is often useful to think of these curves as **interfaces** between two types of evolving media (for example, two fluids). A convenient way to describe a curve in $\mathbb{R}^2$ is as the level set of a function of two variables $\phi_0(x, y)$, that is, the curve given by

$$\Gamma_0 := \left\{(x, y) \in \mathbb{R}^2 \mid \phi_0(x, y) = 0\right\}.$$

To model the evolution of such curves, it suffices to model the evolution of the **level set functions**. To this end, let us now consider a function of three variables $\phi(x, y; t)$ with $\phi(x, y; 0) = \phi_0(x, y)$. At any time $t > 0$, the curve Γ_0 has now moved to the level set of $\phi(x, y; t)$; i.e., the curve

$$\Gamma(t) := \left\{(x, y) \in \mathbb{R}^2 \mid \phi(x, y; t) = 0\right\}.$$

Note the use of the semicolon in ϕ to denote the difference in the use of the independent variables (x, y) vs. t. We think of t as time and $\Gamma(t)$ as the evolution of curves (interfaces) which we will now describe via PDEs for the level set functions $\phi(x, y; t)$.

As a warm-up problem, let $\mathbf{v}$ be a fixed two-dimensional vector and consider the IVP

$$\phi_t + \mathbf{v} \cdot \nabla\phi = 0, \qquad \phi(x, y; 0) = \phi_0. \tag{2.41}$$

This linear PDE is a two-dimensional version of the transport equation discussed in Section 2.2.4. We explicitly solved this equation; however, the readers should convince themselves now that from the perspective of interface motion, the PDE simply dictates that the level set of ϕ_0 is translated in space with **constant velocity** $\mathbf{v}$ (cf. Exercise **2.33**(a)).

We could now let the velocity vector $\mathbf{v}$ depend on space (x, y) (or even time t) and, while we would still arrive at a linear PDE, the shape of the initial curve would now change with time. However, fundamental to many interface problems in physics and geometry are the nonlinear cases when the interfacial velocity is always in the direction normal to the interface (the normal direction $\mathbf{n}$). In terms of ϕ, this means that for some scalar v,

$$\mathbf{v} = v\mathbf{n} = v\left(\pm\frac{\nabla\phi}{|\nabla\phi|}\right). \tag{2.42}$$

Let us choose the positive sign. Note that this velocity depends on the dependent variable ϕ; hence it will depend on space and time but implicitly via ϕ. Placing (2.42) into (2.41), we find

$$\phi_t + \mathbf{v} \cdot \nabla\phi = \phi_t + v\frac{\nabla\phi}{|\nabla\phi|} \cdot \nabla\phi = \phi_t + v|\nabla\phi| = 0.$$

We have arrived at the fully nonlinear PDE known as the **level set equation**:

$$\boxed{\phi_t + v|\nabla\phi| = 0.} \tag{2.43}$$

The level set equation is a particular example of a Hamilton-Jacobi equation in two space variables. It has proven to be surprisingly useful in modeling a wide class of interface problems: crystal growth, flame dynamics (the spread of a forest fire), the shape of oil droplets in water, computer vision, and more. The scalar speed v could be a constant, a function of space and/or time, or a function of ϕ itself.

It is insightful to compare the solution of the level set equation with that of the purely spatial eikonal equation. Recall from Section 2.5.5 the distance function to $\partial\Omega$ as a solution to the eikonal equation. Now, for simplicity, set the speed v to be the constant 1 and let $\phi_0(x, y)$ be the distance function solution to

$$|\nabla\phi_0| = 1 \text{ in } \Omega, \qquad \phi_0 = 0 \text{ on } \partial\Omega.$$

Then

$$\phi(x, y; t) = \phi_0(x, y) - t$$

is a solution to the level set equation

$$\phi_t + |\nabla\phi| = 0, \qquad \phi(x, y; 0) = \phi_0(x, y).$$

Hence, one can view the interface motion of $\partial\Omega$ as motion with unit speed in the direction of the distance function.

A particularly important example of (2.43) wherein v depends on ϕ itself is the case where the speed v is proportional to the **curvature of the interface**. A very important calculation in differential geometry gives the curvature (mean curvature in higher space dimensions) as the **divergence of the unit normal**. In terms of the level set function ϕ, this means that the curvature κ of the level set of $\phi(x, y; t)$ is given by

$$\kappa = \operatorname{div}\left(\frac{\nabla\phi}{|\nabla\phi|}\right).$$

Hence, let us now place

$$v = -\operatorname{div}\left(\frac{\nabla\phi}{|\nabla\phi|}\right) \tag{2.44}$$

into the level set equation (2.43) to obtain the second-order PDE

$$\phi_t - \operatorname{div}\left(\frac{\nabla\phi}{|\nabla\phi|}\right)|\nabla\phi| = 0. \tag{2.45}$$

The associated interfacial flow is known as **motion by mean curvature**[8]. Note that it is no longer a first-order nonlinear PDE; rather it is second order. The choice of minus sign in the speed in (2.44) ensures the flow is inward; for example the flow of a circle of a fix radius will produce circles with decreasing radii.

In 1988, Stanley Osher and James Sethian wrote a tremendously influential paper[9] introducing numerical methods for solving level set equations and motion by mean curvature. The reader is encouraged to research online level set simulations in a variety of physical contexts — many videos of interface dynamics are readily available. Readers may also consult Sethian's book[10] for further details.

2.6. • Some General Questions

In the previous sections, the method of characteristics has been extensively used to derive closed form solutions (i.e., explicit formulas) to many first-order PDEs. Several questions need to be addressed:

- In all our examples, the characteristic ODEs could easily be solved in closed form. Is this generic? The answer is **no**. In general, the system of ODEs would need to either be addressed with more sophisticated analytical techniques or be solved numerically.
- Can at least the simplest case of a linear equation with appropriately chosen data always be solved by the method of characteristics? The answer is **no**. We have already seen cases where the solutions to the ODEs for either the characteristics or the solution along the characteristics blow up. Degeneracy in the linear PDE, i.e., a point (x_0, y_0) where $a(x, y)$ and $b(x, y)$ simultaneously vanish, will significantly affect the nature of the solution around (x_0, y_0). Consider, for example, the case of $xu_x + yu_y = 0$. The characteristics are straight lines passing through the origin and any solution must be constant along each characteristic.

---

[8] In 2D, it is simply motion by curvature.

[9] Fronts Propagating with Curvature-Dependent Speed: Algorithms Based on Hamilton–Jacobi Formulations, S. Osher and J.A. Sethian, Journal of Computational Physics, 79, pp. 12–49, 1988.

[10] J.A. Sethian, Level Set Methods and Fast Marching Methods: Evolving Interfaces in Computational Geometry, Fluid Mechanics, Computer Vision, and Materials Science, Cambridge University Press, 1999.

- In our linear and in some of our quasilinear and fully nonlinear examples, we were able to solve uniquely for the parameters in the characteristics in terms of the independent variables. Is this generic? The answer, again, is **no**. For example, as we saw from Burgers's equation, characteristics can collide and it may not be possible to find a solution in a given domain. As we shall see in a future volume planned for this subject, methods, which are physically and mathematically natural, exist for generalizing the notion of a solution to include discontinuities associated with intersecting characteristics.
- Generically (for most first-order PDEs), a closed form solution is not possible, which begs the question as to whether we can at least use the method of characteristics to **prove the existence** of a solution in some region of the domain which encompasses the data set Γ. The answer is **yes**, with some caveats. Given a first-order PDE in the form of a function F and some data for the solution on a set Γ we would need certain additional assumptions: (i) on the function F defining the PDE in terms of smoothness and nondegeneracy; (ii) on the smoothness on the data defined on Γ, and the set Γ itself; (iii) and finally, the data curve Γ must never be in a characteristic direction. That is, no portion of it can overlap with one of the characteristic curves in the domain. In this way, we say Γ is **noncharacteristic**.

  Even with these assumptions, characteristics can collide and the solution can become overdetermined (cf. Burgers's equation). Hence, we will only be able to prove the existence of a **local** solution in some domain of Ω which contains Γ (a neighborhood of the data Γ). Such a **local existence theorem** is by no means easy to prove; see for example Section 3.2 of Evans [**10**]. Together with an existence and uniqueness for systems of ODEs, a key step (tool) is an inversion to solve for the characteristic parameters. For this step, one needs the **Inverse Function Theorem**.

2.7. • A Touch of Numerics, I: Computing the Solution of the Transport Equation

> *The purpose of computing is insight, not numbers.*
>
> **– Richard Hamming**[a]
>
> [a]Richard Wesley Hamming (1915–1998) was an American mathematician whose work had an enormous impact on computer engineering and telecommunications.

The numerical approximation of solutions to partial differential equations has transformed our ability to gain inference into many, otherwise intractable, PDEs. While in this text we will not address this vast subject in any detail, it is instructive to focus for the moment on computing the solution to the transport equation with one of the most widely used numerical methods, **the finite difference method**. We will focus on the simple, well-posed problem for the transport equation:

$$u_t + cu_x = 0 \quad \text{with} \quad c > 0, \qquad u(x,0) = g(x). \tag{2.46}$$

This problem has an explicit solution; in other words, everything is known. As such, it presents one with a perfectly accessible vehicle to address some basic tenets of the numerical analysis of PDEs. Indeed, our purpose here is to show that, even with simple numerical approximation of derivatives, the following are true:

- One cannot blindly apply a numerical scheme (a method based upon some discrete numerical approximation of derivatives) without knowledge about the structure of the particular PDE.
- There are always certain stability requirements pertaining to our particular numerical approximation which can lead to nonsense unless the proper analysis is first performed.

Thus the following analysis of this elementary PDE gives simple, yet convincing, support to a fundamental warning for all the future practitioners of PDEs:

> Blindly applying numerical recipes to a PDE can be dangerous, without some underlying grasp of the particular character of the PDE.

Based upon this section, we will briefly return to finite difference computations for the wave, diffusion, and Laplace equations in Sections 3.11, 7.10, and 8.6, respectively. However, for more details on the finite difference method the reader is encouraged to consult the book of LeVeque.[11]

2.7.1. • Three Consistent Schemes.

We begin by **discretizing** space and time. This means that we replace the continuums of space and time with a finite, or at most countable, collection of positions and times. Specifically, we do the following:

- Fix values of $\Delta x > 0$ and $\Delta t > 0$ called the spatial and temporal **step size**.
- Consider "grid points" $x_j = j\Delta x$ and $t_n = n\Delta t$ with $j = 0, \pm 1, \pm 2, \ldots$ and $n = 0, 1, 2, \ldots$.

Numerically computing the solution means that we find (approximate) values of the true solution at these particular values of x and t. That is, we then attempt to find the solution at the grid points

$$\boxed{U_j^n := u(j\Delta x, n\Delta t).} \tag{2.47}$$

In the **finite difference method**, we approximate the values of the partial derivatives at a grid point, based upon values at neighboring grid points. For example, a natural way to **approximate** u_t is by a *forward finite difference*

$$u_t(j\Delta x, n\Delta t) \approx \frac{u(j\Delta x, n\Delta t + \Delta t) - u(j\Delta x, n\Delta t)}{\Delta t} = \frac{U_j^{n+1} - U_j^n}{\Delta t}. \tag{2.48}$$

This discretization is often referred to as **forward Euler** and is justified by Taylor series; assuming u is smooth, the error in the approximation tends to zero as $\Delta t \to 0$. For

[11] **Randall LeVeque**, *Finite Difference Methods for Ordinary and Partial Differential Equations: Steady-State and Time-Dependent Problems*, Society for Industrial and Applied Mathematics, 2007.

the approximation of u_x, we have a few choices (all justified by Taylor series):

Forward finite difference:

$$u_x(j\Delta x, n\Delta t) \approx \frac{u(j\Delta x + \Delta x, n\Delta t) - u(j\Delta x, n\Delta t)}{\Delta x} = \frac{U_{j+1}^n - U_j^n}{\Delta x}.$$

Backward finite difference:

$$u_x(j\Delta x, n\Delta t) \approx \frac{u(j\Delta x, n\Delta t) - u(j\Delta x - \Delta x, n\Delta t)}{\Delta x} = \frac{U_j^n - U_{j-1}^n}{\Delta x}.$$

Centered finite difference:

$$u_x(j\Delta x, n\Delta t) \approx \frac{u(j\Delta x + \Delta x, n\Delta t) - u(j\Delta x - \Delta x, n\Delta t)}{2\Delta x} = \frac{U_{j+1}^n - U_{j-1}^n}{2\Delta x}.$$

You should reflect on the choice of names: forward, backward, and centered.

Once we implement one of these choices for the spatial derivatives, solving the PDE numerically boils down to:

(i) solving a system of algebraic equations (which can be nonlinear for nonlinear PDEs) which allows us to compute the discrete solution at time $(n+1)\Delta t$ from information about the discrete solution at time $n\Delta t$;

(ii) iterating step (i).

For example, with the forward difference in space, the PDE (2.46) on the grid amounts to solving

$$\frac{U_j^{n+1} - U_j^n}{\Delta t} + c\,\frac{U_{j+1}^n - U_j^n}{\Delta x} = 0$$

which, solving for U_j^{n+1}, yields

$$U_j^{n+1} = U_j^n - c\frac{\Delta t}{\Delta x}\left(U_{j+1}^n - U_j^n\right).$$

We call this an **explicit scheme** since the value of the solution at time step $n+1$ can be inferred directly from values of the solution at the earlier time step n. There is an important dimensionless quantity which measures the relative size of the grid:

$$r = c\frac{\Delta t}{\Delta x}.$$

In terms of r, our forward difference in space scheme is

$$\boxed{U_j^{n+1} = U_j^n - r\left(U_{j+1}^n - U_j^n\right).} \tag{2.49}$$

On the other hand, the backward difference in space scheme is

$$\boxed{U_j^{n+1} = U_j^n - r\left(U_j^n - U_{j-1}^n\right),} \tag{2.50}$$

while the centered difference in space scheme is

$$\boxed{U_j^{n+1} = U_j^n - \frac{r}{2}\left(U_{j+1}^n - U_{j-1}^n\right).} \tag{2.51}$$

For any fixed value of r, all three schemes are **consistent** with the PDE. This means that we have used an approximation to the derivatives in the PDE which gets more and more accurate as the Δx and Δt tend to 0. Specifically, the definition of a **consistent scheme** is as follows. Let $u(x,t)$ be any smooth function. We can write the continuous PDE by introducing the operator $\mathcal{L}$ where $\mathcal{L}(u) := u_t + cu_x$. On the other hand, we can write any of the schemes by using a discrete operator $\mathcal{S}$ defined over U, the discrete set of numbers given by (2.47), i.e., values of u on the grid. For example, using forward difference in space, we have that

$$\mathcal{S}(U) := \frac{U_j^{n+1} - U_j^n}{\Delta t} + c\,\frac{U_{j+1}^n - U_j^n}{\Delta x}.$$

We say the scheme is consistent if for any smooth function $\phi(x,t)$, we have

$$\mathcal{L}(\phi) - \mathcal{S}(\phi) \longrightarrow 0 \quad \text{as} \quad \Delta x,\ \Delta t \longrightarrow 0,$$

where, in writing $\mathcal{S}(\phi)$, we mean the application of $\mathcal{S}$ to the discrete set of numbers consisting of ϕ evaluated at the fixed grid points.

The fact that all the above schemes are consistent follows from the fact that all the grid approximations to the derivatives were supported by (based upon) Taylor series expansions. Hence, the error made will indeed tend to 0 as the grid size tends to 0. For example, by Taylor series we have

$$\begin{aligned} u(j\Delta x + \Delta x, n\Delta t) &= u(j\Delta x, n\Delta t) + u_x(j\Delta x, n\Delta t)\Delta x \\ &\quad + \frac{1}{2}u_{xx}(j\Delta x, n\Delta t)(\Delta x)^2 + \text{higher-order terms}, \end{aligned}$$

where here higher-order terms means the following: If we divide any of these terms by $(\Delta x)^2$, they tend to zero as $\Delta x \to 0$. Hence the forward difference approximation to the spatial derivative satisfies

$$\frac{u(j\Delta x + \Delta x, n\Delta t) - u(j\Delta x, n\Delta t)}{\Delta x} = u_x(j\Delta x, n\Delta t) + \text{error},$$

where the error is $O(\Delta x)$ as $\Delta x \to 0$. Here, we use standard big O notation[12]. As a result, we often say the forward difference scheme is **first order** as the error made (specifically, the truncation error as described below) is $O(\Delta x)$. One can readily check that the backward difference scheme is also first order, but the centered difference scheme is **second order** as it gives an error which is $O((\Delta x)^2)$. Note that all these statements pertain to a regime where Δx is small; hence the higher the order of the scheme, the higher the accuracy. Note that all three schemes are first order in time.

In implementing a finite difference approximation for a PDE, there is far more subtlety than simply working with consistent schemes. To illustrate this, an excellent computational exercise (cf. Exercise **2.34**) is the following: Suppose we considered a particular initial condition, say

$$g(x) = \begin{cases} e^{-\frac{1}{1-|x|^2}} & \text{if } |x| < 1, \\ 0 & \text{if } |x| \geq 1. \end{cases} \tag{2.52}$$

[12]The error E (which depends on Δx) is $O(\Delta x)$ as $\Delta x \to 0$ means that for Δx small, $E \leq C\Delta x$ for some constant C. In other words, as $\Delta x \to 0$ the error tends to zero at the same rate as Δx.

Fix an $r > 0$ (say, $r = 1/2, 1, 2$), implement all three schemes, and examine the results. For example, take $\Delta t = \frac{1}{100}$, iterate the scheme 300 times, and plot the numerical solution at $t = 1$. You might be surprised by the findings! We now present a little essential theory which will explain the results.

Errors are a basic tenet of any numerical approximation and must be controlled. There are two main types:

1. **Truncation errors** are, roughly speaking, the error made by truncating an infinite sum and approximating with a finite sum. These result in the very essence of our schemes which approximate true derivatives with only a few terms in their Taylor expansions. The fact that these schemes are consistent means these errors will tend to zero as the step sizes tend to zero.
2. **Round-off errors** are also present, as any real number can only be captured by a computer via **finite precision** (a finite number of decimal places). These errors are typically much smaller than truncations errors; they are of order 10^{-15} for double precision arithmetic. However, one should certainly be aware of them and Exercise **2.35** is a great way to appreciate how they can, for example, pop up in the numerical approximation of derivatives.

Thus, any numerical scheme will yield errors. If these errors grow uncontrollably, numerical implementation of the scheme will not yield an acceptable approximation to the true solution. We address this growth of errors in the next subsection.

2.7.2. • von Neumann Stability Analysis. We are interested in the propagation of small errors at different grid points. If these errors are not **controlled** at each stage of the time iteration, they can grow and create a solution completely different from the desired solution at a later time (achieved by many time iterations). So what do we mean when we say "controlled". To this end, we follow **von Neumann**[13] **stability analysis.** We start with an initial pulse which has bounded size but oscillates with a frequency k, for example $\sin kx$ or $\cos kx$. From the perspective of the algebra, it is more convenient to just take a complex exponential (cf. Section 6.1) $e^{ikx} = \cos kx + i \sin kx$, for some $k \in \mathbb{R}$. The size of such a complex-valued pulse is given by its modulus, and note that $|e^{ikx}| = 1$. We then look at the *growth in size* when applying the scheme once to e^{ikx}, captured by an amplification constant called the **growth factor**. Let us illustrate these steps with the backward difference in space scheme (2.50). At time step n, we take

$$U_j^n = e^{ik(j\Delta x)}, \qquad j \in \mathbb{Z},$$

noting that

$$U_{j-1}^n = e^{ik(j\Delta x)}e^{-ik\Delta x}, \qquad j \in \mathbb{Z}.$$

Applying scheme (2.50) we find that

$$\begin{aligned} U_j^{n+1} &= U_j^n - r\left(U_j^n - U_{j-1}^n\right) \\ &= e^{ik(j\Delta x)} - r\left(e^{ik(j\Delta x)} - e^{ik(j\Delta x)}e^{-ik\Delta x}\right) \\ &= \left(1 - r + re^{-ik\Delta x}\right) e^{ik(j\Delta x)}. \end{aligned} \tag{2.53}$$

[13]**John von Neumann** (1903–1957) was a Hungarian-American mathematical scientist who made fundamental contributions to mathematics, physics, and computer science.

The modulus of the expression in the parentheses is called the **growth factor**. In one time step, the amplitude of the oscillatory signal $e^{ik(j\Delta x)}$ increases/decreases by this factor. If it is larger than 1, then as we iterate the time step, small errors will grow uncontrollably: Hence we require that for any k, the growth factor must not be larger than 1. Computing the modulus in (2.53), we find

$$\begin{aligned} |U_j^{n+1}| &= \left|1 - r + re^{-ik\Delta x}\right| \left|e^{ik(j\Delta x)}\right| \\ &= \left|1 - r + re^{-ik\Delta x}\right| \\ &\leq |1 - r| + r. \end{aligned}$$

This estimate holds independently of k. Since by definition $r > 0$ we find that

$$|U_j^{n+1}| \leq 1 \qquad \text{if and only if} \qquad 0 \leq r \leq 1.$$

This is our stability condition for scheme (2.50). It states that the **errors will be controllable**; i.e., they will not grow in size if we choose r in the interval $0 \leq r \leq 1$. So the scheme will work nicely as long as we choose $r \leq 1$.

What about the forward difference in space scheme (2.49)? The analogous steps yield

$$U_j^{n+1} = \left(1 - re^{ik\Delta x} + r\right) e^{ik(j\Delta x)}.$$

This time we find that

$$\begin{aligned} |U_j^{n+1}| &= \left|1 - re^{ik\Delta x} + r\right| \\ &\geq 1 \end{aligned}$$

for all $r > 0$. Moreover it is only 1 when $k\Delta x$ is an integer multiple of 2π. In other words, the growth factor will always have size greater than 1 for certain k. Hence this scheme is **always unstable**[14]. In other words, it will never give us an accurate approximation to the real solution, regardless of the step size.

So, what is going on? Let us think for a moment about the structure of the solution of (2.46), i.e., the characteristics. Characteristics are lines in the upper x- vs. t-plane with slope $c > 0$ upon which the solution in constant. In other words, information moves to the right with speed c. Hence at a given time step $(n + 1)\Delta t$, should we base the value of the solution $U_j^{(n+1)} = u(j\Delta x, (n + 1)\Delta t)$ at position $x = j\Delta x$ on U_{j+1}^n or U_{j-1}^n? Clearly the latter, since information is propagating in time to the right. In this sense we say the backward difference in space scheme (2.50) is the **upwind** scheme, as it travels "with the wind" which is moving to the right. The forward difference in space scheme (2.49) is called the **downwind** scheme. The upwind is stable for $r \leq 1$ while the downwind is always unstable. In fact it is interesting to note for this very particular PDE that if $r = 1$, the upwind scheme is exact — no error is made! This is simply a consequence of the fact that the solution to (2.46) is constant on lines with slope c, and when $r = 1$, such a line passes through the grid points $(j\Delta x, (n + 1)\Delta t)$ and $((j - 1)\Delta x, n\Delta t)$.

[14] One can also check that the centered difference in space scheme (2.51) is also always unstable.

The condition on r, ensuring that the size of the growth factor is never larger than 1, is an example of **the CFL condition**, named after[15] **Richard Courant**, **Kurt Friedrichs**, and **Hans Lewy**. What we seek in finite difference schemes for PDEs is those which are **convergent**. A convergent scheme is one for which successive iteration yields an accurate approximation to the solution, with the error tending to 0 as both the temporal and spatial step sizes tends to 0. A major result in the numerical analysis of **linear** PDEs is the following:

Lax Equivalence Theorem

A consistent numerical scheme converges if and only if it is stable.

The notion of stability varies with the context (PDE and numerical method). This result is named after **Peter Lax**[16].

2.8. The Euler Equations: A Derivation

In this concluding section, we present a derivation of one of the most important **first-order systems** of PDEs: the Euler equations[17]. The Euler equations and their second-order cousin the Navier-Stokes equations govern the motion of a fluid (gases and liquids). Given the prevalence and scope of fluids in our world, it is hardly surprising that:

- they are fundamentally important equations (e.g., they are widely used for weather prediction);
- they are amazingly difficult to solve and analyze.

On the other hand, the point of this section is to show how these formidable systems of equations can be derived from two simple physical principles:

- conservation of mass,
- conservation of linear momentum (Newton's Second Law).

Let us consider a fluid occupying a region of space Ω with our independent variables being space and time. We will use (x, y, z) to denote a spatial position in Ω and, as usual, t to denote time. Our dependent (state) variables will be as follows:

- $\rho(x, y, z, t)$, the **mass density** associated with the fluid position at (x, y, z) and time t. Note that the dimensions of ρ are mass per length cubed.
- $\mathbf{u}(x, y, z, t) = (u_1(x, y, z, t), u_2(x, y, z, t), u_3(x, y, z, t))$, the **velocity of the fluid particle which is moving through the point** (x, y, z) **at time** t. The vector

[15]**Richard Courant** (1888–1972) and **Kurt Otto Friedrichs** (1901–1982) were German-American mathematicians who in 1935 founded a mathematics institute in New York which was later renamed the **Courant Institute of the Mathematical Sciences**. It is part of New York University and since its founding has arguably been the Mecca of applied mathematics. **Hans Lewy** (1904–1988) and Kurt Friedrichs were PhD students of R. Courant at the University of Göttingen. All three made fundamental contributions to PDE, applied mathematics, and analysis.

[16]**Peter Lax** (1926–) is a Hungarian-American mathematician and one of the most influential applied mathematicians of the last 70 years. He was a PhD student of Friedrichs and has spent his career at the Courant Institute.

[17]This fundamental system of PDEs is named after the Swiss mathematician and scientist **Leonhard Euler** (1707–1783) who was, arguably, the greatest mathematician of all time. To quote Laplace: "*Read Euler, read Euler, he is the master of us all.*"

field $\mathbf{u}$ is often called the **spatial velocity field** and each component of $\mathbf{u}$ has dimensions of length time^{-1}.

Note that fixing the point (x, y, z) does not fix a particular fluid particle, but rather a place in space. In other words, we have chosen here to label space with our independent variables and **not** to label individual particles of the fluid. This has an important ramification when one computes the acceleration.

Conservation of mass results in **the continuity equation** for the density ρ. Even though we already derived this equation in Section 2.2.6, we repeat the short derivation here to fix ideas and notation.

2.8.1. Conservation of Mass and the Continuity Equation.

Consider the fluid inside a set fixed piece $W \subseteq \Omega$.

The mass of the fluid in W at time t is

$$m_W(t) := \iiint_W \rho(x, y, z, t)\, dxdydz.$$

Note that $m_W(t)$ has dimensions of mass. We use differentiation under the integral sign (cf. Section A.9) to find that

$$\frac{dm_W(t)}{dt} = \iiint_W \rho_t(x, y, z, t)\, dxdydz. \tag{2.54}$$

Now we ask the fundamental question: How can m_W, the total mass of fluid in W, change with time? The simple answer is that fluid can enter or leave W across the boundary ∂W. At any point (x, y, z) on the boundary and time t, the fluid flux across the boundary is instantaneously $\rho\mathbf{u} \cdot \mathbf{n}$, where $\mathbf{n}$ denotes the outer unit normal to W. Note that the fluid flux $\rho\mathbf{u} \cdot \mathbf{n}$ has dimensions of mass per unit area per unit time. The total (net) amount of fluid leaving W at time t is then

$$\iint_{\partial W} \rho\, \mathbf{u} \cdot \mathbf{n}\, dS. \tag{2.55}$$

Note that the above has dimensions of mass per unit time. Balancing (2.54) with (2.55), we find

$$\iiint_W \rho_t(x, y, z, t)\, dxdydz = -\iint_{\partial W} \rho\, \mathbf{u} \cdot \mathbf{n}\, dS.$$

Note the presence of the minus sign: If (2.55), the net amount of fluid leaving through the boundary, is positive, then the net mass in W will decrease. By the Divergence Theorem (cf. Section A.6.1), we have

$$\iint_{\partial W} \rho\, \mathbf{u} \cdot \mathbf{n}\, dS = \iiint_W \operatorname{div}(\rho\mathbf{u})\, dxdydz,$$

and hence

$$\iiint_W \left(\rho_t(x, y, z, t) + \operatorname{div}(\rho\mathbf{u})\right) dxdydz = 0. \tag{2.56}$$

Since (2.56) is true on any piece W, we may invoke the IPW Theorem, Theorem A.6, to obtain the pointwise law (PDE):

$$\boxed{\rho_t + \operatorname{div}(\rho\mathbf{u}) = 0.} \tag{2.57}$$

This scalar partial differential equation is the continuity equation of Section 2.2.6. While previously the velocity field was given (a prescribed function), now it is an unknown vector field which one needs to find with the help of the next set of equations.

2.8.2. Conservation of Linear Momentum and Pressure. We now turn to the conservation of linear momentum (mass × velocity). This conservation principle is dictated by **Newton's Second Law** which states that the rate of change of linear momentum must be balanced by the net forces. Let us first motivate why computing the time rate of change of linear momentum is far more complicated than simply taking the partial derivative with respect to t. We already implicitly addressed this in deriving the continuity equation but it is useful to focus on this point for spatial descriptions of velocity and acceleration. If you are familiar with the notion of a material derivative (cf. Section 2.8.6), or comfortable with the derivation of the continuity equation, you may skip the following comment (digression).

An Aside Comment on Spatial Descriptions of Velocity and Acceleration: Consider for a moment a 2D fluid which is circling around a point. This scenario should be familiar from your advanced calculus class where you no doubt considered plots of a vector field like that shown in Figure 2.15. What do the arrows mean? They are samples of the spatial velocity field $\mathbf{u}$; that is, at a grid point (x, y), they give the velocity vector of the fluid which is at that point. Should this velocity vector at point (x, y) change with time? No, in this scenario the velocity vector of the fluid at a fixed position (x, y) does not change with time. Thus we have $\mathbf{u}(x, y, t) = \mathbf{u}(x, y)$. Does this mean that there is no acceleration? Certainly not, as the direction of fluid flow continuously changes with time. Indeed, the fluid is circling around the vortex. The underlying issue here is that the velocity field $\mathbf{u}$ is based upon a **fixed position** and **not** on a **fixed fluid particle**.

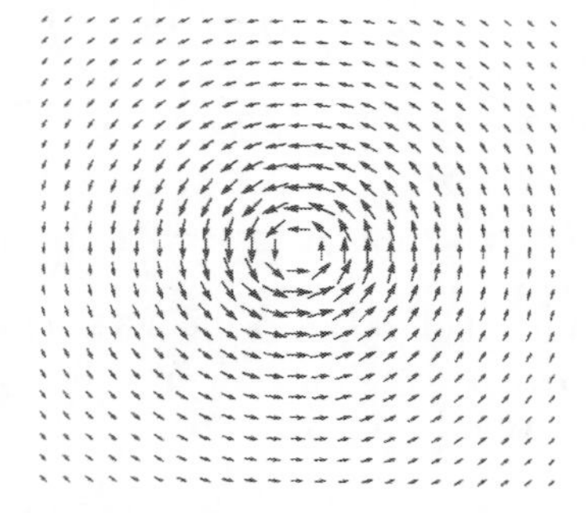

Figure 2.15. Velocity vector field for a 2D fluid around a center vortex.

Let us be more precise by working with a general fluid in 3D and considering the **trajectory of a particular fluid particle**. This trajectory can be represented by the curve $\mathbf{x}(t) = (x(t), y(t), z(t))$ in $\mathbb{R}^3$. The **velocity** and **acceleration** vectors of the fluid particle at time t are given by, respectively,

$$\mathbf{v}(t) := \frac{d\mathbf{x}}{dt} \quad \text{and} \quad \mathbf{A}(t) := \frac{d^2\mathbf{x}(t)}{dt^2} = \frac{d\mathbf{v}(t)}{dt}.$$

We will see shortly why we use above the capital $\mathbf{A}$ as opposed to the lowercase $\mathbf{a}$. How does $\mathbf{v}(t)$ relate to the spatial velocity field $\mathbf{u}(x,y,z,t)$, which we recall gives the velocity of the fluid particle moving through the point (x,y,z) at time t? The relationship is simply $\mathbf{v}(t) = \mathbf{u}(x(t),y(t),z(t),t)$.

Now we consider the spatial description of acceleration which we label $\mathbf{a}(x,y,z,t)$. It gives the acceleration vector of the fluid particle which is traveling through the point (x,y,z) at time t. In terms of the $\mathbf{A}(t)$ above, we have $\mathbf{A}(t) = \mathbf{a}(x(t),y(t),z(t),t)$. The central issue that we wish to make precise is that

$$\mathbf{a}(x,y,z,t) \neq \frac{\partial \mathbf{u}(x,y,z,t)}{\partial t}.$$

To see this, we consider the velocity and acceleration fields **along** a particle trajectory. By the chain rule we find

$$\begin{aligned}
\mathbf{a}(x(t),y(t),z(t),t) = \mathbf{A}(t) &= \frac{d\mathbf{v}(t)}{dt} \\
&= \frac{d\,\mathbf{u}(x(t),y(t),z(t),t)}{dt} \\
&= \frac{\partial \mathbf{u}}{\partial t} + \frac{\partial \mathbf{u}}{\partial x}x'(t) + \frac{\partial \mathbf{u}}{\partial y}y'(t) + \frac{\partial \mathbf{u}}{\partial z}z'(t) \\
&= \frac{\partial \mathbf{u}}{\partial t} + \frac{\partial \mathbf{u}}{\partial x}u_1 + \frac{\partial \mathbf{u}}{\partial y}u_2 + \frac{\partial \mathbf{u}}{\partial z}u_3 \\
&= \frac{\partial \mathbf{u}}{\partial t} + \mathbf{u}\cdot\nabla\mathbf{u}. \qquad (2.58)
\end{aligned}$$

In the last three lines of (2.58), we have omitted the argument $(x(t),y(t),z(t),t)$ from all the functions. As $\mathbf{u}$ is a vector, all the partial derivatives of $\mathbf{u}$ are also vectors and $\nabla\mathbf{u}$ is the matrix (sometimes called the *Jacobian matrix*) whose columns are these partial derivatives.

In conclusion, we note that the acceleration is not simply the time partial derivative of u since the fluid is moving through space. The additional term $\mathbf{u}\cdot\nabla\mathbf{u}$ accounts for the dependence of u on spacial points (x,y,z).

Since ρ is not, in general, a constant, it is difficult to work at the pointwise level (as in the above aside comment) to compute the time rate of change of linear momentum. Rather, we now revert back to a piece W of the fluid and, following the same reasoning as with conservation of mass, examine how the total linear momentum in W can change with time.

As with conservation of mass, we consider **the portion of the fluid that is in** W **at time** t. Let us first consider the linear momentum in the x direction. In the spirit of conservation of mass, we will need to take into consideration that fluid is moving in and out of W. This means that the total rate of change of linear momentum in the x direction is **not** simply

$$\frac{d}{dt}\iiint_W \rho\, u_1\, dxdydz = \iiint_W \frac{\partial(\rho\, u_1)}{\partial t}\, dxdydz.$$

Why? The above integral does not take into account the changes in total linear momentum **as a result of fluid escaping or entering through the boundary of** W. This additional loss or gain of linear momentum through ∂W is exactly the surface integral[18]

$$\iint_{\partial W} (\rho u_1)\mathbf{u}\cdot\mathbf{n}\, dS.$$

The reader should verify that the above surface integral has dimensions of mass $\times$ length $\times$ time^{-2} (the dimensions of force). Hence, the total rate of change of linear momentum in the x direction is given by

$$\iiint_W \frac{\partial(\rho u_1)}{\partial t}\, dxdydz \;+\; \iint_{\partial W} (\rho u_1)\mathbf{u}\cdot\mathbf{n}\, dS.$$

As before, we apply the Divergence Theorem on the second term to obtain

$$\iiint_W \left(\frac{\partial(\rho u_1)}{\partial t} \;+\; \operatorname{div}(\rho u_1 \mathbf{u})\right) dx\,dy\,dz. \tag{2.59}$$

The analogous arguments for the y and z component of the total linear momentum in W yields (2.59) with u_1 replaced with u_2 and u_3, respectively. Note the similarity with (2.56) for conservation of mass. In that instance, the analogous expression was equal to 0 because there were no other facilitators for generating mass. For linear momentum, there will be forces which must balance (Newton's Second Law). Hence we must balance (2.59), the total rate of change of linear momentum in the x direction, by the total net force in the x direction acting on W.

What are the forces pertaining to the fluid? There are two types:

(i) **Body/external** forces which exert force per unit volume. The most natural body force to include would be gravity.

(ii) **Internal forces (stress)** on particles from neighboring particles. These internal forces are assessed by considering two pieces of the material separated by a *surface* and examining the force of one piece on the other through the surface. Stress is thus measured in terms of force per unit area. In a fluid, a standard example of an internal force is pressure.

The defining property of any *material* is the existence of internal stresses, and hence we focus on (ii).

An **ideal fluid** is one of the simplest materials and is characterized by the fact that the forces on any piece of fluid from the surrounding fluid are entirely due to **pressure**. Pressure acts on surfaces and produces a force (which we call *pressure force*) which is always directed in the **normal direction** of the surface. To be more precise, consider a surface $\mathcal{S}$ through a body of fluid occupying region W, and a point (x, y, z) on the surface $\mathcal{S}$ with a normal vector $\mathbf{n}$ (think of $\mathbf{n}$ determining the *upwards* direction). The surface divides the body of fluid in W into upper and lower pieces W_u and W_l, respectively (cf. Figure 2.16). The presence of pressure implies that there exists a positive

[18]This integral may at first seem strange as it involves a product of two terms involving $\mathbf{u}$. The first term ρu_1 is there as it represents the x component of the fluid's linear momentum (the quantity of interest). The second term $\mathbf{u}\cdot\mathbf{n}$ is there as it measures the rate per unit area that fluid leaves the boundary. Since any quantity associated with the fluid (here, the linear momentum) will be transported out of W via this rate, we integrate the product $(\rho u_1)\mathbf{u}\cdot\mathbf{n}$ over ∂W.

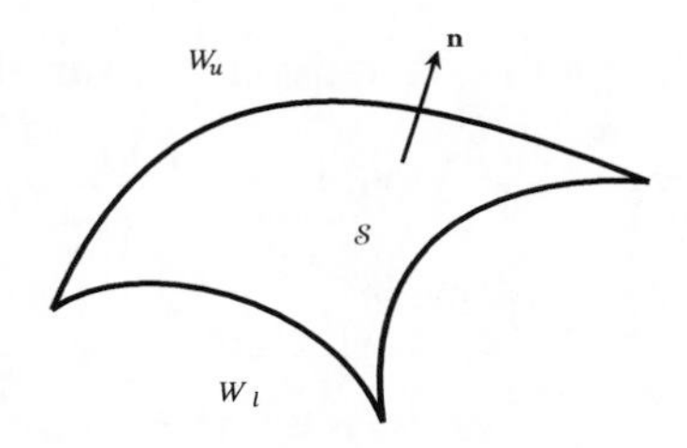

Figure 2.16

scalar function $p(x, y, z, t)$ such that the following is true:

$$\begin{cases}\text{The } \textbf{force} \text{ per unit area at } (x, y, z) \text{ at time } t \textbf{ on} \text{ the fluid in } W_u \\ \textbf{from} \text{ the fluid in } W_l \text{ equals } \; p(x, y, z, t)\,\mathbf{n}.\end{cases} \tag{2.60}$$

The magnitude $p(x, y, z, t)$ of this force density (force per unit area) can have any variation with space and time, but its direction is always dictated by (2.60). The reader should reconcile this fact with their intuition of pressure in a basin of water.

Now let us think of a body of fluid in region W which is under pressure from the surrounding fluid, i.e., the fluid which surrounds the boundary ∂W. The **net** pressure force on W through ∂W is

$$\mathbf{F}_W \; := \; -\iint_{\partial W} p\,\mathbf{n}\,dS.$$

Note that the magnitude of the vector $\mathbf{F}_W$ has dimensions of force. We invoke Theorem A.3 (the Componentwise Divergence Theorem) to obtain

$$\mathbf{F}_W \; = \; -\iint_{\partial W} p\,\mathbf{n}\,dS \; = \; -\iiint_W \nabla p\,dx\,dy\,dz. \tag{2.61}$$

The above is a vector equation. Let us focus on the first component, the x component, of the pressure force:

$$-\iiint_W \frac{\partial p}{\partial x}\,dx\,dy\,dz.$$

Balancing this with the total rate of change in linear momentum from (2.59) gives

$$\iiint_W \left(\frac{(\partial \rho\, u_1)}{\partial t} + \operatorname{div}(\rho\, u_1\, \mathbf{u})\right) dx\,dy\,dz \; = \; -\iiint_W \frac{\partial p}{\partial x}\,dx\,dy\,dz.$$

Since the above holds for all W, we may invoke the IPW Theorem, Theorem A.6, to obtain the pointwise PDE:

$$\frac{\partial(\rho\, u_1)}{\partial t} + \operatorname{div}(\rho\, u_1\, \mathbf{u}) \; = \; -\frac{\partial p}{\partial x}.$$

Carrying out the product differentiation, we find

$$\rho\left(\frac{\partial u_1}{\partial t} + \mathbf{u}\cdot\nabla u_1\right) + u_1\left(\frac{\partial \rho}{\partial t} + \operatorname{div}(\rho\mathbf{u})\right) \; = \; -\frac{\partial p}{\partial x}.$$

But the expression in the second set of parentheses is 0 due to conservation of mass. Thus

$$\rho\left(\frac{\partial u_1}{\partial t} + \mathbf{u}\cdot\nabla u_1\right) \; = \; -\frac{\partial p}{\partial x}. \tag{2.62}$$

The analogous equality holds for the y and z components; that is,

$$\rho\left(\frac{\partial u_2}{\partial t} + \mathbf{u}\cdot\nabla u_2\right) = -\frac{\partial p}{\partial y} \qquad \text{and} \qquad \rho\left(\frac{\partial u_3}{\partial t} + \mathbf{u}\cdot\nabla u_3\right) = -\frac{\partial p}{\partial z}.$$

Thus, in (column) vector form, we obtain

$$\boxed{\rho\left(\frac{\partial \mathbf{u}}{\partial t} + \mathbf{u}\cdot\nabla \mathbf{u}\right) = -\nabla p.} \tag{2.63}$$

Note that in this standard vector notation, $\nabla\mathbf{u}$ is a 3×3 matrix whose respective columns are the vectors ∇u_i, and $\mathbf{u}\cdot\nabla\mathbf{u}$ corresponds to the matrix multiplication of $\mathbf{u}$, interpreted as a row vector (matrix with one row and 3 columns), with the 3×3 matrix $\nabla\mathbf{u}$. Thus, all three terms in (2.63) are $3D$ column vectors.

Remark on the inclusion of body forces like gravity: One can include the effect of a body force on all three sets of equations. For example, including the effects of a constant gravitational force results in the following modification to (2.63):

$$\rho\left(\frac{\partial \mathbf{u}}{\partial t} + \mathbf{u}\cdot\nabla \mathbf{u}\right) = -\nabla p + \rho\mathbf{g},$$

where $\mathbf{g} := \langle 0, 0, g\rangle$ with g the constant gravitational acceleration.

So far this derivation has been generic for any fluid; conservation of mass gave rise to the scalar continuity equation (2.57) and conservation of linear momentum gave rise to the system of the equations (2.63). Recall that the dependent variables are the density ρ and the 3D velocity vector field $\mathbf{u}$. However, there is also the pressure p in (2.63). What do we do with this p? Let us now discuss two separate cases.

2.8.3. Gas Dynamics: The Compressible Euler Equations.

We consider an ideal gas, for example air. A gas is an example of a compressible fluid; as a set region of the fluid (a set collection of fluid particles) moves with time, its volume can change. Alternatively, the density of the fluid can vary with space and time. So what is pressure p in a gas? Is it an additional state variable which we need the PDEs to solve for? No, for a compressible fluid, the pressure at any point and time is a **function of the density**; that is,

$$p(x, y, z, t) = f(\rho(x, y, z, t)), \tag{2.64}$$

for some function f. As you might expect, f is an increasing function as, intuitively, the greater the density the greater the pressure. The particular choice of function will depend on the exact nature of the gas. For many gases, the function f is simply a power law; i.e., $f(\rho) = c_0\,\rho^\gamma$, for some constants $c_0 > 0$ and $\gamma > 0$.

Substituting (2.64) into (2.63), we arrive at the **compressible Euler equations**:

$$\boxed{\begin{cases} \frac{\partial \mathbf{u}}{\partial t} + \mathbf{u}\cdot\nabla\mathbf{u} = -\frac{1}{\rho}\nabla f(\rho), \\ \frac{\partial \rho}{\partial t} + \operatorname{div}(\rho\mathbf{u}) = 0. \end{cases}} \tag{2.65}$$

These are equations for gas dynamics and consist of a system of four partial differential equations in four unknown state variables (ρ, u_1, u_2, u_3). These Euler equations are quasilinear first order. The reader should note the parallel to the inviscid Burgers equation in 1D (uu_x vs. $\mathbf{u} \cdot \nabla \mathbf{u}$). Indeed, discontinuities and shock waves are signature solutions to these equations of gas dynamics.

2.8.4. An Ideal Liquid: The Incompressible Euler Equations. A liquid such as water is also an example of a fluid yet it is different from a gas in that it is **incompressible**; any region (part) of the fluid can change shape but with the same volume. While there are many ways to characterize incompressibility, a simple way is to assert that the density ρ is constant is space and time. In this case, the continuity equation (2.57) becomes

$$\operatorname{div} \mathbf{u} = 0.$$

This equation encapsulates our incompressibility condition.

But how should we address the pressure p in (2.63)? In an ideal incompressible liquid, it is **not** a function of ρ, nor is it constant. Rather, it is a state (dependent) variable which we must find. It must balance the motion of the fluid in such a way that it maintains incompressibility. One can think of the pressure as a *Lagrange multiplier function* associated with the incompressibility constraint.

So we arrive at the **incompressible Euler equations**:

$$\boxed{\begin{cases} \frac{\partial \mathbf{u}}{\partial t} + \mathbf{u} \cdot \nabla \mathbf{u} = -\nabla p, \\ \operatorname{div} \mathbf{u} = 0 \end{cases}} \tag{2.66}$$

with p as a state variable (dependent variable) which we must solve for. Even though p does not appear in the last equation, we are still left with a system of four partial differential equations in four unknown state variables (u_1, u_2, u_3, p). The density ρ is no longer a dependent variable, but the pressure is.

2.8.5. A Viscous Liquid: The Navier-Stokes Equations. In this very short subsection, we will break from first-order equations with the inclusion of the second-order Laplacian. In a true ideal liquid, the only internal force is pressure. However, liquids (like water) that we encounter on a daily basis are **not** ideal fluids. There is another internal force, a frictional force, which loosely speaking is tied to the fluid "rubbing against itself". Thus returning to two pieces of fluid in region W separated by a surface (cf. Figure 2.16), this force is now directed **tangential** to the surface, in contrast to pressure which is **normal** to the surface. This resistive force stems from the relative velocity of different fluid particles. In the language of continuum mechanics, this is **shear stress** in the fluid. It is responsible for "slowing the fluid down" just like friction when you push forward a cart. In the language of fluid mechanics, it is called **viscosity**. Without giving a full treatment and derivation from continuum mechanics (see Chorin and Marsden[19] for a full derivation), the resulting force appears on the right-hand side of (2.63) as

$$\nu \nabla \cdot \nabla \mathbf{u} = \nu \Delta \mathbf{u},$$

where $\nu > 0$ is called the viscosity constant and depends on the particular fluid. Here Δ denotes the vector Laplacian; i.e., $\Delta \mathbf{u}$ is the vector with respective components Δu_i.

[19] **A. Chorin and J. Marsden**, A Mathematical Introduction to Fluid Mechanics, Springer-Verlag, 1993.

While this is the first mention of the Laplacian, it will appear throughout the subsequent chapters (see the short Subsection A.6.3 of the Appendix if this is your first time encountering the Laplacian).

Thus, for a viscous liquid like water, the equations are the system known as the **Navier-Stokes equations**[20]:

$$\boxed{\begin{cases} \frac{\partial \mathbf{u}}{\partial t} + \mathbf{u} \cdot \nabla \mathbf{u} = \nu \Delta \mathbf{u} - \nabla p, \\ \operatorname{div} \mathbf{u} = 0. \end{cases}} \tag{2.67}$$

Unlike the Euler equations, they constitute a **second-order system**.

Like the Euler equations, the Navier-Stokes equations prove to be incredibly complicated. Indeed, a major open problem[21] in mathematics is the well-posedness of the Navier-Stokes equations in suitable function spaces. In particular, it is not known, if we start with nice initial conditions, whether or not the solution will develop singularities. The complicated structure of these solutions is often linked to the phenomenon of **turbulence** in fluids.

2.8.6. Spatial vs. Material Coordinates and the Material Time Derivative.

The Euler and Navier-Stokes equations use **spatial descriptions** of the dependent variables; i.e., the independent variables are positions in space and time. Such spatial descriptions are often referred to as **Eulerian descriptions**. For this subsection, let us switch notation, using $\mathbf{x} = (x_1, x_2, x_3)$ in place of (x, y, z) for the spatial coordinates. There is another point of view which, in some ways, is more natural: **material**, also known as **Lagrangian**, descriptions. Here one does not use spatial points as independent variables but rather material coordinates, in other words, coordinates which label each fluid particle whose position in space varies with time.

Working in spatial coordinates meant that it was necessary for us to account for the fact that fluid particles were moving with the flow. Hence, when considering the time rate change of any relevant *state variable* (e.g., velocity), we did not just take a partial derivative with respect to t but rather we took what is known as **the material time derivative**:

$$\frac{D}{Dt}(*) := \frac{\partial}{\partial t}(*) + \mathbf{u} \cdot \nabla (*). \tag{2.68}$$

It represents taking a time derivative of some quantity with respect to a fixed material (fluid) particle, **not** a fixed position in space. For example, conservation of the x_1 component of the linear momentum (2.62) was exactly the statement

$$\frac{D}{Dt}(\rho u_1) = -\frac{\partial p}{\partial x_1}.$$

On the other hand, we can use the fact that $\operatorname{div}(\rho \mathbf{u}) = (\operatorname{div} \mathbf{u})\rho + \mathbf{u} \cdot \nabla \rho$ to write the continuity equation (2.57) as

$$\frac{D}{Dt}(\rho) = -(\operatorname{div} \mathbf{u})\,\rho.$$

---

[20]**Claude-Louis Navier** (1785–1836) was a French engineer and physicist. **George Gabriel Stokes** (1819–1903) was an English-Irish physicist and mathematician.

[21]In fact it is one of the *Millennium Prize* Problems; see `https://www.claymath.org/millennium-problems`.

Hence, in the case of the incompressible Euler equations wherein $\operatorname{div} \mathbf{u} = 0$, the continuity equation (conservation of mass) states that

$$\frac{D}{Dt}\rho = 0,$$

or, in words, the density of the fluid does not change as it flows.

I reiterate that everything so far has been in spatial coordinates; indeed, the material time derivative (2.68) should be thought of as *a description of the material time derivative in spatial coordinates.* This brings up an obvious point: The fundamental difficulty in either the Euler or Navier-Stokes equations lies in the nonlinear (albeit quasilinear) term $\mathbf{u} \cdot \nabla \mathbf{u}$, a direct consequence of the material derivative in spatial coordinates. **Why not use material coordinates to write these fundamental equations?**

First off, what exactly do we mean by material coordinates; in other words, how do we label fluid particles? Consider taking a snapshot of the fluid at $t = 0$ (we call this the **reference frame**). The snapshot shows the fluid occupying some region of space Ω. Associate to any $\mathbf{a} = (a_1, a_2, a_3) \in \Omega$ the fluid particle which is at $\mathbf{a}$ at $t = 0$. We think of the initial position $\mathbf{a}$ of a fluid particle as tagging, or labeling, it throughout its motion. With this labeling of the continuum of fluid particles, we can write any field variable (e.g., density, velocity) in terms of $\mathbf{a}$ and t. For example, $\rho(\mathbf{a}, t) = \rho(a_1, a_2, a_3, t)$ denotes the density at time t of the fluid particle indexed by $\mathbf{a}$, i.e., the fluid particle which was at position $\mathbf{a}$ in the reference frame. We call $\mathbf{a}$ material coordinates as they label (index) material particles, and systems based upon material coordinates are known as **Lagrangian descriptions**. With material coordinates, the material time derivative is simply $\frac{\partial}{\partial t}$; in other words, in material coordinates the left-hand sides of the Euler and Navier-Stokes equations are a simple (linear) derivative with respect to t!

Now we are in a position to answer the above question. One can certainly use material coordinates but the punchline is: *There is no free lunch.* While the pressure, the fundamental internal force of a fluid, had a simple description in spatial coordinates, how would one write the pressure in material coordinates at time $t > 0$? To do this, one would have to consider how the fluid has deformed from the reference frame at $t = 0$, which would result in a nonlinear and complicated dependence of the pressure function on the material description of velocity. Thus the corresponding system of PDEs would have simple linear left-hand sides (time rate of change of linear momentum) **but** complicated nonlinear right-hand sides (balancing forces). Whereas such systems are not of much theoretical or practical use, material (Lagrangian) formulations of fluid dynamics are useful **but** from the point of view of **the particle trajectories**, a perspective rooted in the method of characteristics.

To provide a few details, we denote $\mathbf{X}(t; \mathbf{a})$ to be the position of the fluid particle $\mathbf{a}$ at time t. If, for example, we were given the spatial description of the velocity vector field $\mathbf{u}(\mathbf{x}, t)$, then these particle trajectories would be the solution to the system of ODEs:

$$\frac{\partial}{\partial t}\mathbf{X}(t; \mathbf{a}) = \mathbf{u}\left(\mathbf{X}(t; \mathbf{a}), t\right), \qquad \mathbf{X}(0; \mathbf{a}) = \mathbf{a}.$$

In the particle trajectory formulation of the Euler equation, we do not try to find the velocity field directly, but rather the conservation of linear momentum takes the form

$$\frac{\partial^2}{\partial t^2}\mathbf{X}(t;\mathbf{a}) = -\nabla p\,(\mathbf{X}(t;\mathbf{a}),t),$$

where p is the pressure function of (2.60).

On the other hand, for (say) an incompressible fluid, the Eulerian condition div $\mathbf{u} = 0$ would be replaced with the condition that for any $t \geq 0$, the map $\mathbf{X}(t;\cdot) : \mathbb{R}^3 \to \mathbb{R}^3$ is *volume preserving*, captured by the requirement that det $\nabla\mathbf{X}(t;\cdot) = 1$.

For more on the Lagrangian (particle trajectory) formulation of fluid dynamics see **Gurtin**[22] and **Majda-Bertozzi**.[23]

2.9. Chapter Summary

- Characteristics are associated with first-order PDEs. The characteristics are curves in the domain of the solution *upon which* the PDE tells us how the solution should evolve. In principle, if we know the value of u at one point on a particular characteristic, then the PDE tells us what u should be at all the other points on that characteristic. One could say that the PDE over the domain *degenerates* into ODEs *along* the characteristics.
- There is a systematic method to solve first-order PDEs in which we exploit this property of having characteristics. It is called the **method of characteristics**. As usual, we assume we have a solution (unknown at this point) and use properties of the PDE (here the characteristic ODEs) to gain insight into its structure. For certain simple first-order PDEs with data appropriately given, this is sufficient to characterize the solution, i.e., find a formula for it. For more complicated equations the method needs to be implemented numerically with the help of a computer.
- We can decompose the method of characteristics into three parts:
    1. Write down the **characteristic system of ODEs**. In general, this means the system (2.36); however, as we have seen in most examples, this system of ODEs becomes quite simple. Only in cases where the PDE is fully nonlinear will you need the $\mathbf{p}$ equation to solve for $\mathbf{x}(s)$ and $z(s)$.
    2. **Solve** these ODEs (if possible) such that at $s = 0$, $\mathbf{x}(0)$ lies on the data curve Γ with intersection point $\mathbf{x}_0$.
    3. **Invert the ODE solutions to solve for the characteristic parameters:** So far, we have equations for *all* the characteristic curves $\mathbf{x}(s)$ and for the solution along any one of them. Hence, given any $\mathbf{x} \in \Omega$, to find what the solution is at $\mathbf{x}$ (i.e., $u(\mathbf{x})$), we simply need to know which characteristic curve the point $\mathbf{x}$ lies on and when this characteristic curve passes through $\mathbf{x}$. The first piece of information is determined by the $\mathbf{x}_0$ and the second by the choice of s. Thus, we need to solve for $\mathbf{x}_0$ and s in terms of $\mathbf{x}$.

[22] **M.E. Gurtin**, An Introduction to Continuum Mechanics, Academic Press, 1981.

[23] **A.J. Majda and A.L Bertozzi**, Vorticity and Incompressible Flow, Cambridge University Press, 2001.

- There is a fundamental difference between linear and nonlinear first-order equations:

  If the first-order PDE is **linear** (or semilinear), then the structure of the characteristics is completely determined by the coefficients in the PDE (i.e., the form of F).

  For **nonlinear** (quasilinear and fully nonlinear) first-order PDEs, there is an interplay between the structure of the characteristics and the value of the solution (and its derivatives) along the characteristics. In particular, the data (the "initial values") will influence the structure of the characteristics.

- An important class of first-order PDEs (a wonderful class for students to digest and gain insight into the notion of characteristics) are known as **transport equations**. The simplest examples are

$$\text{in one space dimension:} \qquad u_t + cu_x = 0, \qquad u(x,0) = g(x),$$

$$\text{in several space dimensions:} \qquad u_t + \mathbf{a} \cdot \nabla u = 0, \qquad u(\mathbf{x},0) = g(\mathbf{x}),$$

  where $c \in \mathbb{R}$ and $\mathbf{a} \in \mathbb{R}^N$. We view these PDEs as equations describing the transport in space of an initial signal (piece of information) g with respective velocities c and $\mathbf{a}$. We may also allow these velocities to depend on space (and/or time), still yielding a linear equation, but one which may distort the shape of the initial signal. On the other hand, an important case (our canonical example of a nonlinear PDE) was, in one space dimension, to allow c to depend on u (specifically "$c = u$"). This gives rise to (the inviscid) Burgers equation $u_t + uu_x = 0$.

Exercises

2.1 (**Characterizations of a general solution can look different**) In Example 2.1.5, we showed that the general solution to $au_x + bu_y + u = 0$ was $u(x,y) = f(bx - ay)\, e^{-\frac{ax+by}{a^2+b^2}}$, for any function (of one variable) f. Assuming $a \neq 0$, show that this is equivalent to saying

$$u(x,y) = f(bx - ay)\, e^{-\frac{x}{a}},$$

for any function (of one variable) f.

2.2 Consider the PDE $yu_x + xu_y = 0$. What are the characteristics for this PDE? Describe the general solution in terms of these characteristics (first in words, then more precisely in a formula).

2.3 Suppose that **any** solution $u(x,y)$ to $au_x + bu_y = 0$ satisfies $u(1,2) = u(3,6)$. What is $\frac{a}{b}$?

2.4 Suppose $u(x,y)$ satisfies $u_x + y^2 u_y = 0$. Suppose further that $u(3,2) = 7$, $u(4,2) = -6$, $u(8,1) = -2$, $u(6,-1) = 3$, $u(6,-2) = 0$, $u(10,7) = 8$, $u(15,-4) = 10$. What are the values of $u\left(\frac{5}{2}, \frac{1}{2}\right)$ and $u\left(8, -\frac{2}{5}\right)$? Can you find the value of $u\left(\frac{9}{2}, 1\right)$? Explain your reasoning.

2.5 Solve $4u_x - 3u_y = 0$, together with the condition $u(x, 5x) = e^x$.

2.6 Use the method of characteristics to solve for $u(x,t)$ where $u_t + xu_x = 0$, $u(x,0) = \frac{1}{x^2+1}$. Plot the profiles for $t = 0, 1, 2, 3, 4$.

2.7 Use the method of characteristics to solve for $u(x,y)$ where $(x-y)u_x + u_y = x$, $u(x,0) = g(x)$.

2.8 Use the method of characteristics to solve for $u(x,y)$ where $xu_y - 2xyu_x = 2yu$, $u(x,0) = x^3$.

2.9 Use the method of characteristics to solve for $u(x,y)$ where $yu_x + 3u_y = -u$, $u(x,0) = \sin x$.

2.10 Show that there is no solution to the PDE $u_x + xu_y = 0$ subject to the initial condition $u(x,0) = e^x$ that is valid in a neighborhood of the origin. Sketch the characteristics to see geometrically what went wrong.

2.11 Use the method of characteristics to solve for $u(x,t)$ where $u_t + (x+t)u_x = t$, $u(x,0) = g(x)$.

2.12 Use the method of characteristics to solve for $u(x_1,x_2,x_3)$ where $u_{x_1} + x_1 u_{x_2} - u_{x_3} = u$, $u(x_1,x_2,1) = x_1 + x_2$.

2.13 Find the solution of $(\sqrt{1-x^2})\, u_x + u_y = 0$, $u(0,y) = y$.

2.14 Use the method of characteristics to solve $y\, u_x - 2xy\, u_y = 2x\, u$, with $u = y^3$ when $x = 0$ and $1 \le y \le 2$. **Hint:** To solve the characteristic ODEs for x and y, multiply the first by $2x$ and then add.

2.15 Consider the transport equation with space varying velocity of Subsection 2.2.5. Under the same assumptions, show that $\iiint_{\mathbb{R}^3} |u(\mathbf{x},t)|^2\, d\mathbf{x}$ is constant in time.

2.16 Use the method of characteristics to solve for $u(x_1,x_2,x_3,x_4)$ where

$$u_{x_1} + 2u_{x_2} + u_{x_3} + x_2 u_{x_4} = x_1 + x_2, \qquad u(x_1,x_2,x_3,0) = x_2 + x_3.$$

2.17 Use the method of characteristics to solve for $u(x,y,z)$ where $xu_x + yu_y + zu_z = 1$ and $u = 0$ on $x + y + z = 1$.

2.18 **(a)** Use the method of characteristics to find the general solution for $u(x,y,z)$ of $u_x + u_y + u_z = 0$. Find a particular solution which satisfies the auxiliary condition $u(x,y,0) = x^2 + y^2$.
(b) Use the method of characteristics to find $u(x,y,z)$ where $u_x + u_y + u_z = 1$ and $u(x,y,0) = x + y$.

2.19 Use the method of characteristics to solve $xu_x - 2yu_y = u^2$, $u(x,x) = x^3$.

2.20 Suppose $u(x,t)$ is a smooth solution to $u_t + uu_x = 0$. Suppose further that $u(1,1) = 5$. At which of the following points can you find the value of u: $(2,1)$, $(2,2)$, $(6,2)$, $(4,3)$, $(10,3)$, $(11,3)$? What is the value at these points? Explain your reasoning.

2.21 Suppose $u(x,y)$ is a smooth function which is constant on any curve of the form $y = x^2 + 2x + C$. The value u takes on different curves (parametrized by C) may of course be different. What PDE does u solve?

2.22 **(a)** Use the method of characteristics to solve for $u(x,t)$ in the upper half-plane $t \ge 0$ where $u_t + uu_x = 2$, with $u(x,0) = 3x$.
(b) Use the method of characteristics to solve for $u(x,y)$ where $(y+u)u_x + yu_x = x - y$ with $u(x,1) = 1 + x$.

2.23 Consider the Burgers equation $u_t + uu_x = 0$ with initial data $u(x, 0) = x^3$. Use the method of characteristics to write down the solution in terms of an x_0 which is implicitly related to (x, t). Show that for each (x, t), there exists a unique x_0. Sketch the characteristics. Using any method you like, sketch the profiles $u(x, 0)$, $u(x, 1)$, $u(x, 3)$.

2.24 Consider the Burgers equation $u_t + uu_x = 0$ with initial data given by

$$u(x,0) = \begin{cases} 1 & \text{if } x \leq 0, \\ 1-x & \text{if } 0 \leq x \leq 1, \\ 0 & \text{if } x \geq 1. \end{cases}$$

Sketch the characteristics in the x- vs. t-plane for $0 \leq t \leq 1$. Sketch $u(x, 0)$, $u(x, 1/2)$, $u(x, 3/4)$, $u(x, 1)$. Write down the explicit solution $u(x, t)$ for all x and $0 \leq t < 1$.

2.25 Use the method of characteristics to solve $u_t + uu_x = 1$, $u(x, 0) = x$.

2.26 Use the method of characteristics to solve $(u + 2y)u_x + uu_y = 0$, $u(x, 1) = \frac{1}{x}$.

2.27 Use the method of characteristics to solve $(y+u)u_x + yu_y = x-y$, $u(x, 1) = 1+x$.

2.28 Consider the initial value problem $u_t - tx^2uu_x = 0$, $u(x, 0) = f(x)$. Use the method of characteristics to show that the following implicit formula holds true for the solution $u(x, t)$:

$$u(x,t) = f\left(\frac{2x}{2 - xt^2u(x,t)}\right).$$

2.29 Show that for the (incompressible) transport equation (2.11), the quantity

$$\iiint_{\mathbb{R}^3} |u(\mathbf{x}, t)|^2 \, d\mathbf{x}$$

is constant in time.

2.30 Recall Example 2.5.2 where we solved the Hamilton-Jacobi equation in the upper half-plane $t \geq 0$. Where does this solution exist? What about a solution in the entire x- vs. t-plane? Can we have a solution to the PDE for $t < -1/4$. How is it connected to the data at $t = 0$?

2.31 Use the method of characteristics to solve for $u(x, y)$ on the unit disk $B(\mathbf{0}, 1)$ where

$$u_x^2 + u_y^2 = 4u \quad \text{on } B(\mathbf{0}, 1) \quad \text{with } u = 1 \text{ on } \partial B(\mathbf{0}, 1).$$

2.32 Write down an explicit solution (i.e., a formula) to the eikonal equation (2.38) in the case where:
(a) $\Omega = B(\mathbf{0}, 1)$, the unit disk centered at the origin.
(b) Ω is the unit square $(-1/2, 1/2) \times (-1/2, 1/2)$.
(c) Ω is the inside of the ellipse $\frac{x^2}{4} + \frac{y^2}{9} = 1$. In each case, clearly describe the set of discontinuities of the solution (the subset of Ω where the solution is not differentiable).

2.33 Consider two examples of initial level set curves in $\mathbb{R}^2$: the unit circle and a dumbbell (with smooth boundary) shown here in the dumbbell figure:

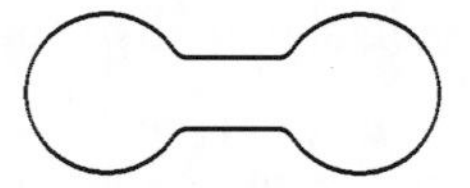

Sketch the resulting shapes at several later times for the following:
(a) Flow by (2.41) with $\mathbf{v} = \langle 1, 1 \rangle$.
(b) Flow by the level set equation (2.43) with $v = 1$ and where the initial function $\phi_0(x, y)$ whose zero level set is the curve is positive inside the curve and negative outside.
(c) Flow by motion by mean curvature (2.45). This question only requires a basic understanding of what these PDEs dictate in terms of interface motion.
(d) Now repeat the three parts above for the sphere and dumbbell in $\mathbb{R}^3$.

2.34 (**Computing the solution to the transport equation**) Using any high-level software (e.g., Matlab of Python), we want to solve numerically for the solution $u(x, t)$ to the transport equation $u_t + u_x = 0$ with initial data $u(x, 0)$ given by (2.52). Fix $\triangle x = 0.1$ and numerically implement all three schemes ((2.49), (2.50), and (2.51)) to find the solution at $t = 1$ when $r = \frac{1}{2}, 1, 2$. Repeat for $\triangle x = 0.01$. Since you are doing this for three schemes, each with three different choices of r and two choices of $\triangle x$, you will end up with $3 \times 3 \times 2$ approximations for $u(x, 1)$. Plot your results against the exact solution and make some conclusions as to the accuracy of these schemes. Discuss why your results are related to the theory presented in Section 2.7.

2.35 (**Round-off errors and the perils of numerical differentiation**)
(a) Let f be a smooth function and consider the centered difference approximation to the first derivative at a point x

$$D_h f(x) := \frac{f(x+h) - f(x-h)}{2h}.$$

Use Taylor's Theorem to show that for any $x \in \mathbb{R}$,

$$f'(x) = D_h f(x) - \frac{f^{(3)}(\zeta^+) + f^{(3)}(\zeta^-)}{12} h^2,$$

where $\zeta^{\pm}$ are some points between x and $x \pm h$, respectively, and $f^{(3)}$ denotes the third derivative of f. Use the Intermediate Value Theorem to show further that one has

$$f'(x) = D_h f(x) - \frac{f^{(3)}(\zeta_0)}{6} h^2, \tag{2.69}$$

for some $\zeta_0 \in (x - h, x + h)$. In particular, note that we have $\lim_{h \to 0} D_h f(x) = f'(x)$ with an error which is $O(h^2)$.
(b) Now let $f(x) = \sin x$ and $x = \pi$. For $h = \frac{1}{10^k}$, $k = 0, \ldots, 12$, compute both $f'(\pi)$ and $D_h f(\pi)$ and use Matlab to make a table recording the absolute value of the errors: $|f'(\pi) - D_h f(\pi)|$. Note that things start off well (errors getting smaller with k) but then errors start to grow!

(c) Now we investigate what is going on: Matlab (or any other computing software system) will only record values of f to a certain number of decimal places, introducing round-off errors. Suppose this error is ϵ; that is, we record $\tilde{f}$ rather than the true f, knowing only that $|\tilde{f}(x+h) - f(x+h)| \leq \epsilon$ and $|\tilde{f}(x-h) - f(x-h)| \leq \epsilon$. Use this together with (2.69) to show that

$$|f'(x) - D_h\tilde{f}(x)| \leq \left|\frac{f^{(3)}(\zeta_0)}{6}h^2\right| + \frac{\epsilon}{h}, \tag{2.70}$$

where $\zeta_0 \in (x-h, x+h)$. Note that the right-hand side does **not** tend to 0 as $h \to 0$.
(d) Now let $f(x) = \sin x$. Noting that $|f^{(3)}(\zeta_0)| \leq 1$, minimize the right-hand side of (2.70), i.e., $\frac{1}{6}h^2 + \frac{\epsilon}{h}$, over $h > 0$. By looking up online the notion of "*machine precision*" and "*eps*" in Matlab, find the relevant choice of ϵ and then compare with your findings in part (b).

2.36 **(a)** Solve the IVP (2.11) with $\mathbf{a}(\mathbf{x}) = \mathbf{A}\mathbf{x} + \mathbf{b}$, where $\mathbf{A}$ is a constant 3×3 matrix and $\mathbf{b}$ is a constant vector in $\mathbb{R}^3$.
(b) Repeat part (a) for the inhomogeneous problem where the right-hand side of the PDE is replaced with a fixed continuous function $f(x,t)$.

2.37 Let $\phi(t)$ be a continuous function for $t \geq 0$ and let $\mathbf{b}$ be a constant vector in $\mathbb{R}^3$.
(a) Solve the IVP for the transport equation

$$u_t + \phi(t)\mathbf{b} \cdot \nabla u = 0, \qquad u(\mathbf{x}, 0) = g(\mathbf{x}), \qquad \mathbf{x} \in \mathbb{R}^3, \ t \geq 0.$$

(b) Repeat part (a) for the inhomogeneous problem where the right-hand side of the PDE is replaced with a fixed continuous function $f(x,t)$.

2.38 **For students interested in more abstract geometry.** There is a more global geometric view point to first-order PDEs and the method of characteristics. This exercise encourages the student to explore this further. By researching online, write a short description of how the notions of the **Monge cone**, **integral surfaces**, and **complete envelopes** are related to the material of this chapter (the method of characteristics). Illustrate these notions in the context of both a linear and nonlinear example in two independent variables.

2.39 **For students interested in history.** By researching online, write a short exposition on the history of first-order PDEs and the method of characteristics. In particular, start with the work of Leonhard Euler and Jean-Baptiste le Rond D'Alembert and then address the contributions of Augustin-Louis Cauchy, Paul Charpit, William Rowan Hamilton, Carl Gustav Jacob Jacobi, Joseph-Louis Lagrange, Sophus Lie, and Gaspard Monge.

Chapter 3

The Wave Equation in One Space Dimension

The previous chapter dealt in some generality with first-order PDEs. Now we begin our study of **second-order PDEs**, **not** in generality, but rather through **three** particular linear second-order equations, each related to an important physical phenomenology. "The big three", **the wave, diffusion, and Laplace equations**, are in some sense prototypes for **all** linear second-order PDEs. We start with the wave equation, which shares certain similarities with the first-order equations of the previous chapter.

The **wave equation** models propagation of a "disturbance" in a medium (a material substance which can propagate waves). For example, in one dimension it models relatively small vibrations of a taut string; in two dimensions it models (relatively small) vibrations of a membrane (drum); finally, in three dimensions it models both the propagation of electromagnetic waves in space and the propagation of sound waves in air. Sound waves are (relatively small) variations in air pressure (or air density) which propagate through space.

We have already seen several first-order PDEs which model a disturbance propagating with time: the transport equations of Examples 2.1.3, 2.2.4, Sections 2.2.4 and 2.2.5, and the nonlinear Burgers equation of Section 2.3. These first-order equations do not capture the physics of many wave propagations. Indeed, the homogeneous transport equation simply preserves the overall structure (shape) of the disturbance, and this is surely not the case for most types of wave propagations. Moreover, to model displacements on a string or membrane, one would expect that Newtonian mechanics should play the driving force. From the perspective of displacement, **Newton's Second Law** involves second derivatives, and hence one would expect a **second-order** PDE to govern the behavior. Indeed, what does model a vast number of wave propagations (including the ones discussed above) is the second-order linear PDE known as the **wave equation**. In one space dimension x, it is given for a function $u(x, t)$ of space and time by

$$u_{tt} = c^2 u_{xx}.$$

In two and three space dimensions ((x, y) and (x, y, z)), it is given for $u(x, y, t)$ and $u(x, y, z, t)$, respectively, by

$$u_{tt} = c^2(u_{xx} + u_{yy}) = c^2 \Delta u, \qquad u_{tt} = c^2(u_{xx} + u_{yy} + u_{zz}) = c^2 \Delta u,$$

where Δ denotes the Laplacian in two and three dimensions, respectively. The constant c denotes a physical parameter which is related to the speed of propagation of the waves.

In this chapter we will focus on the wave equation in space dimension $n = 1$. We will consider the problem either on all of one-dimensional space, the real line $\mathbb{R}$, or on the half-line $[0, \infty)$. As we shall see, in order to obtain a unique solution we will need to supplement the wave equation with two pieces of data: the initial displacement and the initial velocity. In doing so, we will not only have a well-posed problem but one for which there is an explicit formula for the unique solution. The **structure** of this formula will reveal qualitative insight into the propagation of waves/disturbances/information in one space dimension.

3.1. • Derivation: A Vibrating String

We derive the one-dimensional wave equation on $\mathbb{R}$ as a model for the shape of an **infinitely long, flexible, elastic, homogeneous** taut string which undergoes relatively small **transverse** motions (vibrations). Here, one should imagine a very long string of a musical instrument (violin or piano) or a very long elastic band, fixed at both ends and extended so that it is taut. Taut means that there is tension in the string. The tension in the string is responsible for the fact that if we initially disturb the string (for example, pluck it somewhere), this disturbance from equilibrium will propagate along the string as time progresses. Note that it is the disturbance (essentially energy) which propagates along the string.

Taking the string to be infinitely long might seem unphysical. While this is true, we wish to model local disturbances in a string which is sufficiently long so that we can ignore what happens at the two boundary ends; thus, for practical purposes, we can assume the string is infinitely long. As we shall see, this assumption is justified by the fact that the equation of motion dictates that **disturbances propagate at a finite speed**. Hence, the motion of a localized disturbance will only reach and interact with the boundary after a long period of time; therefore, for the time interval before the disturbance reaches the ends, we can ignore boundary effects and conveniently assume the string is infinitely long. We certainly do care about vibrations in a finite string where interactions with the boundary are crucial. This will be discussed later. For the time being, we work with a boundaryless infinite string to focus solely on the propagation of localized disturbances.

Our two independent variables are x and t where (i) x denotes (labels) a position along the string and has dimensions of length and (ii) t denotes time and, not surprisingly, has dimensions of time. We assume the string lies in a plane with the horizontal axis (the x-axis) representing the equilibrium (at rest) position of the taut string. The motion of the string is **only** in the **vertical** direction; i.e., the string moves up or down, not from side to side. We call such motions **transverse** motions since they are in a direction which is perpendicular to the direction in which information (the "waves") will propagate. The vertical axis u measures the vertical (transverse) displacement from equilibrium at a time t. So $u(x, t)$ denotes the displacement of the part of the string at position x and time t (cf. Figure 3.1, left). Note that u has dimensions of length. On the other hand, the spatial derivative $u_x(x, t)$ is dimensionless and plays a central role in our derivation. It represents the **slope of the deformed string** and as such a dimensionless measure of the strength of the deformation.

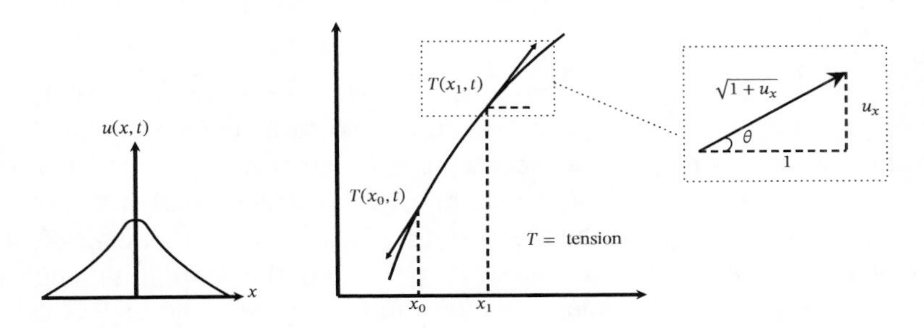

Figure 3.1. Left: The graph of $u(x, t)$ as a function of x for fixed t corresponds to the profile (or picture) of the elastic string at time t, perturbed away from its equilibrium position on the x-axis. **Right:** Here we focus on the piece of the string between $x = x_0$ and $x = x_1$ and the forces on this piece as a result of tension in the string.

We make the following important physical assumptions:

(1) The string is **homogeneous**, meaning that the mass density (mass per unit length) ρ of the string is constant across its length. For example, think of a string made of the same material with constant thickness throughout. The mass density ρ has dimensions of mass $\times$ length^{-1}.

(2) The string is perfectly **flexible and elastic**. Without giving the precise definitions, let us just say that by "flexible" we mean that the string is not a rigid object and is capable of being deformed. By "elastic" we mean the string can be displaced slightly from equilibrium by applying forces; but when these forces are released, it will return to equilibrium. To be more precise, let us focus on the fundamental **internal force**, called **tension**, in the taut string. At any point x and time t, the force **on** the left piece of the string **by** the right piece is due to tension in the string and is directed in the **tangential** direction. The force **on** the right piece of the string **by** the left piece is exactly the opposite force (same magnitude but opposite direction). Let $T(x, t)$ denote the magnitude of this tension force vector. We call $T(x, t)$ the tension in the string, and it has dimensions of force, that is, mass $\times$ length $\times$ time^{-2}.

(3) We assume the tension T does **not depend on time**; thus $T(x, t) = T(x)$[1].

[1] The fact that we assume here that the tension does not depend on time is for simplicity of this derivation. For a more complete derivation see the first few pages of Weinberger's classic book [7].

(4) Lastly, we make a very important assumption that the **disturbances of the string are always small**. In other words, we are only modeling small vibrations. What do we mean by this? We mean the dimensionless quantity $|u_x|$ is small compared to 1. More precisely, we will assume that

$$\sqrt{1+u_x^2} \approx 1. \tag{3.1}$$

This is a strong assumption and, as we will see, together with assumption (1), it will imply that the tension T must be constant throughout the string; i.e., T is independent of x.

The PDE for u will simply be a consequence of **Newton's Second Law**. In particular, we will apply Newton's Second Law (in both the vertical and horizontal directions) to a piece of the string which lies between x_0 to x_1 (see Figure 3.1, right). This amounts to saying that, in either direction, the mass times the acceleration of this piece must be equal to the total (net) force on the piece. In the absence of external forces (e.g., gravity), the only forces on the piece of the string come from the adjoining string itself via the tension. In order to decompose this tension force vector into its vertical (transverse) and horizontal components, we will need to know the angle θ with respect to the positive x-axis at a given position x and time t (see Figure 3.1, right). If we draw a triangle at the point x whose hypotenuse is tangent to the string and horizontal base 1, then the vertical side must be u_x. Thus the hypotenuse is $\sqrt{1+u_x^2}$ and

$$\cos\theta = \frac{1}{\sqrt{1+u_x^2}} \qquad \text{and} \qquad \sin\theta = \frac{u_x}{\sqrt{1+u_x^2}}.$$

Now we invoke the crucial assumption (3.1) so that

$$\cos\theta = 1 \qquad \text{and} \qquad \sin\theta = u_x.$$

Thus, the horizontal and vertical components of the tension force at (x,t) directed by the right part of the string on the left part of the string are

$$T_{\text{horiz}} = T(x)\cos\theta = T(x) \qquad \text{and} \qquad T_{\text{vert}} = T(x)\sin\theta = T(x)\,u_x(x,t).$$

Note, by definition of θ with respect to the positivity convention of the x-axis, that T_{horiz} is always positive whereas T_{vert} could be positive or negative. The total net forces on the piece of the string which lies between x_0 to x_1 are given by

$$F_{\text{horiz}} = T(x_1) - T(x_0) \qquad \text{and} \qquad F_{\text{vert}} = T(x_1)\,u_x(x_1,t) - T(x_0)\,u_x(x_0,t).$$

By assumption (1), there is no movement (hence, no acceleration) in the horizontal directions and, therefore,

$$F_{\text{horiz}} = T(x_1) - T(x_0) = 0. \tag{3.2}$$

Since x_0 and x_1 are arbitrary points, this implies that $T(x)$ is a constant T, and henceforth we may replace $T(x)$ with T.

The total mass times vertical acceleration of the piece is given by

$$\int_{x_0}^{x_1} \rho u_{tt}(s,t)\, ds,$$

where, again, we have used the assumption (3.1) to measure length along the string (arc length) simply by the displacement in x. Invoking Newton's Second Law, $F = ma$, in the vertical direction yields

$$T\, u_x(x_1,t) \;-\; T\, u_x(x_0,t) \;=\; \int_{x_0}^{x_1} \rho u_{tt}(s,t)\, ds. \tag{3.3}$$

Lastly, we *reduce* (3.3) to a PDE. The choices of x_1 and x_0 are arbitrary; in other words, the above holds true for any x_0 and x_1. Hence, fix x_0 but replace x_1 with variable x (this is why we use s as the spatial (dummy) variable of integration). In this notation, (3.3) becomes

$$T\, u_x(x,t) \;-\; T\, u_x(x_0,t) \;=\; \int_{x_0}^{x} \rho u_{tt}(s,t)\, ds,$$

for all x. We can now differentiate with respect to the variable x, using the Fundamental Theorem of Calculus on the right, to find

$$Tu_{xx}(x,t) = \rho u_{tt}(x,t), \qquad \text{or} \qquad \boxed{u_{tt} = c^2 u_{xx}, \qquad \text{where } c^2 = \frac{T}{\rho}.}$$

Since the dimensions of the constants T and ρ are, respectively, mass × length × time^{-2} and mass × length^{-1}, the constant c has dimensions of length × time^{-1}. In other words, it represent a velocity.

We have derived the wave equation which related the acceleration of the string to spatial derivatives. To find the unique trajectory of the string we would need to supplement the PDE with

- the initial displacement of the string, i.e., $u(x,0)$ for $x \in \mathbb{R}$,
- the initial velocity of the string, i.e., $u_t(x,0)$ for $x \in \mathbb{R}$.

Note that if we started with no initial displacement and velocity, i.e., $u(x,0) = 0 = u_t(x,0)$ for all x, then the solution would be $u(x,t) = 0$. In other words, the string stays at its equilibrium position for all time. If one of these initial conditions is not identically zero, then the string will move according to the wave equation. Shortly, we will show how to find this explicit solution but, first, note that the general solution of the PDE alone has a very simple and illuminating form.

Before continuing we make an important remark on what exactly is propagating through space (here one-dimensional). Any particular piece of the string just goes up and down, so the string is not moving with x. What is being transported in the x direction is a "disturbance": More precisely, **energy** is being transported across the string (cf. Section 3.5).

3.2. • The General Solution of the 1D Wave Equation

Let us now find the general solution to the wave equation $u_{tt} = c^2 u_{xx}$. Note that we can rewrite this equation as

$$\partial_{tt} u - c^2 \partial_{xx} u = 0 \quad \text{or} \quad (\partial_{tt} - c^2 \partial_{xx}) u = 0.$$

Think of $(\partial_{tt} - c^2 \partial_{xx})$ as an **operator**, something which operates on functions; it takes a function u and gives back another function involving its second derivatives. This "wave" operator is often called the **D'Alembertian operator** and is denoted with a box $\Box$. Let us assume the solution u is C^2. Hence the temporal and spatial derivatives commute, and we can then "factor" the wave operator $\Box$ into the composition of two other operators:

$$\Box = (\partial_t - c\partial_x)\,(\partial_t + c\partial_x).$$

The above simply means that $(\partial_{tt} - c^2\partial_{xx})\, u = (\partial_t - c\partial_x)\,[(\partial_t + c\partial_x)\, u]$. Each of these two first-order operators leads to a transport equation (cf. Example 2.1.3 of Section 2.1) and, hence, it is not too surprising that the general solution to the wave equations takes the following form:

General Solution to the 1D Wave Equation

Proposition 3.2.1. *The general solution to* $u_{tt} - c^2 u_{xx} = 0$ *is of the form*

$$u(x,t) = f(x+ct) + g(x-ct), \tag{3.4}$$

for arbitrary C^2 *functions* f *and* g *of one variable.*

Proof. Our decomposition (factoring) of the equation suggests that we should change to characteristic coordinates $\zeta = x + ct$ and $\eta = x - ct$, since the wave equation is the composition of two directional derivatives (one in direction ζ and the other in direction η). As discussed in Section 2.1, we abuse notation by using u to denote the unknown function both in terms of (x,t) and in terms of (ζ, η). By the chain rule we have

$$\frac{\partial u}{\partial x} = \frac{\partial u}{\partial \zeta}\frac{\partial \zeta}{\partial x} + \frac{\partial u}{\partial \eta}\frac{\partial \eta}{\partial x} = \frac{\partial u}{\partial \zeta} + \frac{\partial u}{\partial \eta}, \qquad \frac{\partial u}{\partial t} = \frac{\partial u}{\partial \zeta}\frac{\partial \zeta}{\partial t} + \frac{\partial u}{\partial \eta}\frac{\partial \eta}{\partial t} = \frac{\partial u}{\partial \zeta}c + \frac{\partial u}{\partial \eta}(-c).$$

Written in shorthand, we have

$$\partial_x u = \partial_\zeta u + \partial_\eta u \qquad \text{and} \qquad \partial_t u = c\partial_\zeta u - c\partial_\eta u.$$

Combining these equations, we find

$$(\partial_t - c\partial_x) u = -2c\partial_\eta u \qquad \text{and} \qquad (\partial_t + c\partial_x) u = 2c\partial_\zeta u.$$

Thus,

$$\partial_{tt} u - c^2 \partial_{xx} u = (\partial_t + c\partial_x)(\partial_t - c\partial_x)\, u = -4c^2 \partial_\zeta \partial_\eta u = -4c^2 \partial_{\eta\zeta} u.$$

In other words, in the new variables (ζ, η) the wave equation has transformed into the simpler PDE $u_{\eta\zeta} = 0$. Note here that we divided through by the nonzero factor of $-4c^2$.

Now recall from Example 1.5.3 that the general solution to $u_{\zeta\eta} = 0$ in terms of ζ and η is

$$u(\zeta, \eta) = f(\zeta) + g(\eta),$$

for any functions f and g. Returning to our original variables this translates to

$$u(x,t) = f(x+ct) + g(x-ct), \qquad \text{for any functions } f \text{ and } g.$$

□

We make two remarks concerning (3.4):

(i) Note that ct (a velocity times a time) has dimensions of length. Hence, $x-ct$ and $x+ct$ have dimensions of length and denote positions on the string. The reader should get accustomed to this **interplay between space and time**; while slightly confusing at first, it is fundamental to the propagation of waves (recall the transport equation, its solution, and interpretation discussed in Example 2.1.3 of Section 2.1).

(ii) The general solution to the wave equation is the sum of two solutions to transport equations: one which moves to the left with speed c and the other which moves to the right with speed c. Each signal keeps its shape as time changes. However, as we shall see next, the dependence of the two functions (transport shapes) f and g on the initial data is nontrivial.

3.3. • The Initial Value Problem and Its Explicit Solution: D'Alembert's Formula

The general solution of the wave equation involved two arbitrary functions. If we consider the model problem of a vibrating string and the resulting derivation of the wave equation, we expect that we will need both an initial displacement and an initial velocity.[2] Thus, in order to have a unique solution, we consider the following initial value problem:

Initial Value Problem for the 1D Wave Equation

$$\begin{cases} u_{tt} = c^2 u_{xx}, & -\infty < x < \infty,\ t > 0, \\ u(x,0) = \phi(x), & -\infty < x < \infty, \\ u_t(x,0) = \psi(x), & -\infty < x < \infty. \end{cases} \tag{3.5}$$

Unless otherwise specified, we assume throughout this chapter that ϕ and ψ are smooth, say, C^2. The general solution to the wave equation always has the form $f(x+ct) + g(x-ct)$ for functions f and g. By specifying the initial displacement $\phi(x)$ and initial velocity $\psi(x)$, it is natural to ask what the functions f and g should be. While they are not simply $f = \phi$ and $g = \psi$, it is straightforward to determine f and g in terms of ϕ and ψ and, hence, derive the explicit solution to (3.5), first written down by **D'Alembert**[3] in 1746.

[2]Recall from high school physics that if we specify the acceleration of a body, we need to give an initial position and initial velocity in order to determine the position of the body at every time.

[3]**Jean-Baptiste le Rond D'Alembert** (1717–1783) was a French mathematician, physicist, philosopher, and music theorist who is most remembered for the solution formula to the 1D wave equation.

D'Alembert's Formula

Theorem 3.1. *If $u(x,t)$ solves* (3.5), *then*

$$u(x,t) = \frac{1}{2}[\phi(x+ct)+\phi(x-ct)] + \frac{1}{2c}\int_{x-ct}^{x+ct}\psi(s)\,ds. \tag{3.6}$$

One can readily check (cf. Exercise **3.1**) that if u is given as above, then it solves (3.5). Hence, the initial value problem (3.5) has a unique solution given by (3.6), and take note that most of this chapter is based upon consequences of D'Alembert's formula.

Proof. By the Proposition 3.2.1, the general solution of $u_{tt} = c^2 u_{xx}$ takes the form $u(x,t) = f(x+ct) + g(x-ct)$. Keeping in mind that our goal is to determine how the functions f and g are related to ϕ and ψ, we note that

$$u(x,0) = f(x) + g(x) = \phi(x),$$

and, since $u_t(x,t) = cf'(x+ct) - cg'(x-ct)$,

$$u_t(x,0) = cf'(x) - cg'(x) = \psi(x).$$

As we shall soon see, it is useful to replace x with a *neutral* dummy variable α and rewrite these equations as

$$\phi(\alpha) = f(\alpha) + g(\alpha) \qquad \text{and} \qquad \psi(\alpha) = cf'(\alpha) - cg'(\alpha).$$

Differentiating the first equation with respect to α gives us

$$\phi'(\alpha) = f'(\alpha) + g'(\alpha) \qquad \text{and} \qquad \psi(\alpha) = cf'(\alpha) - cg'(\alpha).$$

We can now solve this system for f' and g' as functions of ϕ' and ψ:

$$f'(\alpha) = \frac{1}{2}\left(\phi'(\alpha) + \frac{\psi(\alpha)}{c}\right) \qquad \text{and} \qquad g'(\alpha) = \frac{1}{2}\left(\phi'(\alpha) - \frac{\psi(\alpha)}{c}\right).$$

Integrating with respect to α gives

$$f(\alpha) = \frac{1}{2}\phi(\alpha) + \frac{1}{2c}\int_0^\alpha \psi(s)\,ds + C_1 \qquad \text{and} \qquad g(\alpha) = \frac{1}{2}\phi(\alpha) - \frac{1}{2c}\int_0^\alpha \psi(s)\,ds + C_2,$$

where C_1 and C_1 are arbitrary constants. But what then should we take for C_1 and C_2? Well, we actually only care about $C_1 + C_2$ since the general solution dictates that we add these two functions. Since $\phi(\alpha) = f(\alpha) + g(\alpha)$, we conclude that $C_1 + C_2 = 0$. Therefore, we now know the explicit dependence of the functions f and g on the data ϕ and ψ. Hence, we have

$$\begin{aligned} u(x,t) &= f(x+ct) + g(x-ct) \\ &= \frac{1}{2}[\phi(x+ct) + \phi(x-ct)] + \frac{1}{2c}\int_0^{x+ct} \psi(s)\,ds - \frac{1}{2c}\int_0^{x-ct} \psi(s)\,ds \\ &= \frac{1}{2}[\phi(x+ct) + \phi(x-ct)] + \frac{1}{2c}\int_0^{x+ct} \psi(s)\,ds - \frac{1}{2c}\left(-\int_{x-ct}^{0} \psi(s)\,ds\right) \\ &= \frac{1}{2}[\phi(x+ct) + \phi(x-ct)] + \frac{1}{2c}\left(\int_0^{x+ct} \psi(s)\,ds + \int_{x-ct}^{0} \psi(s)\,ds\right) \\ &= \frac{1}{2}[\phi(x+ct) + \phi(x-ct)] + \frac{1}{2c}\int_{x-ct}^{x+ct} \psi(s)\,ds. \end{aligned}$$ □

D'Alembert's formula gives the solution at position x and time t in terms of the combinations of the initial data. As we shall see, in general this can lead to nonobvious solutions. However, it is insightful here to pause and consider a very special choice of ϕ and ψ.

Example 3.3.1. Let $l > 0$ and suppose $\phi(x) = \sin\frac{x}{l}$ and $\psi(x) = 0$. Then D'Alembert's formula gives

$$u(x,t) = \frac{1}{2}\left(\sin\frac{x+ct}{l} + \sin\frac{x-ct}{l}\right) = \sin\frac{x}{l}\cos\frac{ct}{l},$$

where we used a standard double angle formula for sine. This is a rather special solution, and the fact that it is easy to verify that it solves the wave equation rests upon the special property of sines and cosines, i.e., that their second derivative is, respectively, also a sine and a cosine. This property lies at the heart of Fourier series and the method of separation of variables for boundary value problems (cf. Chapters 11 and 12). The take-away point here is that at any fixed time t, $u(x,t)$ is simply $\phi(x)$ (the initial displacement) multiplied by a damping factor. Hence the basic shape of the initial displacement is kept through time. This will not be the case for most initial data.

3.4. • Consequences of D'Alembert's Formula: Causality

In the previous section we arrived at an explicit form for the solution (Theorem 3.1). What is particularly enlightening about this formula is that it tells us exactly how the displacement of the string at a later time depends on initial displacement and initial velocity. Let us translate the formula into words — D'Alembert's formula says the following:

To find the displacement of the string at a position x and a time $t > 0$, we start by taking the **average of the initial displacements**, not at the same position x, but at positions $x - ct$ and $x + ct$. Then we add this to the **average velocity** over the part of the string which lies between the positions $x - ct$ and $x + ct$ multiplied by t (the time that has elapsed); that is, we add

$$t \times \underbrace{\frac{1}{2ct}\int_{x-ct}^{x+ct} \psi(s)\, ds}_{\text{average of } \psi \text{ over } [x-ct,\, x+ct]} .$$

The relevant points are illustrated in Figure 3.2.

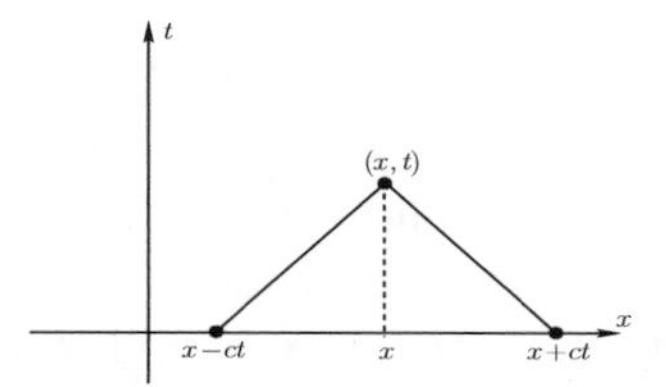

Figure 3.2. The central figure in the x- vs. t-plane (the domain of the solution) which accompanies D'Alembert's formula.

Next we explore the consequences of D'Alembert's formula in two ways: one very general way in terms of regions of space and time and the other via two particular examples.

3.4.1. • The Domain of Dependence and Influence.

The most general way to view **causality** (what causes what) is via the domain of dependence and domain of influence. Both are regions in the x- vs. t-plane (space-time), one based upon the past and the other on the future.

The first is the **domain of dependence**, illustrated in Figure 3.3: Here we fix a point (x_1, t_1) in space-time and **look at the past**. To this end, consider the dependence of the solution $u(x_1, t_1)$ on the initial data. Using D'Alembert's formula, the solution $u(x_1, t_1)$ depends on the initial displacements at positions $x_1 - ct_1$ and $x_1 + ct_1$ and all the initial velocities between them. Hence, it depends on what happened initially (displacement and velocity) in the interval $[x_1 - ct_1, x_1 + ct_1]$. Now let us consider any earlier time t_2, i.e., $0 < t_2 < t_1$, starting our stopwatch here. We can think of t_2 as the initial time and ask what positions are relevant at time t_2 for determining $u(x_1, t_1)$. This is a shorter interval as less time has elapsed from t_2 to t_1 than from 0 to t_1. In fact the interval is exactly $[x_1 - c(t_1 - t_2), x_1 + c(t_1 - t_2)]$. Thus, the entire domain of dependence for a point (x_1, t_1) in space-time is the shaded triangular region shown on the left of Figure 3.3.

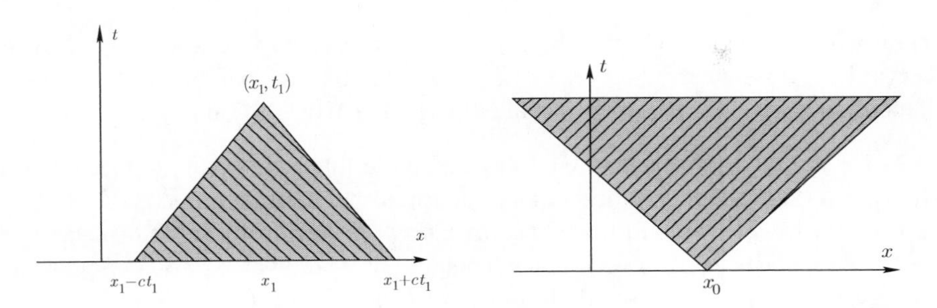

Figure 3.3. Left: Domain of dependence wherein the left and right boundary lines have slope (with respect to t) c and $-c$, respectively. **Right:** The Domain of influence wherein the left and right boundary lines have slope (with respect to t) $-c$ and c, respectively. Note that while we have only indicated a finite region, the domain of influence looks to the future and is, hence, unbounded and extends up for all time.

Alternatively we may look to the future and this will yield the **the domain of influence**. That is, we fix a point on the string x_0 at $t = 0$ and ask: At a later time $t_1 > 0$, what points on the string are effected (influenced) by the displacement/velocity at x_0 for $t = 0$? By D'Alembert, all points in the interval $[x_0 - ct_1, x_0 + ct_1]$ will be influenced by what happens to the point x_0 at $t = 0$. By repeating this for all time $t > 0$, we obtain the domain of influence, partly illustrated by the shaded triangular region shown on the right of Figure 3.3.

These descriptions of the domains of dependence and influence go a long way in explaining the propagation of information in 1D; however, experience has shown that the concept of causality can cause confusion particularly in dealing with the **interplay** between space and time. We will work through an exercise which will hopefully help you digest these concepts.

Example 3.4.1. Consider the 1D wave equation $u_{tt} = u_{xx}$ with initial data

$$u(x,0) = 0, \qquad u_t(x,0) = \begin{cases} 1 & \text{if } |x-3| \le 1 \text{ or } |x+3| \le 1, \\ 0 & \text{otherwise.} \end{cases}$$

(a) At $t = 1$ and $t = 4$, what are the positions on the string where the displacement u is nonzero?

(b) At $t = 10$, what are the positions on the string where the displacement u is maximal?

(c) What is $u(0,t)$ as a function of t?

By D'Alembert, the solution $u(x,t)$ is given by

$$u(x,t) = \frac{1}{2}\int_{x-t}^{x+t} \psi(s)\,ds. \tag{3.7}$$

All the information needed to answer these questions is here. But to do so it would be helpful to absorb what the formula says in words in the context of the particular initial data. At any fixed t, we will have a nonzero displacement at x if the integral is nonzero. The integrand, i.e., $\psi(\cdot)$, is nonzero precisely in the set $(-4,-2) \cup (2,4)$, and, thus, the integral will be nonzero if and only if there is some **overlap** between the **sampling interval** $(x-t, x+t)$ **and** $(-4,-2) \cup (2,4)$. Alternatively, this happens if the domain of dependence for (x,t) includes part of either $(-4,-2)$ or $(2,4)$ at $t=0$.

(a) Fix $t = 1$. We have that $u(10,1) = 0$, since the interval $(9,11)$ intersects neither $(-4,-2)$ nor $(2,4)$. At $t=1$, the same is true for all $x > 10$ and, indeed, also $x < -10$. So let us start at $x = 10$ and trace back in x (lowering x) until the first place where $(x-1, x+1)$ starts to **intersect** $(2,4)$. This will be exactly at $x = 5$. If we continue the process past $x = 5$, we see that after $x = 1$ the interval $(x-1, x+1)$ no longer intersects $(2,4)$, nor does it intersect $(-4,-2)$. Thus, the displacement now becomes zero again. Continuing past $x = 0$ we see that after $x = -1$ the interval $(x-1, x+1)$ intersects $(-4,-2)$ and continues to do so until $x = -5$. Thus at $t = 1$ the displacement is nonzero exactly at points x in the set $(-5,-1) \cup (1,5)$.
Repeating this process at $t = 4$ with positions x from right to left, we see that we begin to have a nonzero disturbance at $x = 8$. As we trace back in x, we now continuously find a nonzero disturbance until we reach $x = -8$. So at $t = 4$ there is no middle region where we do not feel the initial disturbance; this is a consequence of the fact that the sample interval $(x-4, x+4)$ now has a bigger length $2t = 8$. Thus at $t = 4$, the displacement is nonzero exactly at points x in the interval $(-8,8)$.

(b) If we follow the reasoning in part (a), we see that at $t = 10$ the displacement $u(x,10)$ will be maximal if the interval $(x-10, x+10)$ includes the entire set $(-4,-2) \cup (2,4)$; this way the integral in (3.7) is maximal. The set of x with this property is exactly $x \in (-6,6)$.

(c) We have

$$u(0,t) = \frac{1}{2}\int_{-t}^{t} \psi(s)\, ds.$$

From $t = 0$ to $t = 2$, the interval $(-t,t)$ does not intersect the set $(-4,-2) \cup (2,4)$. Hence, $u(0,t) = 0$ for $t \in [0,2]$. This changes at $t = 2$ where the interval $(-t,t)$ begins to intersect the set $(-4,-2) \cup (2,4)$. Indeed, as t increases past 2, the interval $(-t,t)$ intersects more and more of this set until $t = 4$, after which it **encompasses** the entire set. Thus

$$u(0,t) = \begin{cases} 0, & t \in [0,2], \\ t-2, & t \in (2,4], \\ 2, & t > 4. \end{cases}$$

3.4.2. • Two Examples: A Plucked String and a Hammer Blow. We present two examples wherein one of the data functions is identically 0. They correspond, respectively, to the situations of a plucked string and a hammer blow.

Warning: In each case, the illustrated dynamics given by the wave equation will not coincide with your intuition of plucking a guitar string nor hitting a piano string. This is because of our assumption of an infinitely long string. An actual guitar, or piano, string is finite with fixed ends (no displacement at the end points), and as we shall see later, these boundary conditions greatly affect the wave dynamics.

The Plucked String. Let $c = 1$,

$$\phi(x) = \begin{cases} 1 - |x,| & |x| \leq 1, \\ 0, & |x| > 1, \end{cases}$$

and $\psi \equiv 0$. By D'Alembert's formula, the solution is

$$u(x,t) = \frac{1}{2}\big[\phi(x+t) + \phi(x-t)\big].$$

Thus, at any time t, the solution is simply the sum of two *waves* with the same initial shape but half the amplitude, one which has moved to the right by t and one which has moved to the left by t. We plot some of the profiles in Figure 3.4.

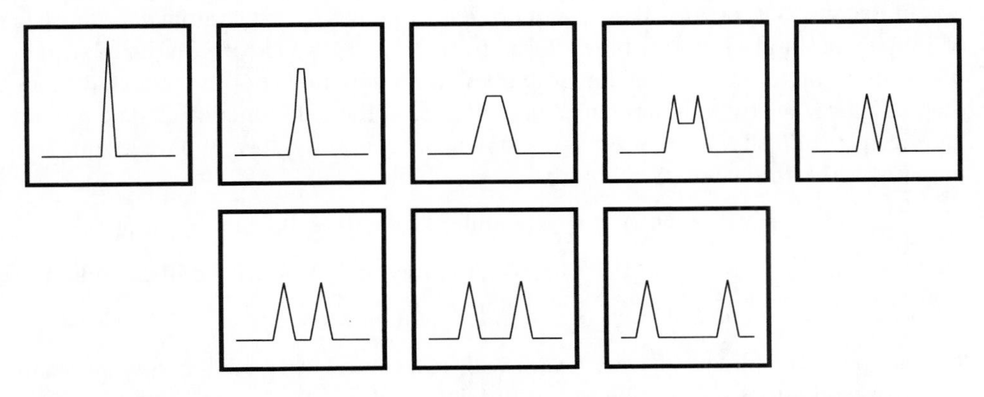

Figure 3.4. Profiles at increasing times showing the propagation of the **plucked string**.

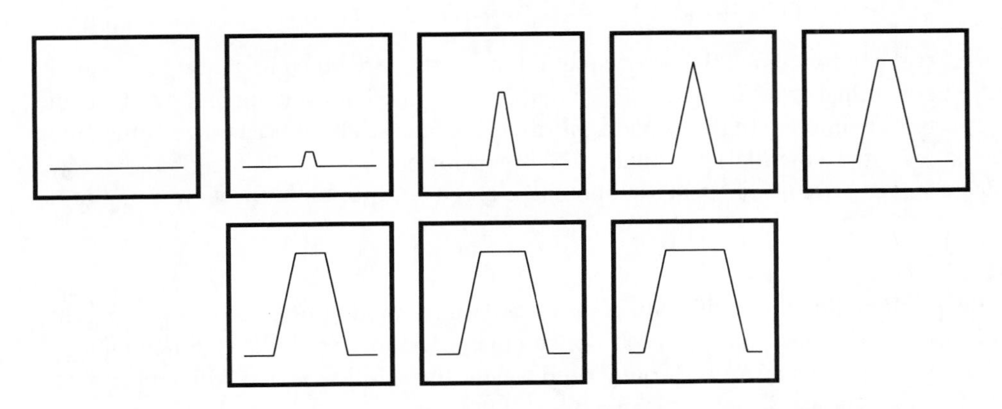

Figure 3.5. Profiles at increasing times showing the propagation of the **hammer blow**.

The Hammer Blow. To model a hammer blow (think of the effect on a piano string by hitting a piano key), we let $c = 1$, $\phi \equiv 0$, and

$$\psi(x) = \begin{cases} 1, & |x| \leq 1, \\ 0, & |x| > 1. \end{cases}$$

By D'Alembert's formula, the solution is

$$u(x,t) = \frac{1}{2}\int_{x-t}^{x+t} \psi(s)\, ds.$$

Following on Example 3.4.1, it is a good exercise in deciphering notation with many "parameters" to *play with* the above formula for this precise ψ and work out the profiles $u(x,t)$ for different t. Basically, fix a time t and then look at different values of x: The formula tells us exactly the value of u at a fixed x and t. Indeed, given the simple structure of ψ, it says the value of $u(x,t)$ is exactly one half the length of the intersection between $[x-t, x+t]$ and $[-1,1]$. From this information alone you should be able to concur with the profiles of Figure 3.5.

In Exercise **3.8** you are asked to plot profiles in cases where the initial data consists of both a pluck and a hammer blow. Recall from Section 1.4.3 **the principle of superposition**, a consequence of the linearity of the wave equation. In this context, it states that we may solve two subproblems, wherein one of the data functions is taken to be identically 0, and add. More precisely, it states that if w and v solve the wave equation with respective initial conditions

$$w(x,0) = \phi(x),\ w_t(x,0) \equiv 0, \qquad \text{and} \qquad v(x,0) \equiv 0,\ v_t(x,0) = \psi(x),$$

then $u = w + v$ solves the full IVP (3.5), i.e., the wave equation with initial conditions

$$u(x,0) = \phi(x) \qquad \text{and} \qquad u_t(x,0) = \psi(x).$$

This simple principle proves very useful with respect to both initial data and certain boundary conditions.

3.5. • Conservation of the Total Energy

After the derivation of the wave equation, we highlighted the fact that it is energy that is being transported along the string; that is, the "wave" (movement of the disturbance) pertains to energy. In this section, we are more precise about the nature of the energy (kinetic and potential) and make the important observation that the total energy is conserved. To this end, let us suppose $u(x,t)$ is a C^2 solution to the wave equation

$$u_{tt} = \frac{T}{\rho} u_{xx}.$$

We assume that the initial data $\phi(x)$ and $\psi(x)$ are smooth functions with compact support. This means that they both vanish outside the interval $[-R, R]$, for some $R > 0$. By D'Alembert's formula, for any fixed time t, the solution $u(x,t)$ will vanish outside the interval $[-R-ct, R+ct]$, where $c = \sqrt{T/\rho}$.

Fix $t > 0$ and let us consider the total kinetic energy of the string at time t. Recall the usual form of kinetic energy: one half the mass times the velocity squared. In our continuum setting, in which velocity changes continuously across the string and mass is given by density ρ times length (dx), this translates as

$$KE(t) = \frac{1}{2}\int_{-\infty}^{\infty} \rho u_t^2\, dx. \tag{3.8}$$

In terms of dimensions, $KE(t)$ has dimensions of energy which is mass × length^2 × time^{-2}. Next we define the total potential energy of the string at time t. It is reasonable to expect the potential energy to be proportional to the tension T. Moreover, it is not hard to convince oneself that it should also increase with $|u_x|$; the higher the slope, the greater the restoring force wishing to return the string to its equilibrium position. Hence, the greater the tendency or potential for the restoration of the string to its equilibrium state. Via dimensional analysis (cf. Section A.11) it is not hard to derive the formula for the total potential energy:

$$PE(t) = \frac{T}{2}\int_{-\infty}^{\infty} u_x^2\,dx.$$

In particular, note that $PE(t)$ has dimensions of energy.

Let us see how the kinetic energy changes with time. We compute the time derivative of (3.8) to find

$$\frac{d\,KE(t)}{dt} = \frac{d}{dt}\left(\frac{1}{2}\int_{-\infty}^{\infty} \rho u_t^2\,dx\right) = \frac{1}{2}\int_{-\infty}^{\infty} \frac{\partial(\rho u_t^2)}{\partial t}\,dx = \rho\int_{-\infty}^{\infty} u_t\,u_{tt}\,dx.$$

Note that the differentiation under the integral sign is legal because of the smoothness of the integrand and its compact support. Since u satisfies the wave equation, we have $\rho u_{tt} = T u_{xx}$. Hence

$$\frac{d\,KE(t)}{dt} = T\int_{-\infty}^{\infty} u_t\,u_{xx}\,dx.$$

Integrating by parts in the above integral, we have

$$\frac{dKE(t)}{dt} = Tu_tu_x|_{-\infty}^{\infty} - T\int_{-\infty}^{\infty} u_{tx}\,u_x\,dx = -T\int_{-\infty}^{\infty} u_{tx}\,u_x\,dx.$$

Here we used the fact that for any fixed t, u_t and u_x will have compact support, and hence we can dismiss the boundary terms at $\pm\infty$. Therefore, we have

$$\begin{aligned}\frac{dKE(t)}{dt} = -T\int_{-\infty}^{\infty} u_{tx}\,u_x\,dx &= -T\int_{-\infty}^{\infty}\left(\frac{1}{2}u_x^2\right)_t\,dx\\ &= -\frac{d}{dt}\left(T\int_{-\infty}^{\infty}\left(\frac{1}{2}u_x^2\right)dx\right) = -\frac{dPE(t)}{dt}.\end{aligned}$$

We have just shown that

$$\frac{dKE(t)}{dt} = -\frac{dPE(t)}{dt}. \tag{3.9}$$

In terms of the total energy $E(t) = KE(t) + PE(t)$, equation (3.9) gives

$$\frac{dE(t)}{dt} = 0,$$

and, hence, $E(t)$ is constant. Thus, the wave equation has the property that the **total energy is conserved**.

We can use *conservation of energy* to show that the initial value problem

$$\begin{cases} u_{tt} = c^2 u_{xx}, & -\infty < x < \infty,\ t > 0, \\ u(x,0) = \phi(x), & -\infty < x < \infty, \\ u_t(x,0) = \psi(x), & -\infty < x < \infty, \end{cases}$$

has a unique (smooth) solution. Note that we have, in principle, already proven this (why?). To see this, let us again assume that the initial data $\phi(x)$ and $\psi(x)$ are smooth functions with compact support. Suppose there were two C^2 solutions u_1 and u_2. Then $u = u_1 - u_2$ would be a solution to the wave equation with initial conditions $u(x,0) = 0 = u_t(x,0)$. Thus, uniqueness (i.e., the fact that $u_1 \equiv u_2$) boils down to whether or not the only solution to this initial problem is the zero solution ($u(x,t) \equiv 0$). Turning to the total energy,

$$E(t) = \frac{1}{2} \int_{-\infty}^{\infty} \left(\rho u_t^2 + T u_x^2\right) dx \geq 0,$$

we see that $E(0) = 0$. But then, by conservation of energy, we must have $E(t) \equiv 0$. The only way this can happen for a smooth solution u is for $u_t \equiv 0$ and $u_x \equiv 0$. Thus, $u(x,t)$ must be constant for all (x,t). Since $u(x,0) = 0$ we must have $u \equiv 0$.

While the total energy across the string is conserved in time, the **distribution of kinetic and potential energy** varies across the string (i.e., in x). Thus in terms of wave propagation, it is exactly the **transfer** of kinetic to potential energy (and vice versa) which is transported across the string. This is a more precise description of the statement that it is energy that propagates through space. The reader is encouraged to reflect on this transfer of energy in the visual profiles of the plucked string and hammer blow of Section 3.4.2. For example in the hammer blow scenario, some of the initial burst of kinetic energy is transferred into potential energy as the string deforms from equilibrium. Try to visualize the propagation of the hammer blow in terms of the transfer of kinetic energy (changes in displacement from one time frame to another) with potential energy (places in a frame where the slope is not 0).

3.6. • Sources

Suppose the infinite string is now subjected to a vertical external **body force** which, at each place x and time t, is given by $f(x,t)$. The body force is understood to be force per unit mass and hence has physical dimensions of length per time^2. For example, in the case of gravity, the body force is simply $f(x,t) \equiv -g$, a constant. If we reconsider the balancing of the vertical component of the forces in Section 3.1, we find (cf. Exercise **3.6**) the modified wave equation

$$u_{tt} - c^2 u_{xx} = f(x,t),$$

where $c^2 = T/\rho$. We call the inhomogeneous right-hand side **a source term**. The associated initial value problem is

$$\boxed{\begin{cases} u_{tt} - c^2 u_{xx} = f(x,t), & -\infty < x < \infty,\ t > 0, \\ u(x,0) = \phi(x), & -\infty < x < \infty, \\ u_t(x,0) = \psi(x), & -\infty < x < \infty. \end{cases}} \tag{3.10}$$

We claim that the solution is given by the following extension of D'Alembert's formula.

Theorem 3.2. *If u solves* (3.10), *then*

$$u(x,t) = \frac{1}{2}[\phi(x+ct) + \phi(x-ct)] + \frac{1}{2c}\int_{x-ct}^{x+ct}\psi(s)\,ds + \frac{1}{2c}\iint_D f(y,\tau)\,dy\,d\tau, \tag{3.11}$$

where D is precisely the domain of dependence associated with (x,t), *i.e., the triangle in the xt-plane with top point* (x,t) *and base points* $(x-ct,0)$ *and* $(x+ct,0)$.

As before, one can readily check that if u is defined by (3.11), then u solves (3.10). Thus, we have found the unique solution to (3.10). Before proving this, let's look at what (3.11) is saying. The first two terms are simply D'Alembert's formula. However, the last term says that to calculate the displacement at a point x and time t, we need to integrate the **body force history** over the domain of dependence for (x,t).

A general approach to solving linear PDEs with inhomogenous terms is known as **Duhamel's Principle**[4]. This principle easily extends to higher-space dimensions and, indeed, to many other linear partial differential equations (cf. Exercise **3.18** and Section 7.8). The principle is related to the method of **variation of parameters** in ODEs (cf. Exercise **3.16**). Next, we will present, and in fact prove, the principle for the wave equation. This approach will yield (cf. Exercise **3.17**) the formula (3.11). We will also present an alternate direct proof of Theorem 3.2 by invoking Green's Theorem.

3.6.1. • Duhamel's Principle.

We state and prove the principle in the context of the wave equation. In Section 7.8.1, we will give a physical explanation for Duhamel's Principle but in the context of the heat (diffusion) equation with a heat source.

First note that, by superposition, it suffices to solve

$$\begin{cases} u_{tt} - c^2 u_{xx} = f(x,t), & -\infty < x < \infty,\ t > 0, \\ u(x,0) = 0, & -\infty < x < \infty, \\ u_t(x,0) = 0, & -\infty < x < \infty. \end{cases} \tag{3.12}$$

Indeed, once this is solved, we can simply add it to D'Alembert's solution to the (homogeneous) wave equation (3.5), with initial data ϕ and ψ. The sum will solve (3.10). To solve (3.12) we introduce an additional temporal parameter $s \in [0,\infty)$. For each **fixed** s, we consider an indexed solution (indexed by s) $w(x,t;s)$ to

$$\begin{cases} w_{tt}(x,t;s) = c^2 w_{xx}(x,t;s), & -\infty < x < \infty,\ t > s, \\ w(x,s;s) = 0 \quad w_t(x,s;s) = f(x,s), & -\infty < x < \infty. \end{cases} \tag{3.13}$$

Effectively we are now solving the wave equation with time starting at $t = s$ (not $t = 0$) and incorporating the source function f into the "initial" data for the velocity at $t = s$. In invoking Duhamel's Principle, we always place the source function f in the initial data corresponding to one time derivative less than the total number of time derivatives. Note that the solution w has physical dimensions of length per time (as

[4]**Jean-Marie Constant Duhamel** (1797–1872) was a French mathematician and physicist who is primarily remembered for this principle.

opposed to u which has dimensions of length). Duhamel's Principle states that the solution to (3.12) is given by

$$\boxed{u(x,t) = \int_0^t w(x,t;s)\,ds.} \tag{3.14}$$

Let us check (effectively prove) Duhamel's Principle for the wave equation by checking that $u(x,t)$ defined by (3.14) does indeed solve (3.12). In Exercise **3.17** you are asked to further show that (3.14) reduces to the exact same formula as in (3.11).

Proof of Duhamel's Principle: We show that $u(x,t)$ defined by (3.14) does indeed satisfy the inhomogeneous wave equation in (3.12). To this end, we note by differentiation under the integral sign (Theorem A.10), we have

$$u_{xx}(x,t) = \int_0^t w_{xx}(x,t;s)\,ds.$$

On the other hand, by the Leibnitz Rule Theorem A.12 applied to $F(t) := u(x,t)$ for x fixed, we have

$$F'(t) = u_t(x,t) = w(x,t;t) + \int_0^t w_t(x,t;s)\,ds. \tag{3.15}$$

Since by (3.13), $w(x,t;t) = 0$, we have

$$F'(t) = u_t(x,t) = \int_0^t w_t(x,t;s)\,ds.$$

Using the Leibnitz rule once more on $F'(t) = u_t(x,t)$, we find

$$F''(t) = u_{tt}(x,t) = w_t(x,t;t) + \int_0^t w_{tt}(x,t;s)\,ds.$$

Again using (3.13), we have $w_t(x,t;t) = f(x,t)$ and

$$u_{tt} - c^2 u_{xx} = f(x,t) + \int_0^t \underbrace{\left(w_{tt}(x,t;s) - c^2 w_{xx}(x,t;s)\right)}_{0}\,ds = f(x,t).$$

Finally, it is clear from the definition of u and (3.13) that $u(x,0) = 0$. Moreover, via (3.15), we see that we also have $u_t(x,0) = 0$.

3.6.2. Derivation via Green's Theorem. As an alternate approach, we show how to prove Theorem 3.2 by using the classical Green's Theorem (Theorem A.2). This provides a nice example of a case where we effectively solve the PDE by integrating both sides! We follow the presentation in Strauss [**6**].

Proof. Since it is convenient to use x and t as dummy variables of integration, let's denote the point at which we want to derive (3.11) by (x_0, t_0). Thus, fix (x_0, t_0) and let D denote its associated domain of dependence in the x- vs. t-plane (see Figure 3.6). We now assume a solution to (3.10) exists. We integrate both sides of the PDE over D; i.e., we have

$$\iint_D f(x,t)\,dx\,dt = \iint_D \left(u_{tt} - c^2 u_{xx}\right)\,dx\,dt.$$

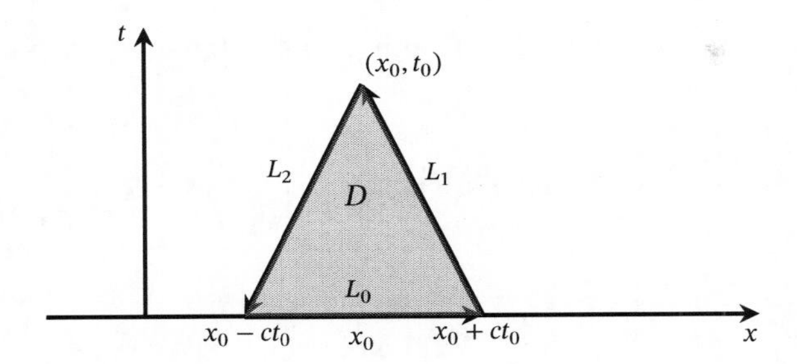

Figure 3.6. The curve $L = L_0 \cup L_1 \cup L_2$.

The left-hand side appears in our solution formula, so our goal is to derive the other terms from the right-hand side. Next, we label and orient the boundary of D with $L = L_0 \cup L_1 \cup L_2$ (shown in Figure 3.6) and invoke Green's Theorem (Theorem A.2) with $P = -u_t$ and $Q = -c^2 u_x$ to yield

$$\begin{aligned} \iint_D (-c^2 u_x)_x - (-u_t)_t \, dx\, dt &= \int_{\partial D} -c^2 u_x \, dt \; - \; u_t \, dx \\ &= \int_{L_0 \cup L_1 \cup L_2} -c^2 u_x \, dt \; - \; u_t \, dx \\ &= \sum_{i=0}^{2} \int_{L_i} -c^2 u_x \, dt \; - \; u_t \, dx. \end{aligned} \tag{3.16}$$

Let us compute each of these line integrals. On L_0, we will use x as our parameter and note that $dt = 0$ and x goes from $x_0 - ct_0$ to $x_0 + ct_0$. Note, also, that $u_t(x, 0) = \psi(x)$ so we have

$$\int_{L_0} -c^2 u_x dt \; - \; u_t \, dx \; = \; - \int_{x_0 - ct_0}^{x_0 + ct_0} \psi(x) \, dx.$$

The line L_1 has equation $x + ct = x_0 + ct_0$. We use t as our parameter so that $dt = dt, dx = -cdt$, and t ranges from 0 to t_0. This gives

$$\begin{aligned} \int_{L_1} -c^2 u_x \, dt \; - \; u_t \, dx &= \int_0^{t_0} (-c^2 u_x \; + \; c u_t) \, dt \\ &= c \int_0^{t_0} \frac{du(x(t), t)}{dt} \, dt \\ &= c u(x_0, t_0) \; - \; c\phi(x_0 + ct_0). \end{aligned}$$

Similarly, on L_2 we find

$$\int_{L_2} -c^2 u_x \, dt \; - \; u_t \, dx \; = \; -c\phi(x_0 - ct_0) \; + \; c u(x_0, t_0).$$

Adding the three line integrals in (3.16), noting that on D we have $u_{tt} - c^2 u_{xx} = f(x,t)$, and dividing by $2c$ yields

$$u(x_0, t_0) = \frac{1}{2}[\phi(x_0 + ct_0) + \phi(x_0 - ct_0)] + \frac{1}{2c}\int_{x_0-ct_0}^{x_0+ct_0} \psi(s)\,ds + \frac{1}{2c}\iint_D f(x,t)\,dx\,dt.$$

□

Note that by setting $f \equiv 0$ in (3.11) we recover D'Alembert's formula; hence, our proof of Theorem 3.2 via Green's Theorem yields another derivation/proof of D'Alembert's formula.

3.7. • Well-Posedness of the Initial Value Problem and Time Reversibility

Recall that there are three core ingredients that make a well-posed problem: (i) existence, (ii) uniqueness, and (iii) stability with respect to small perturbations in the data.

The initial value problem (3.5) is an example of a well-posed problem. We have already shown the existence of a unique solution (why unique?). The explicit nature of D'Alembert's solution allows us to readily address the stability by demonstrating that small changes in either ϕ or ψ lead to small changes in the solution u at any later time t. To address small changes (perturbations) in ϕ and ψ, we need to be precise about a notion of **closeness of functions** as there are many such notions. One simple notion is to say that two functions $\phi_1(x)$ and $\phi_2(x)$ are close if the maximum value of $|\phi_1(x) - \phi_2(x)|$ over all $x \in \mathbb{R}$ is small. Technically, to be precise, one should replace maximum with supremum. We use the notation that for any function $\phi(x)$, $||\phi(x)||_\infty$ denotes the supremum of $|\phi(x)|$ over $x \in \mathbb{R}$. In this case, **a statement of stability** is as follows: Fix a time t, say, $t_* > 0$. Assume ϕ_i, ψ_i $(i = 1, 2)$ are bounded functions on the real line. Let u_i $(i = 1, 2)$ denote the solution to

$$\frac{\partial^2 u_i}{\partial t^2} - c^2 \frac{\partial^2 u_i}{\partial x^2} = 0, \quad u_i(x, 0) = \phi_i(x), \quad \frac{\partial u_i}{\partial t}(x, 0) = \psi_i(x).$$

Then for every $\epsilon > 0$ we can find $\delta > 0$ such that

$$\textbf{if} \quad ||\phi_1(x) - \phi_2(x)||_\infty < \delta \text{ and } ||\psi_1(x) - \psi_2(x)||_\infty < \delta,$$

$$\textbf{then} \qquad ||u_1(x, t_*) - u_2(x, t_*)||_\infty < \epsilon.$$

In Exercise **3.5**, you are asked to use D'Alembert's formula to prove this statement. If you are unfamiliar with the notion of supremum, you may just assume ϕ_i, ψ_i $(i = 1, 2)$ are bounded, continuous functions with compact support. Then **all** functions will attain their maxima and minima. Hence, $||f(x)||_\infty$ just denotes the maximum value of $|f(x)|$ over all $x \in \mathbb{R}$.

As an equation involving time, the initial value problem to the wave equation, whereby we wish to infer information at time $t > 0$ from the information at $t = 0$, makes physical sense. However, the wave equation is also completely **time reversible**. This means that we can prescribe the displacement and velocity at any time t_0 and use the wave equation to determine the solution for all times $t \leq t_0$. More precisely, for any smooth $\phi(x)$ and $\psi(x)$, the **backward problem**

$$\begin{cases} u_{tt} = c^2 u_{xx}, & x \in \mathbb{R},\ t \leq t_0, \\ u(x, t_0) = \phi(x), & x \in \mathbb{R}, \\ u_t(x, t_0) = \psi(x), & x \in \mathbb{R}, \end{cases}$$

is well-posed. Here, we specify the displacement and the velocity at a certain time t_0 and wish to determine the history (the configuration at all earlier times). This is not too surprising based both on (i) physical grounds (what the wave equation models and the conservation of energy) and (ii) the algebraic structure of the equation; if we make a change of independent variable $s = -t$ in the wave equation, we obtain the exact same equation $u_{ss} = c^2 u_{xx}$.

3.8. • The Wave Equation on the Half-Line with a Fixed Boundary: Reflections

We consider our first *boundary value problem*: the equation for a semi-infinite string parametrized by $x \in [0, \infty)$. In this case, we will only have initial data defined for $x \geq 0$. The problem cannot be well-posed unless we specify some additional information about what happens for all time $t > 0$ at the end point $x = 0$. This information is called a boundary condition.

3.8.1. • A Dirichlet (Fixed End) Boundary Condition and Odd Reflections.

We model the situation of a fixed end for the left boundary of the semi-infinite string, i.e., the end labeled by $x = 0$. We prescribe that

$$u(0, t) = 0 \quad \text{for all } t \geq 0.$$

This boundary condition ensures that there is never any vertical displacement at the boundary point. Together, we obtain the following boundary/initial value problem:

$$\begin{cases} v_{tt} = c^2 v_{xx}, & x \geq 0,\ t \geq 0, \\ v(x, 0) = \phi(x), & x \geq 0, \\ v_t(x, 0) = \psi(x), & x \geq 0, \\ v(0, t) = 0, & t \geq 0. \end{cases} \tag{3.17}$$

As we shall see shortly, there is a reason why we use v for the displacement instead of u. What is different from our previous IVP (3.5) is precisely that we now only work on the half-line $x \geq 0$ and have an additional condition at the left end ($x = 0$).

Can we just blindly apply D'Alembert's formula to (3.5)? There are two issues here:

- D'Alembert's formula for $u(x, t)$ requires information about the initial data on the interval $(x - ct, x + ct)$, and even if $x \geq 0$, $x - ct$ might still be negative. We only have data for $x \geq 0$.
- How do we incorporate the fixed boundary condition with the D'Alembert approach?

Both of these complications are solved by **extending ϕ, ψ to all of $\mathbb{R}$ via odd reflections**. This will at first seem unmotivated, but it turns out to be mathematically and physically the right thing to do! Indeed, we will soon see that it **naturally** enforces the fixed boundary condition at $x = 0$.

Define

$$\phi_{\text{odd}}(x) = \begin{cases} \phi(x), & x \geq 0, \\ -\phi(-x), & x < 0, \end{cases} \qquad \psi_{\text{odd}}(x) = \begin{cases} \psi(x), & x \geq 0, \\ -\psi(-x), & x < 0. \end{cases}$$

Since ϕ_{odd} and ψ_{odd} are defined for all x, we can apply D'Alembert's formula to ϕ_{odd} and ψ_{odd} and obtain a solution $u(x,t)$ to the wave equation, valid for all $x \in \mathbb{R}$ and $t > 0$ with initial displacement and velocity ϕ_{odd} and ψ_{odd}, respectively. Now, what if we restrict this solution to $x \geq 0$? That is, for $x \geq 0$ we define

$$v(x,t) \;=\; u(x,t) \;=\; \frac{1}{2}[\phi_{\text{odd}}(x+ct) \;+\; \phi_{\text{odd}}(x-ct)] \;+\; \frac{1}{2c}\int_{x-ct}^{x+ct} \psi_{\text{odd}}(s)\,ds. \tag{3.18}$$

Does this work? It certainly satisfies the PDE and initial conditions in (3.17), but what about the boundary condition? If we place $x = 0$ into the above formula (3.18), we find that for all $t \geq 0$,

$$v(0,t) \;=\; u(0,t) \;=\; \frac{1}{2}[\phi_{\text{odd}}(ct) \;+\; \phi_{\text{odd}}(-ct)] \;+\; \frac{1}{2c}\int_{-ct}^{ct} \psi_{\text{odd}}(s)\,ds.$$

But since by definition both ϕ_{odd} and ψ_{odd} are odd functions, we have $v(0,t) = 0$ for all $t \geq 0$. Hence choosing these odd extensions and applying D'Alembert's formula on the entire real line was, at least mathematically, the right thing to do! To see that it was physically the right thing to do requires some, perhaps empirical, physical intuition about wave reflections.

Next, let us write (3.18) exclusively in terms of ϕ and ψ to unravel exactly what is happening. To this end, we have two cases to consider:

(i) If $x \geq ct$ (i.e., $x - ct \geq 0$), then all the relevant positions, $s \in [x-ct, x+ct]$, are positive and ϕ_{odd} and ψ_{odd} agree exactly with ϕ and ψ. Thus,

$$v(x,t) \;=\; \frac{1}{2}[\phi(x+ct) \;+\; \phi(x-ct)] \;+\; \frac{1}{2c}\int_{x-ct}^{x+ct} \psi(s)\,ds, \tag{3.19}$$

the same as before. In this case, the position x is sufficiently (with respect to time t) away from the end position 0 that the fixed boundary has no effect on the solution.

(ii) If $x \leq ct$ (i.e., $x - ct \leq 0$), then some of the relevant positions, $s \in [x-ct, x+ct]$, will be negative, precisely $s \in [x-ct, 0)$. Using the definition of ϕ_{odd} and ψ_{odd}, we have

$$v(x,t) \;=\; \frac{1}{2}[\phi(x+ct) \;-\; \phi(ct-x)] \;+\; \frac{1}{2c}\left(\int_{0}^{x+ct} \psi(s)\,ds \;+\; \int_{x-ct}^{0} -\psi(-s)\,ds\right).$$

By a change of variables, $s \longmapsto -s$, we note that

$$\int_{x-ct}^{0} -\psi(-s)\,ds = \int_{ct-x}^{0} \psi(s)\,ds = -\int_{0}^{ct-x} \psi(s)ds$$

and, hence, by cancellation in the integral we have

$$v(x,t) = \frac{1}{2}[\phi(x+ct) - \phi(ct-x)] + \frac{1}{2c}\int_{ct-x}^{x+ct} \psi(s)\,ds. \tag{3.20}$$

We have proven the following theorem:

Solution Formula for the Fixed End BVP/IVP

Theorem 3.3. *If v solves (3.17), then*

$$v(x,t) = \begin{cases} \frac{1}{2}[\phi(x+ct) + \phi(x-ct)] + \frac{1}{2c}\int_{x-ct}^{x+ct} \psi(s)\,ds & \text{if } x \geq ct, \\ \frac{1}{2}[\phi(x+ct) - \phi(ct-x)] + \frac{1}{2c}\int_{ct-x}^{x+ct} \psi(s)\,ds & \text{if } 0 < x \leq ct. \end{cases} \tag{3.21}$$

As usual, one can directly check that if v is defined above, then v solves (3.17).

3.8.2. • Causality with Respect to the Fixed Boundary.

We will now explore the preceding formula (3.21) from the perspective of causality and the fixed boundary. In (3.21), we are presented with a dichotomy depending on whether $x < ct$ or $x \geq ct$.

- If $x \geq ct$, then the solution and domains of dependence/influence are **the same as before**. In this instance, for the particular time t, the position x is sufficiently far from the fixed boundary point $x = 0$; consequently the fixed boundary has no influence on the displacement $u(x,t)$.
- If $x < ct$, the situation is different; the solution $u(x,t)$ will now depend on initial displacement and velocity for string position in the interval $[ct - x, x + ct]$. This is illustrated in Figure 3.7, where one notes that what happens initially to the string in the region $(0, ct - x)$ is irrelevant and has been **shadowed** by the fixed boundary. Alternatively, given any point x_0, the domain of influence of displacement/velocity at x_0 and $t = 0$ is illustrated in Figure 3.8.

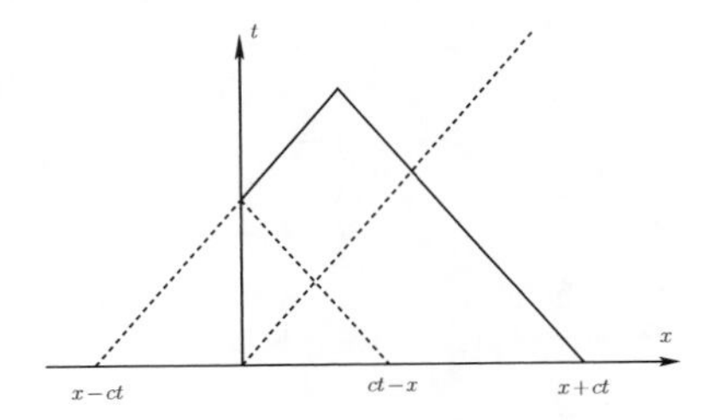

Figure 3.7. Note the shadowing effect of the initial disturbance in the region $(0, ct - x)$.

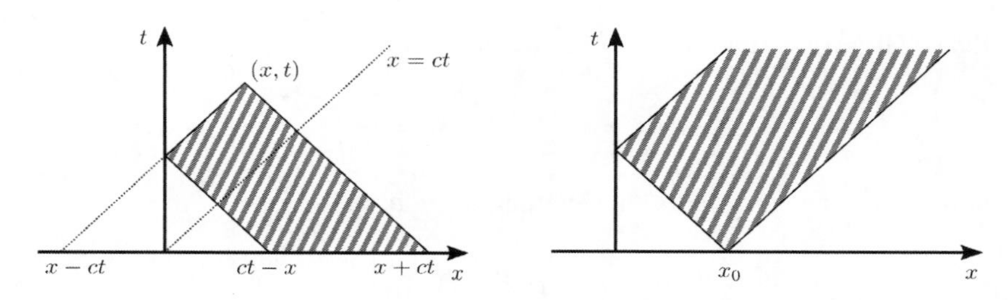

Figure 3.8. The domain of dependence (left) and the domain of influence (right). Note that the domain of influence continues infinitely upwards for all time t.

3.8.3. • The Plucked String and Hammer Blow Examples with a Fixed Left End. We now return to the examples of the plucked string and the hammer blow to observe the effect of including a boundary condition. In this scenario, one can think of a long string which is clamped down at $x = 0$.

Example 3.8.1 (Plucked Semi-Infinite String with a Fixed Left End). We return to the example of a plucked string. Let $c = 1$ and consider the initial data

$$\phi(x) = \begin{cases} 1 - |x-5|, & |x-5| \le 1, \\ 0, & |x-5| > 1, \end{cases}$$

and $\psi \equiv 0$. We plot the profiles in Figure 3.9. These can readily be achieved by inputting the formula into some mathematical software packages. Note that fixing the left boundary has the effect of an **odd reflection** on an incoming wave.

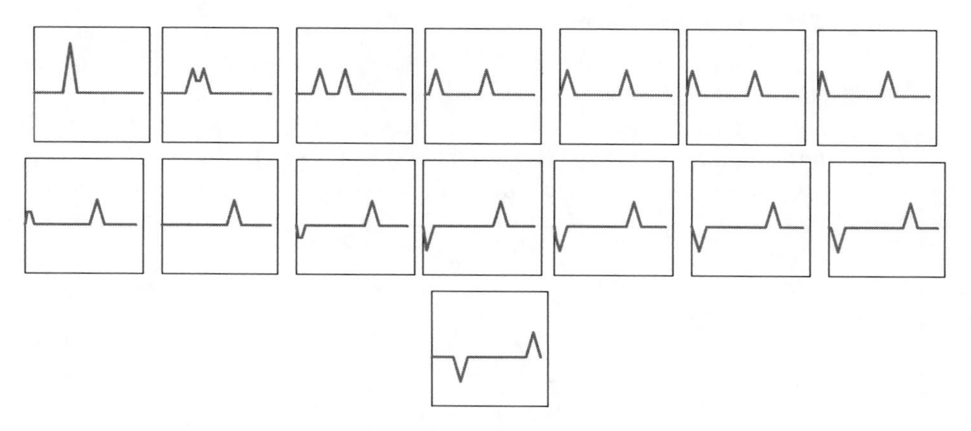

Figure 3.9. Propagation of the plucked string with a **fixed** boundary at $x = 0$. Notice that the initial disturbance (the "pluck") splits in two, with one piece traveling to the left and the other to the right. Once the left-hand tail of the left piece reaches the boundary $x = 0$, it is **absorbed** by the boundary, and then its **odd reflection** (with respect to the horizontal axis) is ejected and moves to the right thereafter.

On the other hand, it is rather instructive to see what is going on close to $x = 0$ by hand using only the simple, yet rich, structure of the solution formula. To this end, let me guide you through the steps which are presented for this initial data (plucked string about $x = 5$). As with the infinite string, the initial pulse (pluck) breaks apart into two pieces, one going to the right and the other to the left. After a certain interval of time, the pulse traveling to the left will reach the fixed boundary point. This will happen at $t = 4$. What happens next? Since we are interested in what occurs close to $x = 0$, our solution formula (3.21) tells us that

$$v(x,t) = \frac{1}{2}\left(\phi(x+t) - \phi(t-x)\right).$$

Fix $t > 0$ and let us first consider the right-hand side for all $x < t$. As a function of x evolving with time past $t = 4$, we can view the first term as transporting the support of ϕ (the pluck) to the left through the origin ($x = 0$) with speed $c = 1$. On the other hand, the second term, again viewed as a function of x evolving with time, is simply the odd reflection of the first (cf. Figure 3.10). That is, if we fix time t and let

$$f_1(x) := \frac{1}{2}\phi(x+t)$$

and

$$f_2(x) := -\frac{1}{2}\phi(t-x),$$

then for all x,

$$f_1(x) = -f_2(-x).$$

Note that since $\phi(x)$ is defined for $x \geq 0$, $f_1(x)$ is defined for $x \geq -t$ and $f_2(x)$ for $x \leq t$. At a fixed time t, we obtain the solution close to $x = 0$ (the profile of the string close to the fixed end) by simply:

- adding $f_1(x)$ and $f_2(x)$ and
- restricting attention to positive x.

This process is illustrated in Figure 3.10 which superimposed the graphs of $\frac{1}{2}\phi(x+t)$, $-\frac{1}{2}\phi(t-x)$ (as functions of x), and their sum at fixed times $t = 4.25$ and $t = 4.75$. Note that at $t = 5$ both parts will exactly cancel out and the string will be completely flat close to the boundary.

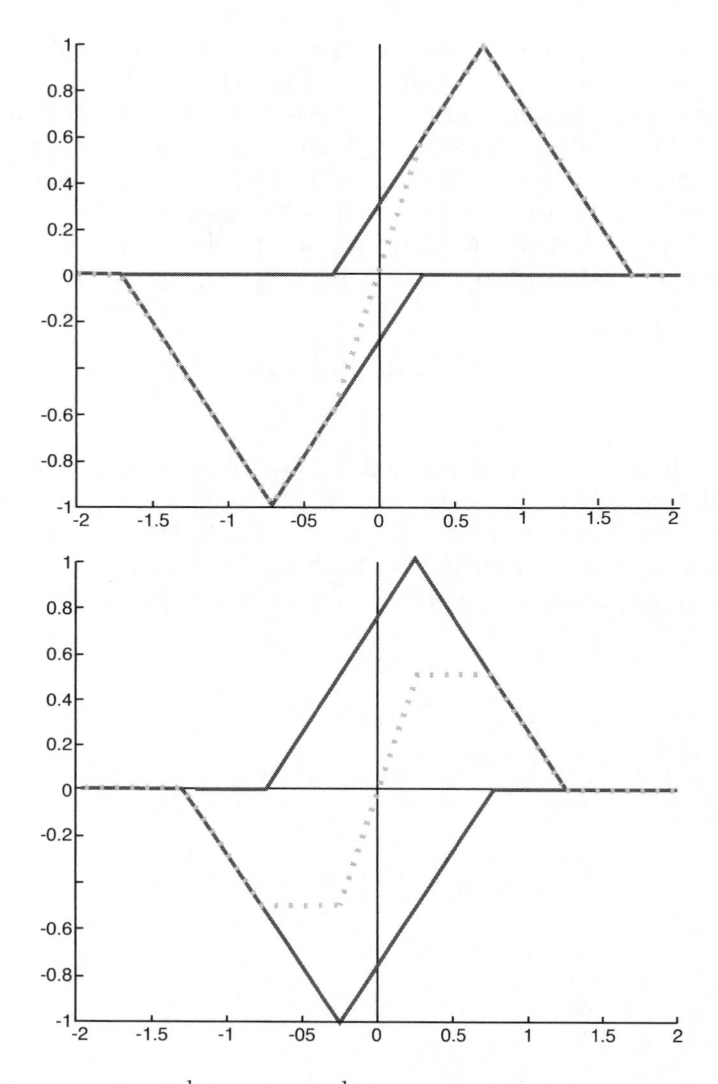

Figure 3.10. Plots of $\frac{1}{2}\phi(x+t)$ and $-\frac{1}{2}\phi(t-x)$ as well as their sum (dotted) as a function of x for fixed $t = 4.25$ and 4.75.

Example 3.8.2 (Hammer Blow for the Semi-Infinite String with a Fixed Left End). We again take $c = 1$, but this time choose the initial data $\phi \equiv 0$ and $\psi = 1$ if $|x-8| \leq 1$ and $\psi = 0$ otherwise. We plot some profiles in Figure 3.11. Note that, in this case, fixing the left boundary has the effect of an **even reflection** of the incoming wave. Moreover, note that the support of the wave (the values of x where the displacement is nonzero) does not grow as it did for the case of the hammer blow on the infinite string. This is a consequence of the restricted domain of influence as can be seen by comparing the right-hand sides of Figure 3.8 and Figure 3.3. As with the plucked string, it is an excellent idea for the readers to convince themselves of these profiles by looking solely

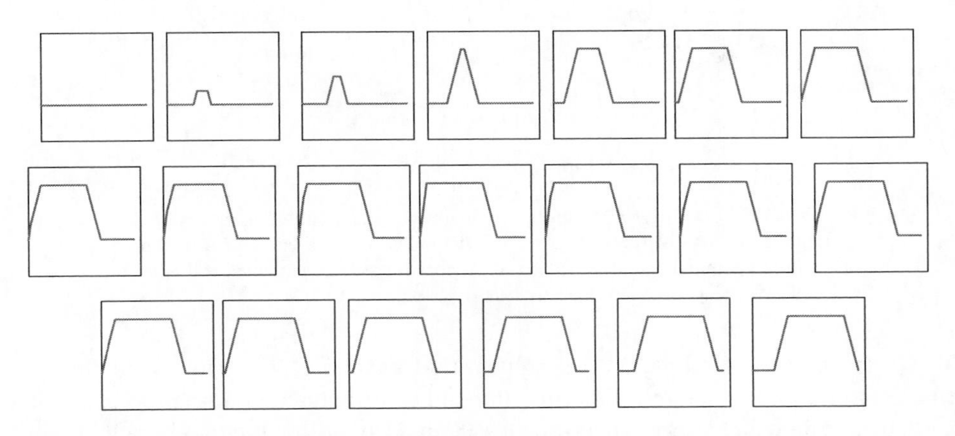

Figure 3.11. Propagation of a **hammer blow** with the $x = 0$ **boundary fixed**.

at the solution formula (3.21); for a collection of fixed times t, look at the resulting integral for different values of x for the cases $x > t$ and $0 \leq x < t$.

3.9. • Neumann and Robin Boundary Conditions

The boundary condition $u(0,t) = 0$ that we have just visited is called a **Dirichlet boundary condition**; it corresponds to fixing the end point of the string. There are many other types of boundary conditions, two of which have direct interpretations for the wave equation.

The **Neumann**[5] **boundary condition** at $x = 0$ takes the form

$$u_x(0,t) = 0.$$

In our scenario, the interpretation (see Figure 3.12, left) is that the end of the string is not fixed but rather attached to a ring on a frictionless bar. This attached end can move freely up and down but cannot exert any vertical force on the string to the right. Hence, at this end point, there must be no vertical component of the tension force. This amounts to enforcing that $u_x(0,t) = 0$ for all $t > 0$.

The associated semi-infinite string BVP/IVP with the Neumann condition at $x = 0$ is

$$\begin{cases} u_{tt} = c^2 u_{xx}, & x \geq 0,\ t \geq 0, \\ u(x,0) = \phi(x),\ u_t(x,0) = \psi(x), & x \geq 0, \\ u_x(0,t) = 0, & t \geq 0. \end{cases}$$

It can be solved in an analogous way to the Dirichlet condition by considering **even extensions** of the data (cf. Exercise **3.10**).

The **Robin**[6] **boundary condition** at $x = 0$ takes the following form for some $a > 0$:

$$u_x(0,t) + au(0,t) = 0. \tag{3.22}$$

[5]Named after the German mathematician Carl Neumann (1832–1925).

[6]Named after the French mathematician Victor Gustave Robin (1855–1897).

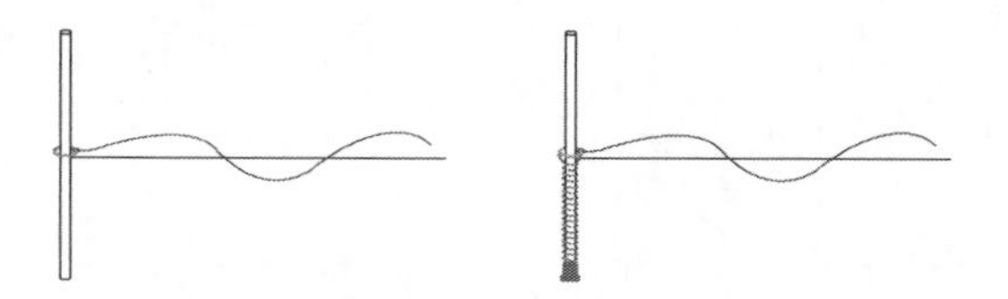

Figure 3.12. **Left:** Illustration of the Neumann boundary condition at $x = 0$: A vibrating string is attached at the left end to a frictionless vertical rod. **Right:** Illustration of the Robin boundary condition at $x = 0$: A vibrating string is attached at the left end to a coiled vertical **spring** enclosing a frictionless vertical rod.

The interpretation (see Figure 3.12, right) is that the end of the string is attached to a ring on a frictionless bar, but this time the end is also attached to the top of a coiled vertical **spring** which exerts a restoring force aimed at pulling the ring back to its equilibrium position. The spring, with spring constant $k > 0$, is at equilibrium when its top is at height $u(0,t) = 0$ (i.e., at the same height where the string is in equilibrium). The left end is free to move, i.e., $u(0,t)$ can change, but in such a way that the force from the spring balances the vertical component of the tension in the string. Recall that the restoring force of the spring is simply k times the displacement from equilibrium. Accounting for the right signs (with positive denoting the upward direction) this means $k(0 - u(0,t))$. On the other hand, the vertical component of the tension in the string at $x = 0$ is given by $Tu_x(0,t)$ where $T > 0$ is the constant tension in the string. Accounting for the right signs (with positive denoting the upward direction), we must have

$$k(0 - u(0,t)) = Tu_x(0,t).$$

This is equivalent to (3.22) with $a = k/T$.

3.10. • Finite String Boundary Value Problems

A more realistic boundary problem is for the scenario of a *finite* vibrating string which is fixed at both ends (Dirichlet boundary conditions). The corresponding BVP/IVP is given by

$$\begin{cases} u_{tt} = c^2 u_{xx}, & 0 \le x \le l, \\ u(x,0) = \phi(x), \quad u_t(x,0) = \psi(x), & 0 \le x \le l, \\ u(0,t) = 0, \quad u(l,t) = 0, & t > 0. \end{cases} \tag{3.23}$$

A natural question is whether or not one can derive the solution with a D'Alembert approach and odd reflections. The answer is twofold:

- Yes, in the sense that for any **fixed** (x,t), we could trace back all the reflections to the initial data and write down a D'Alembert-like formula for $u(x,t)$ entirely in terms of the initial data.
- No, in the sense that there is **no one formula** which would work for all (x,t); for any given $x \in (0,l)$, the number of reflections will increase as t increases.

To give a hint of this complexity, consider the effect (influence) at later times of an initial displacement at a point $x_0 \in (0,l)$. This information would travel on characteristics with speeds c and $-c$ and become successively reflected at the boundaries $x = 0$

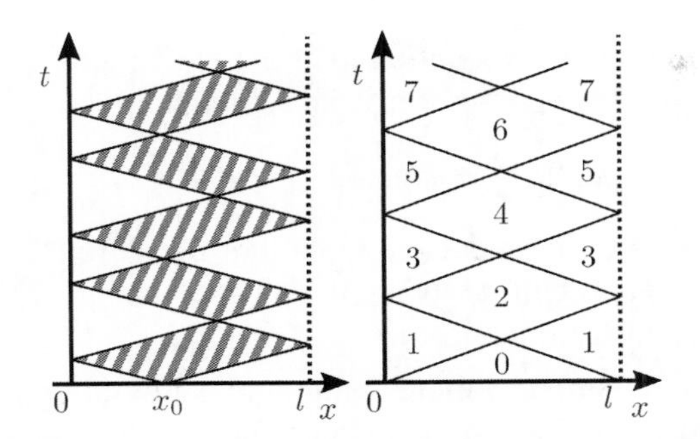

Figure 3.13. Left: The domain of influence of an initial disturbance (either via displacement or velocity) at x_0 and $t = 0$. Both characteristics are reflected back and forth as they hit the boundaries. Right: The division of space-time for the finite string problem labeled with the number of reflections needed to connect back to the initial data; on each region (diamond or triangle), one could write down a *D'Alembert-like* formula based upon the set number of reflections from each of the end points.

and $x = l$ (cf. Figure 3.13, left). Thus, if we wanted to work backward in time and determine $u(x, t)$ from data at $t = 0$, the formula would depend on how many reflections we would need to trace back to the initial x-axis. Hence, unlike for the semi-infinite string where there were only two cases to consider ($x < ct$ and $x > ct$), there would be a separate formula for each region in Figure 3.13, right; in any such region, as we trace back in time, the number of reflections from the left end point and the number of reflections from the right end point are **both** fixed.

Fortunately, there is a simpler, more direct way to solve (3.23) called **separation of variables**, and this method will be the focus of Chapter 12. One drawback to this method will be that, in general, it presents the solution as an infinite series. However, it will expose some fundamental properties of oscillations, including preferred frequencies and modes. To provide a taste of this, let $\phi(x) = \sin(27\pi x/l)$ and $\psi \equiv 0$. One can directly check that the unique solution to (3.23) is given by

$$u(x,t) = \sin\left(\frac{27\pi x}{l}\right)\cos\left(\frac{27c\pi t}{l}\right).$$

The simple, **separated** structure of the solution is most telling. Indeed, the solution at any time t is simply a constant times the initial displacement, i.e., some amplification of the initial displacement. In particular, the basic shape is preserved, certainly not the case for most choices of initial data. Our choice here for $\phi(x)$ was rather special, and we call such choices **normal modes** of the string. They are preferred shapes associated with the material and size parameters of the string, in this case the tension T, density ρ, and length l. On the other hand, if we were to fix a position x on the string and let time evolve, we would see that the displacement oscillates with a frequency equal to $\frac{27\pi c}{l}$. We call these frequencies the **natural frequencies** of the vibrating string. As we shall see in Chapter 12, there is a sequence of such frequencies and a sequence of normal modes:

$$\frac{n\pi}{l} \quad \text{and} \quad \frac{nc\pi}{l}, \quad \text{respectively, for } n = 1, 2, \ldots.$$

Moreover, we will see that **every** solution to (3.23) can be written as an infinite linear combination (a superposition) of these special solutions

$$u_n(x,t) = \sin\Big(\frac{n\pi x}{l}\Big)\cos\Big(\frac{nc\pi t}{l}\Big),$$

associated with the natural frequencies. Note that these frequencies depend on the physical characteristics of the finite string: the physical length l and speed c (which in turn depends on the density and tension).

While we will not go into any more detail here, let us end by quickly mentioning **energy conservation** in a finite string. Consider both the BVP (3.23) as well as the BVP that corresponds to Neumann boundary conditions at both ends:

$$\begin{cases} u_{tt} = c^2 u_{xx}, & 0 \le x \le l, \\ u(x,0) = \phi(x), \quad u_t(x,0) = \psi(x), & 0 \le x \le l, \\ u_x(0,t) = 0, \quad u_x(l,t) = 0, & t > 0. \end{cases} \tag{3.24}$$

The BVPs (3.23) and (3.24) both conserve energy (cf. Exercise **3.15**). This can be proven using analogous steps to those performed for the infinite string, i.e., differentiating the total energy (kinetic plus potential) and applying integration by parts. The reader should also attempt to argue the conservation of energy purely on physical grounds.

3.11. • A Touch of Numerics, II: Numerical Solution to the Wave Equation

This section requires the reading of Section 2.7.

Following the approach of Section 2.7 on the finite difference numerical solution to the transport equation, let us briefly apply the same ideas for finding a numerical solution to the IVP for the wave equation

$$\begin{cases} u_{tt} = c^2 u_{xx}, & -\infty < x < \infty,\ t > 0, \\ u(x,0) = \phi(x),\ u_t(x,0) = \psi(x), & -\infty < x < \infty. \end{cases}$$

We discretize space and time via spatial and temporal steps $\Delta x > 0$ and $\Delta t > 0$, and we consider "grid points" $x_j = j\Delta x$ and $t_n = n\Delta t$ for integer values of j and positive integer values of n. Numerically computing the solution means that we find (approximate) values of the true solution at these particular values of x and t. That is, we attempt to find the solution at the grid points

$$U_j^n := u(j\Delta x, n\Delta t).$$

Of course, any numerical experiment can only store a finite amount of information, i.e., a finite number of j's and n's. We could directly address a BVP; however, the finite propagation speed of the wave equation allows us to argue as follows: Suppose we work with data ϕ and ψ which has compact support and we wish to compute the solution up to a time t_*. Then we choose our domain to be $[-L, L]$ with L sufficiently large so that at time t_*, the domain of influence of the data lies in $(-L, L)$. Of course, the larger the t_*, the larger we need to take our L.

Let us fix such a t_* and interval $[-L, L]$ and consider $x_j = j\Delta x$ for $j = -J, \ldots, J$ where $J\Delta x = L$. One now needs an approximation to the second partial derivatives of u in the wave equation solely in terms of U_j^n, the value u at the grid points. We use what is known as the **centered finite difference approximation** based upon the simple fact that if $f(x)$ is a smooth function, then

$$f''(x) = \lim_{\Delta x \to 0} \frac{f(x + \Delta x) + f(x - \Delta x) - 2f(x)}{(\Delta x)^2}. \tag{3.25}$$

If you are unfamiliar with this, please see the short Subsection 7.3.1. Based upon (3.25), we now approximate the partial derivatives u_{xx} and u_{tt} as follows:

$$u_{xx}(j\Delta x, n\Delta t) \approx \frac{U_{j+1}^n - 2U_j^n + U_{j-1}^n}{(\Delta x)^2},$$

$$u_{tt}(j\Delta x, n\Delta t) \approx \frac{U_j^{n+1} - 2U_j^n + U_j^{n-1}}{(\Delta t)^2}.$$

Hence, the discrete wave equation (the wave equation on the grid) becomes

$$\frac{U_j^{n+1} - 2U_j^n + U_j^{n-1}}{(\Delta t)^2} = c^2 \frac{U_{j+1}^n - 2U_j^n + U_{j-1}^n}{(\Delta x)^2},$$

or

$$\boxed{U_j^{n+1} = r\left(U_{j+1}^n + U_{j-1}^n\right) + 2(1 - r)U_j^n - U_j^{n-1},} \tag{3.26}$$

where

$$r := \frac{c^2(\Delta t)^2}{(\Delta x)^2}$$

is a dimensionless parameter which measures the relative size of the grid. Equation (3.26) is our explicit scheme for the wave equation. It is often called a **leapfrog scheme** and its stencil is illustrated in Figure 3.14. One can readily check that this scheme is **consistent**; in fact the errors made are, respectively, $O((\Delta x)^2)$ as $\Delta x \to 0$ and $O((\Delta t)^2)$ as $\Delta t \to 0$. In Exercise **3.23** you are asked to perform the von Neumann stability analysis of Subsection 2.7.2 to find that the scheme is **stable** if and only if $0 < r \leq 1$.

Now let us choose such an r and attempt to solve the IVP for a given compactly supported ϕ and ψ. As with the transport equation of Section 2.7, our procedure is to use the data to find the discrete solution at $n = 0$; use the explicit scheme to compute the discrete solution at $n = 1$; and then "**march**" along to compute at $n = 2, 3, 4$, etc. However, to implement our scheme (3.26), we require information at two previous time steps, not one. Let us go through a few steps to see how best to proceed.

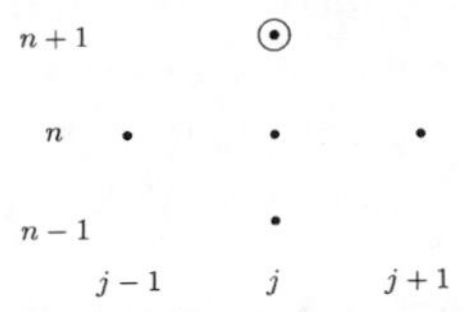

Figure 3.14. The "*Stencil*" associated with the leapfrog scheme (3.26) highlights the dependence of the solution at the circled point in terms of the values at the four other points.

First we note that we can use ϕ to determine U_j^0; i.e., $U_j^0 = \phi(j\Delta x)$. To implement the scheme (3.26) to compute the solution U_j^1, we require information from two previous time steps. To this end, let us introduce an artificial time $t = -\Delta t$, i.e., starting at $n = -1$. How do we determine the values of U_j^{-1}? We use the data for $\psi(j\Delta x) = u_t(j\Delta x, 0)$. To do this, one needs a finite difference approximation for $u_t(x, 0)$. The simplest approximation scheme would be

$$\psi(j\Delta x) = u_t(x,0) \approx \frac{U_j^0 - U_j^{-1}}{\Delta t}. \tag{3.27}$$

Since we already know the values of U_j^0 and $\psi(j\Delta x)$, we can use (3.27) to determine U_j^{-1}. We can then use the scheme to compute U_j^1 and march on to find U_j^2, etc. The problem with this approach is that the error made in (3.27) is $O(\Delta t)$ (first order) whereas our scheme has a higher degree of accuracy (second order) in Δt. In other words, in order to preserve the second-order accuracy of the numerical method, we cannot make large first-order errors in the data. To remedy this situation, we use a second-order finite difference approximation

$$\psi(j\Delta x) = u_t(x,0) \approx \frac{U_j^1 - U_j^{-1}}{2\Delta t}. \tag{3.28}$$

This allows us to write the value of U_j^{-1} in terms of $\psi(j\Delta x)$ and U_j^1. We then use this equation **together with** one application of the scheme (3.26) for $n = 0$ to determine U_j^1. With U_j^0 and U_j^1 known, we simply march on to find U_j^2, U_j^3, etc. In Exercises **3.24** you are asked to go through the details for particular examples.

3.12. Some Final Remarks

A few remarks on the wave equation are in order. Of particular importance are the last two subsections on more general wave equations and the notion of dispersion.

3.12.1. Nonsmooth Solutions. When solving the IVP (3.5) to arrive at D'Alembert's formula, we implicitly made some assumptions on the initial data ϕ and ψ. Alternatively, in order to show that D'Alembert's formula does indeed solve the wave equation, we would need ϕ to be C^2 and ψ to be C^1. On the other hand, we do not need these assumptions to just plug ϕ and ψ into D'Alembert's formula: Indeed, in our examples of the plucked string and the hammer blow, we simply applied D'Alembert's formula to, respectively, a ϕ which was not C^1 and a ψ which was not even continuous. While we discussed the resulting solutions and plotted profiles, in what sense do we have a solution? Do the discontinuities count or can they be ignored? The answer to these questions will be revealed once we introduce the notion of a solution in the sense of distributions (cf. Chapter 9). This more integral-focused notion of a solution will indeed allow us to conclude that our placement of these nonsmooth functions in D'Alembert's formula does indeed yield a solution, albeit a distributional one.

3.12.2. Heterogeneous Media and Scattering. Wave propagation through a heterogeneous string leads to a wave equation in which the speed parameter c is spatially dependent:

$$u_{tt} = c^2(x)\, u_{xx}.$$

This spatial dependence of c is determined by the physical nature of the string. With certain general assumptions on the coefficient function $c(x)$, the resulting IVPs are well-posed. However, closed form solution formulas are not usually available. An interesting feature here is the **scattering** of waves, whereby aspects of the inhomogeneous medium can cause incoming waves to have a change in their outgoing behavior. For example, consider the 1D wave equation with

$$c(x) = \begin{cases} c_1, & x < 0, \\ c_2, & x \geq 0, \end{cases}$$

where $c_1 > 0, c_2 > 0$ with $c_1 \neq c_2$. You can think of this as a model for a vibrating string made of two materials of differing mass densities, one for $x < 0$ and the other for $x > 0$. A signal concentrated at say $x_0 < 0$ will propagate in the positive x direction until it interacts with the change in medium at $x = 0$. This interaction will cause the wave to be scattered and give rise to both a change in its amplitude and speed. In Exercise **3.22**, We give more details and outline the steps in order to find an exact solution.

3.12.3. Finite Propagation Speed, Other "Wave" Equations, and Dispersion. The solution to the wave equation exhibits *finite propagation speed.* This was illustrated in the domain of influence where we saw that information can propagate through space (x) with different speeds, but no speed greater than c. The name "**the** wave equation" is mainly for historical reasons and one often loosely calls any time-dependent PDE in which information (data) propagates at finite speeds *"a wave equation"*. For example, the simple first-order transport equations of Chapter 2 can be thought of as *wave equations.* On the other hand, there are also far more complicated linear and nonlinear PDEs for wave propagation, for example,

- the **inviscid Burgers equation** which we saw in Section 2.3,
- the **Schrödinger, Klein-Gordon, Telegraph, and KdV equations**.

The latter two PDEs exhibit a fundamental process associated with the propagation of waves known as **dispersion**. Dispersion for our vanilla wave equation proves rather trivial. It is true that light propagation (as an electromagnetic wave) satisfies the 3D wave equation with c being the speed of light; however, it does so in a vacuum. This is rather uninteresting as there is literally nothing that "can be seen". On the other hand, when lights hits a glass prism, the waves of different wavenumbers have different velocities and the resulting dispersion produces a spectrum of colors. In general, dispersion describes the way in which an initial disturbance of the wave medium distorts over time, explicitly specifying how composite plane wave components of an initial signal evolve temporally for different wavenumbers.

To illustrate the notion of **wavenumber**, consider solutions to the 1D wave equation $u_{tt} = c^2 u_{xx}$ of the *separated* form

$$u(x, t) = \sin kx \cos ckt.$$

You should check that for each $k \in \mathbb{R}$, the above does indeed solve the wave equation. As we shall see in Chapter 12, all solutions to finite domain BVPs of the 1D wave equation can be written as an infinite sum of such (or similar) separated solutions. The number k is called the **wavenumber**; for any fixed time t, it gives the frequency associated with **spatial** oscillations. On the other hand, **temporal frequency** ω is given by ck and for a fixed x gives the frequency of **temporal** oscillations. An important definition is the following:

The relationship between wavenumber k and the temporal frequency ω of the wave is called the **dispersion relation.**

For our (vanilla) wave equation, the dispersion relation is rather trivial:

$$\omega = ck, \qquad \text{where } c \text{ is the wave speed.}$$

The ratio of ω to k is associated with a velocity (or speed); more precisely it is known as the **phase velocity**. For the wave equation this ratio is always the c in the PDE. One can rephrase this by saying that the phase velocity (or wave speed) is the same for all wavenumbers k. However, for many types of wave motion (e.g., light refraction, surface water waves, internal gravity waves, etc.) the phase velocity varies with k. In other words, waves with different wavenumbers travel at different speeds and the dispersion relation is nontrivial. The study of dispersion of **waves** and their **interactions** is an important and difficult subject in science.

3.12.4. Characterizing Dispersion in PDEs via the Ubiquitous Traveling Wave Solution. We just discussed the relation between the temporal frequency and the wavenumber for the vanilla 1D wave equation. These two notions are central to all impulses which propagate with time and can readily be related to almost any time-dependent linear PDE with constant coefficients. Let us provide a few details here. To do so, however, it is extremely convenient to deal with complex-valued solutions. If the reader wants a short primer on complex numbers, read first the short Section 6.1. Surprisingly, all linear PDEs in space x and time t which have constant coefficients admit complex-valued solutions of the form

$$e^{i(kx-\omega t)}. \tag{3.29}$$

Such a solution is often referred to as a **traveling wave**. Why? Note that we can write this solution as some function of $(x - ct)$ where $c = \frac{\omega}{k}$. Now recall Example 2.1.3 of the simple transport equation, and the basic structure of $f(x - ct)$ in space and time.

In the traveling wave solution (3.29), the parameter k is the wavenumber with dimensions of length^{-1}. It is a real number. The constant ω is the temporal frequency and on occasion can be a complex number. The dimension of $|\omega|$ is time^{-1}. At this early stage, it may be more transparent for the reader to digest the roles and meaning of k and ω by using Euler's formula

$$e^{i(kx-\omega t)} = \cos(kx - \omega t) + i \sin(kx - \omega t)$$

and the two real-valued solutions which are imbedded in the complex-valued solution. Note that assuming ω is real, the ratio

$$\frac{\omega}{k} \tag{3.30}$$

gives the **phase velocity** of the wave (the speed of the wave moving in space).

Now, consider a time-dependent linear PDE with constant coefficients in one space dimension. It is certainly not the case that

$$u(x,t) = e^{i(kx-\omega t)} \tag{3.31}$$

is a complex-valued solution for any ω and k. However, it is a solution for certain values of ω and k, and the resulting relationship between ω and k is the **dispersion relation**.

We present some examples starting with one of the first time-dependent PDEs of this book (the transport equation): $u_t = cu_x$. Placing the potential solution (3.31) into this PDE gives

$$-i\omega e^{i(kx-\omega t)} = c\,ike^{i(kx-\omega t)}.$$

Can we make this work? Yes, by choosing $\omega = -ck$, yielding a linear dispersion relation.

Next, we revisit the wave equation

$$u_{tt} = c^2 u_{xx}, \qquad c > 0.$$

Placing (3.31) into the PDE gives

$$-\omega^2 e^{i(kx-\omega t)} = c^2\,(-k^2)e^{i(kx-\omega t)}.$$

Hence we require

$$\omega^2 = c^2k^2 \quad \text{or} \quad \omega = \pm ck.$$

This is in exact agreement with the previous subsection where we showed a linear dispersion relation. So in both of these examples, the phase velocity (3.30) is c, independent of the wavenumber k.

Let us give one example where there is a nontrivial dispersion relation. The **Klein-Gordon equation**[7] is a relativistic (in the sense of Einstein's Special Theory of Relativity) version of the wave equation and is given by

$$\hbar^2 u_{tt} - \hbar^2 c^2 \Delta u + m^2 c^4 u = 0, \tag{3.32}$$

where $\hbar$ is the reduced Planck constant and m is the mass of the particle. In one space dimension, it is given by

$$u_{tt} - c^2 u_{xx} + M^2 u = 0 \qquad \text{where} \qquad M = \frac{mc^2}{\hbar}.$$

If we place (3.31) into the 1D Klein-Gordon equation, we find

$$-\omega^2 e^{i(kx-\omega t)} - c^2\,(-k^2)e^{i(kx-\omega t)} + M^2 e^{i(kx-\omega t)} = 0.$$

Hence, we require

$$-\omega^2 + c^2k^2 + M^2 = 0 \quad \text{or} \quad \omega = \pm\sqrt{c^2k^2 + M^2}.$$

So here we have a **nonlinear dispersion relation** for a linear PDE; in particular, the speed of propagation (3.30) now depends on k with smaller values of k (slower oscillations) having larger speeds. As we discussed in the previous subsection, this phenomenon is known as dispersion.

[7]Named after theoretical physicists **Oskar Klein** and **Walter Gordon**, who in 1926 proposed the equation to model relativistic electrons.

We can play the same game with other linear PDEs including the diffusion equation wherein ω becomes complex and the Schrödinger equation (cf. Exercise **7.40**. Lastly, we make two remarks:

(i) While our description so far has been for linear PDEs, one can (with more involved tools) address dispersion in **nonlinear PDEs**. A famous example is the **KdV (or Korteweg-de Vries) equation**

$$u_t + u_{xxx} - 6uu_x = 0,$$

which models shallow water waves. Look up the equation online for more details.

(ii) The particular traveling wave solutions (3.31) might seem rather special; however, as we shall see when venturing into the world of Fourier (Chapters 6 and 11), general solutions to these PDEs can be obtained by adding (superimposing) such special solutions.

3.13. Chapter Summary

- The 1D wave equation is a second-order PDE $u_{tt} = c^2 u_{xx}$ which, under certain assumptions, models the propagation of a "disturbance" in a medium (a material substance which can propagate waves). In particular, the wave equation defined for all $x \in \mathbb{R}$ is a model for the shape of an infinitely long, elastic, and homogeneous taut string which undergoes relatively small transverse vibrations.
- When we specify both the initial displacement $\phi(x)$ and the initial velocity $\psi(x)$, the resulting initial value problem

$$\begin{cases} u_{tt} = c^2 u_{xx}, & -\infty < x < \infty,\ t > 0, \\ u(x,0) = \phi(x), & -\infty < x < \infty, \\ u_t(x,0) = \psi(x), & -\infty < x < \infty, \end{cases}$$

  is well-posed. Moreover, there is an explicit formula for the unique solution called D'Alembert's formula:

$$u(x,t) = \frac{1}{2}\left[\phi(x+ct) + \phi(x-ct)\right] + \frac{1}{2c}\int_{x-ct}^{x+ct} \psi(s)\, ds.$$

  The formula implies that to find the displacement of the string at position x and time $t > 0$, we start by taking the average of the **initial displacements** at positions $x - ct$ and $x + ct$. Then we add this to t (the time that has elapsed) multiplied by the average **initial velocity** over the part of the string which lies between the positions $x - ct$ and $x + ct$.
- The best way to digest and appreciate D'Alembert's formula is via **causality** and the domains of dependence and influence.
- One can solve the wave equation with a source term (body force) $u_{tt} - c^2 u_{xx} = f(x,t)$ to find D'Alembert's formula plus a double integral of the source function f over the domain of dependence. One way to arrive at this formula is to invoke Duhamel's Principle.
- The wave equation conserves the total energy (kinetic + potential).

- We considered the semi-infinite string with a Dirichlet boundary condition (no displacement) on the left end. This resulted in the IVP/BVP

$$\begin{cases} v_{tt} = c^2 v_{xx}, & x \geq 0,\ t \geq 0, \\ v(x,0) = \phi(x),\ v_t(x,0) = \psi(x), & x \geq 0, \\ v(0,t) = 0, & t \geq 0. \end{cases}$$

We solved this equation via the notion of an odd reflection, arriving at the following formula:

$$v(x,t) = \begin{cases} \frac{1}{2}[\phi(x+ct) + \phi(x-ct)] + \frac{1}{2c}\int_{x-ct}^{x+ct} \psi(s)\,ds & \text{if } x \geq ct, \\ \frac{1}{2}[\phi(x+ct) - \phi(ct-x)] + \frac{1}{2c}\int_{ct-x}^{x+ct} \psi(s)\,ds & \text{if } 0 < x \leq ct. \end{cases}$$

We noted that the effect of the fixed boundary on an incoming wave (towards the origin) was an odd reflection.

- The physically relevant problem of a finite string with both boundaries fixed has a rather complicated picture from the D'Alembert perspective due to the multitudes of possible reflections. The problem is best solved by the method of *separation of variables* which will be presented in Chapter 12.
- Two other types of boundary conditions are Neumann and Robin, each with a particular physical interpretation for the vibrating string.
- One of the central properties of the wave equation (in any space dimension) is that it enforces a **fixed, finite propagation speed** for any disturbance (signals). There are far more complicated "wave equations" associated with finite propagation speed, many of them nonlinear. For historical reasons, the linear PDE considered in this chapter and the next are dubbed "the wave equation"; however, from the point of view of wave propagation in physical media, these two chapters *only scratch the surface.*

Exercises

3.1 Let $u(x,t)$ be defined by D'Alembert's formula (3.6). Show directly that u solves the wave equation for all x and $t > 0$. Show directly that it satisfies the initial condition of (3.5).

3.2 Let $g(\cdot)$ be a smooth function of one variable.
(a) Show directly that $u(x,t) = g(x-ct)$ solves the wave equation $u_{tt} - c^2 u_{xx} = 0$ on $\mathbb{R} \times (0,\infty)$.
(b) For what initial data ϕ and ψ is $u(x,t) = g(x-ct)$ the unique solution of (3.5).

3.3 Consider the electromagnetic fields $\mathbf{B}(x,y,z,t)$ (magnetic field vector) and $\mathbf{E}(x,y,z,t)$ (electric field vector). As you may know, their evolution is governed by Maxwell's equations which, for a vacuum, can be stated as

$$\nabla \times \mathbf{E} = -\frac{\partial \mathbf{B}}{\partial t}, \qquad \nabla \times \mathbf{B} = \frac{1}{c^2}\frac{\partial \mathbf{E}}{\partial t}, \qquad \nabla \cdot \mathbf{E} = 0, \qquad \nabla \cdot \mathbf{B} = 0,$$

where μ and ϵ are constants. Show that if $\mathbf{B}(x, y, z, t) = (0, B(x, t), 0)$ and $\mathbf{E}(x, y, z, t) = (0, 0, E(x, t))$, then both $E(x, t)$ and $B(x, t)$ satisfy the 1D wave equation $u_{tt} - c^2 u_{xx} = 0$.

3.4 Solve the IVP: $u_{tt} - c^2 u_{xx} = 0, u(x, 0) = \sin x, u_t(x, 0) = x$.

3.5 Fix a time t, say, $t_* > 0$. Assume ϕ_i, ψ_i $(i = 1, 2)$ are bounded functions on the real line. Let u_i $(i = 1, 2)$ denote the solution to $\frac{\partial^2 u_i}{\partial t^2} - c^2 \frac{\partial^2 u_i}{\partial x^2} = 0$ with $u_i(x, 0) = \phi_i(x)$, $\frac{\partial u_i}{\partial t}(x, 0) = \psi_i(x)$. For any function $f(x)$, $\|f(x)\|_\infty$ denotes the supremum of $|f|$ over the real line.

Consider the following statement: For every $\epsilon > 0$ we can find $\delta > 0$ such that **if** $\|\phi_1(x) - \phi_2(x)\|_\infty < \delta$ and $\|\psi_1(x) - \psi_2(x)\|_\infty < \delta$, **then** $\|u_1(x, t_*) - u_2(x, t_*)\|_\infty < \epsilon$.

(a) Explain in words what this statement means in terms of solving the initial value problem for the wave equation.

(b) Prove the statement.

Note: If you are unfamiliar with the notion of supremum, just assume ϕ_i, ψ_i $(i = 1, 2)$ are bounded continuous functions which are identically zero outside the interval $[-1, 1]$. Then **all** functions will attain their maxima and minima. Hence, $\|f(x)\|_\infty$ just denotes the maximum value of $|f|$.

3.6 Consider the vibrating string with only small vibrations in the transverse direction. Suppose now that the string is also subjected to gravity. **Derive the equation** which governs the motion of the string. You may assume that the acceleration due to gravity is constant with magnitude g. **Find the general solution** of the equation.

3.7 Solve $u_{xx} - 3u_{xt} - 4u_{tt} = 0$, $u(x, 0) = x^2$, $u_t(x, 0) = e^x$. Hint: "Factor" the equation.

3.8 **(a)** Consider the 1D wave equation for an infinite piano string which is subject to **both** a "pluck" and a "hammer blow" at $t = 0$. To model this let the speed be 1 and let

$$\phi(x) = \begin{cases} 1 - |x - 1| & \text{if } |x - 1| < 1, \\ 0 & \text{if } |x - 1| \geq 1, \end{cases} \qquad \psi(x) = \begin{cases} 1 & \text{if } 3 < x < 4, \\ 0 & \text{if otherwise.} \end{cases}$$

Plot the profiles $u(x, t_{\text{fixed}})$ as functions of x for times $t_{\text{fixed}} = 0, \frac{1}{2}, 1, 2, 4, 8$. You may use any software to plot these profiles. Have the software generate a movie for the first 20 seconds.

(b) Repeat part (a) for the semi-infinite string with a fixed end point at $x = 0$.

3.9 Consider the semi-infinite string but now with Dirichlet boundary condition at $x = 0$ (fixed end) replaced with a prescribed displacement according to a function $b(t)$. This means that we consider (3.17) with boundary condition $v(0, t) = b(t)$ for all $t \geq 0$.

(a) Solve this problem and give a physical interpretation.

(b) If $\phi(x) = 0 = \psi(x)$ and $b(t) = \sin \pi t$, write down the explicit solution and plot profiles for $t = 0, 1, 3, 5$.

Hint for (a): First off, you can assume for consistency that $\phi(0) = b(0)$ and $\psi(0) = b'(0)$. However, as you will see in part (b), we can break from this assumption at the expense of introducing a singularity along a characteristic line. Now, by the superposition principle, first solve two problems: one with $b \equiv 0$ (this is the classic fixed boundary problem) and the other with the nonzero b but where ϕ and ψ are taken to be identically 0. Then add the two. For the second problem, argue as follows: The general solution of the wave equation has the form $v(x,t) = f(x+ct)+g(x-ct)$. Use the fact that $\phi(x) = \psi(x) = 0$ for all $x \geq 0$ to impose constraints on f and g. Then use the boundary condition $v(0,t) = b(t)$ to determine exactly these unknown functions.

3.10 Consider the semi-infinite string with a **Neumann boundary condition**, that is, (3.17) with the boundary condition replaced with $v_x(0,t) = 0$ for $t \geq 0$.
(a) Solve this problem. Hint: Proceed as we did for the half-line with a fixed end point where we used odd **reflections**. This time an **odd** reflection will not work. What will?
(b) Based upon your solution to part (a), answer the following question with the interpretation that we are modeling vibrations of a semi-infinite string: Suppose an initial disturbance (nonzero values of ϕ and ψ) is made in the interval $[1,2]$. At $t = 5$, what parts of the string will be affected by this initial disturbance?
(c) Plot profiles of the solution for $t = 0, 1, 3, 5, 7$ for both the plucked string and hammer blow initial data used in Section 3.8.1.

3.11 (**Wave Equation with Friction**) Consider the PDE $u_{tt} + \nu u_t = c^2 u_{xx}$, where $\nu > 0$ is the friction coefficient.
(a) Derive this equation as a model for a vibrating string in a medium ("air") which subjects the motion of the string to friction (damping).
(b) Prove that the total energy (kinetic + potential) is no longer conserved but rather decreases with time.

3.12 **(a)** Consider a vibrating string (infinite string) whose displacement is given by the wave equation $u_{tt} = u_{xx}$. Suppose at $t = 0$ the string is subjected to a disturbance **exactly** in the intervals $[1,2]$ and $[4,5]$. At $t = 10$, **describe the positions** on the string which will be affected by the initial disturbance.
(b) Repeat part (a) for the semi-infinite string with fixed boundary at $x = 0$.

3.13 (**The Semigroup Property**) Via D'Alembert's formula, the solution to the IVP of the wave equation (3.5) has the following *semigroup property* which we implicitly assumed in describing the domains of dependence and influence. Suppose we fix two instances $t = t_1$ and $t = t_2$ with $0 < t_1 < t_2$. There are two ways we could compute $u(x,t_2)$:

- On one hand, we could apply D'Alembert's formula directly to the initial displacement ϕ and velocity ψ.
- On the other hand, one could apply D'Alembert's formula to compute $u(x,t_1)$ from the same initial data but only for time t_1 and arrive at $u(x,t_1)$. Then we could take as a new pair of "initial values" $\phi_1(x) = u(x,t_1), \psi_1(x) = u_t(x,t_1)$ and apply D'Alembert's formula with time $t_2 - t_1$.

The semigroup property says that the two resulting functions for $u(x,t_2)$ would be exactly the same.

(a) State precisely what this means and prove the assertion.

(b) Why is this called the semigroup property? You may research online.

3.14 Consider the finite string problem (3.23). Restrict your attention to the lowest diamond region in Figure 3.13, right (i.e., the diamond labeled with 2 for 2 reflections), and derive a *d'Alembert-like* formula for the solution.

3.15 Consider the two boundary value problems for the finite string: (3.23) and (3.24). Show that both problems conserve the total energy (kinetic plus potential). Why is this physically reasonable?

3.16 (**Duhamel's Principle and Variation of Parameters from ODEs**) In this exercise you will address the relationship between Duhamel's Principle and the method of variations of parameters which you probably explored in your ODE class. Fix $a \in \mathbb{R}$ and some smooth function $f(t)$. Let us solve for $t \geq 0$ the following inhomogenous IVP for the ODE:

$$y'(t) - ay(t) = f(t), \qquad y(0) = 0. \tag{3.33}$$

First note that the general solution to the homogenous ODE $y' - ay = 0$ is $y(t) = y_0 e^{at}$.

(a) Variation of parameters entails allowing the coefficient y_0 to vary, say equal to $v(t)$, and looking for a solution to the inhomogeneous problem (3.33) of the form $y(t) = v(t)e^{at}$. You can think of $v(t)$ as an "initial value" not at 0 but at time t. Find such a solution; i.e., find v in terms of f.

(b) Now show that you get the same solution via the following steps: For each $s > 0$ let $y_s(t)$, for $t \geq s$, be the solution to $y_s'(t) - ay_s(t) = 0, y_s(s) = f(s)$. Take $y(t) = \int_0^t y_s(t)\,ds$. Convince yourselves of the similarities between the parts (a) and (b).

3.17 (**Duhamel's Principle for the Wave Equation**) Use (3.14) to prove Theorem 3.2.

3.18 (**Duhamel's Principle for the Transport Equation**) Consider the IVP for the transport equation $u_t + cu_x = f(x,t), u(x,0) = g(x)$.

(a) Interpret Duhamel's Principle for solving this IVP and come up with a solution formula.

(b) Show that this solution formula does indeed solve the IVP.

(c) Show that applying the method of characteristics results in the same solution formula.

3.19 (**The Parallelogram Law**) Let $u(x,t)$ be a solution to the 1D wave equation $u_{tt} = c^2 u_{xx}$. The parallelogram law states that if A, B, C, and D are the vertices of a parallelogram (labeled as in Figure 3.15, left) which lies in the domain of u (a subset of the upper half xt-plane) such that its sides have slopes $\frac{\Delta x}{\Delta t} = \pm c$, then $u(A) + u(B) = u(C) + u(D)$.

(a) Using only the form (3.4), prove the parallelogram law.

(b) Use the parallelogram law to derive the solution formula in Theorem 3.3 for the important case when $x < ct$. Hint: Look at the right-hand side of Figure 3.15.

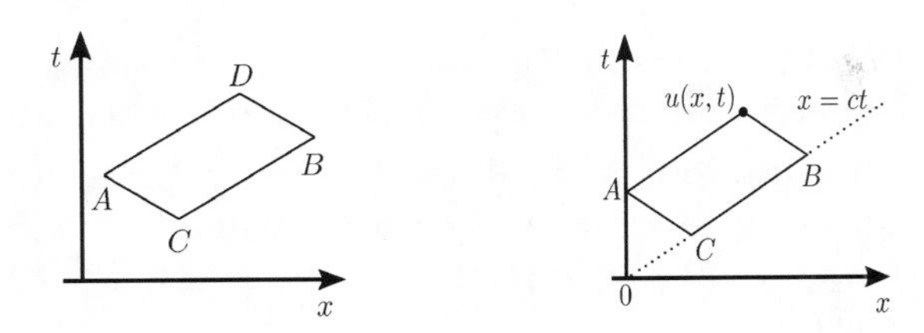

Figure 3.15. Figures for Exercise **3.19**.

(c) Now consider the finite string problem (3.23) with $l = 1$ and $c = 1$ with $\phi \equiv 0$ and $\psi = 1$ if $\frac{1}{4} \leq x \leq \frac{3}{4}$ and 0 otherwise. Use the parallelogram law to find the solution at $x = \frac{1}{2}$ and $t = 1$. Use the parallelogram law to find a solution formula valid in the entire lowest diamond region in Figure 3.13, right (i.e., the diamond labeled with 2 for 2 reflections).

3.20 (**Half-Line with a Robin Boundary Condition**) Solve the IVP/BVP for the semi-infinite string with a Robin boundary condition: $v_{tt} = c^2 v_{xx}$ for $x \geq 0$, $t \geq 0$; $v(x, 0) = \phi(x)$, $v_t(x, 0) = \psi(x)$ for $x \geq 0$; and $v_x(0, t) + v(0, t) = 0$ for $t \geq 0$.

3.21 (**Half-Line with a Different Mixed Boundary Condition**)
(a) Fix $\lambda \neq 1$ and let $\phi(x), \psi(x)$ be data functions with $\phi(0) = \psi(0) = 0$. Solve the initial value problem for the semi-infinite string from the previous question except the boundary condition now reads: $v_t(0, t) + \lambda v_x(0, t) = 0$ for $t \geq 0$.
(b) What happens if $\lambda = 1$? Do you have a solution for all ϕ and ψ?

3.22 (**Inhomogeneous String and Scattering**) Consider here an infinite string made of two materials, one to the left of $x = 0$ and another with a different density to the right. We may model this via the wave equation

$$u_{tt} = c^2(x) u_{xx} \quad \text{where} \quad c(x) := \begin{cases} c_1 & \text{if } x < 0, \\ c_2 & \text{if } x \geq 0, \end{cases} \tag{3.34}$$

with $c_i > 0$ and distinct. Let us say that $u(x, t)$ is a solution to (3.34) on $x \in \mathbb{R}$ and $t > 0$ if (i) it is a solution to the homogeneous wave equation in each of the half-planes $x < 0$, $t > 0$ and $x > 0$, $t > 0$ and (ii) both u and u_x are continuous functions at $x = 0$. The latter condition implies that the string does not break and its slope varies continuously.

Our goal in this question is to explore how an incoming wave $\phi(x - c_1 t)$ on the left of $x < 0$, which is moving towards $x = 0$ with speed c_1, is affected by the change in the density of the string (reflected in the change in speed c). You may think of ϕ as a smooth function with compact support (cf. Section A.2.2) in the interval $[-11, -9]$ (a blip centered at $x = -10$). If we supplement (3.34) with initial displacement $\phi(x)$ and initial velocity $-c_1 \phi'(x)$, then for small times, the solution will simply be $\phi(x - c_1 t)$. However, eventually this blip will hit $x = 0$.
(a) Find the solution for all x and $t > 0$ to the resulting IVP. Hint: Use the general form of the solution (3.4) separately in each of the half-planes and then impose both the continuity conditions at $x = 0$ and the initial data.
(b) Describe in words what happens after the blip (described above) passes through $x = 0$. In this context, interpret the notion of *scattering*?

3.23 (**Numerical Computations 1)** Perform the von Neumann stability analysis of Subsection 2.7.2 to find that the scheme (3.26) is **stable** if and only if $0 < r \leq 1$.

3.24 (**Numerical Computations 2**) In this exercise, you will use the scheme (3.26) to numerically compute the solution to the wave equation with $c = 1$ with discrete initial data: $\phi(j\Delta x) = 2$ if $j = 0$; $\phi(j\Delta x) = 1$ if $j = \pm 1$; $\phi(j\Delta x) = 0$ otherwise; $\psi(j\Delta x) = 0$ for all j.

(a) Let $\Delta x = 0.1$ and compute the solution after 10 time steps using (i) $r = 1$, (ii) $r = 0.5$, and then (iii) $r = 2$. By using linear interpolation between the grid points, graph the solution in each case.

(b) Let $\Delta x = 0.1$ and $r = 1$ compute the approximate solution at time $t = 10$.

(c) Repeat parts (a) and (b) for discrete data: $\phi(j\Delta x) = 0$ for all j; $\psi(j\Delta x) = 2$ if $j = 0$; $\psi(j\Delta x) = 1$ if $j = \pm 1$; $\psi(j\Delta x) = 0$ otherwise.

3.25 (**Numerical Computations 3 — A Maximum Principle**) Let $\phi(x)$ in (3.5) be a bounded and continuous function and set $\psi \equiv 0$. From D'Alembert's formula (3.6) it is clear that the solution to (3.5) has the following property (called a maximum principle):

$$|u(x,t)| \leq \max_{y \in \mathbb{R}} |\phi(y)| \qquad \text{for all } x \in \mathbb{R} \text{ and } t > 0\,.$$

In this problem, we show that the scheme (3.26), using $r = 1$ and (3.28), also shares this property; namely,

$$|U_j^n| \leq \max_i |\phi(i\Delta x)| \qquad \text{for all } j \in \mathbb{Z} \text{ and } n \in \mathbb{N}. \tag{3.35}$$

(a) Show that $U_j^1 = \frac{1}{2}\left(U_{j+1}^0 + U_{j-1}^0\right)$ and conclude that $|U_j^1| \leq \max_i |\phi(i\Delta x)|$.

(b) Now prove that $U_j^2 = \frac{1}{2}\left(U_{j+2}^0 + U_{j-2}^0\right)$ and hence $|U_j^2| \leq \max_i |\phi(i\Delta x)|$.

(c) By mathematical induction prove (3.35).

(d) Here we show that the scheme also preserves area. To this end, first show that if $\phi(x) > 0$ with $\int_{-\infty}^{\infty} \phi(x)dx < \infty$, then for all $t > 0$, the solution $u(x,t)$ satisfies

$$\int_{-\infty}^{\infty} u(x,t)\,dx = \int_{-\infty}^{\infty} \phi(x)\,dx.$$

Next, prove the analogous statement for the scheme:

$$\sum_{j \in \mathbb{Z}} U_j^n = \sum_{j \in \mathbb{Z}} \phi(j\Delta x) \qquad \text{for all } n \in \mathbb{N}.$$

Chapter 4

The Wave Equation in Three and Two Space Dimensions

The reader is encouraged to first review
Sections A.4–A.6 and A.9 from the Appendix.

In this chapter we address the wave equation in space dimensions three and two. In general, the wave equation in N space dimensions is the second-order PDE for $u(\mathbf{x}, t)$, $\mathbf{x} = (x_1, \ldots, x_N) \in \mathbb{R}^N$, and $t \in \mathbb{R}$:

$$u_{tt} = c^2 \Delta u, \tag{4.1}$$

where Δu denotes the **Laplacian operator**; i.e.,

$$\Delta u = \operatorname{div} \nabla u = \sum_{i=1}^{N} u_{x_i x_i}.$$

In the previous chapter we dealt with one space dimension $N = 1$ and found that the solution at any point x and time t depended on initial data on a spatial interval $[x - ct, x + ct]$. In other words, the solution involved an integral over the one-dimensional ball centered at x with radius ct. When the space dimension is larger than one, we would expect that these single spatial integrals would be replaced by multiple integrals. But of what type? We will soon see that these multiple integrals are surface integrals over spheres in 3D, and bulk integrals over balls (in 2D). It is thus very useful to review Section A.5 before embarking any further. We begin with two separate physical derivations of the 3D wave equation.

4.1. • Two Derivations of the 3D Wave Equation

In this section, we denote the coordinates of three-dimensional space by (x, y, z). We give two derivations of the 3D wave equation:

$$\boxed{u_{tt} = c^2(u_{xx} + u_{yy} + u_{zz}) = c^2 \Delta u.} \tag{4.2}$$

The first is for the propagation of electromagnetic waves in a vacuum and is exact in the sense that no approximations are needed for the equation to be valid. The second one is for acoustics — the propagation of sound in air. Here, following the same path as for the vibrating string, we will derive the wave equation by making an assumption that disturbances are relatively small. Neither derivation will be given from "first principles"; rather, each will come from the following respective underlying systems of PDEs which we will take at face value:

- Maxwell's equations for electromagnetic waves,
- the compressible Euler equations for sound waves.

4.1.1. • Derivation 1: Electromagnetic Waves from Maxwell's Equations.

Consider the electromagnetic fields $\mathbf{B}(x, y, z, t)$ (magnetic field vector) and $\mathbf{E}(x, y, z, t)$ (electric field vector). Their evolution is governed by **Maxwell's equations**[1], a system of PDEs. In a vacuum (ignoring charges and currents), these equations[2] can be summarized by

$$\nabla \times \mathbf{E} = -\frac{\partial \mathbf{B}}{\partial t}, \qquad \nabla \times \mathbf{B} = \frac{1}{c^2}\frac{\partial \mathbf{E}}{\partial t}, \qquad \nabla \cdot \mathbf{E} = 0, \qquad \nabla \cdot \mathbf{B} = 0, \tag{4.3}$$

where c is the speed of light. Note that the first two equations are vector equations which are equivalent to stating that the respective scalar components are equal. Here we have used the more descriptive notation (favored by physicists) for the curl and divergence; that is,

$$\text{curl } \mathbf{E} = \nabla \times \mathbf{E} \qquad \text{and} \qquad \text{div } \mathbf{E} = \nabla \cdot \mathbf{E}.$$

We will use the more descriptive notation for the derivation. To uncover the wave equation, we start by differentiating the second equation in (4.3) with respect to t and then apply the first equation in (4.3). This yields

$$\begin{aligned}\frac{\partial^2 \mathbf{E}}{\partial t^2} &= c^2 \frac{\partial}{\partial t}(\nabla \times \mathbf{B}) \\ &= c^2\, \nabla \times \frac{\partial \mathbf{B}}{\partial t} \\ &= c^2\, \nabla \times (-\nabla \times \mathbf{E}).\end{aligned}$$

Note that in the second equality above we interchanged the order of the t derivative with the spatial derivatives. A useful vector identity from advanced calculus states that for any (smooth) vector field $\mathbf{X}$, we have

$$\nabla \times (\nabla \times \mathbf{X}) = \nabla(\nabla \cdot \mathbf{X}) - \nabla \cdot \nabla \mathbf{X} = \nabla(\nabla \cdot \mathbf{X}) - \Delta \mathbf{X},$$

---

[1] **James Clerk Maxwell** (1831–1879) was a Scottish physicist and mathematician well known for these equations. He is also regarded as the founder of the field of *electrical engineering*.

[2] In their original form, Maxwell's equations involve the universal constants ϵ_0 and μ_0, the "permittivity" and "permeability" of free space. These quantities are measured experimentally when working with electric and magnetic fields and have a priori no relationship. As we will show here, the vacuum Maxwell's equations imply is a wave equation with speed c equal to $1/\sqrt{\epsilon_0 \mu_0}$. The fact that this value c equals the experimentally measured speed of light hinted at light being a so-called electromagnetic wave. In contrast to the wave equation in a string say, light then propagates in a vacuum, i.e., without any medium to "support" oscillations. This was an extremely controversial idea at the time and led to Einstein's theory of special relativity.

where $\Delta\mathbf{X}$ is the vector with each component being the Laplacian of the respective component of $\mathbf{X}$. This, combined with the fact that $\nabla \cdot \mathbf{E} = 0$ (third equation in (4.3)), yields

$$\frac{\partial^2 \mathbf{E}}{\partial t^2} = c^2\, \nabla \times (-\nabla \times \mathbf{E}) = c^2\, \Delta\mathbf{E}.$$

One can repeat this argument, replacing $\mathbf{E}$ with $\mathbf{B}$, to obtain

$$\frac{\partial^2 \mathbf{B}}{\partial t^2} = c^2\, \Delta\mathbf{B}.$$

These are vector wave equations which we can interpret componentwise; for example if

$$\mathbf{E}(x, y, z, t) = \big(E_1(x, y, z, t), E_2(x, y, z, t), E_3(x, y, z, t)\big),$$

then for each $i = 1, 2, 3$,

$$\frac{\partial^2 E_i}{\partial t^2} = c^2\, \Delta E_i.$$

Conclusion: We have shown that if $\mathbf{E}(x, y, z, t)$ and $\mathbf{B}(x, y, z, t)$ solve the Maxwell equations (4.3), then all six component functions $E_i, B_i,\ i = 1, 2, 3$, solve the 3D wave equation $u_{tt} - c^2 \Delta u = 0$.

4.1.2. Derivation 2: Acoustics from the Euler Equations. Sound waves are density (alternatively, pressure) disturbances which propagate in a medium. The basis for describing sound waves in air will be the equations of motion for a compressible fluid. These are the **compressible Euler equations**, which we derived in Section 2.8.3:

$$\begin{cases} \frac{\partial \mathbf{u}}{\partial t} + \mathbf{u} \cdot \nabla \mathbf{u} = -\frac{1}{\rho} \nabla f(\rho), \\ \frac{\partial \rho}{\partial t} + \operatorname{div}(\rho \mathbf{u}) = 0. \end{cases} \tag{4.4}$$

The independent variables are space $(x, y, z) \in \mathbb{R}^3$ and time $t \in \mathbb{R}^+$. The dependent variables are what is known as **spatial (or Eulerian)** descriptions of the velocity $\mathbf{u} : \mathbb{R}^3 \times \mathbb{R} \to \mathbb{R}^3$ and the density $\rho : \mathbb{R}^3 \times \mathbb{R} \to \mathbb{R}$. Spatial (Eulerian) descriptions of velocity and density are quite subtle and the reader is encouraged to consult the first part of Section 2.8.3 in order to absorb the consequences of using spatial descriptions of the dependent variables. In a nutshell, $\rho(x, y, z, t)$ denotes the density of air **at position** (x, y, z) at time t, while $\mathbf{u}(x, y, z, t)$ denotes the velocity of the fluid particle **which is at position** (x, y, z) at time t.

The first equation in (4.4) is presented in vector form and constitutes three coupled scalar PDEs (cf. Section 2.8.3). It includes the pressure f which, for an ideal gas, is given by a function of the density ρ. For air, this has the form

$$f(\rho) = p_0 \left(\frac{\rho}{\rho_0}\right)^{\gamma}, \tag{4.5}$$

where γ denotes the *adiabatic index* and where p_0 and ρ_0 denote, respectively, the sea level atmospheric pressure and air density at a reference temperature. For air the *adiabatic index* is 1.4. The second (scalar) equation in (4.4) is the **continuity equation** previously discussed in Section 2.2.6.

The Euler equations constitute a difficult nonlinear (albeit quasilinear) system of equations. Let us now reduce this complicated system of four PDEs in four unknowns to the simple 3D wave equation, under the assumption of **small vibrations**. Precisely, we assume the following:

- The density ρ has small variations with respect to some constant ρ_0. That is, we assume that $|\rho-\rho_0|$ and the absolute value of its derivatives (spatial and temporal) are small.
- The velocity $\mathbf{u}$ has small variations from 0. That is, $|\mathbf{u}|$ and the absolute value of its derivatives (spatial and temporal) are small. Remember that $\mathbf{u}$ is the *speed of the air particles* and **not** the speed of the disturbance; just because air molecules move slowly doesn't mean that the disturbance (the wave propagation through the air) has to also be small. For example, when you hear the voice of your instructors as they speak, you cannot at the same time smell what they had for breakfast! You can also see the analogy with the "domino" effect in a linear sequence.

Dimensional Red Flag (cf. Section A.11): In the current form, the Euler equations are dimensional. That is, ρ has dimensions of mass length^{-3} and u has dimensions of length time^{-1}. Hence, saying that these quantities are small makes no sense without fixing a reference scale for mass, length, and time. Such a process is called nondimensionalizing the problem. While, in principle, we should do this first, we will be a little sloppy here and just agree to use units of kilograms, meters, and seconds, respectively. Roughly speaking, this sets the scale for the value of 1 in these quantities and, hence, smallness can be measured relative to **this** "1".

I will now demonstrate that under these assumptions, ρ satisfies the 3D wave equation with a wave speed equal to the speed of sound. Our basis for extracting the consequences of these assumptions relies on Taylor series. That is, we use a Taylor series to linearize the Euler equations (4.4) around the base states $\rho = \rho_0$ and $\mathbf{u} = 0$. This may be the first time you see such a linearization. In fact, it is a very common procedure, and even though this is not the easiest linearization to start with, it actually proves to be quite short. To this end, we assume that there is some small number ϵ (think of ϵ as $\rho - \rho_0$ or $|\mathbf{u}|$) such that $\rho - \rho_0, u_i$, and their derivatives are, using standard big O notation, all $O(\epsilon)$ as $\epsilon \to 0$. We can then write out (4.4) in terms of $\rho - \rho_0$ and $\mathbf{u}$ and **neglect all terms** which are smaller than ϵ (e.g., terms which are $O(\epsilon^2)$).

Let us begin with the second equation in the system (4.4), which can be rewritten as

$$\rho_t + \operatorname{div}((\rho - \rho_0)\mathbf{u}) + \rho_0 \operatorname{div}(\mathbf{u}) = 0. \tag{4.6}$$

If we linearize this equation about $\rho = \rho_0$, i.e., drop all terms of order $O(\epsilon^2)$, we obtain

$$\rho_t + \rho_0 \operatorname{div}(\mathbf{u}) = 0.$$

Turning to the first (vector) equation in (4.4), we focus on the right-hand side noting that

$$\nabla f(\rho) = f'(\rho)\nabla \rho = f'(\rho)\nabla(\rho - \rho_0).$$

Thus we need two Taylor series expansions in ρ close to $\rho = \rho_0$; one for $f'(\rho)$ and the other for $\frac{1}{\rho}$. To this end, we find that

$$f'(\rho) = \left(f'(\rho_0) + f''(\rho_0)(\rho - \rho_0) + O((\rho - \rho_0)^2)\right)$$

and

$$\frac{1}{\rho} = \left(\frac{1}{\rho_0} - \frac{1}{\rho_0^2}(\rho - \rho_0) + O((\rho-\rho_0)^2)\right).$$

Hence, the first equation can be written as

$$\begin{aligned}\mathbf{u}_t + \mathbf{u}\cdot\nabla\mathbf{u} &= -\left(\frac{1}{\rho_0} - \frac{1}{\rho_0^2}(\rho-\rho_0) + O((\rho-\rho_0)^2)\right)\\ &\quad \cdot \left(f'(\rho_0) + f''(\rho_0)(\rho-\rho_0) + O((\rho-\rho_0)^2)\right)\nabla(\rho-\rho_0).\end{aligned}$$

Note that in our Taylor expansions, we have included **all** the higher-order terms in the single expression $O((\rho-\rho_0)^2)$, which amounts to terms which are all $O(\epsilon^2)$. Dropping all terms which are $O(\epsilon^2)$ (this includes terms like $\mathbf{u}\cdot\nabla\mathbf{u}$ which are products of $O(\epsilon)$ terms) brings us to

$$\mathbf{u}_t = -\frac{f'(\rho_0)}{\rho_0}\nabla\rho.$$

Thus, we have arrived at the linearized Euler equations

$$\rho_t + \rho_0\,\mathrm{div}(\mathbf{u}) = 0 \qquad \text{and} \qquad \mathbf{u}_t = -\frac{f'(\rho_0)}{\rho_0}\nabla\rho. \tag{4.7}$$

So where is the wave equation? If we differentiate the first equation of (4.7) with respect to time and then use the second equation, we find

$$\rho_{tt} = -\rho_0\,\mathrm{div}(\mathbf{u}_t) = f'(\rho_0)\,\mathrm{div}(\nabla\rho) = f'(\rho_0)\Delta\rho.$$

This is precisely the 3D wave equation[3] for ρ:

$$\boxed{\rho_{tt} = c^2\Delta\rho, \qquad \text{with speed of propagation} \qquad c = \sqrt{f'(\rho_0)}.}$$

The reader should verify that c does indeed have dimensions of length time^{-1}. Also, note that it is natural to expect that the pressure f is an increasing function of density ρ; hence, f' is positive. From the adiabatic relation (4.5), we find[4]

$$c = \sqrt{f'(\rho_0)} = \sqrt{\gamma\frac{p_0}{\rho_0}}. \tag{4.8}$$

Finally, let me show how this little bit of theory, which yielded (4.8), can be used to predict the **speed of sound**. At sea level and temperature $288K$, one measures the density of air and atmospheric pressure as $\rho_0 = 1.225\ kg\ m^{-3}$ and $p_0 = 101325\ Pa$ (pascal), where a $Pa = N\ m^{-2} = kg\ m^{-1}\ s^{-2}$. For air, the adiabatic index is $\gamma = 1.4$.

---

[3]One might also ask if the components of $\mathbf{u}$ also satisfy the same 3D wave equation. The answer is *yes* if we make an additional assumption that the curl of the velocity field, $\nabla\times\mathbf{u}$, vanishes (cf. Exercise **4.4**). This quantity measures the rotational properties of the fluid and is known as **vorticity**. Zero vorticity corresponds to an **irrotational** flow. As it turns out, for the system (4.7), vorticity is conserved over time. Hence, one needs only to assume that the vorticity vanishes initially; that is, the air is initially irrotational.

[4]Formula (4.8) has an interesting pre-wave-equation history. In the latter half of the 17th century **Isaac Newton** asserted that speed of sound in a medium (a gas, liquid, or solid) was given by the simple formula

$$c = \sqrt{\frac{p}{\rho}}$$

where ρ was the density of the medium and p was the pressure "surrounding the sound waves". Though dimensionally correct, this formula was found to incorrectly predict the speed of sound. Later in the 17th century, **Pierre-Simon Laplace** realized how to correct Newton's formula via the inclusion of a dimensionless factor γ related to heat capacity. This γ is precisely the adiabatic index.

Placing these values in (4.8), we predict the speed of sound to be $c \approx 340.29\, m/s$. Up to two decimal places, this is exactly the speed of sound at sea level — not bad!

4.2. • Three Space Dimensions: The Initial Value Problem and Its Explicit Solution

Read the review Sections A.4 and A.5 on integrals before proceeding.

We now consider the initial value problem for the 3D wave equation. As in 1D, we must prescribe both the initial displacement and velocity. These are scalar functions of $\mathbb{R}^3$ and will be denoted, respectively, by $\phi(\mathbf{x})$ and $\psi(\mathbf{x})$. We assume throughout this chapter that ϕ and ψ are smooth functions. For simplicity of notation, let us use $\mathbf{x} = (x_1, x_2, x_3)$ for spatial points instead of (x, y, z). Our goal now is to solve the initial value problem:

IVP for the 3D Wave Equation in All of Space

$$\begin{cases} u_{tt} = c^2 \Delta u, & \mathbf{x} \in \mathbb{R}^3,\ t > 0, \\ u(\mathbf{x}, 0) = \phi(\mathbf{x}), & \mathbf{x} \in \mathbb{R}^3, \\ u_t(\mathbf{x}, 0) = \psi(\mathbf{x}), & \mathbf{x} \in \mathbb{R}^3, \end{cases} \tag{4.9}$$

where ϕ and ψ are smooth functions.

As we shall soon see, the key to solving (4.9) is to exploit the fact that the right-hand side of the wave equation (the Laplacian) treats all the spatial directions (i.e., x_1, x_2, and x_3) equally. This is simply a consequence of the fact that the PDE models the propagation of a disturbance in a homogeneous medium wherein there is no preferred spatial direction.

4.2.1. • Kirchhoff's Formula.

As in the 1D case, there is an explicit solution to (4.9) due to Kirchhoff[5].

Kirchhoff's Formula

The solution to (4.9) is given by Kirchhoff's formula:

$$u(\mathbf{x}, t) = \frac{1}{4\pi c^2 t^2} \iint_{\partial B(\mathbf{x}, ct)} \left(\phi(\mathbf{y}) + \nabla\phi(\mathbf{y}) \cdot (\mathbf{y} - \mathbf{x}) + t\psi(\mathbf{y})\right) dS_{\mathbf{y}}. \tag{4.10}$$

As with D'Alembert's formula, Kirchhoff's formula explicitly gives the solution at any place and time **in terms of** the initial data. However, there is a fundamental difference which exemplifies how waves propagate in 3D and this is directly related to the Huygen's Principle, which we will discuss shortly in Section 4.2.2. Before proceeding, we should **ground** ourselves in the **notation** of (4.10). The formula tells us how to find the value of $u(\mathbf{x}, t)$ (the displacement at position $\mathbf{x}$ and time t) in terms of the initial data. Hence, for the purposes of the formula you can think of $\mathbf{x}$ and t as fixed. The initial data is sampled over $\partial B(\mathbf{x}, ct)$, a sphere centered at $\mathbf{x}$ with radius ct. We use the

---

[5] **Gustav Robert Kirchhoff** (1824–1887) was a German physicist who made fundamental contributions to our understanding of electrical circuits, spectroscopy, and the emission of black-body radiation by heated objects.

dummy variable $\mathbf{y}$ as the integration variable over this sphere and $dS_{\mathbf{y}}$ to emphasize that the surface integration over the sphere $\partial B(\mathbf{x}, ct)$ is parametrized by $\mathbf{y}$.

To emphasize the fixed nature of $\mathbf{x}$ in the integral, it is convenient to change the notation and denote this fixed point by $\mathbf{x}_0$. Consequently, we are now free to use $\mathbf{x}$ as the dummy variable of integration over the sphere with center $\mathbf{x}_0$ and radius ct. Thus, we can rewrite Kirchhoff's formula from the perspective of giving the value of the disturbance u at the point $(\mathbf{x}_0, t)$:

$$\boxed{u(\mathbf{x}_0, t) = \frac{1}{4\pi c^2 t^2} \iint_{\partial B(\mathbf{x}_0, ct)} \left(\phi(\mathbf{x}) + \nabla\phi(\mathbf{x}) \cdot (\mathbf{x} - \mathbf{x}_0) + t\psi(\mathbf{x})\right) dS_{\mathbf{x}}.} \tag{4.11}$$

We will derive Kirchhoff's formula in Sections 4.2.3 and 4.2.4. We end this subsection by asking the usual reverse question associated with the solution formula: If u is defined by Kirchhoff's formula (4.11), does it solve (4.9)? One can verify (cf. Exercise **4.7**) that it does indeed solve the 3D wave equation for all $\mathbf{x} \in \mathbb{R}^3$ and $t > 0$. But what about the initial data? Unlike D'Alembert's formula, Kirchhoff's formula does not even make sense when we plug in $t = 0$! This will prove to be typical with many solution formulas that we will encounter. The solution formula $u(\mathbf{x}, t)$ and its temporal derivative $u_t(\mathbf{x}, t)$ are defined for all $t > 0$. The question becomes: If we were to extend them to the larger domain $t \geq 0$ by defining them to be the initial data at $t = 0$, do we generate a continuous function all the way down to $t = 0$? This will be the case if for any $\mathbf{x}_0 \in \mathbb{R}^3$, we have

$$\lim_{t \to 0^+} u(\mathbf{x}_0, t) = \phi(\mathbf{x}_0) \qquad \text{and} \qquad \lim_{t \to 0^+} u_t(\mathbf{x}_0, t) = \psi(\mathbf{x}_0). \tag{4.12}$$

Hence, showing that the solution attains the initial data is equivalent to establishing these limits[6]. Let us show the first limit and leave the second as part of Exercise **4.7**.

Attainment of the First Initial Condition: We have

$$\begin{aligned}
\lim_{t \to 0^+} u(\mathbf{x}_0, t) &= \lim_{t \to 0^+} \frac{1}{4\pi c^2 t^2} \iint_{\partial B(\mathbf{x}_0, ct)} \left(\phi(\mathbf{x}) + \nabla\phi(\mathbf{x}) \cdot (\mathbf{x} - \mathbf{x}_0) + t\psi(\mathbf{x})\right) dS_{\mathbf{x}} \\
&= \lim_{t \to 0^+} \frac{1}{4\pi c^2 t^2} \iint_{\partial B(\mathbf{x}_0, ct)} \phi(\mathbf{x})\, dS_{\mathbf{x}} + \lim_{t \to 0^+} \frac{1}{4\pi c^2 t^2} \iint_{\partial B(\mathbf{x}_0, ct)} \nabla\phi(\mathbf{x}) \cdot (\mathbf{x} - \mathbf{x}_0)\, dS_{\mathbf{x}} \\
&\quad + \lim_{t \to 0^+} \frac{1}{4\pi c^2 t^2} \iint_{\partial B(\mathbf{x}_0, ct)} t\psi(\mathbf{x})\, dS_{\mathbf{x}}.
\end{aligned} \tag{4.13}$$

By the Averaging Lemma (Lemma A.7.1),

$$\lim_{t \to 0^+} \frac{1}{4\pi c^2 t^2} \iint_{\partial B(\mathbf{x}_0, ct)} \phi(\mathbf{x})\, dS_{\mathbf{x}} = \phi(\mathbf{x}_0), \quad \lim_{t \to 0^+} \frac{1}{4\pi c^2 t^2} \iint_{\partial B(\mathbf{x}_0, ct)} \psi(\mathbf{x})\, dS_{\mathbf{x}} = \psi(\mathbf{x}_0).$$

Hence, the first limit in (4.13) gives $\phi(\mathbf{x}_0)$ while the third limit

$$\lim_{t \to 0^+} \frac{1}{4\pi c^2 t^2} \iint_{\partial B(\mathbf{x}_0, ct)} t\psi(\mathbf{x})\, dS_{\mathbf{x}} = \left(\lim_{t \to 0^+} t\right) \lim_{t \to 0^+} \frac{1}{4\pi c^2 t^2} \iint_{\partial B(\mathbf{x}_0, ct)} \psi(\mathbf{x})\, dS_{\mathbf{x}} = 0.$$

[6]Actually, this is not quite true. Here we have fixed $\mathbf{x}_0$ when taking the limit to the $t = 0$ axis. We should really show that

$$\lim_{t \to 0^+,\, \mathbf{x} \to \mathbf{x}_0} u(\mathbf{x}, t) = \phi(\mathbf{x}_0) \qquad \text{and} \qquad \lim_{t \to 0^+,\, \mathbf{x} \to \mathbf{x}_0} u_t(\mathbf{x}, t) = \psi(\mathbf{x}_0).$$

For the second limit in (4.13), we find

$$\begin{aligned}\left|\iint_{\partial B(\mathbf{x}_0,ct)} \nabla\phi(\mathbf{x})\cdot(\mathbf{x}-\mathbf{x}_0)\,dS_{\mathbf{x}}\right| &\le \iint_{\partial B(\mathbf{x}_0,ct)} |\nabla\phi(\mathbf{x})\cdot(\mathbf{x}-\mathbf{x}_0)|\,dS_{\mathbf{x}}\\ &\le \iint_{\partial B(\mathbf{x}_0,ct)} |\nabla\phi(\mathbf{x})|\,|\mathbf{x}-\mathbf{x}_0|\,dS_{\mathbf{x}}\\ &= ct\iint_{\partial B(\mathbf{x}_0,ct)} |\nabla\phi(\mathbf{x})|\,dS_{\mathbf{x}}.\end{aligned}$$

To pass to the second inequality we used the Cauchy-Schwartz inequality[7] on the integrand. Hence,

$$\begin{aligned}&\left|\lim_{t\to0^+}\frac{1}{4\pi c^2t^2}\iint_{\partial B(\mathbf{x}_0,ct)} \nabla\phi(\mathbf{x})\cdot(\mathbf{x}-\mathbf{x}_0)\,dS_{\mathbf{x}}\right|\\ &\qquad\le\left(\lim_{t\to0^+} ct\right)\lim_{t\to0^+}\frac{1}{4\pi c^2t^2}\iint_{\partial B(\mathbf{x}_0,ct)} |\nabla\phi(\mathbf{x})|\,dS_{\mathbf{x}}.\end{aligned}$$

By the Averaging Lemma, the second limit above is $|\nabla\phi(\mathbf{x}_0)|$ while the first is, of course, 0. Thus, the second limit in (4.13) is also 0. Done!

4.2.2. • Consequences of Kirchhoff's Formula: Causality and the Huygens Principle. Kirchhoff's formula reveals many important features about the **causality** of wave propagation in 3D. We begin at $t = 0$ with an initial disturbance $\phi(\mathbf{x})$ having initial velocity $\psi(\mathbf{x})$. Fixing a point $\mathbf{x}_0$ in space and time $t > 0$, we wish to know the disturbance at position $\mathbf{x}_0$ and time t. Kirchhoff's formula tells us that we can obtain this information via the values of the initial data on the **sphere** centered at $\mathbf{x}_0$ with radius ct. In fact, it is obtained by averaging the values of ϕ and its derivatives and the values of $t\psi$ on this sphere. All values of ϕ, $\nabla\phi$, and ψ away from this sphere are irrelevant, including, for example, the values at the point $\mathbf{x}_0$ itself. Note that the radius of the relevant information is ct (which has dimensions of length); therefore the longer we wait the larger the sphere we need to average over. We can rephrase this crucial observation in terms of the domains of dependence and influence.

- **Domain of Dependence**: We fix a position $\mathbf{x}_0$ in space and a time $t > 0$ and then address the dependence of $u(\mathbf{x}_0, t)$ on the initial data $u(\mathbf{x}, 0), u_t(\mathbf{x}, 0)$. More precisely, we want to know the places in space where the initial disturbance can affect the value of $u(\mathbf{x}_0, t)$. From Kirchhoff's formula, we see that the value of $u(\mathbf{x}_0, t)$ depends on the initial values $u(\mathbf{x}, 0)$ **only if** $|\mathbf{x}-\mathbf{x}_0| = ct$, in other words, only for $\mathbf{x}$ which lie on the sphere centered at $\mathbf{x}_0$ with radius ct. More generally, for any two times $t_2 > t_1 \ge 0$, $u(\mathbf{x}_0, t_2)$ depends on the values of $u(\mathbf{x}, t_1)$ only for $\mathbf{x}$ on the sphere centered at $\mathbf{x}_0$ with radius $c(t_2 - t_1)$.
- **Domain of Influence**: Suppose initial information about u and its derivatives is given at a point $\mathbf{x}_0$; i.e., we are provided with the values of $u(\mathbf{x}_0, 0)$ and its derivatives. At any later time t, we are interested in the region of space which will be effected (influenced) by this information. From Kirchhoff's formula, this information will influence the values of $u(\mathbf{x}, t)$ **only if** $|\mathbf{x}-\mathbf{x}_0| = ct$, that is, only for

[7] Namely, the trivial inequality $|ab| \le |a|\,|b|$ for all $a, b \in \mathbb{R}$.

those $\mathbf{x}$ which lie on the sphere centered at $\mathbf{x}_0$ with radius ct. More generally, for any two times $t_2 > t_1 \geq 0$, information at position $\mathbf{x}_0$ and time t_1 will influence the values of $u(\mathbf{x}, t_2)$ only for $\mathbf{x}$ on the sphere centered at $\mathbf{x}_0$ with radius $c(t_2 - t_1)$.

We illustrate the domain of dependence and the domain of influence with the same cartoon in Figure 4.1. In each case, you should interpret the cartoon differently, in one case to the past and the other to the future.

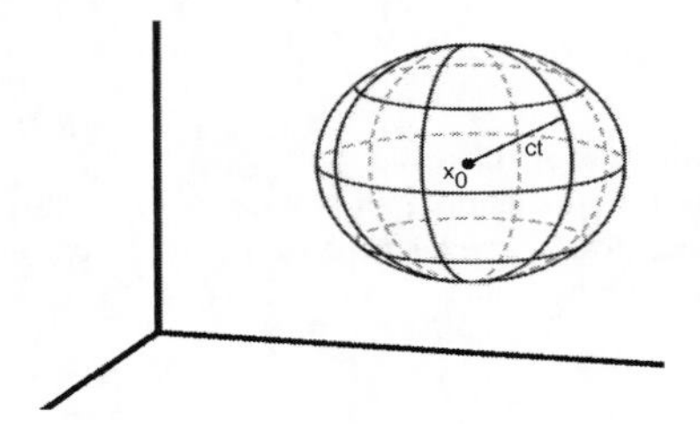

Figure 4.1. Two interpretations of the same figure. (i) **Dependence:** The value of the disturbance at position $\mathbf{x}_0$ and time t depends on the initial disturbance ($t = 0$) on the sphere of radius ct. **Influence:** An initial disturbance ($t = 0$) at position $\mathbf{x}_0$ will influence the value of u at time t on all points which lie on the sphere of radius ct.

The analogous space-time pictures to the 1D wave equation in Section 3.4.1 would now require 4 dimensions (3 for space and 1 for time). However, you can think of either the domain of dependence or influence as a **cone** in four-dimensional space-time. This cone is often called the **light cone**. Here we are referring only to the "surface" of the cone and not the "inside". In other words, these domains are three-dimensional subsets of four-dimensional space-time. For example, the domain of dependence at the point $\mathbf{x}_0, t_0$ is the light cone represented by

$$\mathcal{C} = \left\{(\mathbf{x}, t) \,\middle|\, 0 \leq t \leq t_0,\ |\mathbf{x} - \mathbf{x}_0| = c(t_0 - t)\right\}.$$

The punchline remains that information propagates on surfaces — **a sharp front** in 3D. Perhaps it is easiest to illustrate this in terms of sound waves. Suppose a bomb, say concentrated at the point $\mathbf{x}_0$, detonates at $t = 0$ and we are fixed at a different position $\mathbf{x}_1$ in space. Then we will hear the bomb (feel the disturbance) only for an instance, that is, precisely at time

$$t_1 = \frac{|\mathbf{x}_1 - \mathbf{x}_0|}{c}. \tag{4.14}$$

This is illustrated by the cartoon in Figure 4.2. The observation that disturbances propagate in 3D on **sharp fronts** is part of **the Huygens Principle**[8]. Since the 3D wave equation is a model for the propagation of sound waves in air, the fact that information (i.e., sound) propagates on sharp fronts is the essential ingredient to our notion of sound, music, and speech. This is in contrast to 1D and, as we shall see in the next subsection, in 2D as well. For example, suppose we lived in a 1D universe where sound propagated via the 1D wave equation. Recalling the domain of influence from Section 3.4.1 we see that the following is true: If a bomb detonated at position $\mathbf{x}_0$ at $t = 0$ and

[8]**Christiaan Huygens** (1629–1695) was a Dutch mathematician and scientist.

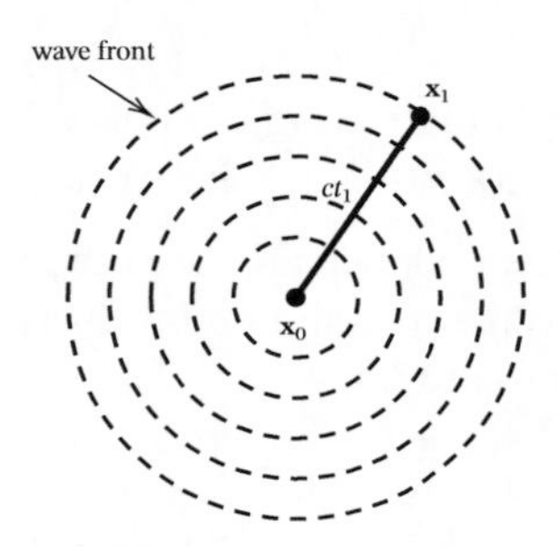

Figure 4.2. A two-dimensional slice of three space wherein a bomb detonates at $\mathbf{x}_0$ at $t = 0$. The sound from the detonation propagates on a sharp wave front. If we are always standing at position $\mathbf{x}_1$, we hear the sound resulting from the detonation only for an instance at t_1 given by (4.14).

we (as a one-dimensional being) were fixed at position $\mathbf{x}_1$, we would initially feel the disturbance (hear the bomb) at time $t_1 = \frac{|\mathbf{x}_1 - \mathbf{x}_0|}{c}$ **yet** continue to hear it at all future times $t \geq t_1$. Hence, if we lived in one- or two-dimensional space, music and speech would be rather muddled; after first hearing a sound, we would continue to hear versions of it for eternity[9].

The Huygens Principle actually elaborates further and often includes the following statement:

"Every point of a wave front may be considered as the source of secondary wavelets that spread out in all directions with a speed equal to the speed of propagation of the waves".

This recognizes that each point of an advancing wave front is, in fact, the center of a fresh disturbance and the source of a new train of waves; additionally, the advancing wave (as a whole) may be regarded as the sum of all the secondary waves arising from points in the medium already traversed. These statements are embedded in Kirchhoff's formula but, at first glance, they may seem **self-contradictory** to the fact that Kirchhoff's formula dictates the propagation on sharp fronts. For example, suppose we are at a fixed point and an initial disturbance (sound), concentrated at some other point, passes us at time t_0. **Question:** Won't some of the new train of waves (wavelets) from the advanced wave front come back to us at a later time $t > t_0$? The answer is yes, but these waves come from all directions and in such a way that they will cancel each other out!

One can view the wave front at any time as an envelope that encloses all of the secondary wavelets. In other words, the new wave front surface is **tangent** to the front surface of the primary and all secondary wave fronts. We illustrate this phenomena with a two-dimensional slice in Figure 4.3.

[9]Do not confuse this with **echo** effects which have to do with wave reflections from boundaries or physical obstacles.

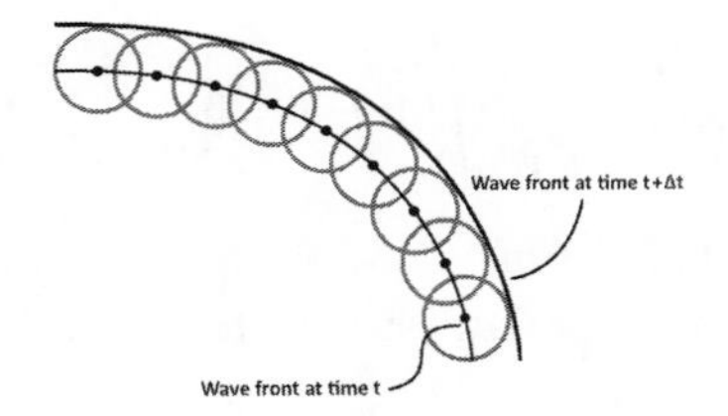

Figure 4.3. The propagating wave front surface is **tangent** to the front surface of the primary and all secondary wave fronts.

4.2.3. • Kirchhoff's Formula via Spherical Means.

The reader should be curious and perhaps a bit suspicious as to several aspects of Kirchhoff's formula: Where did it come from? Why the averaging over spheres? Why is causality different from that in 1D? What is special about three space dimensions? The most direct route to answering these questions lies in deriving Kirchhoff's formula from scratch. We will shortly present this derivation in full but first, let us set up the basic notion of a **spherical mean** (an average over a sphere), outline how the derivation will proceed, and give a hint as to why dimension three is special.

Our first observation is that the Laplacian in the 3D wave equation treats all the spacial variables equally; there is no real difference between x_1, x_2, and x_3. For example, if we were to interchange any two, we would find the exact same PDE. This is reflected in the fact that the PDE models the propagation of a disturbance in a homogeneous medium wherein there is no preferred propagation direction. Hence one would expect an "intrinsic radial symmetry" to the propagation of disturbances in a homogeneous medium. This is **not** to say that the solution to the 3D wave equation will always be radially symmetric.

Now let u be a C^2 (in $\mathbf{x}$ and t) solution to (4.9), and **fix** a point in space $\mathbf{x}_0 \in \mathbb{R}^3$. We wish to exploit the fact that information at $\mathbf{x}_0$ should propagate only in the radial direction. Said another way, from the perspective of determining $u(\mathbf{x}_0, t)$, our wave equation in three space dimensions and time should effectively be captured by a PDE in only the radial direction and time. Working with spherical means is our route to discovering this reduced PDE.

With these comments in mind, we define for $r > 0$ the **spherical mean**

$$\overline{u}(r,t) \quad := \quad \frac{1}{4\pi r^2} \iint_{\partial B(\mathbf{x}_0,r)} \quad u(\mathbf{x},t)\,dS. \tag{4.15}$$

As usual, $\partial B(\mathbf{x}_0, r)$ denotes the boundary of the ball centered at $\mathbf{x}_0$ with radius r, i.e., **the sphere** with radius r and center $\mathbf{x}_0$. Based upon the previous comments about radial propagation in space and the fact that u solves the 3D wave equation, one might expect that $\overline{u}(r,t)$ solves a PDE in one space variable $r \geq 0$, time $t > 0$, and initial values:

$$\overline{u}(r,0) = \frac{1}{4\pi r^2} \iint_{\partial B(\mathbf{x}_0,r)} \phi(\mathbf{x})\,dS =: \overline{\phi}(r) \tag{4.16}$$

and

$$\overline{u}_t(r,0) \;=\; \frac{1}{4\pi r^2}\iint_{\partial B(\mathbf{x}_0,r)} \psi(\mathbf{x})\,dS \;=:\; \overline{\psi}(r). \tag{4.17}$$

Hence, if we can find this PDE for $\overline{u}(r,t)$, we have the initial data (4.16) and (4.17) to solve it and find/derive a formula for $\overline{u}(r,t)$ in terms of $\overline{\phi}(r)$ and $\overline{\psi}(r)$. Then by the Averaging Lemma (Lemma A.7.1),

$$u(\mathbf{x}_0,t) \;=\; \lim_{r\to 0^+} \overline{u}(r,t), \tag{4.18}$$

and we will obtain a formula for $u(\mathbf{x}_0,t)$ solely in terms of the initial data ϕ and ψ. Note that while the mother point $\mathbf{x}_0$ is fixed throughout this process, it can be any point in $\mathbb{R}^3$; hence, we will have completely solved the IVP to the 3D wave equation.

Let us pause to remind ourselves about the notation we have just introduced:

- $\mathbf{x}_0 \in \mathbb{R}^3$ is a **fixed** spatial point; in other words, it will not vary throughout this derivation. It denotes the center of the spheres in **all** the spherical means of u.
- $t \geq 0$ is the time at which we take the solution u to be averaged in the spherical means. It will vary throughout the derivation.
- $r \geq 0$ denotes the radius of the sphere for a spherical mean. It will vary throughout the derivation.
- $\mathbf{x} \in \mathbb{R}^3$ is a dummy variable of integration used for all surface integrals over the spheres.
- The spherical mean $\overline{u}(r,t)$ is the **spatial average** of $u(\mathbf{x},t)$ over $\mathbf{x}$ lying on the sphere centered at $\mathbf{x}_0$ with radius r.
- The spherical means $\overline{\phi}(r)$ and $\overline{\psi}(r)$ are the spatial averages of $u(\mathbf{x},0)$ and $u_t(\mathbf{x},0)$ (respectively) over $\mathbf{x}$ lying on the sphere centered at $\mathbf{x}_0$ with radius r.

The PDE for $\overline{u}(r,t)$ comes from the following:

(a) the fact that the “mother” function u satisfies the 3D wave equation,

(b) some basic integral calculus (the Divergence Theorem and spherical coordinates).

It is not, as one might hope, the 1D wave equation, but rather what is known as the **Euler-Poisson-Darboux equation**

$$\boxed{\overline{u}_{tt} = c^2\left(\overline{u}_{rr} + \frac{2}{r}\overline{u}_r\right).}$$

Next, comes a miracle of sorts which works because the space dimension is three[10]. If we define

$$v(r,t) := r\overline{u}(r,t),$$

then $v(r,t)$ satisfies the 1D wave equation on the half-line $r \geq 0$ with a self-imposed Dirichlet boundary condition at $r = 0$. Hence, we can use the solution formula we derived in Section 3.8 to find $v(r,t)$, and hence $\overline{u}(r,t)$, in terms of $\overline{\phi}(r)$ and $\overline{\psi}(r)$. Using (4.18), we finally recover $u(\mathbf{x}_0,t)$ by taking the limit as $r \to 0^+$. After some “bookkeeping” we arrive at Kirchhoff's formula. The details are presented in the next section.

[10]Actually a “related miracle” works in all **odd** space dimensions except one!

4.2.4. Full Details: The Proof of Kirchhoff's Formula. Here we present the full derivation of Kirchhoff's formula by proving the following theorem. We follow Evans [**10**].

Theorem 4.1 (Kirchhoff's Formula). *If $u(\mathbf{x},t)$ solves (4.9), then at any point $\mathbf{x}_0$ and time $t > 0$, we have*

$$\boxed{u(\mathbf{x}_0,t) = \frac{1}{4\pi c^2 t^2}\iint_{\partial B(\mathbf{x}_0,ct)} \left(\phi(\mathbf{x}) + \nabla\phi(\mathbf{x})\cdot(\mathbf{x}-\mathbf{x}_0) + t\psi(\mathbf{x})\right) dS_{\mathbf{x}}.}$$

One can check (cf. Exercise **4.7**) that if we define u via this formula, then we do indeed solve (4.9); that is, the function solves the 3D wave equation for $t > 0$ and attains the initial data in the sense of (4.12). Therefore, we have found the unique solution.

Proof. Step 1: Spherical means from $u(\mathbf{x},t)$. We begin, as usual, by assuming we have a function $u(\mathbf{x},t)$ that solves (4.9). Fix $\mathbf{x}_0 \in \mathbb{R}^3$ for the remainder of the proof wherein we will derive a formula for the solution at $(\mathbf{x}_0,t)$. Since this formula will hold for any choice of $\mathbf{x}_0 \in \mathbb{R}^3$ and $t > 0$, we will be done. Following (4.15)–(4.17), we write down our spherical means $\overline{u}(r,t), \overline{\phi}(r)$, and $\overline{\psi}(r)$ calculated from the solution $u(\mathbf{x},t)$. Our goal is to obtain a formula for $\overline{u}(r,t)$ in terms of $\overline{\phi}(r)$ and $\overline{\psi}(r)$ and then *recover* $u(\mathbf{x}_0,t)$ by sending $r \to 0^+$, invoking the the Averaging Lemma (Lemma A.7.1) to find

$$u(\mathbf{x}_0,t) = \lim_{r\to 0^+} \overline{u}(r,t).$$

Step 2: Derivation of the PDE for $\overline{u}(r,t)$. Our goal is show that $\overline{u}(r,t)$ satisfies

$$\boxed{\overline{u}_{tt} = c^2\left(\overline{u}_{rr} + \frac{2}{r}\overline{u}_r\right).} \tag{4.19}$$

The only ingredients in the proof of this claim are (i) the fact that u itself satisfies the wave equation and (ii) basic calculus (the Divergence Theorem and integration over spherical shells). To start, we wish to calculate $\overline{u}_r(r,t)$, i.e., differentiate the spherical mean

$$\overline{u}(r,t) = \frac{1}{4\pi r^2}\iint_{\partial B(\mathbf{x}_0,r)} u(\mathbf{x},t)\, dS$$

with respect to r. To this end, it is convenient to rescale and translate variables so that the domain of integration is simply $\partial B(\mathbf{0},1)$. Therefore, let $\mathbf{y} = \frac{\mathbf{x}-\mathbf{x}_0}{r}$ or $\mathbf{x} = \mathbf{x}_0 + r\mathbf{y}$. Noting that (cf. Section A.5.4) $dS_{\mathbf{x}} = r^2\, dS_{\mathbf{y}}$, we find that

$$\overline{u}(r,t) = \frac{1}{4\pi r^2}\iint_{\partial B(\mathbf{x}_0,r)} u(\mathbf{x},t)\, dS_{\mathbf{x}} = \frac{1}{4\pi}\iint_{\partial B(\mathbf{0},1)} u(\mathbf{x}_0 + r\mathbf{y},t)\, dS_{\mathbf{y}}.$$

Now we can differentiate with respect to r under the integral sign to find, using the chain rule, that

$$\overline{u}_r(r,t) = \frac{1}{4\pi}\iint_{\partial B(\mathbf{0},1)} \nabla u(\mathbf{x}_0 + r\mathbf{y},t)\cdot \underbrace{\mathbf{y}}_{=\frac{d}{dr}(\mathbf{x}_0+r\mathbf{y})}\, dS_{\mathbf{y}}.$$

After taking the derivative, we can now transform back to our original variable $\mathbf{x}$ to find

$$\overline{u}_r(r,t) = \frac{1}{4\pi}\iint_{\partial B(\mathbf{0},1)} \nabla u(\mathbf{x}_0 + r\mathbf{y}, t)\cdot \mathbf{y}\, dS_{\mathbf{y}} = \frac{1}{4\pi r^2}\iint_{\partial B(\mathbf{x}_0,r)} \nabla u(\mathbf{x},t)\cdot \frac{\mathbf{x}-\mathbf{x}_0}{r}\, dS_{\mathbf{x}}.$$

Next, we note that for $\mathbf{x} \in \partial B(\mathbf{x}_0, r)$, $\frac{\mathbf{x}-\mathbf{x}_0}{r}$ denotes the unit outer normal. Hence, using the Divergence Theorem we have that

$$\overline{u}_r(r,t) = \frac{1}{4\pi r^2}\iiint_{B(\mathbf{x}_0,r)} \Delta u(\mathbf{x},t)\, d\mathbf{x}.$$

Since u satisfies the wave equation $u_{tt} = c^2\Delta u$, we have

$$\overline{u}_r(r,t) = \frac{1}{4\pi c^2 r^2}\iiint_{B(\mathbf{x}_0,r)} u_{tt}(\mathbf{x},t)\, d\mathbf{x} \quad \text{or} \quad r^2\overline{u}_r(r,t) = \frac{1}{4\pi c^2}\iiint_{B(\mathbf{x}_0,r)} u_{tt}(\mathbf{x},t)\, d\mathbf{x}.$$

By (A.11), we can differentiate with respect to r to find

$$\begin{aligned}\left(r^2\overline{u}_r(r,t)\right)_r &= \frac{1}{4\pi c^2}\iint_{\partial B(\mathbf{x}_0,r)} u_{tt}(\mathbf{x},t)\, dS_{\mathbf{x}} \\ &= \frac{r^2}{c^2}\left(\frac{1}{4\pi r^2}\iint_{\partial B(\mathbf{x}_0,r)} u_{tt}(\mathbf{x},t)\, dS_{\mathbf{x}}\right) \\ &= \frac{r^2}{c^2}\,\overline{u}_{tt}(r,t).\end{aligned}$$

Using the product rule on the left yields (4.19).

Step 3: Reducing (4.19) to the 1D wave equation and recovering $u(\mathbf{x}_0, t)$. Now, define $v(r,t) := r\overline{u}(r,t)$ and ask: what PDE does v satisfy? The answer will justify the definition. To this end, we first note that

$$v_r = r\overline{u}_r + \overline{u}, \qquad v_{rr} = r\overline{u}_{rr} + 2\overline{u}_r, \qquad v_{tt} = r\overline{u}_{tt}.$$

Hence, by (4.19), we have

$$v_{tt} = r\overline{u}_{tt} = c^2(r\overline{u}_{rr} + 2\overline{u}_r) = c^2 v_{rr}.$$

So v satisfies the 1D wave equation! Note that $v(r,t)$ is only defined for $r \geq 0$ and, by definition, $v(0,t) = 0$ for all $t \geq 0$. In terms of initial data we have

$$v(r,0) = r\overline{\phi}(r), \qquad v_t(r,0) = r\overline{\psi}(r).$$

If we define

$$\Phi(r) := r\overline{\phi}(r) \quad \text{and} \quad \Psi(r) := r\overline{\psi}(r),$$

then $v(r,t)$ must be a solution to the fixed boundary, semi-infinite string boundary value problem:

$$\begin{cases} v_{tt} = c^2 v_{rr}, & r \geq 0,\ t \geq 0, \\ v(r,0) = \Phi(r), & x \geq 0, \\ v_t(r,0) = \Psi(r), & x \geq 0, \\ v(0,t) = 0, & t \geq 0. \end{cases}$$

But this problem was solved in Section 3.8. The solution is given by either (3.19) or (3.20) depending on whether $r < ct$ or $r \geq ct$. Do we need both cases? In the end, we want to know $u(\mathbf{x}_0, t)$, which is related to $v(r, t)$ by

$$u(\mathbf{x}_0, t) = \lim_{r \to 0^+} \frac{v(r,t)}{r}. \tag{4.20}$$

Consequently, we only care about the case where $r < ct$. By (3.20), we have

$$v(r,t) = \frac{1}{2}\left[\Phi(r+ct) - \Phi(ct-r)\right] + \frac{1}{2c}\int_{ct-r}^{ct+r} \Psi(s)\,ds.$$

Now, (4.20) becomes

$$\begin{aligned} u(\mathbf{x}_0, t) &= \lim_{r \to 0^+} \left[\frac{\Phi(r+ct) - \Phi(ct-r)}{2r} + \frac{1}{2cr}\int_{ct-r}^{ct+r} \Psi(s)\,ds \right] \\ &= \Phi'(ct) + \frac{1}{c}\Psi(ct). \end{aligned}$$

To pass to the last equality, we used the definition of the derivative for the first term and the Averaging Lemma (1D version) for the second.

Step 4: Writing out the spherical means of the data. Recall the definition of Φ and Ψ:

$$\begin{aligned} \Phi(ct) = ct\overline{\phi}(ct) &= \frac{ct}{4\pi c^2t^2}\iint_{\partial B(\mathbf{x}_0, ct)} \phi(\mathbf{x})\,dS, \\ \Psi(ct) = ct\overline{\psi}(ct) &= \frac{ct}{4\pi c^2t^2}\iint_{\partial B(\mathbf{x}_0, ct)} \psi(\mathbf{x})\,dS. \end{aligned}$$

Noting that by the chain rule, $\Phi'(ct) = \frac{1}{c}\frac{\partial}{\partial t}\Phi(ct)$, we have

$$\begin{aligned} u(\mathbf{x}_0, t) &= \frac{\partial}{\partial t}\left(\frac{t}{4\pi c^2t^2}\iint_{\partial B(\mathbf{x}_0, ct)} \phi(\mathbf{x})\,dS\right) + \frac{1}{c}\left(\frac{ct}{4\pi c^2t^2}\iint_{\partial B(\mathbf{x}_0, ct)} \psi(\mathbf{x})\,dS\right) \\ &= t\frac{\partial}{\partial t}\left(\frac{1}{4\pi c^2t^2}\iint_{\partial B(\mathbf{x}_0, ct)} \phi(\mathbf{x})\,dS\right) + \left(\frac{1}{4\pi c^2t^2}\iint_{\partial B(\mathbf{x}_0, ct)} \phi(\mathbf{x})\,dS\right) \\ &\quad + \frac{1}{c}\left(\frac{ct}{4\pi c^2t^2}\iint_{\partial B(\mathbf{x}_0, ct)} \psi(\mathbf{x})\,dS\right), \end{aligned} \tag{4.21}$$

where we used the product rule on the first term. You should now be able to recognize the second and third terms as two of the three components in Kirchhoff's formula. What remains to be shown is that the derivative in the first term completes Kirchhoff's formula.

Step 5: Carrying out the differentiation in the first term. We perform the differentiation by first changing variables to write the integral over the unit sphere. This removes the t dependence from the region of integration, placing it solely in the integrand. Then we can differentiate under the integral sign and, finally, resort back to the original variables. To this end, we evoke the change of variables $\mathbf{x} = \mathbf{x}_0 + ct\mathbf{z}$ or $\mathbf{z} = \frac{\mathbf{x}-\mathbf{x}_0}{ct}$ to find

$$\frac{1}{4\pi c^2t^2}\iint_{\partial B(\mathbf{x}_0, ct)} \phi(\mathbf{x})\,dS = \frac{1}{4\pi}\iint_{\partial B(\mathbf{0},1)} \phi(\mathbf{x}_0 + ct\mathbf{z})\,dS_{\mathbf{z}}.$$

Hence, by the chain rule we have

$$\begin{aligned}\frac{\partial}{\partial t}\left(\frac{1}{4\pi c^2 t^2}\iint_{\partial B(\mathbf{x}_0,ct)} \phi(\mathbf{x})\,dS\right) &= \frac{1}{4\pi}\iint_{\partial B(\mathbf{0},1)} \nabla\phi(\mathbf{x}_0+ct\mathbf{z})\cdot c\mathbf{z}\,dS_{\mathbf{z}} \\ &= \frac{1}{4\pi c^2 t^2}\iint_{\partial B(\mathbf{x}_0,ct)} \nabla\phi(\mathbf{x})\cdot\left(\frac{\mathbf{x}-\mathbf{x}_0}{ct}\right) c\,dS_{\mathbf{x}}.\end{aligned}$$

Placing this into (4.21), we arrive at Kirchhoff's formula

$$u(\mathbf{x}_0,t) = \frac{1}{4\pi c^2 t^2}\iint_{\partial B(\mathbf{x}_0,ct)} \big(\phi(\mathbf{x}) + \nabla\phi(\mathbf{x})\cdot(\mathbf{x}-\mathbf{x}_0) + t\psi(\mathbf{x})\big)\,dS_{\mathbf{x}}.$$

□

4.3. Two Space Dimensions: The 2D Wave Equation and Its Explicit Solution

The IVP for the 2D wave equation is given by

$$\begin{cases} u_{tt} = c^2\Delta u, & \mathbf{x}=(x_1,x_2)\in\mathbb{R}^2,\ t>0, \\ u(\mathbf{x},0)=\phi(\mathbf{x}), & \mathbf{x}\in\mathbb{R}^2, \\ u_t(\mathbf{x},0)=\psi(\mathbf{x}), & \mathbf{x}\in\mathbb{R}^2. \end{cases} \tag{4.22}$$

A natural setting for the 2D wave equation, one which can yield a physical derivation, is the **vibrations of a flat, stretched, elastic membrane** (cf. Exercise **4.10**). For example, one can model the vibrations in a circular drum by considering the wave equation on a disc and prescribing Dirichlet conditions on the boundary. This would correspond to fixing the displacement at the rim of the drum. We will solve this boundary value problem in Section 12.8.1. Here we simply solve the equation in all of $\mathbb{R}^2$ to reveal a fundamental difference between wave propagation in two and three space dimensions.

4.3.1. The Solution via the Method of Descent.

One could certainly proceed as we did in 3D by considering spherical means. However in 2D, there is no transformation which would lead to the 1D wave equation. On the other hand, there is a simple way of deriving the solution from the 3D solution formula which is dubbed **the method of descent**[11]. It consists of the following steps:

- We assume that we have a solution $u(x_1,x_2,t)$ to (4.22) and *artificially lift* this solution to 3D by prescribing no dependence on an artificial third variable x_3.
- We note that this lifted 3D function trivially solves the 3D wave equation.
- We use Kirchhoff's formula to write down the solution formula for the lifted solution.
- Since Kirchhoff's formula only depends on a sphere (a two-dimensional object in 3D), we can parametrize this sphere via x_1 and x_2 and project the 3D solution back to 2D. This last step will be clearer shortly.

[11]The method of descent is often attributed to the French mathematician **Jacques Hadamard** (1865–1963).

We detail these steps by proving the following theorem.

Theorem 4.2 (Solution Formula for the 2D Wave Equation). *If* $u(\mathbf{x},t) = u(x_1, x_2, t)$ *solves* (4.22), *then*

$$u(\mathbf{x},t) = \frac{2}{4\pi ct} \iint_{B_{2D}(\mathbf{x},ct)} \frac{\phi(\mathbf{y}) + \nabla\phi(\mathbf{y})\cdot(\mathbf{y}-\mathbf{x}) + t\psi(\mathbf{y})}{\sqrt{c^2t^2 - |\mathbf{y}-\mathbf{x}|^2}}\, dy_1\, dy_2. \tag{4.23}$$

We have used the notation $B_{2D}(\mathbf{x}, ct)$ to emphasize that we are integrating over the two-dimensional ball (a disc) with center $\mathbf{x} = (x_1, x_2) \in \mathbb{R}^2$ and radius ct. Note that this region is a (solid) ball and not a boundary as in Kirchhoff's formula and, hence, this is a regular "bulk" 2D integral. The dummy variable of integration which traverses the disc is denoted by $\mathbf{y} = (y_1, y_2)$.

Proof. Let $u(x_1, x_2, t)$ solve (4.22) and define $\tilde{u}, \widetilde{\phi}, \widetilde{\psi}$ by

$$\tilde{u}(x_1,x_2,x_3,t) := u(x_1,x_2,t), \qquad \widetilde{\phi}(x_1,x_2,x_3) := \phi(x_1,x_2),$$
$$\widetilde{\psi}(x_1,x_2,x_3) := \psi(x_1,x_2).$$

We then automatically determine that

$$\tilde{u}_{tt}(x_1,x_2,x_3,t) = u_{tt}(x_1,x_2,t), \qquad \tilde{u}_{x_1x_1}(x_1,x_2,x_3,t) = u_{x_1x_1}(x_1,x_2,t),$$
$$\tilde{u}_{x_2x_2}(x_1,x_2,x_3,t) = u_{x_2x_2}(x_1,x_2,t), \qquad \tilde{u}_{x_3x_3}(x_1,x_2,x_3,t) = 0.$$

Since $u_{tt} = c^2(u_{x_1x_1} + u_{x_2x_2})$ for all $(x_1, x_2) \in \mathbb{R}^2, t \geq 0$, this means $\tilde{u}$ solves the 3D wave equation initial value problem:

$$\begin{cases} \tilde{u}_{tt} = c^2(\tilde{u}_{x_1x_1} + \tilde{u}_{x_2x_2} + \tilde{u}_{x_3x_3}), & (x_1,x_2,x_3) \in \mathbb{R}^3,\ t \geq 0, \\ \tilde{u}(x_1,x_2,x_3,0) = \widetilde{\phi}(x_1,x_2,x_3), & (x_1,x_2,x_3) \in \mathbb{R}^3, \\ \tilde{u}_t(x_1,x_2,x_3,0) = \widetilde{\psi}(x_1,x_2,x_3), & (x_1,x_2,x_3) \in \mathbb{R}^3. \end{cases}$$

By Kirchhoff's formula, written in terms of $\tilde{\mathbf{x}} = (x_1, x_2, x_3) \in \mathbb{R}^3$,

$$\tilde{u}(\tilde{\mathbf{x}},t) = \frac{1}{4\pi c^2t^2} \iint_{\partial B_{3D}(\tilde{\mathbf{x}},ct)} \left(\widetilde{\phi}(\mathbf{y}) + \nabla\widetilde{\phi}(\mathbf{y})\cdot(\mathbf{y}-\tilde{\mathbf{x}}) + t\widetilde{\psi}(\mathbf{y})\right) dS_{\mathbf{y}}. \tag{4.24}$$

Let us consider this formula for $\tilde{\mathbf{x}}$ on the x_1x_2-plane (i.e., $x_3 = 0$). The sphere centered at $\tilde{\mathbf{x}}$ has two parts but by symmetry we may perform the integral over the top hemisphere and double the result. The key observation is that we can easily parametrize the top hemisphere with x_1 and x_2 traversing the disk in the x_1x_2-plane. Recall that if a surface can be represented as the graph of a function of two variables, $y_3 = f(y_1, y_2)$, then

$$dS_{\mathbf{y}} = \sqrt{1 + \left(\frac{\partial f}{\partial y_1}\right)^2 + \left(\frac{\partial f}{\partial y_2}\right)^2}\, dy_1\, dy_2.$$

The top hemisphere of the sphere $\partial B_{3D}(\tilde{\mathbf{x}}, ct)$ has equation

$$y_3 = \sqrt{c^2t^2 - (y_1 - x_1)^2 - (y_2 - x_2)^2},$$

which implies

$$dS = \frac{ct}{\sqrt{c^2t^2 - (y_1 - x_1)^2 - (y_2 - x_2)^2}}\, dy_1\, dy_2.$$

Note that the projection of the top hemisphere on the x_1x_2-plane is the disc centered at (x_1, x_2) with radius ct, or $B_{2D}((x_1, x_2), ct)$. Thus, writing (4.24) as a double integral over this projection gives

$$u(x_1, x_2, t)$$
$$= \frac{2}{4\pi c^2 t^2} \iint_{B_{2D}((x_1,x_2),ct)} \frac{ct\phi(y_1, y_2) + ct\nabla\phi(y_1, y_2) \cdot (y_1 - x_1, y_2 - x_2) + ct^2\psi(y_1, y_2)}{\sqrt{c^2t^2 - (y_1 - x_1)^2 - (y_2 - x_2)^2}} \, dy_1 dy_2.$$

□

4.3.2. Causality in 2D. Note the fundamental difference between (4.23) and Kirchhoff's formula: In the former, the solution depends on a two-dimensional **solid** region of the plane. As we did for D'Alembert's formula, one can draw/picture the analogous domain of dependence and domain of influence. They would be **solid cones** in the x_1, x_2, t space, i.e., cones with the insides filled in. Hence, the Huygens Principle is **false** in two space dimensions. In particular, disturbances do not propagate on sharp fronts, and at any given position, once one feels a disturbance it will be felt for all later times. If it so happened that we lived in a two-dimensional world, as depicted in *Flatland*[12], then sound waves would propagate according to the 2D wave equation. This would be unfortunate as it would be rather difficult for us to communicate with sound; once we hear a sound, we would continue to hear variants of it forever.

On the other hand, it is worth remarking on the factor in the denominator of (4.23). Here we are clearly weighting information which is closer to the sharp front $\partial B_{2D}(\mathbf{x}, ct)$, far more than information in the interior of the disk. Thus, in some sense, one could argue that the Huygens Principle still has some validity but in a very weak sense. Note that this was not the case with D'Alembert's formula in 1D.

4.4. Some Final Remarks and Geometric Optics

4.4.1. The Wave Equation in Space Dimensions Larger Than Three. Following the approaches of Sections 4.2 and 4.3, one can derive formulas for the solution to the wave equation in the other odd space dimensions $5, 7, 9, \ldots$ and the other even space dimensions $4, 6, 8, \ldots$ (see Evans [**10**] and Exercise **4.18**). In doing so, one finds that the Huygens Principle of disturbances propagating on sharp fronts holds in odd dimensions strictly larger than 1 yet fails in even dimensions. As we saw, it also fails in dimension 1, but this dimension is rather special and consequently should be viewed as a unique case; for instance, recall that D'Alembert's formula treats the initial displacements and velocities differently from the point of view of fronts, something not seen in other space dimensions.

4.4.2. Regularity. Thus far we have only dealt with smooth solutions of the wave equation resulting from smooth initial conditions[13]. Two points are in order. First, it is useful to make a quantitative statement about the smoothness (regularity) of the solution based upon the smoothness of the initial data. For example, for which positive

[12] *Flatland: A Romance of Many Dimensions* is an 1884 satirical novella by the English schoolmaster Edwin Abbott Abbott. It is based on the fictional two-dimensional world of Flatland.

[13] Well, actually we have been at times been a bit sloppy and plugged in initial data with singularities (think of the plucked string or hammer blow examples).

integers k_1 and k_2 will $\phi(\mathbf{x}) \in C^{k_1}$ and $\psi(\mathbf{x}) \in C^{k_2}$ ensure that $u(\mathbf{x}, t) \in C^2$? Interestingly, this statement is dimension dependent. For example, if $N = 1$, then we would require $k_1 = 2$ and $k_2 = 1$; however if $N = 3$, we would require $k_1 = 3$ and $k_2 = 2$ (cf. Exercise **4.16**). See Section 2.4 of Evans [**10**] for other space dimensions.

It is also useful and indeed physical to consider **nonsmooth**, perhaps even discontinuous, initial data (as in Exercises **4.2** and **4.6**) and, furthermore, data which is not a function at all, e.g., a delta function. In the case of discontinuous initial data (as in the hammer blow and Exercises **4.2** and **4.6**), one can simply place the data into the respective formula. However, one can also *make precise sense of* and analyze the resulting nonsmooth solutions to the wave equation by venturing into the world of distributions (Chapters 5 and 9).

4.4.3. Nonconstant Coefficients and Inverse Problems.

Wave propagation in heterogeneous media leads to 3D wave equations in which the speed parameter c is spatially dependent: $u_{tt} = c^2(x, y, z)\, \Delta u$. This spatial dependence of c is determined by the medium. With certain general assumptions on the coefficient function $c(\mathbf{x})$, the resulting IVPs are well-posed. However, closed form solution formulas are not usually available. The study of wave equations with nonconstant coefficients is fundamentally important to the fields of seismology, geology, oil exploration, and biomedical imaging. What prevails in these applications is the notion of an **inverse problem**. Here we wish to infer information about the structure of the media (spatially varying properties) by measuring wave signals at different times and places. More precisely, suppose we are given a domain $\Omega \subset \mathbb{R}^3$ and we know a measurable quantity $u(\mathbf{x}, t)$ satisfies a wave equation for $\mathbf{x} \in \Omega$ and $t \geq 0$ with nonconstant coefficients. An example of an inverse problem would be: *Can we determine the coefficients of the wave equation (hence, infer information about the physical medium in Ω) by observing (measuring) $u(\mathbf{x}, t)$ for $\mathbf{x}$ near $\partial\Omega$ over a certain time interval?* The study of such inverse problems is an active research area for both academic and industrial scientists across many disciplines.

4.4.4. Geometric Optics and the Eikonal Equation.

This subsection complements Subsection 2.5.5 on the eikonal equation, as well as Subsections 3.12.3 and 3.12.4 on dispersion. In Subsection 4.1.2, we saw how to derive the 3D wave equation from the Euler equations (the fundamental equations describing a compressible fluid) under an assumption: In air, the relative size of pressure and density variations is small compared to the other scales of the system. Here let us use the 3D wave equation, which encapsulates Maxwell's equations in a vacuum, and see that under an assumption of the relatively small wavelength of visible light, we can reduce the 3D wave equation to the purely spatial eikonal equation. This reduction, dating back to William Rowan Hamilton, is commonly referred to as *geometric optics*. As in the previous subsection, let us start with the 3D wave equation in a possible heterogeneous medium:

$$u_{tt} - c^2(\mathbf{x})\, \Delta u = 0. \tag{4.25}$$

Light, more precisely, visible light is electromagnetic radiation (waves of the electromagnetic field) whose range of frequencies/wavelengths can be perceived by the human eye. As described in Subsection 4.1.1, the 3D wave equation above models the propagation of these waves in a medium. As in Subsection 3.12.4, it is instructive to look for a solution in a particular form[14]. To start, let us consider wave-like solutions of the form

$$u(\mathbf{x},t) = A(\mathbf{x})e^{ik(S(\mathbf{x})-c_0t)} = A(\mathbf{x})e^{ikS(\mathbf{x})}\,e^{-ikc_0t}. \tag{4.26}$$

Here, c_0 denotes an average velocity of the system; k denotes the **wavenumber** with $1/(2\pi k)$ the **wavelength**; and the real-valued function $S(\mathbf{x})$ encodes the spatial variation of the solution and is known as the **phase**. Note that c_0 has dimensions of length per time, k has dimensions of time^{-1}, and S has dimensions of length. The **amplitude** A is a complex-valued function. Taking the amplitude to be the constant 1 will significantly reduce the number of calculations in the reduction of the eikonal equation. However, there is a danger in making this assumption and we will shortly see the consequence.

Plugging the wave form (4.26) with $A \equiv 1$ into (4.25) yields

$$e^{ikS(\mathbf{x})}(-ikc_0)^2e^{-ikc_0t} - c^2(\mathbf{x})e^{-ikc_0t}\Delta\left(e^{ikS(\mathbf{x})}\right) = 0$$

which simplifies to

$$\Delta\left(e^{ikS(\mathbf{x})}\right) + \frac{k^2c_0^2}{c^2(\mathbf{x})}e^{ikS(\mathbf{x})} = 0.$$

Noting that $\Delta = \text{div}\nabla$, we have

$$\text{div}\,\left(ike^{ikS}\,\nabla S\right) + \frac{k^2c_0^2}{c^2(\mathbf{x})}e^{ikS} = 0.$$

Lastly we use the vector identity $\text{div}(\psi\mathbf{A}) = \psi\text{div}\mathbf{A} + \nabla\psi\cdot\mathbf{A}$, for any smooth scalar function ψ and vector field $\mathbf{A}$, to find

$$ike^{ikS(\mathbf{x})}\Delta S(\mathbf{x}) - k^2e^{ikS(\mathbf{x})}\,\nabla S(\mathbf{x})\cdot\nabla S(\mathbf{x}) + \frac{k^2c_0^2}{c^2(\mathbf{x})}e^{ikS(\mathbf{x})} = 0.$$

Canceling $e^{ikS(\mathbf{x})}$ and dividing by k^2, we find

$$\left(\frac{c_0^2}{c^2(\mathbf{x})} - |\nabla S|^2\right) + \frac{i}{k}\Delta S = 0. \tag{4.27}$$

The reader should be alarmed by the above equation stating that a purely real term plus a purely imaginary term is 0. This issue is a consequence of not including the complex-valued amplitude A. The point we wish to make now is that under certain physically justifiable assumptions, the second (complex) term in (4.27) is negligibly small; this fact would not have changed if we had included the amplitude. The assumptions stem from the fact that the wavelength of visible light is on the order of a micron (10^{-6} meters) but all other length scales of the problem are far larger. For example, think of the lens in a pair of eye glasses and how the variations in thickness of the lens compares to a micron. Hence, we can assume that other parameters and fields which depend on space, e.g., c and S, have spatial variations on a much larger scale than the wavelength of visible light. Since our ansatz (4.26) is for visible light, we may thus assume that its wavenumber k is very large (alternatively the wavelength $1/k$ is very small) and ignore terms which tend to zero as $k \to \infty$. Just as in Subsection 4.1.2, our equations are in dimensional form and the parameter $1/k$ has dimensions of length. So what do we mean by $1/k$ small? Again, we assume that the length scales have been normalized by some

[14]In mathematics, the assumption of a particular structural form is often referred to with the German word **ansatz**.

macroscopic length scale associated with the system, say, the length scale associated with the geometry of the lens. The punchline is that we may neglect the last term in (4.27) and arrive at the eikonal equation for the phase S:

$$|\nabla S|^2 = \frac{c_0^2}{c^2(\mathbf{x})} \qquad \text{or} \qquad |\nabla S| = n(\mathbf{x}) \quad \text{where } n(\mathbf{x}) := \frac{c_0}{c(\mathbf{x})}.$$

The dimensionless function $n(\mathbf{x})$ is known as the **refractive index**. In the simple case of homogeneous media, one derives the eikonal equation $|\nabla S| = 1$.

We have not given any real motivation for the ansatz (4.26), nor why the phase function $S(\mathbf{x})$ encapsulates the structure of light rays. For that, one must dive deeper into the field of optics. For example, see the classic book of Whitham.[15]

4.5. Chapter Summary

- The wave equation in N space dimensions is given by $u_{tt} = c^2 \Delta u$. Under certain assumptions, it models the propagation of a "disturbance" in an N-dimensional medium (the ambient environment in which the waves propagate).
- This chapter focused on the 3D wave equation, which models both the propagation of electromagnetic waves in a vacuum and the propagation of sound waves in air.
- We explicitly solved the IVP for the 3D wave equation on all of $\mathbb{R}^3$

$$\begin{cases} u_{tt} = c^2 \Delta u, & \mathbf{x} \in \mathbb{R}^3, \ t > 0, \\ u(\mathbf{x}, 0) = \phi(\mathbf{x}), & \mathbf{x} \in \mathbb{R}^3, \\ u_t(\mathbf{x}, 0) = \psi(\mathbf{x}), & \mathbf{x} \in \mathbb{R}^3, \end{cases}$$

  by the method of spherical means (averaging over spheres). In doing so, we found an explicit formula to this well-posed initial value problem which is known as Kirchhoff's formula:

$$u(\mathbf{x}, t) = \frac{1}{4\pi c^2 t^2} \iint_{\partial B(\mathbf{x}, ct)} \phi(\mathbf{y}) + \nabla \phi(\mathbf{y}) \cdot (\mathbf{y} - \mathbf{x}) + t\psi(\mathbf{y}) \, dS_{\mathbf{y}}.$$

  This formula allowed us to conclude a completely different behavior for the propagation of initial data than in 1D: In 3D, information propagates on sharp fronts. This fact is part of what is known as the Huygens Principle.
- The 2D wave equation can be used to model the vibrations of an elastic membrane. Via the method of descent, we used our 3D solution to solve the 2D wave equation. We saw a completely different behavior for the propagation of waves than in 3D; in 2D, waves do not follow the Huygens Principle, but rather there is a similar qualitative behavior to 1D.
- From the point of view of causality, we *remarked* that the Huygens Principle (the propagation on sharp fronts) was valid in all **odd** space dimensions larger than one, and false in all the other space dimensions.

[15] **G.B. Whitham**, *Linear and Nonlinear Waves*, Wiley-Interscience.

Exercises

4.1 Solve the 3D wave equation with initial data $\phi(x_1, x_2, x_3) = 0, \psi(x_1, x_2, x_3) = x_2$. You may take the speed c to be 1.

4.2 Suppose a pressure disturbance in 3D propagates according to the 3D wave equation $u_{tt} = \Delta u$. Suppose at $t = 0$ a bomb explodes creating a disturbance exactly at position $\mathbf{x} = 0$ and causes an initial displacement and velocity to be, respectively, $\phi(\mathbf{x}) \equiv 0$ and $\psi(\mathbf{x})$, the characteristic function of $B(\mathbf{0}, 1)$.
(a) At $t = 10$, what is the value of u at the point $(10, 0, 0)$? You can leave your answer as an integral.
(b) At $t = 10$, what is the value of u at the point $(20, 8, 17)$? Here, we want a numerical value for the answer.
(c) Suppose we are at the point $(20, 20, 20)$. At what times will we feel the initial disturbance; i.e., at what $t > 0$ will $u((20, 20, 20), t)$ be nonzero?

4.3 Let $u(\mathbf{x}, t)$ solve the 3D wave equation.
(a) Suppose that if $\mathbf{x} \in B(\mathbf{0}, 1)$, then $\phi(\mathbf{x}) = \psi(\mathbf{x}) \equiv 0$. At what points $(\mathbf{x}, t)$ in space-time must $u(\mathbf{x}, t) = 0$?
(b) Suppose that if $\mathbf{x} \in B(\mathbf{0}, 1)^c$ (i.e., $|\mathbf{x}| \geq 1$), then $\phi(\mathbf{x}) = \psi(\mathbf{x}) \equiv 0$. At what points $(\mathbf{x}, t)$ in space-time must $u(\mathbf{x}, t) = 0$?

4.4 This question involves acoustics and the derivation of the wave equation from the Euler equations (cf. Section 4.1.2). Suppose that the velocity field is irrotational; that is, $\nabla \times \mathbf{u} = \mathbf{0}$. Under the same assumptions of "small vibrations", show that each of the three components of $\mathbf{u}$ satisfy the 3D wave equation.

4.5 **(Radially Symmetric Solutions)** Suppose we consider the 3D wave equation with radially symmetric initial conditions:

$$u(\mathbf{x}, 0) = \phi(r), \quad u_t(\mathbf{x}, 0) = \psi(r) \qquad \text{where } r := |\mathbf{x}|.$$

Note that here ϕ and ψ are functions of a single variable $r \geq 0$. You may assume that the derivatives from the right, $\phi'(0)$ and $\psi'(0)$, are both zero. Let $\tilde{\phi}$ and $\tilde{\psi}$ be the even extensions to all of $\mathbb{R}$. Show that the solution to the 3D IVP is radially symmetric and is given by

$$u(\mathbf{x}, t) = \frac{1}{2r}(r + ct)\tilde{\phi}(r + ct) + (r - ct)\tilde{\phi}(r - ct) + \frac{1}{2cr}\int_{r-ct}^{r+ct} s\tilde{\psi}(s)\, ds.$$

Hint: Do not try to show this from Kirchhoff's formula. Rather, look for a radially symmetric solution $u(\mathbf{x}, t) = u(r, t)$ (note the abuse of notation) to the 3D wave equation by working in spherical coordinates and noting that $v(r, t) := ru(r, t)$ solves the 1D wave equation.

4.6 **(a)** Find an explicit solution to the 3D IVP given in Exercise **4.2**.
(b) Find an explicit solution to the 3D IVP for the wave equation with ϕ given by the characteristic function of $B(\mathbf{0}, 1)$ and $\psi \equiv 0$.
Hint: Use the result of the previous question.

4.7 Show that Kirchhoff's formula does indeed solve the initial value problem (4.9) in the sense that it solves the 3D wave equation and matches the initial data as in (4.12). Hint: To show that the formula actually solves the PDE, it is useful to work with the formula in the first line of (4.21), before the calculation of the t derivative. You may find (A.11) useful.

4.8 Repeat Exercise **4.7** in 2D. That is, show that the solution formula given by (4.23) solves the initial value problem (4.22) for the 2D wave equation where the data is again interpreted in terms of extensions and limits.

4.9 Use Kirchhoff's formula to write down the explicit solution to Maxwell's equations (4.3). You can write your answer in vector form.

4.10 (**Derivation of the 2D Wave Equation**) We want to derive the 2D wave equation as a model for a stretched elastic membrane (think of a large drum). Here x and y will denote the positions on a horizontal membrane, and $u(x, y, t)$ denotes the vertical displacement of the membrane from equilibrium. Assume the mass density of the membrane is a constant ρ (units of mass per area) and the tension T in the membrane is fixed (in space and time). Now the tension is a force per unit length and acts on pieces of the membrane through their boundary curves. As with the 1D vibrating string, assume that the vertical displacements are small in the sense that all angles of inclination are small. Apply Newton's Second Law to a piece of the membrane with a corner (x, y) and sides Δx and Δy. For the "mass times vertical acceleration" part you need not write this (exactly) as an integral, but rather approximate it with $\rho \times \Delta x\, \Delta y \times u_{tt}(x, y, t)$. Balance this term with the vertical components of the tension force, and let Δx and Δy tend to 0.

4.11 (**Using Duhamel's Principle to Solve the 3D Wave Equation with a Source**) Let $f(\mathbf{x}, t)$ be any smooth function of $\mathbf{x} \in \mathbb{R}^3$ and $t > 0$. Consider the IVP

$$\begin{cases} u_{tt} - c^2 \Delta u = f(\mathbf{x}, t), & \mathbf{x} \in \mathbb{R}^3,\ t > 0, \\ u(\mathbf{x}, 0) = \phi(\mathbf{x}), & \mathbf{x} \in \mathbb{R}^3, \\ u_t(\mathbf{x}, 0) = \psi(\mathbf{x}), & \mathbf{x} \in \mathbb{R}^3. \end{cases} \tag{4.28}$$

Solve this problem by using Duhamel's Principle to solve

$$\begin{cases} u_{tt} - c^2 \Delta u = f(\mathbf{x}, t), & \mathbf{x} \in \mathbb{R}^3,\ t > 0, \\ u(\mathbf{x}, 0) = 0, & \mathbf{x} \in \mathbb{R}^3, \\ u_t(\mathbf{x}, 0) = 0, & \mathbf{x} \in \mathbb{R}^3. \end{cases} \tag{4.29}$$

Recall from Section 3.6.1 that this is done by introducing an additional temporal parameter $s \in [0, \infty)$. For each **fixed** s we consider a solution $u(\mathbf{x}, t; s)$ to

$$\begin{cases} u_{tt}(\mathbf{x}, t; s) = c^2 \Delta u(\mathbf{x}, t; s), & \mathbf{x} \in \mathbb{R}^3, \quad t > s, \\ u(\mathbf{x}, s; s) = 0, \quad u_t(\mathbf{x}, s; s) = f(\mathbf{x}, s), & \mathbf{x} \in \mathbb{R}^3, \end{cases}$$

and define

$$u(\mathbf{x}, t) = \int_0^t u(\mathbf{x}, t; s)\, ds. \tag{4.30}$$

Show that (4.30) solves (4.29) and use Kirchhoff's formula to explicitly write down the solution. In other words, determine an explicit solution formula for (4.28).

4.12 (**Taken from the First Edition of Evans [10]**) Let u solve (4.9) where ϕ and ψ are smooth functions with compact support on $\mathbb{R}^3$. Show that there exists a constant C such that for all $\mathbf{x} \in \mathbb{R}^3$ and $t > 0$,

$$|u(\mathbf{x}, t)| \leq \frac{C}{t}.$$

Hint: Use the compact support to bound all occurrences of ϕ, $\nabla\phi$, and ψ in Kirchhoff's formula by some constant C. Now derive an estimate by integrating a "constant" over the spheres $\partial B(x, ct)$. Note that it is not the desired (optimal) decay estimate. However ask yourself if we really need to integrate over the **full** $\partial B(x, ct)$ given that the data has compact support.

4.13 Use the method of descent to derive D'Alembert's formula from the solution formula for the 2D wave equation.

4.14 (**Conservation of Energy**) Let $u(\mathbf{x}, t)$ be a solution to (4.9) where ϕ and ψ are smooth functions with compact support. (a) Show that the total energy (kinetic + potential)

$$E(t) := \frac{1}{2} \iiint_{\mathbb{R}^3} \left(u_t^2(\mathbf{x}, t) + c^2 |\nabla u(\mathbf{x}, t)|^2\right) d\mathbf{x}$$

is conserved in time; i.e., $E'(t) = 0$. (b) Use this to prove that there exists at most one solution to (4.9).

4.15 Consider the IVP for the 3D wave equation with friction:

$$u_{tt} + k u_t = c^2 \Delta u, \qquad k > 0.$$

Assume that the initial data ϕ and ψ are smooth functions with compact support. Show that the energy $E(t)$ (defined in the previous question) satisfies $E'(t) \leq 0$. Use this to prove that there exists at most one solution to the IVP.

4.16 Prove that if $\phi \in C^3(\mathbb{R}^3)$ and $\psi \in C^2(\mathbb{R}^3)$, then the solution given by Kirchhoff's formula (4.10) is C^2 in $\mathbf{x}$ and t.

4.17 (**Difficult**) Fix a unit vector $\omega \in \mathbb{R}^3$ and consider the wave equation in three space dimensions where the initial data ϕ, ψ are both constant on all planes with normal ω. This means that there exist functions f and g of one variable such that for all $\mathbf{x} \in \mathbb{R}^3$,

$$\phi(\mathbf{x}) = f(\mathbf{x} \cdot \omega) \qquad \text{and} \qquad \psi(\mathbf{x}) = g(\mathbf{x} \cdot \omega).$$

Since the data just depends on one variable, one might expect that D'Alembert's formula applies. Show that this is indeed the case and the solution is given by

$$u(\mathbf{x}, t) = \frac{1}{2}\left[f(\mathbf{x} \cdot \omega + ct) + f(\mathbf{x} \cdot \omega - ct)\right] + \frac{1}{2c} \int_{-ct}^{ct} g(\mathbf{x} \cdot \omega + s)\, ds.$$

Why does this not contradict Huygen's Principle? **Food for thought and further study:** What is the connection between this exercise and the Radon transform (cf. Section 6.13)?

4.18 (**Difficult**) In this exercise we want to find explicit formulas for the solution to the initial value problem of the 4D and 5D wave equation. Note that formulas are derived in Evans [**10**] for all odd and even dimensions N. However, it is best to consult [**10**] only after you have undertaken this exercise.

First we work in dimension $N = 5$ to solve

$$\begin{cases} u_{tt} = c^2 \Delta u, & \mathbf{x} \in \mathbb{R}^5,\ t > 0, \\ u(\mathbf{x}, 0) = \phi(\mathbf{x}), & \mathbf{x} \in \mathbb{R}^5, \\ u_t(\mathbf{x}, 0) = \psi(\mathbf{x}), & \mathbf{x} \in \mathbb{R}^5. \end{cases} \tag{4.31}$$

(i) Define for any fixed $x \in \mathbb{R}^5$ the spherical mean $\overline{u}(r,t)$ analogous to what we did in three dimensions. Show that it solves

$$\overline{u}_{tt} = c^2 \left(\overline{u}_{rr} + \frac{4}{r} \overline{u}_r \right).$$

(ii) Show that

$$v(r,t) := \left(\frac{1}{r} \frac{\partial}{\partial r} \right) (r^3 \overline{u}(r,t))$$

solves the 1D wave equation on the half-line with fixed boundary at $r = 0$.
(iii) Show that

$$u(x,t) = \lim_{r \to 0+} \frac{v(r,t)}{3r}.$$

(iv) As in 3D, use the explicit solution for the 1D wave equation on the half-line to solve (4.31).
(v) Use your answer from part (iv) and the method of descent to solve the IVP for the 4D wave equation.

Chapter 5

The Delta "Function" and Distributions in One Space Dimension

"Thus, physicists lived in a fantastic universe which they knew how to manipulate admirably, and almost faultlessly, without ever being able to justify anything. They deserved the same reproaches as Heaviside: their computations were insane by standards of mathematical rigor, but they gave absolutely correct results, so one could think that a proper mathematical justification must exist. Almost no one looked for one, though.
I believe I heard of the Dirac function for the first time in my second year at the ENS. I remember taking a course, together with my friend Marrot, which absolutely disgusted us, but it is true that those formulas were so crazy from the mathematical point of view that there was simply no question of accepting them. It didn't even seem possible to conceive of a justification. These reflections date back to 1935, and in 1944, nine years later, I discovered distributions. This at least can be deduced from the whole story: it's a good thing that theoretical physicists do not wait for mathematical justifications before going ahead with their theories!"

- **Laurent Schwartz**[a]

[a]**Laurent Schwartz (1915–2002)** was a French mathematician who made fundamental contributions to the mathematical analysis of PDEs, particularly the theory of distributions. While several aspects of the theory were previously presented by the great Russian mathematician **Sergei Sobolev (1908–1989)**, it was Schwartz whose full development led to his winning the Fields Medal in 1950. His published two-volume treatise is simply titled "*Théorie des Distributions*". This quote is taken from pages 217–218 of his book *A Mathematician Grappling with His Century*, Birkhäuser, 2001. The translation from the original French was by Leila Schneps.

Throughout most of this text we are interested in finding solutions to PDEs which are inarguably functions. However, one of the key objects central to studying a large class of PDEs is *something* which is actually not a function and which encapsulates a rather central phenomenon: **concentration**. This *object* was not officially introduced by a mathematician, but rather by the Nobel Prize winning physicist **Paul Dirac**[1]. Dirac clearly understood what this object was and, more importantly, how to use it. However, for us easily confused mortals, it is most unfortunate that it has come to be known as the **Dirac delta function** because as we shall see, it is **not** a function or, at least, not a classical function as we know them. In the 1940s, the mathematician **Laurent Schwartz** developed a beautiful mathematical theory which gave such objects precise meaning by generalizing the notion of a function. These **generalized functions** are called **distributions** and are the focus of this chapter. The *Theory of Distributions* is extensive and, incidentally, Schwartz received the Fields Medal ("the Nobel Prize of mathematics") in 1950, in part for the development and advancement of this theory.

In this text it is our hope to convince you, the reader, that the **basic** rudiments of the theory of distributions are accessible, digestible, and tremendously useful and insightful. To this end, we will primarily address the following two types of distributions:

- distributions associated with point concentrations (delta "functions"),
- distributions which are generated by classical functions with discontinuities

from two points of view: **differentiation** and **limits**.

Much of this chapter (cf. sections with bullets) is fundamental to material we will cover in the subsequent chapters. Spending time and effort to appreciate and digest the delta "function" and, further, what it means to interpret a function as a distribution will serve all readers well, regardless of which scientific discipline they endeavor to pursue.

We begin by recalling what exactly a function is, with the underlying goal of a fundamental shift in mindset away from **pointwise values** to **integral values (averages)**. This shift in mindset will lead us naturally into the spirit of distributions.

5.1. • Real-Valued Functions

5.1.1. • What Is a Function?

We recall that a **function** f is a well-defined map or **input-output machine** from one set of numbers into another. In other words, a function f takes an input number and outputs another number:

$$f: \text{(an input number)} \longrightarrow \text{the output number.}$$

If f is a function of several variables, the input is a vector of numbers. Real numbers, or vectors of real numbers, are ideal for presenting either the input or the output variables. They are ideal for parametrizing either positions in physical space and/or time. Hence, throughout this text we will deal with functions defined on values in $\mathbb{R}^N$ (or some domain in $\mathbb{R}^N$) with N usually 1, 2, or 3. Specifically, if we **input** a value for $\mathbf{x} \in \mathbb{R}^N$, the function f **outputs** a real (or complex) number $f(\mathbf{x})$. For example, here are two

[1]**Paul Dirac (1902–1984)** was an English physicist who made fundamental contributions to both quantum mechanics and quantum electrodynamics. In 1932, Paul Dirac published his treatise *The Principles of Quantum Mechanics*, in which the delta function was officially singled out and widely used. However, the notion of such an object can be traced back much earlier to several mathematicians, in particular **Cauchy**.

functions with $N = 2$ and $N = 1$, respectively:

$$f_1(x_1, x_2) = x_1 + 4(x_2)^2, \qquad f_2(x) = \begin{cases} x & \text{if } x < 0, \\ 2 & \text{if } x = 0, \\ 4 & \text{if } x > 0. \end{cases}$$

One often associates a geometric picture to such a function by considering its **graph**, the locus (set) of points in $\mathbb{R}^{N+1}$ of the form $(\mathbf{x}, f(\mathbf{x}))$. In the case $N = 1$, the graph is a curve in $\mathbb{R}^2$. At this point in your studies you have spent a good deal of time learning tools for *analyzing* functions of one or more real variables based upon differential and integral calculus.

5.1.2. • Why Integrals (or Averages) of a Function Trump Pointwise Values.

In what follows, we loosely use the terms **integrals** and **averages** as synonyms.

From a physical (and mathematical) point of view, functions often denote **densities**. The density at any given point has no significance; rather, it is the value of its **integrals** (or **averages**) over parts of the domain which have physical significance, and it is these values which can be physically observed and measured. We give four examples.

(i) Consider the **mass density** of a certain quantity, such as water, in the atmosphere. For the length scales of which we are interested, we assume that the mass density is a function which can continuously change over space. We cannot directly measure mass density at a particular point; after all, any single point should have zero mass. What one can measure is the mass of regions with nonzero volume; in other words, compute integrals of the mass density over regions of space. In fact even the physical dimensions of mass density, namely mass per volume, suggest the underlying notion of a region of space, rather than a point.

(ii) Consider the **temperature** (as a measure of the amount of heat) in a room or in the atmosphere. Can one measure the temperature (heat) at exactly one particular point in space? No, one measures temperature over a small region of space where some measure of heat energy can be registered. Moreover, temperature by its very definition entails **averaging** over velocities of a particle and over the particles themselves.

So while one might consider a temperature function $T(\mathbf{x})$ defined over points in space, we should actually focus our attention not on its pointwise values, but on its integral over regions of space. Indeed, we should think of the temperature function $T(\mathbf{x})$ as giving a temperature density, with physical dimensions of temperature per unit volume.

(iii) You may recall from a preliminary course in probability and statistics the notion of a **continuous probability density function** $p(x)$ defined for $x \in \mathbb{R}$, for example, the normal probability density function associated with the infamous bell-shaped curve. This is directly connected with the probability of **events**. However, note that the probability of x being any given value, say, $x = 3$, is not $p(3)$; rather it is zero! What has nonzero probability is that x lies in some interval I, and this probability is given by

$$\int_I p(x)\,dx.$$

(iv) Here is an example of a function which you probably would not refer to as a density. Consider the **velocity** of a falling ball. Because of gravity, this velocity will continuously change with time. How would we measure the velocity v at a given time t_1? We would fix a small time interval I including t_1 of size Δt, measure how far the ball traveled during this time interval, and then divide by Δt. This approximation to the actual velocity at t_1 is an **average** velocity over the time interval chosen:

$$\frac{1}{\Delta t}\int_I v(t)\,dt.$$

An Important Remark: The premise of these examples, demonstrating that integral values "trump" pointwise values, might seem rather contradictory and indeed a bit hypocritical, given that the focus of this text is on PDEs — pointwise equations relating the values of partial derivatives. Phrasing basic laws of physics/mathematics in the form of pointwise equations (PDEs) can indeed be very useful. However, the reader will soon see that many PDEs, especially those modeling some aspect of **conservation**, are derived from integral laws: equations relating integrals of an unknown function. These include[2] the diffusion equation, Laplace's equation, and Burgers's equation. Via integral to pointwise (IPW) theorems (cf. Section A.7.1), we can reduce these integral laws to pointwise laws (PDEs); see, for example, Section 7.1. This reduction is based upon the assumption of an underlying smooth solution. On the other hand, the ability to incorporate **discontinuities/singularities** into solutions (or initial values) to these PDEs is tremendously useful. These singularities in solutions/initial values can only be captured and analyzed at the level of the integrals, not at the level of the pointwise values. The purpose of this chapter, as well as Chapter 9, is to provide a new way of interpreting the solution to these PDEs in a manner focused on integrals (or averages). The new idea here will be "distributions" and as we shall eventually see, interpreting these PDEs in the "sense of distributions" amounts to returning to, or rather being **faithful to**, their original integral-based derivation.

5.1.3. • Singularities of Functions from the Point of View of Averages.

We just made a case that in dealing with functions describing a physical variable which depends on space and/or time, **averages (integrals)** of the function are more relevant than their **pointwise values**. These averages will characterize the essence of the function. In particular one could ask, if we know the value of

$$\int_S f(x)\,dx,$$

for all sets S, do we know exactly the function? The answer is yes modulo a "small" set which has negligible size, where "negligible" is from the point of view of integration. This means a set sufficiently small enough such that any function integrated over it is zero. An example of such a set would be a finite set of points; however, there are infinite sets which also have this property. While we will not need them in this text,

[2]In fact, even the wave equation was derived based upon the calculation of integrals.

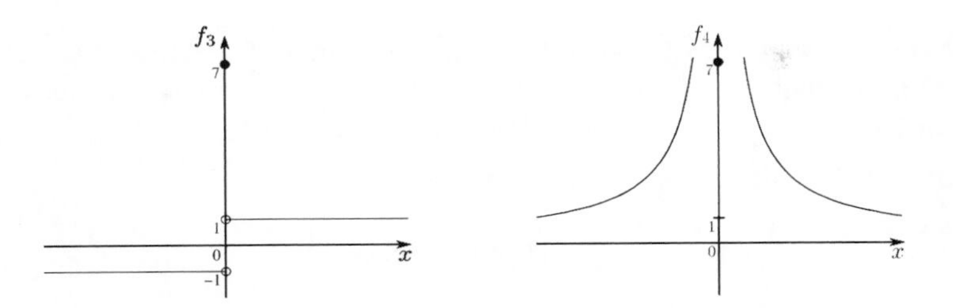

Figure 5.1. Graphs of the functions defined in (5.1) with, respectively, a jump and a blow-up singularity at $x = 0$.

it should be noted that there is a beautiful mathematical theory called *measure theory* which makes such notions very precise.

Thus, changing the values of a function at a finite number of inputs has no effect on any of its integrals. For example, from the point of view of integral calculus and its applications, the following two functions are the same:

$$f_1(x) = x^2 \qquad \text{and} \qquad f_2(x) = \begin{cases} x^2, & x \neq 2, \\ 7, & x = 2. \end{cases}$$

Certainly the values of their integrals over any interval are equal. On the other hand, there are other "singularities" that a function can possess which do have effects on integrals.

A **singularity** of a function is simply an input point x_0 where the function fails to be well-behaved (in terms of continuity or differentiability). Certain singularities, often labeled as **removable**, like that of $f_2(x)$ above, are for the most part irrelevant to the structure of the function. Other singularities, often labeled as **essential**, like those of

$$f_3(x) = \begin{cases} -1, & x < 0, \\ 1, & x > 0, \\ 7, & x = 0, \end{cases} \qquad \text{or} \qquad f_4(x) = \begin{cases} \frac{1}{|x|}, & x \neq 0, \\ 7, & x = 0, \end{cases} \tag{5.1}$$

are indeed essential to the structure of the function. Both of these functions have a singularity at $x = 0$. From the perspective of integrals and averages, the value of the function at $x = 0$ is indeed irrelevant. However, the behavior around the singularity — a jump and blow-up discontinuity, respectively — is relevant. Note that there is no redefinition of the value at $x = 0$ which would change the essential behavior around the singularity. As you will come to appreciate, both of these two types of singularities are something one would "*detect or see*" via integrals. They are central to the function's character. **Green's functions**, the subject of Chapter 10, are functions which exhibit such a singular behavior.

In this chapter we will consider piecewise smooth functions of **one variable** which have either jump or blow-up discontinuities. However, we will analyze these functions from the point of view of **integrals/averages** and, in doing so, unlock the true effect that these singularities have from the important perspective of **differentiation**.

With our focus on integration and averages of a function, we introduce (cf. Section A.4.3) two large classes of functions which will frequently appear in this and subsequent chapters.

A function $f(x)$ defined on $\mathbb{R}$ is **integrable** if

$$\int_{-\infty}^{\infty} |f(x)|\, dx \;<\; \infty.$$

This means that the improper integral exists and is a finite number. A function $f(x)$ defined on $\mathbb{R}$ is **locally integrable** if the integral of the absolute value over any finite interval is finite; i.e., for any $a < b$ with $a, b \in \mathbb{R}$,

$$\int_{a}^{b} |f(x)|\, dx \;<\; \infty.$$

5.2. • The Delta "Function" and Why It Is Not a Function. Motivation for Generalizing the Notion of a Function

You may have already encountered the **Dirac delta "function"** $\delta_0(x)$ which can be loosely defined as

$$\delta_0(x) = \begin{cases} 0, & x \neq 0, \\ \text{"suitably infinite"}, & x = 0, \end{cases}$$

where **"suitably infinite"** means that for any $a > 0$,

$$\text{“}\int_{-a}^{a} \delta_0(x)\, dx \;=\; 1\text{”}. \tag{5.2}$$

Alternatively, we can reformulate the above as follows: Given any function ϕ which is continuous on $[-a, a]$,

$$\text{“}\int_{-a}^{a} \delta_0(x)\, \phi(x)\, dx \;=\; \phi(0)\text{”}. \tag{5.3}$$

Thus, somehow the infinite density at $x = 0$ has resulted in the fact that multiplying the function $\phi(x)$ by the delta "function" and integrating *picks out* the value of ϕ at $x = 0$. **There is no function, that is, no input-output machine defined on the real numbers, that can do this.** This is why we make use of quotation marks in (5.2) and (5.3). Even if we decided to allow $+\infty$ as a possible output value for the function and consider a definition like

$$\text{“}\;\delta_0(x) = \begin{cases} 0, & x \neq 0, \\ +\infty, & x = 0, \end{cases}\text{”} \tag{5.4}$$

there is no unambiguous way to interpret its integral and, hence, enforce either (5.2) or (5.3). To achieve this one would have to make sense of the ambiguous product: $0 \times +\infty$. Moreover, we think you would agree that $2 \times +\infty$ should equal $+\infty$, but then by definition (5.4), should we conclude that

$$\text{“}\,\delta_0(x) \;=\; 2\delta_0(x)\,\text{”?}$$

In fact, the same reasoning would suggest that $\delta_0(x) = C\delta_0(x)$, for any $C > 0$.

From these observations, one might be tempted to conclude that this delta "function" is an abstract fabrication[3] with little connection to reality which, up to this point, seems to be well described by the standard calculus of *honest-to-goodness* functions. Well, nothing could be further from the truth; the delta "function" comes up naturally, even when one just deals with functions. As we shall see, it plays a **central role** in two fundamental areas:

- linear second-order partial differential equations,
- Fourier analysis (Fourier series and the Fourier transform).

Indeed, we stress that it is one of the most important "*players*" in this book. To give a small hint of this, we present in the next subsection two simple examples involving honest-to-goodness functions where this, as yet ill-defined, delta "function" arises. These examples will help provide the intuition behind both the essential character of the delta "function" and its eventual definition as a distribution. The key phenomenon prevalent in both examples is **concentration**.

5.2.1. • The Delta "Function" and the Derivative of the Heaviside Function. Consider the Heaviside[4] function defined by

$$H(x) := \begin{cases} 0, & x < 0, \\ 1, & x \geq 0. \end{cases} \tag{5.5}$$

What is its derivative? If we look *pointwise*, i.e., fix a point x and ask what $H'(x)$ is, we see that

$$H'(x) = \begin{cases} 0, & x \neq 0, \\ undefined, & x = 0. \end{cases}$$

Can we ignore the fact that $H'(0)$ is undefined? Since $H'(x) = 0$ except at one point, our previous discussion about values at a single point might suggest that the derivative H' behaves like the zero function. But then $H(x)$ should be a constant, which is not the case: $H(x)$ changes from 0 to 1 but does so *instantaneously*. Hence perhaps we should assert that the derivative is *entirely concentrated at* $x = 0$ and define $H'(0)$ to be $+\infty$, associated with the instantaneous change from 0 up to 1 at $x = 0$. In other words, take

$$H'(x) = \begin{cases} 0, & x \neq 0, \\ +\infty, & x = 0. \end{cases} \tag{5.6}$$

Unfortunately there is ambiguity in this $+\infty$; for example, in order to reconstruct H (up to a constant) from its derivative we would need to make sense of the recurring product "$0 \times +\infty$". Further to this point, suppose we considered a larger jump, say

$$H_7(x) = \begin{cases} 0, & x < 0, \\ 7, & x \geq 0, \end{cases}$$

---

[3]In fact even one of the greatest mathematicians / computer scientists of the 20th century, **John von Neumann (1903–1957)**, was so adamant that Dirac's delta function was "*mathematical fiction*" that he wrote his monumental monograph, published in 1932, *Mathematische Grundlagen der Quantenmechanik* (Mathematical Foundations of Quantum Mechanics) in part to explain quantum mechanics with absolutely no mention of the delta "function". John von Neumann described this artifact in the preface as an "*improper function with self-contradictory properties*". This counterpart to Dirac's treatise was based on functional analysis and Hilbert spaces.

[4]Named after the British self-taught scientist and mathematician **Oliver Heaviside (1850–1925)**, a rather interesting person who was quite critical about contemporary mathematical education. Look up online his "Letter to Nature".

would its derivative still be (5.6)? The functions $H(x)$ and $H_7(x)$ are clearly different functions whose structure is **lost** in the derivative definition (5.6).

Let us pursue this further by focusing on **integration**. Fix a smooth (C^1) function ϕ which is identically 0 if $|x| \geq 1$. Consider the function $f(x)$ (which, as we will later see, is called a convolution) defined by

$$f(x) := \int_{-\infty}^{\infty} H(x-y)\,\phi(y)\,dy,$$

where H is the Heaviside function of (5.5). While the definition of $f(x)$ is rather "convoluted", it presents a perfectly well-defined function. Indeed, despite the discontinuity in H, f is continuous and, in fact, differentiable at all x. While, as we shall see, it is easy to prove these assertions (cf. (5.8) and (5.9) below), you should try to convince yourself of this before proceeding on.

Next, let us attempt to compute $f'(x) = \frac{df(x)}{dx}$ in two different ways:

First way: On one hand, we could "attempt" to bring the x-derivative inside[5] the integral (with respect to y) to find

$$f'(x) = \frac{df(x)}{dx} = \int_{-\infty}^{\infty} \frac{dH(x-y)}{dx}\,\phi(y)\,dy. \tag{5.7}$$

Should we take $\frac{dH(x-y)}{dx}$ as a function of y to be identically 0? After all, it is 0 except at the one point $y = x$. But then this would imply that $f'(x) = 0$ at all points x. If we try to incorporate the behavior at $y = x$ and claim

$$\frac{dH(x-y)}{dx} \text{ is infinity when } y = x,$$

then we are back to the same question on how to compute the integral.

Second way: On the other hand, by definition of the Heaviside function note that

$$H(x-y) = \begin{cases} 0 & \text{if } y > x, \\ 1 & \text{if } y \leq x. \end{cases}$$

Hence $f(x)$ can be conveniently written as

$$f(x) = \int_{-\infty}^{\infty} H(x-y)\,\phi(y)\,dy = \int_{-\infty}^{x} \phi(y)\,dy. \tag{5.8}$$

But now things are more transparent. By the Fundamental Theorem of Calculus, f is continuous and differentiable with

$$f'(x) = \phi(x). \tag{5.9}$$

Now there is absolutely no ambiguity, and the answer is not, in general, 0.

So what happened in (5.7)? What occurred was the derivative of the translated Heaviside function,

$$\frac{dH(x-y)}{dx} \quad \text{viewed as a function of } y,$$

[5]Note here that the hypotheses for differentiation under the integral sign (Theorem A.10) fail to hold true.

did what a delta "function" (concentrated at $y = x$) was supposed to do; i.e., when integrated against $\phi(y)$, it picked out the value of $\phi(y)$ at $y = x$. Note that, because of the translation, the concentration was moved from $y = 0$ to $y = x$.

5.2.2. • The Delta "Function" as a Limit of a Sequence of Functions Which Concentrate.

The delta "function" also appears when we look at what happens to **sequences of functions which concentrate.** Suppose that f_n is defined by

$$f_n(x) = \begin{cases} n - n^2x, & 0 < x < 1/n, \\ n + n^2x, & -1/n < x \leq 0, \\ 0, & |x| \geq 1/n. \end{cases} \tag{5.10}$$

Do not be put off by the formula, as these are simply steeper and steeper spike functions as illustrated in Figure 5.2. Note that for all $n \in \mathbb{N}$,

$$\int_{-\infty}^{\infty} f_n(x)\, dx = 1. \tag{5.11}$$

However, for all $x \neq 0$, $f_n(x)$ (as a sequence in n) is eventually 0. So what happens to these functions as $n \to \infty$? If we look at this from a pointwise perspective, it would seem as if $f_n(x)$ tends to 0 as $n \to \infty$ except at $x = 0$. On the other hand, $f_n(0)$ tends to $+\infty$. If we include $+\infty$ as a possible output and write the pointwise limit as the function

$$f(x) = \begin{cases} 0, & x \neq 0, \\ +\infty, & x = 0, \end{cases} \tag{5.12}$$

then what is

$$\int_{-\infty}^{\infty} f(x)\, dx?$$

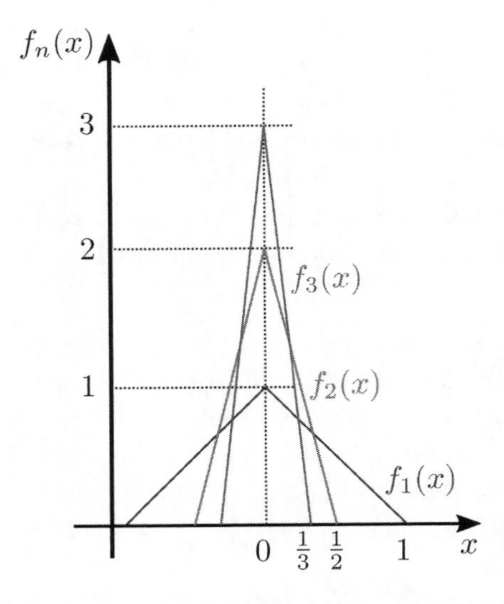

Figure 5.2. Plots of functions in the sequence $\{f_n(x)\}_{n\in\mathbb{N}}$ defined by (5.10) for $n = 1, 2, 3$.

Given that for every n, $\int_{-\infty}^{\infty} f_n(x)\,dx = 1$, one would expect it to be 1. But again, there is no way to make sense of this integral without having a nonambiguous value for $0 \times +\infty$. Indeed, the **same pointwise limit** (5.12) would occur with the sequence of functions $2f_n(x)$ or $7f_n(x)$, which differ from $f_n(x)$ in that the area under the curve is no longer always 1, but rather 2 and 7, respectively. This area information is **lost** by simply assigning the value of $+\infty$ at $x = 0$ in (5.12).

These points should sound familiar! As we just did for the derivative of the Heaviside function, let us investigate this from the perspective of integration or, more precisely, integration *against* another function. Suppose that ϕ is a continuous function on $\mathbb{R}$ and then consider what happens to

$$\int_{-\infty}^{\infty} f_n(x)\phi(x)\,dx \quad \text{as } n \to \infty.$$

Due to the concentration property of this sequence, we have

$$\int_{-\infty}^{\infty} f_n(x)\phi(x)\,dx = \int_{-\frac{1}{n}}^{\frac{1}{n}} f_n(x)\phi(x)\,dx.$$

When n is very large, the continuous function ϕ is approximately constant on the tiny interval $\left[-\frac{1}{n}, \frac{1}{n}\right]$; i.e., for $x \in \left[-\frac{1}{n}, \frac{1}{n}\right]$, $\phi(x) \sim \phi(0)$, where the notation $\sim$ simply means "close to" or "approximately equal to". Thus, when n is very large we have

$$\int_{-\frac{1}{n}}^{\frac{1}{n}} f_n(x)\phi(x)\,dx \quad \sim \quad \int_{-\frac{1}{n}}^{\frac{1}{n}} f_n(x)\phi(0)\,dx = \phi(0)\int_{-\frac{1}{n}}^{\frac{1}{n}} f_n(x)\,dx = \phi(0),$$

where we use (5.11) as well as the concentration property of the sequence for the last equality. Thus, it would seem that

$$\lim_{n\to\infty} \int_{-\infty}^{\infty} f_n(x)\phi(x)\,dx = \phi(0),$$

and as before, what the f_n seem to be converging to is this, as yet ill-defined, delta "function".

The sequence of functions represented in (5.10) is a very good way to *visualize* the delta "function" in terms of honest-to-goodness functions. In fact, this is exactly how Dirac thought of it[6]:

> *"To get a picture of $\delta(x)$, take a function of the real variable x which vanishes everywhere except inside a small domain, of length ϵ say, surrounding the origin $x = 0$, and which is so large inside this domain that its integral over this domain is unity. The exact shape of the function inside this domain does not matter, provided there are no unnecessarily wild variations (for example provided the function is always of order ϵ^{-1}). Then in the limit $\epsilon \to 0$ this function will go over to $\delta(x)$."*

[6]From page 58 of **P. Dirac**, *The Principles of Quantum Mechanics*, Oxford at the Clarendon Press, third edition, 1947.

Conclusion/Punchline of This Section

- Dirac's delta "function" pops up in instances when (i) differentiating the Heaviside function to capture a derivative which concentrates at one point and (ii) when considering limits of certain functions which seem to concentrate.
- In both cases, we were able to make sense of what was happening, not via pointwise values of x, but by **focusing on integration** involving a generic function $\phi(x)$.

True functions have significance without any notion of integration **but** the delta "function" **does not**. In the next section we will provide a precise definition of this delta "function" as a generalized function or **distribution**. After this, we will agree to keep the name and dispense with quotation marks in "function".

5.3. • Distributions (Generalized Functions)

Hopefully the previous section has given you some motivation for *generalizing* the notion of a function to include the delta "function" and to emphasize the importance of averages of classical "honest-to-goodness" functions. This generalization should preserve the important character of a classical function from the point of view of calculus. The basic object of our generalization will be a **distribution**.

Unlike with classical functions where we **understand/capture/characterize** the object as an input/output machine on the real numbers, we will **understand/capture/characterize** a distribution by "what it does" to **test functions**. In other words, a distribution will remain an input/output machine, but the difference is we now input a test function and output a real number. In this scenario, where the inputs are functions themselves, we often call the input/output machine a **functional**. Once we have a precise notion of a distribution, we can readily provide a precise meaning to the delta "function", transforming it into a proper mathematical object upon which we have a firm and solid foundation to explore its presence and uses.

5.3.1. • The Class of Test Functions $C_c^\infty(\mathbb{R})$.

The class of test functions will consist of **localized smooth functions** where here, by *smooth*, we will mean infinitely differentiable (that is, very smooth!). By *localized* we mean that they are only nonzero on a bounded set. Precisely, we consider $\phi : \mathbb{R} \to \mathbb{R}$ in the class $C_c^\infty(\mathbb{R})$, where the "c" in the subindex position denotes **compact support**. Recall that a continuous function of one variable has compact support if it is identically zero outside of some closed finite interval.

The definition of $C_c^\infty(\mathbb{R})$ may seem very restrictive and, indeed, you might wonder whether or not such a function even exists; remember, we want a perfectly smooth function (no singularities in any derivative) which is eventually flat in either direction. In other words the function becomes identically flat (zero) at some finite value in a perfectly smooth fashion (cf. Figure 5.3). While no polynomial function can do this, here is the generic example which is called the **bump function** (illustrated in Figure

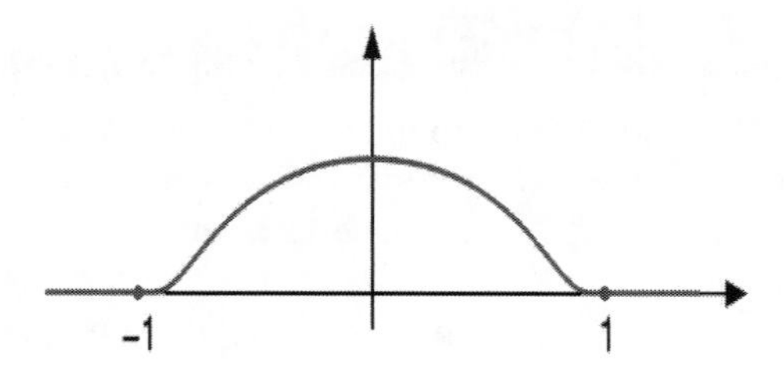

Figure 5.3. A canonical function in $C_c^\infty(\mathbb{R})$: graph of the *bump function* defined by (5.13) with $a = 1$.

5.3): For any $a > 0$, let

$$\phi_a(x) := \begin{cases} e^{-\frac{1}{a^2-x^2}} & \text{if } |x| < a, \\ 0 & \text{if } |x| \geq a. \end{cases} \tag{5.13}$$

Clearly the support of this function is contained in the interval $[-a, a]$. In Exercise **5.1**, you are asked to further show that this function is $C^\infty(\mathbb{R})$; the only obstacle is at the joining points $x = \pm a$. You may view this example as rather contrived or special. However, even though we may not have simple explicit formulas for them, there are many more functions in $C_c^\infty(\mathbb{R})$. Indeed, in Section 6.3.4 we will show how the notion of convolution of functions enables one to use the bump function to "infinitely smooth out" any function with compact support, generating designer functions in $C_c^\infty(\mathbb{R})$. This will provide a way to "approximate" any function with compact support with a function in $C_c^\infty(\mathbb{R})$. In mathematical analysis one can phrase and prove precise statements about such approximations and, in doing so, establish that the set of $C_c^\infty(\mathbb{R})$ comprises a **dense** subset of many larger function spaces. **Conclusion: There are "lots and lots" of test functions.**

Lastly, note that the sum of any two test functions in $C_c^\infty(\mathbb{R})$ is also a test function in $C_c^\infty(\mathbb{R})$. The same is true for the product of a constant multiplied by a test function. Moreover,

$$\text{if } \phi \in C_c^\infty(\mathbb{R}), \text{ then } \phi' = \frac{d\phi}{dx} \in C_c^\infty(\mathbb{R}).$$

Consequently, derivatives of all orders of test functions are also test functions; this was one of the reasons why we imposed the C^∞ smoothness criterion.

5.3.2. • The Definition of a Distribution.

We are now at a point where we can define the basic object of this chapter — a distribution. To this end, let us first state the definition and then explain several of the terms used.

Definition of a Distribution

Definition 5.3.1. A **distribution** F (also known as a **generalized function**) is a rule, assigning to each test function $\phi \in C_c^\infty(\mathbb{R})$ a real number, which is linear and continuous.

By a rule, we simply mean a map (or functional) from the space of test functions $C_c^\infty(\mathbb{R})$ to $\mathbb{R}$. We denote by

$$F : \phi \in C_c^\infty(\mathbb{R}) \longrightarrow \mathbb{R}$$

such a functional on $C_c^\infty(\mathbb{R})$ and adopt the notation

$$\langle F, \phi \rangle$$

for the action of the distribution F on ϕ. That is, for each $\phi \in C_c^\infty(\mathbb{R})$, $\langle F, \phi \rangle$ is a real number. We require this functional F to be linear and continuous in the following sense:

Linearity means

$$\langle F, a\phi + b\psi \rangle = a \langle F, \phi \rangle + b \langle F, \psi \rangle \qquad \text{for all } a, b \in \mathbb{R} \text{ and } \phi, \psi \in C_c^\infty(\mathbb{R}).$$

Continuity can be phrased in two ways; one "loose" and informal, the other precise.

Loose and informal description of continuity: One way is to assert that if two test functions are "close", then the associated respective actions of the distribution (two real numbers) are also "close". Closeness of two real numbers is clear and unambiguous, but we need to specify what closeness means for two test functions. In the context of the test functions, we (loosely) say two test functions ϕ_1 and ϕ_2 in $C_c^\infty(\mathbb{R})$ are close if

$$\max_{x \in \mathbb{R}} |\phi_1(x) - \phi_2(x)| \qquad \text{is small,}$$

and, further, the same is deemed true for their derivative functions of any order; i.e., for any $k = 1, 2, \ldots$

$$\max_{x \in \mathbb{R}} |\phi_1^{(k)}(x) - \phi_2^{(k)}(x)| \qquad \text{is small,}$$

where $\phi_i^{(k)}$ is the k-th derivative of ϕ_i. Geometrically, two test functions ϕ_i would satisfy the first criterion of closeness if the graph of ϕ_2 can fit in a small **bar-neighborhood** of the graph of ϕ_1, as depicted in Figure 5.4. They would satisfy the second criterion of closeness if the same holds true for the respective functions obtained by taking the k-th derivative.

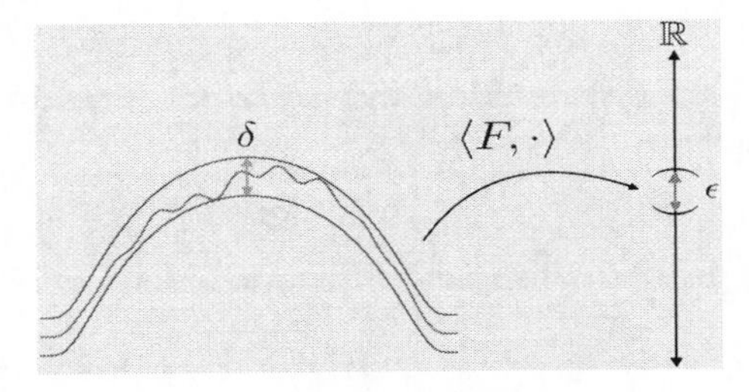

Figure 5.4. On the left is a δ bar-neighborhood in $C_c^\infty(\mathbb{R})$. For δ small, any two functions lying in the δ bar-neighborhood are deemed "close". Note, however, that for closeness in $C_c^\infty(\mathbb{R})$, we require each of the respective derivative functions to also lie in a δ bar-neighborhood. Any distribution F should map a function in this bar-neighborhood to an interval of width ϵ.

Precise description of continuity: Alternatively, we can rephrase continuity in the following sequential statement: If ϕ_n "converges to" ϕ in $C_c^\infty(\mathbb{R})$, then

$$\langle F, \phi_n \rangle \xrightarrow{n\to\infty} \langle F, \phi \rangle. \tag{5.14}$$

Since the second convergence relates to real numbers, there should not be any ambiguity here. But what about convergence of the test functions? While we were informal with the notion of closeness of functions, let us give here a precise definition of convergence of test functions. It relies on the notion of uniform convergence (cf. Section 11.7). We say

$$\phi_n \xrightarrow{n\to\infty} \phi \qquad \text{in } C_c^\infty(\mathbb{R}) \qquad \text{if}$$

(i) there exists an $a > 0$ such that all the functions $\phi_n(x)$ vanish for $|x| \geq a$ and

(ii) for any $k = 0, 1, 2, \ldots,$

$$\phi_n^{(k)}(x) \text{ converges uniformly to } \phi^{(k)}(x), \tag{5.15}$$

where $\phi_n^{(k)}(x)$ denotes the k-th derivative of ϕ_n (with the understanding that $\phi_n^{(0)}(x) = \phi_n(x)$). In the context of test functions in $C_c^\infty(\mathbb{R})$, the uniform convergence, i.e., (5.15), is equivalent to

$$\max_{x\in\mathbb{R}} |\phi_n^{(k)}(x) - \phi^{(k)}(x)| \xrightarrow{n\to\infty} 0, \qquad \text{for all } k = 0, 1, 2, \ldots.$$

Hence the precise definition of continuity is the following condition: **If** $\phi_n \xrightarrow{n\to\infty} \phi$ in $C_c^\infty(\mathbb{R})$, **then** (5.14) holds true.

Before continuing, let us provide one example of a distribution.

Example 5.3.1. Consider the functional F_1 on $C_c^\infty(\mathbb{R})$ which assigns to each $\phi \in C_c^\infty(\mathbb{R})$ its integral over $\mathbb{R}$; that is, the rule is given by

$$\langle F_1, \phi \rangle = \int_{\mathbb{R}} \phi(x)\, dx.$$

We claim that F_1 is a distribution. It should be clear that this rule (or functional) is linear (check this). It should also be intuitively clear that if the functions are close (in the above sense), then their integrals are close.[7]

The set (class) of distributions[8] forms a vector space over $\mathbb{R}$. This essentially means:

(i) We can multiply any distribution F by a scalar $a \in \mathbb{R}$ and generate a new distribution aF defined by

$$\langle aF, \phi \rangle := a\langle F, \phi \rangle \qquad \text{for all } \phi \in C_c^\infty.$$

(ii) We can add two distributions F and G to form a new distribution $F + G$ defined by

$$\langle F + G, \phi \rangle := \langle F, \phi \rangle + \langle G, \phi \rangle \qquad \text{for all } \phi \in C_c^\infty.$$

[7] A precise argument can be achieved via Theorem A.8.

[8] Readers familiar with the theory of linear algebra may recognize this formalism in the context of **dual** vector spaces. Indeed, one may view the space of distributions over the class $C_c^\infty(\mathbb{R})$ as the dual space of $C_c^\infty(\mathbb{R})$. A finite-dimensional vector space X is coupled with an additional vector space of the same dimension which is a vector space of linear functionals on X. The study of duality on function spaces (infinite-dimensional vector spaces) is far more involved and this lies in the realm of functional analysis.

To conclude, a distribution is characterized by how it **acts** on test functions. One should never ask what the *value* of a distribution F is at a point $x \in \mathbb{R}$, but rather what the value of F is "at" a test function ϕ. Put another way, you may think of the action (value) of F on a test function as a way of **sampling the distribution**. However, be warned to not directly identify these notions of sampling and distributions with their common uses in statistics; while they are related, they are not exactly the same.

In the next two subsections, we will present two fundamental types/examples of distributions. There are many other far more complicated distributions beyond these two types. While we will discuss one such class in Section 5.8, for most purposes in this book, these two types will suffice. The first type is indicative of a distribution **generalizing the notion of a function** and encapsulates the statement that every function can be regarded (or captured) as a distribution in a **natural way**.

5.3.3. • Functions as Distributions.

Any Locally Integrable Function as a Distribution

Let $f(x)$ be **a locally integrable function** on $\mathbb{R}$. Then $f(x)$ can be interpreted as the distribution F_f where

$$\langle F_f, \phi \rangle := \int_{-\infty}^{\infty} f(x)\,\phi(x)\,dx \quad \text{for any } \phi \in C_c^\infty(\mathbb{R}). \tag{5.16}$$

Think of F_f as **the distribution generated** by the function $f(x)$.

Why do we use the phrase "in a natural way"? Well, it is based upon the paradigm, argued in Section 5.1.2, that integrals (averages) of a function are fundamentally more important than its pointwise values. Indeed, the definition is only based upon the integration properties of the function f (averaging properties); hence, if we were to change the value of $f(x)$ at a finite number of points, we would still generate the **same** distribution[9]. For example, the two functions

$$f_1(x) = x^3 \qquad \text{and} \qquad f_2(x) = \begin{cases} x^3 & \text{if } x \neq 7, \\ 5 & \text{if } x = 7 \end{cases}$$

generate the same distribution. One could ask if, by only "recording" these weighted integrals of $f(x)$ (i.e., $\langle F_f, \phi \rangle$), do we lose information about f as a function? Stated a different way, do all the values of $\langle F_f, \phi \rangle$ uniquely determine the function $f(x)$ at all points $x \in \mathbb{R}$? The answer is yes modulo a negligible set of real numbers which has no effect on integration, i.e., zero measure in the sense of measure theory **[17]**.

Lastly, note that our first example of a distribution (Example 5.3.1) was exactly the distribution generated by the function which was identically 1. This distribution assigns to every test function $\phi \in C_c^\infty(\mathbb{R})$ its integral over all $\mathbb{R}$.

Notation: The notation F_f becomes cumbersome over time so we will usually dispense with it, just using f **but** making it clear in words that we are "thinking" of f

[9] In fact, you can change the value of the functions on any set *which will have no effect on integration*. This is made precise if one studies a bit of *measure theory* **[17]**.

as a distribution. Therefore, when one makes a statement about a function f and adds the phrase *in the sense of distributions*, one is speaking about F_f.

5.3.4. • The Precise Definition of the Delta "Function" as a Distribution.

We now give the delta "function" δ_0 **official status**. Indeed, while it is **not** a function, it most certainly **is** a distribution. Whereas part of our motivation in Section 5.1.2 was to capture functions from their "averages" rather than their pointwise values, here it will be crucial that **our test functions** are continuous and their value at a point (say $x = 0$) represents the local environment of the test function. Indeed, the following definition simply states that δ_0 is a distribution whose action on any test function picks out its value at 0.

Definition of the Delta Function

Definition 5.3.2. The delta "function" δ_0 is the distribution defined as follows:

$$\langle \delta_0, \phi \rangle = \phi(0) \qquad \text{for any } \phi \in C_c^\infty(\mathbb{R}).$$

We can also consider delta functions *concentrated at* point $x_0 \in \mathbb{R}$; that is,

$$\langle \delta_{x_0}, \phi \rangle = \phi(x_0) \qquad \text{for any } \phi \in C_c^\infty(\mathbb{R}).$$

With Definition 5.3.2 in hand, let us now agree to **abandon the quotation marks** around the word *function* and yield to tradition, partially admitting defeat, by simply referring to the distribution δ_0 as **the delta function**. However, we will usually suppress the explicit functional dependence on x and not write "$\delta_0(x)$".

Two important remarks are in order:

(i) **A delicate and subtle point on integration and the delta function:** There is **no integral** in the above definition of the delta function; we did **not** ever write (or need to write)

$$\text{“} \int \delta_0 \, \phi(x)\, dx \text{”}.$$

On the other hand, recall the punchline of Section 5.2.2 where we concluded by claiming that while functions have significance without any notion of integration, the delta function does not. So where is the notion of integration in Definition 5.3.2? The subtle answer is that integration is implicitly embedded in the definition because of the way in which we envision a function as a distribution. More precisely, to address δ_0 in any scenario involving functions, we must view the functions as objects to be integrated against test functions.

(ii) **Functional notation and the use of quotation marks in the sequel:** Occasionally, we may use the functional notation for the delta function to focus directly on the intuition; however, in these cases, we will also use quotation marks when viewing an object as a function which is the **delta function in disguise**, for example, "$H'(x)$", the derivative of the Heaviside function. Another important example of the delta function in disguise is the following indefinite integral:

$$\text{“}\ \frac{1}{2\pi}\int_{-\infty}^{\infty} e^{-ixy}\,dy.\ \text{”} \tag{5.17}$$

Here, i is the complex number $\sqrt{-1}$ (cf. Section 6.1) and x is a real parameter. It is natural to consider (5.17) as a function of x. However, this function of x makes no sense; for any fixed x, how do you make sense of the indefinite integral which can be thought of as periodically traversing the unit circle in the complex plane infinitely many times! On the other hand we shall see later, when studying the Fourier transform, that (5.17) is one of the most important and famous *disguises* for the delta function. That is, this improper integral can be interpreted in the sense of distributions and is the same distribution as the delta function. In fact, (5.17) is often referred to as an *integral representation of the delta function.*

To conclude in a nutshell: δ_0 ~~(x)~~. Well, at least for now! As just mentioned, sometimes it is useful to guide us by informally treating the delta function as a function and manipulating the argument of the function, i.e., the $(\cdot)$. We will on occasion also do this but will always use the quotation marks. See Section 5.6 for more on this.

5.4. • Derivative of a Distribution

5.4.1. • Motivation via Integration by Parts of Differentiable Functions.

Given a function $f \in C^1$, we can find the derivative $f'(x)$ at any $x \in \mathbb{R}$ as

$$\lim_{h\to 0}\frac{f(x+h)-f(x)}{h}.$$

The definition highlights that this pointwise derivative is an *instantaneously local* quantity associated with the function f. We can think of $f'(x)$ itself as a continuous function on $\mathbb{R}$.

Let us explore the role that this pointwise derivative (or classical derivative) function plays from the perspective of integrating (or averaging) against a test function $\phi \in C_c^\infty(\mathbb{R})$. The key notion here is **integration by parts**. Indeed, since $\phi(x)$ has compact support, there exists some finite interval, say, $[-L, L]$, such that $\phi(x) = 0$ for all x with $|x| \geq L$. Hence, via integration by parts, we have

$$\begin{aligned}
\int_{-\infty}^{\infty} f'(x)\,\phi(x)\,dx &= \int_{-L}^{L} f'(x)\,\phi(x)\,dx \\
&\overset{\text{integrations by parts}}{=} -\int_{-L}^{L} f(x)\,\phi'(x)\,dx + \Big[f(x)\phi(x)\Big]\Big|_{-L}^{L} \\
&= -\int_{-L}^{L} f(x)\,\phi'(x)\,dx + f(L)\,\phi(L) - f(-L)\,\phi(-L) \\
&= -\int_{-\infty}^{\infty} f(x)\,\phi'(x)\,dx.
\end{aligned}$$

Note, since $\phi(-L) = 0 = \phi(L)$, the boundary terms after the integration by parts are zero. The choice of L depends on the particular test function ϕ but, regardless, we still have

$$\text{for all } \phi \in C_c^\infty(\mathbb{R}), \qquad \int_{-\infty}^{\infty} f'(x)\,\phi(x)\,dx = -\int_{-\infty}^{\infty} f(x)\,\phi'(x)\,dx. \tag{5.18}$$

This simple formula tells us that, with respect to integration, we can place the derivative on either f or ϕ, at the expense of a minus sign.

5.4.2. • The Definition of the Derivative of a Distribution.

With (5.18) as motivation, we now define the derivative for **any** distribution.

Definition of the Derivative of a Distribution

Definition 5.4.1. Let F be any distribution. Then F' (its derivative) is also a distribution defined by

$$\langle F', \phi \rangle := -\langle F, \phi' \rangle \qquad \text{for } \phi \in C_c^\infty(\mathbb{R}). \tag{5.19}$$

Note that the definition makes sense since if $\phi \in C_c^\infty(\mathbb{R})$, then $\phi' \in C_c^\infty(\mathbb{R})$.

Thus every distribution has a distributional derivative. Since every locally integrable function generates a distribution, it follows that every locally integrable function, even those which are not classically differentiable, has a distributional derivative. But now let me pose the essential question with respect to **consistency**. Above we have given a definition for the distributional derivative. But is it a "good" definition in the sense that it is **consistent** with our classical notion of differentiability? To address this, consider a classically differentiable function $f \in C^1$. Its pointwise derivative f' exists and is a continuous function on $\mathbb{R}$. Hence, both f and f' are locally integrable and generate (can be thought of as) distributions via (5.16). For consistency to hold, we require that

$$(F_f)' = F_{f'}. \tag{5.20}$$

This says the following:

The distributional derivative of the distribution generated by a smooth function is simply the distribution generated by the classical derivative.

You should read this sentence over and over again until you have fully digested its meaning (this might take some time so please be patient!).

It is easy to see that it is precisely (5.18) which makes (5.20) hold true. To this end, the equality of distributions means that for any $\phi \in C_c^\infty(\mathbb{R})$,

$$\langle (F_f)', \phi \rangle = \langle F_{f'}, \phi \rangle.$$

But this follows directly from (i) the definition of the distributional derivative (Definition 5.4.1), (ii) the interpretation of a function as a distribution (5.16), and (iii) the integration by parts formula (5.18). Indeed,

$$\langle (F_f)', \phi \rangle = -\langle F_f, \phi' \rangle = -\int_{-\infty}^{\infty} f(x)\,\phi'(x)\,dx = \int_{-\infty}^{\infty} f'(x)\,\phi(x)\,dx = \langle F_{f'}, \phi \rangle.$$

Hence, the classical notion of differentiation is **carried over** to our generalized setting of distributions. However, Definition 5.4.1 also applies to **any** distribution. We now give a few words on the two important types of distributions (distributions generated by functions and the delta function).

Derivative of a Locally Integrable Function in the Sense of Distributions. Even if a locally integrable function f is not differentiable in the classical pointwise sense, we have a derivative in the sense of distributions. This derivative in the sense of distributions is a distribution G which may, or may not, be generated by another locally integrable function (cf. the next subsection). In terms of language, if f is a locally integrable function and G is a distribution, we say the derivative of f equals G **in the sense of distributions** if $(F_f)' = G$. To repeat, when we speak of a derivative of a function f in the sense of distributions, we always mean the distributional derivative of F_f.

The Derivative of the Delta Function. Consider the delta function δ_0. As we have discussed at length, it is not a function but it certainly is a distribution, and hence we can find its distributional derivative. This distributional derivative, conveniently denoted as δ_0', is the distribution defined by

$$\langle \delta_0', \phi \rangle := -\langle \delta_0, \phi' \rangle = -\phi'(0) \qquad \text{for } \phi \in C_c^\infty(\mathbb{R}).$$

In other words, it is the distribution which assigns to each test function ϕ the negative value of its derivative at the point $x = 0$.

Finally, let us note that differentiation of distributions immediately carries over to **higher-order derivatives.** One can repeat this process of distributional differentiation (5.19) to define any number of derivatives of a distribution; this was the advantage of choosing C^∞ test functions. The sign on the right-hand side needs to be adjusted in accordance with the number of derivatives. The definition is again a consequence of the integration by parts formula which motivated definition (5.19). Indeed, if n is any positive integer, we define the n-th derivative of a distribution F to be the new distribution (denoted by $F^{(n)}$) defined by

$$\boxed{\langle F^{(n)}, \phi \rangle = (-1)^n \langle F, \phi^{(n)} \rangle \qquad \text{for } \phi \in C_c^\infty(\mathbb{R}).}$$

Here $\phi^{(n)}$ denotes the n-th derivative of the function $\phi(x)$.

5.4.3. • Examples of Derivatives of Piecewise Smooth Functions in the Sense of Distributions.

Let us begin by differentiating some piecewise smooth functions in the sense of distributions. Why? Because we shall now see the effect of jump discontinuities in the distributional derivatives.

Example 5.4.1 (the Heaviside Function). Let us return to the Heaviside function (5.5) and differentiate it as a distribution; i.e., differentiate in the sense of distributions. We have for any test function ϕ

$$\begin{aligned} \langle (F_H)', \phi \rangle = -\langle F_H, \phi' \rangle &= -\int_{-\infty}^{\infty} H(x)\,\phi'(x)\,dx \\ &= -\int_{0}^{\infty} \phi'(x)\,dx \\ &= -(\phi(+\infty) - \phi(0)) \\ &= \phi(0) \\ &= \langle \delta_0, \phi \rangle . \end{aligned}$$

Note that the term $\phi(+\infty)$ vanished because of compact support. Thus, $(F_H)' = \delta_0$ or, in other words, the derivative of the Heaviside function in the sense of distributions is the Dirac delta function.

Note: We stress that at no point did we write, or need to write,

$$\text{“} \int_{-\infty}^{\infty} H'(x)\,\phi(x)\,dx \text{ ”.}$$

This does not make sense since $H'(x)$ is a distribution which cannot be generated by a function. Only distributions which are generated by a function have the property that their application to a test function can be written as a regular integral. However, it is common practice to write down expressions like

$$\text{“} \int_{-\infty}^{\infty} \underbrace{H'(x)}_{\delta_0}\ \phi(x)\,dx\text{”.}$$

Even though such expressions are technically incorrect, we now know how to make sense of them using the theory of distributions. In the rest of this text, we will sometimes write expressions like this for intuition and motivation. However, in these cases, we will always write them within parentheses "...".

Example 5.4.2. Let $f(x) = |x|$. First, let us ask what $f'(x)$ is. Pointwise, we would say

$$f'(x) = \begin{cases} -1, & x < 0, \\ undefined, & x = 0, \\ 1, & x > 0. \end{cases}$$

So, again, we have the issue of undefined at $x = 0$. Is this "undefined" important? In other words, if we set the value of $f'(0)$ to be some number (say, 45), would we miss something? Let us do this differentiation in the sense of distributions. We have

$$\langle (F_f)', \phi \rangle = -\langle F_f, \phi' \rangle = -\int_{-\infty}^{\infty} |x|\,\phi'(x)dx = -\int_{-\infty}^{0} -x\,\phi'(x)dx - \int_{0}^{\infty} x\,\phi'(x)dx$$

and integrating by parts in each integral, we find that

$$\begin{aligned}
\langle (F_f)', \phi \rangle &= -\int_{-\infty}^{0} \phi(x)\, dx + [x\phi(x)]\Big|_{-\infty}^{0} + \int_{0}^{\infty} \phi(x)\, dx - [x\phi(x)]\Big|_{0}^{+\infty} \\
&= -\int_{-\infty}^{0} \phi(x)\, dx + \int_{0}^{\infty} \phi(x)\, dx \\
&= \int_{-\infty}^{\infty} g(x)\,\phi(x)\, dx = \langle F_g, \phi \rangle,
\end{aligned}$$

where g is the function

$$g(x) = \begin{cases} -1, & x < 0, \\ 1, & x \geq 0. \end{cases}$$

Note here how we dispensed with the boundary terms. They were trivially zero at $x = 0$ and were zero at $\pm\infty$ because ϕ has compact support; for example,

$$[x\,\phi(x)]\Big|_{-\infty}^{0} := \lim_{L\to\infty} [x\,\phi(x)]\Big|_{-L}^{0} = \lim_{L\to\infty} L\,\phi(-L) = 0,$$

since $\phi \equiv 0$ outside some fixed interval. In conclusion, f' is simply g *in the sense of distributions.* The function g is known as the **signum function** (usually abbreviated by sgn) since it effectively gives the sign (either ± 1) of x; that is,

$$\mathrm{sgn}(x) := \begin{cases} -1, & x < 0, \\ 1, & x > 0, \\ 0, & x = 0. \end{cases}$$

Note that in this instance we took the value at $x = 0$ to be 0 but could just as well have taken it to be 45. The value at one point has no effect on *sgn as a distribution.*

Now, suppose we want to find f''? From our previous example of the Heaviside function, we know that doing this pointwise will **not** give us the right answer. So let us work in the sense of distributions. We have

$$\begin{aligned}
\langle (F_f)'', \phi \rangle &= \langle F_f, \phi'' \rangle \\
&= \int_{-\infty}^{\infty} |x|\,\phi''(x)\, dx \\
&= \int_{-\infty}^{0} (-x)\,\phi''(x)\, dx + \int_{0}^{\infty} x\,\phi''(x)\, dx \\
&\overset{\substack{\text{integration} \\ \text{by parts}}}{=} \int_{-\infty}^{0} \phi'(x)\, dx - [x\,\phi'(x)]\Big|_{-\infty}^{0} - \int_{0}^{\infty} \phi'(x)\, dx + [x\,\phi(x)]\Big|_{0}^{\infty} \\
&= \int_{-\infty}^{0} \phi'(x)\, dx - \int_{0}^{\infty} \phi'(x)\, dx \\
&= 2\phi(0).
\end{aligned}$$

Thus, $f'' = g' = 2\delta_0$ *in the sense of distributions.*

Example 5.4.3. Let

$$f(x) = \begin{cases} x^2, & x \geq 0, \\ 2x+3, & x < 0. \end{cases}$$

We find f' in the sense of distributions. Let $\phi \in C_c^\infty(\mathbb{R})$ be a test function; then we have

$$\begin{aligned}
\langle (F_f)', \phi \rangle &= -\langle F_f, \phi' \rangle \\
&= -\int_{-\infty}^{\infty} f(x)\,\phi'(x)\,dx \\
&= -\int_{-\infty}^{0} (2x+3)\,\phi'(x)\,dx \;-\; \int_0^\infty x^2\,\phi'(x)\,dx \\
&\overset{\substack{\text{integration}\\ \text{by parts}}}{=} \int_{-\infty}^{0} 2\,\phi(x)\,dx \;-\; \left[(2x+3)\,\phi(x)\right]\Big|_{-\infty}^{0} \\
&\qquad + \int_0^\infty 2x\,\phi(x)\,dx \;-\; \left[x^2\,\phi(x)\right]\Big|_0^\infty \\
&= \int_{-\infty}^{0} 2\,\phi(x)\,dx \;-\; 3\phi(0) \;+\; \int_0^\infty 2x\,\phi(x)\,dx \\
&= \int_{-\infty}^{\infty} g(x)\,\phi(x)\,dx - 3\phi(0),
\end{aligned}$$

where

$$g(x) = \begin{cases} 2x, & x \geq 0, \\ 2, & x < 0. \end{cases}$$

We used, as before, the fact that ϕ had compact support to dismiss the boundary terms at $\pm\infty$. Thus the derivative of f in the sense of distributions is the distribution $g - 3\delta_0$, or more precisely the sum of the two distributions, $F_g - 3\delta_0$. The $3\delta_0$ is a result of the jump discontinuity in f at $x = 0$.

To summarize this section:

- The notion of the derivative of a distribution allows us to differentiate any distribution to arrive at another distribution.
- We are primarily concerned with distributions which are either generated by functions or are constructed using delta functions. Whereas we can now consider the derivative of a distribution generated by **any** integrable function, we focused on distributions generated by piecewise smooth functions. In certain cases, this distributional derivative was simply the distribution generated by the pointwise derivative function, wherein we ignored the points at which the derivative function was undefined. In these cases, the function was either continuous or had a removable discontinuity at the singularity. The true singularity was in the first derivative and, hence, played no role in the distributional first derivative. In cases where the function had an essential jump discontinuity, the singularity was important and gave rise to a delta function in the distributional derivative.

In the examples presented here we have only addressed differentiating piecewise smooth functions in the sense of distributions.[10] One might thus be tempted to conclude that this new machinery of distributions was a bit of overkill. From the perspective of differentiating functions with discontinuities, the power and richness of this new machinery is really only seen in functions of several variables. We will begin to explore this in Chapter 9.

5.5. • Convergence in the Sense of Distributions

5.5.1. • The Definition.

We now define a notion of convergence for distributions and then apply this definition to sequences of functions interpreted as distributions.

Definition of Convergence of a Sequence of Distributions

Definition 5.5.1. A sequence F_n of distributions converges to a distribution F if

$$\langle F_n, \phi \rangle \xrightarrow{n\to\infty} \langle F, \phi \rangle \quad \text{for all } \phi \in C_c^\infty(\mathbb{R}).$$

In other words, for every $\phi \in C_c^\infty(\mathbb{R})$, the sequence of real numbers $\langle F_n, \phi \rangle$ converges to the real number $\langle F, \phi \rangle$. This convergence is written $F_n \to F$ in the sense of distributions.

Given a sequence of functions f_n, we may consider these functions as distributions and then apply the above definition. This leads to the following:

Definition 5.5.2. We say a sequence of locally integrable functions $f_n(x)$ **converges in the sense of distributions** to a distribution F if

$$\langle F_{f_n}, \phi \rangle \xrightarrow{n\to\infty} \langle F, \phi \rangle \quad \text{for all } \phi \in C_c^\infty(\mathbb{R}),$$

that is, if

$$\boxed{\int_{-\infty}^{\infty} f_n(x)\,\phi(x)\,dx \xrightarrow{n\to\infty} \langle F, \phi \rangle \quad \text{for all } \phi \in C_c^\infty(\mathbb{R}).} \tag{5.21}$$

Please note that occasionally we will leave out the "$n \to \infty$". It is understood!

A sequence of functions may converge in the sense of distributions to either another function or a more general distribution not generated by a function, e.g., a delta function. Specifically, $f_n(x)$ converging in the sense of distributions to a function $f(x)$ means that

$$\int_{-\infty}^{\infty} f_n(x)\,\phi(x)\,dx \xrightarrow{n\to\infty} \int_{-\infty}^{\infty} f(x)\,\phi(x)\,dx \quad \text{for all } \phi \in C_c^\infty(\mathbb{R}).$$

On the other hand, $f_n(x)$ converging in the sense of distributions to δ_0 means that

$$\int_{-\infty}^{\infty} f_n(x)\,\phi(x)\,dx \xrightarrow{n\to\infty} \phi(0) \quad \text{for all } \phi \in C_c^\infty(\mathbb{R}).$$

[10]**Warning:** In general, the distributional derivative of an integrable function can have a very complex structure, for example, a sample path of *Brownian motion* (cf. Section 7.3.4) which is continuous but nowhere differentiable, or what is called *the devil's staircase or the Cantor function*. These functions have a distributional derivative **but** this distributional derivative is neither generated by a function nor a combination of delta functions.

Automatic Distributional Convergence of Derivatives. The following result is a direct consequence of the definitions of differentiation and convergence in the sense of distributions (cf. Exercise **5.5**).

Proposition 5.5.1. *Suppose $\{F_n\}$ is a sequence of distributions which converges to a distribution F. Then*

$$F_n' \to F' \quad \text{in the sense of distributions.}$$

The same holds true for higher derivatives.

5.5.2. • Comparisons of Distributional versus Pointwise Convergence of Functions. Suppose we are given a sequence of functions $f_n(x)$ and a limit function $f(x)$. In this scenario, **pointwise convergence** would mean the following: For every $x \in \mathbb{R}$, $f_n(x)$ as a sequence of numbers converges to the number $f(x)$. Pointwise convergence is, in principle, a stronger notion of convergence in the following sense:

Proposition 5.5.2. *Suppose $f_n(x)$ is a sequence of locally integrable functions which converges pointwise to a locally integrable function f. Further, suppose that there exists a locally integrable function g such that*

$$|f_n(x)| \le g(x) \qquad \text{for all } n \text{ and for all } x \in \mathbb{R}.$$

Then f_n converges to f in the sense of distributions.

The proof is a simple consequence of Theorem A.8. As an example, the sequence of functions

$$f_n(x) = \frac{1}{n} e^{-\frac{x^2}{4n}}$$

converges pointwise to 0, and hence by Proposition 5.5.2, in the sense of distributions to the zero function $f(x) \equiv 0$. On the other hand, it is possible for a sequence of functions $f_n(x)$ to converge in the sense of distributions to function $f(x)$, but not in **any** pointwise sense. In fact, as we will discuss shortly in Section 5.5.5, the sequence $f_n(x) = \sin(nx)$ converges in the sense of distributions to the zero function $f(x) \equiv 0$; yet, for any fixed $x \in \mathbb{R}$ (different from integer multiples of π), the sequence $f_n(x) = \sin(nx)$ does not converge to any number.

5.5.3. • The Distributional Convergence of a Sequence of Functions to the Delta Function: Four Examples. The sequence of spikes, defined by (5.10), was previously used in Section 5.2.2 to motivate the appearance of the delta function. In this and the next subsection, we address the distributional convergence of (5.10) and three other sequences of functions to δ_0, summarizing the essential properties shared by all of them and providing several proofs. For convenience of notation, let $\sigma_n := \frac{1}{n}$. Consider the following four sequences of functions:

1. **The sequence of hats**

$$f_n(x) = \begin{cases} \frac{n}{2} & \text{if} \quad |x| \leq \sigma_n, \\ 0 & \text{otherwise.} \end{cases} \tag{5.22}$$

2. **The sequence of spikes**

$$f_n(x) = (n - n^2|x|)\,\chi_{[-\sigma_n,\sigma_n]} = \begin{cases} n - n^2 x & \text{if} \quad 0 < x < \sigma_n, \\ n + n^2 x & \text{if} \quad -\sigma_n < x \leq 0, \\ 0 & \text{if} \quad |x| \geq \sigma_n. \end{cases} \tag{5.23}$$

3. **The sequence involving the derivative of arctangent**

$$f_n(x) = \frac{1}{\pi}\frac{\sigma_n}{x^2 + \sigma_n^2}. \tag{5.24}$$

4. **The sequence of Gaussians**

$$f_n(x) = \frac{1}{\sqrt{2\pi\,\sigma_n}} e^{-\frac{x^2}{2\sigma_n}}. \tag{5.25}$$

Figure 5.5 shows plots of these functions for increasing n.

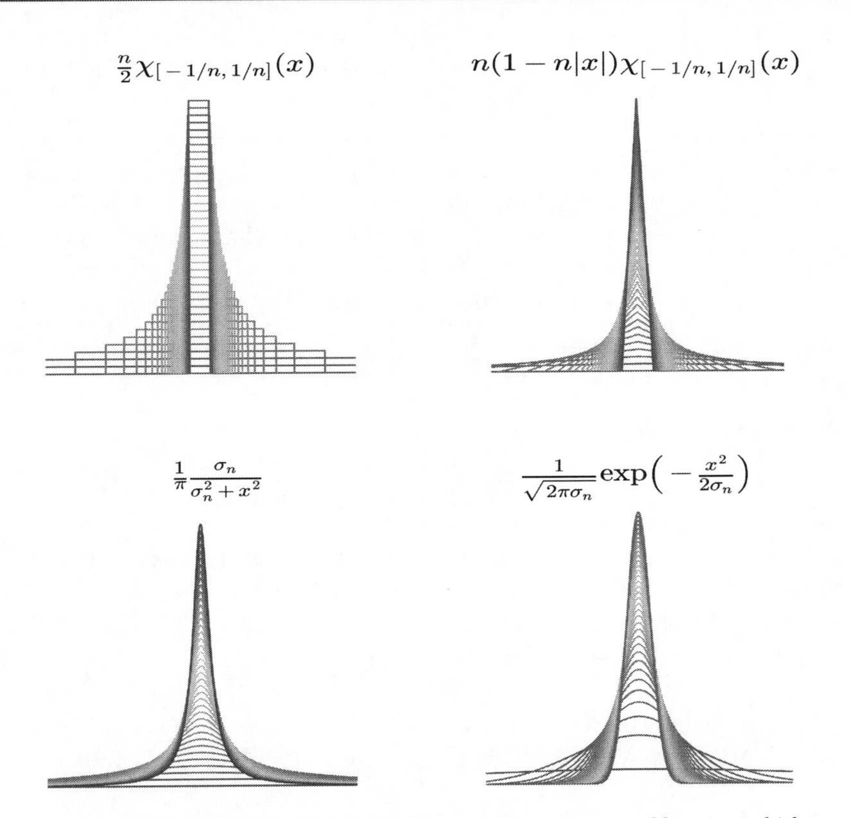

Figure 5.5. Plots for increasing n of the respective four sequences of functions which converge in the sense of distributions to δ_0.

All four sequences of functions converge in the sense of distributions to the distribution δ_0 (the delta function). This means, for any $\phi \in C_c^\infty(\mathbb{R})$, we have

$$\int_{-\infty}^{\infty} f_n(x)\,\phi(x)\,dx \;\to\; \phi(0) \qquad \text{as } n \to \infty. \tag{5.26}$$

Mathematicians sometimes call such a sequence **an approximation to the identity** and, as we previously pointed out (cf. the quote at the end of Section 5.2.2), such sequences were central to Dirac's vision of his delta function. There are **three properties** shared by all of these sequences that are responsible for their distributional convergence to δ_0:

Properties 5.5.1. *The three properties for $f_n(x)$ are:*

(1) ***Nonnegativity:*** The functions f_n are always nonnegative; i.e.,

$$\text{for all } x \in \mathbb{R} \text{ and } n \in \mathbb{N}, \text{ we have } f_n(x) \geq 0.$$

(2) ***Unit mass (area under the curve):*** The functions f_n all integrate to 1. That is, for each $n = 1, 2, 3, \ldots$

$$\int_{-\infty}^{\infty} f_n(x)\,dx = 1.$$

(3) ***Concentration at*** 0: As $n \to \infty$, the functions concentrate their **mass** (area under the curve) around $x = 0$. This loose condition can be made precise (see Exercise **5.15**).

Observe from Figure 5.5 that each of our four sequences appears to satisfy these three properties. We claim that Properties 5.5.1 are sufficient for (5.26) to hold. Let us first intuitively argue why. Properties (1) and (3) imply that if we consider $f_n(x)$ for very large n, there is a very small $\delta > 0$ such that the mass of f_n (i.e., the area under the curve) is concentrated in $[-\delta, \delta]$. Hence, by the second property, for large n we have

$$\int_{-\infty}^{\infty} f_n(x)\,dx \sim \int_{-\delta}^{\delta} f_n(x)\,dx \sim 1,$$

and for any continuous function ϕ,

$$\int_{-\infty}^{\infty} f_n(x)\,\phi(x)\,dx \sim \int_{-\delta}^{\delta} f_n(x)\,\phi(x)\,dx.$$

Here, again we use $\sim$ to loosely mean *close to* or *approximately equal to*. Thus for large n, we may focus our attention entirely on the small interval $x \in [-\delta, \delta]$. Since ϕ is continuous, it is essentially constant on this small interval; in particular, $\phi(x)$ on the interval $[-\delta, \delta]$ is close to $\phi(0)$. Thus, for n large,

$$\begin{aligned}
\int_{-\infty}^{\infty} f_n(x)\,\phi(x)\,dx \sim \int_{-\delta}^{\delta} f_n(x)\,\phi(x)\,dx &\sim \int_{-\delta}^{\delta} f_n(x)\,\phi(0)\,dx \\
&\sim \phi(0)\int_{-\delta}^{\delta} f_n(x)\,dx \\
&\sim \phi(0).
\end{aligned} \tag{5.27}$$

For any particular sequence with these properties, one can make these statements precise (i.e., formulate a proof) via an "ϵ vs. N" argument. In fact, a proof for the hat sequence (5.22) follows directly from the Averaging Lemma (Exercise **5.6**). We will now present the "ϵ vs. N" proofs for two sequences: the spike functions (defined by (5.23)) and sequence involving the derivative of arctangent (defined by (5.24)). While the above heuristics for this convergence are indeed rather convincing, we recommend all students read the first proof — even if you have not taken a class in mathematical analysis. Both proofs are actually relatively straightforward and serve as excellent examples to digest and appreciate the estimates behind a convergence proof.

5.5.4. ϵ vs. N Proofs for the Sequences (5.23) and (5.24).

We begin with the sequence of spikes (5.23); here, the proof is straightforward since the $f_n(x)$ are identically zero on $|x| \geq \sigma_n = \frac{1}{n}$.

Theorem 5.1. *Let $f_n(x)$ be defined by (5.23). Then $f_n \to \delta_0$ in the sense of distributions.*

Proof. Fix any $\phi \in C_c^\infty$. We need to prove that

$$\int_{-\infty}^{\infty} f_n(x)\,\phi(x)\,dx \xrightarrow{n\to\infty} \phi(0). \tag{5.28}$$

In other words, the sequence of real numbers

$$\int_{-\infty}^{\infty} f_n(x)\,\phi(x)\,dx$$

converges to $\phi(0)$. To be precise, this means for any $\epsilon > 0$, there exists $N \in \mathbb{N}$ such that

$$\text{if } n > N, \qquad \text{then} \qquad \left| \left(\int_{-\infty}^{\infty} f_n\,\phi\,dx \right) - \phi(0) \right| < \epsilon.$$

To this end, we first note that since the functions f_n all integrate to 1, we have (rather trivially) that

$$\phi(0) = \phi(0)\,1 = \phi(0) \int_{-\infty}^{\infty} f_n(x)\,dx = \int_{-\infty}^{\infty} f_n(x)\,\phi(0)\,dx.$$

Thus,

$$\begin{aligned} \left| \int_{-\infty}^{\infty} f_n(x)\,\phi(x)\,dx - \phi(0) \right| &= \left| \int_{-\infty}^{\infty} f_n(x)\,\phi(x)\,dx - \int_{-\infty}^{\infty} f_n(x)\,\phi(0)\,dx \right| \\ &= \left| \int_{-\infty}^{\infty} f_n(x)\,(\phi(x) - \phi(0))\,dx \right| \\ &\leq \int_{-\infty}^{\infty} f_n(x)\,|\phi(x) - \phi(0)|\,dx. \end{aligned} \tag{5.29}$$

Since ϕ is continuous, **there exists** $\delta > 0$ **such that** if $|x| < \delta$, then $|\phi(x) - \phi(0)| < \epsilon$. Now **choose** $N \in \mathbb{N}$ sufficiently large such that $1/N < \delta$. Then if $n > N$, we have

$$\sigma_n = \frac{1}{n} < \frac{1}{N} < \delta$$

and

$$\begin{aligned}
\int_{-\infty}^{\infty} f_n(x)|\phi(x) - \phi(0)|dx &= \int_{-\frac{1}{n}}^{\frac{1}{n}} f_n(x)|\phi(x) - \phi(0)|\,dx \\
&< \epsilon \underbrace{\int_{-\frac{1}{n}}^{\frac{1}{n}} f_n(x)\,dx}_{=1} \\
&= \epsilon.
\end{aligned}$$

Combining this with (5.29), we have for all $n > N$

$$\left| \left(\int_{-\infty}^{\infty} f_n(x)\phi(x)\,dx \right) - \phi(0) \right| < \epsilon. \qquad \square$$

In the previous sequence of functions, the functions become identically zero on more and more of the domain as n tends to infinity: As a result, the proof never used the precise formula for f_n; that is, the exact shape of the spike was irrelevant. This is a rather strong form of the concentration property (3). Many sequences of functions still concentrate their mass around zero but are never actually equal to zero at a particular x. This is the case with (5.24) and (5.25) where $f_n(x) > 0$ for any n and any $x \in \mathbb{R}$. However, just by plotting $f_5(x)$, $f_{10}(x)$, and $f_{50}(x)$ for either case, one can readily see the concentration of mass (area under the curve). We now present a proof for (5.24).

First, note that property (2) holds; indeed,[11] for any n we have

$$\int_{-\infty}^{\infty} \frac{\sigma_n}{x^2 + \sigma_n^2}\,dx = \sigma_n \int_{-\infty}^{\infty} \frac{1}{x^2 + \sigma_n^2}\,dx = \sigma_n \frac{1}{\sigma_n} \tan^{-1}\left(\frac{x}{\sigma_n}\right)\Big|_{-\infty}^{\infty} = \frac{\pi}{2} - (-\frac{\pi}{2}) = \pi.$$

Hence, the same heuristics behind (5.27) suggest the distributional convergence to δ_0. The following theorem and proof present the precise argument which, as you will observe, is only slightly more involved than the proof of Theorem 5.1.

Theorem 5.2. *Let $f_n(x)$ be defined by (5.24). Then $f_n \to \delta_0$ in the sense of distributions.*

Proof. Fix any $\phi \in C_c^\infty$. We need to prove that for any $\epsilon > 0$, there exists $N \in \mathbb{N}$ such that

$$\text{if } n > N, \qquad \text{then} \qquad \left| \int_{-\infty}^{\infty} f_n(x)\,\phi(x)\,dx - \phi(0) \right| < \epsilon.$$

Let $\epsilon > 0$. Due to the fact that the functions f_n are nonnegative and always integrate to 1 (i.e., satisfy properties (1) and (2)), we can repeat the initial steps leading up to (5.29) verbatim to find

$$\left| \int_{-\infty}^{\infty} f_n(x)\,\phi(x)\,dx - \phi(0) \right| \le \int_{-\infty}^{\infty} f_n(x)\,|\phi(x) - \phi(0)|\,dx.$$

[11] Recall from calculus the fact that for any $a > 0$,

$$\int \frac{1}{x^2 + a^2}\,dx = \frac{1}{a} \tan^{-1}\left(\frac{x}{a}\right) + C.$$

Now comes the extra work associated with the fact that $f_n(x)$ is not identically zero on most of $\mathbb{R}$. Since ϕ is continuous, **there exists** $\delta > 0$ **such that** if $|x| < \delta$, then

$$|\phi(x) - \phi(0)| < \frac{\epsilon}{2}.$$

You will see shortly why we need the one-half on the right.

Splitting the integral into two pieces, we have

$$\int_{-\infty}^{\infty} f_n(x)\,|\phi(x)-\phi(0)|\,dx = \int_{-\delta}^{\delta} f_n(x)\,|\phi(x)-\phi(0)|\,dx + \int_{\{|x|\geq\delta\}} f_n(x)\,|\phi(x)-\phi(0)|\,dx. \tag{5.30}$$

For the first integral, we note that with our choice of δ we have

$$\int_{-\delta}^{\delta} f_n(x)\,|\phi(x) - \phi(0)|\,dx < \frac{\epsilon}{2}\int_{-\delta}^{\delta} f_n(x)\,dx \leq \frac{\epsilon}{2}. \tag{5.31}$$

In the last inequality, we used the fact that f_n was positive and integrated over $\mathbb{R}$ to 1; hence, the integral of f_n on any subinterval must be less than or equal to 1.

For the second integral, we need to work a little harder. First, note that since ϕ has compact support and is continuous, it must be bounded. This means there exists a constant C such that $|\phi(x)| \leq C$ for all $x \in \mathbb{R}$. Hence,

$$|\phi(x) - \phi(0)| \leq |\phi(x)| + |\phi(0)| \leq 2C,$$

and

$$\int_{\{|x|\geq\delta\}} f_n(x)\,|\phi(x) - \phi(0)|\,dx \leq 2C\int_{\{|x|\geq\delta\}} f_n(x)\,dx. \tag{5.32}$$

Since for each n the function $f_n(x)$ is an even function,

$$2C\int_{\{|x|\geq\delta\}} f_n(x)\,dx = 4C\int_{\delta}^{\infty} f_n(x)\,dx = \frac{4C}{\pi}\int_{\delta}^{\infty}\frac{\sigma_n}{x^2+\sigma_n^2}\,dx. \tag{5.33}$$

Note that this is the first time we are actually writing down the precise formula for f_n. Until now, all we required were properties (1) and (2). Now we need to interpret property (3) by means of choosing n large enough that this tail integral in (5.33) can be made small. How small? Less than $\epsilon/2$. To this end, note that

$$\int_{\delta}^{\infty}\frac{\sigma_n}{x^2+\sigma_n^2}\,dx = \frac{\pi}{2} - \tan^{-1}\left(\frac{\delta}{\sigma_n}\right).$$

Now comes a slightly delicate argument. We have fixed an ϵ and, consequently, have established the existence of a δ (depending on ϵ) which gave the estimate (5.31). With this fixed δ, we note that since

$$\lim_{\theta\to\infty}\tan^{-1}\theta = \frac{\pi}{2} \qquad\text{and}\qquad \lim_{n\to\infty}\sigma_n = 0,$$

we can make $\tan^{-1}\left(\frac{\delta}{\sigma_n}\right)$ as close as we like to $\frac{\pi}{2}$ by choosing n sufficiently large. In our present context, this means we can **choose** N **such that** if $n > N$, then

$$\frac{\pi}{2} - \tan^{-1}\left(\frac{\delta}{\sigma_n}\right) < \frac{\pi}{4C}\,\frac{\epsilon}{2}. \tag{5.34}$$

Hence, combining (5.32)–(5.34), we find that for $n > N$,

$$\int_{\{|x|\geq\delta\}} f_n(x)\,|\phi(x) - \phi(0)|\,dx \;<\; \frac{\epsilon}{2}.$$

Bringing this together with (5.30) and (5.31), we have shown the following: For every $\epsilon > 0$, we can find (there exits) N such that

$$\text{if } n > N, \text{ then } \quad \left| \int_{-\infty}^{\infty} f_n(x)\,\phi(x)\,dx \;-\; \phi(0) \right| < \epsilon.$$

Note that the parameter δ was needed only as an intermediate variable in the course of the proof. □

This technique of splitting an integral into multiple parts and controlling each part to be less than the appropriate fraction of epsilon is quite common in analysis. The method of control (i.e., ensuring smallness) for each term can be distinct but must be mutually compatible. In this proof, our two methods of control were (a) continuity of the test function at $x = 0$ and (b) the precise notion that as n gets arbitrarily large, more and more of the unit mass of our functions gets pulled into the origin; in the previous proof, this entailed controlling the residual mass of the tails, $\pi/2 - \tan^{-1}(\delta/\sigma_n)$. These two methods were mutually compatible because after fixing a neighborhood of the origin of a given length 2δ in which we controlled the fluctuations of ϕ, we were able to control the residual mass at the tail determining the cutoff index, N, in terms of our window length δ.

In a very similar fashion (cf. Exercise **5.14**) one can prove that the sequence of Gaussians (5.25) converges to δ_0 in the sense of distributions as $n \to \infty$. One may ask why it is necessary to provide individual proofs for each particular sequence; why not just prove that the three properties of Properties 5.5.1 are sufficient for distributional convergence to the delta function? The issue here is that the third condition (concentration at 0) needs to be made precise. Exercise **5.15** addresses a general result in this direction. Another general result is the following theorem whose proof is left for Exercise **5.16**.

Theorem 5.3. *Let $f(x)$ be **any** nonnegative integrable function on $\mathbb{R}$ which integrates to one. For $n = 1, 2, \ldots$ define*

$$f_n(x) := n\,f(nx).$$

Then

$$f_n \longrightarrow \delta_0 \quad \textit{in the sense of distributions.}$$

5.5.5. • The Distributional Convergence of $\sin nx$. We will now present an example of a sequence of functions which converge in the sense of distributions to another function (in fact the function which is identically zero). What is new here is that this sequence of functions does **not** converge in any pointwise or regular sense that we have seen before. This is an important example and it is worth spending some time to digest it. Consider the sequence of functions

$$\boxed{f_n(x) = \sin(nx).}$$

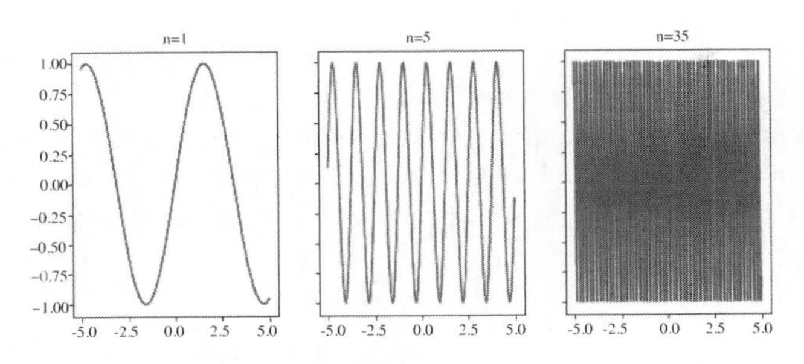

Figure 5.6. Plots of $\sin nx$ for $n = 1, 5, 35$.

What happens to f_n as $n \to \infty$? The functions oscillate between 1 and -1, increasingly as n gets larger (i.e., their period of oscillation tends to 0). In particular, except when x is an integer multiple of π, the sequence of numbers $f_n(x)$ does not converge[12]. So there exists no pointwise limit. On the other hand, what happens in the sense of distributions? In other words, if ϕ is a test function, what happens to

$$\int_{-\infty}^{\infty} \sin(nx)\, \phi(x)\, dx? \tag{5.35}$$

Plots of $\sin nx$ for $n = 1, 5, 35$ are given in Figure 5.6. What is important to note is that there are two fundamental properties of $\sin(nx)$ which stem from the mother function $\sin x$:

- **Periodicity:** The function $\sin x$ is periodic with period 2π. Hence, the function $\sin(nx)$ has period $\frac{2\pi}{n}$. As $n \to \infty$, this period tends to 0 and, hence, the larger the value of n the more oscillations of $\sin(nx)$ per unit length in x.
- **Averaging out over a period:** The function $\sin x$ integrates to 0 ("averages out") over any interval of length 2π. Note that this is not only for, say, the interval $[0, 2\pi]$, but for any interval $[a, b]$ with $b - a = 2\pi$. Hence, the function $\sin(nx)$ integrates to 0 (averages out) over any interval of size $\frac{2\pi}{n}$.

So in (5.35), we are multiplying a function, which oscillates more and more rapidly between 1 and -1, by ϕ and then integrating. For very large n, think of blowing up on the scale of the period $\frac{2\pi}{n}$, and then focus for a moment on one such interval. The continuous function ϕ is essentially constant on this tiny interval (period); hence, if we integrate $\phi(x)\sin nx$ over this interval, we are effectively integrating a constant times $\sin nx$ over a period interval, where the end result is "almost" *zero*. The full integral (5.35) can be viewed as a sum over all the period intervals and, consequently, as $n \to \infty$, (5.35) "should" tend to 0.

Does this sound convincing? Well, be careful! Yes, it is true that the integrals over the smaller and smaller periods will get closer to 0 but there are more and more of them. If this argument convinced you that the sum of all of them should also tend to 0, then we fear that we have convinced you that every Riemann integral is zero! In the

[12]Test this on your computer: Compute $\sin n$ (so $x = 1$) for $n = 1, \ldots, 1{,}000$ and see if you detect convergence.

present case, with ϕ smooth and compactly supported, one can quantify the analysis more precisely by estimating the size of the integral over each period subinterval of size $\frac{2\pi}{n}$ versus the number of such small intervals (essentially, a constant times n). While this calculation would indeed yield that the full integral (5.35) tends to 0 as $n \to \infty$, here we provide a very quick and slick proof which only uses integration by parts!

Theorem 5.4.

$$\boxed{\sin(nx) \xrightarrow{n\to\infty} 0 \quad \textit{in the sense of distributions.}}$$

Note that the same distributional convergence holds for the sequence $\cos(nx)$.

Proof. Fix $\phi \in C_c^\infty(\mathbb{R})$. Our goal is to prove that for every $\epsilon > 0$, there exists N such that if $n \geq N$, then

$$\left| \int_{-\infty}^{\infty} \sin(nx)\, \phi(x)\, dx \right| < \epsilon.$$

We first note that since ϕ has compact support, there exists some $L > 0$ such that

$$\left| \int_{-\infty}^{\infty} \sin(nx)\, \phi(x)\, dx \right| = \left| \int_{-L}^{L} \sin(nx)\, \phi(x)\, dx \right|,$$

with $\phi(\pm L) = 0$. Let $C = \max_{x\in[-L,L]} |\phi'(x)|$. Then we have,

$$\begin{aligned}
\left| \int_{-L}^{L} \sin(nx)\, \phi(x)\, dx \right| &= \left| \int_{-L}^{L} \left(-\frac{\cos nx}{n} \right)' \phi(x)\, dx \right| \\
&\overset{\text{integrations by parts}}{=} \left| \frac{1}{n} \int_{-L}^{L} \cos nx\, \phi'(x)\, dx \right| \\
&\leq \frac{1}{n} \int_{-L}^{L} |\cos nx\, \phi'(x)|\, dx \\
&\leq \frac{1}{n} \int_{-L}^{L} |\phi'(x)|\, dx \leq \frac{2LC}{n}.
\end{aligned}$$

Thus we simply need to choose $N > \frac{2LC}{\epsilon}$ and we are done. □

In this proof, we conveniently exploited two facts about ϕ: compact support and C^1. However, the result is true for ϕ with far less restrictive assumptions and goes by the name of **the Riemann-Lebesgue Lemma**. It is connected with Fourier analysis (the Fourier transform and Fourier series) and will resurface in Chapters 6 and 11.

5.5.6. • The Distributional Convergence of the Sinc Functions and the Dirichlet Kernel: Two Sequences Directly Related to Fourier Analysis. Here we look at two more examples, without delving into their proofs, which are fundamentally related to the Fourier transform (Chapter 6) and to Fourier series (Chapter 11), respectively. They are, again, examples of a sequence of functions converging in the sense of distributions to the delta function; however, the functions now take both negative and positive values. They are, respectively, the (rescaled and normalized) **Sinc functions** and what is known as **the Dirichlet kernel**:

$$S_n(x) := \frac{\sin(nx)}{\pi x}, \qquad K_n(x) := \frac{\sin\left(\left(n+\frac{1}{2}\right)x\right)}{\sin\left(\frac{x}{2}\right)}.$$

While it is not immediately obvious, we will see later in Section 11.5.2 that for each n, $K_n(x)$ is a periodic function with period 2π. Also note that both S_n and K_n have a removable discontinuity at $x = 0$ (in the case of K_n, at all integer multiples of 2π). Hence redefining $S_n(0) = \frac{n}{\pi}$ and K_n at any integer multiple of 2π to be $1 + 2n$ yields smooth functions of $x \in \mathbb{R}$.

Let us begin with S_n, plots of which are presented in Figure 5.7. Again, we ask what happens as $n \to \infty$? With a bit of work, one can verify that for each n, the following holds:

$$\int_{-\infty}^{\infty} S_n(x)\,dx = 1. \tag{5.36}$$

However, in this instance, there is a major difference with our previous examples (5.22)–(5.25). The functions S_n are **no longer nonnegative** and, consequently, there are cancellation effects in (5.36). Figure 5.7 suggests that there is concentration near $x = 0$. Moving away from zero, the functions oscillate more and more but with an amplitude which tends to 0. With some mathematical analysis, one can indeed prove that

$$S_n(x) \xrightarrow{n\to\infty} \delta_0 \qquad \text{in the sense of distributions.}$$

We will encounter these sinc functions again in the next chapter, Chapter 6, on the Fourier transform. Moreover, Exercise **6.25** asks for a direct proof of the distributional convergence of the sinc functions S_n.

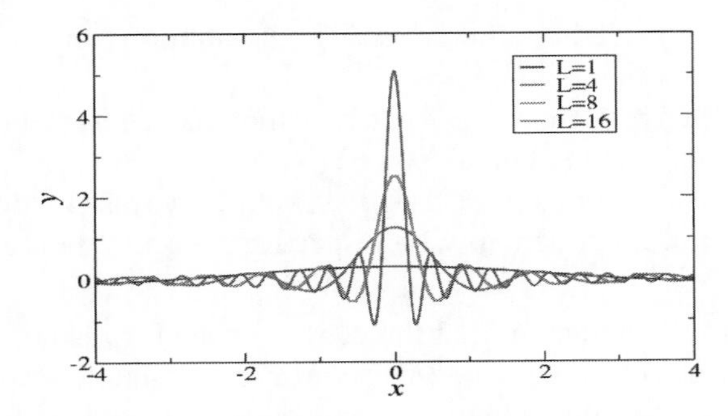

Figure 5.7. Sinc functions $S_n(x)$ for $n = L = 1, 4, 8, 16$. Note that the values at $x = 0$ are taken to be the $\lim_{x\to 0} S_n(x) = \frac{n}{\pi}$.

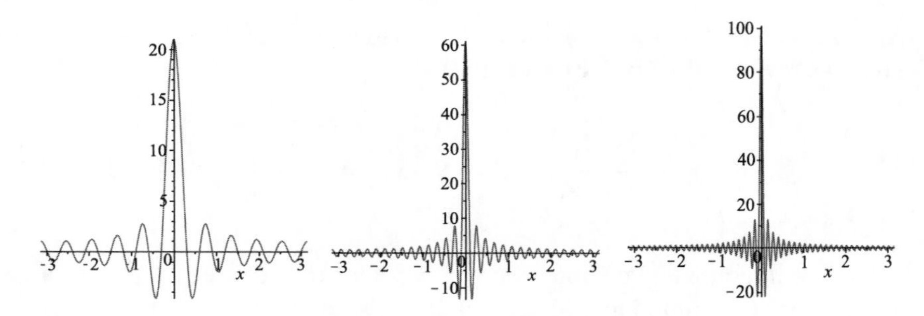

Figure 5.8. Plots of the Dirichlet kernel $K_n(x)$ on $[-\pi, \pi]$ for $n = 10, n = 30$, and $n = 50$ from left to right. Note that the values at $x = 0$ are taken to be the $\lim_{x\to 0} K_n(x) = 1 + 2n$.

The functions K_n are also no longer nonnegative and their convergence is even more subtle, entailing the focus of Section 11.5. First off, note that these functions are all periodic with period 2π, and hence we may restrict our attention to $x \in [-\pi, \pi]$. With a bit of simple trigonometry and algebra (cf. Section 11.5), one can verify that over any period, for each n the functions K_n integrate to 2π; i.e.,

$$\int_{-\pi}^{\pi} K_n(x)\, dx \;=\; 2\pi.$$

Figure 5.8 shows plots of K_n on $[-\pi, \pi]$ and, again, there appears to be concentration around $x = 0$ as n gets larger. As with S_n, there are increasing oscillations in the (here, finite) tails. However, the plots for K_n are deceiving, in that at first sight they seem to suggest that as $n \to \infty$, the amplitude of oscillation in the finite tails is tending to 0. This was the case for the infinite tails of S_n, but **not** for K_n. For K_n, the tail amplitudes appear to be tending to 0 because of the scaled y-axis. In fact, as n increases, the kernel's tail oscillates more and more frequently but at a fixed positive amplitude. As $n \to \infty$, these tails **do** tend to 0 in the sense of distributions (similar to the distributional convergence of $\sin(nx)$) but **not** in any pointwise sense. In the end, we have

$$\boxed{K_n(x) \xrightarrow{n\to\infty} 2\pi\delta_0 \quad \text{in the sense of distributions on } (-\pi, \pi),}$$

and this will be proven in Section 11.5. Note here that we are restricting our attention to test functions defined on the interval $(-\pi, \pi)$ (cf. Section 5.7.1). Instead, if we wanted the statement for test functions $\phi \in C_c^\infty(\mathbb{R})$, we would have distributional convergence to an infinite sum of delta functions with respective concentrations at $x = 2\pi m$ for all $m \in \mathbb{Z}$.

It is important to emphasize that the convergence of K_n is far more subtle than in **any** of the previous examples. Indeed, in the earlier examples, we only needed the test functions to be continuous. For K_n, continuity alone is insufficient for the test functions, and this fact is responsible for making the pointwise convergence analysis of Fourier series an amazingly difficult and delicate subject.

5.6. Dirac's Intuition: Algebraic Manipulations with the Delta Function

"Dirac is of course fully aware that the δ function is not a well-defined expression. But he is not troubled by this for two reasons. First, as long as one follows the rules governing the function (such as using the δ function only under an integral sign, meaning in part not asking the value of a δ function at a given point), then no inconsistencies will arise. Second, the δ function can be eliminated, meaning that it can be replaced with a well-defined mathematical expression. However, the drawback in that case is, according to Dirac, that the substitution leads to a more cumbersome expression that obscures the argument. In short, when pragmatics and rigor lead to the same conclusion, pragmatics trumps rigor due to the resulting simplicity, efficiency, and increase in understanding."

From the article by Fred Kronz and Tracy Lupher: *Quantum Theory and Mathematical Rigor*, The Stanford Encyclopedia of Philosophy (Fall 2019 Edition), Edward N. Zalta (ed.), `https://plato.stanford.edu/archives/fall2019/entries/qt-nvd/`.

We have gone to great lengths to emphasize why the delta function is not a function and agreed not to write expressions like "$\delta_0(x-a)$" to represent a delta function concentrated at $x = a$, but rather δ_a. However, we strongly feel that it is not contradictory to also claim that there is merit and insight in Dirac's informal approach, and there are times when "pragmatics trumps rigor". In the physics, engineering, and even mathematics literature, one often encounters informal calculations which treat the delta function as a true function with argument "(x)" of real numbers. With the correct guidance, such informal calculations will steer us to make correct conclusions. This will be the case, for example, with the higher-dimensional delta function and its use in finding the Green's function with Neumann boundary conditions for a ball (cf. Section 10.5.2) and in formulating Green's functions for the wave equation (cf. Sections 13.2.3 and 13.2.4). See also Section 9.7.3 and Exercise **10.22**.

In this section, we review a few informal algebraic calculations with delta functions based upon manipulations of the argument. We will show that, with the right guidance, these manipulations are justifiable and, hence, acceptable (or, at the very least, tolerable). We repeat that, in this text, **we will place within quotation marks all such informal expressions**.

5.6.1. Rescaling and Composition with Polynomials.

Example 1: Consider first the following:

$$\text{“}\delta_0(2x)\text{”}.$$

While we will **not** give this expression **precise meaning as a distribution**, let us at least try to make some sense of it. The delta function is basically unitary mass concentration at $x = 0$. Since x and $2x$ are both 0 at $x = 0$, does that mean "$\delta_0(2x) = \delta_0(x)$"?

No, this type of pointwise reasoning will get us nowhere with the delta function. In fact, we will now argue that

$$\text{“}\delta_0(2x) = \frac{1}{2}\delta_0(x)\text{”}. \tag{5.37}$$

What is important to remember is that the delta function concentrates *unit mass* at a point and **mass** is captured by **integration**. The best way to proceed is to follow Dirac's lead (cf. the quote at the end of Section 5.2.2) and to view the delta function as a limit (in the sense of distributions) of a sequence of nonnegative functions f_n such that the following hold:

- For all n, $\int_{-\infty}^{\infty} f_n(x)\,dx = 1$.
- The supports of the f_n (i.e., the set of x in the domain for which $f(x) > 0$) get smaller and smaller as $n \to \infty$ and concentrate about 0.

Fix one of these sequences, say, (5.10), and let us **think of** the delta function "$\delta_0(x)$" as the true function $f_n(x)$ for n very large. We can then address "$\delta_0(2x)$" by considering $f_n(2x)$ for n very large; more precisely, what does $f_n(2x)$ converge to in the sense of distributions as $n \to \infty$? Let $\phi \in C_c^\infty(\mathbb{R})$ be a test function. Then by letting $y = 2x$, we have

$$\int_{-\infty}^{\infty} f_n(2x)\,\phi(x)\,dx = \frac{1}{2}\int_{-\infty}^{\infty} f_n(y)\,\phi(y/2)\,dy.$$

Since $f_n \to \delta_0$ in the sense of distributions and noting that $\psi(y) := \phi(y/2)$ is also a test function in $C_c^\infty(\mathbb{R})$, we have

$$\frac{1}{2}\int_{-\infty}^{\infty} f_n(y)\,\psi(y)\,dy \longrightarrow \frac{1}{2}\psi(0) \quad \text{as} \quad n \to \infty.$$

Since $\frac{1}{2}\psi(0) = \frac{1}{2}\phi(0)$, we have shown that for any test function ϕ,

$$\int_{-\infty}^{\infty} f_n(2x)\,\phi(x)\,dx \longrightarrow \frac{1}{2}\phi(0) \quad \text{as} \quad n \to \infty.$$

But this means that $f_n(2x)$ for a very large n behaves like $\frac{1}{2}\delta_0$, and so informally (5.37) holds true.

We can generalize this observation with respect to the composition of a delta function with certain C^1 functions. To this end, let $g(x)$ be a C^1 function which is zero exactly at $x = a$ with $g'(a) \neq 0$. Then we claim,

$$\text{“}\delta_0(g(x)) = \frac{1}{|g'(a)|}\delta_0(x-a)\text{”}. \tag{5.38}$$

The details follow exactly the same steps as above for $g(x) = 2x$ and are left as an exercise (Exercise **5.22**). The assumption that $g'(a) \neq 0$, which insured that g was one-to-one in a neighborhood of a, is critical in order to make sense of "$\delta_0(g(x))$". For example, we **cannot** make sense of "$\delta_0(x^2)$". Try to!

Example 2: Let us now justify the following informal equality for $a < b$:

$$\text{“}\delta_0[(x-a)(x-b)] = \frac{1}{|a-b|}\left(\delta_0(x-a) + \delta_0(x-b)\right)\text{”}.$$

To this end, consider any sequence $f_n(\cdot)$ which converges to δ_0 in the sense of distributions. Our goal is to determine what $f_n((x-a)(x-b))$ converges to in the sense of distributions. This means we need to determine what happens to

$$\int_{-\infty}^{\infty} f_n((x-a)(x-b))\,\phi(x)\,dx,$$

as $n \to \infty$ for any test function ϕ. In particular, we want to show it tends to

$$\frac{\phi(a)}{|a-b|} + \frac{\phi(b)}{|a-b|}.$$

If we let $y = (x-a)(x-b)$, we have an issue here because unlike with the example "$\delta_0(2x)$", we cannot solve uniquely for x as a function of y. Let $x = c$ be the point at which the curve

$$y(x) = (x-a)(x-b)$$

attains its minimum (sketch this curve). On either interval $x \in (-\infty, c)$ or (c, ∞), we can solve uniquely for x as a function of y. Let us call these two branches $x_1(y)$ and $x_2(y)$, respectively. Since $f_n(y)$ will concentrate its support (mass) when y is close to 0, $f_n(y(x))$ on $x \in (-\infty, c)$ will concentrate around $x = a$. Similarly, $f_n(y(x))$ on $x \in (c, \infty)$ will concentrate around $x = b$.

We now have

$$\begin{aligned}\int_{-\infty}^{\infty} f_n((x-a)(x-b))\,\phi(x)\,dx &= \int_{-\infty}^{c} f_n((x-a)(x-b))\,\phi(x)\,dx \\ &\quad + \int_{c}^{\infty} f_n((x-a)(x-b))\,\phi(x)\,dx.\end{aligned}$$

With the substitution $y = (x-a)(x-b)$, $dy = (2x-(a+b))dx$ in each of the two integrals, we focus on the first integral to find that

$$\begin{aligned}\int_{-\infty}^{c} f_n((x-a)(x-b))\,\phi(x)\,dx &= \int_{+\infty}^{y(c)} f_n(y)\frac{\phi(x_1(y))}{2x_1(y)-(a+b)}\,dy \\ &= -\int_{y(c)}^{+\infty} f_n(y)\frac{\phi(x_1(y))}{2x_1(y)-(a+b)}\,dy.\end{aligned}$$

Noting that $y(c) < 0$ and, hence, viewing

$$\frac{\phi(x_1(y))}{2x_1(y)-(a+b)}$$

as a test function in y, the effect of the integral as $n \to \infty$ will be to pick out the value of the test function at $y = 0$. Since by definition $x_1(0) = a$, this indicates that the integral will approach

$$-\frac{\phi(a)}{a-b} = \frac{\phi(a)}{b-a} = \frac{\phi(a)}{|a-b|}.$$

The analogous argument on the second integral yields the convergence to $\frac{\phi(b)}{|a-b|}$.

5.6.2. Products of Delta Functions in Different Variables. It is also commonplace to use **products** of delta functions in different variables[13] in order to denote the delta function in more than one variable. For example, in two dimensions we can define (cf. Chapter 9) the delta function $\delta_{\mathbf{0}}$ where $\mathbf{0} = (0,0)$ over test functions ϕ on $\mathbb{R}^2$ by

$$\langle \delta_{\mathbf{0}}, \phi(x,y)\rangle = \phi(0,0) \qquad \text{for all } \phi \in C_c^\infty(\mathbb{R}^2).$$

The distribution $\delta_{\mathbf{0}}$ is often written as

$$\text{“}\delta_{\mathbf{0}} = \delta_0(x)\,\delta_0(y)\text{”}. \tag{5.39}$$

In Exercise **9.2** we will present a precise definition of such **direct products of distributions**. Here, let us just argue that (5.39) is *intuitively correct.*

Let $f_n(x)$ be the particular approximation of δ_0 given by (5.10). We also consider $f_n(y)$ as an approximation of δ_0 in the y variable. We trivially extend these two sequences of functions on $\mathbb{R}$ to sequence of functions on $\mathbb{R}^2$ by

$$g_n(x,y) := f_n(x) \qquad \text{and} \qquad h_n(x,y) := f_n(y).$$

Now, consider the product sequence

$$g_n(x,y)h_n(x,y).$$

By plotting this product function for $n = 10, 50, 100$, you should be able to convince yourself that $g_n(x,y)h_n(x,y) \longrightarrow \delta_{\mathbf{0}}$ in the sense of distributions. Hence, the product is a good approximation to $\delta_{\mathbf{0}}$, the 2D delta function.

In Section 9.7.3, we will solve the 1D wave equation with an instantaneous source concentrated at $x = 0$ and $t = 0$. This equation is often written in the informal, but customary, form

$$\text{“}u_{tt} - cu_{xx} = \delta(x)\delta(t)\text{”}.$$

5.6.3. Symmetry in the Argument. It is also common, and often quite useful, to invoke intuitive symmetry properties of the delta function. For example, one often encounters equations such as

$$\text{“}\delta_0(x) = \delta_0(-x)\text{”} \qquad \text{and} \qquad \text{“}\delta_0(x-y) = \delta_0(y-x)\text{”}.$$

Since, loosely speaking, the delta function is 0 except when the argument is 0, these two equations seem plausible. Let us focus on the first equation, "$\delta_0(x) = \delta_0(-x)$". First note that this equation is justified by viewing the delta function as a limit of functions f_n (for example (5.10)) since each of the functions f_n is itself symmetric. Hence we would expect that symmetry is preserved in any reasonable limit. However, unlike in the previous examples, we can make complete sense of this equality in the sense of distributions. To this end, given any test function ϕ, note that

$$\phi^-(x) := \phi(-x)$$

is also a test function. For any distribution F, we define the distribution F^- as

$$\langle F^-, \phi\rangle := \langle F, \phi^-\rangle.$$

[13]Note that we are **not** considering products of distributions over **the same** underlying independent variable. In general, we **cannot** make sense of such products.

As with differentiation, this presents a reasonable definition if it is the case that for any integrable function f,

$$F_{f^-} = (F_f)^-.$$

You can readily verify that this is indeed the case.

With these definitions, one immediately sees that the informal equation "$\delta_0(x) = \delta_0(-x)$" is simply a way of expressing the fact that

$$\text{if } F = \delta_0, \text{ then } \quad F = F^-.$$

5.7. • Distributions Defined on an Open Interval and Larger Classes of Test Functions

In this section we briefly discuss two natural extensions to the theory of distributions.

5.7.1. • Distributions Defined over a Domain. We often encounter problems in which the independent variables do not lie in all of $\mathbb{R}$ but rather some domain Ω in $\mathbb{R}$, i.e., an open interval $(-r, r)$ for some $r > 0$. To this end it is useful to consider distributions where the underlying space of the test functions is not $\mathbb{R}$ but, rather, some open interval $(-r, r)$. In these cases the test functions ϕ must have compact support in $(-r, r)$, and we denote this space as $C_c^\infty((-r, r))$.

But what exactly does compact support in $(-r, r)$ mean? We mean a C^∞ function defined on $\mathbb{R}$ (yes, $\mathbb{R}$) for which there exists a (proper) subset $K \subset (-r, r)$ which is bounded and closed in $\mathbb{R}$ such that

$$\phi(x) = 0 \qquad \text{for all } x \notin K.$$

So certainly such a test function must vanish (i.e., be 0) outside of the interval $(-r, r)$ and at the two boundary points $x = \pm r$. However more is true; as one approaches any boundary point, ϕ must be zero before reaching the boundary, that is, at some nonzero distance from the boundary. So if ϕ_a is the blip function defined by (5.13), then $\phi_a \in C_c^\infty((-r, r))$ if $a < r$ (but not if $a = r$). Loosely speaking, think of an end point as a point which we never "reach" or "touch" with our test functions.

We give one example without proof for $\Omega = (-\pi, \pi)$.

Example 5.7.1 (An Infinite Series). Recall that an infinite series of functions is precisely a limit of partial (finite) sums of the functions. Let $x \in \Omega = (-\pi, \pi)$. Then

$$\sum_{k \text{ is odd}} \frac{2}{\pi} \cos(kx) = \delta_0 \quad \text{in the sense of distributions on } \Omega. \tag{5.40}$$

This means that if we define the sequence of functions

$$f_n(x) = \sum_{k=0}^{n} \frac{2}{\pi} \cos\big((2k+1)x\big),$$

then as $n \to \infty$, we have

$$\int_{-\pi}^{\pi} f_n(x)\,\phi(x)\,dx \longrightarrow \phi(0) \qquad \text{for all } \phi \in C_c^\infty((-\pi, \pi)).$$

5.7.2. • Larger Classes of Test Functions. A statement in the sense of distributions involves a class of test functions which, so far, have been conveniently taken to be infinitely smooth localized functions. However, it is not always the case that we require the infinite smoothness or even the compactness of the support and, therefore, we can often **extend the statement** to a **wider (larger)** class of test functions. This means that the statement (i.e., the equality) holds for more general test functions than just C_c^∞. Note that the **larger** the class of test functions, the **stronger** (meaning the more general) the statement becomes. We give a few examples:

(i) We can define the delta function with test functions ϕ which are simply continuous functions, not necessarily smooth and not necessarily with compact support. That is, $\langle \delta_0, \phi \rangle = \phi(0)$ for all continuous functions ϕ.

(ii) The statement that $H'(x) = \delta_0$ in the sense of distributions (cf. Example 5.4.1) means that for every $\phi \in C_c^\infty(\mathbb{R})$,

$$-\int_{-\infty}^{\infty} H(x)\,\phi'(x)\,dx \;=\; \phi(0).$$

This actually holds true for all ϕ which are simply C^1 with compact support. In fact, it would also hold for any C^1 function which is integrable.

(iii) The statement that f_n, defined by (5.10), converges to δ_0 in the sense of distributions holds for all continuous functions ϕ; i.e., (5.28) holds for all continuous functions ϕ.

(iv) The sequence of functions $\sin nx$ converges to 0 in the sense of distributions. However, it turns out that the test functions can be taken to be just **integrable** (not necessarily continuous or possessing any derivatives). Precisely, we have

$$\int_{-\infty}^{\infty} \sin(nx)\,\phi(x)\,dx \;\longrightarrow\; 0,$$

for any function ϕ, such that $\int_{-\infty}^{\infty} |\phi(x)|\,dx \;<\; \infty$. This is an important result in Fourier analysis and is known as the **Riemann-Lebesgue Lemma**.

(v) In the next chapter we will see that, in order to extend the Fourier transform to distributions, we need a larger class of distributions stemming from a larger class of test functions. These test functions will still be C^∞ but we will relax the condition on compact support to the point where we only require the functions to decay *rapidly* to 0 as $|x| \to \infty$. This larger class of test functions is called the **Schwartz class** and the distributions, which can act not only on C_c^∞ test functions but also on Schwartz functions, are called **tempered distributions**.

5.8. Nonlocally Integrable Functions as Distributions: The Distribution PV $\frac{1}{x}$

This section is optional and should be skipped on first reading.

Throughout this text we will primarily be interested in distributions that are either generated by locally integrable functions or involve delta functions. Yet, there are important distributions which are neither. As we shall see, their Fourier transforms are very useful in solving many PDEs and one such distribution, called the principle value of $\frac{1}{x}$, is the building block of the **Hilbert transform**, which plays a central role in harmonic analysis. Consider the function

$$f(x) = \frac{1}{x}$$

which is **not** locally integrable around $x = 0$; that is, $\int_I \frac{1}{x}\, dx$ diverges for any interval I whose closure contains 0. We cannot directly interpret the function f as a distribution; however, there are several distributions which encapsulate the essence of this function. The key here is the idea of approximating the nonlocally integrable function $1/x$ by a sequence of functions in such a way as to exploit cancellation effects; we call this process **regularization**.

5.8.1. Three Distributions Associated with the Function $\frac{1}{x}$.

We discuss three ways to regularize $1/x$, each giving rise to a slightly different distribution.

1. **The principle value (PV)**. Because the function $\frac{1}{x}$ is odd, one can **exploit the cancellation effects** and define the distribution PV $\frac{1}{x}$ as follows: For any $\phi \in C_c^\infty(\mathbb{R})$,

$$\boxed{\left\langle \text{PV}\,\frac{1}{x}, \phi \right\rangle := \lim_{\epsilon \to 0+} \int_{\{x \,|\, |x| > \epsilon\}} \frac{1}{x}\,\phi(x)\,dx.} \tag{5.41}$$

One can readily verify that this definition gives PV $\frac{1}{x}$ the status of a distribution. Note that we are defining this distribution as a limit of distributions stemming from the truncated functions:

$$\left\langle \text{PV}\,\frac{1}{x}, \phi \right\rangle = \lim_{n \to \infty} \int_{-\infty}^{\infty} f_n(x)\,\phi(x)\,dx,$$

where for $\sigma_n = \frac{1}{n}$,

$$f_n = \begin{cases} \frac{1}{x} & \text{if } |x| > \sigma_n, \\ 0 & \text{if } |x| \le \sigma_n. \end{cases} \tag{5.42}$$

In other words, PV $\frac{1}{x}$ is the distribution limit of the functions f_n viewed as distributions.

2. The distribution

$$\frac{1}{x + i0}$$

is defined by

$$\boxed{\left\langle \frac{1}{x + i0}, \phi \right\rangle := \lim_{\epsilon \to 0+} \int_{-\infty}^{\infty} \frac{1}{x + i\epsilon}\,\phi(x)\,dx.} \tag{5.43}$$

3. The distribution
$$\frac{1}{x-i0}$$
is defined by
$$\boxed{\left\langle \frac{1}{x-i0}, \phi \right\rangle := \lim_{\epsilon\to 0+} \int_{-\infty}^{\infty} \frac{1}{x-i\epsilon}\,\phi(x)\,dx.} \tag{5.44}$$

Note the appearance of the complex number $i = \sqrt{-1}$ in the above distributions 2 and 3. This imaginary number demonstrates its utility by translating the singularity away from the real axis of integration. These two distributions are hence complex-valued distributions and, as we will see, important in the Fourier transform of certain functions.

While all three distributions
$$\mathrm{PV}\,\frac{1}{x}, \qquad \frac{1}{x+i0}, \qquad \frac{1}{x-i0}$$
are related to the function $1/x$, they should all be understood only as distributions and **not** as functions in any pointwise sense. They represent different distributions; indeed, we will prove shortly that in the sense of distributions[14],
$$\boxed{\frac{1}{x+i0} = \mathrm{PV}\,\frac{1}{x} - i\pi\delta_0 \quad \text{and} \quad \frac{1}{x-i0} = \mathrm{PV}\,\frac{1}{x} + i\pi\delta_0,} \tag{5.45}$$
and, hence,
$$\mathrm{PV}\,\frac{1}{x} = \frac{1}{2}\left(\frac{1}{x+i0} + \frac{1}{x-i0}\right).$$
Surely you are curious about the appearance of the delta function in (5.45). It comes from the behavior of the (nonintegrable) function $1/x$ around $x = 0$. We will prove (5.45) shortly; it will be a simple consequence of another possible way (sequence) to regularize $1/x$ and define the same distribution $\mathrm{PV}\,\frac{1}{x}$.

5.8.2. A Second Way to Write $\mathrm{PV}\,\frac{1}{x}$.

Consider the function
$$g(x) = \begin{cases} \log|x|, & x \neq 0, \\ 0, & x = 0. \end{cases}$$
This function is locally integrable (indeed, it has a finite integral about 0) and, therefore, one can consider it in the sense of distributions (i.e., as the distribution F_g). What is its derivative in the sense of distributions? The answer cannot be the distribution generated by $1/x$; the function is not locally integrable and, hence, cannot be directly viewed as a distribution.

We shall now see that
$$(\log|x|)' = \mathrm{PV}\,\frac{1}{x} \quad \text{in the sense of distributions} \tag{5.46}$$

[14]These equations are related to what is known as the Sokhotski-Plemelj Theorem in complex analysis.

and, in doing so, we will provide an alternate but equivalent definition of PV $\frac{1}{x}$. Consider the sequence of functions, where as usual $\sigma_n = \frac{1}{n}$,

$$g_n(x) := \begin{cases} \log|x| & \text{if } |x| > \sigma_n, \\ 0 & \text{if } |x| \le \sigma_n. \end{cases}$$

Then by Proposition 5.5.2, we have

$$g = \lim_{n\to\infty} g_n \quad \text{in the sense of distributions.} \tag{5.47}$$

This means that for all $\phi \in C_c^\infty(\mathbb{R})$,

$$\int_{-\infty}^{\infty} g_n(x)\,\phi(x)\,dx \longrightarrow \int_{-\infty}^{\infty} g(x)\,\phi(x)\,dx \qquad \text{as } n\to\infty. \tag{5.48}$$

With (5.47) in hand, it follows from Proposition 5.5.1 that g_n' converges in the sense of distributions to g'. However, g_n' as a distribution is simply f_n, defined in (5.42), as a distribution. Thus, for any test function ϕ,

$$\langle (\log|x|)', \phi\rangle = \lim_{n\to\infty} \int_{-\infty}^{\infty} f_n(x)\,\phi(x)\,dx.$$

Hence, by definition of PV $\frac{1}{x}$, (5.46) holds true.

Now here is a trick: We can repeat the previous argument using a different approximating sequence:

$$g_n(x) = \log\sqrt{x^2+\sigma_n^2}.$$

Again, by Proposition 5.5.2 , this sequence converges in the sense of distributions to $\log|x|$. Hence, its derivative

$$\frac{x}{x^2+\sigma_n^2}$$

must also converge in the sense of distributions to $(\log|x|)'$. We have thus given a different way of writing PV $\frac{1}{x}$; namely it is the distributional limit

$$\mathrm{PV}\,\frac{1}{x} = \lim_{n\to\infty} \frac{x}{x^2+\sigma_n^2}.$$

This means that

$$\boxed{\left\langle \mathrm{PV}\,\frac{1}{x}, \phi\right\rangle := \lim_{n\to\infty} \int_{-\infty}^{\infty} \frac{x}{x^2+\sigma_n^2}\,\phi(x)\,dx.} \tag{5.49}$$

With (5.49) in hand, we can now prove (5.45).

Proof of (5.45). First note that for any $x\in\mathbb{R}$ and $\sigma_n = \frac{1}{n}$,

$$\frac{1}{x+i\sigma_n} = \left(\frac{1}{x+i\sigma_n}\right)\left(\frac{x-i\sigma_n}{x-i\sigma_n}\right) = \frac{x}{x^2+\sigma_n^2} - \frac{i\sigma_n}{x^2+\sigma_n^2}.$$

Hence, by (5.43) and (5.49), for any test function ϕ we have that

$$\begin{aligned}\left\langle \frac{1}{x+i0}, \phi \right\rangle &:= \lim_{n\to\infty} \int_{-\infty}^{\infty} \frac{1}{x+i\sigma_n}\, \phi(x)\, dx \\ &= \lim_{n\to\infty} \int_{-\infty}^{\infty} \frac{x}{x^2+\sigma_n^2}\, \phi(x)\, dx \;-\; \lim_{n\to\infty} \int_{-\infty}^{\infty} \frac{i\sigma_n}{x^2+\sigma_n^2}\, \phi(x)\, dx \\ &= \left\langle \mathrm{PV}\,\frac{1}{x}, \phi \right\rangle \;-\; \lim_{n\to\infty} \int_{-\infty}^{\infty} \frac{i\sigma_n}{x^2+\sigma_n^2}\, \phi(x)\, dx.\end{aligned}$$

Now recall the sequence (5.24) from Section 5.5.3

$$\frac{1}{\pi}\, \frac{\sigma_n}{x^2+\sigma_n^2}$$

which converged in the sense of distributions to δ_0 as $n \to \infty$. Since

$$\frac{i\sigma_n}{x^2+\sigma_n^2} = i\pi \left(\frac{1}{\pi}\, \frac{\sigma_n}{x^2+\sigma_n^2} \right),$$

we have

$$\left\langle \frac{1}{x+i0}, \phi \right\rangle = \left\langle \mathrm{PV}\,\frac{1}{x}, \phi \right\rangle \;-\; i\pi\phi(0) \qquad \text{for all } \phi \in C_c^\infty.$$

This proves the first equation in (5.45). In a similar fashion, one proves the second equation.

5.8.3. A Third Way to Write PV $\frac{1}{x}$.

There is yet another way to write the distribution PV $\frac{1}{x}$, which highlights the issue surrounding the loss of integrability at $x = 0$. First, we show that if the test function ϕ happened to satisfy $\phi(0) = 0$, then we would simply have

$$\left\langle \mathrm{PV}\,\frac{1}{x}, \phi \right\rangle = \int_{-\infty}^{\infty} \frac{1}{x}\, \phi(x)\, dx.$$

To see this, recall from calculus the Mean Value Theorem, which applied to ϕ says that

$$\phi(x) = \phi(x) - \phi(0) = \phi'(\eta)x, \qquad \text{for some } \eta \in (0, x).$$

Moreover, since $\phi \in C_c^\infty$, there exists an L such that ϕ and ϕ' vanish outside the interval $[-L, L]$, and ϕ' is bounded by some constant C on $[-L, L]$. Then the integrand in the definition

$$\left\langle \mathrm{PV}\,\frac{1}{x}, \phi \right\rangle := \lim_{\epsilon\to 0+} \int_{\{x\,|\,|x|>\epsilon\}} \frac{1}{x}\, \phi(x)\, dx$$

is bounded; that is,

$$\frac{1}{x}\phi(x) = \frac{1}{x}\phi'(\eta)x = \phi'(\eta) \le C.$$

Thus, the integrand is integrable around 0 and there is no need to take the limit over the truncations on the set $\{x \,|\, |x| > \epsilon\}$.

The issue with turning $1/x$ into a distribution pertains to the behavior of the test function at $x = 0$. Hence, one might wonder what happens if we replace the test function ϕ with $\phi(x) - \phi(0)$. We now show that we indeed get the same distribution; that is,

$$\boxed{\left\langle \text{PV}\,\frac{1}{x}, \phi \right\rangle = \int_{-\infty}^{\infty} \frac{1}{x}\,(\phi(x) - \phi(0))\,dx \quad \text{for all } \phi \in C_c^\infty.} \tag{5.50}$$

So we have yet another way to write the distribution PV $\frac{1}{x}$.

To prove (5.50), fix $\phi \in C_c^\infty$. For any $\epsilon > 0$ and any $L > \epsilon$, we have

$$\begin{aligned}\int_{\{x\,|\,|x|>\epsilon\}} \frac{1}{x}\,\phi(x)\,dx &= \int_{\{x\,|\,|x|>L\}} \frac{1}{x}\,\phi(x)\,dx + \int_{\{x\,|\,\epsilon<|x|<L\}} \frac{1}{x}\,(\phi(x)-\phi(0))\,dx \\ &\quad + \int_{\{x\,|\,\epsilon<|x|<L\}} \frac{1}{x}\,\phi(0)\,dx.\end{aligned}$$

For any ϵ and L, the third integral on the right is always 0 since $1/x$ is an odd function. The previous argument using the Mean Value Theorem implies that

$$\lim_{\epsilon\to 0+} \int_{\{x\,|\,\epsilon<|x|<L\}} \frac{1}{x}\,(\phi(x)-\phi(0))\,dx = \int_{-L}^{L} \frac{1}{x}\,(\phi(x)-\phi(0))\,dx.$$

Thus we have for any $L > 0$,

$$\begin{aligned}\left\langle \text{PV}\,\frac{1}{x}, \phi \right\rangle &:= \lim_{\epsilon\to 0+} \int_{\{x\,|\,|x|>\epsilon\}} \frac{1}{x}\,\phi(x)\,dx \\ &= \int_{\{x\,|\,|x|>L\}} \frac{1}{x}\,\phi(x)\,dx + \int_{-L}^{L} \frac{1}{x}\,(\phi(x)-\phi(0))\,dx.\end{aligned}$$

If we choose L large enough so that $[-L, L]$ encompasses the support of ϕ, then the first integral above vanishes and we arrive at (5.50).

The presence of $\phi(0)$ in (5.50) is suggestive of a delta function. Informally this is the case, but with the caveat that the delta function comes with an infinite constant factor! Indeed, if we let

$$\text{“}C_\infty = \int_{-\infty}^{\infty} \frac{1}{x}\,dx\text{”}$$

denote the “infinite” constant, then informally (5.50) says

$$\text{“PV}\,\frac{1}{x} = \frac{1}{x} - C_\infty \delta_0\text{”}.$$

What we are witnessing here is a particular cancellation of the infinite behavior (surrounding $x = 0$) and this is made precise by the right-hand side of (5.50).

5.8.4. The Distributional Derivative of PV $\frac{1}{x}$.

We end this section by computing

$$\left(\text{PV}\,\frac{1}{x}\right)' \quad \text{in the sense of distributions.}$$

One would expect that the distributional derivative has something to do with the function $-\frac{1}{x^2}$. However, we should be careful. As with the function $\frac{1}{x}$, $\frac{1}{x^2}$ is not locally integrable and hence does not directly generate a distribution. Moreover, unlike with the odd function $\frac{1}{x}$, there is no cancellation effect for x negative and positive and hence, we cannot define a principal value in a similar way to the regularizations used to define PV $\frac{1}{x}$ in (5.41) and (5.49); they were all based on the fact that the function in question was odd. The key is to cancel off the infinite behavior around $x = 0$, as was seen in the third definition (5.50).

Consider the approximating sequence of truncated functions f_n defined in (5.42), which converge in the sense of distributions to PV $\frac{1}{x}$. By Proposition 5.5.1, we must have $f_n' \longrightarrow \left(\text{PV}\,\frac{1}{x}\right)'$ in the sense of distributions. In other words, for every $\phi \in C_c^\infty(\mathbb{R})$,

$$\left\langle \left(\text{PV}\,\frac{1}{x}\right)', \phi \right\rangle = \lim_{n\to\infty} \langle f_n', \phi\rangle.$$

We find via integration by parts that

$$\begin{aligned} \langle f_n', \phi\rangle &= -\langle f_n, \phi'\rangle \\ &= -\int_{\{x\,|\,|x|>\sigma_n\}} \frac{1}{x}\,\phi'(x)\,dx \\ &\overset{\substack{\text{integration}\\ \text{by parts}}}{=} \int_{\{x\,|\,|x|>\sigma_n\}} -\frac{1}{x^2}\,\phi(x)\,dx + \frac{\phi(\sigma_n)+\phi(-\sigma_n)}{\sigma_n} \\ &= \int_{\{x\,|\,|x|>\sigma_n\}} -\frac{1}{x^2}\,\phi(x)\,dx + \frac{2\phi(0)}{\sigma_n} \\ &\qquad + \left(\frac{\phi(\sigma_n)+\phi(-\sigma_n)-2\phi(0)}{\sigma_n}\right). \end{aligned}$$

Now, we take the limit as $n \to \infty$, to find

$$\begin{aligned} \left\langle \left(\text{PV}\,\frac{1}{x}\right)', \phi \right\rangle &= \lim_{n\to\infty} \langle f_n', \phi\rangle \\ &= \lim_{n\to\infty} \left(\int_{\{x\,|\,|x|>\sigma_n\}} -\frac{1}{x^2}\,\phi(x)\,dx + \frac{2\phi(0)}{\sigma_n} \right) \\ &\qquad + \lim_{n\to\infty} \left(\frac{\phi(\sigma_n)+\phi(-\sigma_n)-2\phi(0)}{\sigma_n}\right). \end{aligned}$$

But, by Taylor's Theorem applied to the C^∞ function ϕ, the second limit[15] is 0. For the first limit, note that

$$\begin{aligned} \int_{\{x\,|\,|x|>\sigma_n\}} \frac{1}{x^2}\phi(0)\,dx &= \phi(0)\left(\int_{-\infty}^{-\sigma_n} \frac{1}{x^2}\,dx + \int_{\sigma_n}^{\infty} \frac{1}{x^2}\,dx \right) \\ &= \phi(0)\left(\frac{1}{\sigma_n} + \frac{1}{\sigma_n}\right) = \frac{2\phi(0)}{\sigma_n}. \end{aligned}$$

Thus

$$\boxed{\left\langle \left(\text{PV}\,\frac{1}{x}\right)', \phi \right\rangle = \lim_{n\to\infty} \int_{\{x\,|\,|x|>\sigma_n\}} -\frac{1}{x^2}\left(\phi(x) - \phi(0)\right) dx.}$$

---

[15] By Taylor's Theorem, we have $\phi(\sigma_n)-\phi(0) = \sigma_n\phi'(0) + \frac{\sigma_n^2}{2}\phi''(0) + O(\sigma_n^3)$ and $\phi(-\sigma_n)-\phi(0) = -\sigma_n\phi'(0) + \frac{\sigma_n^2}{2}\phi''(0) + O(\sigma_n^3)$. Add these, divide by σ_n, and let $n \to \infty$.

This defines $\left(\text{PV}\,\frac{1}{x}\right)'$ as a distribution. Informally, we can write this as

$$\text{``}\left(\text{PV}\,\frac{1}{x}\right)' = -\frac{1}{x^2} + C_\infty \delta_0\text{''},$$

where C_∞ is the "infinite constant" "$C_\infty = \int_{-\infty}^{\infty} \frac{1}{x^2}\,dx$".

5.9. Chapter Summary

- Distributions are defined by **their effect (or action) on test functions**. While the choice of test functions can vary, we have focused on the convenient class of **infinitely smooth functions with compact support**. In this chapter we only considered test functions which depended on one variable and, as such, only addressed distributions in one space dimension.
- **The theory of distributions** allows us to both
  - interpret and analyze (i.e., do calculus with) classical functions, not in a pointwise input/output sense, but from the point of view of averaging and integration;
  - give precise meaning to, and find occurrences of, objects like δ_0, which cannot be captured by a pointwise-defined classical function.
- In general, there are many types of distributions with very complicated structures. For our purposes, we are concerned with three important classes:
  (1) Distributions **generated by locally integrable functions**: If $f(x)$ is a locally integrable function on $\mathbb{R}$, then f can be thought of as the distribution F_f where

  $$\langle F_f, \phi\rangle := \int_{-\infty}^{\infty} f(x)\,\phi(x)\,dx \quad \text{for any } \phi \in C_c^\infty(\mathbb{R}).$$

  When we speak of the function f in the sense of distributions we mean F_f.
  (2) Distributions which **concentrate at points**; for example the **delta function** δ_0, which is defined by $\langle \delta_0, \phi\rangle = \phi(0)$ for any test function ϕ.
  (3) Distributions related to **nonlocally integrable functions**; for example PV $\frac{1}{x}$ which was the focus of Section 5.8.

  The first two classes of distributions will prevail throughout the sequel of this text. The PV $\frac{1}{x}$ will be important in using the Fourier transform.
- **Convergence in the sense of distributions** allows us to capture and calculate the limiting behavior of functions which do the following:
  - **Concentrate** their "mass" at a point; for example, f_n defined by (5.10), which concentrates at $x = 0$ and converges in the sense of distributions to δ_0.
  - **Oscillate** more and more about a fixed number; for example, $f_n(x) = \sin nx$, whose values oscillate more and more about 0.

  As Dirac demonstrated, it is useful to *visualize* the delta function δ_0 with one of these sequences which concentrate at $x = 0$; one can think of δ_0 as f_n, for some very large value of n. With $\sigma_n = 1/n$, we discussed six different examples of such sequences of functions. They are, roughly ordered in terms of increasing complexity,

$$\text{(i) } f_n(x) = \begin{cases} \frac{n}{2}, & |x| \le \sigma_n, \\ 0, & \text{otherwise,} \end{cases} \qquad \text{(ii) } f_n(x) = \begin{cases} n - n^2 x & \text{if } \quad 0 < x < \sigma_n, \\ n - n^2 x & \text{if } \; -\sigma_n < x < 0, \\ 0 & \text{if } \quad |x| \ge \sigma_n, \end{cases}$$

$$\text{(iii) } f_n(x) = \frac{1}{\pi} \frac{\sigma_n}{x^2 + \sigma_n^2}, \qquad \text{(iv) } f_n(x) = \frac{1}{\sqrt{4\pi \sigma_n}} e^{-\frac{x^2}{4\sigma_n}},$$

$$\text{(v) } f_n(x) = \frac{\sin nx}{\pi x}, \qquad \text{(vi) } f_n(x) = \frac{\sin\left(n + \frac{1}{2}\right)x}{\sin\left(\frac{x}{2}\right)}.$$

The first two denote hats and spikes, respectively, and are best viewed in terms of their simple graphs (Figure 5.5). As we shall see, the sequence of Gaussians (iv) are key to solving the diffusion equation, and the sequences (v) and (vi) play a fundamental role in the Fourier transform and Fourier series, respectively. Sequence (vi) consists of periodic functions with period 2π, and hence their distributional convergence to δ_0 is restricted to test functions with compact support in $(-\pi, \pi)$.

- We can take the **derivative** of any distribution by considering the action of the respective derivative on the test function. Such a derivative is also a distribution over the same class of test functions. For our purposes, we are concerned with derivatives of **distributions generated by functions**, that is, derivatives of functions in the sense of distributions. Moreover, we are primarily concerned with differentiating functions which are smooth except for certain **singularities**. We focused on the differentiation of a piecewise smooth function of one variable with a jump discontinuity. In doing so, we saw **two possible components** for the distributional derivatives of these functions: functions and delta functions. Important examples where distributional derivatives of functions (derivatives in the sense of distributions) yield the delta function are (in 1D)

$$\frac{d}{dx} H(x) = \delta_0 \quad \text{and} \quad \frac{d^2}{dx^2}\left(\frac{|x|}{2}\right) = \delta_0, \quad \text{in the sense of distributions.}$$

In these examples there is no nonzero functional component to the distributional derivatives but, rather, only the delta function.

- In order to **find** derivatives in the sense of distributions of these functions with singularities, we do the following:
  (i) Fix a test function and write down the integral of the function multiplied by the derivative of the test function.
  (ii) **Isolate** the singularity (singularities) by breaking up the domain in $\mathbb{R}$ into disjoint intervals.
  (iii) On regions of integration where the original function is smooth, we perform **integration by parts** to place the derivatives back on the original function, at the expense of additional boundary terms.

Exercises

5.1 (a) Show that the test function (5.13) is in $C_c^\infty(\mathbb{R})$.
(b) Give an explicit function in $C_c^\infty(\mathbb{R})$ whose support is equal to a generic bounded interval $[a, b]$.

5.2 Let $f(x) = x$ if $x \geq 0$ and $f(x) = 0$ if $x < 0$. Find f' and f'' in the sense of distributions.

5.3 Let $f(x) = |x|$ for $-1 < x \leq 1$. Extend f to all of $\mathbb{R}$ by periodicity. In other words, repeat the function from 1 to 3, from 3 to 5, from -3 to -1, and so forth. Note that the extension f will satisfy $f(x + 2) = f(x)$ for all $x \in \mathbb{R}$. Find f' and f'' in the sense of distributions.

5.4 Give an example of a function $f(x)$ whose first and second derivatives in the sense of distributions are both distributions generated by functions, but f''' in the sense of distributions is δ_0.

5.5 Prove Proposition 5.5.1.

5.6 Prove that the sequence of hat functions defined by (5.22) converges *in the sense of distributions* to δ_0.

5.7 Consider the following function:

$$f(x) = \begin{cases} 0 & \text{if } x \leq 0, \\ e^x & \text{if } 0 < x \leq 1, \\ x & \text{if } 1 < x \leq 2, \\ 0 & \text{if } x > 2. \end{cases}$$

What is its derivative in the sense of distributions? What is its second derivative in the sense of distributions?

5.8 Let $f(x) = e^{-x}$ if $x > 0$ and $f(x) = -e^x$ if $x \leq 0$. Find f' in the sense of distributions and show that $f'' = 2\delta_0' + f$ in the sense of distributions.

5.9 Find a function $f(x)$ whose derivative in the sense of distributions is the distribution $x^2 + 4x + \delta_2$. Note that this means a function f such that $(F_f)' = (F_g) + \delta_2$, where $g(x) = x^2 + 4x$ and δ_2 is the delta function with concentration at $x = 2$.

5.10 (**Summarizing the Distributional Derivative of a Piecewise Smooth Function**) This exercise generalizes (and summarizes) many of the previous exercises and examples. Suppose f is a C^1 (continuously differentiable) function except at a point x_0. Define the function

$$g(x) = \begin{cases} f'(x) & \text{if } x \neq x_0, \\ 0 & \text{if } x = x_0. \end{cases}$$

(a) If f is continuous at $x = x_0$, prove (show) that $f' = g$ in the sense of distributions.
(b) Suppose f has a jump discontinuity at x_0 with $a = \lim_{x \to x_0^-} f(x)$ and $b = \lim_{x \to x_0^+} f(x)$. Show that $f' = g + (b - a)\delta_{x_0}$ in the sense of distributions.

5.11 Is $\sum_{n=1}^{\infty} \delta_n$ a well-defined distribution? Note that to be a well-defined distribution, its action on any test function must be a finite number. Provide an example of a function $f(x)$ whose derivative in the sense of distributions is $\sum_{n=1}^{\infty} \delta_n$.

5.12 (a) Consider the function $f(x)$ defined on $[0, 1)$ to be -1 if $0 \le x \le \frac{1}{2}$ and 1 if $\frac{1}{2} < x < 1$. Extend f to the entire real line periodically (with period 1). For each $n = 1, 2, \ldots$, define $f_n(x) := f(nx)$. Now look at f_n as a sequence of distributions. Does $f_n(x)$ converge as $n \to \infty$ **in the sense of distributions**? If so, to what? If not, why? You do not need to prove your answer but, instead, provide some explanation. You should start by sketching a few of the f_n: say, $f_1(x)$, $f_2(x)$, $f_4(x)$.
(b) Repeat part (a) with $f(x)$ now defined on $[0, 1)$ to be -1 if $0 \le x \le \frac{1}{2}$ and 5 if $\frac{1}{2} < x < 1$.

The next five exercises are geared towards students with some exposure to mathematical analysis.

5.13 While we discussed sequential limits of functions, one can also consider continuum limits of functions, for example, the distributional limit of f_ϵ as $\epsilon \to 0^+$. Let f be any integrable function such that $\int_{\mathbb{R}} f(x)\,dx = 1$. For any fixed $x_0 \in \mathbb{R}$, prove that

$$\frac{1}{\epsilon} f\left(\frac{x - x_0}{\epsilon}\right) \xrightarrow{\epsilon \to 0^+} \delta_{x_0} \qquad \text{in the sense of distributions.}$$

Hint: Use the continuum analogue of Theorem A.8 in the Appendix.

5.14 Prove that the sequence $f_n(x)$ of Gaussians (5.25) converges to δ_0 in the sense of distributions.

5.15 From the basic structure of the proofs of Theorems 5.1 and 5.2 and Exercise **5.14**, it is natural to ask for one *mother* proof which shows that the three properties of Properties 5.5.1 are sufficient for distributional convergence to the delta function. The issue here is that the third property (concentration at 0) needs to be made precise. Use the following two conditions for property (3): (i) For any $a > 0$, $f_n(x)$ converges uniformly to 0 on the interval $|x| \ge a$; (ii) for any $a > 0$,

$$\lim_{n \to \infty} \int_{|x| \ge a} f_n(x)\,dx = 0,$$

to prove that the sequence of functions $f_n(x)$ converges to the delta function in the sense of distributions. Prove that the sequences (5.23), (5.24), and (5.25) all satisfy this precise concentration property.

5.16 Prove Theorem 5.3. Extra: Is the hypothesis on non-negativity necessary?

5.17 Let $f(x)$ be any integrable, periodic function with period l such that its integral over any interval of length l (i.e., over a period) is A for some $A \in \mathbb{R}$. Define $f_n(x) := f(nx)$, and prove that $f_n(x)$ converges to $f(x) \equiv \frac{A}{l}$ (the average) in the sense of distributions.

5.18 Find the distributional limit of the sequence of distributions $F_n = n\delta_{-\frac{1}{n}} - n\delta_{\frac{1}{n}}$. Hint: F_n is the distributional derivative of some function.

5.19 Consider the distribution δ_0', that is, the derivative of the 1D delta function in the sense of distributions. Find a sequence of functions $f_n(x)$ such that f_n converges to δ_0' in the sense of distributions. Sketch $f_n(x)$ for $n = 1, 5, 10, 20$.

5.20 Consider the equation (5.40). (a) By looking up references online, give either a partial or full (i.e., a proof) justification for this equation. (b) Find a situation in mathematics or physics where this sum turns up. (c) Suppose we considered the sum not on $\Omega = (-\pi, \pi)$, but on the whole real line $\mathbb{R}$. In that case, what would it converge to in the sense of distributions?

5.21 Let $y \in \mathbb{R}$ and consider the distribution δ_y (the delta function with concentration at $x = y$). As usual, x will denote the underlying independent variable of this distribution (i.e., the variable of the test functions). Suppose we wish to differentiate δ_y with respect to the source point y (not as we have previously done with respect to the underlying independent variable x). Rather than trying to make a precise distributional definition, simply justify the informal equation:

$$\text{“}\frac{\partial}{\partial y}\delta_y = -\delta_y' \left(= -\frac{\partial}{\partial x}\delta_y\right)\text{”}.$$

5.22 (a) Justify (5.38).
(b) Suppose $P(x)$ is a cubic polynomial with distinct real roots a, b, c such that the derivative $P'(x)$ is not zero at any of these roots. Justify the informal equation

$$\text{“}\delta_0[P(x)] = \frac{1}{|P'(a)|}\delta_0(x-a) + \frac{1}{|P'(b)|}\delta_0(x-b) + \frac{1}{|P'(c)|}\delta_0(x-c)\text{”}.$$

(c) (**Change of Variables in the Delta Function**) Let $y(x)$ be a smooth function. Justify the following informal change of variables statement: For all $a \in \mathbb{R}$, “$\delta_0(x-a) = y'(a)\,\delta_0(y - y(a))$”.
(d) Justify the informal statement “$\delta_0(\sin \pi x) = \sum_{n=-\infty}^{\infty} \delta_0(x-n)$”. The right-hand side is actually the distribution $\sum_{n=-\infty}^{\infty} \delta_n$ which is known as the **Dirac comb**. See Exercise **6.29** for more on the Dirac comb.

5.23 Justify the informal equations

$$\text{“}\int_{-\infty}^{a} \delta_0(x)\,dx = \begin{cases} 1 & \text{if } a > 0, \\ 0 & \text{if } a < 0 \end{cases}\text{”} \quad \text{and} \quad \text{“}\int_{a}^{\infty} \delta_0(x)\,dx = \begin{cases} 0 & \text{if } a > 0, \\ 1 & \text{if } a < 0. \end{cases}\text{”}$$

5.24 Let $g \in C^\infty(\mathbb{R})$. If F is any distribution, we can define the product of g with F as a new distribution defined by

$$\langle gF, \phi\rangle = \langle F, g\phi\rangle \qquad \text{for any } \phi \in \mathcal{D}.$$

(a) Why is this a reasonable definition?
(b) Prove (show) that in the sense of distributions, $x\,\delta_0 = 0$ and $x\,(\delta_0)' = -\delta_0$.
(c) If for any $n = 1, 2, \ldots$, $\delta_0^{(n)}$ denotes the n-th distributional derivative of δ_0, prove (show) that

$$x^n\,\delta_0^{(n)} = (-1)^n\, n!\; \delta_0.$$

(d) In the context of this question, show that the informal equation (written by Dirac on page 60 of *The Principles of Quantum Mechanics*)

$$\text{“}f(x)\delta(x-a) = f(a)\delta(x-a)\text{”}$$

has full justification in the sense of distributions.

5.25 Justify the informal equation written by Dirac on page 60 of *The Principles of Quantum Mechanics*:

$$\text{“}\left(\int \delta_0(x-a)\right)\delta_0(x-b) = \delta_0(a-b)\text{”}.$$

5.26 Consider the function $f(x) = x^{1/3}$. Show that $f(x)$ is locally integrable on $\mathbb{R}$ and, hence, can be considered as a distribution. Find f' in the sense of distributions. Note that the pointwise derivative of f is not locally integrable.

5.27 Consider the informal equation written by Dirac on page 61 of *The Principles of Quantum Mechanics*:

$$\text{“}\frac{d}{dx}\log x = \frac{1}{x} - i\pi\delta(x)\text{”}. \tag{5.51}$$

By $\frac{1}{x}$ above, Dirac is talking about PV $\frac{1}{x}$. However, recall from (5.46) that the distributional derivative of $\log|x|$ was simply PV $\frac{1}{x}$. In (5.51), Dirac considers the logarithm for all $x \in \mathbb{R}$, interpreting the log via its extension to the complex plane. With this interpretation of log, show (5.51) as a precise statement in the sense of distributions.

Chapter 6

The Fourier Transform

> *"Fourier's theorem is not only one of the most beautiful results of modern analysis, but it may be said to furnish an indispensable instrument in the treatment of nearly every recondite question in modern physics."*
>
> \- **Lord Kelvin** (William Thomson)[a]
>
> [a]William Thomson and Peter Guthrie Tait, Treatise on Natural Philosophy, Vol. 1, Section 75, Oxford Univ. Press, 1867.

The Fourier transform and its predecessor, Fourier series, are tools which have revolutionized the physical sciences, engineering, medicine, and, of course, mathematics. Despite the fact that Fourier series came first, appearing in the 1807 treatise of **Jean-Baptiste Joseph Fourier** on heat flow, it will be convenient to devote this chapter fully to the Fourier transform. Fourier series will be the topic of Chapter 11, and only at the end of that chapter will we link the two. We hope to impress upon the reader that these are truly groundbreaking tools which every self-respecting scientist, engineer, or mathematician must have in their toolbox.

While we will present the Fourier transform in a manner slanted towards its application in solving PDEs, it is far more than just a tool for PDEs. Indeed, it lies at the heart of signal processing and, as such, has tremendous physical significance. Roughly speaking, the Fourier transform takes a signal and expresses it in terms of the frequencies of the waves that make up that signal. We will discuss this interpretation in Section 6.12 but, for now, let us present the Fourier transform and its properties from the perspective of solving PDEs. On the other hand, it may be insightful to keep in mind from the start the following wonderful loose description (though we are unfortunately unaware of its originator):

"The Fourier transform takes a smoothie and gives you back the recipe".
" The inverse Fourier transform takes the recipe and makes the smoothie".

Complex numbers are very useful in the presentation and analysis of the Fourier transform, and hence it is beneficial to present a quick summary.

6.1. • Complex Numbers

Complex numbers are so named because they are composed of two types of more fundamental numbers: real numbers and imaginary numbers. You should view the adjective **complex** not as "complicated" or "difficult", but rather, describing certain numbers which have a more complex structure than real numbers. Complex numbers have proven themselves to be exceptionally useful when addressing many physical phenomena and mathematical structures. They encode two pieces of information and can be considered as ordered pairs (x, y), where each component is a real number. What makes these entities different from vectors in $\mathbb{R}^2$ is an operation (multiplication) in which one component, the second component, can **interact/transfer** to the first component. More precisely, we have a multiplication of these numbers for which

$$(0,1)\times(0,1) = (-1,0). \tag{6.1}$$

Just as with vectors in $\mathbb{R}^2$, we can add these ordered pairs and multiply them by real scalars to form new ordered pairs but, further, we now have a way of multiplying them to form a new ordered pair based upon (6.1). This new multiplication, defined below in (6.2), is very different from the dot product of two vectors in $\mathbb{R}^2$.

For a complex number $z = (x, y)$, we say $x \in \mathbb{R}$ is the real part of z and $y \in \mathbb{R}$ is the imaginary part of z. We say z is a real number if and only if it has 0 as its imaginary part; in other words, for $x \in \mathbb{R}$, we identify x with $(x, 0) = x(1, 0)$, and we regard the real numbers as a subset of the complex numbers. We denote the complex number $(0, 1)$ by i. In this case, we can write z as

$$z = (x,y) = (x,0) + (0,y) = x(1,0) + y(0,1) = x + iy,$$

and the fundamental relation (6.1) can be written as

$$i^2 = -1 \qquad \text{or} \qquad i = \sqrt{-1}.$$

Henceforth we will not use the bracket notation for a complex number but rather denote the complex numbers by

$$\boxed{z = x + iy.}$$

Geometrically we still view z as living in a plane, albeit the complex plane, with the horizontal component being the real component x and the vertical component being the imaginary component y. We thus speak of the real and imaginary axes. The modulus of a complex number z, denoted by $|z|$, is defined to be $|z| := \sqrt{x^2 + y^2}$. It represents the length of the position vector of z in the complex plane.

Following the fundamental relation (6.1), we can multiply two complex numbers and obtain another complex number; if $z_1 = x_1 + iy_1$ and $z_2 = x_2 + iy_2$, then we define their product as the complex number

$$z_1 z_2 := (x_1 + iy_1)(x_2 + iy_2) = (x_1x_2 - y_1y_2) + i(x_1y_2 + x_2y_1). \tag{6.2}$$

We also define the complex conjugate of a complex number $z = x + iy$ to be

$$\bar{z} := x - iy.$$

Geometrically, this conjugation operation reflects a complex number about the real axis — the axis spanned by $(1, 0)$. Note that $|z|^2 = z\bar{z}$.

We will denote the set of complex numbers by $\mathbb{C}$. While a course in complex variables (or analysis) is primarily concerned with functions of a complex variable, for our purposes we will only be concerned with **functions of a real variable**[1] but whose values are **complex**. Such a function $f : \mathbb{R} \to \mathbb{C}$ can be written as

$$f(x) = f_1(x) + if_2(x),$$

where f_i are functions from $\mathbb{R}$ to $\mathbb{R}$ and denote the real and imaginary parts of the function. In this case, the derivative and integral of f can be obtained by differentiating (resp. integrating) separately the real and imaginary parts of f; i.e.,

$$f'(x) = f_1'(x) + if_2'(x) \qquad \text{and} \qquad \int f(x)\,dx = \int f_1(x)\,dx + i\int f_2(x)\,dx.$$

The **exponential function**, *the mother function of mathematics*, can readily be defined on a complex variable via its Taylor series; namely if $z \in \mathbb{C}$, then

$$e^z := \sum_{n=0}^{\infty} \frac{z^n}{n!}. \tag{6.3}$$

With this definition in hand, all the properties enjoyed by the exponential function of a real variable carry over; in particular, we have $e^{z_1+z_2} = e^{z_1}e^{z_2}$.

The building block of the Fourier transform is the exponential function of a purely imaginary argument: $e^{ix}, x \in \mathbb{R}$. **Euler's formula** tells us that the real and imaginary parts of this function are, respectively, cosine and sine; i.e.,

$$\boxed{e^{ix} = \cos x + i \sin x.} \tag{6.4}$$

Note that one can easily derive Euler's formula from the definition of the complex exponential (6.3). In words, Euler's formula tells us that the exponential will map purely imaginary numbers to rotations around the unit circle in the complex plane. One could say it wraps a vertical real axis around the unit circle in the complex plane (cf. Figure 6.1). Note that by Euler's formula,

$$\overline{e^{ix}} = e^{-ix} \qquad \text{and} \qquad \frac{d}{dx}e^{ix} = ie^{ix}.$$

[1]This is not to say that functions of a complex variable are not important in PDEs (see for example [7]). In addition many Fourier transform computations are nicely facilitated via complex integration (integration with respect to a complex variable).

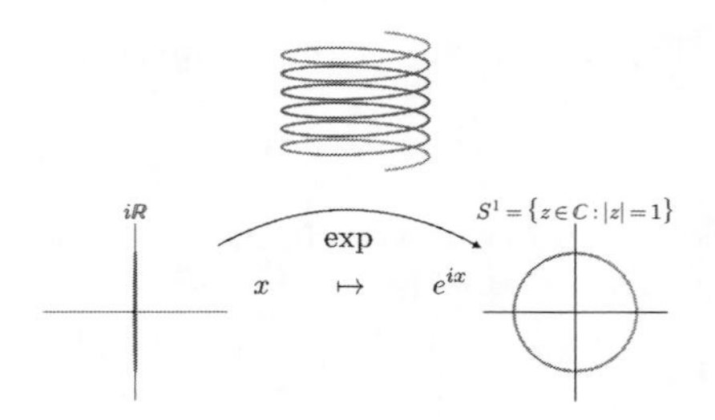

Figure 6.1. Cartoon to illustrate Euler's formula.

6.2. • Definition of the Fourier Transform and Its Fundamental Properties

6.2.1. • The Definition. Let $f(x)$ be an integrable real-valued[2] function on $\mathbb{R}$; that is, $\int_{-\infty}^{\infty} |f(x)|\, dx < \infty$.

The Definition of the Fourier Transform

Definition 6.2.1. For each $k \in \mathbb{R}$, we define **the Fourier transform** of the function f at k to be

$$\widehat{f}(k) := \int_{-\infty}^{\infty} f(x)\, e^{-ikx}\, dx. \tag{6.5}$$

We often say the Fourier transform has *transformed* the function of x into Fourier space (k-space) as opposed to the original "real" space where x lives. In this regard, "real" does not mean real numbers (both x and k are real numbers) but rather refers to the space in which the original problem is posed. Following convention, we mostly use k as the variable of a Fourier transform; in other words, k is the independent variable of the Fourier transform of a function. Symbolically, we view the Fourier transform as the following process:

$$\text{a function of } x \xrightarrow{\text{Fourier transform}} \text{a function of } k.$$

Rather than writing $\widehat{f(x)}\,(k)$, we simply write $\widehat{f}(k)$. However, it is sometimes convenient to have another notation for the Fourier transform involving $\mathcal{F}$:

$$\mathcal{F}\{f\} = \widehat{f} \qquad \text{or} \qquad \mathcal{F}\{f(x)\}(k) = \widehat{f}(k).$$

Before continuing, a few comments about definition (6.5) are in order.

- The Fourier transform is related to the original function in an awfully *convoluted and nonlocal* way, meaning that to find $\widehat{f}$ at any point k (say, $k = 3$), we need to incorporate (integrate) the values of the original function f everywhere.
- For each k, it follows that $\widehat{f}(k)$ is a complex number. So while x, k, and the values of $f(x)$ are real, in general, the values of $\widehat{f}(k)$ are complex. However, as we shall see, the Fourier transform of several important functions has no imaginary part; i.e., it is real valued.

[2]The definition naturally extends to complex-valued functions (cf. Section 6.2.5).

- By Euler's formula

$$e^{-ikx} = \cos kx - i \sin kx, \tag{6.6}$$

meaning that e^{-ikx} lies on the unit circle in the complex plane. However, note that in this chapter the integration is, and will always be, with respect to a real variable. This "real" integration of the complex-valued integrand can be viewed as the separate integration of its real and complex parts.

- As a function of k, $\widehat{f}$ is bounded. Indeed, since $|e^{-ikx}| = 1$, we have for all $k \in \mathbb{R}$,

$$|\widehat{f}(k)| \leq \int_{-\infty}^{\infty} |f(x)|\,|e^{-ikx}|\,dx = \int_{-\infty}^{\infty} |f|\,dx < \infty.$$

In fact, one can show that $\widehat{f}$ is also a continuous function of k (see Section 6.5).

- The Fourier transform is a **linear** operation on functions. That is, if f_1 and f_2 are integrable functions and $a, b \in \mathbb{R}$ (or even $\mathbb{C}$), then

$$(\widehat{af_1 + bf_2})(k) = a\widehat{f_1}(k) + b\widehat{f_2}(k).$$

We give a simple yet important example.

Example 6.2.1. Let $a > 0$ be fixed and consider the characteristic function of the interval $[-a, a]$:

$$f(x) = \begin{cases} 1 & \text{if } |x| < a, \\ 0 & \text{if } |x| \geq a. \end{cases}$$

We find that

$$\widehat{f}(k) = \int_{-\infty}^{\infty} f(x)\,e^{-ikx}\,dx = \int_{-a}^{a} e^{-ikx}\,dx = \frac{e^{-iak} - e^{iak}}{-ik} = 2\frac{\sin ak}{k},$$

where we used Euler's formula (6.6) in the last equality. Note that this Fourier transform has the following properties:

- It has no imaginary part (i.e., it is always real valued) — a consequence of the function being even.
- While f had compact support and took on only the two values 0 or 1, its Fourier transform is not compactly supported and takes on a continuum of values.
- It is **not** integrable over $\mathbb{R}$ (cf. Exercise **6.3**).

6.2.2. • Differentiation and the Fourier Transform.

Surely you are curious as to **why** one would make such a convoluted definition in (6.5). We obtain partial motivation by looking at how the Fourier transform behaves with derivatives. Suppose $f(x)$ is a differentiable, integrable function with derivative $\frac{df}{dx}$. Assume that $\frac{df}{dx}$, as a function of x, is also integrable. What happens if we take the Fourier transform of $\frac{df}{dx}$? Is it related to the Fourier transform of $f(x)$?

By integration by parts, we find

$$\begin{aligned}
\widehat{\frac{df}{dx}}(k) &= \int_{-\infty}^{\infty} \frac{df}{dx}(x)e^{-ikx}dx \\
&= \lim_{L\to\infty} \int_{-L}^{L} \frac{df}{dx}(x)e^{-ikx}dx \\
&\overset{\substack{\text{integration}\\ \text{by parts}}}{=} \lim_{L\to\infty} \left(-\int_{-L}^{L} f(x)e^{-ikx}(-ik)dx \;+\; f(L)e^{-ikL} \;-\; f(-L)e^{ikL} \right) \\
&= \lim_{L\to\infty} -\int_{-L}^{L} f(x)e^{-ikx}(-ik)dx \\
&= ik\int_{-\infty}^{\infty} f(x)e^{-ikx}dx \\
&= ik\hat{f}(k).
\end{aligned}$$

Note that in the integration by parts, the boundary terms vanished; the moduli (sizes) of these complex numbers are, respectively, $|f(L)|$ and $|f(-L)|$, which, since f and f' are integrable over the entire real line, must[3] tend to 0 as $L \to \infty$. We summarize this important calculation:

Differentiation and the Fourier Transform

The Fourier transform has the following property:

$$\widehat{\frac{df}{dx}}(k) = ik\hat{f}(k). \tag{6.7}$$

In words, the Fourier transform **transforms differentiation into complex polynomial multiplication!** This is truly an amazing fact; surely differentiation is more complicated than polynomial multiplication.

Under the assumption of integrability of all the relevant derivatives, we can repeat this argument to find

$$\boxed{\widehat{\frac{df^{(n)}}{dx^n}}(k) = i^n\, k^n\, \hat{f}(k),} \tag{6.8}$$

where $\frac{df^{(n)}}{dx^n}$ denotes the n-th derivative of f.

6.2.3. • Fourier Inversion via the Delta Function. We have just witnessed that, by **transforming** into Fourier space (i.e., taking the Fourier transform), *nice*

[3]This should sound very believable but many students might want a proof. To this end, write

$$f(L) - f(0) = \int_0^L f'(x)\,dx,$$

and let $L \to \infty$, noting that f' is integrable. This will prove that the limit must exist. Now use the integrability of f to conclude that this limit must be 0.

things happen from the perspective of differentiation. However, should we be concerned that in transforming to Fourier space, we have lost information about the original function $f(x)$? In other words, is it possible to **go back** by **inverting** the Fourier transform? Fortunately, for reasonably nice functions, we can completely invert the transformation, and, hence, no information has been lost. Indeed, the following inversion formula holds true.

Fourier Inversion Theorem/Formula

Theorem 6.1. *Suppose f and $\hat{f}$ are integrable and continuous*[a]. *Then*

$$f(x) = \frac{1}{2\pi}\int_{-\infty}^{\infty} \hat{f}(k)\, e^{ikx}\, dk. \tag{6.9}$$

[a] As we shall see, the continuity is redundant since it will follow from the integrability of both f and $\hat{f}$.

Note that in the integrand we are dealing with the product of two complex numbers, $\hat{f}(k)$ and e^{ixk}. However, if we start with a real-valued function $f(x)$, the theorem states that the integral of the complex-valued function must be real. In other words, the complex part of the integral on the right of (6.9) must be identically 0.

Formula (6.9) is often called the **Fourier inversion formula**. We can rephrase the Fourier inversion formula by introducing the inverse Fourier transform:

The Definition of the Inverse Fourier Transform

Definition 6.2.2. The inverse Fourier transform of a function $g(k)$ is defined to be

$$\check{g}(x) := \frac{1}{2\pi}\int_{-\infty}^{\infty} g(k)\, e^{ikx}\, dk. \tag{6.10}$$

When notationally convenient, we use the notation $\mathcal{F}^{-1}$ for the inverse Fourier transform; i.e.,

$$\mathcal{F}^{-1}\{g\} = \check{g} \quad \text{or} \quad \mathcal{F}^{-1}\{g(k)\}(x) = \check{g}(x).$$

With this definition in hand, the Fourier inversion formula (6.9) states that

$$\boxed{\widecheck{\hat{f}(k)}(x) = f(x) \quad \text{or simply } \check{\hat{f}} = f,}$$

justifying the use of the word "inverse".

While we will not present here a direct proof[4] of the inversion formula (Theorem 6.1), clearly some explanation is in order. Why on earth would this work? Let us *place* the definition of the Fourier transform into the right-hand side of the inversion formula (6.9), rearrange the terms, and see that it is **again the devious delta function at work!**

[4] See Folland [**14**] for a direct proof.

To this end, **fix** x and **suppose we can** interchange the order of integration (cf. Section A.10 of the Appendix) to find

$$\frac{1}{2\pi}\int_{-\infty}^{\infty} \hat{f}(k)\, e^{ixk}\, dk = \frac{1}{2\pi}\int_{-\infty}^{\infty}\left[\int_{-\infty}^{\infty} f(y)\, e^{-iky}\, dy\right] e^{ixk}\, dk$$
$$= \frac{1}{2\pi}\int_{-\infty}^{\infty}\int_{-\infty}^{\infty} f(y)\, e^{-iky}\, e^{ixk}\, dy\, dk$$
$$\overset{\substack{\textbf{illegal}\text{ change in}\\ \text{order of integration}}}{=} \text{“}\frac{1}{2\pi}\int_{-\infty}^{\infty}\int_{-\infty}^{\infty} f(y)\, e^{ik(x-y)}\, dk\, dy\text{”} \tag{6.11}$$
$$= \int_{-\infty}^{\infty} f(y)\,\text{“}\left[\frac{1}{2\pi}\int_{-\infty}^{\infty} e^{ik(x-y)}\, dk\right]\text{”}\, dy. \tag{6.12}$$

If the integral in the last line above is going to give us back $f(x)$, it must be that *somehow*

$$\text{“}\frac{1}{2\pi}\int_{-\infty}^{\infty} e^{ik(x-y)}\, dk\text{”} \qquad \text{as a function of } y \text{ behaves like } \delta_x, \tag{6.13}$$

a delta function with concentration at $y = x$. In particular, somehow we must have

$$\boxed{\text{“}\int_{-\infty}^{\infty} e^{-iky}\, dk\text{”} = 2\pi\,\delta_0 \qquad \text{in the sense of distributions.}} \tag{6.14}$$

Note above the four occurrences of quotation marks.

- In (6.12), (6.13), and (6.14), they surround an improper integral, which viewed as a function of y makes no sense; we are integrating with respect to k, a complex-valued function infinitely many times around the unit circle in the complex plane with no signs of *decay*.
- In (6.11) they were used because of an illegal change of order in the integration. Indeed, in the lines above (6.11), e^{-iky} was multiplied by $f(y)$ **before** being integrated with respect to y. The function $f(y)$ had decaying properties which **tamed** the tails of e^{-iky} as $y \to \pm\infty$, rendering the **product** integrable; hence, there was no need for quotation marks in the line above (6.11). The illegal change in the order of integration left $e^{-ik(x-y)}$ all by itself to be integrated with respect to k, giving rise to the issue discussed in the first bullet.

At this point, we have enough experience with distributions and the delta function to make sense of (6.14), and hence the above calculation ending with (6.12). An improper integral is defined as a limit over larger and larger domains, and in our case this means

$$\text{“}\frac{1}{2\pi}\int_{-\infty}^{\infty} e^{-iky}\, dk\text{”} = \lim_{n\to\infty}\frac{1}{2\pi}\int_{-n}^{n} e^{-iky}\, dk. \tag{6.15}$$

This limit does not exist in the classical sense but it does in the sense of distributions, and it is exactly a distributional sense that we want for (6.14); note the presence of the *test function* f in (6.12). To find this limit in the sense of distributions, we note that for fixed n, Euler's formula implies that

$$\frac{1}{2\pi}\int_{-n}^{n} e^{-iky}\,dk \;=\; \frac{1}{2\pi}\frac{e^{-iny}-e^{iny}}{-iy} \;=\; \frac{\sin ny}{\pi y}.$$

These are the **sinc functions**, previously discussed in Section 5.5.6, which we stated (but did not prove) converge to δ_0 in the sense of distributions.

Let us now return to the calculation surrounding (6.12), incorporating the limit on the right-hand side of (6.15). To this end, we find that

$$\begin{aligned}
\frac{1}{2\pi}\int_{-\infty}^{\infty} \hat{f}(k)\,e^{ixk}\,dk &= \lim_{n\to\infty}\left(\frac{1}{2\pi}\int_{-n}^{n}\left[\int_{-\infty}^{\infty} f(y)\,e^{-iky}\,dy\right] e^{ixk}\,dk\right)\\
&= \lim_{n\to\infty}\left(\frac{1}{2\pi}\int_{-n}^{n}\int_{-\infty}^{\infty} f(y)\,e^{-iky}\,e^{ixk}\,dy\,dk\right)\\
&\overset{\substack{\textbf{legal}\text{ change in}\\ \text{order of integration}}}{=} \lim_{n\to\infty}\left(\frac{1}{2\pi}\int_{-\infty}^{\infty}\int_{-n}^{n} f(y)\,e^{ik(x-y)}\,dk\,dy\right)\\
&= \lim_{n\to\infty}\left(\int_{-\infty}^{\infty} f(y)\underbrace{\left[\frac{1}{2\pi}\int_{-n}^{n} e^{ik(x-y)}\,dk\right]}_{\frac{\sin n(y-x)}{\pi(y-x)}} dy\right).
\end{aligned}\tag{6.16}$$

Note that the change in the order of integration is legal for any finite n (cf. Theorem A.13); hence there is no need for quotation marks. Moreover, assuming

- $\frac{\sin n(y-x)}{\pi(y-x)}$ as a function of y converges to δ_x (as $n \to \infty$) in the sense of distributions,
- we can view $f(y)$ as a test function,

it is clear what happens when we take the limit as $n \to \infty$ in (6.16): All integral signs *disappear* and we are simply left with $f(x)$.

One can in fact directly prove the distributional convergence of the sinc functions to δ_0 (cf. Exercise **6.25**). Moreover, one can prove the statement in the case where the test functions are simply continuous and integrable (this would prove Theorem 6.1). However, in Section 6.9 we will take a far more general, systematic, and versatile approach by **extending the notion of the Fourier transform to distributions**. Note that the undefined integral on the left of (6.14) can be informally thought of as

"the Fourier transform of the constant function $f(x) \equiv 1$".

Physicists have no issues proclaiming that the Fourier transform of 1 is (2π times) the delta function. They are of course correct and this is really not so hard to believe; intuitively, if we took the Fourier transform of the delta function, we would indeed obtain the constant function 1. Hence, if there is any justice in the world, at least the inverse Fourier transform of 1 *ought to be*, up to a constant, the delta function.

By viewing the function 1 as a distribution, we will soon make precise that its Fourier transform is $2\pi\delta_0$. This will not exactly prove the inversion formula (Theorem 6.1), but it will yield a simple proof for functions in a certain class (the Schwartz class; cf. Theorem 6.3), which in particular includes functions in $C_c^\infty(\mathbb{R})$. With this result in hand, extending the inversion formula to larger classes of functions (for example, those stated in Theorem 6.1) is actually not that hard with the right tools from analysis (tools beyond the scope of this book).

6.2.4. Finding the Fourier and Inverse Fourier Transforms of Particular Functions. Finding the Fourier and inverse Fourier transforms of particular functions turns out to be very useful, and examples of many functions appear throughout the science literature. Indeed, historically, entire books were devoted to presenting Fourier and other transforms of particular functions. Many functions have explicit Fourier transforms which can be written with our vast repertoire of named functions. Often complex integration, that is, integration with respect to a complex variable, proves to be a valuable tool.

Here, let us just remark that one can exploit the Fourier inversion formula to determine Fourier transforms, basically by going back and forth in the following sense. Note that, besides the factor of 2π, the only difference between the Fourier transform and the inverse Fourier transform is a sign change in the exponential of the integrand. Using the neutral independent variable y, this means that for any nice function g,

$$\mathcal{F}\{g\}(y) = \hat{g}(y) = 2\pi\check{g}(-y) = 2\pi\mathcal{F}^{-1}\{g\}(-y).$$

Thus, suppose we know the Fourier transform of $f(x)$ is some function and let's call it $g(k)$. Now suppose we want to find the Fourier transform of g. It is not exactly f, but close: Indeed,

$$\hat{g}(y) = 2\pi\check{g}(-y) = 2\pi\left(\check{\hat{f}}(-y)\right) = 2\pi f(-y).$$

So, for example, for any $a > 0$,

$$\mathcal{F}\left\{e^{-a|x|}\right\} = \frac{2a}{k^2 + a^2}; \qquad \text{hence,} \quad \mathcal{F}\left\{\frac{2a}{x^2 + a^2}\right\} = 2\pi e^{-a|k|}.$$

In general, many functions do not have explicit Fourier transforms in terms of known functions. While there are analytical tools to approximate these, the computer has now provided a revolutionary tool. Through the **discrete Fourier transform** and the **fast Fourier transform**, one can quickly and efficiently compute Fourier and inverse Fourier transforms. In particular, we can compute Fourier transforms of functions for which only a finite number of values are given, for example, in instances where we may only know the values of the function on a finite grid in the domain.

6.2.5. The Fourier Transform of a Complex-Valued Function. Since, in general, the Fourier transform of a real-valued function can be complex valued, we may as well extend the definition of the Fourier transform to **integrable complex-valued functions** on $\mathbb{R}$. These are complex-valued functions $f(x)$ for which

$$\int_{-\infty}^{\infty} |f(x)|\, dx < \infty.$$

Remember that here $|f(x)|$ denotes the modulus of the complex number $f(x)$. If f is an integrable complex-valued function on $\mathbb{R}$, we can define the Fourier transform by the same formula as before with

$$\widehat{f}(k) := \int_{-\infty}^{\infty} f(x)e^{-ikx}\,dx, \tag{6.17}$$

but now with the understanding that the multiplication in the integrand is the product of two complex numbers.

Note that this is equivalent to taking the Fourier transform of the real and imaginary parts of f. That is, if we let

$$f(x) = f_1(x) + if_2(x),$$

where each f_i is a real-valued function, the integrability of the complex-valued f is equivalent to each of the f_i being integrable in the usual sense; i.e.,

$$\int_{-\infty}^{\infty} |f_i(x)|\,dx < \infty, \qquad i = 1, 2.$$

Hence, one can take the Fourier transform of the real and complex parts f_i, as we did in the previous sections, and find that

$$\widehat{f}(k) = \widehat{f_1}(k) + i\widehat{f_2}(k).$$

6.3. • Convolution of Functions and the Fourier Transform

6.3.1. • The Definition of Convolution. An important form of "multiplication of functions" is called convolution. If $f(x)$ and $g(x)$ are functions over $\mathbb{R}$, then the convolution $f * g$ is a new function defined over $\mathbb{R}$ by

$$(f * g)(x) := \int_{-\infty}^{\infty} f(x-y)\,g(y)\,dy. \tag{6.18}$$

Note that for the purpose of the integral, x is fixed.

Solutions to many important PDEs (for example, the diffusion equation and PDEs involving the Laplacian) are expressed as convolutions; in other words, nature favors convolutions! Hence it is best to start now and make this operation your friend, contemplating the structure of the formula (6.18). It is indeed an awfully "convoluted" way to multiply functions; take two of your favorite functions and try to compute the convolution.

The first question[5] to ask is whether or not this type of multiplication is commutative; i.e., is $(f * g)(x) = (g * f)(x)$? Although this may not be immediately obvious,

[5] Actually, to some mathematics students, the first question will be, *Is the convolution well-defined*? That is, for $x \in \mathbb{R}$ is $f(x-y)\,g(y)$ as a function of y integrable? It turns out that if f and g are both integrable, this convolution is defined for all but a "negligible" number of values of x. As before, "negligible" is with respect to integration and mathematicians call this set a set of measure zero. With this in hand, one can show that the convolution of two integrable functions (which is defined almost everywhere) is also integrable with respect to x. Students who will pursue mathematical analysis will eventually see a basic duality structure in the definition of the convolution. A fundamental inequality, called Hölder's inequality, will state that if $f \in L^p$ (meaning that $|f|^p$ is integrable) with $p > 1$ and if g is in L^q where $\frac{1}{p} + \frac{1}{q} = 1$, then the convolution is defined for all x and is a continuous function of x. See, for example, Lieb and Loss [**18**]. As we shall see next, if **one of** the two functions f or g happens to be in C_c^∞, then the convolution is a smooth (C^∞) function, regardless of the continuity or differentiability of the other function.

the answer is yes. This proves to be just a change of variable, keeping track of some minus signs. Indeed, letting $z = x - y$, we have that

$$\begin{aligned}(f * g)(x) = \int_{-\infty}^{\infty} f(x-y)\, g(y)\, dy &= -\int_{\infty}^{-\infty} f(z)\, g(x-z)\, dz \\ &= \int_{-\infty}^{\infty} f(z)\, g(x-z)\, dz \\ &= (g * f)(x).\end{aligned}$$

Hence, we arrive at

$$\boxed{(f * g)(x) = (g * f)(x).}$$

To gain some insight into convolution, we consider a special example. Let f be any function and for any $a > 0$ (think of a as a small number), let $g_a(x)$ be defined by

$$g_a(x) := \begin{cases} \frac{1}{2a} & \text{if } x \in [-a, a], \\ 0 & \text{otherwise}. \end{cases} \tag{6.19}$$

Note that for any $a > 0$, g_a integrates to 1, and as a gets smaller, $g_a(x)$ concentrates around 0 (sound familiar?). Now fix such an a, and let us investigate the convolution $(f * g_a)(x)$. We compute

$$(f * g_a)(x) = \int_{-\infty}^{\infty} f(x-y)\, g_a(y)\, dy = \int_{-\infty}^{\infty} f(y)\, g_a(x-y)\, dy = \frac{1}{2a}\int_{x-a}^{x+a} f(y)\, dy.$$

Thus, $(f * g_a)(x)$ is the **average** of the function f over the interval with center x and size $2a$. In particular, note that if f had a jump discontinuity at some point x_0, its average $(f * g_a)(x)$ would be continuous at x_0.

It is instructive to further work out two specific examples: Let $a = 1/2$ and consider both

$$(g_{\frac{1}{2}} * g_{\frac{1}{2}})(x) \qquad \text{and} \qquad (g_{\frac{1}{2}} * g_{\frac{1}{2}} * g_{\frac{1}{2}})(x) = g_{\frac{1}{2}} * (g_{\frac{1}{2}} * g_{\frac{1}{2}})(x).$$

Note that convolving any function f with $g_{\frac{1}{2}}$ gives a new function which at any point x is the average of f over the interval $(x - 0.5, x + 0.5)$. Let us now take f to also be $g_{\frac{1}{2}}$. We imagine traversing the x-axis from left to right and at any point x taking the local average of $g_{\frac{1}{2}}$ over an interval of size 1. We would obtain a piecewise linear continuous function with maximum at $x = 0$. This spike function is illustrated in the middle, wedge-shaped, curve in Figure 6.2. Note that the derivative of this function is singular at three points. Now suppose we want to convolute this spike function again with $g_{\frac{1}{2}}$. We would repeat the procedure of traversing the x-axis and computing the local average. Diagrammatically, this would give the bottom, rounded, curve in Figure 6.2. Note that this triple convolution is C^1. Is it C^2?

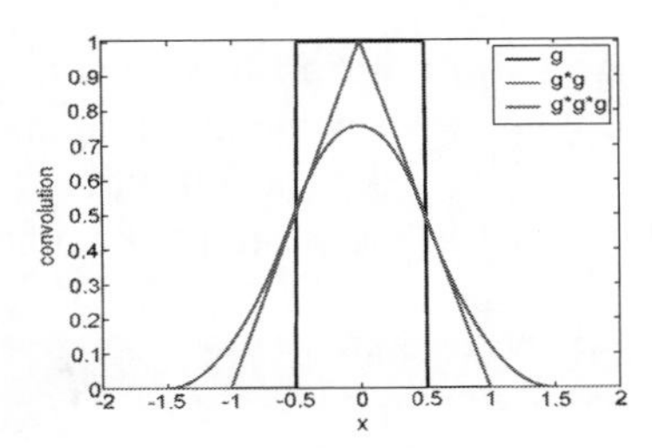

Figure 6.2. Plots of $g_{\frac{1}{2}}(x)$ (top), $(g_{\frac{1}{2}} * g_{\frac{1}{2}})(x)$ (middle), and $(g_{\frac{1}{2}} * g_{\frac{1}{2}} * g_{\frac{1}{2}})(x)$ (bottom).

6.3.2. • Differentiation of Convolutions.

There is a remarkable property concerning derivatives of convolutions which is at the heart of the solution formulas to many PDEs, IVPs, and BVPs, in particular, Poisson's equation, the IVP for the diffusion equation, and the Dirichlet problem for the Laplacian. For these problems, we will see that the solution formula will be in the form of a **convolution** between

- a certain function associated with the PDE (called the fundamental solution or Green's function) and
- the data (e.g., source terms, initial values, or boundary values).

Let f and g be two integrable functions and consider their convolution

$$F(x) \;:=\; (f*g)(x) \;=\; \int_{-\infty}^{\infty} f(x-y)\,g(y)\,dy.$$

Suppose we wish to compute $F'(x) = \frac{dF}{dx}$. By definition, this amounts to **differentiation under the integral sign** (cf. Section A.9). Under certain additional assumptions on f, we can apply Theorem A.10 (with $N = 1$, $\mathbf{x} = y$, and $t = x$) to conclude that F is continuous and differentiable with

$$F'(x) \;=\; (f*g)'(x) \;=\; \frac{d}{dx}\int_{-\infty}^{\infty} f(x-y)\,g(y)\,dy \;=\; \int_{-\infty}^{\infty} \frac{\partial f(x-y)}{\partial x}\,g(y)\,dy.$$

While f is a function of one variable, $f(x-y)$ can be interpreted as a function of two variables, hence the use of the partial derivative notation in the integrand. If for example $f \in C_c^1(\mathbb{R})$, one can verify that the hypotheses of Theorem A.10 hold true and F is $C^1(\mathbb{R})$. Moreover, if $f \in C_c^\infty(\mathbb{R})$, then F is $C^\infty(\mathbb{R})$. Note that the only requirement on the function g is integrability; g need not be differentiable at any point and in fact need not even be continuous on $\mathbb{R}$.

On the other hand, the commutative property of convolution means that we can also compute the derivative of F as

$$F'(x) \;=\; (g*f)'(x) \;=\; \frac{d}{dx}\int_{-\infty}^{\infty} g(x-y)\,f(y)\,dy \;=\; \int_{-\infty}^{\infty} \frac{\partial g(x-y)}{\partial x}\,f(y)\,dy.$$

Hence we could alternatively rely on g to provide the required smoothness and decay, with f only integrable. We can summarize the above observations as follows: Under certain smoothness and decay assumptions on, respectively, f and g, we have

$$\boxed{(f*g)'(x) \;=\; (f'*g)(x) \;=\; (f*g')(x).}$$

6.3.3. • Convolution and the Fourier Transform. One of the amazing properties of the Fourier transform concerns convolution. Following the remark closing the last subsection, one can verify that the convolution of two integrable functions is integrable, and hence we may take the Fourier transform of the convolution to find the following:

Convolution and the Fourier Transform

$$\widehat{(f * g)}(k) = \hat{f}(k)\,\hat{g}(k). \tag{6.20}$$

Thus convolution, a rather complicated and "convoluted" form of multiplication in real space, is **transformed into ordinary pointwise multiplication** in Fourier space. This is another indication of the potential power of the Fourier transform.

It is straightforward to show/prove (6.20). To this end, by definition of the Fourier transform

$$\begin{aligned}
\widehat{(f * g)}(k) &= \int_{-\infty}^{\infty} (f * g)(x) e^{-ikx}\, dx \\
&= \int_{-\infty}^{\infty} \left(\int_{-\infty}^{\infty} f(x-y) g(y)\, dy \right) e^{-ikx}\, dx \\
&= \int_{-\infty}^{\infty} \int_{-\infty}^{\infty} f(x-y) e^{-ik(x-y)} g(y)\, e^{-iky}\, dy\, dx \\
&= \int_{-\infty}^{\infty} \left(\int_{-\infty}^{\infty} f(x-y)\, e^{-ik(x-y)}\, dx \right) g(y)\, e^{-iky}\, dy.
\end{aligned}$$

Note that to pass to the third line we multiplied the integrand by 1 in the form of $e^{iky}e^{-iky}$. To pass to the last line we made a legal change in the order of integration.

A change of variable $z = x - y$ $(dz = dx)$ in the inner integral with respect to x yields

$$\int_{-\infty}^{\infty} f(x-y)\, e^{-ik(x-y)}\, dx = \int_{-\infty}^{\infty} f(z)\, e^{-ikz}\, dz.$$

Hence, the integral above is independent of y and

$$\begin{aligned}
\widehat{(f * g)}(k) &= \int_{-\infty}^{\infty} \left(\int_{-\infty}^{\infty} f(z)\, e^{-ikz}\, dz \right) g(y)\, e^{-iky}\, dy \\
&= \left(\int_{-\infty}^{\infty} f(z)\, e^{-ikz}\, dz \right) \left(\int_{-\infty}^{\infty} g(y)\, e^{-iky}\, dy \right) \\
&= \hat{f}(k)\, \hat{g}(k).
\end{aligned}$$

In terms of the inverse Fourier transform, one can write (6.20) as

$$\boxed{\left(\hat{f}(k)\hat{g}(k)\right)^{\vee} = (f * g)(x).} \tag{6.21}$$

One can also check that if f and g are integrable functions such that fg is also integrable, then

$$\boxed{\widehat{f(x)g(x)}(k) = \frac{1}{2\pi}(\hat{f} * \hat{g})(k).} \tag{6.22}$$

6.3.4. Convolution as a Way of Smoothing Out Functions and Generating Test Functions in C_c^∞. We pause for a moment, breaking away from the Fourier transform, to note that convolution is a natural way to smooth out functions, and, in doing so, generate many functions in $C_c^\infty(\mathbb{R})$. Indeed, the skeptical reader might have a bone of contention with the previous **long** chapter on distributions. Our theory of distributions was presented with the underlying premise that one can capture distributions (more general objects than functions) by sampling them on test functions in $C_c^\infty(\mathbb{R})$. Hence there better be "*lots and lots*" of such test functions! We gave one example, the *bump function*:

$$\phi_1(x) := \begin{cases} e^{-\frac{1}{1-x^2}} & \text{if } |x| < 1, \\ 0 & \text{if } |x| \geq 1. \end{cases}$$

This begs the question: **What are other examples?**

While other examples may not have a nice closed-form formula, let us now use convolution and our **one** specific example ϕ_1 to show that given any continuous function f with compact support (this is easy to find), we can generate a function in $C_c^\infty(\mathbb{R})$ which is "arbitrarily close" to f. In fact this algorithm works even if f is not continuous.

To this end, let

$$c_* := \int_{-\infty}^{\infty} \phi_1(x)\, dx,$$

which is some finite positive number (we do not really care about its exact value). Next, we normalize ϕ_1 by defining

$$\phi^*(x) = \frac{1}{c_*}\phi_1(x),$$

and we define for any $n = 1, 2, \ldots$ the functions $\Phi_n(x) := n\,\phi^*(nx)$. Note that these definitions are motivated by the following two facts:

- For each n
$$\int_{-\infty}^{\infty} \Phi_n(x)\, dx = 1.$$
- The support of $\Phi_n(x)$ is the interval $\left[-\frac{1}{n}, \frac{1}{n}\right]$.

In particular, Φ_n will converge to δ_0 in the sense of distributions (the proof is almost identical to that of Theorem 5.1). These facts have nice consequences from the perspective of convolution.

Now consider **any** function f with compact support (the function need not even be continuous!). Define for each n, the new function

$$\boxed{f_n(x) := (\Phi_n * f)(x) = \int_{-\infty}^{\infty} \Phi_n(x-y)\, f(y)\, dy.}$$

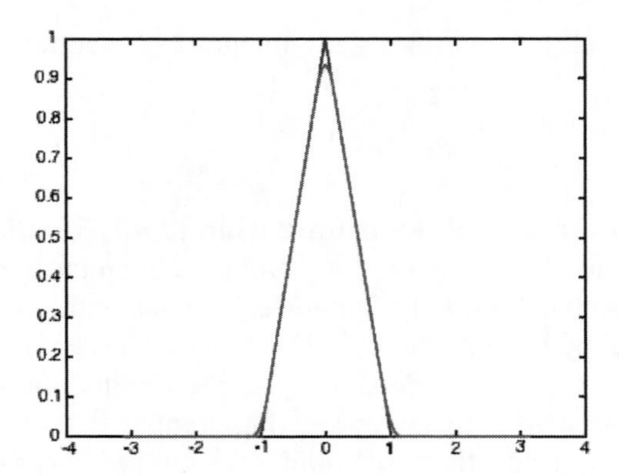

Figure 6.3. Plot of the spike $f(x)$ (taller) and $(\Phi_5 * f)(x) \in C_c^\infty(\mathbb{R})$ (shorter).

Then, for each n, the following hold:

- $f_n(x) \in C_c^\infty(\mathbb{R})$.
- For n large, f_n is a good approximation (is "close to") $f(x)$.

It is straightforward to check that f_n must have compact support. The fact that f_n is C^∞ directly follows from Section 6.3.2. You can also see intuitively why f_n must be smooth by viewing the convolution $f_n(x)$ as giving a weighted average of f over the interval $\left[x - \frac{1}{n}, x + \frac{1}{n}\right]$. For any fixed n, this weighted average will "average out" any singularities in f. Moreover, as n becomes larger we obtain a better and better approximation to f.

While all these statements can be made precise and proven, here we present a simple, yet rather convincing, graphical illustration. Let f be the spike function

$$f(x) = \begin{cases} 1 - x & \text{if } 0 \leq x < 1, \\ 1 + x & \text{if } -1 < x \leq 0, \\ 0 & \text{if } |x| \geq 1. \end{cases}$$

Note that this was the first element of the sequence of spikes defined by (5.23) in the previous chapter. Figure 6.3 plots both $f(x)$ and $f_5(x) := (\Phi_5 * f)(x)$. While we have not been precise here in terms of defining what "close" means, we think you will agree that the two functions are visually close.

6.4. • Other Important Properties of the Fourier Transform

The following are a few important properties of the Fourier transform. It is a straightforward exercise to show the first three (cf. Exercise **6.2**).

(i) Translation and frequency shifting:

$$\boxed{\widehat{f(x-a)} = e^{-iak}\hat{f}(k) \qquad \text{and} \qquad \widehat{e^{iax}f(x)} = \hat{f}(k-a).} \tag{6.23}$$

Note that for the second equality, we are taking the Fourier transform of a complex-valued function. However, this can readily be done by taking the Fourier transform of its real and complex parts separately.

(ii) Scaling: For any $a > 0$, let

$$f_a(x) = \frac{1}{a} f\left(\frac{x}{a}\right) \quad \text{and} \quad \widehat{f}_a(k) = \frac{1}{a}\widehat{f}\left(\frac{k}{a}\right).$$

Then

$$\boxed{\widehat{f_a(x)} = \widehat{f}(ak) \quad \text{and} \quad \widehat{f(ax)} = \widehat{f}_a(k).} \tag{6.24}$$

(iii) Polynomial multiplication in real space and differentiation in Fourier space: This is the dual statement of the fact that the Fourier transform "transforms" differentiation in real space into polynomial multiplication in Fourier space.

$$\boxed{\widehat{xf(x)} = i\frac{d\widehat{f}(k)}{dk}.} \tag{6.25}$$

(iv) Invariance of a Gaussian: The familiar bell-shaped curve $e^{-\frac{x^2}{2}}$, often dubbed a Gaussian, plays an important role in mathematics and physics. Indeed, you should recognize it from probability and the normal distribution. Gaussians are special functions for the Fourier transform in that they are essentially left-invariant. That is, the Fourier transform of a Gaussian is also (essentially) a Gaussian. For example,

$$\boxed{\text{if } f(x) = e^{-\frac{x^2}{2}}, \quad \text{then } \widehat{f}(k) = \sqrt{2\pi}e^{-\frac{k^2}{2}}.} \tag{6.26}$$

One can readily show (6.26) by direct computation of the Fourier transform integral. However, this does require a tiny bit of complex (contour) integration. So instead, we give a slick indirect approach which foreshadows the use of the Fourier transform to solve differential equations.

It is straightforward to see that both $f(x)$ and $xf(x)$ are integrable. Note that the tails of both functions (values for large $|x|$) decay very rapidly as $|x| \to \infty$. For notational simplicity, let us denote $\widehat{f}(k)$ by $F(k)$. Note that $f(x)$ satisfies the ODE

$$f'(x) + xf(x) = 0.$$

Now, we take the Fourier transform of both sides of this equation and use properties (6.7) and (6.25) to find that

$$ikF(k) + iF'(k) = 0 \quad \text{or} \quad \frac{F'(k)}{F(k)} = -k.$$

This is easy to solve by integrating with respect to k: We find

$$\log F(k) = -\frac{k^2}{2} + C \quad \text{or} \quad F(k) = C_1 e^{-\frac{k^2}{2}},$$

for some constant C with $C_1 := e^C$. To find the positive constant C_1, we evaluate the definition of the Fourier transform at $k = 0$:

$$C_1 = F(0) = \int_{-\infty}^{\infty} e^{-\frac{x^2}{2}}\, dx = \sqrt{2\pi}.$$

The last integral is a famous integral in probability; see Exercise **6.4** for a quick way to show it is $\sqrt{2\pi}$. This yields (6.26).

More generally, one has the following for any $a > 0$:

$$\boxed{\text{If } f(x) = e^{-ax^2}, \quad \text{then} \quad \widehat{f}(k) = \frac{\sqrt{\pi}}{\sqrt{a}} e^{-\frac{k^2}{4a}}.} \tag{6.27}$$

6.5. Duality: Decay at Infinity vs. Smoothness

When comparing functions with their Fourier transforms, two important properties present an interesting duality:

(i) Smoothness of the function is based upon continuity and differentiability. One way in which to **quantify** the level of smoothness is via the spaces $C^0(\mathbb{R}) = C(\mathbb{R})$, $C^1(\mathbb{R})$, $C^2(\mathbb{R})$, $\ldots$, $C^\infty(\mathbb{R})$ (cf. Section A.2.1). Note that $f \in C^m(\mathbb{R})$ is its m-th derivative and is continuous; i.e., $f^{(m)}(x) \in C(\mathbb{R})$.

(ii) Decay at infinity is based upon how fast the function tends to 0 as $x \to \pm\infty$. By example, note that the following functions all decay to 0 at infinity but with progressively faster rates: $\frac{1}{x}, \frac{1}{x^2}, \frac{1}{x^3}, \ldots, e^{-|x|}$. The most direct way to **quantify** the rate of decay of $f(x)$ is that for some $m = 0, 1, 2, \ldots,$

$$|x^m f(x)| \to 0 \quad \text{as} \quad |x| \to \infty.$$

An alternative way to quantify the rate of decay of $f(x)$ is via integrability of $x^m f(x)$; i.e.,

$$\int_{-\infty}^{\infty} |x^m f(x)|\, dx < \infty.$$

The smoothness of a function f is directly related to the decay of its Fourier transform, and vice versa. In particular, the following statements between the function and its Fourier transform hold true:

Duality: Smoothness and Decay

$$\underbrace{\int_{-\infty}^{\infty} |x^m f(x)|\, dx < \infty}_{\text{decay of } f \text{ of order } m} \quad \Longrightarrow \quad \underbrace{\widehat{f}(k) \in C^m(\mathbb{R})}_{\text{smoothness of } \widehat{f} \text{ of order } m},$$

$$\underbrace{f \in C^m(\mathbb{R}), f^{(m)} \text{ integrable}}_{\text{smoothness of } f \text{ of order } m} \quad \Longrightarrow \quad \underbrace{\widehat{f} \in C(\mathbb{R}),\ |k^m \widehat{f}(k)| \to 0 \text{ as } |k| \to \infty}_{\text{decay of } \widehat{f} \text{ of order } m}.$$

The analogous statements hold true when reversing the roles of f and $\widehat{f}$. We can summarize in words: **(i) The smoother the function, the faster its Fourier transform decays at infinity; (ii) the faster the function decays at infinity, the smoother is its Fourier transform.**

While we will neither state nor prove a full if and only if theorem[6], surely some explanation is needed for the proof of the above implications. The essentials follow from the definition of the Fourier transform, the Fourier inversion formula, and the differentiation property.

First, let us argue that if f is integrable, then the Fourier transform $\widehat{f}(k)$ is a continuous functions of k. To see this, let us recall the definition,

$$\widehat{f}(k) = \int_{-\infty}^{\infty} f(x)\, e^{-ikx}\, dx.$$

We want to show that a small perturbation in the value of k results in a controlled small perturbation in the value of $\widehat{f}(k)$. That is, we want to show that if $|k_1 - k_2|$ is small, then $|\widehat{f}(k_1) - \widehat{f}(k_2)|$ is small, where $|\cdot|$ denotes the modulus of a complex number. The dependence on k in the definition of $\widehat{f}$ lies entirely in the function e^{-ikx} inside the integral. This function is a continuous complex-valued function; i.e., it has continuous real and imaginary parts. The subtlety here is the presence and dependence on $x \in \mathbb{R}$ (the variable of integration) in e^{-ikx}. To overcome this, we will require the interchange of limits and integrals result, Theorem A.8, our watered-down version of the famous Lebesgue Dominated Convergence Theorem from measure theory. For completeness, we will provide some details. Fix $k_0 \in \mathbb{R}$ with a goal of proving the continuity of $\widehat{f}(k)$ at $k = k_0$. This amounts to proving that $\lim_{k\to k_0} \widehat{f}(k) = \widehat{f}(k_0)$ or $\lim_{k\to k_0} |\widehat{f}(k) - \widehat{f}(k_0)| = 0$. Since

$$\lim_{k\to k_0} |\widehat{f}(k) - \widehat{f}(k_0)| = \lim_{k\to k_0} \left| \int_{-\infty}^{\infty} f(x)\,(e^{-ikx} - e^{-ik_0x})\, dx \right|,$$

it suffices to prove

$$\lim_{k\to k_0} \int_{-\infty}^{\infty} |f(x)|\, |e^{-ikx} - e^{-ik_0x}|\, dx = 0.$$

The issue we need to address is bringing the limit into the integral sign. This is the realm of Theorem A.8 with the caveat that there we took a limit as $n \to \infty$ of a sequence of functions, while here we are taking a limit $k \to k_0$ of a continuum of functions indexed by k. The same theorem with such a continuum limit does hold true. Setting $F_k(x) := |f(x)||e^{-ikx} - e^{-ik_0x}|$, we note that for each $x \in \mathbb{R}$, $\lim_{k\to k_0} F_k(x) = 0$, and for all $x, k \in \mathbb{R}$, $|F_k(x)| \leq 2|f(x)|$. Since $f(x)$ is integrable, Theorem A.8 (in its continuum-limit version) implies

$$\lim_{k\to k_0} \int_{-\infty}^{\infty} F_k(x) = \int_{-\infty}^{\infty} \left(\lim_{k\to k_0} F_k(x) \right) = 0.$$

One can reverse this argument via the Fourier inversion formula and come to the conclusion that if $\widehat{f}(k)$ is integrable, then f must have been continuous. So if f is not continuous, for example the discontinuous function of Example 6.2.1, then its Fourier transform will not be integrable.

[6]The framework of **Sobolev spaces** (see Section 9.8 for a very brief introduction) provides the right setting for a complete statement.

Next, we suppose $\widehat{f}$ is integrable and continuous. In this scenario, consider the Fourier inversion formula

$$f(x) = \frac{1}{2\pi}\int_{-\infty}^{\infty} \widehat{f}(k)\, e^{ikx}\, dk.$$

From this formula, we wish to determine if f is differentiable and if its derivative is integrable. The dependence on x is entirely confined to the function e^{ikx} in the integrand. It has C^∞ real and imaginary parts. Hence, we will be able to differentiate under the integral sign (cf. Section A.9) and find that

$$f'(x) = \frac{1}{2\pi}\int_{-\infty}^{\infty} (ik)\, \widehat{f}(k)\, e^{ikx}\, dk,$$

as long as $k\widehat{f}(k)$ is integrable; i.e., $\int_{-\infty}^{\infty} |k\widehat{f}(k)|\, dk < \infty$, where here $|\cdot|$ denotes the complex modulus. Moreover, following the previous argument for the continuity of $\widehat{f}$, we find that if $k\widehat{f}(k)$ is integrable, then $f'(x)$ is continuous; i.e., $f \in C^1(\mathbb{R})$. Similarly, continuity of the m-th derivative $f^{(m)}$, i.e., whether $f \in C^m(\mathbb{R})$, reduces to integrability of $k^m \widehat{f}$.

The reader is encouraged to **test** the previous duality statement by examining Fourier transforms of particular functions. For instance, note that the function $\frac{1}{x^2+a^2}$ for fixed $a > 0$ is an infinitely smooth function, i.e., in $C^\infty(\mathbb{R})$, and its Fourier transform $\frac{\pi}{a}e^{-a|k|}$ decays very fast, indeed of degree m for every m. On the other hand, the function $\frac{1}{x^2+a^2}$ has only 0 rate of decay and hence its Fourier transform is not even C^1; note the "corner effect" associated with $|k|$.

6.6. Plancherel's Theorem and the Riemann-Lebesgue Lemma

In this section we present two important and useful results on the Fourier transform. We begin with a word on two important classes of functions and the Riemann vs. Lebesgue integral.

6.6.1. The Spaces $L^1(\mathbb{R})$ and $L^2(\mathbb{R})$. Recall the class of **integrable functions**, namely functions f with $\int_{-\infty}^{\infty} |f(x)|\, dx < \infty$, which is often denoted by $L^1(\mathbb{R})$. Here, the "1" refers to the power in the integrand. Another important class of functions is those which are **square integrable**; namely

$$\int_{-\infty}^{\infty} |f(x)|^2\, dx < \infty.$$

This class is usually denoted by $L^2(\mathbb{R})$. In order to make these classes "complete" spaces, we need a more sophisticated notion of integration, namely the Lebesgue integral. In this text, all integrals are defined as the usual Riemann integral from your calculus class. While this rather rigid notion of integration suffices for the functions considered in this text, proofs of many results, including Plancherel's Theorem below, are not easily presented without the more robust and indeed beautifully complete Lebesgue theory.

On $\mathbb{R}$, neither class is contained in the other; i.e., $L^1(\mathbb{R}) \not\subset L^2(\mathbb{R})$ and $L^2(\mathbb{R}) \not\subset L^1(\mathbb{R})$. For example, the function $\frac{1}{1+|x|}$ is in $L^2(\mathbb{R})$ but not in $L^1(\mathbb{R})$. On the other hand, the function

$$\begin{cases} \frac{1}{\sqrt{|x|}}, & -1 < x < 0 \text{ or } 0 < x < 1, \\ 0, & \text{otherwise}, \end{cases} \qquad \text{is in } L^1(\mathbb{R}) \text{ but not in } L^2(\mathbb{R}).$$

Note here that the necessary properties that entail entrance to one space and not the other are either the decay rate at infinity **or** the rate of blow-up around a singularity (here at $x = 0$).

6.6.2. Extending the Fourier Transform to Square-Integrable Functions and Plancherel's Theorem. Thus far we have defined the Fourier transform for any integrable complex-valued function $f(x)$ on $\mathbb{R}$, that is, for any function satisfying $\int_{-\infty}^{\infty} |f(x)|\, dx < \infty$. Indeed, complex-valued $L^1(\mathbb{R})$ functions provide a natural space upon which to define the Fourier transform. However, recall from Example 6.2.1 that the Fourier transform of an integrable function need not be integrable. Given the back and forth nature of the Fourier and inverse Fourier transforms, it is highly desirable to seek a large space of functions with the property that the Fourier (or inverse Fourier) transform of any member of the space is also a member of the space. There is such a space and it is $L^2(\mathbb{R})$, the space of all square-integrable complex-valued functions. Since, in general, a function in $L^2(\mathbb{R})$ may not be integrable, it is not clear how to define the Fourier transform on L^2. The key to extending the Fourier transform to L^2 is a fundamental result known as Plancherel's Theorem[7].

Plancherel's Theorem

Theorem 6.2. *If $f \in L^1(\mathbb{R}) \cap L^2(\mathbb{R})$, then*

$$\int_{-\infty}^{\infty} |f(x)|^2\, dx = \frac{1}{2\pi} \int_{-\infty}^{\infty} |\widehat{f}(k)|^2\, dk. \tag{6.28}$$

With (6.28) in hand, one can **extend** the Fourier transform in "a natural way" **to all** complex-valued functions in $L^2(\mathbb{R})$ (even ones which are not $L^1(\mathbb{R})$) so that their Fourier transforms are also square integrable, and (6.28) holds true for all $f \in L^2(\mathbb{R})$. For a variety of reasons, one often refers to the integral of the squares as the **energy** associated with the function. Thus Plancherel's Theorem can be viewed as a statement about conservation of energy in real and Fourier space (up to a factor of 2π[8]). Mathematicians often phrase the theorem by saying that *the Fourier transform extends to become a* ***bijective isometry*** *on the space of square-integrable functions.* By a bijective isometry we mean a map from $L^2(\mathbb{R})$ to $L^2(\mathbb{R})$ which is both one-to-one and onto and preserves the "length" of the function (up to a factor of 2π), where the length (or size) of a function $f(x) \in L^2(\mathbb{R})$ is defined to be $\|f\| = \left(\int_{-\infty}^{\infty} |f(x)|^2\, dx\right)^{1/2}$.

[7] **Michel Plancherel** (1885–1967) was a Swiss mathematician who is best known for this result.

[8] An alternate definition of the Fourier transform is given by (6.31) and presented in the next subsection. With this definition, one obtains (6.28) without any prefactor on the right-hand side, that is, pure equality for the integral of the squares.

This extension of the Fourier transform allows us to assign Fourier transforms to square-integrable functions which are not integrable, for example $\frac{\sin ax}{x}$. Recall from Example 6.2.1 that the Fourier transform of the characteristic function of the interval $[-a, a]$ was $2\frac{\sin ak}{k}$, which was not integrable. However, you can check that it is square integrable and hence this extension of the Fourier transform to $L^2(\mathbb{R})$ allows us, in a natural way, to conclude that the Fourier transform of $\frac{\sin ax}{x}$ is $\pi\chi_a(k)$, where χ_a is the characteristic function of the interval $(-a, a)$.

To construct this extension to all of L^2 and indeed present a proof of Plancherel's Theorem, one needs more advanced notions and tools based upon the *Lebesgue integral*, specifically the notions of *dense sets* and *completeness* in the *space* L^2. These concepts are usually presented in a graduate-level analysis course, and one can find such a complete rigorous treatment of Plancherel and this extension in [**10**]. Nevetheless, we will end this section with an intuitive explanation of (6.28) via the delta function.

An Explanation of Plancherel's Theorem (i.e., (6.28)) via the Delta Function. While we will neither state precisely nor prove Plancherel's Theorem, it is not hard to justify (6.28) via our devious delta function. Indeed, let us show that for suitably nice complex-valued functions f and g,

$$\int_{-\infty}^{\infty} f(x)\,\overline{g}(x)\,dx \;=\; \frac{1}{2\pi}\int_{-\infty}^{\infty} \widehat{f}(k)\,\overline{\widehat{g}}(k)\,dk. \tag{6.29}$$

Then taking $f = g$ will yield (6.28). To this end, we use the Fourier inversion formula for $f(x)$ and $g(x)$, and for the latter note that

$$\overline{g}(x) \;=\; \overline{\frac{1}{2\pi}\int_{-\infty}^{\infty} \widehat{g}(w)\,e^{iwx}\,dw} \;=\; \frac{1}{2\pi}\int_{-\infty}^{\infty} \overline{\widehat{g}(w)}\,e^{-iwx}\,dw,$$

since $\overline{e^{iwx}} = e^{-iwx}$. Now, suppose we make an illegal change in the order of integration to find

$$\int_{-\infty}^{\infty} f(x)\,\overline{g}(x)\,dx = \frac{1}{(2\pi)^2}\int_{-\infty}^{\infty}\left(\int_{-\infty}^{\infty}\widehat{f}(k)\,e^{ikx}\,dk\right)\left(\int_{-\infty}^{\infty}\overline{\widehat{g}(w)}\,e^{-iwx}\,dw\right)dx$$

$$\overset{\substack{\textbf{illegal}\text{ change in}\\ \text{order of integration}}}{=} \text{“}\frac{1}{(2\pi)^2}\int_{-\infty}^{\infty}\widehat{f}(k)\left[\int_{-\infty}^{\infty}\overline{\widehat{g}(w)}\underbrace{\left(\int_{-\infty}^{\infty}e^{-ix(w-k)}\,dx\right)}_{=\,2\pi\delta_{w=k}}dw\right]dk\text{”}$$

$$= \frac{1}{2\pi}\int_{-\infty}^{\infty}\widehat{f}(k)\,\overline{\widehat{g}}(k)\,dk.$$

Recall from Section 6.2.3 that the ill-defined x integral in the parentheses (of the second line) with w as the variable is, if interpreted in the right way, a delta function with concentration at $w = k$. One can view $\overline{\widehat{g}(w)}$ as the test function and, hence, the effect of "integrating" this delta function against $\overline{\widehat{g}(w)}$ with respect to w gives $\overline{\widehat{g}(k)}$. As in Section 6.2.3 one can address this ill-defined improper integral via truncated limits.

6.6.3. The Riemann-Lebesgue Lemma.

The Riemann-Lebesgue Lemma

For every complex-valued ϕ which is integrable, we have

$$|\widehat{\phi}(k)| \longrightarrow 0 \qquad \text{as } |k| \to \infty.$$

In particular, if ϕ is a real-valued integrable function, we have

$$\int_{-\infty}^{\infty} \phi(x)\, \sin(kx)\, dx \longrightarrow 0 \qquad \text{as } k \to \infty. \tag{6.30}$$

Note that (6.30) holds true with $\sin(kx)$ replaced with either $\cos(kx)$ or e^{ikx}. In the context of (6.30), the Riemann Lebesgue Lemma is a continuum version (i.e., $k \to \infty$ with $k \in \mathbb{R}$ vs. $n \to \infty$) of the limit considered in Section 5.5.5; however, it now holds not only for $\phi \in C_c^\infty(\mathbb{R})$, but for any integrable function.

Whereas the Lebesgue integral furnishes a very simple proof of the Riemann-Lebesgue Lemma, one can also prove it directly with the Riemann integral (cf. Exercise **6.23**).

6.7. The 2π Issue and Other Possible Definitions of the Fourier Transform

There are several other ways to define the Fourier transform depending on where one includes the factor of 2π. However there is no way of completely eliminating a factor of 2π; it will always be present somewhere. In our definition, this factor appeared in the inverse Fourier transform. Conversely, many mathematicians would argue quite strongly that the "best" place to include the 2π is via the exponent of the exponentials; i.e., they would define the Fourier transform by

$$\widehat{f}(k) := \int_{-\infty}^{\infty} f(x)\, e^{-2\pi ikx}\, dx. \tag{6.31}$$

In this case, the inverse Fourier transform can simply be written as

$$\check{g}(x) := \int_{-\infty}^{\infty} g(k)\, e^{2\pi ikx}\, dk. \tag{6.32}$$

Observe now the perfect symmetry in the two formulas. Moreover, this definition would yield Plancherel's equality (6.28) without any factor of 2π; i.e., the Fourier transform would indeed become a perfect isometry on square-integrable functions.

While we agree with this preference, the reason why we will adopt the earlier definition (6.5) is that it is widely adopted in the applied mathematics, partial differential equations, physics, and engineering literature. Hence, for the majority of readers, it will prove consistent with the definition they are most likely to encounter in other courses and in the future.

In addition to the placement of the 2π constant, some authors differ as to the placement of the minus sign in the exponent (the one before kx); it can appear in the definition of the inverse Fourier transform (the Fourier inversion formula) rather than in the definition of the Fourier transform.

As a rule of thumb, whenever you find a reference to the Fourier transform online, in a book, in an article, in code, or in a software package, check first which definition they use. It may be necessary to adjust the constants if their definition is different from the one you are expecting.

6.8. • Using the Fourier Transform to Solve Linear PDEs, I: The Diffusion Equation

With the inversion formula in hand and the fact that differentiation is transformed into polynomial multiplication, a **strategy** comes into play for solving linear PDEs involving space x and time t with data at $t = 0$:

(1) We Fourier transform both sides of the PDE in the space variable and also Fourier transform the data at $t = 0$.

(2) Solve for $\hat{u}$ using the fact that in the transformed variables, the differentiation is now algebraic. For each k, this results in an ordinary differential equation (ODE) for $\hat{u}(k,t)$ with initial condition given by the Fourier transform of the data at k.

(3) Solve the ODE initial value problem for $\hat{u}(k,t)$ at any time t.

(4) Take the inverse Fourier transform in the spatial variables to obtain $u(x,t)$. Here you will often need the fact that multiplication in Fourier space is convolution in real space.

6.8.1. • Prelude: Using the Fourier Transform to Solve a Linear ODE.

We begin with a purely spatial example, in order to first illustrate Steps (1), (2), and (4) above. For any fixed continuous and integrable function $f(x)$ on $\mathbb{R}$, let us use the Fourier transform to find a solution $y(x)$ to the ODE

$$y''(x) - y(x) = f(x), \tag{6.33}$$

on all of $\mathbb{R}$.

We start by taking the Fourier transform in the variable x of both sides of the ODE. By using the linearity of the Fourier transform (the Fourier transform of a sum is the sum of the Fourier transforms), we find

$$\widehat{y''}(k) - \hat{y}(k) = \hat{f}(k).$$

Note that

$$\widehat{y''}(k) = (ik)^2\hat{y}(k).$$

Thus, we find

$$-\left(1+k^2\right)\hat{y}(k) = \hat{f}(k).$$

This is an algebraic equation in k and, hence,

$$\hat{y}(k) = -\frac{\hat{f}(k)}{(1+k^2)}.$$

We now take the inverse Fourier transform of both sides to find

$$y(x) = \left(-\frac{\widehat{f}(k)}{(1+k^2)}\right)^{\vee}(x) = -\left(\widehat{f}(k)\,\frac{1}{(1+k^2)}\right)^{\vee}(x).$$

The right-hand side involves a product of two terms in Fourier space, and consequently the inverse Fourier transform of this product is the convolution of the inverse Fourier transforms of each term. In other words, if

$$g(x) = \left(\frac{1}{(1+k^2)}\right)^{\vee},$$

then

$$\left(\widehat{f}(k)\,\frac{1}{(1+k^2)}\right)^{\vee}(x) = (f * g)(x).$$

One can check that g is given by

$$g(x) = \frac{e^{-|x|}}{2}.$$

Hence, a solution is

$$y(x) = -(f * g)(x) = -\frac{1}{2}\int_{-\infty}^{\infty} f(x-z)\,e^{-|z|}\,dz. \tag{6.34}$$

Stopping here, we have just found **one** solution to the ODE on all of $\mathbb{R}$. But are we sure this formula for $y(x)$ does indeed yield a solution? Check this (cf. Exercise **6.14**).

6.8.2. • Using the Fourier Transform to Solve the Diffusion Equation.

We use the Fourier transform to solve the initial value problem for the diffusion equation:

$$\begin{cases} u_t = \alpha u_{xx}, \\ u(x,0) = g(x). \end{cases} \tag{6.35}$$

This is a fundamental equation and, hence, an extremely important calculation which will be the basis for Chapter 7. Here, let us not discuss the origin of the PDE nor its interpretations, but rather just use the Fourier transform to solve it (i.e., derive a solution formula). We add here that we will spend much time discussing the implications of this solution formula in Section 7.2.

Let us take $\alpha = 1$ for notational simplicity and rewrite the diffusion equation as $u_t - u_{xx} = 0$. We assume that we have a solution $u(x,t)$ which, for every t, is an integrable function of x. Once we have derived a solution formula for u, we can circle back to check that it is indeed a solution and, further, satisfies the integrability condition (under suitable assumptions on the initial data).

Fix $t > 0$ for the moment, and Fourier transform both sides of the equation $u_t - u_{xx} = 0$ with respect to x. Using the linearity of the Fourier transform and the fact that the Fourier transform of 0 is 0, we have

$$(\widehat{u_t - u_{xx}})(k,t) = (\widehat{u_t})(k,t) - (\widehat{u_{xx}})(k,t) = 0.$$

For the first term, we note that the Fourier transform in x of u_t is simply the derivative with respect to t of the Fourier transform in x of u. Indeed, by definition of the Fourier transform we have

$$\begin{aligned}\widehat{(u_t)}(k,t) = \int_{-\infty}^{\infty} u_t(x,t)\,e^{-ikx}\,dx &= \frac{\partial}{\partial t}\left(\int_{-\infty}^{\infty} u(x,t)\,e^{-ikx}\,dx\right)\\ &= \frac{\partial\,\hat{u}(k,t)}{\partial t} = \hat{u}_t(k,t).\end{aligned}$$

Note that on the left the hat is over u_t whereas on the right it is just over u. Here, we used the fact that the t derivative can be inside or outside the integral (cf. Section A.9). In other words, the operators of time differentiation and space Fourier transformation commute. For the second term, we use the property of the Fourier transform and differentiation to find

$$\widehat{u_{xx}}(k,t) = (ik)^2\hat{u}(k,t) = -k^2\hat{u}(k,t).$$

Thus, we have

$$\hat{u}_t(k,t) + k^2\hat{u}(k,t) = 0$$

with

$$\hat{u}(k,0) = \hat{g}(k).$$

For each k, this is an ODE for $f(t) := \hat{u}(k,t)$:

$$f'(t) = -k^2 f(t) \quad \text{with} \quad f(0) = \hat{g}(k).$$

The solution is $f(t) = f(0)\,e^{-k^2t}$ and, hence,

$$\hat{u}(k,t) = \hat{g}(k)\,e^{-k^2t}.$$

Taking the inverse Fourier transform and noting how the transform behaves with multiplication/convolution (cf. (6.21)), we find

$$u(x,t) = (\hat{g}(k)e^{-k^2t})^{\vee} = (g*F)(x,t) = (F*g)(x,t)$$

where

$$F = (e^{-k^2t})^{\vee} = \mathcal{F}^{-1}(e^{-k^2t}).$$

By (6.27), incorporating an additional constant factor, we have

$$F(x) = \frac{1}{\sqrt{4\pi t}}\,e^{-\frac{x^2}{4t}}.$$

Thus

$$\boxed{u(x,t) = \frac{1}{\sqrt{4\pi t}}\int_{-\infty}^{\infty} e^{-\frac{(x-y)^2}{4t}}\,g(y)\,dy.}$$

One can readily check that if we include a constant α in the IVP (6.35) with numerical value not necessarily 1, we arrive at

$$\boxed{u(x,t) = \frac{1}{\sqrt{4\pi\alpha t}}\int_{-\infty}^{\infty} e^{-\frac{(x-y)^2}{4\alpha t}}\,g(y)\,dy.} \tag{6.36}$$

6.9. The Fourier Transform of a Tempered Distribution

Note to the reader: It is possible to skip this section, but this would be at the expense of having to take at face value the Fourier transforms of certain non-integrable functions. However, if the reader is comfortable with the notion of a distribution, then this section will provide a natural extension.

Our justification for both the Fourier inversion formula and Plancherel's equality was based upon the informal fact that

$$\text{“}\hat{1}\text{”} = 2\pi\delta_0. \tag{6.37}$$

Recall from Section 6.2.3 that we justified (6.37) via the distributional convergence of $\frac{\sin nx}{\pi x}$ to δ_0, without actually proving this convergence. One of the purposes of this section is to provide a precise statement for (6.37) with proof. However, the fruits of this section extend far beyond this task. Fourier transforms of functions which are neither integrable nor square integrable turn out to be fundamental tools in physics, engineering, and mathematics. Examples include the following:

- the function $f(x) \equiv 1$,
- the Heaviside and Sgn functions,
- sines, cosines, and e^{-iax},
- the function $f(x) = 1/x$,
- polynomials in x.

For these functions, the pointwise definition of the Fourier and inverse Fourier transforms makes no sense. In fact, many of the most useful Fourier transforms in physics and engineering are those for which the definition (6.5) makes no sense! Although one can give intuitive justifications for the Fourier transforms of all these functions, it is just as easy and, in our opinion, far more satisfying, to simply generalize the notion of the Fourier transform (and inverse Fourier transform) to any distribution and then view these functions as distributions.

6.9.1. Can We Extend the Fourier Transform to Distributions?

Recall how we defined the derivative of a distribution by “placing” the derivative on the test function. Will this work for the Fourier transform? In other words, if F is a distribution, suppose we define $\widehat{F}$ to be a new distribution whose action on any test function ϕ is given by

$$\boxed{\langle \widehat{F}, \phi \rangle := \langle F, \widehat{\phi} \rangle.} \tag{6.38}$$

There are two immediate issues that arise:

- If $\phi \in C_c^\infty(\mathbb{R})$, then, in general, it is **not true** that $\widehat{\phi} \in C_c^\infty(\mathbb{R})$ since the Fourier transform need not have compact support and will in general be complex valued. This suggests that the real-valued space of test functions $C_c^\infty(\mathbb{R})$ is too restrictive, and we need to consider a larger class.
- This definition will **generalize** the Fourier transform of an integrable function **only if the following is true:** If f is an integrable function on $\mathbb{R}$ and we consider

f as the distribution F_f (the distribution generated by function f), then with the definition (6.38), is it true that

$$\widehat{F_f} = F_{\hat{f}} \ ? \tag{6.39}$$

By (6.38) and our interpretation of a function as a distribution, this amounts to whether or not[9]

$$\int_{-\infty}^{\infty} f(y)\,\hat{\phi}(y)\,dy = \int_{-\infty}^{\infty} \hat{f}(y)\,\phi(y)\,dy \qquad \text{for all test functions } \phi. \tag{6.40}$$

But this is clearly true since

$$\begin{aligned} \int_{-\infty}^{\infty} f(y)\,\hat{\phi}(y)\,dy &= \int_{-\infty}^{\infty} f(y) \int_{-\infty}^{\infty} \phi(z) e^{-iyz}\,dz\,dy \\ &= \int_{-\infty}^{\infty} \phi(z) \int_{-\infty}^{\infty} f(y) e^{-iyz}\,dy\,dz. \end{aligned}$$

Note that changing the order of integration above was legal since f and ϕ are integrable. So it appears the definition (6.38) will indeed generalize the Fourier transform. Good! But we still need to address the class of test functions.

A minor notational issue/possible confusion: You are likely accustomed to thinking of ϕ and $\hat{\phi}$ as functions of two different variables, x and k, respectively. In (6.40), we use both ϕ and $\hat{\phi}$ as test functions of the same underlying independent variable. It is important here to not get fixated on their arguments. In this context, they both represent test functions of the same underlying independent variable.

It is a good idea to first review Section 6.5 before proceeding further.

6.9.2. The Schwartz Class of Test Functions and Tempered Distributions. In order to address the class of test functions, we must relax the constraint of compact support. Our goal is to find a class **perfectly suited (or tempered to)** the Fourier transform in the sense that the Fourier transform and the inverse Fourier transform are well-defined on this class. In other words, the Fourier transform should map functions in this class to the same class. Recall from Section 6.5 that smoothness of a function is directly related to the decay of the Fourier transform and vice versa. This would suggest the following properties for our class of test functions on $\mathbb{R}$:

- complex-valued;
- very smooth in the sense that their real and imaginary parts are in $C^\infty(\mathbb{R})$;
- the function and all its derivatives decay very rapidly as $x \to \pm\infty$.

These properties are summarized in the following definition.

[9] Notice that we have used the variable y for both real and Fourier space.

The Definition of the Schwartz Class

Definition 6.9.1. The Schwartz class, denoted by $\mathcal{S}(\mathbb{R})$, consists of complex-valued C^∞ functions whose real and imaginary parts (and all their derivatives) **decay**[a] to 0 (as $x \to \pm\infty$) **faster** than any function of the form $\frac{1}{x^m}$ with $m \in \{1, 2, \dots\}$. More precisely, if $\phi = \phi_1 + i\phi_2$, we have for $i = 1, 2$,

$$\lim_{|x|\to\infty} \left| x^m \frac{d^k \phi_i}{dx^k}(x) \right| = 0 \quad \text{for all } m, k \in \mathbb{N}. \tag{6.41}$$

[a]Since we are dealing with C^∞ functions (i.e., continuous derivatives of all orders), the decay condition (6.41) can be phrased in terms of the boundedness condition:

$$\max_{x\in\mathbb{R}} \left(|x|^m \left| \frac{d^k \phi_i}{dx^k}(x) \right| \right) < \infty, \quad \text{for all } m, k \in \mathbb{N}.$$

First note that any real-valued (or complex-valued) C^∞ function with compact support is automatically in the Schwartz class. However, there is now room for functions which do not have compact support. A good case in point is the Gaussian $\phi(x) = e^{-x^2}$. Despite the fact that it does not have compact support, it is still in the Schwartz class since the tails of this Gaussian decay very rapidly; in particular, for any $m > 0$, $\lim_{|x|\to\infty} x^m e^{-x^2} = 0$. Moreover we readily see that the same holds true for any derivative of e^{-x^2}. On the other hand, C^∞ functions which either do not decay to 0 at infinity, e.g., $\phi(x) = x^2$, or do not decay fast enough to 0 at infinity, e.g., $\phi(x) = \frac{1}{1+x^2}$, are **not** in the Schwartz class.

The power gained by enforcing (6.41) to hold for all m and k is that the Fourier transform of any function in the Schwartz class is also in the Schwartz class. In fact, the Fourier transform can be viewed as a **one-to-one** map of $\mathcal{S}(\mathbb{R})$ **onto** itself: If $\phi \in \mathcal{S}(\mathbb{R})$, then $\hat{\phi} \in \mathcal{S}(\mathbb{R})$. Moreover, any $\phi \in \mathcal{S}(\mathbb{R})$ is the Fourier transform of some function in $\mathcal{S}(\mathbb{R})$ and, thus, the inverse Fourier transform can be applied to functions in $\mathcal{S}(\mathbb{R})$. This should be intuitively clear from our earlier discussion around differentiation and decay of the Fourier transform in Section 6.5 and is actually straightforward to prove (see Folland [**17**]).

The Definition of a Tempered Distribution

Definition 6.9.2. A distribution which is defined on the larger Schwartz class $\mathcal{S}(\mathbb{R})$ is called a **tempered distribution**. Precisely, F is a tempered distribution if it acts on functions in $\mathcal{S}(\mathbb{R})$ and produces a complex number: $\langle F, \phi \rangle \in \mathbb{C}$, and this action is linear and continuous[a] on $\mathcal{S}(\mathbb{R})$. The associated notion of convergence is called **convergence in the sense of tempered distributions**.

[a]Following our precise definition of continuity for regular distributions, continuous means that if $\phi_n \to \phi$ in $\mathcal{S}(\mathbb{R})$, then $\langle F, \phi_n \rangle \to \langle F, \phi \rangle$. However now we must define what it means for $\phi_n \to \phi$ in $\mathcal{S}(\mathbb{R})$. To this end, for any $m, k \in \mathbb{N}$, define for $\phi \in \mathcal{S}$, $\|\phi\|_{m,k} := \sup_{x\in\mathbb{R}} \left| x^m \frac{d\phi^{(k)}}{dx}(x) \right|$. Then $\phi_n \to \phi$ in $\mathcal{S}(\mathbb{R})$ if $\|\phi_n - \phi\|_{m,k} \to 0$ for all $m, k \in \mathbb{N}$.

Think of the adjective "tempered" as referring to a distribution which has been **tempered to** (i.e., given a specific disposition for) the Fourier transform. Since[10] $C_c^\infty(\mathbb{R}) \subset \mathcal{S}(\mathbb{R})$, any real-valued tempered distribution is automatically a distribution as defined in the previous chapter; that is,

$$\Big\{\text{real-valued tempered distributions}\Big\} \subset \Big\{\text{real-valued distribution}\Big\}.$$

Thus, many definitions and examples of (regular) distributions naturally carry over to tempered distributions. For example, any complex-valued locally integrable function, which does not grow too fast as $x \to \pm\infty$, naturally generates (or can be viewed as) a distribution F_f where for any $\phi \in \mathcal{S}(\mathbb{R})$,

$$\langle F_f, \phi \rangle = \int_{-\infty}^{\infty} \overline{f(x)} \phi(x)\, dx,$$

where $\overline{f(x)}$ denotes the complex conjugate of $f(x)$. The key here is that we need $f(x)\,\phi(x)$ to be integrable; thus, for example, the function x^n would generate a tempered distribution but the function e^x would not. On the other hand, the delta function is also a tempered distribution.

One can readily show that the notion of the derivative of a distribution from Section 5.4.2 directly carries over to tempered distributions: in particular, note that if $\phi \in \mathcal{S}(\mathbb{R})$, then so are all its derivatives. But unlike for distributions (where the test functions have compact support), we can now extend the notion of the Fourier transform to any tempered distribution.

The Fourier Transform of a Tempered Distribution

Definition 6.9.3. If F is a tempered distribution, the Fourier transform of F is the tempered distribution $\widehat{F}$ defined by

$$\langle \widehat{F}, \phi \rangle := \langle F, \widehat{\phi} \rangle \qquad \text{for any } \phi \in \mathcal{S}(\mathbb{R}).$$

Similarly, we define the inverse Fourier transform $\check{F}$ by

$$\langle \check{F}, \phi \rangle := \langle F, \check{\phi} \rangle \qquad \text{for any } \phi \in \mathcal{S}(\mathbb{R}).$$

So, for example, we can now legitimately compute the Fourier transform of the delta function. To this end, for any $\phi \in \mathcal{S}$, we have

$$\langle \widehat{\delta_0}, \phi \rangle = \langle \delta_0, \widehat{\phi} \rangle = \widehat{\phi}(0) = \int_{-\infty}^{\infty} \phi(x)\, dx = \langle 1, \phi \rangle.$$

Thus in the sense of tempered distributions, $\widehat{\delta_0} = 1$.

It is straightforward to see that the following Fourier inversion formula holds in the sense of tempered distributions: For any tempered distribution F, we have that

$$\boxed{\check{\widehat{F}} = F.} \tag{6.42}$$

[10]Actually, to be a distribution entails a continuity requirement with respect to the test functions and the two notions of continuity were slightly different. Hence in order for tempered distributions to be distributions, we do require more then the simple embedding $C_c^\infty(\mathbb{R}) \subset \mathcal{S}(\mathbb{R})$; rather we require that this is a "continuous embedding". One can check that this is indeed the case.

Indeed, for any $\phi \in \mathcal{S}$ we have

$$\left\langle \check{\widehat{F}}, \phi \right\rangle = \left\langle \widehat{F}, \check{\phi} \right\rangle = \langle F, \widehat{\check{\phi}} \rangle = \langle F, \phi \rangle,$$

where in the last equality we used the Fourier inversion formula for integrable functions.

For the analogous differentiation property of the Fourier transform — transformation of differentiation into polynomial multiplication — we need to define the product of a polynomial with a tempered distribution F. This is easy: Suppose F is a tempered distribution and $p(x)$ is any complex-valued polynomial in $x \in \mathbb{R}$. Then $p(x)F$ is the tempered distribution defined by

$$\langle p(x)F, \phi \rangle \ := \ \langle F, p(x)\phi(x) \rangle.$$

Then it is immediately apparent from the definitions that, in the sense of tempered distributions, we have

$$\boxed{\widehat{F'} \ = \ ik\widehat{F},} \tag{6.43}$$

where k is the underlying real variable of the distribution $\widehat{F}$, i.e., the independent variable of the test functions. What is critical here is the fact that if $\phi \in \mathcal{S}$, then so is $k^m \phi$ for any $m \in \mathbb{N}$.

One main reason for introducing tempered distributions is to enable us to take the Fourier transform of functions which are neither integrable nor square integrable. Now, for any locally integrable function $f(x)$ with moderate growth as $|x| \to \infty$, we can take its Fourier transform by taking the Fourier transform in the sense of tempered distributions of F_f. A useful way of taking (i.e., finding) such a Fourier transform is via **regularization**, whereby the object in question is taken to be the distributional limit of a sequence of functions. We can then use the following result (the sister result for Proposition 5.5.1).

Proposition 6.9.1. *Suppose $\{F_n\}$ is a sequence of tempered distributions of functions which converge in the sense of tempered distributions to a tempered distribution F. Then*

$$\widehat{F_n} \to \widehat{F} \quad \text{in the sense of tempered distributions.}$$

The proof follows directly from the definitions of both convergence and the Fourier transform in the sense of tempered distributions. In the next subsections we will apply Proposition 6.9.1 to find Fourier transforms in the sense of tempered distributions of certain functions f which are neither integrable nor square integrable. In a nutshell, we consider a sequence of integrable functions f_n which converge in the sense of tempered distributions to f. We then find the Fourier transform of f in the sense of tempered distributions by investigating the distributional limit of $\widehat{f_n}$. Such a sequence of integrable functions $f_n(x)$ is called a **regularization sequence**.

6.9.3. The Fourier Transform of $f(x) \equiv 1$ and the Delta Function.

In this subsection, we consider the tempered distribution generated by the function $f(x) \equiv 1$, and we present two ways of proving that

$$\boxed{\hat{1} \ = \ 2\pi\delta_0 \quad \text{in the sense of tempered distributions;}} \tag{6.44}$$

that is,

$$\langle \hat{1}, \phi \rangle = 2\pi\phi(0) \qquad \text{for all } \phi \in \mathcal{S}(\mathbb{R}).$$

The first is complete, while the second requires an additional step which is left as an exercise.

Via Regularization Sequence 1: Let

$$f_n(x) = e^{-\frac{|x|}{n}}. \tag{6.45}$$

Note that for every fixed $x \in \mathbb{R}$, the numbers $f_n(x)$ converge to 1 as n tends to ∞. By Theorem A.9[11], the sequence f_n converges in the sense of tempered distributions to $f \equiv 1$. This means that for every test function $\phi \in \mathcal{S}(\mathbb{R})$,

$$\int_{-\infty}^{\infty} e^{-\frac{|x|}{n}}\, \phi(x)\, dx \longrightarrow \int_{-\infty}^{\infty} \phi(x)\, dx.$$

Thus, by Proposition 6.9.1, $\widehat{f_n} \longrightarrow \hat{1}$ in the sense of tempered distributions. The functions f_n are integrable and, hence, they have a Fourier transform $\widehat{f_n}(k)$ in the usual functional sense. This functional Fourier transform generates the distributional Fourier transform. Moreover, one can check via direct integration (cf. Exercise **6.5**) that

$$\widehat{f_n}(k) = \frac{2\sigma_n}{\sigma_n^2 + k^2} = 2\pi \left(\frac{1}{\pi} \frac{\sigma_n}{\sigma_n^2 + k^2} \right) \quad \text{where } \sigma_n = \frac{1}{n}.$$

Recall from Section 5.5.3 that as a function of k,

$$\widehat{f_n}(k) = 2\pi \left(\frac{1}{\pi} \frac{\sigma_n}{\sigma_n^2 + k^2} \right) \longrightarrow 2\pi\delta_0 \quad \text{in the sense of distributions.}$$

We proved this in Theorem 5.2, wherein the test functions were $C^\infty(\mathbb{R})$ with compact support. However, in that instance, we only used the compact support to infer that the functions were bounded. Hence the same conclusions can be drawn for test functions in $\mathcal{S}(\mathbb{R})$, and the proof actually yields convergence to $2\pi\delta_0$ **in the sense of tempered distributions**. In other words, we do not just have

$$\int_{-\infty}^{\infty} \widehat{f_n}(k)\, \phi(k)\, dk \longrightarrow 2\pi\phi(0) \qquad \text{for any } \phi \in C_c^\infty(\mathbb{R}),$$

but also for any $\phi \in \mathcal{S}(\mathbb{R})$. This proves (6.44).

Via Regularization Sequence 2: We can also prove (6.44) via a different regularization sequence, one which is more attuned to unraveling the ill-defined integral associated with $\hat{1}$,

$$\text{“}\int_{-\infty}^{\infty} e^{-ixk}\, dx\text{”}. \tag{6.46}$$

Consider the simple sequence

$$f_n(x) = \begin{cases} 1, & |x| \leq n, \\ 0, & \text{otherwise.} \end{cases}$$

[11] Here we are applying Theorem A.9 to the real and imaginary parts of the sequence of functions $e^{-\frac{|x|}{n}}\, \phi(x)$ for some fixed $\phi \in \mathcal{S}(\mathbb{R})$. These real and imaginary parts need not be nonnegative; however, one can break them up into their positive and negative parts. For example, denoting ϕ_1 as the real part of ϕ, we would decompose ϕ_1 into ϕ_1^+ and ϕ_1^-, both nonnegative functions such that $\phi_1 = \phi_1^+ - \phi_1^-$. Now we can apply the Theorem A.9 separately to $e^{-\frac{|x|}{n}}\, \phi_1^\pm(x)$.

Note that truncating the improper integral (6.46) and taking the limit as $n \to \infty$ would seem to be a natural way to make sense of it. Then again, f_n converges in the sense of tempered distributions to $f \equiv 1$. In terms of the Fourier transform,

$$\widehat{f_n}(k) = \int_{-n}^{n} e^{-ixk}\, dx = 2\pi\left(\frac{\sin nk}{\pi k}\right),$$

which are the (rescaled and renormalized) **sinc functions** previously discussed in Section 5.5.6. In that subsection, we stated without proof that

$$\frac{\sin nk}{\pi k} \longrightarrow \delta_0 \qquad \text{in the sense of distributions.}$$

One can directly prove this. In fact one can prove the statement in the sense of tempered distributions (cf. Exercise **6.25**). This yields another way to prove (6.44). Recall from Section 6.2.3 that this sequence was also a path to proving the Fourier inversion formula (Theorem 6.1).

6.9.4. The Fourier Transform of $f(x) = e^{iax}$ and the Delta Function δ_a. Let $a \in \mathbb{R}$. Using property (6.23), which states that multiplying an integrable function by e^{iax} and then taking the Fourier transform results in a shift of length a, it is not too far-fetched to expect that

$$\boxed{\widehat{e^{iax}} = \widehat{e^{iax}\,1} = 2\pi\delta_a \quad \text{in the sense of tempered distributions.}} \tag{6.47}$$

Indeed, this is the case and we can readily prove (6.47) via the regularization sequence f_n defined by (6.45). To this end, first let us recall this sequence:

$$f_n(x) = e^{-\frac{x}{n}} \qquad \text{with} \qquad \widehat{f_n}(k) = \frac{2\sigma_n}{\sigma_n^2 + k^2} \qquad \left(\sigma_n = \frac{1}{n}\right).$$

Again, by Theorem A.9, we find that

$$e^{iax}\, f_n(x) \xrightarrow{n\to\infty} e^{iax} \qquad \text{in the sense of tempered distributions.}$$

By property (6.23) we have

$$(\widehat{e^{iax} f_n(x)}) = 2\pi\left(\frac{1}{\pi}\,\frac{\sigma_n}{\sigma_n^2 + (k-a)^2}\right).$$

The right-hand side converges in the sense of tempered distributions to $2\pi\delta_a$. This proves (6.47).

On the other hand, since

$$\langle \widehat{\delta_a}, \phi\rangle = \langle \delta_a, \widehat{\phi}\rangle = \widehat{\phi}(a) = \int_{-\infty}^{\infty} \phi(w)\, e^{-iaw}\, dw \qquad \text{for any } a \in \mathbb{R},$$

one has

$$\boxed{\widehat{\delta_a} = e^{-iak} \quad \text{in the sense of tempered distributions.}} \tag{6.48}$$

Let us note here that with (6.47) in hand, one can easily prove the Fourier inversion formula for functions in $\mathcal{S}(\mathbb{R})$. For completeness, let us state this as an official theorem:

Theorem 6.3 (Fourier Inversion Formula for Schwartz Functions). *Let $f \in \mathcal{S}(\mathbb{R})$. Then for each $x \in \mathbb{R}$, we have*

$$f(x) = \frac{1}{2\pi}\int_{-\infty}^{\infty} \widehat{f}(k)\, e^{ikx}\, dk.$$

Proof. Fix $x \in \mathbb{R}$ and let us use k as the variable of the test functions. Then by (6.47), we have

$$\begin{aligned} f(x) &= \langle \delta_x\,, f(k)\rangle \\ &= \frac{1}{2\pi}\langle \widehat{e^{ikx}}\,, f(k)\rangle \\ &= \frac{1}{2\pi}\langle e^{ikx}\,, \widehat{f}(k)\rangle \\ &= \frac{1}{2\pi}\int_{-\infty}^{\infty} \widehat{f}(k)\, e^{ikx}\, dk. \end{aligned}$$ □

6.9.5. The Fourier Transform of Sine, Cosine, and Sums of Delta Functions.

By Euler's formula, we have

$$e^{i\theta} = \cos\theta + i\sin\theta \qquad \text{and} \qquad e^{-i\theta} = \cos\theta - i\sin\theta.$$

Hence, we can write

$$\cos\theta = \frac{e^{i\theta} + e^{-i\theta}}{2} \qquad \text{and} \qquad \sin\theta = \frac{e^{i\theta} - e^{-i\theta}}{2i}.$$

Combining this with (6.48), we find that

$$\boxed{\widehat{\left(\frac{\delta_a + \delta_{-a}}{2}\right)} = \cos ak \qquad \text{and} \qquad \widehat{\left(\frac{\delta_{-a} - \delta_a}{2i}\right)} = \sin ak} \tag{6.49}$$

in the sense of tempered distributions.

On the other hand, either via the Fourier inversion formula (or by direct integration), one has for any $a \in \mathbb{R}$,

$$\boxed{\widehat{\cos ax} = \pi(\delta_a + \delta_{-a}) \qquad \text{and} \qquad \widehat{\sin ax} = \frac{\pi}{i}(\delta_a - \delta_{-a})} \tag{6.50}$$

in the sense of tempered distributions.

6.9.6. The Fourier Transform of x.

We consider the function $f(x) = x$, which is clearly not integrable on $\mathbb{R}$ since it tends to $+\infty$ as $x \to \infty$. We show here that the Fourier transform of $f(x) = x$ in the sense of tempered distributions is $2\pi i\delta_0'$; i.e.,

$$\boxed{\hat{x} = 2\pi i \delta_0' \quad \text{in the sense of tempered distributions.}} \tag{6.51}$$

To show this we will not use a regularization sequence but simply the Fourier inversion formula on the test function ϕ. To this end, for any $\phi \in \mathcal{S}(\mathbb{R})$ we have by definition

$$\langle \hat{x}, \phi\rangle = \langle x, \widehat{\phi}\rangle = \int_{-\infty}^{\infty} k\,\widehat{\phi}(k)\, dk. \tag{6.52}$$

Note that in the integral we have chosen to use k as the dummy variable of integration (we could have any variable). On the other hand, by the Fourier inversion formula

$$\phi(x) = \frac{1}{2\pi}\int_{-\infty}^{\infty} \widehat{\phi}(k)e^{ixk}\,dk.$$

Hence, differentiating with respect to x we have

$$\phi'(x) = \frac{1}{2\pi}\int_{-\infty}^{\infty} ik\,\widehat{\phi}(k)e^{ixk}\,dk.$$

Evaluating the above at $x = 0$, we find

$$\phi'(0) = \frac{1}{2\pi}\int_{-\infty}^{\infty} ik\,\widehat{\phi}(k)\,dk.$$

Thus, combining this with (6.52), we have

$$\langle \hat{x}, \phi\rangle = \int_{-\infty}^{\infty} k\,\widehat{\phi}(k)\,dk = \frac{2\pi}{i}\,\phi'(0) = -2\pi i\,\phi'(0) = \langle 2\pi i\delta_0', \phi\rangle.$$

One can similarly derive expressions for the Fourier transform of any polynomial in the sense of tempered distributions (cf. Exercise **6.26**).

6.9.7. The Fourier Transform of the Heaviside and Sgn Functions and PV $\frac{1}{x}$. This subsection requires a reading of Section 5.8. Consider the Sign and Heaviside functions $\operatorname{sgn}(x)$ and $H(x)$ defined, respectively, by

$$\operatorname{sgn}(x) = \begin{cases} 1 & \text{if } x \geq 0, \\ -1 & \text{if } x < 0, \end{cases} \qquad H(x) = \begin{cases} 1 & \text{if } x \geq 0, \\ 0 & \text{if } x < 0. \end{cases}$$

Note that

$$H(x) = \frac{1 + \operatorname{sgn}(x)}{2}.$$

These functions are not integrable, yet their Fourier transforms play an important role, for example, in the wave equation. Just as with the constant function 1, putting either $\operatorname{sgn}(x)$ or $H(x)$ into the definition (6.5) of the Fourier transform gets us nowhere. However, let us now consider $\operatorname{sgn}(x)$ as a tempered distribution and compute its Fourier transform in the sense of distributions via a regularization sequence and Proposition 6.9.1. Note that $\operatorname{sgn}(x)$ is the distributional limit of the sequence of functions (interpreted as tempered distributions)

$$f_n(x) := \begin{cases} e^{-\frac{1}{n}x}, & x \geq 0, \\ -e^{\frac{1}{n}x}, & x < 0. \end{cases}$$

Now, for each n, f_n is integrable and hence has a functional Fourier transform $\widehat{f_n}$. The limit of $\widehat{f_n}$, in the sense of tempered distributions, will generate the distributional

Fourier transform. Computing $\widehat{f_n}$ we find

$$\begin{aligned}\widehat{f_n}(k) &= \int_{-\infty}^{\infty} f_n(x)\, e^{-ikx}\, dx \\ &= \int_0^{\infty} e^{-\frac{1}{n}x}\, e^{-ikx}\, dx + \int_{-\infty}^{0} -e^{-\frac{1}{n}x}\, e^{-ikx}\, dx \\ &= \int_0^{\infty} e^{-\frac{1}{n}x}\, e^{-ikx}\, dx - \int_{-\infty}^{0} e^{-\frac{1}{n}x}\, e^{-ikx}\, dx \\ &= \frac{1}{\frac{1}{n}+ik} - \frac{1}{\frac{1}{n}-ik} \\ &= \frac{-2ik}{\frac{1}{n^2}+k^2} \\ &= \frac{2}{i}\left(\frac{k}{\frac{1}{n^2}+k^2}\right).\end{aligned}$$

Recall (5.49) from Section 5.8 that

$$\frac{2}{i}\left(\frac{k}{\frac{1}{n^2}+k^2}\right) \longrightarrow \frac{2}{i}\,\mathrm{PV}\,\frac{1}{k},$$

in the sense of distributions. The exact same reasoning implies that the convergence also holds in the sense of tempered distributions. Thus,

$$\boxed{\widehat{\mathrm{sgn}}\,(k) = \frac{2}{i}\,\mathrm{PV}\,\frac{1}{k} \quad \text{in the sense of tempered distributions.}} \tag{6.53}$$

Finally, we can use the linearity of the Fourier transform and (6.44) to find that in the sense of tempered distributions,

$$\widehat{H}(k) = \left(\widehat{\frac{1+\mathrm{sgn}}{2}}\right)(k) = \frac{1}{2}\hat{1} + \frac{1}{2}\widehat{\mathrm{sgn}}(k) = \pi\delta_0 + \frac{1}{i}\mathrm{PV}\,\frac{1}{k}.$$

Hence,

$$\boxed{\widehat{H}\,(k) = \pi\delta_0 + \frac{1}{i}\mathrm{PV}\,\frac{1}{k} \quad \text{in the sense of tempered distributions.}} \tag{6.54}$$

By (5.45) we can equivalently write

$$\boxed{\widehat{H}\,(k) = \frac{1}{i}\,\frac{1}{k-i0} \quad \text{in the sense of tempered distributions.}}$$

6.9.8. Convolution of a Tempered Distribution and a Function. One often encounters convolution with distributions. While it is possible under certain circumstances to define the convolution of a distribution with another distribution, we focus here on the convolution of a function with a tempered distribution and its Fourier transform. Let $\psi \in \mathcal{S}(\mathbb{R})$ and define $\psi^-(x) := \psi(-x)$. For any tempered distribution F, we define the distribution $\psi * F$ by

$$\boxed{\langle \psi * F\, ,\, \phi\rangle := \langle F\, ,\, \psi^- * \phi\rangle \qquad \text{for any } \phi \in \mathcal{S}(\mathbb{R}).}$$

As with all our distributional definitions, this definition needs to generalize the notion of convolution of functions; hence, it must be the case that if f is any function which generates a tempered distribution, then $\psi * F_f = F_{\psi * f}$ in the sense of tempered distributions. So let us check that this is indeed the case; for any $\phi \in \mathcal{S}(\mathbb{R})$ we have

$$\begin{aligned}
\langle F_{\psi * f}, \phi \rangle &= \int_{-\infty}^{\infty} \phi(y)\,(\psi * f)(y)\,dy \\
&= \int_{-\infty}^{\infty} \phi(y) \left(\int_{-\infty}^{\infty} f(x)\psi(y-x)\,dx \right) dy \\
&= \int_{-\infty}^{\infty}\int_{-\infty}^{\infty} f(x)\,\psi(y-x)\,\phi(y)\,dx\,dy \\
&= \int_{-\infty}^{\infty} f(x) \left(\int_{-\infty}^{\infty} \psi^-(x-y)\,\phi(y)\,dy \right) dx \\
&= \int_{-\infty}^{\infty} f(x)\,(\psi^- * \phi)(x)\,dx \\
&= \langle F_f, \psi^- * \phi \rangle \\
&= \langle \psi * F_f, \phi \rangle.
\end{aligned}$$

Convolution of a function with the Delta function. It is now straightforward to show that

$$\boxed{\psi * \delta_0 = \psi \quad \text{in the sense of tempered distributions.}} \tag{6.55}$$

Indeed, for any $\phi \in \mathcal{S}(\mathbb{R})$ we have

$$\begin{aligned}
\langle \psi * \delta_0, \phi \rangle &= \langle \delta_0, \psi^- * \phi \rangle \\
&= (\psi^- * \phi)(0) \\
&= \int_{-\infty}^{\infty} \psi^-(y)\,\phi(-y)\,dy \\
&= \int_{-\infty}^{\infty} \psi(-y)\,\phi(-y)\,dy \\
&= \int_{-\infty}^{\infty} \psi(y)\,\phi(y)\,dy \\
&= \langle F_\psi, \phi \rangle.
\end{aligned}$$

So, in words, convolving with the delta function returns us back to the original function. This actually makes complete sense. Consider, for example, the sequence of hat functions

$$\chi_n := \begin{cases} \frac{n}{2} & \text{if } |x| \le \frac{1}{n}, \\ 0 & \text{otherwise.} \end{cases}$$

Then $\chi_n \longrightarrow \delta_0$ in the sense of tempered distributions. On the other hand, if we convolute a function ψ with χ_n, we get the function whose value at x is the average of ψ on the interval from $x - \frac{1}{n}$ to $x + \frac{1}{n}$; i.e.,

$$\psi * \chi_n(x) = \frac{1}{\left(\frac{2}{n}\right)} \int_{x-\frac{1}{n}}^{x+\frac{1}{n}} \psi(y)\, dy.$$

Since ψ is continuous, the Averaging Lemma implies that this average value at x converges as $n \to \infty$ to $\psi(x)$.

Similarly, one can show that for any fixed $x_0 \in \mathbb{R}$,

$$\boxed{\psi * \delta_{x_0} = \psi(x - x_0),}$$

in the sense of tempered distributions with x as the underlying variable.

Convolution of a Schwartz function with a tempered distribution is actually a C^∞ function. We have just shown that the convolution of a Schwartz function ψ with the delta function gives the distribution generated by ψ itself. One could naturally ask whether or not the convolution of ψ with a general tempered distribution is a distribution generated by a function. The answer is yes and, indeed, the function is C^∞. In fact, one can show that $\psi * F$ is the distribution generated by the function

$$(\psi * F)(x) := \langle F_y, \psi(x - y) \rangle,$$

where we use F_y to denote the fact that, in this scenario, y is the underlying variable of the distribution F. One can prove (see Folland **[17]**) that this defines a C^∞ function.

Fourier transform of the convolution of a function with a tempered distribution. Recall that the Fourier transform of the convolution of two functions is simply the product of the individual Fourier transforms. We have the analogous property for the convolution of a Schwartz function with a tempered distribution:

$$\boxed{\widehat{\psi * F} = \widehat{\psi}\widehat{F} \quad \text{in the sense of tempered distributions.}} \tag{6.56}$$

Note that since $\widehat{\psi} \in \mathcal{S}(\mathbb{R})$, the product of the function times the distribution on the right defines a new tempered distribution. To show (6.56) note that, by definition, for all $\phi \in \mathcal{S}(\mathbb{R})$ we have

$$\langle \widehat{\psi * F}, \phi \rangle = \langle \psi * F, \widehat{\phi} \rangle = \langle F, \psi^- * \widehat{\phi} \rangle.$$

Next, we compute

$$\begin{aligned}
(\psi^- * \widehat{\phi})(k) &= \int_{-\infty}^{\infty} \psi^-(y)\,\widehat{\phi}(k-y)\,dy \\
&= \int_{-\infty}^{\infty} \psi^-(y)\left(\int_{-\infty}^{\infty} \phi(x)\,e^{-ix(k-y)}\,dx\right)dy \\
&= \int_{-\infty}^{\infty}\left(\int_{-\infty}^{\infty} \psi(-y)\,e^{-ix(-y)}\,dy\right)\phi(x)\,e^{-ixk}\,dx \\
&= \int_{-\infty}^{\infty}\left(\int_{-\infty}^{\infty} \psi(y)\,e^{-ixy}\,dy\right)\phi(x)\,e^{-ixk}\,dx \\
&= \int_{-\infty}^{\infty} \left(\widehat{\psi}(x)\phi(x)\right)\,e^{-ixk}\,dx \\
&= \widehat{\widehat{\psi}\phi}\,(k).
\end{aligned}$$

Thus, we have for all $\phi \in \mathcal{S}(\mathbb{R})$,

$$\langle \widehat{\psi * F}, \phi\rangle = \langle F, \psi^- * \widehat{\phi}\rangle = \langle F, \widehat{\widehat{\psi}\phi}\,\rangle = \langle \widehat{F}, \widehat{\psi}\phi\rangle = \langle \widehat{\psi}\widehat{F}, \phi\rangle,$$

and (6.56) holds true.

6.10. Using the Fourier Transform to Solve PDEs, II: The Wave Equation

We now revisit the wave equation as a simple illustration of using the Fourier transform of nonintegrable functions in the sense of tempered distributions.

Recall the initial value problem for the wave equation with $c = 1$:

$$\begin{cases} u_{tt} = u_{xx}, & -\infty < x < \infty,\ t > 0, \\ u(x,0) = \phi(x), & -\infty < x < \infty, \\ u_t(x,0) = \psi(x), & -\infty < x < \infty. \end{cases}$$

Let us assume that we have a solution $u(x,t)$ which for every t is an integrable function. We fix t and Fourier transform both sides of the equation $u_{tt} - u_{xx} = 0$ with respect to x. Using the linearity of the Fourier transform and the fact that the Fourier transform of 0 is 0, we have

$$(\widehat{u_{tt} - u_{xx}})(k,t) = (\widehat{u_{tt}})(k,t) - (\widehat{u_{xx}})(k,t) = 0.$$

As we saw with the diffusion equation, we have

$$(\widehat{u_{tt}})\,(k,t) = \hat{u}_{tt}(k,t).$$

By the property of the Fourier transform and differentiation, we have

$$\widehat{u_{xx}}(k,t) = (ik)^2\hat{u}(k,t) = -k^2\hat{u}(k,t).$$

Thus, we have

$$\hat{u}_{tt}(k,t) + k^2\hat{u}(k,t) = 0. \tag{6.57}$$

We also Fourier transform the initial data with

$$\hat{u}(k,0) = \widehat{\phi}(k), \qquad \hat{u}_t(k,0) = \widehat{\psi}(k). \tag{6.58}$$

Now fix $k \in \mathbb{R}$ for the moment and define $f(t) := \hat{u}(k,t)$. By (6.57) and (6.58), we obtain the following ODE with initial conditions for $f(t)$:

$$f''(t) = -k^2 f(t) \qquad \text{with} \qquad f(0) = \widehat{\phi}(k) \text{ and } f'(0) = \widehat{\psi}(k).$$

The solution is

$$f(t) = \widehat{\phi}(k)\cos kt + \frac{\widehat{\psi}(k)}{k}\sin kt,$$

and, hence,

$$\hat{u}(k,t) = \widehat{\phi}(k)\cos kt + \frac{\widehat{\psi}(k)}{k}\sin kt. \tag{6.59}$$

In solving the ODE we fixed k, but now note that (6.59) holds true for all $k \in \mathbb{R}$ and $t \geq 0$.

Next, we use the inverse Fourier transform to get a formula for $u(x,t)$. Let us use the notation of $\mathcal{F}^{-1}$ for the inverse Fourier transform instead of the upside-down hat. By the property (6.20) — multiplication in Fourier space is convolution in real space — we have

$$u(x,t) = \phi(x) * \mathcal{F}^{-1}\big(\cos kt\big) + \psi(x) * \mathcal{F}^{-1}\left(\frac{\sin kt}{k}\right).$$

All that remains is to evaluate these inverse Fourier transforms. Recall from Example 6.2.1 that

$$\mathcal{F}^{-1}\left(\frac{\sin kt}{k}\right) = \begin{cases} \frac{1}{2}, & |k| < t, \\ 0, & |k| \geq t. \end{cases}$$

The function $\cos kt$ is not integrable on $\mathbb{R}$ and, hence, the inverse Fourier transform as defined by (6.10) does not exist. However, we have already seen how to interpret these Fourier transforms in the sense of tempered distributions and, indeed, this function was chosen as one of the examples in Section 6.9. We determined that in the sense of tempered distributions,

$$\mathcal{F}^{-1}\big(\cos kt\big) = \frac{1}{2}(\delta_t + \delta_{-t}).$$

Thus, applying this convolution and noting (6.55), we arrive at D'Alembert's formula

$$u(x,t) = \frac{1}{2}\big[\phi(x+t) + \phi(x-t)\big] + \frac{1}{2}\int_{x-t}^{x+t} \psi(s)ds.$$

While this approach to D'Alembert's formula may seem far more complicated (especially as it requires far more machinery) than the simple method of characteristics taken in Section 3.3, the method via the Fourier transform is very robust for linear extensions of the wave equation. Three cases in point follow:

- The **wave equation with friction** (cf. Exercise **6.20**)

$$u_{tt} + \alpha u_t = c^2 u_{xx} \qquad \text{where} \quad \alpha > 0.$$

- The **Telegraph equation** (cf. Exercise **6.15**)

$$u_{tt} + (\alpha+\beta)u_t + \alpha\beta u = c^2 u_{xx} \qquad \text{where} \quad \alpha, \beta > 0.$$

- The **Klein-Gordon** equation (cf. Exercise **6.22**)

$$\hbar^2 u_{tt} - \hbar^2 c^2 u_{xx} + m^2 c^4 u = 0.$$

6.11. The Fourier Transform in Higher Space Dimensions

6.11.1. Definition and Analogous Properties. The Fourier transform and its inverse Fourier transform can be extended to functions of several variables, say, $f(\mathbf{x})$ for $\mathbf{x} \in \mathbb{R}^N$. Essentially, this is achieved by successively taking the Fourier transform in each variable x_i. Noting that the exponential of a sum is simply the product of the exponentials, we arrive at the following definition.

The N-Dimensional Fourier Transform

Let f be an integrable function on $\mathbb{R}^N$. We define its Fourier transform to be the complex-valued function on $\mathbb{R}^N$:

$$\begin{aligned}\widehat{f}(\mathbf{k}) &:= \int_{-\infty}^{\infty} \cdots \int_{-\infty}^{\infty} \cdots \int_{-\infty}^{\infty} f(\mathbf{x})\, e^{-i\mathbf{k}\cdot\mathbf{x}}\, d\mathbf{x} \\ &= \int_{-\infty}^{\infty} \cdots \int_{-\infty}^{\infty} \cdots \int_{-\infty}^{\infty} f(x_1, \ldots, x_n)\, e^{-ik_1 x_1} \ldots e^{-ik_n x_n}\, dx_1 \cdots dx_n.\end{aligned}$$

So for example in dimension $N = 3$,

$$\widehat{f}(\mathbf{k}) := \int_{-\infty}^{\infty} \int_{-\infty}^{\infty} \int_{-\infty}^{\infty} f(x_1, x_2, x_3)\, e^{-ik_1 x_1}\, e^{-ik_2 x_2}\, e^{-ik_3 x_3}\, dx_1\, dx_2\, dx_3.$$

Note that the multiplication kx in the exponential of the 1D Fourier transform is now replaced with the dot product $\mathbf{k} \cdot \mathbf{x}$.

Partial differentiation and the Fourier transform. One can readily verify that the analogous property of the transformation of partial derivatives into polynomial multiplication holds true. To conveniently label a general partial derivative of a function of several variables, we adopt the notation of Laurent Schwartz. Let α denote a vector with N components whose entries are nonnegative integers. We agree that a nonnegative i-th entry α_i means α_i derivatives with respect to x_i, with the understanding that if $\alpha_i = 0$, no derivatives are taken. We call such an α a **multi-index**. Thus, for any such index α we have an associated partial derivative ∂^α. So for example in $N = 3$,

$$\partial^{(1,0,0)} \phi(x_1, x_2, x_3) = \frac{\partial}{\partial x_1} \phi(x_1, x_2, x_3), \qquad \partial^{(0,2,0)} \phi(x_1, x_2, x_3) = \frac{\partial^2}{\partial x_2^2} \phi(x_1, x_2, x_3),$$

$$\partial^{(2,3,2)} \phi(x_1, x_2, x_3) = \frac{\partial^7}{\partial x_1^2\, \partial x_2^3\, \partial x_3^2} \phi(x_1, x_2, x_3).$$

We denote by $|\alpha|$ the total number of derivatives taken which, by definition, is simply the sum of the components α_i. So, in full generality, if $\alpha = (\alpha_1, \alpha_2, \dots, \alpha_N)$, then

$$\partial^\alpha \phi(x_1, x_2, \dots, x_N) = \frac{\partial^{|\alpha|}}{\partial x_1^{\alpha_1} \partial x_2^{\alpha_2}, \dots, \partial x_N^{\alpha_N}} \phi(x_1, x_2, \dots, x_N).$$

Let ∂^α denote the partial derivative associated with the multi-index $\alpha = (\alpha_1, \dots, \alpha_N)$. Then for any $\mathbf{k} = (k_1, \dots, k_N) \in \mathbb{R}^N$, we define the polynomial $\mathbf{k}^\alpha$ to be

$$\mathbf{k}^\alpha = k_1^{\alpha_1} k_2^{\alpha_2} \cdots k_N^{\alpha_2}.$$

One can directly check the following:

Partial Differentiation and the N-dimensional Fourier Transform

If α is any multi-index, then

$$\widehat{\partial^\alpha f}(\mathbf{k}) = i^{|\alpha|}\, \mathbf{k}^\alpha\, \hat{f}(\mathbf{k}).$$

So for example if $N = 3$ and $\alpha = (0, 1, 0)$, i.e., $\partial^\alpha f = \frac{\partial f}{\partial x_2}$, then

$$\widehat{\partial^\alpha f}(\mathbf{k}) = \widehat{\frac{\partial f}{\partial x_2}}(\mathbf{k}) = ik_2\, \hat{f}(\mathbf{k}).$$

If $N = 3$ and $\alpha = (1, 1, 1)$, i.e., $\partial^\alpha f = \frac{\partial^3 f}{\partial x_1 \partial x_2 \partial x_3}$, then

$$\widehat{\partial^\alpha f}(\mathbf{k}) = \widehat{\frac{\partial^3 f}{\partial x_1 \partial x_2 \partial x_3}}(\mathbf{k}) = -ik_1 k_2 k_3\, \hat{f}(\mathbf{k}).$$

If $N = 3$ and $\alpha = (1, 1, 2)$, i.e., $\partial^\alpha f = \frac{\partial^4 f}{\partial x_1 \partial x_2 \partial x_3^2}$, then

$$\widehat{\partial^\alpha f}(\mathbf{k}) = \widehat{\frac{\partial^4 f}{\partial x_1 \partial x_2 \partial x_3^2}}(\mathbf{k}) = k_1 k_2 k_3^2\, \hat{f}(\mathbf{k}).$$

Since the Fourier transform is a linear operation, we can readily take the Fourier transform of any linear combinations of partial derivatives. For example, in terms of the Laplacian, we have

$$\widehat{\Delta f}(\mathbf{k}) = \mathcal{F}\left(\frac{\partial^2 f}{\partial x_1^2} + \frac{\partial^2 f}{\partial x_2^2} + \cdots + \frac{\partial^2 f}{\partial x_N^2}\right)(\mathbf{k}) = -(k_1^2 + \cdots + k_N^2)\, \hat{f}(\mathbf{k}) = -|\mathbf{k}|^2\, \hat{f}(\mathbf{k}).$$

For the **inverse Fourier transform**, one needs to adjust the normalizing factor:

N-dimensional Fourier Inversion Theorem

Theorem 6.4. *Suppose both f and $\hat{f}$ are integrable on $\mathbb{R}^N$ (hence both are continuous). Then*

$$f(\mathbf{x}) = \frac{1}{(2\pi)^N} \int_{-\infty}^{\infty} \cdots \int_{-\infty}^{\infty} \cdots \int_{-\infty}^{\infty} \hat{f}(\mathbf{k})\, e^{i\mathbf{k}\cdot\mathbf{x}}\, d\mathbf{k}.$$

Alternatively, as we did in Section 6.2.3, we can define the inverse Fourier transform for possibly complex function $g(\mathbf{k})$ on $\mathbb{R}^N$ as

$$\check{g}(\mathbf{x}) := \frac{1}{(2\pi)^N} \int_{-\infty}^{\infty} \cdots \int_{-\infty}^{\infty} \cdots \int_{-\infty}^{\infty} g(\mathbf{k})\, e^{i\mathbf{k}\cdot\mathbf{x}}\, d\mathbf{k}.$$

Then the Fourier Inversion Theorem states that under certain conditions on f, we have

$$\widecheck{\hat{f}(\mathbf{k})}(\mathbf{x}) = f(\mathbf{x}).$$

The notions of convolution and its Fourier transform directly carry over to N dimensions. Indeed, we have the following:

N-Dimensional Convolution and Its Fourier Transform.

Convolution of functions f and g on $\mathbb{R}^N$ can be defined in a similar way:

$$(f * g)(\mathbf{x}) = \int_{-\infty}^{\infty} \cdots \int_{-\infty}^{\infty} \cdots \int_{-\infty}^{\infty} f(\mathbf{x}-\mathbf{y})\, g(\mathbf{y})\, d\mathbf{y}.$$

By the analogous computation performed in one dimension, one can show that the Fourier transform will *transform* convolution in $\mathbb{R}^N$ into pointwise multiplication; that is,

$$\widehat{(f * g)}(\mathbf{k}) = \hat{f}(\mathbf{k})\, \hat{g}(\mathbf{k}).$$

We conclude by noting the analogous properties of translation, scaling, and invariance of the Gaussian.

Translation. Let $\mathbf{a} \in \mathbb{R}^N$ and let f be integrable on $\mathbb{R}^N$. Then we have

$$\boxed{\widehat{f(\mathbf{x}-\mathbf{a})} = e^{-i\mathbf{a}\cdot\mathbf{k}}\, \hat{f}(\mathbf{k}) \qquad \text{and} \qquad \widehat{(e^{i\mathbf{a}\cdot\mathbf{x}} f(\mathbf{x}))} = \hat{f}(\mathbf{k}-\mathbf{a}).}$$

Scaling. Let $g(\mathbf{y})$ be integrable on $\mathbb{R}^N$, and for any $c > 0$ define

$$g_c(\mathbf{y}) := \frac{1}{c^N}\, g\left(\frac{\mathbf{y}}{c}\right).$$

Then for any f integrable on $\mathbb{R}^N$, we have

$$\boxed{\widehat{f(c\mathbf{x})} = \left(\hat{f}\right)_c(\mathbf{k}) \qquad \text{and} \qquad \widehat{f_c}(\mathbf{k}) = \hat{f}(c\mathbf{k}).}$$

Separable functions and some examples. Suppose a function f *separates* in terms of its independent variables as follows:

$$f(\mathbf{x}) = f_1(x_1)\, f_2(x_2) \cdots f_N(x_N),$$

where the f_i are functions of one variable. Then one can readily check that

$$\widehat{f}(\mathbf{k}) = \widehat{f}_1(k_1)\, \widehat{f}_2(k_2) \cdots \widehat{f}_n(k_N).$$

Here are two examples:

- Consider the **N-dimensional Gaussian**

$$f(\mathbf{x}) = e^{-a|\mathbf{x}|^2} = e^{-a(x_1^2+x_2^2+\cdots+x_N^2)} = e^{-ax_1^2}\, e^{-ax_2^2} \cdots e^{-ax_N^2},$$

  where $a > 0$. Then its Fourier transform is given by

$$\widehat{f}(\mathbf{k}) = \frac{\pi^{N/2}}{a^{N/2}}\, e^{-\frac{|\mathbf{k}|^2}{4a}}.$$

- Consider the **N-dimensional step function**

$$f(\mathbf{x}) = \begin{cases} 1 & \text{if for all } i = 1 \ldots N,\ |x_i| \le a, \\ 0 & \text{otherwise.} \end{cases}$$

  This function is simply the product of the one-dimensional step functions

$$f_i(x_i) = \begin{cases} 1 & \text{if } |x_i| \le a, \\ 0 & \text{if } |x_i| > a. \end{cases}$$

  Hence, by Example 6.2.1, we have

$$\widehat{f}(\mathbf{k}) = 2^N \frac{\sin ak_1}{k_1} \cdots \frac{\sin ak_N}{k_N}.$$

6.11.2. The Fourier Transform of a Radially Symmetric Function and Bessel Functions. Many functions of several variables that we encounter are radially symmetric. This is a reflection of the simple fact that, when modeling certain dependent variables defined over a homogeneous space, it is often natural to treat all directions equally; i.e., there is no bias for one direction (say, the x direction in $\mathbb{R}^3$) as opposed to another (say, the z direction in $\mathbb{R}^3$). By a radially symmetric function, we mean one with the form $f(\mathbf{x}) = f_0(|\mathbf{x}|)$, for some function f_0 of one variable. Note that it is often common to abuse notation and give the same notation for both f and f_0.

If one looks at the definition of the Fourier transform (which indeed treats all variables equally), one would expect the Fourier transform of a radially symmetric function to also be radially symmetric. This turns out to be the case and, as such, we say the Fourier transform is **rotationally symmetric**. Precisely, this means that if $\mathbf{R}$ denotes an $N \times N$ matrix which corresponds to a rotation in $\mathbb{R}^N$, then for any integrable function f on $\mathbb{R}^N$ we have $\widehat{f(\mathbf{Rx})} = \widehat{f}(\mathbf{Rk})$.

We can obtain a specific form for the Fourier transform of a radial function and this particular form introduces us to very important class of C^∞ functions which are ubiquitous when dealing with spherical symmetry. They are called **Bessel**[12] **functions.** Presenting this calculation in two dimensions, let f be an integrable function of $\mathbb{R}^2$ which is radial; i.e, $f(\mathbf{x}) = f_0(|x|)$, for some function $f_0(r)$ of one variable defined for $r \geq 0$. We calculate the Fourier transform

$$\widehat{f}(\mathbf{k}) = \int_{-\infty}^{\infty}\int_{-\infty}^{\infty} f(\mathbf{x})e^{-i\mathbf{x}\cdot\mathbf{k}}\,d\mathbf{x}$$

by changing to polar coordinates. We have two variables in the plane $\mathbf{x}$ and $\mathbf{k}$ but note that, for the purposes of the integration, $\mathbf{x}$ varies but $\mathbf{k}$ is fixed. Let us denote by (r, θ) the polar coordinates for $\mathbf{x}$ and by (ρ, ϕ) the polar coordinates for $\mathbf{k}$. Then, by the elementary identity for cosine of a sum of angles, we have $\mathbf{x}\cdot\mathbf{k} = r\rho\cos(\theta-\phi)$. Hence

$$\begin{aligned}\widehat{f}(\mathbf{k}) = \widehat{f}(\rho,\phi) &= \int_0^\infty\int_0^{2\pi} f_0(r)e^{-ir\rho\cos(\theta-\phi)}\,r\,d\theta\,dr\\ &= \int_0^\infty rf_0(r)\left(\int_0^{2\pi} e^{-ir\rho\cos(\theta-\phi)}d\theta\right)dr\\ &= \int_0^\infty rf_0(r)\left(\int_0^{2\pi} e^{-ir\rho\cos\theta}d\theta\right)dr,\end{aligned}$$

where we passed to the last line by noting the fact that the inner integrand was a 2π periodic function of θ; therefore its integral over $\theta \in [\phi, 2\pi+\phi]$ is the same as its integral over $[0, 2\pi]$. Note that the independence of ϕ tells us that the Fourier transform must be radial, i.e., only a function of ρ. Now, for any $z \in [0,\infty)$ define

$$J_0(z) := \frac{1}{2\pi}\int_0^{2\pi} e^{-iz\cos\theta}d\theta. \tag{6.60}$$

This defines a C^∞ function which is called the **zero-th order Bessel function of the first kind**. Then

$$\boxed{\widehat{f}(\mathbf{k}) = \widehat{f}(\rho) = 2\pi\int_0^\infty f_0(r)J_0(r\rho)\,r\,dr.} \tag{6.61}$$

The right-hand side of (6.61) is known as **the Hankel transform of order** 0 of the function $f_0(r)$.

There exists a similar formula for the Fourier transform of a radial function in any dimension $n = 3, 4, \ldots$. However, these involve Bessel functions of higher order and things get messy pretty quickly! In general, **Bessel functions of the first kind of order** m can be defined for $m = 0, 1, 2, \ldots$ by the power series

$$J_m(z) := \sum_{j=0}^{\infty}(-1)^j\frac{\left(\frac{z}{2}\right)^{m+2j}}{j!\,(m+j)!}. \tag{6.62}$$

[12]**Friedrich Wilhelm Bessel** (1784–1846) was a German astronomer, mathematician, and scientist most known for this class of functions. However, Bessel functions were in fact first discovered by the Swiss mathematician and physicist **Daniel Bernoulli** (1700–1782) and were later generalized by Bessel.

This power series converges for all $z \in \mathbb{R}$ and, in the case $m = 0$, it has the integral representation (6.60). For higher orders there are analogous integral representations. However, one can also define the **Bessel function of order** ν for any real number $\nu \geq 0$ in terms of the power series, but in order to do so one needs the **Gamma function** which is, roughly speaking, a continuum version of the factorial. Specifically, the Gamma function, defined for any real number $t > 0$, is

$$\Gamma(t) := \int_0^\infty x^{t-1} e^{-x}\, dx. \tag{6.63}$$

In fact, its definition can be extended to a large portion of the complex plane. It might seem like a crazy definition but it turns out that for any positive integer $n = 1, 2, \ldots$, $\Gamma(n) = (n-1)!$ With the Gamma function in hand, the **Bessel function of order** ν with $\nu \geq 0$ is defined by

$$J_\nu(z) = \left(\frac{z}{2}\right)^\nu \sum_{j=0}^\infty (-1)^j \frac{\left(\frac{z}{2}\right)^{2j}}{\Gamma(j+1)\,\Gamma(\nu+j+1)}. \tag{6.64}$$

For the radial Fourier transform in higher dimensions, Bessel functions of integer multiples of $1/2$ come into play; that is, the radial Fourier transform in dimension $N \geq 3$ involves the Bessel function of order $\frac{N-2}{2}$. For example in dimension $N = 3$, if $f(\mathbf{x}) = f_0(|\mathbf{x}|)$ is a function on $\mathbb{R}^3$, then

$$\boxed{\hat{f}(\mathbf{k}) = \hat{f}(\rho) = (2\pi)^{\frac{3}{2}} \int_0^\infty f_0(r) J_{\frac{1}{2}}(r\rho)\, r^{\frac{1}{2}}\, r\, dr.} \tag{6.65}$$

Note that the definition (6.64) yields the simple formula for $J_{\frac{1}{2}}$:

$$J_{\frac{1}{2}} = \sqrt{\frac{2}{\pi x}}\, \sin x.$$

An important radial function in $\mathbb{R}^3$ is

$$f(\mathbf{x}) = \frac{1}{|\mathbf{x}|^2 + a^2}, \qquad \text{where} \quad a > 0.$$

The relationship of the power in the denominator (here 2) to the dimension of the space (here 3) is crucial. Note that in dimension 3, this function is locally integrable but not integrable over $\mathbb{R}^3$. However, it is square integrable on $\mathbb{R}^3$ (check this) and, hence, it does have a Fourier transform which is also square integrable. After some calculations, one finds that

$$\hat{f}(\mathbf{k}) = \frac{(2\pi)^3 e^{-a|\mathbf{k}|^2}}{4\pi|\mathbf{k}|}.$$

On the other hand, if

$$g(\mathbf{x}) = \frac{e^{-a|\mathbf{x}|^2}}{4\pi|\mathbf{x}|}, \qquad \text{then} \quad \hat{g}(\mathbf{k}) = \frac{1}{|\mathbf{k}|^2 + a^2}. \tag{6.66}$$

The latter will prove useful in solving the Helmholtz equations.

6.12. Frequency, Harmonics, and the Physical Meaning of the Fourier Transform

> *"To calculate a [Fourier] transform, all you have to do is listen.*
> *The ear formulates a mathematical transform by converting sound -*
> *the waves of pressure traveling through time and the atmosphere -*
> *into a spectrum, a description of the sound as a series of volumes*
> *at distinct pitches. The brain turns this information into perceived sound."*
> – R. N. Bracewell[a]
>
> [a] R.N. Bracewell, The Fourier Transform, Scientific American, June 1989, pp. 62–69.

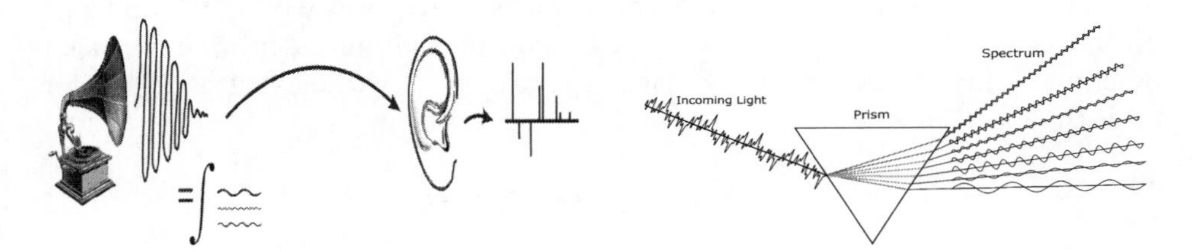

Figure 6.4. Left: The **ear** as a **Fourier transformer**: A cartoon to illustrate Bracewell's quote. Right: The **prism** as a **Fourier transformer** — similar cartoon appearing in Bracewell's article.

We have introduced the Fourier transform, discussed its properties, and illustrated why it is so amenable to solving linear PDEs. One might be tempted to conclude that the transform and the role of the Fourier space variable k exist for analytical convenience only, divorced from any direct physical reality. This perception would be completely false and purely a result of our presentation thus far! In Figure 6.4 we see both the ear and the prism acting as a Fourier transformer. In these settings, it is instructive to think of the independent variable as time. For the case of the ear, Bracewell's referenced quote says it all. In the case of a prism, sunlight enters the prism and separates in space as it exits into pure colors at different frequencies of different amplitude. Hence, in both cases, one can view the Fourier transform as taking a function of **strength versus time** and transposing it into a function of **amplitude versus frequency**. In this section, we further explore these interpretations by revisiting previous material from the more physical perspective of waves, frequencies, and harmonics. For further interpretations of the Fourier transform, more from the perspective of electrical engineering, see the recent text of Osgood [**13**].

6.12.1. Plane Waves in One Dimension.

We begin with a brief review of the basic terminology used when describing cosine and sine waves. We consider a (right moving) spatiotemporal **plane wave** in the form of

$$u_{PW}(x,t) = A\cos(kx - \omega t - \phi).$$

If the problem at hand does not involve time, spatial waves are described by the same form with $t = 0$. The parameter A, known as the **amplitude**, is the maximum value the plane wave will achieve; hence it describes the strength of the wave. The parameter k is known as the **wavenumber** and counts the number of waves per unit distance, describing the spatial periodicity of the wave. It is related to the wavelength λ of the wave as $k = 2\pi/\lambda$. On the other hand, ω is known as the **(temporal) frequency** and describes the temporal periodicity of the wave. The last parameter ϕ is known as the **phase** of the wave and encodes the delay of the plane wave relative to other plane waves. These quantities, describing undamped vibrations of a medium, are assumed to be real numbers. It should also be noted that by modifying the definition of our phase parameter $\pm\pi/2$, we can equivalently define plane waves via sine curves rather than cosines. Consequently, they are equivalent mathematical descriptions.

As we have seen throughout this chapter, is is tremendously insightful and algebraically useful to elevate our discussion to the **complex plane** and provide equivalent descriptions in terms of complex exponentials. To this end, we note that via Euler's formula (6.4), we can write

$$u_{PW}(x,t) = A\Re e^{i(kx-\omega t-\phi)}, \tag{6.67}$$

where $\Re$ denotes the **real part** of the complex argument.

6.12.2. Interpreting the Fourier Inversion Formula. Recall that if $f(x)$ is integrable, then its Fourier transform $\hat{f}(k) = \int_{-\infty}^{\infty} e^{-ikx} f(x)dx$ is well-defined. If, in addition, $\hat{f}(k)$ is integrable, we can invert and represent $f(x)$ in terms of $\hat{f}(k)$ by the Fourier inversion formula:

$$f(x) = \frac{1}{2\pi}\int_{-\infty}^{\infty} e^{ikx}\hat{f}(k)dk. \tag{6.68}$$

What is the physical interpretation of this formula? Specifically, what does the variable k represent? Clearly, by dimensional analysis, if x relates to position, then k will have dimensions of inverse length. Moreover, how is it guaranteed that by summing over the complex exponentials weighted by the complex-valued Fourier transform $\hat{f}(k)$ we will even get back a real number? To answer this, let us write $\hat{f}(k)$ in complex polar coordinates; that is,

$$\hat{f}(k) = \mathcal{A}(k)e^{-i\theta(k)} \tag{6.69}$$

where $\mathcal{A}(k)$ is the real-valued amplitude and $\theta(k)$ is the clockwise phase of the complex number $\hat{f}(k)$. One can show that in order to ensure that $f(x)$ is real, the Fourier transform should obey $\overline{\hat{f}(k)} = \hat{f}(-k)$. In the language of polar coordinates, one concludes that the amplitude and phase functions must have even and odd symmetry, respectively[13]. We can now reexamine (6.68) by summing over only positive k:

$$
\begin{aligned}
f(x) = \frac{1}{2\pi}\int_{-\infty}^{\infty} e^{ikx}\hat{f}(k)\,dk &= \frac{1}{2\pi}\int_{0}^{\infty} e^{ikx}\hat{f}(k)\,dk + \frac{1}{2\pi}\int_{-\infty}^{0} e^{ikx}\hat{f}(k)\,dk \\
&= \frac{1}{2\pi}\int_{0}^{\infty}\left(e^{ikx}\hat{f}(k) + e^{-ikx}\hat{f}(-k)\right)dk \\
&= \frac{1}{2\pi}\int_{0}^{\infty}\left(e^{ikx}\hat{f}(k) + e^{-ikx}\overline{\hat{f}(k)}\right)dk \\
&= \frac{1}{2\pi}\int_{0}^{\infty}\left(e^{ikx}\mathcal{A}(k)e^{-i\theta(k)} + e^{-ikx}\mathcal{A}(k)e^{i\theta(k)}\right)dk \\
&= \frac{1}{2\pi}\int_{0}^{\infty}\mathcal{A}(k)\left(e^{i(kx-\theta(k))} + e^{-i(kx-\theta(k))}\right)dk \\
&= \frac{1}{\pi}\int_{0}^{\infty}\mathcal{A}(k)\cos(kx-\theta(k))\,dk.
\end{aligned}
$$

Thus, there is a direct correlation between the amplitudes and phases of the complex-valued transform and the amplitudes and phases of composite plane waves that interfere to give our original function, $f(x)$ (modulo a factor of π). Now we see that the variable k of the Fourier transform represents the wavenumber (or temporal frequency if the real variable is time) representing the spatial periodicity of $f(x)$'s composite waves. Curious to know how much of a wave with wavenumber (spatial frequency) k makes up f? Examine the amplitude and phase of the complex number $\hat{f}(k)$. It is for this reason that the domain of the variable k is often called frequency space. For periodic functions (those that will be investigated in Chapter 11), the Fourier transform will be strongly peaked at integer multiples of the fundamental periodicity of the function (cf. Section 11.8.2). These integer multiples of the fundamental frequency are known as **harmonics**. In fact, most musical instruments give off sound waves resonating at multiple harmonics of the fundamental frequency that is being perceived. Further, the nature of these additional peaks in Fourier space distinguish different instruments from each other, yielding different **timbres**.

6.12.3. Revisiting the Wave Equation.

Given that we are talking about "waves", it would seem logical at this point to connect with the wave equation. For simplicity, let us revisit the one-dimensional wave equation, derived and solved in Chapter 3, through the **lens** of the Fourier transform. Recall that the solution to the IVP (3.5) for the 1D wave equation was given by D'Alembert's formula

$$u(x,t) = \frac{1}{2}[\phi(x+ct)+\phi(x-ct)] + \frac{1}{2c}\int_{x-ct}^{x+ct}\psi(s)\,ds. \tag{6.70}$$

In Section 3.4, we thoroughly investigated the consequences and interpretations of this formula. Now we will use the Fourier transform to reinterpret this solution. Assuming the required integrability on the initial data (and its transform), we can apply the

[13] Note that this encompasses the fact that even functions have purely real Fourier transforms, while odd functions have purely imaginary Fourier transforms (cf. Exercise **6.1**)

Fourier inversion formula and the Fourier translation property in (6.70) to find

$$u(x,t) = \frac{1}{4\pi}\int_{-\infty}^{\infty} e^{ikx}\hat{\phi}(k)(e^{ikct}+e^{-ikct})\,dk + \frac{1}{4\pi c}\int_{x-ct}^{x+ct}\int_{-\infty}^{\infty}\frac{1}{2\pi}e^{iks}\hat{\psi}(k)\,dk\,ds$$

$$= \frac{1}{4\pi}\int_{-\infty}^{\infty} e^{ikx}\left(\hat{\phi}(k)(e^{ikct}+e^{-ikct}) + \frac{1}{c}\hat{\psi}(k)\frac{e^{ikct}-e^{-ikct}}{ik}\right)dk. \tag{6.71}$$

Now let us rewrite the superpositions of complex exponentials back into sine and cosine functions, converting this expression into a statement about real numbers. However, we must remember that the Fourier transforms of our real-valued initial data are complex valued. For any fixed k, imagine parametrizing these complex numbers in polar coordinates as $\hat{\phi}(k) = \rho_1(k)e^{-i\theta_1(k)}$ and $\hat{\psi}(k) = \rho_2(k)e^{-i\theta_2(k)}$ where $\rho_1(k) = |\hat{\phi}(k)|$ and $\theta_1(k) = \arg(\hat{\phi}(k))$ are the real-valued amplitudes and clockwise phases of the complex number $\hat{\phi}(k)$, likewise, for the transformed initial velocity data. Note that because the initial data is real, this actually places a constraint on the properties of the Fourier transform. Indeed, as before, for a real square-integrable function f, $f(x) = \bar{f}(x)$ implies that $\bar{\hat{f}}(k) = \hat{f}(-k)$, and this translates to our amplitude and phase functions by constraining them to be even and odd functions of k, respectively. Now we will simplify equation (6.71), integrating only over positive k by exploiting the symmetry of the transformed data:

$$\begin{aligned}
u(x,t) &= \frac{1}{4\pi}\int_0^{\infty}(e^{ikct}+e^{-ikct})(\rho_1(k)e^{i(kx-\theta_1(k))} + \rho_1(-k)e^{-i(kx+\theta_1(-k))})\,dk \\
&\quad + \frac{1}{4\pi c}\int_0^{\infty}\frac{e^{ikct}-e^{-ikct}}{ik}(\rho_2(k)e^{i(kx-\theta_2(k))} + \rho_2(-k)e^{-i(kx+\theta_2(-k))})\,dk \\
&= \frac{1}{4\pi}\int_0^{\infty}\rho_1(k)(e^{ikct}+e^{-ikct})(e^{i(kx-\theta_1(k))} + e^{-i(kx-\theta_1(k))})\,dk \\
&\quad + \frac{1}{4\pi c}\int_0^{\infty}\rho_2(k)\frac{e^{ikct}-e^{-ikct}}{ik}(e^{i(kx-\theta_2(k))} + e^{-i(kx-\theta_2(k))})\,dk.
\end{aligned}$$

By Euler's formula, this yields

$$\begin{aligned}
u(x,t) = &\frac{1}{\pi}\int_0^{\infty}\rho_1(k)\cos(kct)\cos(kx-\theta_1(k))\,dk \\
&+ \frac{1}{\pi}\int_0^{\infty}\rho_2(k)\frac{\sin(kct)}{kc}\cos(kx-\theta_2(k))\,dk.
\end{aligned}$$

Finally, we can apply the product-to-sum trigonometric identities to eliminate the trigonometric products and yield

$$\begin{aligned}
u(x,t) = &\frac{1}{2\pi}\int_0^{\infty}\rho_1(k)\left(\cos(k(x+ct)-\theta_1(k)) + \cos(k(x-ct)+\theta_1(k))\right)dk \\
&+ \frac{1}{2\pi}\int_0^{\infty}\frac{\rho_2(k)}{ck}\left(\sin(k(x+ct)-\theta_2(k)) + \sin(k(x-ct)+\theta_2(k))\right)dk.
\end{aligned}$$

Hence, via the Fourier transform we can see that the solution of the wave equation is an infinite superposition of right and left moving plane waves. The temporal frequency of each plane wave is proportional to the wavenumber of the wave, scaled by the wave speed. Crucially, **the amplitude of each wave is determined by the amplitude of the complex Fourier transform $\hat{\phi}(k)$ and $\hat{\psi}(k)/ck$, with phases determined by the phases of these complex numbers!** Hence, the Fourier transform of a function evaluated at a particular wavenumber k is a complex number encoding the amplitude and phase of the plane wave nested within the function. Integrating or summing over all plane waves (all positive wavenumbers) reconstructs the desired function! Recall that these wavenumbers are the spatial frequencies of the wave and are often called harmonics.

An astute reader may wonder about the appearance of the factor $\frac{1}{ck}$ in the second integral over the initial velocity data and why sines appear there instead of cosines. We will come back to this in the next subsection, as it can be explained by the physical interpretation of the incredible property of the Fourier transforms to transform derivatives into complex polynomial multiplication.

Before proceeding, it is worthwhile to reflect on a few other aspects that arise from this formula. First of all, the notion of summing infinitely many plane waves is, in fact, in complete harmony (no pun intended) with our general solution to the wave equation: $u(x,t) = f(x-ct) + g(x+ct)$, for functions f and g. Using the Fourier transform, we can write these functions as

$$f(s) = \frac{1}{2\pi}\int_0^\infty \left(\rho_1(k)\cos(ks-\theta_1(k)) + \frac{\rho_2(k)}{ck}\sin(ks-\theta_2(k))\right) dk,$$

$$g(s) = \frac{1}{2\pi}\int_0^\infty \left(\rho_1(k)\cos(ks+\theta_1(k)) + \frac{\rho_2(k)}{ck}\sin(ks+\theta_2(k))\right) dk.$$

Indeed these forward and backward moving signals explicitly arise from the wave decomposition of the initial data.

Secondly, recall from Section 3.12.3 that there exist many other "wave equations" both linear and nonlinear. We remark here that the solution to a wide class of **linear** "wave equations" can be written as

$$\begin{aligned} u(x,t) = {} & \frac{1}{2\pi}\int_0^\infty \rho_1(k)\left(\cos(kx+\omega(k)t-\theta_1(k)) + \cos(kx-\omega(k)t+\theta_1(k))\right) dk \\ & + \frac{1}{2\pi}\int_0^\infty \frac{\rho_2(k)}{\omega(k)}\left(\sin(kx+\omega(k)t-\theta_2(k)) + \sin(kx-\omega(k)t+\theta_2(k))\right) dk \end{aligned}$$

where $\omega(k)$ is the **dispersion relation** which we introduced in Section 3.12.3. Dispersion describes the way in which an initial disturbance of the wave medium distorts over time, explicitly specifying how composite plane wave components of an initial signal evolve temporally for different wavenumbers. We previously observed that for the classical wave equation, the dispersion relation was trivial: $\omega(k) = ck$. On the other hand, in Section 3.12.4 we saw that the dispersion relation for the **Klein-Gordon equation** (3.32) (the PDE that describes the quantum mechanical wave-like nature of a particle of mass m traveling near the speed of light c) was $\omega(k) = \sqrt{c^2k^2 + (mc^2/\hbar)^2}$, where $\hbar$ is the reduced Planck constant.

6.12.4. Properties of the Fourier Transform, Revisited. We will now revisit the fundamental mathematical properties of the Fourier transform noted in Section 6.4 from the perspective of composite plane waves. In this setting, many of the properties will seem obvious; indeed, for example, the statement that translations of functions in real space map onto phase dilations in Fourier space is simply the observation that translations of cosine functions are (by definition) the phases of the waves. This is solely the physical interpretation of

$$\mathcal{F}\{f(x-a)\}(k) = e^{-ika}\mathcal{F}\{f(x)\}(k).$$

The remarkable property whereby differentiation is mapped into complex polynomial multiplication is really just a statement about the derivatives of sine and cosine functions. Namely, they are each other, scaled by the frequency of the wave due to the chain rule. Since cosines and sines are out of phase by $-\pi/2$, a factor of $-i = e^{-i\pi/2}$ necessarily encapsulates this fact in the realm of complex multiplication. Hence, when you Fourier transform a derivative in real space, in actuality you are really just kicking back the phase of the composite cosine waves and dilating the amplitude of each wave corresponding to its frequency. Not so mysterious now, is it?

The issue of invariance of the Gaussians under the Fourier transform is addressed in the following subsection on the uncertainty principle.

6.12.5. The Uncertainty Principle. Consider the family of Gaussian functions parametrized by their standard deviations, σ_x, which quantifies the narrowness of the peak (i.e., σ_x parametrizes the concentration of the function in real space):

$$G(x;\sigma_x) := \frac{1}{\sqrt{2\pi\sigma_x^2}}e^{-x^2/2\sigma_x^2}.$$

Upon computing the Fourier transform of this function, we obtain

$$\mathcal{F}\{G(x;\sigma_x)\}(k) = \frac{1}{\sqrt{2\pi}}e^{-k^2\sigma_x^2/2} = \sigma_k G(k;\sigma_k)$$

where the Fourier standard deviation is $\sigma_k = 1/\sigma_x$. This acknowledgment that the product of real space and frequency space concentration lengths is of order unity is known as the **uncertainty principle**

$$\sigma_x\,\sigma_k = O(1). \tag{6.72}$$

Readers will probably have heard of the **Heisenberg uncertainty principle** in quantum mechanics and be wondering if this is the same principle. It is indeed: In the context of quantum mechanics the Fourier variable k represents momentum p and Heisenberg's uncertainty principle is often phrased by saying that the standard deviations of position and momentum (σ_x and σ_p) satisfy

$$\sigma_x\,\sigma_p \geq \frac{\hbar}{2},$$

where $\hbar$ is the reduced Planck constant.

The uncertainly principle asserts that the narrower the peak in real space, the wider the support of the Fourier transform. This is why a delta function ($\sigma_x \to 0$)

will necessarily incorporate all possible wavenumbers ($\sigma_k \to \infty$) in its plane wave decomposition. Recall that loosely speaking, all plane waves $\cos(kx)$ take on the same value at $x = 0$ but will cancel each other out at other values of x summing to get a delta function. We find that the more frequencies we integrate over, the stronger this peak will concentrate. The fact that the Gaussian with $\sigma_x = 1$ is really an eigenfunction of the Fourier transform is actually a statement that the Gaussian with this width is a function with minimal uncertainty in both its position and frequency (momentum) simultaneously. It is mapped back onto itself by the transform, hence achieving the perfect balance of position spread and frequency (momentum) spread.

One final, perhaps crude, example will show just how natural/obvious the existence of this relationship is in the context of periodic functions. Suppose you encounter a "periodic looking" function, $f(x)$, whose support is limited to some range $\sigma_x \approx n\lambda$, where $\lambda = 2\pi/k$ is your guess at the wavelength of the wave. However, because the support of this periodic structure is limited, so is your estimate on this spatial frequency/ wavelength of the underlying signal. More precisely, you know the "true" wavelength lies somewhere between

$$\frac{\sigma_x}{n + 1/2} < \lambda < \frac{\sigma_x}{n - 1/2}.$$

Since the wavenumber behaves like the inverse of wavelength, we have

$$\frac{2\pi}{\sigma_x}(n - 1/2) < k < \frac{2\pi}{\sigma_x}(n + 1/2),$$

yielding an uncertainty $\sigma_k \approx \frac{2\pi}{\sigma_x}$. This again reaffirms the uncertainty relation (6.72), without the machinery of the Fourier transform; indeed, we only used the simple idea that measuring the period of an oscillation is limited solely by the length of measurement in real space. This explains why, even if you have perfect pitch, you will not be able to determine the frequency of a clap due to the fact that the duration in time (real) space is too short to have a small uncertainty in frequency (Fourier) space.

6.13. A Few Words on Other Transforms

There are many other transforms which are related to the Fourier transform. While we will not utilize them in this text, they all have proven importance in many fields. For example, (i) the Laplace transform is useful in solving many linear ODEs and PDEs, (ii) the Hilbert transform is a fundamental operation in the mathematical field of harmonic analysis, and (iii) the Radon transform has been proven to be revolutionary in medical imaging. We very briefly describe these three cousins of the Fourier transform.

The Laplace transform: The Laplace transform is similar in spirit to the Fourier transform but is often taken in a time variable t, where $t \geq 0$. For any function $f(t), t \geq 0$, locally integrable on $[0, \infty)$, we define the Laplace transform of f to be the new function of a, in general, complex variable s:

$$\mathcal{L}\{f\}(s) := \int_0^\infty f(t)e^{-st}\,dt = \lim_{R\to\infty} \int_0^R f(t)e^{-st}\,dt,$$

provided the limit exists. If the function f is defined for $t \in \mathbb{R}$, then the analogous definition, sometimes called the *bilateral Laplace transform* is

$$\mathcal{L}\{f\}(s) := \int_{-\infty}^{\infty} f(t)e^{-st}\,dt.$$

Note that, unlike in the Fourier transform, the exponent in the exponential is real. We immediately see the relationship between the Laplace and Fourier transforms; namely,

$$\widehat{f}(k) = \mathcal{L}\{f\}(ik).$$

Also note that, in general, the Laplace transform of a function $f(t)$ will only be defined in a subset of the complex plane. For example, one can readily check that the Laplace transform of the function $f(t) \equiv 1$ is only defined for s real with $s > 0$.

In terms of differentiation, one has (cf. Exercise **6.16**) that

$$\mathcal{L}\left\{\frac{df}{dt}\right\}(s) = s\mathcal{L}\{f\}(s) - f(0) \quad \text{and} \quad \mathcal{L}\left\{\frac{d^2f}{dt^2}\right\}(s) = s^2\mathcal{L}\{f\}(s) - sf(0) - f'(0).$$

One can also show that the Laplace transform behaves in a very similar way to the Fourier transform with respect to convolutions. Indeed, if $g(t)$ is a function defined on all of $\mathbb{R}$, $f(t)$ is a function defined on $[0,\infty)$, and $F(s) = \mathcal{L}\{f\}(s)$, $G(s) = \mathcal{L}\{g\}(s)$, then we can define a convolution $H(t)$ for $t \geq 0$ by

$$H(t) := \int_0^t g(t-t')\,f(t')\,dt'.$$

One can readily verify (cf. Exercise **6.16**) that $\mathcal{L}\{H\}(s) = F(s)\,G(s)$.

The Hilbert transform: The Hilbert transform is named after **David Hilbert**[14]. Besides its importance in signal processing, it plays a fundamental role in the mathematical field of harmonic analysis, in particular an area known as **singular integrals**. Key to Hilbert transform is the distribution PV $\frac{1}{x}$ (cf. Section 5.8). For a function $f(x)$, the Hilbert transform of f is a new function defined by

$$\mathcal{H}(f)(t) := \frac{1}{\pi}\,\mathrm{PV}\int_{-\infty}^{\infty} \frac{f(\tau)}{t-\tau}\,d\tau = \frac{1}{\pi}\lim_{\epsilon\to 0}\int_{\{\tau\,|\,|\tau|\geq\epsilon\}} \frac{f(\tau)}{t-\tau}\,d\tau.$$

Note that the Hilbert transform of f is simply the convolution of f with the (tempered) distribution PV $\frac{1}{x}$. While it might not be obvious that the defining limit always exists, one can put very general assumptions on the function f which implies existence of the limit (for almost every t).

The Hilbert transform is essentially its own inverse: $\mathcal{H}(\mathcal{H}(f))(t) = -f(t)$, or alternatively, $\mathcal{H}^{-1} = -\mathcal{H}$. There is also a beautiful relationship between the Hilbert and Fourier transforms which states that

$$\mathcal{F}(\mathcal{H}(f))(k) = (-i\,\mathrm{sgn}(k))\,\mathcal{F}(f)(k),$$

where sgn is the signum function which gives 1 if the argument is positive and -1 if it is negative.

---

[14]**David Hilbert** (1862–1943) was a German mathematician who had a pioneering role in the development of modern mathematics.

The Radon transform: The Radon transform is named after **Johann Radon**[15] It is defined for functions of more than one variable (the definition degenerated in dimension $N = 1$). Because of its importance in two-dimensional medical imaging, it is natural to work in space dimension $N = 2$. Let $f : \mathbb{R}^2 \to \mathbb{R}$ be a function with *suitable* decay properties as $|\mathbf{x}| \to \infty$. Roughly speaking the Radon transform of f is obtained by integrating f over parallel lines. More precisely, we can write any line in $\mathbb{R}^2$ via a unit normal vector $\mathbf{n}$ and the distance t to the origin; precisely the line is the set

$$\mathcal{L}_{t,\mathbf{n}} := \{\mathbf{x} \in \mathbb{R}^2 \,|\, \mathbf{x}\cdot\mathbf{n} = t\}.$$

Then we define the Radon transform of f as a function of t and $\mathbf{n}$ as the line integral:

$$(\mathcal{R}f)(t,\mathbf{n}) := \int_{\mathcal{L}_{t,\mathbf{n}}} f(\mathbf{x})\,ds.$$

First note that this definition easily extends to functions on $\mathbb{R}^N$, $N > 2$. For example, in 3D, we would have a surface integral over planes of the form $\{\mathbf{x} \in \mathbb{R}^3 \,|\, \mathbf{x}\cdot\mathbf{n} = t\}$. In dimension 2, it is convenient to set α to be the angle made by $\mathbf{n}$ and the positive x-axis, i.e., $\mathbf{n} = \langle\cos\alpha, \sin\alpha\rangle$, and parametrize each $\mathcal{L}_{t,\mathbf{n}}$ via

$$(x(z), y(z)) := (z\sin\alpha + t\cos\alpha, -z\cos\alpha + t\sin\alpha).$$

Note that in this parametrization, z denotes arclength along the line; hence conveniently we have $ds = dz$ and we may write $\mathcal{R}f$ in terms of t and α as

$$(\mathcal{R}f)(t,\alpha) = \int_{-\infty}^{\infty} f(z\sin\alpha + t\cos\alpha, -z\cos\alpha + t\sin\alpha)\,dz.$$

Note the slight abuse of notation here; initially, $\mathcal{R}f$ is a function of $\mathbb{R}\times$(unit vectors in $\mathbb{R}^2$) while in the parametrization it becomes a function of $\mathbb{R}\times\mathbb{R}$.

As with the Fourier transform, the Radon transform is invertible; that is, one can recover a function from its Radon transform. The details are slightly more complicated than for the Fourier transform and hence are not presented here. There is also a relationship between the 2D Radon transform and the 1D and 2D Fourier transforms. To this end, if we define

$$(\mathcal{R}_\alpha f)(t) := (\mathcal{R}f)(t,\alpha),$$

then we have

$$\widehat{\mathcal{R}_\alpha f}(s) = \hat{f}(s\cos\alpha, s\sin\alpha). \tag{6.73}$$

Note that on the left is the 1D Fourier transform and on the right is the 2D Fourier transform. The relation (6.73) is often referred to as the *Fourier Slicing Theorem* or the *Projection Slice Theorem* and it is not difficult to prove (cf. Exercise **6.30**). In words, it states that the 2D Fourier transform of f along lines through the origin with inclination angle α is the same as the 1D Fourier transform of the Radon transform at angle α. Alternatively, we can use the Fourier inversion formula to write (6.73) as

$$(\mathcal{R}_\alpha f)(t) = \frac{1}{2\pi}\int_{-\infty}^{\infty} \hat{f}(s\cos\alpha, s\sin\alpha)\,e^{its}\,ds;$$

[15] **Johann Karl August Radon** (1887–1956) was an Austrian mathematician famous for his transform and several notions in measure theory.

in other words, one can use only the Fourier and inverse Fourier transforms to compute the Radon transform. For more on the Radon transform and its fundamental role in medical imaging see the book of Epstein[16].

6.14. Chapter Summary

- If $f(x)$ is an integrable real-valued function on $\mathbb{R}$, then its Fourier transform is the complex-valued function of $k \in \mathbb{R}$, defined by
$$\hat{f}(k) := \int_{-\infty}^{\infty} f(x)\, e^{-ikx}\, dx.$$
- We define the inverse Fourier transform of a function $g(k)$ to be
$$\check{g}(x) := \frac{1}{2\pi}\int_{-\infty}^{\infty} g(k)\, e^{ikx}\, dk.$$
We use the variable x to describe the original space and the variable k to describe the transformed space, Fourier space. Both are real variables.
- The Fourier transform is a very useful tool for solving linear PDE. This is primarily due to the following three properties.
  1. It transforms **differentiation** into polynomial multiplication; for example,
  $$\widehat{\frac{df}{dx}}(k) = ik\hat{f}(k).$$
  Note that the polynomial multiplication is in the complex plane.
  2. It transforms **convolution** into pointwise multiplication; that is,
  $$\widehat{(f * g)}(k) = \hat{f}(k)\,\hat{g}(k), \quad \text{where} \quad (f * g)(x) := \int_{-\infty}^{\infty} f(x-y)\, g(y)\, dy.$$
  3. The Fourier transform is **invertible** in the sense that we can recover a function from its Fourier transform by taking the inverse Fourier transform. More precisely, the **Fourier inversion formula** states that $\left(\hat{f}(k)\right)^{\vee}(x) = f(x)$.
- Other important properties of the Fourier transform include the following:
  - The invariance of a **Gaussian**: The Fourier transform of a function of Gaussian type is also of Gaussian type.
  - **Plancherel's Theorem**: Up to a constant factor, we can compute the squared integral of a function, $\int_{-\infty}^{\infty} |f(x)|^2\, dx$, by computing the squared integral of the Fourier transform, $\int_{-\infty}^{\infty} |\hat{f}(k)|^2\, dk$.
- Key to the inversion formula is that
$$\text{“}\int_{-\infty}^{\infty} e^{-iky}\, dk\text{”} = 2\pi\,\delta_0 \qquad \text{in the sense of distributions.}$$
Alternatively, the Fourier transform of 1 is $2\pi\delta_0$. This is understood as the statement
$$\int_{-n}^{n} e^{-iky}\, dk \xrightarrow{n\to\infty} 2\pi\delta_0 \qquad \text{in the sense of distributions.}$$

[16] **C.L. Epstein**, *Introduction to the Mathematics of Medical Imaging*, second edition, SIAM, 2007.

Our approach to proving the above was more general and was based upon the extension of the Fourier transform to **tempered distributions**. These are distributions defined over test functions in the **Schwartz class**, the larger class consisting of C^∞ test functions which rapidly decay. Extending the notion of the Fourier transform to tempered distributions allows us to Fourier transform **non-integrable functions**, e.g., constants, trigonometric functions, polynomials, and the Heaviside function.

- **Higher dimensions:** One can analogously define the Fourier transform of any integrable function on $\mathbb{R}^N$. All the properties and results carry over.
- The algorithm for using the Fourier transform to **solve linear differential equations** can roughly be summarized as follows:
    1. Take the Fourier transform in one or more of the variables of both sides of the differential equation (and the initial data as well if solving an IVP).
    2. Solve for $\hat{u}$, the Fourier transform of the unknown solution. This often just entails algebra and/or solving an ODE.
    3. Take the inverse Fourier transform of $\hat{u}$ to obtain a formula for u.

6.15. Summary Tables

Basic Properties of the Fourier Transform

The following properties hold under suitable assumptions of the function f:

Function on $\mathbb{R}$	**Fourier Transform**
$f(x)$	$\widehat{f}(k)$
$f'(x)$	$ik\widehat{f}(k)$
$xf(x)$	$i\widehat{f}'(k)$
$f(x-c)$	$e^{-ick}\widehat{f}(k)$
$e^{icx}f(x)$	$\widehat{f}(k-c)$
$f(ax)$	$\frac{1}{a}\widehat{f}\left(\frac{k}{a}\right)$
$(f*g)(x)$	$\widehat{f}(k)\widehat{g}(k)$
$f(x)g(x)$	$\frac{1}{2\pi}(\widehat{f}*\widehat{g})(k)$

Function on $\mathbb{R}^N$	**Fourier Transform**
$f(\mathbf{x})$	$\widehat{f}(\mathbf{k})$
$\partial^\alpha f(\mathbf{x})$	$i^{\lvert\alpha\rvert}\,\mathbf{k}^\alpha\widehat{f}(\mathbf{k})$
$(i\mathbf{x})^\alpha f(\mathbf{x})$	$\partial^\alpha\widehat{f}(\mathbf{k})$
$f(\mathbf{x}-\mathbf{a})$	$e^{-i\mathbf{a}\cdot\mathbf{k}}\widehat{f}(\mathbf{k})$
$e^{i\mathbf{a}\cdot\mathbf{x}}f(\mathbf{x})$	$\widehat{f}(\mathbf{k}-\mathbf{a})$
$f(c\mathbf{x})$	$\frac{1}{c^N}\widehat{f}\left(\frac{\mathbf{k}}{c}\right)$
$(f*g)(\mathbf{x})$	$\widehat{f}(\mathbf{k})\widehat{g}(\mathbf{k})$
$f(\mathbf{x})g(\mathbf{x})$	$\frac{1}{(2\pi)^N}(\widehat{f}*\widehat{g})(\mathbf{k})$

Fourier Transforms of a Few Particular Functions

Function on $\mathbb{R}$	**Fourier Transform**
$e^{-ax^2}, \quad a > 0$	$\frac{\sqrt{\pi}}{\sqrt{a}} e^{-\frac{k^2}{4a}}$
$e^{-a\|x\|}, \quad a > 0$	$\frac{2a}{k^2+a^2}$
$\frac{1}{x^2+a^2}, \quad a > 0$	$\frac{\pi}{a} e^{-a\|k\|}$
$\chi_a(x) := \begin{cases} 1 & \|x\| < a, \\ 0 & \|x\| \geq a \end{cases}$	$2\frac{\sin ak}{k}$
$\frac{\sin ax}{x}$	$\pi\chi_a(k) = \begin{cases} \pi & \|k\| < a, \\ 0 & \|x\| \geq a \end{cases}$

Function on $\mathbb{R}^N$	**Fourier Transform**
$e^{-a\|\mathbf{x}\|^2} \quad a > 0, \mathbf{x} \in \mathbb{R}^N$	$\frac{\pi^{N/2}}{a^{N/2}} e^{-\frac{\|\mathbf{k}\|^2}{4a}}, \mathbf{k} \in \mathbb{R}^N$
$\frac{1}{\|\mathbf{x}\|^2+a^2} \quad a > 0, \mathbf{x} \in \mathbb{R}^3$	$\frac{(2\pi)^3 e^{-a\|\mathbf{k}\|^2}}{4\pi\|\mathbf{k}\|} \quad \mathbf{k} \in \mathbb{R}^3$
$\frac{e^{-a\|\mathbf{x}\|^2}}{4\pi\|\mathbf{x}\|} \quad a > 0, \mathbf{x} \in \mathbb{R}^3$	$\frac{1}{\|\mathbf{k}\|^2+a^2} \quad \mathbf{k} \in \mathbb{R}^3$

Fourier Transforms of a Few Tempered Distributions in One Variable

Distribution	**Fourier Transform**
1	$2\pi\delta_0$
δ_0	1
e^{iax}	$2\pi\delta_a$
δ_a	e^{iak}
$\frac{\delta_a+\delta_{-a}}{2}$	$\cos ak$
$\frac{\delta_{-a}-\delta_a}{2i}$	$\sin ak$
$\cos ax$	$\pi(\delta_a + \delta_{-a})$
$\sin ax$	$\frac{\pi}{i}(\delta_a - \delta_{-a})$

Distribution	**Fourier Transform**
x	$2\pi i\, \delta_0'$
$\mathrm{sgn}(x)$	$\frac{2}{i}\mathrm{PV}\frac{1}{k}$
$\mathrm{PV}\frac{1}{k}$	$\frac{\pi}{i}\mathrm{sgn}(k)$
$H(x)$	$\pi\delta_0 + \frac{1}{i}\mathrm{PV}\frac{1}{k} = \frac{1}{i}\frac{1}{k-i0}$
$\frac{1}{x+i0}$	$-2\pi i\, H(k)$

Exercises

6.1 Show that the Fourier transform of any **even** real-valued function is real valued (i.e., has no imaginary part). Show that the Fourier transform of any **odd** real-valued function is purely imaginary (i.e., has no real part).

6.2 Prove directly, using the definition of the Fourier transform, Properties (6.23), (6.24), and (6.25).

6.3 Prove that $\frac{\sin ak}{k}$ is not integrable on $\mathbb{R}$.
Hint: You can assume without loss of generality that $a = 1$. Show that

$$\int_{-\infty}^{\infty} \left|\frac{\sin k}{k}\right| dk \geq \int_{0}^{\infty} \left|\frac{\sin k}{k}\right| dk \ \geq \ \sum_{n=1}^{\infty} \int_{n\pi}^{(n+1)\pi} \left|\frac{\sin k}{k}\right| dk$$

$$\geq \ \sum_{n=1}^{\infty} \frac{1}{\pi} \frac{1}{n+1} \int_{n\pi}^{(n+1)\pi} |\sin k|\, dk.$$

Compute the integral on the right and use the fact that the harmonic series diverges.

6.4 Show that $\int_{-\infty}^{\infty} e^{-x^2}\, dx = \sqrt{\pi}$. Hint: You will not be able to find an antiderivative of e^{-x^2}. Hence, argue as follows: Use polar coordinates to compute the double integral over $\mathbb{R}^2$

$$\int_{-\infty}^{\infty} \int_{-\infty}^{\infty} e^{-(x^2+y^2)}\, dx\, dy,$$

and then note that

$$\int_{-\infty}^{\infty} \int_{-\infty}^{\infty} e^{-(x^2+y^2)}\, dx\, dy = \left(\int_{-\infty}^{\infty} e^{-x^2}\, dx\right)\left(\int_{-\infty}^{\infty} e^{-y^2}\, dy\right).$$

6.5 Show that the Fourier transform of $\frac{e^{-|x|}}{2}$ is $\frac{1}{k^2+1}$.

6.6 Find the Fourier transform of each the functions

$$f(x) = \begin{cases} \sin x & \text{if } |x| \leq \pi, \\ 0 & \text{otherwise,} \end{cases} \qquad f(x) = \begin{cases} 1 - |x| & \text{if } |x| \leq 1, \\ 0 & \text{otherwise.} \end{cases}$$

Hint: For the first function, recall the integration formulas for products of sines with different angles.

6.7 (a) Let

$$g(x) = \begin{cases} 1 & \text{if } |x| < \frac{1}{2}, \\ 0 & \text{otherwise.} \end{cases}$$

Compute explicitly the function $(g*g)(x)$. Also, compute explicitly $(g*g*g)(x)$. Finally, for an integrable function f, describe in words how the convolution function $f * g$ is related to f.
(b) Now suppose g is any integrable function which need not be continuous (not to mention differentiable). Suppose f is a C^∞ function with compact support. Show that $(f * g)(x)$ is a C^∞ function on $\mathbb{R}$. Must it have compact support?

(c) If $f \in C_c^\infty(\mathbb{R})$, what would you expect if you convolute f with δ, the Dirac delta "function" concentrated at $x = 0$? In other words, what can you say about $f * \delta$? Is it a function? If so, what function? Here we just want an intuitive answer (no proof or rigorous treatment). Indeed, you cannot actually place the delta "function" inside the integral as it is not a function. However, we will treat this rigorously in proving Theorem 7.1 of the next chapter.

6.8 Let $f(x) = e^{-x^2}$ be a Gaussian. Compute explicitly $(f * f)(x)$.

6.9 Use (6.26) and rescaling to show that for fixed t, the inverse Fourier transform of $F(k) = e^{-k^2 t}$ is $f(x) = \frac{1}{\sqrt{4\pi t}} e^{-\frac{x^2}{4t}}$.

6.10 Use the properties of the Fourier transform from Section 6.4 to find the Fourier transform of $f(x) = x^n e^{-x^2}$ where $n \in \mathbb{N}$.

6.11 Let $n \in \mathbb{N}$ and $a \in \mathbb{R}$ be positive. Find the Fourier transform of the following two functions:

$$f(x) = \begin{cases} x^n e^{-ax}, & x > 0, \\ 0, & x \le 0, \end{cases} \qquad \text{and} \qquad f(x) = \begin{cases} x^n e^{-ax} \cos x, & x > 0, \\ 0 & x \le 0. \end{cases}$$

6.12 First note that the definition of convolution (6.18) easily carries over if one (or both) of the functions are complex valued.
(a) Let $k \in \mathbb{R}$, $f(x)$ be a real-valued integrable function, and $g(x) = e^{ikx}$. Show that $(g * f)(x) = \sqrt{2\pi} e^{ikx} \hat{f}(k)$.
(b) Use part (a) to prove that if $f(x) = e^{-|x|}$ and $g(x) = \cos x$, then $(g * f)(x) = \cos x$; in other words, as functions we have $\cos x * e^{-|x|} = \cos x$.

6.13 Suppose $y(x)$ is a function which solves the ODE $\frac{d^4 y}{dx^4} + 4\frac{d^2 y}{dx^2} + y = f(x)$, where $f(x)$ is some given function which is continuous and has compact support. Use the Fourier transform to write down an expression for $\hat{y}(k)$, the Fourier transform of the solution. Note that your answer will involve $\hat{f}(k)$.

6.14 We used the Fourier transform to find a solution to the ODE $y''(x) - y(x) = f(x)$ on all of $\mathbb{R}$. Prove that (6.34) is indeed a solution. You may assume f is continuous and has compact support. How will you do this? Recall how we showed the derivative of the Heaviside function was a delta function in the sense of distributions. The proof here is similar. Compute y'' by taking the derivative with respect to x into the integral of (6.34). Now break up the z integral into two pieces: $z < x$ and $z \ge x$. Integrate by parts twice on each piece.

6.15 **(Telegraph Equation)**
(a) Consider the telegraph equation:

$$u_{tt} + (\alpha + \beta) u_t + \alpha\beta u = c^2 u_{xx},$$

where α and β are fixed positive parameters. By doing some research online find a physical situation which is modeled by the telegraph equation. Explain all the variables and parameters.
(b) Use the Fourier transform to find the general solution to the telegraph equation.

(c) Explain the notion of **dispersion** in the context of the solutions. You may do some research online.
(d) Let $\alpha = \beta$. Find an explicit solution with initial conditions $u(x, 0) = \phi(x)$ and $u_t(x, 0) = \psi(x)$. What can you say here about dispersion?

6.16 (The Laplace Transform)
(a) Find the Laplace transform of $f(t) \equiv 1$ and of $f(t) = \sin \omega t$, where ω is any constant.
(b) Show that

$$\mathcal{L}\left\{\frac{df}{dt}\right\}(s) = s\mathcal{L}\{f\}(s) - f(0) \quad \text{and} \quad \mathcal{L}\left\{\frac{d^2 f}{dt^2}\right\}(s) = s^2\mathcal{L}\{f\}(s) - sf(0) - f'(0).$$

(c) Let $g(t)$ be a function defined on all of $\mathbb{R}$ and let $f(t)$ be a function defined on $[0, \infty)$. Let $F(s) := \mathcal{L}\{f\}(s)$ and $G(s) := \mathcal{L}\{g\}(s)$. Then we can define a convolution $H(t)$ by

$$H(t) := \int_0^t g(t - t')\, f(t')\, dt'.$$

Show that $\mathcal{L}\{H\}(s) = F(s)\, G(s)$.
(d) Following the basic strategy we used with the Fourier transform in solving the ODE (6.33); use the Laplace transform to solve the IVP for $y(t)$: $y''(t) + \omega^2 y(t) = f(t)$, $y(0) = y'(0) = 0$, where f is a given function on $[0, \infty)$.

6.17 Recall Example 6.2.1. As $a \to +\infty$, the step functions

$$\chi_a(x) := \begin{cases} 1 & \text{if } |x| \le a, \\ 0 & \text{if } |x| > a \end{cases}$$

seem to tend to the constant function $f(x) = 1$. Since, in the sense of tempered distributions, the Fourier transform of 1 is $2\pi\delta_0$, we would expect that the Fourier transforms of the step function χ_a, i.e., $2\frac{\sin ak}{k}$, to tend to $2\pi\delta_0$ as $a \to +\infty$. Plot these functions for $a = 1, 4, 10$, etc., to get some hint of what is happening. This foreshadows the **Dirichlet kernel** which will play an important role in Fourier series (Chapter 11).

6.18 Recall from Section 6.6 that we can extend the notion of the Fourier transform to all square-integrable functions. In doing so, we can take the Fourier transform of $\frac{\sin ax}{x}$. While we did not provide the details, here we show the following related fact. Consider the function $\frac{\sin ax}{x}$ as a tempered distribution and find its Fourier transform in the sense of tempered distributions using the obvious regularization sequence

$$f_n(x) = \chi_n(x)\frac{\sin ax}{x} \qquad \text{where} \quad \chi_n(x) := \begin{cases} 1 & \text{if } |x| \le n, \\ 0 & \text{otherwise.} \end{cases}$$

6.19 Consider the function $f(x) = \frac{1}{\sqrt{|x|}}$ which is neither integrable nor square integrable on $\mathbb{R}$. Note the function is locally integrable (integrable close to the blow-up at $x = 0$) but fails to be integrable because of its tails; for example, $\int_1^\infty f(x)\, dx = +\infty$. While we could take its Fourier transform in the sense of tempered distributions, we claim the pointwise definition (6.5) does make sense except at $k = 0$. This is because of the cancellation effects of e^{-ikx}.

(a) Show that except for $k = 0$, $\widehat{f}(k) = \frac{C}{\sqrt{|k|}}$ for some constant C.

(b) By using integral tables, find C explicitly.

(c) What is $\widehat{f}$ in the sense of tempered distributions?

6.20 Use the Fourier transform to solve the IVP for the wave equation with friction: $u_{tt} + \alpha u_t = c^2 u_{xx}$ where $\alpha > 0$ and $u(x,0) = \phi(x)$, $u_t(x,0) = \psi(x)$. Make any necessary assumptions on the initial value functions ϕ and ψ.

6.21 Use the Fourier transform as another route to deriving the formula (3.11) for the solution to the IVP for the wave equation with source term:

$$\begin{cases} u_{tt} - c^2 u_{xx} = f(x,t), & -\infty < x < \infty,\ t > 0, \\ u(x,0) = \phi(x), & -\infty < x < \infty, \\ u_t(x,0) = \psi(x), & -\infty < x < \infty. \end{cases}$$

You may make appropriate assumptions on the functions f, ϕ, and ψ.

6.22 (**Klein-Gordon Equation**)

(a) Use the 1D Fourier transform to solve the 1D Klein-Gordon equation $\hbar^2 u_{tt} - \hbar^2 c^2 u_{xx} + m^2 c^4 u = 0$ where $c, m > 0$ and $\hbar$ is the reduced Planck constant (Planck's constant h divided by 2π).

(b) Use the 3D Fourier transform to solve the 3D Klein-Gordon equation $\hbar^2 u_{tt} - \hbar^2 c^2 \Delta u + m^2 c^4 u = 0$.

6.23 (**A Proof of the Riemann-Lebesgue Lemma with Riemann Integration**) In this exercise, we prove (6.30) for every complex-valued ϕ which is Riemann integrable. For convenience, you may assume ϕ is a continuous function.

(a) First assume the ϕ has support on the interval $[-L, L]$ for some fixed L. Let $\epsilon > 0$ and recall that ϕ being Riemann integrable on $[-L, L]$ means that there exists a partition of $[-L, L]$, $-L = x_0 < x_1 < \cdots < x_n = L$, such that **if**

$$m_i := \min_{x \in [x_{i-1}, x_i]} \phi(x) \qquad \text{and} \qquad \psi(x) := \sum_{i=1}^{n} m_i \chi_{[x_{i-1}, x_i]},$$

where $\chi_{[x_{i-1}, x_i]}$ is the characteristic function of the interval $[x_{i-1}, x_i]$, **then**

$$0 \leq \int_{-L}^{L} \phi(x) - \psi(x)\, dx < \frac{\epsilon}{2}.$$

Note that $\phi(x) - \psi(x) \geq 0$ for all $x \in [-L, L]$. Now write

$$\left| \int_{-L}^{L} \phi(x) \sin(kx)\, dx \right| = \left| \int_{-L}^{L} (\phi(x) - \psi(x)) \sin(kx)\, dx + \int_{-L}^{L} \psi(x) \sin(kx)\, dx \right|,$$

use the triangle inequality, and compute the second integral on the right with the step function ψ. Finally let k be sufficiently large so that the second integral becomes less than $\epsilon/2$.

(b) Let $L \to \infty$ to arrive at (6.30).

6.24 Prove that for an integrable function f on $\mathbb{R}^N$ and an invertible $N \times N$ matrix A, we have

$$\widehat{f(A\mathbf{x})} = \frac{1}{\det A} \hat{f}((A^{-1})^T \mathbf{k})$$

where T denotes the matrix transpose. This provides further generalization to the scaling property of Fourier transforms.

6.25 (**Hard: For Students Interested in Mathematical Analysis**) Prove that the sequence of functions of k, $\frac{\sin nk}{\pi k}$, converges to δ_0 in the sense of tempered distributions. **Hint:** Use (6.30) of the Riemann-Lebesgue Lemma, which certainly holds for Schwartz functions, and the fact that for any n, $\int_0^\infty \frac{\sin nk}{\pi k}\, dk = \int_{-\infty}^0 \frac{\sin nk}{\pi k}\, dk = \frac{1}{2}$.

6.26 Prove that $\widehat{x^n} = (-1)^{n-1} i^n \delta_0^{(n)}$ in the sense of tempered distributions.

6.27 Find the Fourier transform of $|x|$ in the sense of tempered distributions.

6.28 Give an example of a sequence of functions $f_n(x)$ such that f_n converges to 0 in the sense of distributions but **not** in the sense of tempered distributions.

6.29 (**The Dirac Comb and the Poisson Summation Formula**) Part (b) of this exercise is challenging and for students interested in mathematical analysis. The Dirac comb is informally written as[17] "$\text{III}(x) := \sum_{n=-\infty}^{\infty} \delta(x-n)$". More precisely, it is the tempered distribution $\text{III} := \sum_{n=-\infty}^{\infty} \delta_n$. On the other hand, one can consider "$\sum_{n=-\infty}^{\infty} e^{-inx}$" in the sense of tempered distribution; that is, the partial sums are well-defined functions of x, and we can consider the limit in the sense of tempered distributions. In other words we define the tempered distribution by

$$\sum_{n=-\infty}^{\infty} e^{-inx} := \lim_{N\to\infty} \sum_{n=-N}^{N} e^{-inx}.$$

It turns out that this distribution is the same tempered distribution as the Dirac comb; that is,

$$\text{III} = \sum_{n=-\infty}^{\infty} e^{-inx} \quad \text{in the sense of tempered distributions.} \tag{6.74}$$

(a) Show that (6.74) immediately yields the following: For any $\phi \in \mathcal{S}(\mathbb{R})$,

$$\sum_{n=-\infty}^{\infty} \phi(n) = \sum_{n=-\infty}^{\infty} \hat{\phi}(n).$$

This equality is known as the Poisson summation formula. Note that both the left- and the right-hand sides are finite (complex) numbers. The formula also holds for more general functions than Schwartz functions.

(b) (Hard) Both distributions above are known as *periodic distributions*. Look up the definition of a periodic distribution online and show that this is the case. By consulting references online, prove (6.74). You will need to be familiar with Fourier series (the material of Chapter 11). Note that you may view (6.74) as the Fourier series of the Dirac comb.

---

[17] It is also known in signal processing as the Shah function because its "graph" resembles the Cyrillic letter sha (III).

6.30 (**Projection-Slice Theorem**) We work in dimension $N = 2$ but a similar result holds true in higher dimensions. Given a function $f(x, y)$ and a line $\mathcal{L}$ in the xy-plane, we define the **projection** (denoted by $P_{\mathcal{L}} f$) of f onto the line $\mathcal{L}$ as follows: For each point $\mathbf{p}$ on $\mathcal{L}$, $P_{\mathcal{L}} f$ is the 1D integral of f over the line through $\mathbf{p}$ which is perpendicular to $\mathcal{L}$. Note that by parametrizing $\mathcal{L}$, the projection $P_{\mathcal{L}} f$ is a function of one variable. For example, if $\mathcal{L}$ is the x-axis, then

$$(P_{\mathcal{L}_x} f)(x) = \int_{-\infty}^{\infty} f(x, y)\, dy.$$

On the other hand, given a function g of two variables, we define a **slice** of g, $S_{\mathcal{L}} g$, along a line through the origin as simply the function restricted to that line. So for example, a slice of $g(x, y)$ along the y-axis is simply $(S_{\mathcal{L}_x} g)(y) = g(0, y)$.

The Projection-Slice Theorem states that under certain assumptions on f, we have, for any line $\mathcal{L}$,

$$\widehat{(P_{\mathcal{L}} f)}(k) = \left(S_{\mathcal{L}^\perp} \hat{f}\right)(k), \tag{6.75}$$

where $\mathcal{L}^\perp$ is the line through the origin which is perpendicular to $\mathcal{L}$. Note that on the left of (6.75) the hat denotes the 1D Fourier transform, while on the right it is the 2D Fourier transform. In words, (6.75) states that we obtain the same function of one variable if **either** we take a function $f(x, y)$, project it onto a line $\mathcal{L}$, and take the 1D Fourier transform of the projection **or** we take the 2D Fourier transform of f and then slice it through its origin in the direction perpendicular to $\mathcal{L}$.

(a) Prove (6.75) — it is surprisingly easy! You may assume f is integrable and continuous. Hint: First note that without loss of generality, one can assume $\mathcal{L}$ is the x-axis.

(b) Use this to verify (6.73), the relationship between the Radon and Fourier transforms.

(c) State in words the analogous result in 3D. By researching online, explain why the result is important for **computed tomography** in medical imaging.

Chapter 7

The Diffusion Equation

Perhaps the most ubiquitous time-dependent PDE is **the diffusion equation**. It is also referred to as **the heat equation** but we find this name can be slightly misleading as it gives the impression that this is an equation explicitly for heat propagation. Nothing could be further from the truth: The equation forms one of the cornerstones for the study of *stochastic processes and statistical mechanics*. It is also widely used in financial engineering, image processing, and, yes!, it can be used to describe the propagation of heat.

Earlier when studying first-order equations (involving time) and the wave equation, we were introduced to the notion of characteristics, and information at a point, propagating in space with a finite speed. While these PDEs do represent continuum descriptions, they model systems for which we can, and want to, capture the **exact dynamics** of either particles or regions of particles. For example, the wave equation was derived by applying Newton's Second Law to a piece of a string (regions of string particles). However, many systems in nature are far too complex to analyze directly by solving for the dynamics of either each particle or even fixed regions of particles. Rather, one seeks to describe the system of particles by their macroscopic, coherent behavior, that is, the collective behavior on a large scale. In fact, one may never be able to know precisely the dynamics of any fixed set of particles, but in a certain average (statistical) sense, one can ascertain the **collective (*ensemble*) dynamics**. As a simple illustration[1], consider a swarm of millions of bees. Each bee has a particular distinct motion, but from the perspective of swarm tracking it would be overwhelming, and moreover unnecessary, to track each trajectory. Rather one focuses on the overall motion of the swarm and its density variation.

[1]The swarm of bees is to illustrate our focus on collective behavior, **not** as an example of a system which would be modeled by the diffusion equation.

It is when describing **a certain ubiquitous collective behavior of many systems** that the diffusion equation proves invaluable. To quote the excellent text on statistical mechanics by Sethna[2]:

You may think that Newton's law of gravitation, or Einstein's refinement to it, is more fundamental than the diffusion equation. You would be correct; gravitation applies to everything. But the simple macroscopic law of gravitation emerges, presumably, from a quantum exchange of immense numbers of virtual gravitons just as the diffusion equation emerges from large numbers of long random walks. The diffusion equation and other continuum statistical mechanics laws are special to particular systems, but they emerge from the microscopic theory in much the same way as gravitation and the other fundamental laws of nature do. This is the source of many of the surprisingly simple mathematical laws describing nature.

Roughly speaking, the diffusion equation models the concentration, or ensemble average, of some quantity which is **diffusing** through a medium. The meaning of **diffusing** will be made precise shortly. As **Albert Einstein** noted in one of his seminal papers of 1905 (his *Annus Mirabilis*[3]), a natural way to interpret the diffusion equation is in terms of the probabilities for the position of a particle undergoing a certain random motion called **Brownian motion**. We will address this interpretation in Sections 7.3 and 7.4.

In one space dimension the diffusion equation is given for $u(x,t)$ by

$$\boxed{u_t = \alpha u_{xx}.}$$

In N space dimensions it is given for $u(\mathbf{x},t)$ by

$$\boxed{u_t = \alpha \Delta u.}$$

In both cases α denotes a fixed **positive** constant. Note that the only difference between the diffusion equation and the wave equation lies in the fact that, in the former, there is only one time derivative. As we shall see, this difference has profound consequences giving rise to completely different *behaviors* of the respective solutions.

One immediate difference lies in the number of auxiliary conditions needed for a well-posed initial value problem. The diffusion equation is second order in that there are two spatial derivatives; hence, from our knowledge of the wave equation, one might be tempted to assume that the general solution involves two arbitrary functions. We shall soon see that this is not the case. In fact, we need only supplement the PDE with the initial concentration density at $t = 0$ (i.e., $u(\mathbf{x},0) = g(\mathbf{x})$) in order to obtain a well-posed initial value problem:

$$\begin{cases} u_t = \alpha u_{xx} & \text{for } -\infty < x < \infty,\ t > 0, \\ u(x,0) = g(x) & \text{for } -\infty < x < \infty. \end{cases} \tag{7.1}$$

---

[2]**James P. Sethna,** *Statistical Mechanics: Entropy, Order Parameters, and Complexity*, Oxford University Press, 2006.

[3]The Annus Mirabilis papers (Latin for "extraordinary year") were four groundbreaking papers which Einstein published in the *Annalen der Physik* in 1905.

The reader may have wondered about the placement of the previous two chapters which were **not** based on any particular type of PDE. As we will soon see, the delta function, convergence in the sense of distributions, and the Fourier transform all play a role in the study of the diffusion equation.

7.1. • Derivation 1: Fourier's/Fick's Law

One can derive the diffusion equation in any space dimension. Let us choose dimension three.

The Variables and Their Significance. Let $u(\mathbf{x}, t)$ be the concentration of *a quantity* A (e.g., heat in a medium, ink in a liquid, a chemical, ...) at spatial point $\mathbf{x}$ and time t. The basic assumption, to be made precise shortly, is that the quantity A is *diffusing* through some medium. Let V denote a bounded region of the medium (i.e., space). Then the total amount of quantity A in V at time t is given by

$$\iiint_V u(\mathbf{x}, t)\, d\mathbf{x}.$$

This quantity is a function of time, but how and why should it change with time?

Flux and the Generic Flow Equation: Assuming there are no sources or sinks for the quantity A, A can only enter or leave V through its boundary ∂V according to a **flux density** $\mathbf{F}(\mathbf{x}, t)$. At any point $\mathbf{x} \in \partial V$, the magnitude of $\mathbf{F}(\mathbf{x}, t)$ denotes the amount of A that is flowing per unit time per unit surface area. The direction of $\mathbf{F}(\mathbf{x}, t)$ is the direction of the flow. Thus,

$$\frac{d}{dt} \iiint_V u(\mathbf{x}, t)\, d\mathbf{x} \;=\; - \iint_{\partial V} \mathbf{F} \cdot \mathbf{n}\, dS.$$

Note that by convention $\mathbf{n}$ is the outer normal and, hence, if $\mathbf{F} \cdot \mathbf{n}$ is always positive, there is a net loss of A. This accounts for the minus sign. Now using the Divergence Theorem, we have

$$\iiint_V u_t(\mathbf{x}, t)\, d\mathbf{x} = - \iiint_V \operatorname{div} \mathbf{F}\, d\mathbf{x}.$$

Since this holds for any V, we have, by the IPW Theorem (Theorem A.6), that

$$\boxed{u_t \;=\; -\operatorname{div} \mathbf{F}} \tag{7.2}$$

for all $\mathbf{x}$ and $t > 0$.

The Relationship Between Flux and u: The Essence of Diffusion. The above formula is **generic** for the dynamics of **any** concentration as it is simply based upon the principle of conservation. What uniquely characterizes **diffusion** is the law which relates $\mathbf{F}$ to u. If you think about a distribution of heat in a room (or ink in a liquid), why would the concentration density change with time? Loosely speaking, heat (or ink) in places of higher concentration would "move" to places of lower concentration; in other words, the concentration of A changes with time because of **spatial concentration gradients**.

Now recall that the gradient ∇u points in the direction in which u changes the fastest. Thus, one would expect the flow of concentration u (i.e., the flux $\mathbf{F}$) to be in the opposite direction. This is summarized in a law commonly referred to as **Fick's law** or **Fourier's law** (also, other names — depending on the context) which states that

$$\boxed{\mathbf{F} = -\alpha \nabla u,} \tag{7.3}$$

for some constant $\alpha > 0$. Note that the constant of proportionality will adjust for the different dimensions of $|\mathbf{F}|$ vs. $|\nabla u|$. It depends on the precise nature of both, what exactly is diffusing, and the medium in which it is diffusing.

Combining (7.3) with (7.2) gives the **diffusion equation**

$$\boxed{u_t = \alpha \operatorname{div} \nabla u = \alpha \Delta u.} \tag{7.4}$$

In one space dimension, the diffusion equation is simply $u_t = \alpha u_{xx}$. It is instructive to redo the derivation in 1D, specifically for the temperature in an infinitely long thin bar made of some homogeneous material which is insulated along its sides (cf. Exercise **7.1**). Here $u(x,t)$ is the temperature at position x in the bar at time t. The use of the Divergence Theorem when passing from a boundary integral to a bulk integral is now replaced by the vanilla Fundamental Theorem of Calculus. What is particularly instructive about this derivation is to see how α (which now represents thermal diffusivity) relates to the mass density and, further, the specific heat and thermal conductivity of the material. All of these constants, as well as u and α, have physical dimensions (physical units).

7.2. • Solution in One Space Dimension and Properties

We now consider the following:

The IVP for the Diffusion Equation

$$\begin{cases} u_t = \alpha u_{xx} & \text{on } \mathbb{R} \times (0,\infty), \\ u(x,0) = g(x) & \text{on } \mathbb{R}. \end{cases} \tag{7.5}$$

We assume here that g is an integrable function. In Section 6.8.2, we used the Fourier transform to derive[4] a solution formula for (7.5):

The Solution Formula to the IVP for the Diffusion Equation

$$u(x,t) = \frac{1}{\sqrt{4\pi\alpha t}} \int_{-\infty}^{\infty} e^{-\frac{(x-y)^2}{4\alpha t}} g(y)\, dy. \tag{7.6}$$

Recall that (7.6) was derived under the assumption that for any fixed t, we could take the Fourier transform of $u(\cdot,t)$: Hence, we made the implicit assumption that $u(\cdot,t)$ is integrable, and this will have consequences in determining uniqueness (cf.

[4]Recall the basic logic, first discussed in Section 1.6, behind such formula derivations.

Section 7.5.1). An alternate derivation (not using the Fourier transform) of our solution formula (7.6) is outlined in Exercise **7.5** via a scaling argument and the notion of similarity solutions.

We now study the **implications and interpretations** of this solution formula, and in doing so, we will prove that it is indeed a solution to (7.5).

7.2.1. • The Fundamental Solution/Heat Kernel and Its Properties.

Let us define the following:

The Fundamental Solution of the Diffusion Equation

$$\Phi(x,t) := \frac{1}{\sqrt{4\pi\alpha t}}\, e^{-\frac{x^2}{4\alpha t}}, \qquad x \in \mathbb{R},\ t > 0. \tag{7.7}$$

We can then write the solution formula (7.6) in terms of Φ as follows:

$$u(x,t) = \big(\Phi(\cdot,t) * g\big)(x) = \int_{-\infty}^{\infty} \Phi(x-y,t)\, g(y)\, dy. \tag{7.8}$$

This says that in order to obtain the solution u at a given time t, we convolute the initial data with $\Phi(x,t)$. We examine this further in order to gain insight into the solution formula from the perspective of Φ. The function $\Phi(x,t)$ goes by many names including:

- the **fundamental solution** of the diffusion (or heat) equation on all of $\mathbb{R}$,
- the **Green's function** for the diffusion (or heat) equation on all of $\mathbb{R}$,
- the **source function** for the diffusion (or heat) equation on all of $\mathbb{R}$,
- the **heat kernel**.

First, let us observe the graph of $\Phi(x-y,t)$, for fixed $t > 0$ and $x \in \mathbb{R}$, as a function of y. Note that we are using the independent variable y because it is what we used to denote the free variable in the solution formula (7.8). Thus, in the formula we are effectively translating the origin to x via the dependence on $(x-y)$. We plot $\Phi(y,t)$ (with $\alpha = 1$) as a function of y for three values of t in Figure 7.1.

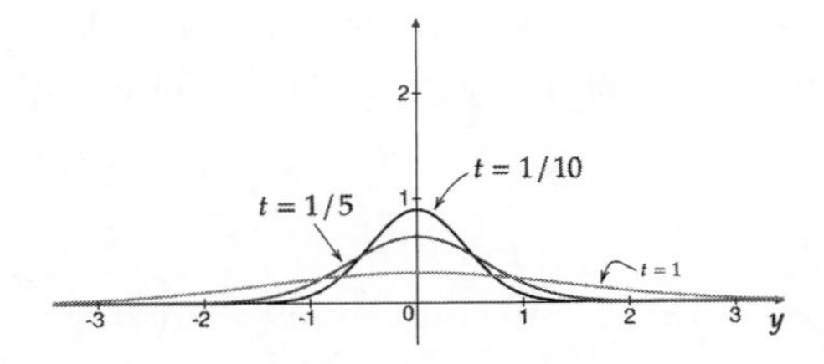

Figure 7.1. Plots of $\Phi(y,t)$ as a function of y for $\alpha = 1$ and $t = 1/10, 1/5, 1$, respectively.

Listed below are four basic properties of Φ. In stating these, we often use x as the spatial independent variable of Φ.

Property 1

For each $t > 0$, $\Phi(x,t)$ is always strictly positive and integrates to one.

For each $t > 0$, $\Phi(x,t)$ is clearly a strictly positive symmetric function of x. To show that

$$\int_{-\infty}^{\infty} \Phi(x,t)\,dx = 1,$$

we change to variable $z = x/\sqrt{4\alpha t}$:

$$\frac{1}{\sqrt{4\pi\alpha t}}\int_{-\infty}^{\infty} e^{-\frac{x^2}{4\alpha t}}\,dx = \frac{1}{\sqrt{\pi}}\int_{-\infty}^{\infty} e^{-z^2}\,dz = 1.$$

Now, we observe that the integral on the right (which is a very important integral in probability theory) can be sneakily calculated by *lifting up* into two dimensions. That is,

$$\int_{-\infty}^{\infty} e^{-z^2}dz = \sqrt{\int_{-\infty}^{\infty} e^{-x^2}dx \int_{-\infty}^{\infty} e^{-y^2}dy} = \sqrt{\int_{-\infty}^{\infty}\int_{-\infty}^{\infty} e^{-(x^2+y^2)}dxdy},$$

and in polar coordinates,

$$\int_{-\infty}^{\infty}\int_{-\infty}^{\infty} e^{-(x^2+y^2)}dxdy = \int_{0}^{2\pi}\int_{0}^{\infty} e^{-r^2}\,r\,dr\,d\theta = 2\pi\int_{0}^{\infty} e^{-r^2}\,r\,dr = \pi.$$

Property 2

$\Phi(x,t)$ is the normal probability density function.

Assuming the reader has been exposed to an introductory course in statistics, one should recognize that for a fixed t, $\Phi(x,t)$ is simply the **probability density function** associated with a **normal distribution** of mean 0 and variance $2\alpha t$. This is the famous **bell-shaped curve** (which students often reference when requesting a better grade). As we shall see in the Section 7.3, this is no coincidence.

From a probabilistic perspective the solution formula may be viewed as follows. For any fixed x and time t, $\Phi(x-y,t)$, as a function of y, is the probability density associated with a normal distribution with mean x and variance $2\alpha t$. The solution formula

$$u(x,t) = \int_{-\infty}^{\infty} \Phi(x-y,t)\,g(y)\,dy$$

states that **to find** u **at position** x **and time** t we must **average** the initial density $g(y)$ with **a certain weight**. The weight is this bell-shaped curve centered at $y = x$ and with "fatness" dictated by t. Note that when t is small, more weight is given to the values of the initial density close to x. As t gets larger, the effect of the initial densities far from x become more relevant (as greater diffusion has had time to occur). In the language of probability and statistics, the solution formula (7.8) says that the value of $u(x,t)$ is the **expectation of the initial data** $g(y)$ under a normal probability distribution (in y) with mean x and variance $2\alpha t$.

Property 3

$\Phi(x,t)$ solves the diffusion equation.

We note that $\Phi(x,t)$ is a C^∞ function of $x \in \mathbb{R}$ and $t > 0$. Moreover, one can directly check (cf. Exercise **7.2**), by computing both the partial with respect to t and the second partial with respect to x, that $\Phi(x,t)$ satisfies the diffusion equation; i.e., $\Phi_t(x,t) = \alpha\Phi_{xx}(x,t)$.

Property 4

$\Phi(x,t)$, as a function of x, converges to δ_0 as $t \to 0^+$ in the sense of distributions.

The plots of $\Phi(x,t)$ for different times t are suggestive of the following observation; while the integral under the curve is always 1, the graphs concentrate their mass close to 0 as t approaches 0. Thus, all the three properties listed in Section 5.5.3 for distributional convergence to the delta function seem to hold true. However, note that in this scenario we are considering a continuum limit of functions indexed by t which is tending to 0 from above, as opposed to a sequential limit of functions indexed by n which is tending to ∞.

It is indeed the case that, as a function of x,

$$\Phi(x,t) \longrightarrow \delta_0 \qquad \text{in the sense of distributions as } t \to 0^+.$$

From the perspective of the solution formula (7.8), we consider x as fixed and view y as the variable in $\Phi(x-y,t)$, thereby concluding that

$$\Phi(x-y,t) \longrightarrow \delta_x \qquad \text{in the sense of distributions as } t \to 0^+.$$

We will actually prove this shortly in Theorem 7.1. At this point, just take note that this property is directly related to the **initial values** of our solution (7.8). Indeed, we make an important observation: **According to** (7.8) **the solution is not defined at** $t = 0$**!** So what happened? We assumed that we had a solution for which we could Fourier transform, and we did some analysis with the transform in which we definitely included the initial data g. But then, where did these “initial values” go? The simple answer is that they are still there but only in the limiting sense: For any fixed $x \in \mathbb{R}$, we have

$$\lim_{t\to 0} u(x,t) = g(x).$$

In fact, more is true as we can approach the data axis in any direction (in x- vs. t-space); i.e., for any $x \in \mathbb{R}$,

$$\lim_{\substack{y\to x \\ t\to 0^+}} u(y,t) = g(x). \tag{7.9}$$

This fact means that the function defined by

$$u(x,t) = \begin{cases} \int_{-\infty}^{\infty} \Phi(x-y,t)\, g(y)\, dy, & t > 0, \\ g(x), & t = 0, \end{cases} \tag{7.10}$$

is **continuous** in the closed upper half-plane $\{(x,t)\,|\, t \geq 0\}$. This is what we require[5] for a solution to the initial value problem (7.5).

7.2.2. • Properties of the Solution Formula.

We derived the solution formula,

$$u(x,t) = \int_{-\infty}^{\infty} \Phi(x-y,t)\, g(y)\, dy, \tag{7.11}$$

under the assumption that a solution exists and is amenable to taking its Fourier transform in x. However, as is typical, one can readily check that the formula does indeed solve the PDE for $x \in \mathbb{R}$ and $t > 0$. This follows from the fact that for any $x \in \mathbb{R}$ and $t > 0$, $\Phi(x,t)$ is C^∞ and solves the diffusion equation (Property 3 of the previous subsection). Indeed, by differentiation under the integral sign (cf. Section A.9), we find that

$$u(x,t) = \int_{-\infty}^{\infty} \Phi(x-y,t)\, g(y)\, dy$$

is a C^∞ function on $\mathbb{R} \times (0,\infty)$. Moreover, for any $x \in \mathbb{R}$, $t > 0$, we have

$$\begin{aligned} u_t(x,t) - \alpha u_{xx}(x,t) &= \int_{-\infty}^{\infty} \left[\Phi_t(x-y,t)\, g(y) \; - \; \alpha\, \Phi_{xx}(x-y,t)\, g(y)\right] dy \\ &= \int_{-\infty}^{\infty} \left[\Phi_t(x-y,t) \; - \; \alpha\, \Phi_{xx}(x-y,t)\right]\, g(y)\, dy \\ &= 0. \end{aligned}$$

Next, we discuss the following properties of the solution formula: **(1)** infinite smoothness of $u(x,t)$ for $t > 0$; **(2)** infinite propagation speed; **(3)** a maximum principle; **(4)** the decay of $u(x,t)$ as $t \to \infty$. All these properties should seem reasonable given some basic intuition on the phenomenon of **diffusion**.

(1) Infinite smoothness: In Property 3 of Φ, we noted that the function $\Phi(x,t)$ is C^∞ for all $t > 0$, $x \in \mathbb{R}$. This implies that for any $t > 0$, $u(x,t)$ (as defined by (7.11)) is a C^∞ function of x and t, regardless of the smoothness of g. This follows from the fact that we can legitimately differentiate u, defined by (7.11), with respect to x or t by differentiating under the integral sign (cf. Section A.9). In doing so, the x to t derivatives **hit** $\Phi(x-y,t)$ but **not** $g(y)$. Thus, even if the function g is discontinuous, for any $t > 0$ the solution $u(x,t)$ is immediately a very smooth function; all discontinuities in g, whether in the function itself or in its derivatives, are smoothed out by the convolution with $\Phi(x-y,t)$.

[5]Note that if we did not care about continuity up to the $t = 0$ axis, we could generate ridiculous solutions to (7.5) such as

$$u(x,t) = \begin{cases} 45, & t > 0, \\ g(x), & t = 0. \end{cases}$$

(2) Infinite propagation speed: Recall the idea of *causality* in studying the wave equation. While we observed different behavior in 1D and 3D, in all cases information propagated no faster than a certain speed (the c in the wave equation). This is **not** the case with the diffusion equation. In fact there are no characteristics here to speak of; to find the solution at any point x and time $t > 0$, it is necessary to **sample** (in fact, average) the initial data **everywhere** on $\mathbb{R}$. Thus, we have a very different picture of the **domain of dependence** from the one we observed for either the wave equation or first-order dynamic equations.

We can make a similar remark from the perspective of the **domain of influence**. Suppose we have an initial density g which is localized in the sense that it is identically 0 except very close to a fixed point, say, $x = 0$, where it is slightly positive. What can you say about the value of $u(x,t)$ for any x and time $t > 0$? It will always be strictly larger than 0; i.e.,

$$u(x,t) > 0 \qquad \forall\, x \in \mathbb{R} \text{ and } t > 0.$$

Hence, that initial piece of localized information at position $x = 0$ has, after any instant of time, propagated to the entire real line. We call this phenomena **infinite propagation speed**. It is a consequence of the basic law (7.3) (Fick's law) which was one of the essential ingredients in deriving the diffusion equation.

Infinite propagation speed may raise a few alarm bells which, for good reason (it is subtle and to a certain extent philosophical), most PDE books do not address: Is it really *physical*? Doesn't **Einstein's special theory of relativity** dictate that *"nothing can travel faster than the speed of light"*. How do we reconcile this fact with the diffusion equation? The key is that our derivation, which was oblivious to any notion of mass or inertia, simply did not account for this physical requirement. One must pause to consider exactly *what* is propagating infinitely fast. It is not particles nor anything which possesses mass and inertia. As we shall soon see, Einstein's 1905 derivation of the diffusion equation via the limit of random walks provides a probabilistic interpretation. From this point of view, exact behavior is not sought, but rather only the statistics of certain possibilities. This approach is one of the foundations of **statistical physics**. Hence we can think of the diffusion equation as modeling the motion of microscopic particles but only in the sense that it **approximates**, on a macroscopic scale, **the statistics of the particle motion**. Since $\Phi(x,t)$ decays very fast as $x \to \pm\infty$, the probability that a particle obeying the diffusion equation has traveled extremely far in time span t is very small. In turn, the probability that a particle's velocity exceeds the speed of light is very small.[6]

(3) A maximum principle: Suppose the initial data g is bounded; i.e., there exists a constant $B > 0$ such that $|g(x)| \leq B$ for all $x \in \mathbb{R}$. Then for all x and $t > 0$,

$$|u(x,t)| \leq \int_{-\infty}^{\infty} \Phi(x-y,t)\,|g(y)|\,dy \;\leq\; B\int_{-\infty}^{\infty} \Phi(x-y,t)\,dy \;=\; B.$$

Note that the last equation on the right follows from Property 1 of Φ. So *if the data is initially bounded by B, then so is the solution at any time t*. Alternatively, the maximum value (over all space) of u initially will automatically set an upper bound for the values

[6]This breaks down when α is large and, hence, systems with extremely high diffusion constants should be given the relativistic treatment. There exist refinements to the diffusion equation with this relativistic treatment.

of $u(x,t)$. This fact is known as **the Maximum Principle**. If you recall the interpretation/derivation of the diffusion equation via the concentration of some quantity diffusing in space, this should seem completely reasonable.

(4) Decay: Let

$$C := \int_{-\infty}^{\infty} |g(y)|\, dy < \infty$$

and note that for all x, y, and $t > 0$, we have $0 < e^{-\frac{(x-y)^2}{4\alpha t}} \leq 1$. Hence, it follows that

$$\begin{aligned} |u(x,t)| \leq \int_{-\infty}^{\infty} \Phi(x-y,t)\,|g(y)|\, dy &= \frac{1}{\sqrt{4\pi\alpha t}} \int_{-\infty}^{\infty} e^{-\frac{(x-y)^2}{4\alpha t}}\, |g(y)|\, dy \\ &\leq \frac{1}{\sqrt{4\pi\alpha t}} \int_{-\infty}^{\infty} |g(y)|\, dy \\ &= \frac{C}{\sqrt{4\pi\alpha t}}. \end{aligned}$$

In particular, for any $x \in \mathbb{R}$

$$\lim_{t\to\infty} |u(x,t)| = 0.$$

In words, at any fixed position x the value of u decays to 0 as time gets larger. Again, this resonates with our interpretation/derivation of the diffusion equation.

7.2.3. The Proof for the Solution to the Initial Value Problem: The Initial Conditions and the Delta Function. Recall that as per usual, we derived our solution formula under the assumption that a solution existed. We now show that u defined by (7.10) does indeed solve the IVP. As noted in the previous subsection, it does solve the diffusion equation for $x \in \mathbb{R}$ and $t > 0$; hence the only issue which remains is that it "attains" the initial conditions. To this end, it is convenient to assume that the initial data g is continuous. We then naturally interpret the "attainment" of the initial conditions to mean that u defined by (7.10) for $t > 0$ is continuous up to $t = 0$, where this continuous function equals the initial data g. Proving this statement is directly related to Property 4 of Φ, asserting that $\Phi(\cdot,t)$ converges to δ_0 in the sense of distributions as $t \to 0^+$. While the plots of $\Phi(\cdot,t)$ for t small are suggestive of this fact, the reader should be curious as to a firmer explanation. We now state and prove a precise result under the assumptions that g is **bounded, continuous, and integrable**. All these hypotheses on the initial data g can be weakened (see the remark following the proof) but they are sufficiently general for our purposes.

Theorem 7.1. *Let $g(x)$ be a bounded, continuous, integrable function on $\mathbb{R}$. Define $u(x,t)$ by (7.10). Then $u(x,t)$ is a C^∞ solution to the diffusion equation on $\{(x,t)\,|\,t > 0\}$, and for any $x \in \mathbb{R}$*

$$\lim_{t\to 0} u(x,t) = g(x). \tag{7.12}$$

In fact, u is continuous on the set $\{(x,t)\,|\,t \geq 0\}$ in the sense that (7.9) holds true. Hence, the function defined by (7.10) constitutes a sensible solution to the full initial value problem (7.5). Here we will only prove the weaker statement (7.12) and leave the full statement (7.9) as an exercise.

Proof. By the previous discussion, all that is left to prove is assertion (7.12). This entails showing that **for every** $\epsilon > 0$, **there exists** $\delta > 0$ such that

$$\text{if } 0 < t < \delta, \qquad \text{then} \qquad \left| \int_{-\infty}^{\infty} \Phi(x-y,t)\, g(y)\, dy \; - \; g(x) \right| < \epsilon. \tag{7.13}$$

To this end, we first use Property 1 to write

$$g(x) = \int_{-\infty}^{\infty} \Phi(x-y,t)\, g(x)\, dy,$$

so that

$$\begin{aligned} \left| \int_{-\infty}^{\infty} \Phi(x-y,t)\, g(y)\, dy \; - \; g(x) \right| &= \left| \int_{-\infty}^{\infty} \Phi(x-y,t)\,(g(y)-g(x))\, dy \right| \\ &\le \int_{-\infty}^{\infty} \Phi(x-y,t)\, |g(y)-g(x)|\, dy. \end{aligned}$$

Note that we used the fact that $\Phi(x-y,t)$ is always positive. Since g is bounded, there exists $B > 0$ such that $|g(x)| \le B$ for all $x \in \mathbb{R}$. Since g is continuous, there exists δ_I such that $|y-x| < \delta_I$ implies $|g(y)-g(x)| < \frac{\epsilon}{2}$. We call this δ_I where I denotes intermediate. It is **not** the δ which is going to complete this proof. As we shall see, that δ will depend on δ_I. Note that δ_I depends on ϵ, the function g, and the point x, but g and x are fixed in this proof. We now break up the integral into two pieces:

$$\begin{aligned} &\int_{-\infty}^{\infty} \Phi(x-y,t)\, |g(y)-g(x)|\, dy \\ &\quad = \int_{|y-x|<\delta_I} \Phi(x-y,t)\, |g(y)-g(x)|\, dy + \int_{|y-x|>\delta_I} \Phi(x-y,t)\, |g(y)-g(x)|\, dy \\ &\quad \le \frac{\epsilon}{2} \int_{|y-x|<\delta_I} \Phi(x-y,t)\, dy + \int_{|y-x|>\delta_I} \Phi(x-y,t)\, |g(y)-g(x)|\, dy \\ &\quad < \frac{\epsilon}{2} + \int_{|y-x|>\delta_I} \Phi(x-y,t)\, |g(y)-g(x)|\, dy \\ &\quad = \frac{\epsilon}{2} + \frac{1}{\sqrt{4\pi\alpha t}} \int_{|y-x|>\delta_I} e^{-\frac{(x-y)^2}{4\alpha t}}\, |g(y)-g(x)|\, dy \\ &\quad \le \frac{\epsilon}{2} + \frac{2B}{\sqrt{4\pi\alpha t}} \int_{|y-x|>\delta_I} e^{-\frac{(x-y)^2}{4\alpha t}}\, dy. \end{aligned}$$

We change variables by setting $z = (y-x)/\sqrt{4\alpha t}$; hence $dz = dy/\sqrt{4\alpha t}$. Together with the symmetry of the integrand, this gives

$$\frac{2B}{\sqrt{4\pi\alpha t}} \int_{|y-x|>\delta_I} e^{-\frac{(x-y)^2}{4\alpha t}}\, dy \;=\; \frac{4B}{\sqrt{\pi}} \int_{\frac{\delta_I}{\sqrt{4\alpha t}}}^{\infty} e^{-z^2}\, dz.$$

As we saw before, $\int_0^\infty e^{-z^2} dz = \sqrt{\pi}/2$; hence the *tails* of this finite integral, $\int_a^\infty e^{-z^2} dz$, must tend to zero as $a \to \infty$. Since $\delta_I/\sqrt{4\alpha t} \to \infty$ as $t \to 0^+$, we are actually done!

However to be more precise on exactly how we can choose a $\delta > 0$ so that (7.13) holds true, we argue as follows:

$$\begin{aligned}\frac{4B}{\sqrt{\pi}}\int_{\frac{\delta_I}{\sqrt{4\alpha t}}}^{\infty} e^{-z^2}\,dz &= \frac{4B}{\sqrt{\pi}}\int_{\frac{\delta_I}{\sqrt{4\alpha t}}}^{\infty} \frac{1}{z}\,z\,e^{-z^2}\,dz\\ &\leq \frac{4B}{\sqrt{\pi}}\frac{\sqrt{4\alpha t}}{\delta_I}\int_{\frac{\delta_I}{\sqrt{4\alpha t}}}^{\infty} z\,e^{-z^2}\,dz\\ &\leq \frac{4B}{\sqrt{\pi}}\frac{\sqrt{4\alpha t}}{\delta_I}\int_{0}^{\infty} z\,e^{-z^2}\,dz = \frac{2B}{\sqrt{\pi}}\frac{\sqrt{4\alpha t}}{\delta_I}.\end{aligned}$$

Now it is clear what to do: Given $\epsilon > 0$, we can choose a δ such that

$$\delta < \frac{\pi\delta_I^2}{64\alpha B^2}\,\epsilon^2.$$

Then, if $0 < t < \delta$, we have

$$\frac{2B}{\sqrt{\pi}}\frac{\sqrt{4\alpha t}}{\delta_I} < \frac{\epsilon}{2},$$

and (7.13) holds true. Note that this choice of δ depends also on δ_I which in turn depended on ϵ. Hence, dispensing with quantities which are fixed in this proof (the point x, function g, and α), this choice of δ only depends on ϵ. □

A few remarks are in order:

(i) With the other two assumptions in place, **integrability** was not needed in this proof. Since the heat kernel $\Phi(\cdot, t)$ boasts a rather strong (exponential) decay as $\cdot \to \pm\infty$, one can relax the integrability assumption on g, replacing it with a growth condition for $g(y)$ as $y \to \pm\infty$; for instance boundedness, which we also assumed, would suffice.

(ii) The assumption on **continuity** can be weakened with the caveat of interpreting the $\lim_{\substack{y\to x\\ t\to 0^+}} u(y,t)$ as some average of values of g close to x (Exercise **7.23**).

(iii) The **boundedness** assumption on g can also can be relaxed and replaced with a growth condition on $g(y)$ as $y \to \pm\infty$.

(iv) Assuming the reader is familiar with the notion of **uniform continuity and uniform convergence**, one can quickly ascertain that if g is uniformly continuous on $\mathbb{R}$, then the convergence in (7.12) is uniform; that is, for any $\epsilon > 0$, there exists a $\delta > 0$ such that (7.13) holds true **for all** $x \in \mathbb{R}$. This fact can be used to fashion (cf. Exercise **7.30**) a simple proof of a famous and fundamental result in analysis: the **Weierstrass Approximation Theorem**, which states that every continuous function defined on a closed interval $[a, b]$ can be uniformly approximated to any desired accuracy by a **polynomial**. In fact, this simple proof is almost identical to the original proof of Weierstrass[7] from 1885[8].

[7] **Karl Theodor Wilhelm Weierstrass** (1815–1897) was a German mathematician and one of the pioneers of modern analysis.

[8] **Weierstrass**, Über die analytische Darstellbarkeit sogenannter willkürlicher Functionen einer reellen Veränderlichen, *Verlag der Königlichen Akademie der Wissenschaften Berlin*, Vol. 2, 1885, 633–639.

7.3. • Derivation 2: Limit of Random Walks

We (re)derive the diffusion equation via the notion of a **random walk**. The PDE will appear as the equation that characterizes (determines) the probability distribution of a particular limit of random walks. Before we begin, we need a simple fact very familiar to those who have taken some numerical analysis.

7.3.1. • Numerical Approximation of Second Derivatives.

First note that if f is a C^1 function, then as $\Delta x \to 0$, we have

$$\frac{f(x + \Delta x) - f(x)}{\Delta x} \to f'(x).$$

This is, after all, the definition of the first derivative! But now, suppose f is a C^2 function and we compute for any Δx,

$$\frac{f(x + \Delta x) + f(x - \Delta x) - 2f(x)}{(\Delta x)^2}. \tag{7.14}$$

What happens as $\Delta x \to 0$? Naively, perform the following calculation: (i) Choose a particular nonlinear smooth function f (e.g., $f(x) = x^2$); (ii) fix a particular small value of Δx (though not too small! — see Exercise **2.35**) and value of x; and (iii) compute the ratio in (7.14) first with denominator replaced with Δx and then with $(\Delta x)^2$. For the former, you would obtain a number very close to zero, while for the latter you would obtain a number close to $f''(x)$. Indeed, we have that

$$\boxed{\lim_{\Delta x \to 0} \frac{f(x + \Delta x) + f(x - \Delta x) - 2f(x)}{(\Delta x)^2} = f''(x).} \tag{7.15}$$

To see this, compute $f(x + \Delta x)$ using a Taylor series:

$$f(x + \Delta x) = f(x) + f'(x)\Delta x + \frac{f''(x)}{2}(\Delta x)^2 + \text{a sum of higher-order terms}$$

where all we need to know about this sum of higher-order terms (denoted by h.o.) is that they are $o((\Delta x)^2)$; that is,

$$\frac{\text{h.o.}}{(\Delta x)^2} \longrightarrow 0 \quad \text{as} \quad \Delta x \longrightarrow 0.$$

Here, we employ the standard little o notation. Also note that

$$f(x - \Delta x) = f(x) - f'(x)\Delta x + \frac{f''(x)}{2}(\Delta x)^2 + o((\Delta x)^2).$$

If we add these two equations and bring $2f(x)$ to the left-hand side, we find that

$$f(x + \Delta x) + f(x - \Delta x) - 2f(x) = f''(x)(\Delta x)^2 + o((\Delta x)^2).$$

Thus, dividing by $(\Delta x)^2$ and sending $\Delta x \to 0$ leads to (7.15). The ratio (7.14) is called the **centered finite difference** approximation to the second derivative.

7.3.2. • Random Walks. Let us first consider a (finite) random walk on a line. We discretize the line into intervals of length Δx and discretize time into intervals of size Δt. Now we play the following game with a fair coin. A fair (or unbiased) coin is one for which a coin toss ends up in either heads or tails with equal probability. The random walk consists of the following steps:

A Random Walk

- We initially start at the origin and start by flipping a fair coin.
- If the coin turns up heads, we move Δx to the right; alternatively, if it turns up tails, then we move Δx to the left. This happens in a time step Δt.
- At the next time step Δt, we repeat the process.
- We continue repeating this process n times; hence $t = n\Delta t$ units of time have elapsed.

The central question is: **Where will we be at the final time $t = n\Delta t$?** Well, we can't say exactly; we could be at any of the positions $k\Delta x$ where k could be any integer between $-n$ and n. What we can do is compute the probability that we will be at any of these positions. If we did so, we would find that the position with the highest probability is $x = 0$ (assuming n is even). You should be able to convince yourself that this makes sense due to the symmetry of the problem. Computing all the probabilities of the likelihood of being situated at the other points would generate a **probability mass function** at time t — a function $p(x,t)$ defined on grid points $x = k\Delta x$ and $t = n\Delta t$ which gives the probability of being at a position $x = k\Delta x$ at time $t = n\Delta t$.

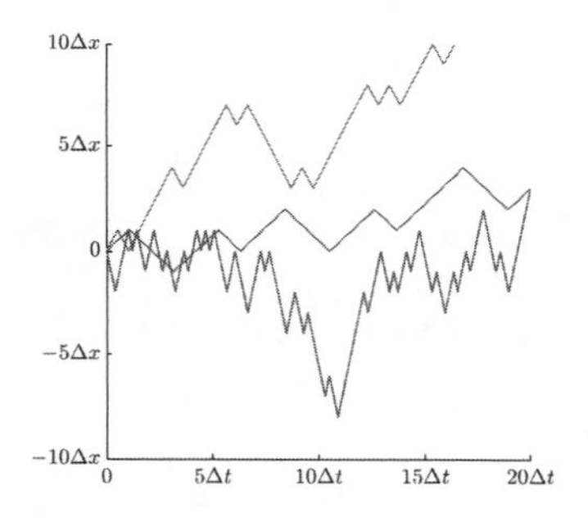

Figure 7.2. Three random walks with step size Δx and $n = 20$ time steps Δt starting at the origin. Here we have linearly interpolated between the discrete position points.

In order to visualize the probability mass function $p(x,t)$, we could perform the following **experiment**. Suppose we have 10,000 fair coins in our pocket! Fix a value of Δt, Δx, and a large positive integer n. We then do the following:

- Start at $x = 0$ and perform an n-step random walk (using one of the coins for the flips) and then place the coin at the position where we end up.

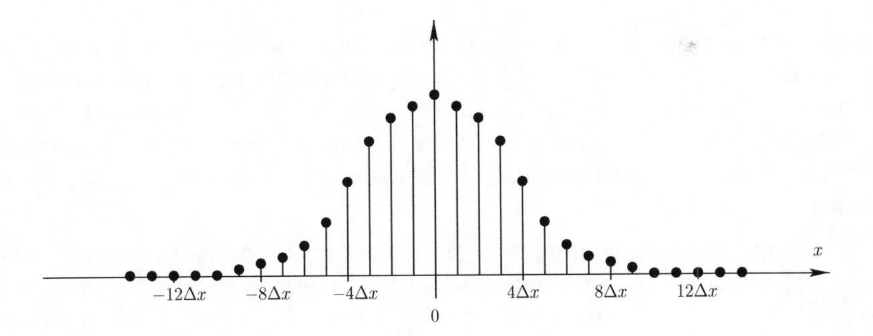

Figure 7.3. Experiment: Perform an n-step random walk and place the coin where we end up, and then repeat this many, many times with different fair coins. If you end up at a place which already has a coin, place the coin on top of the other coin(s), creating a tower. The above bell shape gives a schematic of what you would observe for the distribution of coins (i.e., the heights of the coin towers).

- Return to the origin and then repeat the random walk with a different fair coin, placing the coin where we end up. If we end up at a place which already has a coin, we place the coin directly on top of the other coin.
- Repeat the previous step until we have used up all the 10,000 coins.

Now we consider the **distribution of coins**. This distribution gives us an idea of the probability mass function $p(x, n\Delta t)$ for the random walk at time $t = n\Delta t$. If n is large, you would observe something close to a **bell-shaped curve** like the one in Figure 7.3. The bell-shaped curve would have its center and highest height at the origin. What would change with both the step size Δx and n is the fatness or spread of the bell-shaped curve. You may recall the concept of **variance** (the square of the standard deviation) from elementary statistics, as a measure of the spread of a density/distribution. If we measure the variance for this bell-shaped curve[9], we would find that it is approximately

$$n(\Delta x)^2.$$

This observation is supported, in fact fully validated, by one of the most important theorems in all of mathematics (arguably, one of the most important facts in all of science), the Central Limit Theorem (cf. Section 7.4.2). Note that to actually observe the dependence of the variance on Δx, one would have to repeat this experiment many times, each time with a different value for Δx.

7.3.3. • The Fundamental Limit and the Diffusion Equation.

We consider the fundamental question: **What happens when** $\Delta t \to 0$, $\Delta x \to 0$, **and** $n \to \infty$**?** That is, we want to understand random walks where the spatial and temporal step sizes are very small, but where we perform the random walk more and more times.

[9]Precisely, let C_k be the number of coins at $k\Delta x$ at time $n\Delta$. Then the variance is

$$\frac{1}{10{,}000}\sum_{k=-n}^{n}(k\Delta x)^2 C_k.$$

This question was studied[10] in an extraordinary paper of Albert Einstein in 1905, his *Annus Mirabilis*, in which he wrote four groundbreaking papers. The paper is titled "Über die von der molekular-kinetischen Theorie der Wärme geforderte Bewegung von in ruhenden Flüssigkeiten suspendierten Teilchen"[11], which roughly translates as "Concerning the motion, as required by molecular-kinetic theory of heat, of particles suspended in a liquid at rest".

First off, we reiterate the important (in fact, crucial) observation from the results of our experiment: The variance of the bell-shaped curve in Figure 7.3 is proportional to $n(\Delta x)^2$. What does this mean from the point of letting Δx tend to zero and n (the number of time steps in the random walk) tend to ∞? The bell-shaped curve gives a sense of the probability mass function, the distribution of probabilities at time

$$t = n\Delta t.$$

As $\Delta t \to 0$ and $n \to \infty$, we want to observe what is happening at **a fixed time** $t > 0$. For such a fixed t, we have

$$n = \frac{t}{\Delta t},$$

and hence, the variance of the distribution of coins will be of the order

$$n(\Delta x)^2 = t\frac{(\Delta x)^2}{\Delta t}.$$

What happens to the spread (variance) as Δt and Δx get smaller and smaller? If $(\Delta x)^2$ and Δt do **not** tend to zero at the same rate, then

$$\frac{(\Delta x)^2}{\Delta t} \tag{7.16}$$

will either tend to 0 or to $+\infty$. This means the distribution of coins of Figure 7.3 will either concentrate at the origin or spread out without bound. On the other hand, if we keep the ratio in (7.16) **fixed** as $\Delta t \to 0$ and $\Delta x \to 0$, we will **retain** a nontrivial distribution in the limit.

Now we confirm the importance of fixing the ratio (7.16) by considering the **evolution of the probability mass function** for small Δx and Δt; this is where the diffusion equation will pop up. We start at the discrete level, fixing for the moment values for $\Delta x > 0$ and $\Delta t > 0$ and considering the probability mass function

$$p(x,t) = \text{the probability of being at place } x \text{ at time } t.$$

Note that at this discrete level (i.e., before any limit), the spatial variable x takes on values which are integer multiples of Δx, and t takes on values which are nonnegative integer multiples of Δt. The essence of the random walk lies in the fact that if we are at position x at time $t + \Delta t$, then, at the earlier time t, we had to be either at position $x - \Delta x$ or at position $x + \Delta x$. Since for any coin toss we have equal probability of tails and heads, we must have

$$\boxed{p(x, t + \Delta t) = \frac{1}{2}\, p(x - \Delta x, t) + \frac{1}{2}\, p(x + \Delta x, t).}$$

[10]Interestingly, Einstein was not the first to study this limit. It appears in an article of **L. Bachelier** titled *Théore de la spéculation* which was published in 1900 in *Annales Scientifiques de l'É.N.S series, tomm 17.*

[11]Ann. Phys. (Leipzig) **17**, 549 (1905).

In words, there is a 50% chance that we are at position x at time $t + \Delta t$ because we moved right from position $x - \Delta x$, but there is also a 50% chance we came to x by moving left from position $x + \Delta x$. This simple equation is our **master equation** and everything will follow from it!

As before, we are interested in investigating the limit as Δx and Δt both tend to 0, and n tends to ∞ with time $t = n\Delta t$ fixed. To this end, we algebraically manipulate this master equation. First, we subtract $p(x,t)$ from both sides,

$$p(x,t+\Delta t) - p(x,t) = \frac{1}{2}\left[p(x-\Delta x,t) + p(x+\Delta x,t) - 2p(x,t)\right]$$

and divide by Δt to obtain

$$\frac{p(x,t+\Delta t) - p(x,t)}{\Delta t} = \frac{1}{2\Delta t}\left[p(x-\Delta x,t) + p(x+\Delta x,t) - 2p(x,t)\right].$$

Finally, we multiply by 1 in a clever way to arrive at

$$\frac{p(x,t+\Delta t) - p(x,t)}{\Delta t} = \frac{(\Delta x)^2}{2\Delta t}\left[\frac{p(x-\Delta x,t) + p(x+\Delta x,t) - 2p(x,t)}{(\Delta x)^2}\right] \tag{7.17}$$

Now we are in a better position to address what happens as $\Delta t \to 0$ and $\Delta x \to 0$. In this limit, the spatial and temporal grid points (for which p is defined over) "fill up" the entire real line (half-line for t). We implicitly assume that there exists a limiting function, also called $p(x,t)$, which is smooth in x and t. Then as $\Delta t \to 0$, we have

$$\frac{p(x,t+\Delta t) - p(x,t)}{\Delta t} \longrightarrow p_t(x,t).$$

By (7.15), as $\Delta x \to 0$ we have

$$\frac{p(x-\Delta x,t) + p(x+\Delta x,t) - 2p(x,t)}{(\Delta x)^2} \longrightarrow p_{xx}(x,t).$$

However, we are still left with $(\Delta x)^2/(2\Delta t)$. There are many things that can happen to this ratio as $\Delta t \to 0$ and $\Delta x \to 0$. To get a nontrivial limit (i.e., neither zero nor infinite), we require a particular **slaving** (relationship) of Δx to Δt: For some fixed constant $\sigma > 0$,

$$\frac{(\Delta x)^2}{\Delta t} = \sigma^2 \qquad \text{or} \qquad \frac{(\Delta x)^2}{2\Delta t} = \frac{\sigma^2}{2}.$$

In other words, we require that as Δx and Δt are scaled down to zero, this seemingly peculiar ratio must be kept constant. Note that this relative scaling between space and time is not just apparent in numerical implementation of the heat equation, but it is actually a symmetry of the diffusion equation itself (cf. Exercise **7.5**). Imposing this particular limit in (7.17) leads us to

$$\boxed{p_t = \frac{\sigma^2}{2}\, p_{xx}}$$

which is simply the diffusion equation with $\alpha = \frac{\sigma^2}{2}$. The reason for using σ^2 instead of σ will be made clear shortly, as σ will correspond to a standard deviation and σ^2 to a variance. In the limit, x and t take on a continuum of values, and hence, the probability mass function $p(x,t)$ should now be interpreted as a **probability density** associated with a position x and time t. Moreover, with this continuous probability density, there is zero probability of being at any exact position but rather nonzero probability of being in some $I \subset \mathbb{R}$, calculated by integrating the probability density over interval I; i.e.,

$$\text{the probability of being in the interval } I \text{ at time } t \;=\; \int_I p(x,t)\,dx.$$

We will return to this probability density in the next few sections.

Note that if we followed the previous calculation but this time adopting the relation $\Delta x \sim \Delta t$, i.e., $\Delta x = C\Delta t$ for some constant $C > 0$, then we would obtain $p_t = 0$. In particular, if we started at the origin $x = 0$, then with probability 1, we would not move. The same would be true for any scaling $\Delta x = C(\Delta t)^\beta$ for $\beta > 1/2$. On the other hand, the scaling $\Delta x = C(\Delta t)^\beta$ for $\beta < 1/2$ would lead to $p_t = \infty$, which we could interpret as follows: Starting at the origin $x = 0$ with probability 1, we immediately go off to $\pm\infty$. In conclusion, the only slaving of Δx to Δt which gives a nontrivial limiting behavior is $\Delta x = C(\Delta t)^{1/2}$.

7.3.4. • The Limiting Dynamics: Brownian Motion. Suppose we consider **a particular realization** of a random walk, for example the one depicted on the left of Figure 7.4. We call such a particular realization, a **sample path** of the random walk. In the previous subsection, we found nontrivial behavior in the limit

$$\Delta t \to 0 \quad \text{and} \quad \Delta x \to 0 \quad \text{with} \quad \frac{(\Delta x)^2}{\Delta t} \quad \text{fixed.} \tag{7.18}$$

Precisely, we showed that there was a nontrivial probability density for the position of the particle at time t, and this density evolved according to the diffusion equation. This **continuum limit of the random walks** is known as **Brownian Motion**. It is named after botanist **Robert Brown** who, in 1827, made an observation of highly irregular motion of small grains of pollen suspended in water. Later, in 1905, **Albert Einstein** made the necessary calculations to show that the probability density function associated with this Brownian motion satisfies the diffusion equation. But there is far more one can say about Brownian motion than just this relationship to the diffusion equation. For example, one can consider the **sample paths** as the limit of the sample paths of a random walk as depicted on the left of Figure 7.4. As it turns out, they are **continuous** paths but are highly **jagged**, like the path depicted on the right of Figure 7.4, in fact so much so, that with probability 1 the path of a Brownian motion will be **nowhere differentiable**. A simple way to see this is via "the graph" (on a discrete x- vs. t-plane) of a random walk (left side of Figure 7.4) which approximates a Brownian motion sample path. While this graph is simply a discrete set of points, we linearly interpolate between these discrete points, obtaining a piecewise linear curve with the tangent slopes $\pm(\Delta x)/(\Delta t)$. In our limit (7.18), the ratio $(\Delta x)^2/(\Delta t)$ remains fixed. Hence $(\Delta x)/(\Delta t)$ must blow up (i.e., tend to infinity) which suggests that, generically, continuous curves associated with Brownian paths will be nowhere differentiable.

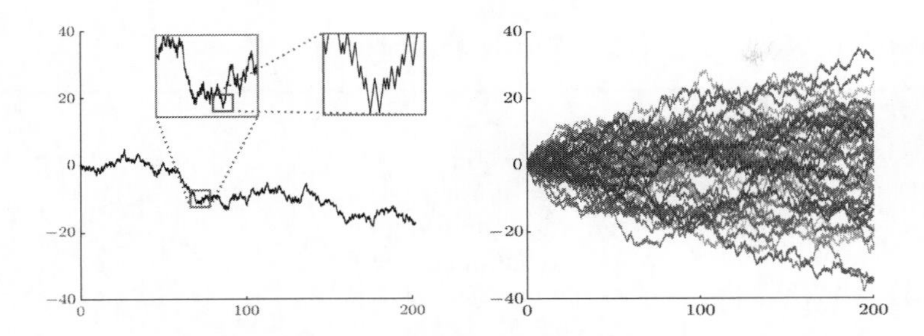

Figure 7.4. Left: A particular sample path of a random walk with step size Δx and time step Δt. Here we have linearly interpolated between the discrete position points. Right: Approximate sample paths of the limiting random walk (the Brownian motion).

Brownian motion is an important example of a **stochastic process**. Its precise structure and related questions are difficult and involve a considerable amount of sophisticated mathematical language and machinery; see, for example, the book of Mörters and Peres[12]. Note that while the process of Brownian motion is stochastic (probabilistic) in nature, the probability density satisfies the diffusion equation, a deterministic equation with no randomness. To be precise, let $g(x)$ be the initial probability density function for the position of the particle at $t = 0$. This means that we do not know exactly where we start; rather, we only know $g(x)$, a starting probability density associated with position x, where $g(x) \geq 0$ for all $x \in \mathbb{R}$ and $\int_{-\infty}^{\infty} g(x)\,dx = 1$. Then, for any fixed time $t_1 > 0$, the probability density for the position of the particle at time t_1 is given by $p(x, t_1)$, where $p(x,t)$ solves

$$\begin{cases} p_t = \frac{\sigma^2}{2} p_{xx} & \text{on } \mathbb{R} \times (0, \infty), \\ p(x, 0) = g(x) & \text{on } \mathbb{R}. \end{cases}$$

If we were to model a situation where, with probability 1, we started at the origin, then we would not use a function $g(x)$ for the initial probability density, but rather the distribution δ_0 (the delta function). The resulting sample paths of the (approximate) Brownian motion are illustrated in Figure 7.4. In this case, we would find that for any later time $t_1 > 0$, the continuous probability density $p(x, t_1)$ for the position of the particle is given by

$$p(x, t_1) = \Phi(x, t_1) = \frac{1}{\sqrt{2\pi\sigma^2 t_1}}\, e^{-\frac{x^2}{2t_1\sigma^2}}, \tag{7.19}$$

where $\Phi(x, t_1)$ is the fundamental solution of the diffusion equation with $\alpha = \frac{\sigma^2}{2}$. You can picture this on the right side of Figure 7.4 where if we fix a vertical line at $t = t_1$, then the probability density of being at any point x (here indexed by the vertical axis) is given by (7.19).

---

[12]**P. Mörters and Y. Peres,** *Brownian Motion*, Cambridge Series in Statistical and Probabilistic Mathematics, 2010.

7.4. Solution via the Central Limit Theorem

★
★ The ★
★ normal ★
★ law of error ★
★ stands out in the ★
★ experience of mankind ★
★ as one of the broadest ★
★ generalisations of natural ★
★ philosophy. It serves as the ★
★ guiding instrument in researches ★
★ in the physical and social sciences ★
★ and in medicine, agriculture and engineering. ★
★ It is an indispensable tool for the analysis and the ★
★ interpretation of the basic data obtained by observation and experiment. ★

– William J. Youden[a]

[a]William J. Youden was an American statistician who made important contributions to statistics and is also known for the way he typeset his anthem to the normal distribution.

In the previous section we derived the diffusion equation for the probability density associated with a certain limit of random walks (i.e., Brownian motion):

$$p_t = \frac{\sigma^2}{2} p_{xx} \tag{7.20}$$

where $p(x,t)$ is the probability density associated with being at x at time t. In Section 6.8.2 we solved the IVP for (7.20) using the Fourier transform. Let us now **resolve** (7.20) in the context of random walks using the Central Limit Theorem. The presentation will not be fully rigorous but, rather, we focus on the intuition and basic ideas. The Central Limit Theorem tells us about the statistics (the distribution of probabilities) of certain weighted sums of **independent, identically distributed random variables**. Assuming the reader is unfamiliar with these (important) terms, let us pause in the next short subsection to present a few intuitive definitions.

7.4.1. Random Variables, Probability Densities and Distributions, and the Normal Distribution.

Important: In this section relating to probability, do not confuse the use of the word "distribution" with our previous *theory of distributions*: Think of it now as a completely separate notion.

A **random variable** X is a variable which can take on many values (a finite or an infinite number) but does so according to a certain probability weight. For example, in tossing a fair coin we could say the pay-off to a player is 1 if the coin comes up heads and -1 (meaning the player owes you a dollar) if the outcome is tails. The **value of the pay-off** is a random variable X taking on two values, 1 and -1, with equal probability (1/2 associated with each value). We call the occurrence of X taking on one of these values ± 1 **events**. In this example, the two events of X are distributed according to a probability density which assigns 1/2 to each. This example of a coin toss is demonstrative of a random variable which took on two values. In general, random variables may take on any number of values, for example, rolling a 6-sided dice with the random variable being exactly the number of the roll. If the number of possible values is finite, or countably infinite, we call it a **discrete random variable**.

A particular important class of random variables, which we call **continuous random variables**, takes on a continuum of values, say, $\mathbb{R}$, with a distribution of probabilities dictated by a **probability density function** $p(x)$ defined for $x \in \mathbb{R}$. To be more precise, the events of such continuous random variables include (possibly infinite) intervals (a, b) of $\mathbb{R}$ and are assigned probabilities according to their probability density function $p(x)$:

$$\text{the probability of the event } \{X \in (a,b)\} \;=\; \int_a^b p(x)\,dx.$$

Thus, for such continuous random variables the probability of the event $X = a$, for any $a \in \mathbb{R}$, is 0: Rather, we can only assign $x = a$ an "instantaneous" probability density value $p(a)$ (note the parallel with integration of a function). Alternatively, we could say the probability of "seeing" X take on values between a and $a + \Delta x$ is approximately $p(a)\Delta x$ when Δx is small. As an illustrative example, think of throwing a dart at a dartboard. The random variable is the position on the dartboard where the dart lands; hence, in this example the random variable takes values in $\mathbb{R}^2$, rather than $\mathbb{R}$.

Another function describing a continuous random variable X is called the **probability distribution function** (also known as the cumulative distribution function) $F(x)$ defined as

$$F(x) \;:=\; \text{probability of the event } \{X \in (-\infty, x)\} \;=\; \text{probability of } \{X \le x\}.$$

Assuming $p(x)$ is a reasonable function, the Fundamental Theorem of Calculus implies that $F'(x) = p(x)$. For any random variable with probability density function $p(x)$, we define the **mean** μ and **variance**[13] σ^2, respectively, as

$$\mu := \int_{-\infty}^{\infty} x\, p(x)\,dx, \qquad \sigma^2 := \int_{-\infty}^{\infty} (x-\mu)^2\, p(x)\,dx.$$

One can analogously define the probability density and distribution functions, as well as their mean and variance, for a discrete random variable. A particularly important continuous random variable is a **normal random variable** whose **normal distribution** has mean 0 and variance σ^2. We will denote such random variables with Z instead of X.

[13]The square notation is motivated by that fact that the variance is the square of the **standard deviation** σ.

The Normal Distribution and a Normal Random Variable

Definition 7.4.1. We say a continuous random variable Z has a **normal distribution** (or equivalently is a **normal random variable**) with mean 0 and variance σ^2 if its probability distribution function is given by

$$F(x) = \text{probability of } \{Z \leq x\} = \frac{1}{\sqrt{2\pi\sigma^2}} \int_{-\infty}^{x} e^{-\frac{y^2}{2\sigma^2}}\, dy. \tag{7.21}$$

We adopt the notation $Z \sim N(0, \sigma^2)$ to mean exactly that (7.21) holds true. Another way to phrase this is to say that the probability density function associated with Z is

$$p(x) = \frac{1}{\sqrt{2\pi\sigma^2}} e^{-\frac{x^2}{2\sigma^2}}.$$

This function is known as a **Gaussian** and its graph is the famous bell-shaped curve.

Two random variables are called **identically distributed** if they take on the same values with the same probability weight. More precisely, they have the same probability density functions or, equivalently, the same probability distribution functions. Returning to our coin example, performing the coin toss experiment again with the same coin, or even another unbiased coin, would lead to a pay-off random variable which was identically distributed. On the other hand, this would not be the case if we used an unfair coin.

Independence of random variables is more subtle. In words, two random variables are independent if knowing the value or values of one of the random variables in no way affects the statistics of the other. Rather than give a precise definition, let us immediately return to the coin toss example. If we toss the coin again, the new pay-off would be another random variable which is independent from the first pay-off. This is simply a consequence of the fact that if the first coin toss turned up heads, this does not make the heads any more likely to be the outcome of the second toss. On the other hand, let us give an example of two random variables which are not independent (i.e., dependent). Consider the pay-off from the first toss to be the random variable X_1 and the pay-off from the second toss to be the random variable X_2. These two random variables are independent and identically distributed. However, suppose we consider the variable Y defined by

$$Y := X_1 + X_2;$$

that is, we add the two pay-offs. There are three events for the random variable Y: $Y = 0, Y = 2$, and $Y = -2$ distributed with probabilities $1/2, 1/4$, and $1/4$, respectively. This random variable is not independent from X_1. Indeed, if we knew $X_1 = 1$, this would mean that the probability that $Y = -2$ is now 0 whereas the probability of being either 0 or 2 is $1/2$ each. In other words, knowing information about X_1 leads to a different probability distribution for the values of Y.

7.4.2. The Central Limit Theorem. The Central Limit Theorem tells us that a certain weighted sum of a family of independent, identically distributed (i.i.d. for short) random variables will eventually have a normal distribution. Even if these random variables take on a discrete number of values, the limiting weighted sum will take on a continuum of values. Here is a precise statement of the theorem:

The Central Limit Theorem

Let $\{X_i, i \geq 1\}$ be a family of independent, identically distributed random variables. Since they are identically distributed, they will (in particular) have the same mean and variance. For simplicity, let's assume the mean is 0 and the variance is $\sigma^2 \in (0, \infty)$. Now, define

$$S_n := \sum_{i=1}^{n} X_i.$$

The Central Limit Theorem states that

$$\frac{S_n}{\sqrt{n}} \xrightarrow[n\to\infty]{\text{in distribution}} Z \sim N(0, \sigma^2). \tag{7.22}$$

While the notion of **convergence in distribution** in (7.22) is a "weak notion" of convergence, it should **not** be confused with **convergence in the sense of distributions.** Rather, (7.22) means that for all $x \in \mathbb{R}$,

$$\text{the probability of } \left\{\frac{S_n}{\sqrt{n}} \leq x\right\} \xrightarrow{n\to\infty} \frac{1}{\sqrt{2\pi\sigma^2}} \int_{-\infty}^{x} e^{-\frac{y^2}{2\sigma^2}}\, dy.$$

In other words, the limiting probability density function of the random variable $\frac{S_n}{\sqrt{n}}$ is given by

$$\frac{1}{\sqrt{2\pi\sigma^2}}\, e^{-\frac{x^2}{2\sigma^2}}.$$

7.4.3. Application to Our Limit of Random Walks and the Solution to the Diffusion Equation. Let us now apply the Central Limit Theorem to our limit of random walks. We will assume that all random walks start at the origin, effectively imposing a delta function as the initial conditions for (7.20). To this end, we fix a time $t > 0$ and keep it fixed for the entire calculation. For the moment, we also fix a small value of Δx and a small value of Δt, such that, for some large integer n, we have

$$t = n\Delta t. \tag{7.23}$$

Step 1: Defining the Random Variables X_i and S_n. Define

$$X_i = \begin{cases} \Delta x & \text{with probability } p = 1/2, \\ -\Delta x & \text{with probability } p = 1/2. \end{cases}$$

Note that the X_i are independent and identically distributed. Indeed, their mean is 0 and their variance is $(\Delta x)^2$. Now for n such that $t = n\Delta t$, define the random variable

$$S_n = \sum_{i=1}^{n} X_i.$$

The value of S_n indicates exactly where we are after performing n random walks with time step Δt. Hence, S_n indicates where we are at time $t = n\Delta t$. Since t will be fixed as $n \to \infty$ and $\Delta x, \Delta t \to 0$, let us **relabel** S_n with S_n^t.

Thus, to repeat: The random variable S_n^t gives our position at time t after n steps of the random walk (with time step size Δt).

Step 2: Invoking the Central Limit Theorem for Finite, but Large, n and Rescaling. Now assuming n is large, we can invoke the Central Limit Theorem to conclude that $\frac{S_n^t}{\sqrt{n}}$ is approximately in distribution $N\left(0, (\Delta x)^2\right)$. Let us write this as

$$\frac{S_n^t}{\sqrt{n}} \overset{\text{approximately}}{\sim} N\left(0, (\Delta x)^2\right).$$

Recall that it means, for n large and $x \in \mathbb{R}$,

$$\text{the probability of } \left\{\frac{S_n^t}{\sqrt{n}} \le x\right\} \quad \text{is approximately equal to} \quad \frac{1}{\sqrt{2\pi\sigma^2}} \int_{-\infty}^{x} e^{-\frac{y^2}{2\sigma^2}}\, dy,$$

with an error tending to 0 as $n \to \infty$. Now we rephrase the above in terms of Δt and t, recalling from (7.23) that $n = \frac{t}{\Delta t}$. This gives that for Δt small (hence n large),

$$\frac{S_n^t \sqrt{\Delta t}}{\sqrt{t}} \overset{\text{approximately}}{\sim} N\left(0, (\Delta x)^2\right).$$

Lastly we use the simple scaling fact that for any constant $a > 0$

$$aZ \sim N(0, \sigma^2) \qquad \text{if and only if} \qquad Z \sim N\left(0, \frac{\sigma^2}{a^2}\right).$$

The reader should pause to verify this fact. Thus, for n large and Δt small so that $t = n\Delta t$, we have

$$S_n^t \overset{\text{approximately}}{\sim} N\left(0, \frac{(\Delta x)^2}{\Delta t}\, t\right). \tag{7.24}$$

Step 3: Taking the Limit as $n \to \infty$. Let

$$\sigma^2 := \frac{(\Delta x)^2}{\Delta t}.$$

Let $\Delta t \to 0$ with σ^2 fixed; this will imply $\Delta x \to 0$ and $n \to \infty$ with $t = n\Delta t$ fixed. In this limit, S_n^t becomes a random variable S^t taking on a continuum of values. Moreover, by passing to the limit in n, we can dispense with the "approximately" in describing the probability distributions, directly applying the full Central Limit Theorem. Precisely, the theorem says that the random variables S_n^t will converge **in distribution** as $n \to \infty$ to a random variable S^t, taking on a continuum of values, which has a normal distribution with mean 0 and variance $\sigma^2 t$; i.e.,

$$S^t \sim N\left(0, \sigma^2\, t\right).$$

This means that for any $x \in \mathbb{R}$

$$\text{the probability of } \{S^t \leq x\} = \frac{1}{\sqrt{2\pi\sigma^2 t}} \int_{-\infty}^{x} e^{-\frac{y^2}{2\sigma^2 t}}\, dy.$$

Alternatively, if we let $p(x,t)$ denote the probability density function of S^t (where we are at time t), we have

$$p(x,t) = \frac{1}{\sqrt{2\pi\sigma^2 t}}\, e^{-\frac{x^2}{2\sigma^2 t}}. \tag{7.25}$$

Again, note that in (7.24), the only way to have a nontrivial distribution of values in the limit, where Δt and Δx tend to 0, is to keep $(\Delta x)^2/\Delta t$ fixed.

But have we solved the diffusion equation? Recall that in our old language $\alpha = \frac{\sigma^2}{2}$, and hence what we have arrived at in (7.25) is the fundamental solution $\Phi(x,t)$ of the diffusion equation, i.e., (7.7), previously derived via the Fourier transform. Is this what we wanted? The answer is yes because our derivation was based upon the premise that we started at $x = 0$. In the continuum limit, this would mean that initially all the probability was at $x = 0$. In other words, we started with a **delta function** with concentration at 0 and, in this case, the solution to the IVP for the diffusion equation is exactly $\Phi(x,t)$. We could take these probabilistic arguments one step further and ask: Supposing that initially all we knew was a probability density $g(x)$ associated with our position, what would be the probability density function at a later $t > 0$? Note here that we would start with an initial density defined only at the discrete step points $n\Delta x$, which in the limit would "tend to" the continuous density g defined over $\mathbb{R}$. With a considerably more sophisticated analysis, similar arguments would indeed yield the limiting probability density as

$$p(x,t) = \frac{1}{\sqrt{2\pi\sigma^2 t}} \int_{-\infty}^{\infty} e^{-\frac{(x-y)^2}{2\sigma^2 t}}\, g(y)\, dy.$$

7.5. • Well-Posedness of the IVP and Ill-Posedness of the Backward Diffusion Equation

As with the wave equation, a natural question to ask is whether or not the initial value problem

$$\begin{cases} u_t = \alpha u_{xx} & \text{on } \mathbb{R} \times (0,\infty), \\ u(x,0) = g(x) & \text{on } \mathbb{R} \end{cases} \tag{7.26}$$

is well-posed. Let us first summarize **existence**. Via the Fourier transform we arrived at a solution formula for (7.26) which entailed convolution of g with the fundamental solution $\Phi(x,t)$. This was derived under certain assumptions on g, in particular decay assumptions as $x \to \pm\infty$, and we presented a proof of existence in Theorem 7.1 (i.e., a proof of the fact that the solution formula actually solves (7.26)). As we remarked, the assumptions of Theorem 7.1 can be weakened. Indeed, note that the solution formula makes sense even when $g(x)$ is a polynomial (cf. Exercise **7.27**). The solution formula would also make sense for g which grows very fast as $x \to \pm\infty$, for example $g(x) = e^{\frac{x^2}{4}}$

of Exercise **7.28**. However, be forewarned that in these cases, the formula may only make sense up to some finite time T; that is, the solution can blow up at time T.

One the other hand, one could fix a class of initial data g (say, those which satisfy the assumptions of Theorem 7.1) and ask about **uniqueness and stability**. **Stability** is straightforward (see Exercise **7.10**). **Uniqueness**, however, is surprisingly **subtle**.

7.5.1. • Nonuniqueness of the IVP of the Diffusion Equation.

Recall that we derived a unique solution formula under the assumption that for any fixed t, we could take the Fourier transform of $u(\cdot, t)$. Hence, we made an assumption about integrability of $u(\cdot, t)$. The property of uniqueness fails since there exist nontrivial (i.e., not identically 0) solutions to

$$\begin{cases} u_t = \alpha u_{xx} & \text{on } \mathbb{R} \times (0, \infty), \\ u(x, 0) \equiv 0 & \text{on } \mathbb{R}, \end{cases} \tag{7.27}$$

which grow very rapidly as $x \to \pm\infty$. This rather startling observation was made by **Andrey Tikhonov**[14] who constructed a nontrivial C^∞ solution to (7.27). The construction is rather involved and not presented here (see Exercise **7.12** for an interesting example on the half-line).

It is, however, true that the initial value problem (7.26) has a unique solution within the class of bounded solutions (not just bounded initial data). In fact, one also has uniqueness if we were to replace the boundedness assumption with an exponential growth restriction. More precisely, the following result holds:

A Uniqueness Assertion for the IVP of the Diffusion Equation

For each $T > 0$, there exists at most one solution to (7.26) on $x \in \mathbb{R}$ and $t \in [0, T]$ among all functions $u(x, t)$ which are C^2 in space and C^1 in time and satisfy, for some constants $C > 0$ and $a > 0$,

$$|u(x, t)| \leq Ce^{ax^2} \qquad \text{for all } x \in \mathbb{R},\ t \in [0, T].$$

On the other hand, if one considers the diffusion equation on a bounded domain (say, an interval) with certain boundary conditions, then uniqueness holds without this caveat. This is because there is a general maximum principle for **any** solution to the diffusion equation on a bounded domain (see Section 7.7). Note that in Section 7.2 we showed that **the** solution of the form (7.6) satisfied a maximum principle; we did not show that **any** solution to (7.26) satisfied this maximum principle.

7.5.2. • Ill-Posedness of the Backward Diffusion Equation.

In contrast to the wave equation, the backward diffusion equation is a notorious example of an ill-posed (not well-posed) problem. What exactly do we mean by the *backward* diffusion equation? To illustrate, we present two equivalent formulations.

[14] **Andrey Tikhonov** (also written Tychonov) was an important Russian mathematician who is perhaps best known for his fundamental contributions to the regularization of ill-posed inverse problems.

1 Suppose we know that $u(x,t)$ solves the diffusion equation and we are given $u(x,t_1)$ for some time $t_1 > 0$. Can we determine the initial values $u(x,0) = g(x)$ which produced $u(x,t_1)$ at $t = t_1$? This type of problem is called **an inverse problem**.

2 Alternatively, we can make the transformation $t \longrightarrow -t$ in the PDE in which case we arrive at

$$u_t = -\alpha u_{xx}, \qquad \alpha > 0.$$

This equation is called the **backward diffusion equation**. Note that the backward diffusion equation is different from the diffusion equation due to the minus sign on the right. However, we can solve the **diffusion equation backward in time** by solving the **backward diffusion equation forward in time**. In other words, the first formulation (the inverse problem) is equivalent to asking whether we can solve the IVP for the backward diffusion equation

$$\begin{cases} u_t = -\alpha u_{xx}, & t > 0, x \in \mathbb{R}, \\ u(x,0) \text{ given.} \end{cases} \tag{7.28}$$

The IVP (7.28) is **ill-posed**, not for lack of existence, but rather due to the question of **stability**. For existence we can find a solution to the backward diffusion equation via **deconvolution**. Let us illustrate deconvolution from the first perspective (the inverse problem) of solving the diffusion equation backward in time (for a more direct example of instability via the second perspective see Exercise **7.11**). Suppose we know the values of $u(x,t_1)$ and wish to infer the initial data $g(x)$. Note that our solution formula for the diffusion equation states that

$$u(x,t_1) = \big[\Phi(y,t_1) * g(y)\big](x).$$

Taking the Fourier transform in x, we find $\hat{u}(k,t_1) = \widehat{\Phi}(k,t_1)\,\hat{g}(k)$. Hence if we know $u(x,t_1)$, we can Fourier transform in space and compute

$$\hat{g}(k) = \frac{\widehat{\Phi}(k,t_1)}{\hat{u}(k,t_1)} \tag{7.29}$$

and then take the inverse Fourier transform in order to recover g. The problem with this process of **deconvolution**, which takes us from $u(x,t_1)$ to $g(x)$, is that it is **highly unstable**; the slightest change in $u(x,t_1)$ could result in a drastically different initial data g. This is hardly surprising given the effect diffusion has on a signal; it spatially averages out the signal and as a result acts as a smoothing process. This smoothing process is stable but the inverse, de-smoothing or de-averaging, is not. An excellent and timely way to appreciate this is via **blurring and deblurring in image processing**.

7.5.3. Deblurring in Image Processing. In our digital world, an image is composed of picture elements called pixels.[15] Each pixel represents a particular part (or sample) of the image and is assigned an intensity value associated with the darkness and/or color of that part of the image. When an image is captured by a camera or scanner, it will be **blurry** as it is unavoidable that information at neighboring pixels will distort the information recorded at a given pixel. For example, a camera lens may be

[15]The number of pixels can range from thousands (low resolution images) to millions (higher resolution images).

out of focus or a scanner may be implemented with a shaky hand. One often experiences the latter at the supermarket, when the cashier needs to rescan a UPC barcode (a particularly simple image) several times before it can be read. Cameras and scanners also produce **noise**, random variations in the recorded intensity at each pixel. Thus, the presence of **blurring and noise** is ubiquitous in the recording (capturing) of images.

The goal of image processing is to attempt to recover the original "ground truth" image via some sort of processing of the measured and recorded data. Note that in general, it is impossible to recover exactly the ground truth image; vital information is simply lost with the blurring and presence of noise. However many possible processing techniques can be implemented to produce a "sufficiently good" approximation of the ground truth. These techniques are quite mathematical and, indeed, PDEs can play an important role[16].

To present a few details in the context of the diffusion equation, let us model the blurring process via **convolution** with a fixed function called a blurring kernel (or point spread function). For a variety of reasons, a natural blurring kernel to consider is a Gaussian[17]. Consequently, we have the following direct analogy of blurring and the **diffusion equation**: Taking a signal g as initial data and evolving it according to the diffusion equation for a small time interval amounts to blurring the signal g. Since most images are two dimensional, one should be thinking here of the 2D diffusion equation; however, there are essentially 1D images like UPC barcodes (see Figure 7.5) for which the 1D diffusion equation is relevant. You can think of signal g as a continuum description of a back and white image where $g(x, y)$ denotes the intensity at (x, y). From the point of view of image processing, blurring an image might not seem very useful. However, the signal g contains noise and an application of the diffusion equation for a very small amount of time (slightly blurring the image) will remove some of the random pixel variations and hence **denoise** the image. Of course, this will also change the basic structure of the image (for example, shapes and edges); hence there are anisotropic versions of the diffusion equation which are preferable denoisers (cf. Exercise **7.43**).

But now comes the point of this section. In image processing, we also want to address the opposite direction — going from a blurred image to the original signal g. This process of **deblurring** amounts to addressing the **inverse problem** by solving the backward diffusion equation, or equivalently performing a **deconvolution**. But as we have just mentioned, deconvolution/solving the backward diffusion equation is highly unstable. If you reflect a little on the deblurring process, you will appreciate why it is so unstable. So whereas the process of blurring, which takes us from the clean image to the blurred image, is **stable** (with respect to noise), the process of deblurring, which takes us from the blurred image to the clean image, is **unstable**. In fact, the **noise** inherent in any measurement can have a drastic effect when recovering all the details of the pure clean image. An active field of contemporary image processing is built upon

[16]For more on the role of PDEs in image processing see either of the excellent books **Tony Chan and Jianhong Shen**, *Image Processing and Analysis*, Society for Industrial and Applied Mathematics, PA, 2005, or **Guillermo Sapiro,** *Geometric Partial Differential Equations and Image Analysis*, Cambridge University Press.

[17]Note that, in general, the blurring kernel may not be Gaussian; in fact, all details about its shape may be unknown.

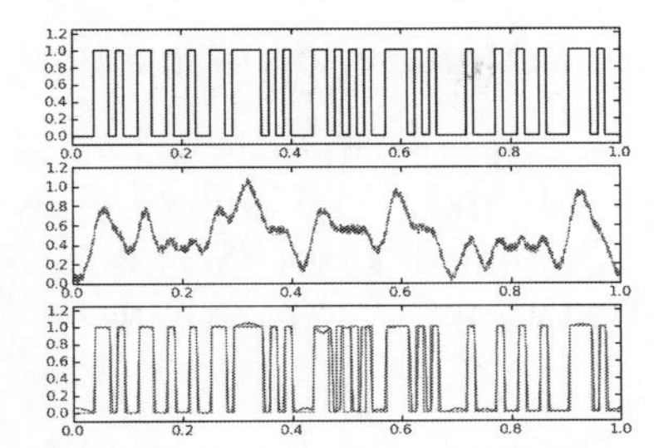

Figure 7.5. Left: A UPC barcode taken from Wikipedia. Right: Deblurring in action. From R. Choksi and Y. Gennip, Deblurring of one dimensional bar codes via total variation energy minimization, SIAM Journal on Imaging Sciences, Vol. 3-4 (2010), pp. 735–764. Copyright (c) 2010 Society for Industrial and Applied Mathematics. Reprinted with permission. All rights reserved.

ways to derive and test methods for **regularizing** such ill-posed inverse problems in order to "solve" them in a "satisfactory" way. In fact, one of the central goals in the contemporary fields of **machine learning** and **deep learning** is to find ways to solve ill-posed problems, many of which are ill-posed because of the lack of uniqueness.

We illustrate these comments with one of the simplest images, a UPC (universal product code) barcode, an example of which is shown on the left side of Figure 7.5. UPC barcodes are currently ubiquitous in the supermarket where a barcode scanner uses, for example, light detectors or photoreceptors (similar to those used by a digital camera) to pick up light and dark areas of the barcode, thereby producing a continuous signal associated with the darkness distribution across the span of the barcode. This continuous signal is an approximation of the actual barcode which, depending on the scanner and the manner in which the scan was captured (distance, vibrations, etc.), will be blurry and noisy. Our goal is to reconstruct our best approximation of the clean barcode from this continuous signal.

One such attempt is illustrated on the right side of Figure 7.5 where a clean barcode is represented by a discontinuous step function g taking only the values of 1 and 0. Its associated graph is depicted at the top. We then convolute this step function with a particular Gaussian; i.e., we evolve g for a small set time by a particular 1D diffusion equation. To this convolution we add some small perturbations on a very fine scale (noise) arriving at a test signal:

$$G(x) := \big(\Phi(\cdot, t_0) * g\big)(x) + \text{noise}, \qquad \text{for some small } t_0.$$

The resulting signal G is depicted in the middle line of Figure 7.5 (right). Our goal is, **given** G and Φ, to **reconstruct** the clean barcode g. Because of the presence of noise, deconvolution as in (7.29) will not yield an accurate approximation of g. While there are many **regularization methods** to perform this task, here we show one such method to extract from G and Φ, the "best" approximation (the deblurred and denoised version) of the clean barcode g; this best approximation is shown in the bottom curve which, for comparison, is superimposed on the clean barcode g. Note that here we assumed knowledge of the blurring kernel Φ; one often knows nothing about its structure yielding a problem which is commonly referred to as **blind deblurring**.

7.6. • Some Boundary Value Problems in the Context of Heat Flow

One often considers the diffusion equation on a finite or semi-infinite interval where we supplement the initial data with boundary conditions. These boundary conditions have a particular interpretation from the perspective of heat transfer. Therefore, let us illustrate them for a finite bar parametrized by the interval $x \in [0, l]$.

7.6.1. • Dirichlet and Neumann Boundary Conditions.

The two types of boundary conditions (Dirichlet and Neumann) considered in Chapter 3 on the wave equation have natural interpretations for the 1D diffusion equation as a model for the temperature in a thin bar (insulated everywhere except at the ends) which lies on the interval $[0, l]$ of the x-axis. Without giving precise definitions, we will view temperature as a measure of heat. A Dirichlet boundary condition (more precisely the homogeneous Dirichlet boundary condition) at the left end would entail fixing the temperature to be 0 at $x = 0$; i.e., $u(0, t) = 0$ for $t > 0$. In general, for any constant T_0, the inhomogeneous Dirichlet boundary condition fixes the temperature at the left end to be T_0; i.e., $u(0, t) = T_0$ for $t > 0$. We have deliberately not included $t = 0$ because it will be useful to deal with initial temperature distributions which are not necessarily T_0 at the location $x = 0$. How do we interpret this? Let us suppose that we attach the left end of the bar to an initially closed (insulated) window to a universe so large that its temperature is always a constant T_0 (cf. Figure 7.6, left). Initially, the part of the bar at the window can be any temperature; however suppose that at $t = 0$, we open the window. We assume that the outside environment is so vast that any excess heat at the left end of the bar immediately dissipates without changing the outside temperature T_0. For the same reason, a lower initial temperature at $x = 0$ would immediately adjust to T_0. Thus after the window is opened, the temperature at $x = 0$ is always T_0. Note here that we are making the assumption that the outside universe is so large that any addition (or loss) of heat via the bar is negligible. Naturally this is an idealization (or rather an approximation) to what really occurs, but for a variety of problems it is sufficient.

For a Neumann boundary condition (cf. Figure 7.6, right right), we set the flux at the left end $x = 0$ to be zero for all later times; i.e., $u_x(0, t) = 0$ for $t > 0$. Thus, there is no transfer of heat at $x = 0$ and one can view this as insulating the side end for $t > 0$.

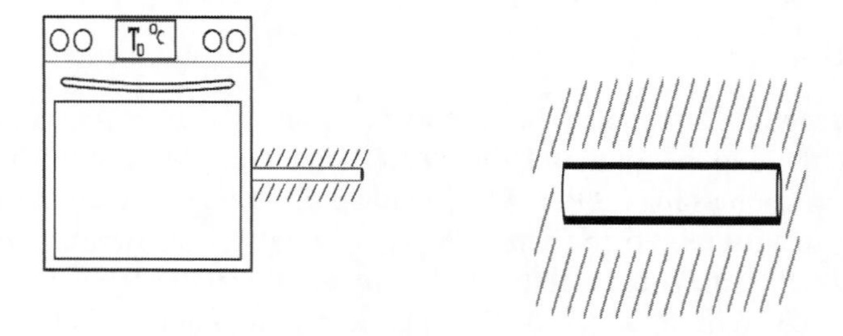

Figure 7.6. Left: Cartoon for a heat flow in an insulated bar with an inhomogeneous Dirichlet boundary condition at the left end of the bar. **Right:** Cartoon for a heat flow in a **fully** insulated bar: Neumann boundary conditions at the ends.

To see some of the consequences of these boundary conditions, it is convenient to consider the long-time behavior (eventual temperature distribution) for the following three IVPs/BVPs:

$$\begin{cases} u_t = u_{xx} & \text{for } \ 0 \le x \le l, \ t > 0, \\ u(0,t) = T_0, \quad u(l,t) = T_1 & \text{for } \ t > 0, \\ u(x,0) = g(x) & \text{for } \ 0 \le x \le l, \end{cases} \tag{7.30}$$

$$\begin{cases} u_t = u_{xx} & \text{for } \ 0 \le x \le l, \ t > 0, \\ u_x(0,t) = 0, \quad u_x(l,t) = 0 & \text{for } \ t > 0, \\ u(x,0) = g(x) & \text{for } \ 0 \le x \le l, \end{cases} \tag{7.31}$$

$$\begin{cases} u_t = u_{xx} & \text{for } \ 0 \le x \le l, \ t > 0, \\ u(0,t) = T_0, \quad u_x(l,t) = 0 & \text{for } \ t > 0, \\ u(x,0) = g(x) & \text{for } \ 0 \le x \le l. \end{cases} \tag{7.32}$$

These problems are called, respectively, nonhomogeneous Dirichlet, (homogeneous) Neumann, and mixed boundary value problems. As we will see in Chapter 12, they can all be readily solved via the techniques of separation of variables. Here, let us just consider the eventual (steady-state) temperature distribution

$$v(x) := \lim_{t \to \infty} u(x,t).$$

We consider (7.32) first. In this case, one of the ends is insulated so no heat can escape. The other end is exposed to a universe kept at temperature T_0. Thus regardless of the initial temperature distribution, in the end ($t \to \infty$), the temperature everywhere in the bar will approach T_0. Any perturbation from T_0 will escape to the universe via the left-hand side. So $v(x) \equiv T_0$.

Next, we consider (7.31) where both ends are insulted. In this instance, there is no contact with the outside. Intuitively, we would expect that the initial heat (or temperature) will just equilibrate to a constant throughout the bar. Indeed, the constant is none other than the average temperature

$$v(x) \equiv \frac{1}{l} \int_0^l g(x)\, dx.$$

Finally, we consider (7.30). If $T_0 = T_1$, one would again expect that all heat would eventually equilibrate to the outside and, hence, $v(x) = T_0 = T_1$. If $T_0 \neq T_1$, then we have a temperature gradient. As we shall see in Chapter 12, the eventual temperature distribution will be a linear function which takes on the values T_0 at $x = 0$ and T_1 at $x = l$; i.e., $v(x) = T_0 + (T_1 - T_0)\frac{x}{l}$.

Let us conclude by remarking that for the semi-infinite interval ($x \in [0, \infty)$) and the fixed boundary $u(0,t) = 0$ for all $t > 0$, we can use the analogous method with odd reflections which was previously adopted with the wave equation (see Exercise **7.8**).

7.6.2. • The Robin Condition and Heat Transfer.

The Robin boundary condition combines both Neumann and Dirichlet. It states that at $x = 0$, we have for all $t > 0$

$$c_1 u(0,t) + u_x(0,t) = c_2, \tag{7.33}$$

for some constants c_1 and c_2. We interpret this in the context of heat flow in the bar. In the case of the nonhomogeneous Dirichlet boundary condition at $x = 0$, we argued that after $t > 0$, the temperature at the end point instantaneously adjust to T_0, the ambient temperature of the outside. A different approach would be to allow for heat transfer. Indeed, if we expose the left end of the bar to a (large) reservoir of a medium (air or liquid) with temperature T_0, one would expect a transfer of heat between the left end of the bar and the reservoir. This transfer of heat can be modeled by **Newton's Law of Cooling** which asserts: *The temperature flux at $x = 0$ is proportional to the difference between the temperature of the bar at $x = 0$ (i.e., $u(0,t)$) and the ambient temperature of its surroundings* (i.e., T_0). This means

$$u_x(0,t) = c_1\big(T_0 - u(0,t)\big),$$

where $c_1 > 0$ and $T_0 \geq 0$ are constants. Note that we have chosen the signs correctly: If $T_0 > u(0,t)$, then heat will flow to the right and $u_x(0,t) > 0$. This is equivalent to (7.33) for some nonnegative constants c_1 and c_2. We again highlight the assumption that the outside reservoir is so large that any addition (or loss) of heat via the bar is negligible.

Boundary value problems with one or two Robin boundary conditions can also be solved via the technique of separation of variables in Section 12.3.5.

7.7. • The Maximum Principle on a Finite Interval

In Section 7.2.2, we noted that the solution formula for the diffusion equation on the entire line exhibited a maximum principle: For any $t > 0$ and $x \in \mathbb{R}$, $|u(x,t)|$ is never greater than the maximum value at $t = 0$, i.e., the maximum of $u(x,0)$ over all $x \in \mathbb{R}$. Actually, more can be said by focusing attention on a finite interval $0 \leq x \leq l$. Here the boundary conditions at $x = 0, l$ are **irrelevant**; all we need is for u to solve the diffusion equation for $x \in (0,l)$ and $t > 0$. To state the maximum principle, fix $T > 0$, and denote by Ω_T the rectangle $[0,l] \times [0,T]$ (or tube) in the space-time plane. Let $\mathcal{S}$ (for sides) be the part of $\partial\Omega_T$ except for the top; precisely,

$$\mathcal{S} = \{(x,0) \in \mathbb{R}^2 \,|\, 0 \leq x \leq l\} \cup \{(0,t) \in \mathbb{R}^2 \,|\, 0 \leq t \leq T\} \cup \{(l,t) \in \mathbb{R}^2 \,|\, 0 \leq t \leq T\}.$$

Theorem 7.2 (The Maximum Principle). *Let $u(x,t)$, continuous on Ω_T, be a smooth solution to the diffusion equation $u_t = \alpha u_{xx}$ $(\alpha > 0)$ for $0 < x < l$ and $0 < t < T$. Then*

$$\max_{\mathcal{S}} u(x,t) = \max_{\Omega_T} u(x,t). \tag{7.34}$$

In words, the Maximum Principle states that the maximum value of u over the finite domain Ω_T must occur at **either time** $t = 0$ **or** at the boundary points $x = 0$ and $x = l$.

All we need for the proof is the following:

- A few simple calculus facts about extrema of functions of one variable: If $f(x)$ is a smooth function on an interval (a, b) which attains its maximum at $x_0 \in (a, b)$, then $f'(x_0) = 0$ and $f''(x_0) \leq 0$.
- A trick.

Proof. Let us begin with the trick which, at first, seems unmotivated. Instead of working with u, we let $\epsilon > 0$ and consider

$$v(x,t) \ := \ u(x,t) \ + \ \epsilon\, x^2.$$

Since u solves the diffusion equation, we have

$$v_t - \alpha v_{xx} \ = \ -2\alpha\epsilon \ < \ 0. \tag{7.35}$$

The point here is that the right-hand side is strictly negative — we will need this. We break the proof up into three steps.

Step 1: The maximum of v over Ω_T can only occur on its boundary $\partial\Omega_T$. Suppose the maximum of v over Ω_T occurred at an interior point (x_0, t_0) where $x_0 \in (0, l)$ and $t_0 \in (0, T)$. Then considering separately v as a function of t and of x, we must have $v_t(x_0, t_0) = 0$ and $v_{xx}(x_0, t_0) \leq 0$. But this means that

$$v_t(x_0, t_0) - \alpha v_{xx}(x_0, t_0) \ \geq \ 0,$$

contradicting (7.35). Since v is continuous, it must attain its maximum on Ω_T; hence it must occur only on the boundary $\partial\Omega_T$.

Step 2: The maximum of v over Ω_T must occur on $\mathcal{S}$. We now show that the maximum cannot occur at a point (x_0, T) on the top side $\{(x,T) \,|\, 0 < x < l\}$. Suppose it did. Then $v(x, T)$ as a function of x would have a maximum at x_0 and hence $v_{xx}(x_0, T) \leq 0$. On the other hand, v as a function of t is only defined for $t \leq T$; hence at $t = T$ we can only compute the t derivative "from below" and conclude that $v_t(x_0, T) \geq 0$. To see this, note that by assumption, for any $h > 0$ we must have $v(x_0, T) > v(x_0, T - h)$; hence

$$v_t(x_0, T) \ = \ \lim_{h \to 0+} \frac{v(x_0, T) - v(x_0, T-h)}{h} \ \geq \ 0.$$

However, this implies that

$$v_t(x_0, T) - \alpha v_{xx}(x_0, T) \geq 0,$$

contradicting (7.35). Hence the maximum of v over Ω_T must occur on $\mathcal{S}$.

Step 3: Relating back to u. By definition of v we have

$$\begin{aligned} \max_{\mathcal{S}} v(x,t) \ &\leq \ \max_{\mathcal{S}} u(x,t) \ + \ \max_{\mathcal{S}} \epsilon x^2 \\ &= \ \max_{\mathcal{S}} u(x,t) \ + \ \epsilon l^2. \end{aligned}$$

Since $u(x,t) = v(x,t) - \epsilon x^2$, we have

$$\max_{\Omega_T} u(x,t) \leq \max_{\Omega_T} v(x,t).$$

Combining this with the previous steps, we have that

$$\max_{\Omega_T} u(x,t) \leq \max_{\Omega_T} v(x,t) \overset{\text{Steps 1 and 2}}{=} \max_{\mathcal{S}} v(x,t) \leq \max_{\mathcal{S}} u(x,t) + \epsilon l^2.$$

Since the above holds for every $\epsilon > 0$, we must have

$$\max_{\Omega_T} u(x,t) \leq \max_{\mathcal{S}} u(x,t).$$

Clearly the reverse inequality holds true and hence we have shown (7.34). □

A few comments are in order:

- By replacing u with $-u$ we obtain the same result for the minimum of u, i.e., the Minimum Principle.
- The readers may ask if u can **only** attain its maximum on $\mathcal{S}$ and not inside Ω_T. We proved that this was the case for v which satisfied $v_t - \alpha v_{xx} < 0$ but our proof does not work for u satisfying $u_t - \alpha u_{xx} = 0$. With a slightly deeper analysis, one can prove the **Strong Maximum Principle** (cf. Section 2.3.3 of [**10**]). This states that if the maximum of u is attained at some point $(x_0, t_0) \in (0,l) \times (0,T]$, then $u(x,t)$ must be identically constant for $(x,t) \in [0,l] \times [0,t_0]$. Note here that we **cannot** make this conclusion in all of Ω_T but only for $t \leq t_0$; that is, the conclusion is false for $t \in (t_0, T]$. The readers should try to reckon this fact with their intuition of heat flow.
- Theorem 7.2 allows one to prove **uniqueness** and **stability** to the IVP/BVP of the diffusion equation on a finite domain $x \in [0,l]$ associated with any of the standard boundary conditions, for example, (7.30)–(7.32) (cf. Exercise **7.25**).

7.8. Source Terms and Duhamel's Principle Revisited

Just as with the wave equation, we can introduce a source term into the diffusion equation. This leads us to the IVP:

$$\begin{cases} u_t - \alpha u_{xx} = f(x,t), & x \in \mathbb{R}, t > 0, \\ u(x,0) = g(x), & x \in \mathbb{R}. \end{cases} \tag{7.36}$$

The addition of a source term can be viewed, for example, as modeling the heat in an infinitely long bar with an external heat source $f(x,t)$ (cf. Exercise **7.16**). The explicit solution is given by

$$u(x,t) = \int_{-\infty}^{\infty} \Phi(x-y,t)\, g(y)dy + \int_0^t \int_{-\infty}^{\infty} f(y,s)\, \Phi(x-y,t-s)\, dy\, ds, \tag{7.37}$$

where as usual, Φ is the heat kernel given by (7.7). Two ways to derive this formula are via

- Duhamel's Principle,
- the Fourier transform (cf. Exercise **7.3**).

For the former, we would follow the same steps as we did in Section 3.6.1. By superposition, we add to the solution of the homogeneous IVP (7.5) the solution of

$$\begin{cases} u_t - \alpha u_{xx} = f(x,t), & x \in \mathbb{R}, t > 0, \\ u(x,0) = 0, & x \in \mathbb{R}. \end{cases} \tag{7.38}$$

Note that the sum indeed solves (7.36). To solve (7.38) we introduce an additional temporal parameter $s \in [0, \infty)$. For each **fixed** s, we consider a solution $w(x,t;s)$ to

$$\begin{cases} w_t(x,t;s) - \alpha w_{xx}(x,t;s) = 0, & x \in \mathbb{R},\ t > s, \\ w(x,s;s) = f(x,s), & x \in \mathbb{R}. \end{cases} \tag{7.39}$$

Note that by a translation of time, for each (fixed) $s > 0$ the solution to (7.39) is simply

$$w(x,t;s) = \int_{-\infty}^{\infty} f(y,s)\,\Phi(x-y,t-s)\,dy.$$

Duhamel's Principle states that the solution to (7.38) is given by

$$u(x,t) = \int_0^t w(x,t;s)\,ds = \int_0^t \int_{-\infty}^{\infty} f(y,s)\,\Phi(x-y,t-s)\,dy\,ds. \tag{7.40}$$

Proving Duhamel's Principle, i.e., that (7.40) solves (7.38), requires more work than for the 1D wave equation (Section 3.6.1) where it entailed a simple application of the Leibnitz formula (Theorem A.12). This is a consequence of the fact that the integral in (7.40) "feels" the singularity of $\Phi(y,t)$ near $t = 0$; hence differentiation under the integral sign requires some care. The details are presented in Theorem 2 of Section 2.3 in [**10**]. There one will note that the essential feature of the proof is still the distributional convergence of $\Phi(\cdot,t)$ to δ_0 as $t \to 0^+$.

7.8.1. An Intuitive and Physical Explanation of Duhamel's Principle for Heat Flow with a Source.

Surely, there must be some physical explanation of why one can *transfer* the source term *to the initial data*. We first saw the following simple explanation in the book by Bleecker and Csordas[18].

Let us interpret the IVP (7.38) as the equation for the temperature distribution $u(x,t)$ across an infinitely long bar which has been subjected to a **heat source** $f(x,t)$. Since there is no heat in the initial conditions, all the heat (temperature) must come from the source. Note that at any time $t > 0$, the heat source has been active since time 0; hence, $u(x,t)$ should depend on the source $f(x,t)$ during the entire interval $[0,t]$ (the source "history"). Also note that while u has physical dimensions of temperature, the heat source $f(x,t)$ has dimensions of temperature per unit time.

We proceed by **fixing** a small time step $\Delta s > 0$, considering a time $t = n\Delta s$, for some integer n, and time subintervals

$$[0,\Delta s], [\Delta s, 2\Delta s], [2\Delta s, 3\Delta s], \ldots, [(n-1)\Delta s, t].$$

We separate the effects of having the heat source **only active on one of these time subintervals** $[(i-1)\Delta s, i\Delta s]$. Precisely, we consider sources of the form

$$f_i(x,t) = f(x,t)\,\chi_{[(i-1)\Delta s, i\Delta s]}(t), \qquad i = 1, \ldots, n,$$

[18]**Basic Partial Differential Equations** by David Bleecker and George Csordas, Van Nostrand Reinhold, 1992.

where $\chi_I(t)$ denotes a characteristic function of $t \in I$, and the resulting temperature distribution $u_i(x,t)$ at time $t = n\Delta s$, i.e., the solution $u_i(x,t)$ to

$$u_t - \alpha u_{xx} = f_i(x,t).$$

Since **the PDE is linear** and

$$f(x,t) = \sum_{i=1}^{n} f_i(x,t),$$

we can obtain the solution $u(x,t)$ to (7.38) by simply adding the $u_i(x,t)$. In the derivation below we will **approximate** each $u_i(x,t)$ by $\tilde{u}_i(x,t)$ with an error which will be $o(\Delta s)$. Since there are essentially $1/\Delta s$ subintervals, the total error will tend to 0 as $\Delta s \to 0$, resulting in the exact solution to (7.38).

On each of the small time subintervals, the approximation $\tilde{u}_i(x,t)$ will stem from assuming $f_i(x,t)$ to be identically equal to its value at the right end point of the interval. Hence, it is convenient here to include an artificial negative time interval $[-\Delta s, 0]$ and end with the source active on $[(n-2)\Delta s, (n-1)\Delta s]$. In other words, we translate the source activity interval from $[0,t]$ to $[-\Delta s, (n-1)\Delta s]$. Since the additional error in this translation will be at most $O(\Delta s)$, the result will be unchanged after we take the limit $\Delta s \to 0$. The reader should recall the analogy with using a Riemann sum to approximate an integral.

We now present the details.

Step 1: Source active only from $[-\Delta s, 0]$**.** First, we assume the source has been active **only** from time $-\Delta s$ to time 0. Our goal is to approximate the resulting temperature distribution, called $\tilde{u}_0(x,t)$, at time t. To do this, we assume the source f is constant in time on $[-\Delta s, 0]$, taking the constant to be the value at the **right end point**. In other words, for any $x \in \mathbb{R}$, we assume that $f(x,t) \approx f(x,0)$ for $t \in [-\Delta s, 0]$. In this case, we can approximate temperature at position x and time 0 by simply multiplying the heat source by the time elapsed; i.e.,

$$\tilde{u}_0(x,0) = f(x,0)\Delta s. \tag{7.41}$$

Now at $t = 0$, the heat source is turned off. Hence to find the resulting temperature distribution at time t, we solve the **homogeneous** diffusion equation with data given by (7.41); i.e.,

$$\begin{cases} (\tilde{u}_0)_t - \alpha(\tilde{u}_0)_{xx} = 0, & x \in \mathbb{R},\ t > 0, \\ \tilde{u}_0(x,0) = f(x,0)\Delta s, & x \in \mathbb{R}. \end{cases}$$

Since Δs is a constant and the diffusion equation is linear, we can **factor out this constant** to find that

$$\tilde{u}_0(x,t) = w(x,t;0)\,\Delta s, \tag{7.42}$$

where $w(x,t;0)$ is defined as the solution to

$$\begin{cases} w_t - \alpha w_{xx} = 0, & x \in \mathbb{R},\ t > 0, \\ w(x,0;0) = f(x,0), & x \in \mathbb{R}. \end{cases}$$

Note that w has physical dimensions of temperature per unit time.

In summary, the function $\tilde{u}_0(x,t)$, given by (7.42), is our approximation of the temperature distribution at time t resulting from the source active only on the time interval $[-\Delta s, 0]$.

Step 2: Source active only from $[(i-1)\Delta s, i\Delta s]$ **for** $i = 1, \dots, n-1$. We now fix an $i = 1, \dots, n-1$ and turn on the heat source **only** during the interval $[(i-1)\Delta s, i\Delta s]$. We will call the resulting approximate temperature distribution at time t, $\tilde{u}_i(x,t)$. Hence, prior to time $(i-1)\Delta s$ the temperature is 0. Following the same reasoning as in Step 1, the approximate temperature at position x at time $i\Delta s$ is

$$\tilde{u}_i(x, i\Delta s) = f(x, i\Delta s)\,\Delta s.$$

In other words, we are assuming the source f is the constant $f(x, i\Delta s)$ for times in $[(i-1)\Delta s, i\Delta s]$.

To find the resulting temperature distribution at time t, we follow the homogeneous diffusion equation, but as before we factor out the constant Δs. To this end, we have

$$\tilde{u}_i(x,t) = w(x,t; i\Delta s)\Delta s \tag{7.43}$$

where $w(x,t;i\Delta s)$ is defined as the solution to

$$\begin{cases} w_t(x,t;i\Delta s) - \alpha w_{xx}(x,t;i\Delta s) = 0, & x \in \mathbb{R},\ t > i\Delta s, \\ w(x, i\Delta s; i\Delta s) = f(x, i\Delta s), & x \in \mathbb{R}. \end{cases}$$

In summary, the function $\tilde{u}_i(x,t)$, given by (7.43), is our approximation of the temperature distribution at time t resulting from the source active only on the time interval $[(i-1)\Delta s, i\Delta s]$.

Step 3: Source active on the full interval. Now we address the solution $u(x,t)$ to (7.38) wherein there is no initial temperature **but** the heat source is active for the **full interval** $[0,t]$. The key here is that the **PDE in** (7.38) **is linear** and hence we can find the approximate temperature distribution at time t by **summing** $\tilde{u}_i(x,t)$. This gives

$$u(x,t) \approx \sum_{i=0}^{n-1} \tilde{u}_i(x,t) = \sum_{i=0}^{n-1} w(x,t;i\Delta s)\,\Delta s. \tag{7.44}$$

Moreover, the error in this approximation tends to 0 as $\Delta s \to 0^+$.

Step 4: Sending $\Delta s \to 0^+$.

We assume that $w(x,t;s)$, defined as the solution to

$$\begin{cases} w_t(x,t;s) - \alpha w_{xx}(x,t;s) = 0, & x \in \mathbb{R},\ t > s, \\ w(x,s;s) = f(x,s), & x \in \mathbb{R}, \end{cases}$$

is continuous in s. Hence, we recognize the right-hand side of (7.44) as a **Riemann sum** of $w(x,t;s)$ with respect to the s variable. Letting $\Delta s \to 0^+$ in this Riemann sum we find

$$u(x,t) = \int_0^t w(x,t;s)\,ds.$$

In words, we can consider $w(x,t;s)\,ds$ as the resulting temperature at time t from the heat source f **instantaneously injected** at time s. We obtain the full effect of the source active on $[0,t]$ by integrating over $s \in [0,t]$.

A similar treatment can be adapted to physically explain Duhamel's Principle for the wave equation (Section 3.6.1). In Exercise **7.33**, you are asked to provide the details.

7.9. The Diffusion Equation in Higher Space Dimensions

Recall that the solution to the wave equation behaved very differently in odd and even dimensions. This is **not** the case for the diffusion equation. The basic structure of the solution formula remains the same, and hence diffusion has a similar behavior in all space dimensions. This is not to say, however, that related objects such as random walks do not have different properties in different space dimensions.

Via the Fourier transform in $\mathbb{R}^N$, one can show that for continuous function $g(\mathbf{x})$ with

$$\int \cdots \int_{\mathbb{R}^N} |g(\mathbf{x})| \, d\mathbf{x} \; < \; \infty,$$

a solution to

$$\begin{cases} u_t = \alpha \Delta u, & \mathbf{x} \in \mathbb{R}^N, \; t > 0, \\ u(\mathbf{x}, 0) = g(\mathbf{x}), & \mathbf{x} \in \mathbb{R}^N, \end{cases} \tag{7.45}$$

is given by

$$\boxed{u(\mathbf{x}, t) \;=\; \frac{1}{(4\pi\alpha t)^{N/2}} \int \cdots \int_{\mathbb{R}^N} \exp\left(-\frac{|\mathbf{x}-\mathbf{y}|^2}{4\alpha t}\right) g(\mathbf{y}) \, d\mathbf{y}.}$$

Here we use the common notation of $\exp(\cdot)$ for $e^{(\cdot)}$. Note that the solution $u(\mathbf{x}, t)$ is given by the N-dimensional convolution of the data g with the N-dimensional fundamental solution/heat kernel:

$$u(\mathbf{x}, t) \;=\; (\Phi(\cdot, t) * g)(\mathbf{x}) \;=\; \int \cdots \int_{\mathbb{R}^N} \Phi(\mathbf{x}-\mathbf{y}, t) \, g(\mathbf{y}) \, d\mathbf{y},$$

where

$$\Phi(\mathbf{x}, t) \;:=\; \frac{1}{(4\pi\alpha t)^{N/2}} \exp\left(-\frac{|\mathbf{x}|^2}{4\alpha t}\right). \tag{7.46}$$

For example, in 3D the solution formula to the diffusion equation is

$$u(\mathbf{x}, t) = \frac{1}{(4\pi\alpha t)^{3/2}} \int_{-\infty}^{\infty} \int_{-\infty}^{\infty} \int_{-\infty}^{\infty} \exp\left(-\frac{|\mathbf{x}-\mathbf{y}|^2}{4\alpha t}\right) g(\mathbf{y}) \, d\mathbf{y}. \tag{7.47}$$

All the properties for Φ and the solution u carry over from 1D. In Exercise **7.19** you are asked to use the Fourier transform to derive (7.47) and verify that Φ satisfies the analogous properties of the 1D heat kernel.

7.10. • A Touch of Numerics, III: Numerical Solution to the Diffusion Equation

This section requires the reading of Section 2.7.

Following the approach of Section 2.7 on the finite difference numerical solution to the transport equation, let us briefly apply the same ideas for finding a numerical solution to the diffusion equation IVP/BVP:

$$\begin{cases} u_t = \alpha u_{xx} & \text{for } 0 \le x \le 1,\ t > 0, \\ u(0,t) = 0 = u(1,t) & \text{for } t > 0, \\ u(x,0) = g(x) & \text{for } 0 \le x \le 1. \end{cases} \tag{7.48}$$

Note that, unlike for the transport equation, the diffusion equation has infinite propagation speed; hence, we need to enforce some boundary conditions.

We start by recalling the notation from Section 2.7. The spatial and temporal step sizes are $\Delta x > 0$ and $\Delta t > 0$ with grid points

$$x_j = j\Delta x, \qquad t_n = n\Delta t, \qquad \text{for } j = 0, 1, \ldots, J,\ n = 0, 1, 2, \ldots.$$

Note that since we are now working on the bounded spatial domain, we have grid points at each of the spatial boundaries; thus

$$1 = J\Delta x.$$

We then attempt to find the solution at the grid points

$$U_j^n := u(j\Delta x, n\Delta t).$$

The finite difference for the single time derivative is the forward Euler

$$u_t(j\Delta x, n\Delta t) \approx \frac{U_j^{n+1} - U_j^n}{\Delta t}.$$

For the second derivative in space, we use the centered difference approximation of Section 7.3.1:

$$u_{xx}(j\Delta x, n\Delta t) \approx \frac{U_{j+1}^n - 2U_j^n + U_{j-1}^n}{(\Delta x)^2}.$$

Thus the diffusion equation on the grid becomes

$$\frac{U_j^{n+1} - U_j^n}{\Delta t} = \alpha \frac{U_{j+1}^n - 2U_j^n + U_{j-1}^n}{(\Delta x)^2},$$

or simply the **explicit scheme**:

$$\boxed{U_j^{n+1} = U_j^n + r\left(U_{j+1}^n - 2U_j^n + U_{j-1}^n\right) \qquad \text{where} \qquad r := \frac{\alpha\Delta t}{(\Delta x)^2}.} \tag{7.49}$$

The dimensionless parameter r measures the relative size of the grid. One can check that the scheme is **consistent**. But is it **stable**? To this end, let us perform the von Neumann stability analysis of Subsection 2.7.2. We choose

$$U_j^n = e^{ik(j\Delta x)}$$

noting that

$$U_{j-1}^n = e^{ik(j\Delta x)}e^{-ik\Delta x} \quad \text{and} \quad U_{j+1}^n = e^{ik(j\Delta x)}e^{ik\Delta x}.$$

Applying scheme (7.49) we find that in one time step

$$\begin{aligned} U_j^{n+1} &= e^{ik(j\Delta x)} + r\left(e^{ik(j\Delta)}e^{ik\Delta x} - 2e^{ik(j\Delta x)} + e^{ik(j\Delta x)}e^{-ik\Delta x}\right) \\ &= e^{ik(j\Delta x)}\left(1 + r\left(e^{ik\Delta x} + e^{-ik\Delta x}\right) - 2r\right) \\ &= e^{ik(j\Delta x)}\left(1 + 2r(\cos k\Delta x - 1)\right). \end{aligned}$$

The scheme (7.49) will be stable if and only if the growth factor satisfies

$$|(1 - 2r(1 - \cos k\Delta x))| \leq 1, \tag{7.50}$$

for all frequencies k. Since for any θ, $0 \leq 1 - \cos\theta \leq 2$, we will ensure (7.50) if and only if

$$-1 \leq 1 - 4r \leq (1 - 2r(1 - \cos k\Delta x)) \leq 1.$$

This yields $1 - 4r \geq -1$ or $r \leq \frac{1}{2}$. Thus the scheme is stable and, hence, by the Lax Equivalence Theorem, convergent if and only if $r \leq \frac{1}{2}$.

In Exercise **7.34** you are asked to illustrate this stability by numerically solving (7.48) with a particular g. Here you will see how the boundary conditions translate to conditions at the left and right end grid points. Essentially, the scheme (7.49) tells us how to compute the discrete solution U_j^1 explicitly from values $g(j\Delta x) = U_j^0$. We repeat the scheme to compute U_j^2 from the values U_j^1 and then "**march**" along to compute the discrete solution U_j^n for $n = 2, 3, \ldots$.

The explicit finite difference scheme (7.49) is one of many possible schemes for the diffusion equation. A famous **implicit** scheme, known as the **Crank-Nicolson scheme**, is derived from using the following finite difference approximation for u_{xx}:

$$u_{xx}(j\Delta x, n\Delta t) \approx \frac{1}{2}\frac{U_{j+1}^n - U_j^n + U_{j-1}^n}{(\Delta x)^2} + \frac{1}{2}\frac{U_{j+1}^{n+1} - U_j^{n+1} + U_{j-1}^{n+1}}{(\Delta x)^2}.$$

Here, we are using centered difference approximations to the second derivative both at time $n\Delta t$ and time $(n+1)\Delta t$. After a little algebra, the diffusion equation on the grid becomes the implicit finite difference scheme:

$$-\frac{r}{2}U_{j+1}^{n+1} + (1+r)U_j^{n+1} - \frac{r}{2}U_{j-1}^{n+1} = \frac{r}{2}U_{j+1}^n + (1-r)U_j^n + \frac{r}{2}U_{j-1}^n,$$

or

$$\begin{aligned} U_j^{n+1} &= \frac{r}{2(1+r)}U_{j+1}^n + \frac{1-r}{1+r}U_j^n + \frac{r}{2(1+r)}U_{j-1}^n \\ &\quad + \frac{r}{2(1+r)}U_{j-1}^{n+1} + \frac{r}{2(1+r)}U_{j+1}^{n+1}, \end{aligned} \tag{7.51}$$

with r defined in (7.49). This scheme is implicit (cf. Figure 7.7) as the values of the solution at time $(n+1)\Delta t$ and position $j\Delta x$ are not explicitly a function of values of the solution at the previous time $n\Delta t$; hence, unlike with an explicit scheme for which we may simply march along in time, to compute the discrete solution at the n-th time step from the solution at the previous step, one must now solve **simultaneously** a large system of coupled linear equations. This is indeed far more complicated than with the explicit scheme (7.49): So why have we used this complicated way of discretizing the second derivative via averaging information at both times $n\Delta t$ and $(n+1)\Delta t$? The answer is that the Crank-Nicolson scheme is stable **for all** r; in this way, we say the

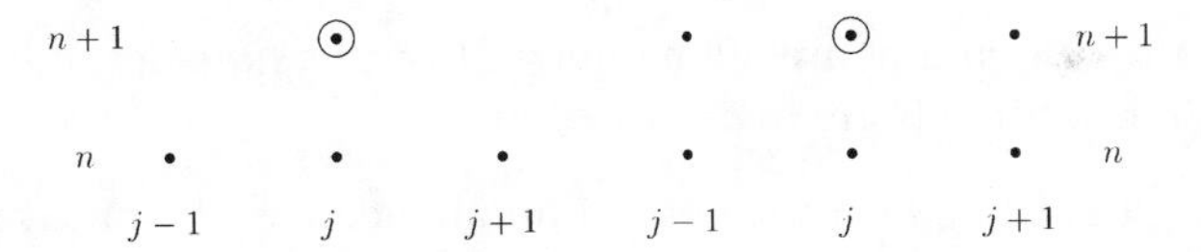

Figure 7.7. Two stencils associated with schemes (7.49) and (7.51), respectively. Discrete time is vertical and space is horizontal. The scheme gives the value at the circled point in terms of values at the other points. The first scheme is explicit as the value at the circled point only depends on values at the previous time. The second scheme (Crank-Nicolson) is implicit, as the solution value at point j at time $n+1$ also depends on the solution at the neighboring points $j-1, j+1$ at time $n+1$.

scheme is **unconditionally stable**. In Exercise **7.35** you are asked to verify this and also implement the scheme for the example of Exercise **7.34**. This time the implementation is not as straightforward and will require solving a linear system of equations.

7.11. Addendum: The Schrödinger Equation

"Where did we get that from?
Nowhere. It is not possible to derive it from anything you know.
It came out of the mind of Schrödinger,
invented in his struggle to find an understanding of the
experimental observations of the real world."
- Richard Feynman[a]

[a]The iconic, Nobel Prize winning, American theoretical physicist **Richard Feynman** (1918–1988) is known for his pioneering work in the path integral formulation of quantum mechanics, quantum electrodynamics, and particle physics. He is equally known as an expositor and this quote is taken from the third volume of *The Feynman Lectures on Physics*, Addison Wesley.

It would be impossible for a book on PDEs to not at least mention the Schrödinger[19] equation; it is **the** equation which governs almost all phenomena at the microscopic (atomic) level. So why do we present it as an addendum to the chapter on the diffusion equation? First, it is beyond the scope of this text to include sufficient material to afford the equation its own (thoroughly deserving) chapter. Second, it bears an uncanny resemblance to the diffusion equation. In one space dimension, the Schrödinger equation for $u(x,t)$ is

$$i\hbar\, u_t \;=\; -\frac{\hbar^2}{2m}u_{xx} \;+\; V(x)u, \tag{7.52}$$

while in 3D, the Schrödinger equation for $u(\mathbf{x},t)$ is

$$i\hbar\, u_t \;=\; -\frac{\hbar^2}{2m}\Delta u \;+\; V(\mathbf{x})u. \tag{7.53}$$

[19]**Erwin Rudolf Josef Alexander Schrödinger** (1887–1961) was an Austrian physicist who was one of the founding fathers of quantum mechanics. He was awarded the Nobel Prize for physics in 1933.

More precisely, these equations are often referred to as the *time-dependent* Schrödinger equations. Several comments are in order:

- Here the i is the complex number $\sqrt{-1}$ and hence u is complex valued. Without the presence of this i, the equation (modulo the V term and constants) is simply the diffusion equation.
- Physically, $u(\mathbf{x}, t)$ represents the complex-valued **wave function** associated with the motion of an elementary particle with its complex modulus squared giving the instantaneous probability density of being at position $\mathbf{x}$ at time t. More precisely, given a region Ω of 3D space,

$$\iiint_{\Omega} |u(\mathbf{x}, t)|^2 \, d\mathbf{x}$$

represents **the probability that at time** t **the particle is located in** Ω. Hence, the **expected position** of the particle at time t is the vector-valued integral

$$\iiint_{\Omega} \mathbf{x}|u(\mathbf{x}, t)|^2 \, d\mathbf{x}.$$

Given this interpretation of $u(\mathbf{x}, t)$, one would expect that (cf. Exercise **7.36**) if initially $\iiint_{\mathbb{R}^3} |u(\mathbf{x}, 0)|^2 \, d\mathbf{x} = 1$, then for all $t > 0$, $\iiint_{\mathbb{R}^3} |u(\mathbf{x}, t)|^2 \, d\mathbf{x} = 1$.
- The parameter m is the **mass** of the elementary particle. The fundamental constant $\hbar$ is the reduced **Planck constant**[20] (Planck's constant h divided by 2π) and relates the energy of the particle to its frequency.
- The function V is called the potential and represents the effect on the particle from another particle or collection of particles. For example, in the **hydrogen atom** there is one electron with charge e moving around a nucleus comprising exactly one proton. If we consider the center of the nucleus (i.e., the proton) to be fixed at the origin, Coulomb's Law (cf. Sections 10.6.1 and 10.6.2) gives the potential resulting from the electron of change e at $\mathbf{x}$ (with $r = |\mathbf{x}|$) to be $V(r) = -k_e \frac{e^2}{r}$, where k_e is Coulomb's constant defined in (10.55). The resulting Schrödinger equation for the motion of the electron is

$$i\hbar u_t = -\frac{\hbar^2}{2m} \Delta u - k_e \frac{e^2}{r} u,$$

where m denotes the reduced mass of the atom; that is, $m = \frac{m_e m_n}{m_2 + m_n}$ with m_e, m_n denoting, respectively, the mass of the electron and nucleus (i.e., the proton).
- If we set $V \equiv 0$, then the equation is often referred to as the **free particle** Schrödinger equation.
- In Newtonian physics, the equation of motion for a particle of mass m under a potential V is given via Newton's Second Law and results in the ODE $m\ddot{\mathbf{x}} = -\nabla V(\mathbf{x})$ for the position $\mathbf{x}(t)$ of the particle at time t. In quantum mechanics, we never know exactly the position of the particle in space; rather, we can only determine

[20]**Max Karl Ernst Ludwig Planck** (1858–1947) was a German physicist whose discovery of energy quanta set the foundations for quantum mechanics and earned him the Nobel Prize. Planck's constant has dimensions of energy times time and appears in the famous simple relation (sometimes called the Planck-Einstein relation) $E = h\nu$ relating the particle energy E to the particle frequency ν. In the SI units Planck's constant h is $6.62607015 \times 10^{-34}$ joule-seconds.

its probability density of being at position $\mathbf{x}$ at time t. Hence the equation which describes the "motion" of the particle changes from a simple ODE to a PDE involving both space and time as independent variables: This PDE is the Schrödinger equation.

Given these remarks, it should not come as a surprise that the Schrödinger equation behaves very differently from the diffusion equation. Indeed, solutions behave more like solutions to a wave equation and this is a direct consequence of the presence of the complex number i. This is actually not that hard to see as the exponential $e^{-(\cdot)^2}$, which is paramount to diffusion, behaves very different from $e^{\pm i(\cdot)^2}$: The former is associated with an (infinite) decay while the latter is oscillatory, traversing the unit circle in the complex plane. One can be more precise by using the Fourier transform to solve the 1D IVP for the free particle Schrödinger equation. This IVP can be conveniently written as

$$u_t = iku_{xx}, \qquad u(x,0) = g(x), \tag{7.54}$$

for some constant $k > 0$. The solution for $t > 0$ (cf. Exercise **7.37**) is

$$u(x,t) = \frac{1}{\sqrt{4\pi ikt}} \int_{-\infty}^{\infty} e^{i\frac{(x-y)^2}{4kt}} g(y)\, dy. \tag{7.55}$$

This is identical to the solution to the diffusion equation modulo the presence of an i in the radical and a $-i$ in the exponential. The 3D Fourier transform can be used to solve the 3D free particle Schrödinger equation (cf. Exercise **7.38**).

Even from these explicit solution formulas, it is hard to get any sense of why the equation proves so fundamental in quantum mechanics. In Section 12.11 on the method of separation of variables, we will show how the Schrödinger equation is able to predict one of the fundamental results of quantum mechanics: the quantized energy levels of the hydrogen atom. The curious reader is encouraged to explore either the online *Feynman Lectures on Physics*[21] or the excellent undergraduate text of **Griffiths**.[22]

7.12. Chapter Summary

- The diffusion equation (also known as the heat equation) is a ubiquitous time-dependent second-order PDE, which in one space dimension takes the form $u_t = \alpha u_{xx}$. In N space dimensions, it is given by $u_t = \alpha\Delta u$.
- The PDE models the diffusion of a quantity (for example, heat) in some medium. It is also the equation for the evolution of the probability density function associated with a Brownian motion, and as such, forms one of the cornerstones of probability and stochastic processes.
- In order for an initial value problem to be well posed, one needs to supplement the PDE with **one** initial condition which we can interpret as the initial distribution, concentration, or probability. The initial value problem can be solved via

[21] See Volume 3 of `https://www.feynmanlectures.caltech.edu/`.

[22] **David J. Griffiths**, Introduction to Quantum Mechanics, second edition, Pearson, 2004.

the Fourier transform to yield the formula

$$u(x,t) = \frac{1}{\sqrt{4\pi\alpha t}} \int_{-\infty}^{\infty} e^{-\frac{(x-y)^2}{4\alpha t}}\, g(y)\, dy$$

which can be written as a convolution

$$u(x,t) = \big(\Phi(\cdot,t) * g\big)(x) = \int_{-\infty}^{\infty} \Phi(x-y,t)\, g(y)\, dy$$

of the initial data with the fundamental solution $\Phi(x,t)$ where

$$\Phi(x,t) = \frac{1}{\sqrt{4\pi\alpha t}} e^{-\frac{x^2}{4\alpha t}}.$$

The function $\Phi(x,t)$ is itself a solution to the diffusion equation for all x and $t > 0$. As $t \to 0^+$, $\Phi(x,t)$ (as a function of x) tends to δ_0 in the sense of distributions. This fact allows us to prove that the solution formula remains continuous up to the $t = 0$ axis and attains the right boundary values.

- From this solution formula, we immediately see some basic properties associated with diffusion, in particular, infinite propagation speed and the Maximum Principle. We also see the natural connection with probability. In dimension $N = 1$, for any fixed x and time t, $\Phi(x-y,t)$, as a function of y, is the probability density associated with a **normal distribution** with mean x and variance $2\alpha t$. Thus, the solution

$$u(x,t) = \int_{-\infty}^{\infty} \Phi(x-y,t)\, g(y)\, dy$$

states that **to find** u **at position** x **and time** t we must **average** the initial density $g(y)$ with **a certain weight**. The weight is this bell-shaped curve centered at $y = x$ and with "fatness" proportional to t, scaled by the diffusion constant. An analogous interpretation can be made in higher space dimensions.

- A natural derivation and interpretation of the diffusion equation is via a particular **limit of random walks, called Brownian motion**. More precisely, the diffusion equation is the evolution equation for $p(x,t)$, the probability density function at position x and time t of Brownian motion. Brownian motion can be realized as a limit of Δx, Δt random walks wherein the step size Δx and time step Δt both tend to zero with $\frac{(\Delta x)^2}{\Delta t}$ fixed, but the number of walks n tends to infinity with $n\Delta t$ remaining finite. In this respect, one can derive the fundamental solution formula from one of the basic results of probability, the Central Limit Theorem.

- In N space dimensions the IVP problem for the diffusion equation can be solved by the Fourier transform to yield

$$u(\mathbf{x},t) = \Big(\Phi(\cdot,t) * g(\cdot)\Big)(\mathbf{x}) = \frac{1}{(4\pi\alpha t)^{N/2}} \int \cdots \int_{\mathbb{R}^N} \exp\left(-\frac{|\mathbf{x}-\mathbf{y}|^2}{4\alpha t}\right) g(\mathbf{y})\, d\mathbf{y},$$

where

$$\Phi(\mathbf{x},t) := \frac{1}{(4\pi\alpha t)^{N/2}} \exp\left(-\frac{|\mathbf{x}|^2}{4\alpha t}\right).$$

Exercises

7.1 Redo the derivation of the diffusion equation specifically for the temperature in an infinitely long thin bar made of some homogeneous material, which is insulated along its top and bottom. You may model the bar as one dimensional with $u(x,t)$ denoting its temperature at position x and time t. You may assume no heat may enter or escape along the bar. The use of the Divergence Theorem in passing from a boundary integral to a bulk integral will now be replaced by the vanilla Fundamental Theorem of Calculus. What is particularly instructive about this derivation is to see how α relates to the mass density, the specific heat, and the thermal conductivity of the material.[23] Give the SI physical units for all these quantities as well as for u and α. What is the physical name for α and what is its value for glass, gold, and copper?

7.2 Show by direct differentiation that $\Phi(x,t)$ solves the diffusion equation $u_t = \alpha u_{xx}$ for $x \in \mathbb{R}$ and $t > 0$.

7.3 Use the Fourier transform to solve the diffusion equation with a source: $u_t - u_{xx} = f(x,t)$, $u(x,0) = g(x)$, on $x \in (-\infty, \infty)$. Here we assume f and g are given integrable functions. We also assume the hidden assumption (why hidden?) that $u(x,t) \to 0$ and $u_x(x,t) \to 0$ as $|x| \to \infty$. Make sure your solution agrees with (7.37).

7.4 Here you will present another proof of the fact that $\lim_{t\to 0^+} \Phi(x,t) = \delta_0$ *in the sense of distributions*, where δ_0 is the delta function with concentration at $x = 0$. Precisely, you will prove that

$$\lim_{t\to 0^+} \int_{-\infty}^{\infty} \Phi(x-y,t)\,\phi(y)\,dy = \phi(x), \qquad \text{for any } \phi \in C_c^\infty(\mathbb{R}).$$

To this end, consider the function

$$Q(x,t) = \frac{1}{2} + \frac{1}{\sqrt{\pi}} \int_0^{\frac{x}{\sqrt{4\alpha t}}} e^{-y^2}\,dy.$$

Compute $\frac{\partial Q}{\partial x}$ and $Q(x,0)$. Then go from there.

7.5 **(Similarity Solutions and Another Route to the Fundamental Solution Φ)** Many important linear and nonlinear PDEs have **similarity solutions**. These are classes of solutions which depend on some **particular grouping** of the independent variables (rather than depending separately on each independent variable). One can often infer these particular groupings via **symmetries (or invariances)** in the PDE structure. A canonical PDE to explore these notions is the diffusion equation. This exercise, essentially taken from Evans [**10**], will result in an alternate derivation of the fundamental solution.

(a) Show that if $u(x,t)$ is a smooth solution to the diffusion equation $u_t = u_{xx}$, then so is $u_\lambda(x,t) := u(\lambda x, \lambda^2 t)$ for each $\lambda \in \mathbb{R}$.

[23] Look up the latter two online.

(b) Based upon this scaling invariance discuss why the ratio $\frac{x^2}{t}$ might play an important role.
(c) Let us now look for a solution of the form

$$u(x,t) = v\left(\frac{x^2}{t}\right), \tag{7.56}$$

for some function of one variable $v(s)$ defined for $s > 0$. Start by showing that if u has the form (7.56), then $v(s)$ must solve the ODE $4sv''(s) + (2+s)v'(s) = 0$.
(d) Find $v(s)$, the general solution to this ODE.
(e) Next, show that if $u(x,t)$ is a C^∞ solution to the diffusion equation, so is $u_x(x,t)$. Use this fact to differentiate (7.56) with respect to x and discover a new solution to the diffusion equation. What solution is it? Note the connection with the previous exercise.

7.6 Consider a random walk where at each integer multiple of Δt we stay where we are with probability 1/2, we move Δx to the right with probability 1/4, and we move Δx to the left with probability 1/4. Write down the equation for the probability of being at any place at time $t + \Delta t$ in terms of probabilities at the earlier time. Now let Δx and Δt go to zero in an appropriate way and derive a PDE for the probability of being at x at time t. How is this different from the diffusion equation derived in Section 7.3? Explain why this makes sense.

7.7 Consider a random walk with $\Delta x = 6$ and $\Delta t = 2$, where at each time step, we move Δx to the left or right, each with probability 1/2. Suppose we implement this random walk 10,000 times. Can we use the solution to a diffusion equation to estimate the probability that where we end up is to the left of $x = 100$? If so, which diffusion equation? Write down an integral which estimates this probability and explain your reasoning.

7.8 Consider the initial value problem for the diffusion equation on the semi-infinite line $[0,\infty)$ with a fixed end point: $u_t = \alpha u_{xx}$ with $u(0,t) = 0$ for $t > 0$ and $u(x,0) = g(x)$ for $x \geq 0$. Here, g is a continuous function on $[0,\infty)$ such that $g(0) = 0$. Find the solution $u(x,t)$. **Hint:** Recall what we did for the wave equation. The same trick will work here.

7.9 **(Conservation and Dissipation)**
(a) Let $u(x,t)$ be a C^∞ solution to (7.1) with $g \geq 0$ integrable. Prove the conservation law that for all $t > 0$,

$$\int_{-\infty}^{\infty} u(x,t)\,dx = \int_{-\infty}^{\infty} g(x)\,dx.$$

You may assume that for any $t > 0$, $\lim_{x\to-\infty} u_x(x,t) = \lim_{x\to\infty} u_x(x,t) = 0$. Thus, in the context of heat flow, we can conclude that the total heat is conserved.
(b) Now consider a notion of "energy" at time t, the total amount of flux

$$E(t) = \int_{-\infty}^{\infty} u_x^2(x,t)\,dx.$$

Prove that E is not conserved but rather it decreases with time; i.e., $E'(t) < 0$. Thus, energy is not conserved, but rather, the diffusion equation **dissipates** energy. Here, you may assume that for any $t > 0$, all partial derivatives of u also tend to 0 as $x \to \pm\infty$.

7.10 Prove the following **stability** result for the initial value problem (7.5): Fix $t > 0$ and $\epsilon > 0$. Let g_1 and g_2 be two continuous, bounded, integrable functions such that $\max_{x\in\mathbb{R}} |g_1(x) - g_2(x)| < \epsilon$. Let u_1, u_2 be the solutions to (7.5) with, respectively, $g(x) = g_1(x)$ and $g_2(x)$. Prove that

$$\max_{x\in\mathbb{R}} |u_1(x,t) - u_2(x,t)| < \epsilon.$$

7.11 (**Instability for the Backward Diffusion Equation**) For each $n \in \mathbb{N}$, let $g_n(x) = \frac{\sin nx}{n}$. For each n, find a solution to the IVP for the backward diffusion equation

$$\begin{cases} u_t = -\alpha u_{xx} & \text{on } \mathbb{R}\times(0,\infty), \\ u(x,0) = g_n(x) & \text{on } \mathbb{R}. \end{cases}$$

Hint: Look for a separated solution $u_n(x,t)$ which is some function of x times some other function of t. Show that this class of solutions gives instability: That is, for n large $g_n(x)$ is uniformly close to $g(x) := 0$ but the associated solutions at time $t = 1$, $u_n(x,1)$, are (uniformly) "not close" to $u \equiv 0$.

7.12 (**A Hint at Nonuniqueness**) Consider the diffusion equation on the half-line with Dirichlet boundary condition

$$\begin{cases} u_t = u_{xx} & \text{for } 0 < x < \infty,\ t > 0, \\ u(0,t) = 0 & \text{for } t > 0, \\ u(x,0) = 0 & \text{for } x \geq 0. \end{cases}$$

Clearly $u(x,t) \equiv 0$ is a solution to this problem, but show that $u(x,t) = \frac{x}{t^{3/2}}e^{-\frac{x^2}{4t}}$ is also a solution except at $x = 0, t = 0$. What type of discontinuity does it have at $(0,0)$?

7.13 Consider the 1D diffusion equation with a drift: $v_t = v_{xx} + dv_x$, $v(x,0) = g(x)$, where $d \in \mathbb{R}$. Show that if u solves the diffusion equation $u_t = u_{xx}$, then $v(x,t) := u(x+dt,t)$ solves the above diffusion equation with a drift. Use this to write down the solution formula.

7.14 Consider a random walk with a "small" drift: Fix a grid size Δx and a time step Δt. Let $\Delta > 0$ be some small number. Then consider the random walk where at each multiple of Δt, **we move Δx to the right with probability $\frac{1}{2} + \delta$ and Δx to the left with probability $\frac{1}{2} - \delta$.** Suppose at $t = 0$ we are at the origin $x = 0$. Suppose we **enforce** that for some **fixed** $\sigma > 0$ and $\alpha > 0$, $\sigma^2 = c\frac{(\Delta x)^2}{2\Delta t}$ and $\delta = \alpha\Delta x$. By considering the limit in which $\Delta x \to 0$, derive the PDE for $p(x,t)$, the probability density function associated with being at point x at time t.

7.15 Consider a random walk with a "finite" drift. For example, in the previous question, take $\delta = 1/4$. This means we move to the right with probability $3/4$ and to the left with probability $1/4$. What happens if we take the limit with $\frac{(\Delta x)^2}{\Delta t}$ fixed?

What happens if we take the limit with $\frac{\Delta x}{\Delta t}$ fixed? What is your conclusion: Can we capture a Brownian motion with a finite drift? For those who have taken a course in probability, make an analogy with the Law of Large Numbers versus the Central Limit Theorem.

7.16 (a) Derive the 1D diffusion equation with source term (7.36) by repeating the derivation from Exercise **7.1**, in this case where there is an internal heat source given by $f(x, t)$. What are the units and physical meaning of $f(x, t)$?
(b) Verify Duhamel's Principle for the diffusion equation by showing that (7.40) satisfies (7.38). Then use this to derive the solution formula (7.37).

7.17 Consider a 2D random walk on a rectangular lattice. That is, we decide on grid sizes Δx and Δy and then divide the plane into discrete points $(n\Delta x, m\Delta y)$ where n and m are integers. Now we perform the following experiment: Suppose at $t = 0$ we are at the grid point $(0, 0)$. At time Δt, we move with equal probability to the right **or** to the left **or** straight up **or** straight down (these are four possibilities here so the probability of any one is 1/4). We repeat this at $2\Delta t$ and so forth. Now take $\Delta x = \Delta y$. By following the same steps as we did for a 1D random walk, show that there is a natural relationship between Δt and $\Delta x = \Delta y$ for which we can let Δt tend to 0 and obtain a 2D diffusion equation for the probability density function $p(x, y, t)$ associated with a point (x, y) at time t.

7.18 Assuming you have done Exercise **6.16**, use the Laplace transform to solve the following BVP for the diffusion equation on $[0, l]$: $u_t = \alpha u_{xx}$, $u(0, t) = u(1, t)$, $t \geq 0$, and $u(x, 0) = \sin \frac{\pi x}{l}$, $x \in [0, l]$.

7.19 Use the 3D Fourier transform to solve the diffusion equation in 3D. Show that the always positive 3D fundamental solution/heat kernel Φ integrates to 1 over $\mathbb{R}^3$ and for any $t > 0$ satisfies the diffusion equation.

7.20 For students who have taken a first course in analysis, let $g(x)$ be a bounded, continuous, integrable function on $\mathbb{R}$ and define $u(x, t)$ by (7.10). Prove that (7.9) holds true.

7.21 Consider the IVP for the diffusion equation (7.26) with the unbounded, nonintegrable initial data $u(x, 0) = e^{-x}$ for $x \in \mathbb{R}$. Technically speaking, our derivation of the solution formula via the Fourier transform should not hold true. However, plug this data into the solution formula, explicitly perform the integration, and show that the simple function of $u(x, t)$ does indeed solve the IVP. What can you say about the structure of this solution? Plot several profiles for different t and make some conclusions.

7.22 (**An Illustration of the Ill-Posedness of the Backward Diffusion Equation**) For each n consider the functions $u_n(x, t) = \frac{\sin nx}{n} e^{-n^2 t}$. Check that for each n, u_n solves the diffusion equation ($\alpha = 1$) for all $x \in \mathbb{R}$ **and all** $t \in \mathbb{R}$. What happens to the values of $u_n(x, 0)$ as $n \to \infty$? What happens to the values of $u_n(x, -1)$ as $n \to \infty$. Make some conclusions with regard to stability.

7.23 Let g satisfy all the hypotheses of Theorem 7.1 except for the fact that g has one jump discontinuity at x_0. Modify slightly the proof of Theorem 7.1 to show

$$\lim_{t\to 0+} u(x_0,t) = \frac{1}{2}[g(x_0+) + g(x_0-)],$$

where $g(x_0+) := \lim_{x\to x_0^+} g(x)$ and $g(x_0-) := \lim_{x\to x_0^-} g(x)$.

7.24 Under the same assumptions as Theorem 7.1, prove the stronger result (7.9).

7.25 Use the Maximum Principle Theorem, Theorem 7.2, to prove **uniqueness** to the IVP/BVP (7.30). Formulate a precise statement of stability for the problem (7.30) with respect to small changes in the initial data g. Prove your statement.

7.26 (**Comparison Principle**) Use the Maximum Principle Theorem, Theorem 7.2, to prove the following comparison principle for the diffusion equation. Suppose $u(x,t)$ and $v(x,t)$ are solutions to the 1D diffusion equation for $0 < x < l$ and $t > 0$ such that $u(x,0) \le v(x,0)$ for all $x \in [0,l]$ and $u(0,t) \le v(0,t)$, $u(l,t) \le v(l,t)$ for all $t > 0$. Prove that $u(x,t) \le v(x,t)$ for all $x \in [0,l]$ and $t \ge 0$. Interpret the result in the context of temperature in a bar.

7.27 Let us find a solution to $u_t = u_{xx}$ on $\mathbb{R}\times(0,\infty)$ with $u(x,0) = x^2$ without using the solution formula (7.6) (i.e., convolution with $\Phi(x,t)$). Argue as follows. Suppose u is a C^∞ solution on $\mathbb{R}\times(0,\infty)$. Show that $v(x,t) := u_{xxx}(x,t)$ is also a solution to the diffusion equation with initial data $v(x,0) \equiv 0$. Let us take $v(x,t) \equiv 0$ as the unique solution (up to satisfying the growth assumption). This means that $u_{xxx}(x,t) \equiv 0$. Show that this means that $u(x,t) = a(t)x^2 + b(t)x + c(t)$, for some smooth functions $a(t), b(t), c(t)$ defined on $t \ge 0$ with $a(0) = 1, b(0) = 0$, and $c(0) = 0$. Plug this form of the solution into the diffusion equation to determine $a(t), b(t)$, and $c(t)$.

7.28 Consider the IVP $u_t = u_{xx}$ on $\mathbb{R}\times(0,\infty)$ with $u(x,0) = e^{\frac{x^2}{4}}$.

(a) Show directly that $u(x,t) = \frac{1}{\sqrt{1-t}}\, e^{\frac{x^2}{4(1-t)}}$ is a solution which blows up at $t = 1$.

(b) Reconcile the above with the solution formula (7.6). In particular, why does the solution formula blow up at $t = 1$?

7.29 Consider the IVP (7.5) where g is continuous and bounded on $\mathbb{R}$ with, for some $a, b \in \mathbb{R}$, $\lim_{x\to-\infty} g(x) = a$ and $\lim_{x\to\infty} g(x) = b$. First show that the solution formula (7.6) is defined for all $x \in \mathbb{R}$ and $t > 0$. Then show that for any $x \in \mathbb{R}$, $\lim_{t\to\infty} u(x,t)$ exists, and find its value.

7.30 (**Weierstrass Approximation Theorem**) Let f be a continuous function on an interval $[a,b]$. Via the following steps, prove that for every $\epsilon > 0$ there exists a polynomial $P(x)$ such that

$$\max_{x\in[a,b]} |f(x) - P(x)| < \epsilon.$$

(a) Extend f to all of $\mathbb{R}$ so that the extension is a bounded continuous function with compact support. Note that the extended f is uniformly continuous on $\mathbb{R}$.

(b) Note that the convergence of Theorem 7.1 is uniform in x.

(c) Note that the Taylor series for e^{-z^2} converges uniformly on **any** closed, finite interval.

7.31 Let Ω be a bounded domain in $\mathbb{R}^3$ and consider the BVP/IVP

$$\begin{cases} u_t = \alpha \Delta u, & \mathbf{x} \in \Omega,\ t \in (0, T], \\ u(\mathbf{x}, t) = 0, & \mathbf{x} \in \partial\Omega,\ t \in (0, T], \\ u(\mathbf{x}, 0) = g(\mathbf{x}), & \mathbf{x} \in \Omega. \end{cases} \tag{7.57}$$

Prove that for all $t \in [0, T]$, $\frac{d}{dt} \iiint_\Omega u^2(\mathbf{x}, t)\, d\mathbf{x} \leq 0$. Conclude that there exists at most one solution to (7.57). Hint: Multiply the PDE by u, integrate over Ω, and use Green's First Identity.

7.32 (**Maximum Principle in 3D**) Let $u(\mathbf{x}, t)$ be a smooth solution to the BVP/IVP (7.57). For $T > 0$ consider the tube in space-time $\Omega_T = \Omega \times [0, T]$ and let $\mathcal{S}$ denotes the parts of $\partial\Omega_T$ except for the top: $\Omega \times \{t = T\}$. Prove that

$$\max_{\mathcal{S}} u(\mathbf{x}, t) = \max_{\Omega_T} u(\mathbf{x}, t).$$

Conclude that there exists at most one solution to (7.57).

7.33 (**Duhamel's Principle for the 1D Wave Equation**) Follow the reasoning of Section 7.8.1 to give a physical explanation for Duhamel's Principle for the 1D wave equation (cf. Section 3.6.1).

7.34 Here you will numerically approximate the solution to (7.48) with $\alpha = 1$ and $g(x) = \sin \pi x$ for $0 \leq x \leq 1$. Use the explicit scheme (7.49) with $\Delta x = 0.1$ and $r = 4/9$. Use 10 time steps to approximate the solution at a certain time (what time?). Now repeat with $r = 2/3$ and explain your findings in terms of the stability condition.

7.35 **(a)** Perform the von Neumann stability analysis on the Crank-Nicolson scheme (7.51) to show that it is stable for all r.
(b) Repeat Exercise **7.34** using the (7.51). You may use any software to invert the matrix.

7.36 Show that for the 1D Schrödinger equation, the integral $\int_{-\infty}^{\infty} |u(x, t)|^2\, dx$ is constant in time.

7.37 Use the 1D Fourier transform to solve (7.54), the 1D free particle Schrödinger equation, and arrive at (7.55).

7.38 Use the 3D Fourier transform to solve the IVP for the 3D free particle Schrödinger equation $i\hbar u_t = -\frac{\hbar^2}{2m}\Delta u$ with $u(\mathbf{x}, 0) = g(\mathbf{x})$.

7.39 (**Probability Current Density and the Continuity Equation from the Schrödinger Equation**) Let $u(\mathbf{x}, t)$ solve the 3D Schrödinger equation (7.53). Define the probability density and the probability current density (or probability flux) by

$$\rho(\mathbf{x}, t) := |u|^2 = u\overline{u} \qquad \text{and} \qquad \mathbf{j}(\mathbf{x}, t) := \frac{\hbar}{2mi}\left(\overline{u}\nabla u - u\nabla\overline{u}\right),$$

respectively. Following Section 6.1, $\overline{u}$ denotes the complex conjugate of u.
(a) Show that $\mathbf{j}$ is real valued by showing that it is precisely the real part of $\frac{\hbar}{mi}(\overline{u}\nabla u)$. Use this to motivate why $\mathbf{j}$ is called the probability current density (or probability flux).

(b) Show that ρ and $\mathbf{j}$ satisfy the continuity equation (discussed in Sections 2.2.6 and 2.8)

$$\rho_t + \operatorname{div}\mathbf{j} = 0.$$

Hint: Derive two new PDEs by (i) multiplying both sides of the Schrödinger equation by $\overline{u}$ and (ii) taking the complex conjugate of both sides of the Schrödinger equation and then multiplying both sides by u. Add these two new equations.

7.40 Following the short discussion of Sections 3.12.3 and 3.12.4, find the dispersion relation for both the 1D diffusion equation and the 1D Schrödinger equation (7.54).

7.41 (**The Hopf-Cole Transformation**) In some situations, one is able to make a change of dependent variable to transform a nonlinear PDE into a linear PDE. The Hopf-Cole transformation and its variants apply to many such cases.
(a) Show that Hopf-Cole transformation $v(x,t) := e^{-\frac{a}{\alpha}u(x,t)}$ transforms the diffusion equation with nonlinear term

$$u_t - \alpha u_{xx} + a u_x^2 = 0, \qquad \alpha > 0, a \neq 0,$$

into the diffusion equation $v_t - \alpha v_{xx} = 0$. Use this fact to explicitly solve the IVP problem for the nonlinear equation with initial data $u(x,0) = g(x)$.
(b) Research online to find a similar change of variables which transforms the Burgers equation with viscosity $u_t + uu_x = \epsilon u_{xx}$ into the diffusion equation.

7.42 (**Nonlinear Diffusion and the Porous Medium Equation**) Let $m > 1$ and consider nonnegative solutions ($u \geq 0$) of the porous medium equation

$$u_t = \Delta(u^m).$$

(a) Show that the equation can be rewritten as $u_t = \operatorname{div}\left(mu^{m-1}\nabla u\right)$. Hence it can be thought of as a diffusion equation with a nonlinear diffusion constant, that is, a diffusion constant which depends on the solution itself.
(b) Let the space dimension be one and show that

$$u(x,t) = t^{-\frac{1}{m+1}}\left(\max\left\{0,\, 1 - \frac{m-1}{2(m+1)m}\frac{|x|^2}{t^{\frac{2}{m+1}}}\right\}\right)^{\frac{1}{m-1}}$$

is a compactly supported (in x) solution to $u_t = (mu^{m-1}u_x)_x$ for $x \in \mathbb{R}$ and $t > 0$. This important solution is known as **Barenblatt's similarity solution**. What happens to this solution as $t \to 0^+$?
(c) Using any software, plot the solution for increasing times. Now observe the following amazing fact: Unlike the diffusion equation wherein $m = 1$, the porous medium equation exhibits **finite** propagation speed.

7.43 (**Anisotropic Diffusion in Image Processing**)
(a) Revisit the derivation of the diffusion equation (say, in 3D) to find that in an inhomogeneous medium the diffusion equation becomes $u_t = \operatorname{div}(\alpha(\mathbf{x},t)\nabla u)$, where $\alpha(\mathbf{x},t)$ is some function representing the medium. Describe a few physical examples for heat flow wherein α varies with $\mathbf{x}$ and t. Explain the term *anisotropy* in these contexts.

(b) As in the previous exercise, an important **nonlinear** case occurs when α depends on u and/or ∇u. Suppose $\alpha(u, \nabla u) = \frac{1}{|\nabla u|}$ which gives rise to the nonlinear anisotropic diffusion equation

$$u_t = \operatorname{div}\left(\frac{\nabla u}{|\nabla u|}\right).$$

This PDE is important in image processing and note, from Section 2.5.7, the relationship with curvature of level sets of u. Suppose we work in 2D. Explore qualitatively the effect of the associated flow on an initial black and white image $g(x, y)$ (where $g(x, y)$ denotes intensity of an instantaneous pixel at (x, y)). In particular, note the diffusive effects on noice removal. Also note that that sharp changes in u (places in the image where $|\nabla u|$ is large) corresponds to *edges* in the image. Explain the flow (via the PDE) in terms of **edge preservation** and compare qualitatively with the application of the regular (isotropic) diffusion equation to initial g.
(c) Consider the following two choices of α:

$$\alpha(u, \nabla u) = e^{-\left(\frac{|\nabla u|}{K}\right)^2} \quad \text{and} \quad \alpha(u, \nabla u) = \left(1 + \left(\frac{|\nabla u|}{K}\right)^2\right)^{-1}.$$

They give rise to a PDE known as the **Perona-Malik equation**. Give heuristic explanations for the effects of these two choices of α on an initial image in terms of edge preservation and image sharpening. Explain the role of the constant $K > 0$. By researching online, view simulations of the PDE highlighting Perona-Malik diffusion.

7.44 (**Anisotropy Diffusion in Medical Imaging: Diffusion MRI**) By researching online first explain the basic mechanism behind an MRI (magnetic resonance imaging) sequence. Then explain the process of diffusion MRI and the role played by the anisotropic diffusion equation.

7.45 (**The Black-Scholes equation**) A famous variant of the diffusion equation from option pricing in mathematical finance is the Black-Scholes equation. Here $u(S, t)$ solves

$$u_t + \frac{1}{2}\sigma^2 S^2 u_{SS} + rSu_S - ru = 0$$

wherein the independent variables are the price S of an underlying asset (security or commodity) and time t. The constants σ and r are, respectively, the *volatility* of the asset price and the *risk-free interest rate*. The dependent variable $u(S, t)$ is the price of a *European option* based upon an underlying asset with price S at time t. Note that this is a modified (ill-posed) backward equation; however, we solve the equation for $t < T$ with a final condition at $t = T$. The final time T is the *expiration date* of the option and for a *call option*, for example, the final condition is $u(S, T) = \max\{S - K, 0\}$, where K is the *strike price*, the fixed price at which the owner of the option can buy the underlying asset.
(a) First research online the definition of a European put or call option, explain all the italicized terms above, and explain why we work backwards in time specifying the final condition $u(S, T) = \max\{S - K, 0\}$.
(b) Solve the Black-Scholes equation via the following steps to transform it into the regular (forward) diffusion equation.

(i) (Removing the "ru" term in the equation) Multiply the equation by the integrating factor e^{-rt} and find a PDE for $\tilde{u} := e^{-rt}u$.

(ii) (Change of variables) Make the change of variables $x = \log S$ and $s = T - t$ to transform the equation into $\tilde{u}_s = \frac{1}{2}\sigma^2\tilde{u}_{xx} + \left(r - \frac{1}{2}\sigma^2\right)\tilde{u}_x$.

(iii) (One more integrating factor) Choose values of α and β for which the PDE for $v(x,s) := e^{\alpha x+\beta s}u(x,s)$ becomes the regular diffusion equation $v_s = \frac{1}{2}\sigma^2 v_{xx}$.

(iv) Now solve the Black-Scholes equation.

7.46 (**A Discrete Version of the Black-Scholes Equation and Random Walks**) This exercise is for those students who wish to venture further into the world of mathematical finance by exploring a discrete approximation to the Black-Scholes equation for option pricing. Let us assume a random walk model for the stock (asset) price. Fix an expiration date T and consider time step size $h = \frac{T}{N}$, for some integer N and discrete times $t = nh$ for $n = 0, \dots, N$. A nondividend paying stock at time 0, whose value is denoted by S_0, evolves from value S_t at time t to value S_{t+h} at time $t + h$ via a coin flip. If the coin lands on heads, $S_{t+h} = c_1S_t$, and if it lands on tails, $S_{t+h} = c_2S_t$, where $c_1 = e^{(rh+\sigma\sqrt{h})}$ and $c_2 = e^{(rh-\sigma\sqrt{h})}$.

(a) First explain the financial interpretations of the parameters c_1 and c_2, and then read about the **Cox-Ross-Rubinstein binomial model**.[24] Explain how the assumption of *risk neutrality* leads to the discrete scheme for the call option price:

$$u(S_t, t) = e^{-rh}\left(\tilde{p}\,u(c_1S_t, t+h) + \tilde{q}\,u(c_2S_t, t+h)\right) \tag{7.58}$$

where $\tilde{p} = \frac{e^{rh}-c_2}{c_1-c_2}$ and $\tilde{q} = 1 - \tilde{p}$ and we work backwards from the known option value $u(S_T, T) = \max\{S_T - K, 0\}$.

(b) Now we begin to address what happens as $N \to \infty$ (equivalently, $h \to 0$) in this binomial model. Use the recursive evolution formula above for the underlying stock price and the scheme (7.58) to price an *at-the-money* call option; i.e., $K = S_0$, with the parameters $K = S_0 = 25$, $T = 1$, $\sigma = 0.30$, $r = 0.01$, and $N = 1, 10, 100, 1{,}000$, and $10{,}000$. Use any computational method of your choice (this could be anything from a spreadsheet to code written in a modern computer language). Do the call option values appear to converge as N approaches 10,000? Now, compute the option value directly using the Black-Scholes equation and the parameters given above.

(c) What can be said about convergence as $N \to \infty$ of this binomial model to the Black-Scholes equation? Feel free to consult references or research online. If you were a "*quant*" tasked with computing an option value, which approach, binomial or PDE, would you prefer to implement?

---

[24]The reader should look this up online or in the book **Steven E. Shreve**, *Stochastic Calculus for Finance I: The Binomial Asset Pricing Model*, Springer, 2004.

Chapter 8

The Laplacian, Laplace's Equation, and Harmonic Functions

For the next three chapters, the reader is encouraged to first review Sections A.4–A.6 and A.9 from the Appendix.

"Physics has its own Rosetta Stones. They're ciphers, used to translate seemingly disparate regimes of the universe. They tie pure math to any branch of physics your heart might desire. And this is one of them:

$$\Delta u = 0.$$

It's in electricity. It's in magnetism. It's in fluid mechanics. It's in gravity. It's in heat. It's in soap films. It's called Laplace's equation. It's everywhere."

- **Brendan Cole** writing in **wired.com**[a]

[a]The opening of Cole's "popular science" article, appearing in June 2016 of the online tech-oriented news magazine **WIRED**, provides a perfect opener to this chapter. His short article is available online for free and is entitled *The Beauty of Laplace's Equation, Mathematical Key to ...Everything.*

We have already encountered the Laplacian several times, appearing, for example, as the spatial derivatives in the multidimensional wave and diffusion equations. What makes the Laplacian so ubiquitous is the coupling of two fundamental concepts/ideas:

the gradient vector field of a function u **and** the Divergence Theorem.

The coupling of the two (*the match made in heaven*) is the divergence of the gradient of a function $u : \mathbb{R}^N \to \mathbb{R}$:

$$\operatorname{div} \nabla u = \operatorname{div}(u_{x_1}, \ldots, u_{x_N}) = \nabla \cdot \nabla u = \sum_{i=1}^{N} u_{x_i x_i}.$$

We call this combination of second-order partial derivatives the **Laplacian** of u. Mathematicians denote it by Δu whereas physicists tend to use the more suggestive notation $\nabla^2 u$, which really stands for $\nabla \cdot \nabla u$. Even though we have already seen the Laplacian, let us formally present its own box definition:

The Laplacian

The Laplacian of a function u defined on a domain in $\mathbb{R}^N$ is defined as

$$\Delta u = \nabla^2 u = \sum_{i=1}^{N} u_{x_i x_i}.$$

We will be primarily interested in space dimensions $N = 2$ and $N = 3$:

$$2\text{D}: \ \Delta u(x_1, x_2) = u_{x_1 x_1} + u_{x_2 x_2}, \qquad 3\text{D}: \ \Delta u(x_1, x_2, x_3) = u_{x_1 x_1} + u_{x_2 x_2} + u_{x_3 x_3}.$$

The primary focus of this relatively short chapter will be on functions whose Laplacian vanishes, and their special properties. We make the following important definition:

The Definition of a Harmonic Function

Definition 8.0.1. A function whose Laplacian is identically zero is said to be **harmonic**.

We will also see that in each space dimension $N \geq 2$, there is, up to additive and multiplicative constants, exactly one radial function whose Laplacian vanishes at all points except the origin. Via a suitable choice of constants, we will define **the fundamental solution**, which will play the key role (cf. Chapter 10) in solving PDEs and BVPs involving the Laplacian. To begin with, let us first document some of these problems involving the Laplacian.

8.1. • The Dirichlet and Neumann Boundary Value Problems for Laplace's and Poisson's Equations

There are many fundamental problems in mathematics, physics, and engineering which are purely spatial and involve the Laplacian. Here are a few:

(i) **Laplace's equations** for a function $u(\mathbf{x})$ is given by $\Delta u = 0$. If this PDE is to be solved, not in all of space $\mathbb{R}^N$ but in a domain $\Omega \subset \mathbb{R}^N$, then we will require a supplementary boundary condition. Two important types are **Dirichlet** and **Neumann** boundary conditions, in which the former gives set boundary values for the solution u, while the latter gives the normal derivative of the solution on $\partial\Omega$. Precisely, the **Dirichlet problem** for Laplace's equation is

$$\begin{cases} \Delta u = 0 & \text{on } \Omega, \\ u = g & \text{on } \partial\Omega, \end{cases} \tag{8.1}$$

while the **Neumann problem** for Laplace's equation is

$$\begin{cases} \Delta u = 0 & \text{on } \Omega, \\ \frac{\partial u}{\partial \mathbf{n}} = g & \text{on } \partial\Omega, \end{cases} \qquad \text{where} \quad \iint_{\partial\Omega} g\, dS = 0. \tag{8.2}$$

In both problems we are looking for **designer harmonic functions** in a domain which have certain prescribed boundary values.

(ii) Alternatively, we can look for a function with prescribed Laplacian; that is, for a given $f(\mathbf{x})$ defined for $\mathbf{x} \in \mathbb{R}^N$, we solve

$$\Delta u = f \quad \text{on } \mathbf{x} \in \mathbb{R}^N.$$

This is called a **Poisson**[1] **equation** or simply Poisson's equation. We may also consider a BVP for Poisson's equation on a domain Ω by prescribing either Dirichlet of Neumann boundary conditions.

In the next section, we discuss the interpretation of harmonic functions in terms of concentrations in equilibrium. As such, the reader should be rather astonished to know that one can indeed solve the Dirichlet and Neumann problems for Laplace's equation under minimal assumptions on Ω and g. However, the reader will have to wait until Chapter 10 for any discussion on how one actually solves these problems.

8.2. • Derivation and Physical Interpretations 1: Concentrations in Equilibrium

Functions whose Laplacian vanishes (is identically zero) can be thought of as **concentrations** (densities) of some quantity[2] **in equilibrium**. We can think of them as steady-state solutions to the diffusion equation[3] $u_t = c^2 \Delta u$, that is, particular solutions which are independent of time ($u_t \equiv 0$). To reinforce this, we revisit the derivation of the diffusion equation (Section 7.1) but, this time, in the context of a **steady-state** density configuration.

Let Ω be a domain in $\mathbb{R}^3$ and let u denote the concentration (density or distribution) of some quantity which is in equilibrium (no change with time). Then, on any subdomain $V \subset \Omega$, we have

$$u_t = 0 = \frac{\partial}{\partial t} \iiint_V u \, d\mathbf{x} = \iint_{\partial V} \mathbf{F} \cdot \mathbf{n} \, dS,$$

where $\mathbf{F}$ denotes the flux density associated with the quantity. As in Section 7.1, we assume $\mathbf{F} = -a\nabla u$ for some $a > 0$. Hence, via the Divergence Theorem,

$$0 = \iint_{\partial V} -a\nabla u \cdot \mathbf{n} \, dS = \iiint_V -a\Delta u \, d\mathbf{x}.$$

[1] **Siméon Denis Poisson** (1781–1840) was a French mathematician, engineer, and scientist who made profound contributions to mathematics and engineering.

[2] One such quantity, directly intertwined with the study of Laplace's equation, is the **electrostatic potential** discussed in Section 10.6.

[3] They are also steady-state solutions to the wave equation.

The above must hold for **all** subdomains $V \subset \Omega$, and hence the IPW Theorem, Theorem A.6, implies that

$$\boxed{\Delta u = 0 \qquad \text{for all } \mathbf{x} \in \Omega.}$$

Note here that we are not saying that in order to be in equilibrium the flux $\mathbf{F}$ must vanish, and hence u must be constant. Rather, there can be nonzero flux at all points in V, but somehow their effects **all cancel out** producing no net flow out of **any** subdomain V. The above shows that this cancellation property is encapsulated in Laplace's equation.

Based upon our derivation, it should be clear that harmonic functions are rather special and denote concentration densities/distributions of quantities which are in equilibrium. One can naturally ask how many such functions exist. Clearly, any constant is harmonic and represents an equilibrium concentration distribution. Any linear function is also a harmonic function. What else? In one space dimension, the answer is nothing else; all 1D harmonic functions are linear. However, in two space dimensions and higher there are many, many more! There are so many, in fact, that one can be greedy and ask the following question: Can one find a **designer** harmonic function with **prescribed** boundary values, i.e., solve **the Dirichlet problem** (8.1)? Rephrasing in physical terms, given the concentration distribution g on the boundary, can one find a continuous equilibrium concentration defined everywhere in $\overline{\Omega}$ which is equal to g on $\partial\Omega$? Remarkably, the answer is yes for any "reasonable" domain Ω and "reasonable" function g. The reader should pause to contemplate this truly amazing fact; how on earth does one find a concentration density which **both** matches the prescribed values on the boundary **and** has the property that inside Ω all concentration gradients *balance out* (in the previous sense).

While we will have to wait until Chapter 10 to address how to find such solutions, let us note here that there must be something fundamentally **nonlocal** about the process. Suppose we were given a solution to the Dirichlet problem (8.1) in some Ω with data g prescribed on $\partial\Omega$. If we were to **slightly change** the values of g **near some** point on the boundary, the solution (i.e., the equilibrium concentration) would have to adjust **everywhere** in Ω. Put another way, the solution at **any** point in Ω must depend on the values of g at **all** points on the boundary. The same will be true with the solution to the Poisson equation and the function f.

The above observations are suggestive of what a potential solution formula for these BVPs is going to look like. At any point $\mathbf{x} \in \Omega$, the solution $u(\mathbf{x})$ must depend on the data (g and/or f) everywhere, and hence one might anticipate formulas (cf. Chapter 10) for $u(\mathbf{x})$ which involve integrals of this data. In this respect, these PDEs are closer to the diffusion equation than to first-order equations and the wave equation.

Given the generality of our derivation, it is hardly surprising that Laplace's equation is rather **ubiquitous** in physics, appearing, for example, throughout electrostatics and fluid mechanics. The Laplacian and Laplace's equation also turn up in many other areas of mathematics, for example, in complex analysis, geometry, and probability. In the next subsection, we briefly describe their connection to probability.

8.3. Derivation and Physical Interpretations 2: The Dirichlet Problem and Poisson's Equation via 2D Random Walks/Brownian Motion

Recall the notion of a random walk from Section 7.3. In Exercise **7.17**, we saw that this notion easily extends to two (or N) dimensions. In this subsection, we will show a connection between 2D random walks and the Dirichlet problem for both Laplace's equation and Poisson's equation. We only present an intuitive connection via some simple calculations.

For simplicity let us work in dimension 2 and consider a domain $\Omega \in \mathbb{R}^2$, the unit square $[0, 1]^2$ with the bottom left corner at the origin. We perform a 2D random walk on a rectangular lattice. In order to avoid confusion with the Laplacian Δ, we denote small spatial and time steps by δ instead of Δ. Let $h := \delta x = \delta y$, and then divide the plane into discrete points (nh, mh) where n and m are integers. At any point, we have $1/4$ probability of moving up or down or left or right.

The role of time in the following derivations is subtle. By its very definition, the notion of a random walk, or the limit of Brownian motion, involves time. However, we will now derive two PDEs, Laplace's and Poisson's equations, which are independent of time. In both cases, our physical interpretation of u will be purely a function of space. In the first derivation, we will not even explicitly mention time t. In the second, t will appear initially but will eventually disappear because of its slaving to h.

Dirichlet Problem for Laplace's equation. Let Γ_1 denote the boundary of the square except for the part on the x-axis, and let Γ_2 denote the part of the x-axis between 0 and 1. Let us think of Γ_2 as an open door and Γ_1 as walls which have been freshly painted. A bug at a grid position inside the room undergoes a 2D random walk (with step size h) **until** it either hits the door and escapes (hence, lives) **or** it hits the wall, sticks, and dies[4]. We let $u(x, y)$ **be the probability that a bug lives if it starts at the point** (x, y). Our goal is to show that in the limit of $h \to 0$, u must solve Laplace's equation in Ω. We first start on the finite h-grid and hence assume (x, y) is a grid point. The master equation will be an equation which relates $u(x, y)$ to the value of u at the four neighboring grid points. In this context, one can readily check that it is simply

$$u(x, y) = \frac{u(x-h, y) + u(x+h, y) + u(x, y-h) + u(x, y+h)}{4}.$$

We can rewrite this equation as

$$\frac{u(x-h, y) + u(x+h, y) - 2u(x, y)}{4} + \frac{u(x, y-h) + u(x, y+h) - 2u(x, y)}{4} = 0.$$

Dividing both sides by h^2 we have

$$\frac{u(x-h, y) + u(x+h, y) - 2u(x, y)}{4h^2} + \frac{u(x, y-h) + u(x, y+h) - 2u(x, y)}{4h^2} = 0.$$

Now we let $h \to 0$, recalling (7.15), to obtain Laplace's equation $u_{xx} + u_{yy} = 0$.

The boundary conditions for u are simple. If $(x, y) \in \Gamma_2$, then we are at the door and have escaped; hence $u(x, y) = 1$. If $(x, y) \in \Gamma_1$, then we have hit the painted wall

[4] It can be shown that with probability 1, the insect will hit the boundary in finite time.

and hence die; so $u(x, y) = 0$. We have shown that $u(x, y)$ must solve

$$\begin{cases} \Delta u = 0 & \text{in } \Omega, \\ u = g & \text{on } \partial\Omega, \end{cases} \quad \text{where} \quad g(x, y) = \begin{cases} 1 & \text{if } (x, y) \in \Gamma_2, \\ 0 & \text{otherwise.} \end{cases}$$

The Dirichlet Problem for Poisson's Equation. Now consider a new scenario in which a bug starts off at a grid point (x, y) and undergoes a random walk. It *lives* until the first time it hits $\partial\Omega$ (the boundary of the square) and then it *dies*. It can be shown that with probability 1, the insect will hit the boundary in finite time. Hence, we are interested in determining

$$u(x, y) = \text{the expected time to arrival at the boundary (the "life expectancy")}$$

of a bug which starts at (x, y). Note that right away we are assuming that the history of the bug is irrelevant when computing its "life expectancy". In other words, the life expectancy of the bug which is currently at (x, y) only depends on (x, y) and not on its past, i.e., how it got to (x, y). This is a consequence of what is called the **Markov property** of the random walk and of its limit, Brownian motion.

It is clear that $u = 0$ on $\partial\Omega$. To find u inside the square, we first consider the discrete random walk and the basic rule for connecting $u(x, y)$ to its neighboring values. As in Section 7.3, this is straightforward: In one time cycle δt, a bug which is at (x, y) can only arrive there in a time step δt from one of its four neighbors with equal probability. Hence, its expected time to arrival at the boundary must equal δt plus the average of u at its four neighbors; i.e.,

$$u(x, y) = \delta t + \frac{1}{4}\Big(u(x-h, y) + u(x+h, y) + u(x, y-h) + u(x, y+h)\Big). \tag{8.3}$$

This is the **master equation** so be sure you are convinced of its validity before continuing. We supplement the master equation with the boundary condition that $u(x, y) = 0$ for $(x, y) \in \partial\Omega$.

We can rewrite (8.3) as

$$\frac{u(x-h, y) + u(x+h, y) - 2u(x, y)}{4} + \frac{u(x, y-h) + u(x, y+h) - 2u(x, y)}{4} = -\delta t.$$

Setting $\sigma^2 = \frac{h^2}{\delta t}$, we divide both sides by h^2 to obtain

$$\frac{u(x-h, y) + u(x+h, y) - 2u(x, y)}{h^2} + \frac{u(x, y-h) + u(x, y+h) - 2u(x, y)}{h^2}$$
$$= -4\frac{\delta t}{h^2} = \frac{-4}{\sigma^2}.$$

Recalling (7.15), we let $h \to 0, \delta t \to 0$ with σ fixed to find that u solves

$$\begin{cases} \Delta u & = -4/\sigma^2 \quad \text{in } \Omega, \\ u & = 0 \quad \text{on } \partial\Omega. \end{cases}$$

In this simple scenario, the right-hand side is constant. We obtain the general Poisson equation if we allow the ratio σ to vary over the domain Ω.

8.4. • Basic Properties of Harmonic Functions

We address four important properties of harmonic functions:

(i) The Mean Value Property.

(ii) The Maximum Principle.

(iii) The Dirichlet Principle.

(iv) Regularity: Infinite smoothness.

From the derivation of Section 8.2, all four properties should seem reasonable given the "equilibrium" interpretation of a harmonic function.

8.4.1. • The Mean Value Property.

The mean value property states that the value of a harmonic function at a given point equals its average on any sphere centered at the point (its **spherical mean**). More precisely, we have the following theorem stated in dimension 3.

The Mean Value Property

Theorem 8.1. *Let u be a C^2 harmonic function on a domain $\Omega \subset \mathbb{R}^3$. Let $\mathbf{x}_0 \in \Omega$ and $r > 0$ such that $B(\mathbf{x}_0, r) \subset \Omega$. Then*

$$u(\mathbf{x}_0) = \frac{1}{4\pi r^2} \iint_{\partial B(\mathbf{x}_0, r)} u(\mathbf{x})\, dS_{\mathbf{x}}. \tag{8.4}$$

The analogous statement holds in 2D (in fact, in any space dimension). In 2D we would replace $4\pi r^2$ with the corresponding "size" of $\partial B(\mathbf{x}_0, r)$ which, in 2D, is $2\pi r$. While we state the proof in 3D, the same proof with minor modifications works in any space dimension. The key step in the proof is the differentiation of the spherical mean. While the same step was previously done in the proof of Kirchhoff's formula (Section 4.2.4), it may not be a bad idea to first review Sections A.5.2 and A.5.4. Whereas the proof is standard, we follow closely Evans [**10**].

Proof. Fix $\mathbf{x}_0 \in \Omega$. For any $r > 0$ such that $B(\mathbf{x}_0, r) \subset \Omega$, we define the spherical mean

$$\phi(r) = \frac{1}{4\pi r^2} \iint_{\partial B(\mathbf{x}_0, r)} u(\mathbf{x})\, dS_{\mathbf{x}}.$$

Our goal is to show that ϕ is a constant; i.e., $\phi'(r) = 0$. Once we have established this, we can invoke the Averaging Lemma (Lemma A.7.1) to conclude that this constant must be equal to

$$\lim_{r \to 0^+} \frac{1}{4\pi r^2} \iint_{\partial B(\mathbf{x}_0, r)} u(\mathbf{x})\, dS_{\mathbf{x}} = u(\mathbf{x}_0).$$

This will yield (8.4).

How do we calculate $\phi'(r)$? We change variables with a translation and dilation in such a way that the new domain of integration becomes $\partial B(\mathbf{0}, 1)$, which is independent of r. To this end, let

$$\mathbf{y} = \frac{\mathbf{x} - \mathbf{x}_0}{r} \qquad \text{or} \qquad \mathbf{x} = \mathbf{x}_0 + r\mathbf{y}.$$

Recall, from spherical coordinates (Section A.5.2), that in the new variable $\mathbf{y} \in \partial B(\mathbf{0}, 1)$, we have

$$r^2 dS_{\mathbf{y}} = dS_{\mathbf{x}}.$$

Thus

$$\phi(r) = \frac{1}{4\pi r^2} \iint_{\partial B(\mathbf{x}_0, r)} u(\mathbf{x})\, dS_{\mathbf{x}} = \frac{1}{4\pi} \iint_{\partial B(\mathbf{0},1)} u(\mathbf{x}_0 + r\mathbf{y})\, dS_{\mathbf{y}}.$$

Note that changing to the new variable **y** still yields an average (a spherical mean). Now we can take the derivative with respect to r by bringing it into the integral (cf. Section A.9) and applying the chain rule:

$$\phi'(r) = \frac{1}{4\pi} \iint_{\partial B(\mathbf{0},1)} \frac{d}{dr} u(\mathbf{x}_0 + r\mathbf{y})\, dS_{\mathbf{y}} = \frac{1}{4\pi} \iint_{\partial B(\mathbf{0},1)} \nabla u(\mathbf{x}_0 + r\mathbf{y}) \cdot \mathbf{y}\, dS_{\mathbf{y}}.$$

Reverting back to our original coordinates of **x**, we find

$$\phi'(r) = \frac{1}{4\pi r^2} \iint_{\partial B(\mathbf{x}_0, r)} \nabla u(\mathbf{x}) \cdot \frac{\mathbf{x} - \mathbf{x}_0}{r}\, dS_{\mathbf{x}}.$$

For $\mathbf{x} \in \partial B(\mathbf{x}_0, r)$, $\frac{\mathbf{x}-\mathbf{x}_0}{r}$ denotes the outer unit normal to $\partial B(\mathbf{x}_0, r)$. Thus, we may apply the Divergence Theorem to conclude that

$$\phi'(r) = \frac{1}{4\pi r^2} \iint_{\partial B(\mathbf{x}_0, r)} \nabla u(\mathbf{x}) \cdot \frac{\mathbf{x} - \mathbf{x}_0}{r}\, dS_{\mathbf{x}} = \frac{1}{4\pi r^2} \iiint_{B(\mathbf{x}_0, r)} \Delta u\, d\mathbf{x}.$$

Since u is harmonic, i.e., $\Delta u = 0$, we have $\phi'(r) = 0$. □

Not surprisingly, we would obtain the same result if we were to average the values over the solid ball; that is,

$$u(\mathbf{x}_0) = \frac{1}{\frac{4\pi}{3} r^3} \iiint_{B(\mathbf{x}_0, r)} u(\mathbf{x})\, d\mathbf{x}. \tag{8.5}$$

This directly follows (cf. Exercise **8.7**) by first integrating over spheres of radius ρ and then integrating from $\rho = 0$ to $\rho = r$. On each spherical integral, we apply the mean value property.

In turns out that the mean value property is **equivalent** to being harmonic in the sense that the converse of Theorem 8.1 holds true (cf. Exercise **8.19**): If $u \in C^2(\Omega)$ and satisfies the mean value property (8.4) for all $B(\mathbf{x}_0, r) \subset \Omega$, then $\Delta u = 0$ in Ω. Hence, one could say the mean value property is the **essence** of being harmonic.

8.4.2. • The Maximum Principle. While we again present the maximum principle in dimension 3, it holds in any space dimension. Loosely speaking, the maximum principle states that *harmonic functions do not tolerate local extrema (maxima and minima)*. More precisely, the principle asserts that if u is a C^2 harmonic function on a **bounded** domain $\Omega \subset \mathbb{R}^3$ which is continuous up to the boundary, i.e., $u \in C(\overline{\Omega})$, then the maximum value of u must occur on the boundary. This means that if we denote M as the maximum value of u over $\overline{\Omega} = \Omega \cup \partial\Omega$, then there must be a point on the boundary for which u takes on M. In fact, a stronger statement is true since, by assumption (cf. Definition A.1.1), our domain Ω is connected. It states that the maximum of u can **only** occur on the boundary **unless** u is identically constant:

The Maximum Principle

Theorem 8.2. *Let u be a C^2 harmonic function on a bounded domain $\Omega \subset \mathbb{R}^3$ which is continuous up to the boundary; i.e., $u \in C(\overline{\Omega})$. If u attains its maximum over $\overline{\Omega} = \Omega \cup \partial\Omega$ at a point in Ω, then u must be identically constant inside $\overline{\Omega}$.*

We give a proof which ends in an informal, yet convincing, step that can readily be made precise with a bit of point set topology (open and closed sets).

Proof. Since u is continuous on $\overline{\Omega}$ (a compact set), u must attain its maximum somewhere in $\overline{\Omega}$. Let M be the maximum value of u over $\overline{\Omega}$. By assumption $u(\mathbf{x}_0) = M$ for some $\mathbf{x}_0 \in \Omega$. Choose a sphere centered at $\mathbf{x}_0$ which lies entirely inside Ω. Then, by the mean value property, the value of the center (i.e., $u(\mathbf{x}_0)$) equals the average of the values on the sphere. Quickly, we arrive at the punch line:

> The only way for M to be **both** the maximum value of u and the average value of u over the sphere is for u to be identically equal to M everywhere on the sphere.

Since the above is true for **any** sphere with center $\mathbf{x}_0$ contained in Ω, we have

$$u(\mathbf{x}) \equiv M \qquad \text{on } \textbf{any} \text{ solid ball with center } \mathbf{x}_0 \text{ which is contained in } \Omega.$$

Fix such a ball and now repeat this argument by choosing a new center point (say, $\mathbf{x}_1$) in the ball. Since $u(\mathbf{x}_1) = M$, we conclude $u(\mathbf{x}) \equiv M$ on any solid ball with center $\mathbf{x}_1$ which lies in Ω. Repeating on successive balls (perhaps of very small radii), we can "*fill up*" all the connected domain Ω (see Figure 8.1). We conclude that $u(\mathbf{x}) \equiv M$ for **all** $\mathbf{x} \in \Omega$. One only needs a bit of point set topology (open and closed sets) to make this last point rigorous. Here one shows that the set of all points upon which u is M is both an open and a closed subset of Ω. The only such subset is Ω itself. □

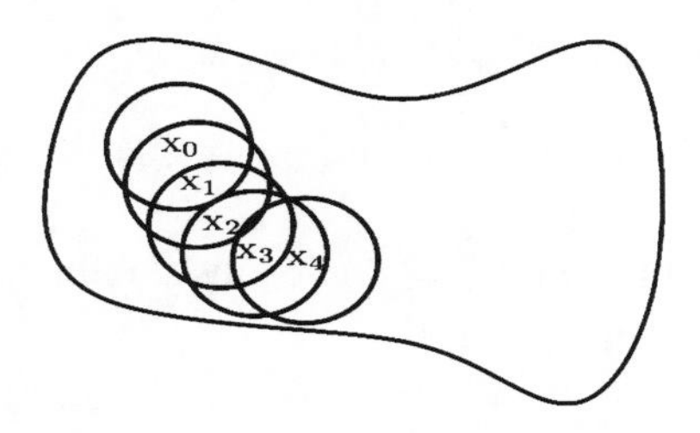

Figure 8.1. Proving the Maximum Principle by successively applying the mean value property.

The Maximum Principle Is Also a Minimum Principle. If u is harmonic, so is $-u$. Now apply the Maximum Principle to $-u$. This gives a Minimum Principle for u; that is, the previous statements hold true with *maximum* replaced by *minimum*.

Combining the two principles, we have the following immediate corollaries:

Corollary 8.4.1. *If u is a $C^2(\Omega) \cap C(\overline{\Omega})$ harmonic function on a bounded domain Ω and the values of u on the boundary are bounded between m and M, then the values of u everywhere are bounded between m and M.*

Maximum principles, in general, are powerful tools. We now show how our Maximum Principle immediately yields two of three conditions needed for the **well-posedness** of the Dirichlet problem: namely **uniqueness and stability**.

Corollary 8.4.2 (Uniqueness). *Let Ω be bounded. There exists at most one $C^2(\Omega) \cap C(\overline{\Omega})$ solution to the Dirichlet problem* (8.1).

Proof. Suppose u_1 and u_2 are both solutions of (8.1). Let $v = u_1 - u_2$. Then v solves (8.1) with $g \equiv 0$ on $\partial\Omega$. By the Maximum Principle, the maximum and minimum of v must be attained on the boundary. Hence $v \equiv 0$ (or $u_1 \equiv u_2$). □

Corollary 8.4.3 (Stability). *Let Ω be bounded and let g_1 and g_2 be continuous functions on $\partial\Omega$. Suppose u_1 and u_2 are $C^2(\Omega) \cap C(\overline{\Omega})$ solutions to the respective Dirichlet problems*

$$\begin{cases} \Delta u = 0 & \text{on } \Omega, \\ u = g_1 & \text{on } \partial\Omega \end{cases} \quad \text{and} \quad \begin{cases} \Delta u = 0 & \text{on } \Omega, \\ u = g_2 & \text{on } \partial\Omega. \end{cases}$$

Then

$$\max_{\mathbf{x}\in\Omega} |u_1(\mathbf{x}) - u_2(\mathbf{x})| \leq \max_{\mathbf{x}\in\partial\Omega} |g_1(\mathbf{x}) - g_2(\mathbf{x})|. \tag{8.6}$$

Proof. Let $v = u_1 - u_2$. Then v solves (8.1) with boundary data $g_1 - g_2$ on $\partial\Omega$. By the Maximum Principle, the maximum and minimum of v must be attained on the boundary. This immediately yields (8.6). □

Note that the inequality (8.6) immediately implies that if the data are close, then so are the respective solutions. Here closeness is measured in terms of distance prescribed by the maximum of pointwise differences over the domain. There are of course other notions of closeness of (distance between) functions; for example, we could measure the distance between two functions u_1 and u_2 on Ω by

$$\iiint_\Omega |u_1(\mathbf{x}) - u_2(\mathbf{x})| \, d\mathbf{x}.$$

For many of these other notions one has analogous stability estimates for the solution to the Dirichlet problem.

The relationship between the Maximum Principle and the Laplacian is rather robust in that the principle holds not only for harmonic functions but also for what are known as **subharmonic functions** (cf. Exercise **8.20**). For the definition of subharmonic we must work with the operator $-\Delta$ (part (c) of Exercise **8.20** will help explain why). Precisely, let Ω be a bounded, connected domain in $\mathbb{R}^3$. A C^2 function u on Ω is called **subharmonic** if

$$-\Delta u(\mathbf{x}) \leq 0 \quad \text{for all } \mathbf{x} \in \Omega. \tag{8.7}$$

This more general maximum principle yields what are known as **comparison principles** (cf. Exercises **8.21–8.22**). These types of comparison principles prove to be a *life line* for many mathematicians working on wide classes of PDEs.

Lastly, let us remark that the Maximum Principle and the uniqueness and stability of the Dirichlet problem are false if the domain is **unbounded** (cf. Exercise **8.14**).

8.4.3. • The Dirichlet Principle. Here is a completely different way to see that harmonic functions are rather special. As usual we illustrate in dimension 3. Consider the class of functions defined on $\overline{\Omega} = \Omega \cup \partial\Omega$ which, in particular, are C^2 in $\Omega \subset \mathbb{R}^3$, continuous on $\overline{\Omega}$, and agree on $\partial\Omega$. That is, we define $w \in \mathcal{A}_h$ where

$$\mathcal{A}_h := \left\{ w \in C^2(\Omega) \cap C^1(\overline{\Omega}) \;\middle|\; w = h \text{ on } \partial\Omega \right\},$$

for some function h. Consider w to be the concentration of some quantity in Ω. It is free to change in Ω but is prescribed (fixed) on the boundary. Recall from the first derivation of Laplace's equation the notion of the flux density wherein $\mathbf{F} = -c\nabla w$. We take $c = 1$ and consider the magnitude of the flux density across Ω. We define one half the **total amount** of this magnitude inside Ω,

$$E(w) := \frac{1}{2} \iiint_\Omega |\nabla w|^2 \, d\mathbf{x}, \tag{8.8}$$

to be the **potential energy** of $w \in \mathcal{A}_h$. It is the energy caused by the spatial variations in concentration: If the concentration w is constant inside Ω, then there is zero potential energy whereas if w oscillates widely, there is a large amount of energy. You will soon see why we include the factor of one half. A natural question is: **At which w do we have the lowest potential energy?** In other words, among all such $w \in \mathcal{A}_h$, find the one which minimizes $E(w)$. Dirichlet's Principle states that it is the one which solves

$$\begin{cases} \Delta u = 0 & \text{in } \Omega, \\ u = h & \text{on } \partial\Omega. \end{cases} \tag{8.9}$$

More precisely, we have the following:

Dirichlet's Principle

Theorem 8.3. *$u \in \mathcal{A}_h$ minimizes E over all $w \in \mathcal{A}_h$, i.e.,*

$$E(u) \leq E(w) \qquad \textit{for all } w \in \mathcal{A}_h,$$

if and only if $\Delta u = 0$ in Ω, i.e., u solves (8.9).

Proof. (i) We first prove that if u is a minimizer, then u solves (8.9). Let u minimize E over $w \in \mathcal{A}_h$; i.e., for any $w \in \mathcal{A}_h$, we have $E(w) \geq E(u)$. Consider for the moment a fixed function $v \in C^2(\Omega) \cap C^1(\overline{\Omega})$ such that $v = 0$ on $\partial\Omega$. Then with our minimizer u and this fixed v, we note that for any $\epsilon > 0$, $u + \epsilon v \in \mathcal{A}_h$ and consider the function from $\mathbb{R}$ to $\mathbb{R}$ defined by

$$f(\epsilon) := E(u + \epsilon v) = \frac{1}{2} \iiint_\Omega |\nabla(u + \epsilon v)|^2 \, d\mathbf{x}.$$

Now comes the key observation: Suppose f is differentiable (we will in fact prove this), then the fact that u is a minimizer implies that

$$f'(0) = 0. \tag{8.10}$$

Convince yourself of this before continuing. To unlock the consequences of (8.10), we first use the fact that for any vector $\mathbf{a}$, $|\mathbf{a}|^2 = \mathbf{a} \cdot \mathbf{a}$. To this end, we find

$$\begin{aligned} f(\epsilon) &= \frac{1}{2} \iiint_\Omega \nabla(u + \epsilon v) \cdot \nabla(u + \epsilon v)\, d\mathbf{x} \\ &= \frac{1}{2} \iiint_\Omega (\nabla u + \epsilon \nabla v) \cdot (\nabla u + \epsilon \nabla v)\, d\mathbf{x} \\ &= \frac{1}{2} \iiint_\Omega \left(\nabla u \cdot \nabla u + 2\epsilon \nabla u \cdot \nabla v + \epsilon^2 \nabla v \cdot \nabla v\right) d\mathbf{x} \\ &= \frac{1}{2} \iiint_\Omega \left(|\nabla u|^2 + 2\epsilon \nabla u \cdot \nabla v + \epsilon^2 |\nabla v|^2\right) d\mathbf{x}. \end{aligned}$$

Differentiation under the integral sign (cf. Theorem A.10) implies that f is indeed C^1 and

$$f'(\epsilon) = \frac{1}{2} \iiint_\Omega \left(2\nabla u \cdot \nabla v + 2\epsilon |\nabla v|^2\right) d\mathbf{x}, \qquad \text{in particular } f'(0) = \iiint_\Omega \nabla u \cdot \nabla v\, d\mathbf{x}.$$

Next we use Green's First Identity (A.16) to find that

$$0 = f'(0) = \iiint_\Omega \nabla u \cdot \nabla v\, d\mathbf{x} = -\iiint_\Omega (\Delta u) v\, d\mathbf{x}. \tag{8.11}$$

Note that v vanishes on $\partial\Omega$, and hence there are no boundary terms. Since (8.11) holds for **all** $v \in C^2(\Omega) \cap C^1(\overline{\Omega})$ with $v = 0$ on $\partial\Omega$, we may invoke the Integral to Pointwise (IPW) Theorem, Theorem A.6, to conclude that

$$\Delta u = 0 \quad \text{for all } \mathbf{x} \in \Omega.$$

(ii) We now prove the opposite direction. Suppose $u \in \mathcal{A}_h$ solves (8.9). We want to show that for any $w \in A_h$, we have

$$E(w) \geq E(u).$$

Let $w \in \mathcal{A}_h$ and define $v = u - w$. Then

$$E(w) = E(u - v) = \frac{1}{2} \iiint_\Omega |\nabla(u - v)|^2\, dx = E(u) - \iiint_\Omega \nabla u \cdot \nabla v\, dx + E(v).$$

Note that, by definition, $v = 0$ on $\partial\Omega$. Hence, by Green's First Identity (A.16), we find

$$0 = \iint_{\partial\Omega} v \frac{\partial u}{\partial \mathbf{n}} dS = \iiint_\Omega \nabla u \cdot \nabla v\, d\mathbf{x} + \iiint_\Omega v \Delta u\, d\mathbf{x} = \iiint_\Omega \nabla u \cdot \nabla v\, d\mathbf{x} + 0.$$

We have proved that $E(w) = E(u) + E(v)$ and since $E(v) \geq 0$, we are done. □

We make two remarks, the first of which is particularly important.

- **The Euler-Lagrange equation and convexity:** Recall from calculus the notion of a critical point when optimizing a smooth function of one variable. We sought critical points which were values of x where the derivative vanished. Criticality was a **necessary condition** for minimality which in general was not **sufficient**; i.e., there existed critical points which were neither local nor global minima (maxima). However, if the function was convex, then criticality (being a critical point) was sufficient for yielding a (global) minimizer.

  In our setting we are minimizing a functional over a class of functions, not points. This is the realm of the **calculus of variations**, which we will discuss in some detail in a future volume planned for this subject. In this theory, the analogue of a critical point is a function which satisfies a PDE (or ODE if there is only one independent variable) called the **Euler-Lagrange equation**. For the Dirichlet energy E above, the Euler-Lagrange equation is simply Laplace's equation.

  As with functions of one (or several) variables, it is rare that criticality alone entails a minimizer (or maximizer). However, the Dirichlet energy has a **convex structure**, and this structure allowed us above to show that satisfying Laplace's equation was sufficient to be a minimizer.

- **The smoothness assumptions:** The assumption of

$$w \in C^2(\Omega) \cap C^1(\overline{\Omega})$$

  in our admissible class was for convenience only in that it fostered a simple proof. Note that in order to define the energy E we only need that the trial functions in $\mathcal{A}_h$ be $C^1(\Omega)$ (not necessarily C^2). Indeed, it turns out that one could minimize E over this wider class and still find that the minimizer would be a C^2 harmonic function. So, whereas the competitors need not have continuous second derivatives, the minimizer would. This enhanced regularity is a consequence of being a minimizer. As we note in the next subsection, not only would the minimizer be C^2, but C^∞. You could also have asked why we required w to also have first derivatives which were continuous up to the boundary. This was necessary in order to apply Green's First Identity to u and v, wherein we required the normal derivative of u on the boundary. Again this is not needed; harmonic functions in domains with smooth boundaries will automatically have derivatives that are continuous up to the boundary. To proceed further with these claims and relax the regularity assumptions, one needs a more sophisticated mathematical treatment which can be found, for example, in [**10**].

8.4.4. Smoothness (Regularity). To speak about a classical solution to $\Delta u = 0$, one would need the particular second derivatives $u_{x_i x_i}$, $i = 1, \ldots, n$, to exist. A natural question to ask is how smooth is this solution? Mathematicians refer to smoothness properties as **regularity**. The particular combination of the derivatives in the Laplacian have a tremendous power in "controlling" all derivatives. In fact, a harmonic function is not only C^2 (in other words, all second derivatives exist and are continuous) but, also, C^∞. Precisely, if $u \in C^2(\Omega)$ solves $\Delta u = 0$ in Ω, then $u \in C^\infty(\Omega)$. This should be intuitive by either the equilibrium interpretation or the mean value property: The averaging out of values of u prohibits any singularities in any derivative. In Exercise **8.18**, you are asked to prove this by using both the convolution with a smooth multivariable *bump function* (cf. Section 6.3.4) and the mean value property.

In fact, even the assumption that $u \in C^2$ turns out to be redundant in the following sense. In the next chapter we will extend distributions to several independent variables and see how to interpret Laplace's equation in the sense of distributions. With this in hand, the following assertion holds true: If F is **any** distribution such that $\Delta F = 0$ in the sense of distributions, then $F = F_u$, a distribution generated by a C^∞ function u. This result is known as **Weyl's**[5] **Lemma**.

A broad class of PDEs which have a similar smoothness property are called **elliptic**. Solutions to elliptic PDEs have an increased measure of smoothness, often referred to as **elliptic regularity**, beyond the derivatives which appear in the PDE. The Laplacian is the canonical example of an elliptic PDE.

8.5. • Rotational Invariance and the Fundamental Solution

The Laplacian treats all spatial variables equally: Interchanging the roles of x_i with x_j ($j \neq i$) would result in the same operator. A more precise and descriptive way of saying this is that the Laplacian is **invariant under rigid motions**. This means that if we change our coordinate system via a rotation and/or a translation, then the Laplacian in the new variables is the same. For example, in 3D, suppose we rotated our $\mathbf{x} = (x_1, x_2, x_3)$ coordinate system by performing a rotation $\mathbf{R}$. Such an $\mathbf{R}$ is captured by a 3×3 orthogonal matrix and the new (rotated) coordinates are $\mathbf{x}' = \mathbf{R}\mathbf{x}$. In Exercise **8.4** you are asked to check that the Laplacian transforms in the new coordinate system to

$$\Delta = u_{x_1'x_1'} + u_{x_2'x_2'} + u_{x_3'x_3'};$$

in other words, it remains unchanged!

Given this invariance, it would seem natural to look for a radial solution to Laplace's equation. That is, we look for a solution to $\Delta u = 0$ of the form

$$u(\mathbf{x}) = v(|\mathbf{x}|) = v(r) \qquad \text{where } r := |\mathbf{x}| = \left(\sum_{i=1}^{N} x_i^2\right)^{1/2}.$$

To this end, suppose $N \geq 2$ and note that for any $i = 1, \ldots, N$ and any $\mathbf{x} \neq 0$, we have

$$\frac{\partial r}{\partial x_i} = \frac{1}{2}\left(\sum_{i=1}^{N} x_i^2\right)^{-1/2} 2x_i = \frac{x_i}{r}.$$

Hence by the chain rule

$$u_{x_i} = v'(r)\frac{x_i}{r} \qquad \text{and} \qquad u_{x_i x_i} = v''(r)\frac{x_i^2}{r^2} + v'(r)\left(\frac{1}{r} - \frac{x_i^2}{r^3}\right).$$

[5]**Herman Weyl** (1885–1955) was a German mathematician and theoretical physicist who made major contributions to mathematical physics, analysis, and number theory.

Summing over $i = 1$ to N, we have

$$\Delta u \;=\; v''(r)\frac{\sum_{i=1}^N x_i^2}{r^2} \;+\; v'(r)\left(\frac{N}{r} - \frac{\sum_{i=1}^N x_i^2}{r^3}\right) \;=\; v''(r) \;+\; v'(r)\left(\frac{N}{r} - \frac{1}{r}\right).$$

Thus we have shown that for a radial function $u(\mathbf{x}) = v(r)$

$$\Delta u \;=\; 0 \qquad \text{iff} \qquad v'' \;+\; \frac{N-1}{r}v' \;=\; 0.$$

Let us assume that $v' \neq 0$ (this eliminates the trivial constant solutions). Then we can rewrite the ODE as

$$\log(|v'|)' \;=\; \frac{v''}{v'} \;=\; \frac{1-N}{r}.$$

Integrating once we have

$$\log(|v'|) \;=\; \log r^{1-N} + \text{constant} \qquad \text{or} \qquad |v'| \;=\; \frac{C}{r^{N-1}},$$

for any constant C. Integrating once more, we find

$$v(r) \;=\; \begin{cases} C_1 \log r \;+\; C_2, & N = 2, \\ \frac{C_1}{r^{N-2}} \;+\; C_2, & N \geq 3, \end{cases} \qquad \text{for any constants } C_1 \text{ and } C_2.$$

Lastly one can check that if v' vanished at any point, then we would be lead to the conclusion that v is constant. Thus we have found all radial solutions. Except for the trivial constant solutions, all solutions blow up at the origin.

Based upon this calculation, let us *assign* certain choices of the constants (which depend on the dimension N) and define the fundamental solution for the Laplacian in different space dimensions. The reasons for this choice will be made clear shortly:

Definition of the Fundamental Solution of the Laplacian

Definition 8.5.1. The **fundamental solution** for Δ in dimension N ($N \geq 2$) is defined to be

$$\Phi(\mathbf{x}) \;:=\; \begin{cases} \frac{1}{2\pi}\log|\mathbf{x}| & \text{in dimension } N = 2, \\ -\frac{1}{4\pi|\mathbf{x}|} & \text{in dimension } N = 3, \\ -\frac{1}{N(N-2)\omega_N|\mathbf{x}|^{N-2}} & \text{in dimensions } N > 3, \end{cases}$$

where ω_N is the (hyper)volume of the unit sphere in $\mathbb{R}^N$.

Sign Warning: For a variety of reasons, some of which will become clear shortly (cf. Section 8.7 and Chapter 11), it is convenient to work with the differential operator $-\Delta$ instead of Δ. In 3D, this has the effect of taking the minus sign from the fundamental solution and putting it on the operator; i.e., the fundamental solution for $-\Delta$ is $\frac{1}{4\pi|\mathbf{x}|}$. Most mathematics texts adopt this tradition wherein the associated Green's functions are also defined for $-\Delta$ and, hence, have the opposite sign. In this text, we will base the definition of the fundamental solution and associated Green's functions **on** Δ rather than $-\Delta$.

In any space dimension $N \geq 2$, the fundamental solution is not defined at the origin; there it can be defined as any number, as this single value will be irrelevant to its character and use. The fundamental solution is a **radial** function which **blows up** at the origin and which is harmonic for all $\mathbf{x} \neq \mathbf{0}$. While the pointwise Laplacian is 0 except at one point, the blow-up at the origin has a profound effect on the distributional Laplacian: In all dimensions, we have

$$\boxed{\Delta\,\Phi(\mathbf{x}) \;=\; \delta_{\mathbf{0}} \qquad \text{in the sense of distributions.}} \tag{8.12}$$

The reason for the choice of constants in Definition 8.5.1 is precisely to ensure that its distributional Laplacian of Φ is exactly $\delta_{\mathbf{0}}$, rather than some constant times the delta function. Of course, we have yet to define distributions in higher space dimensions and the reader will have to wait until Chapter 10 for a proof in 3D.

The reader will surely be curious as to why we call Φ the "fundamental solution". What is it the fundamental solution to? Because of the singularity, it is certainly not a solution to Laplace's equation everywhere; rather we will show in Chapter 10 that it satisfies (8.12). We will then see that this property enables us to use the fundamental solution Φ to **generate solutions** to many linear PDEs involving the Laplacian, particularly Poisson's equation and the Dirichlet and Neumann problems for Laplace's equation. Thus, Φ is the key object needed to solve these problems. Physically, Φ (in 3D) is known as either the **electrostatic potential** (cf. Section 10.6) or the **Newtonian potential** generated by a unit source charge (mass) at the origin.

8.6. • The Discrete Form of Laplace's Equation

It in instructive to consider the discrete form of the Laplacian. Recall from Section 7.3 that Taylor's Theorem implies that if f is a smooth function, then

$$f''(x) \;=\; \lim_{h\to 0} \frac{f(x+h) \,+\, f(x-h) - 2f(x)}{h^2}.$$

In other words, for h small, we can approximate the $f''(x)$ with the discrete finite difference

$$\frac{f(x+h) \,+\, f(x-h) - 2f(x)}{h^2}.$$

Let us now apply this to Laplace's equation $\Delta u = 0$ on a 2D square $\Omega = (0,a) \times (0,a)$. To this end, divide up the rectangle into grid points as follows (cf. Figure 8.2, left). Fix an integer n (think of it as large) and define $h \coloneqq \frac{a}{n-1}$. Then for $i, j = 0, 1, \ldots, n-1$, define the grid points in the square $(x_i, y_j) \;:=\; (ih, jh)$, and let $u_{i,j} \;:=\; u(x_i, y_j)$. Then at any interior grid point (x_i, y_j) where $i, j = 1, \ldots, n-2$, we can approximate the second derivatives with central finite differences:

$$u_{xx}(x_i, y_j) \approx \frac{u_{i+1,j} - 2u_{i,j} + u_{i-1,j}}{h^2} \qquad \text{and} \qquad u_{yy}(x_i, y_j) \approx \frac{u_{i,j+1} - 2u_{i,j} + u_{i,j-1}}{h^2}.$$

Thus, the discrete version of our statement that the Laplacian of u is 0 at an interior grid point (x_i, y_j) is

$$\frac{u_{i+1,j} - 2u_{i,j} + u_{i-1,j}}{h^2} + \frac{u_{i,j+1} - 2u_{i,j} + u_{i,j-1}}{h^2} \;=\; 0,$$

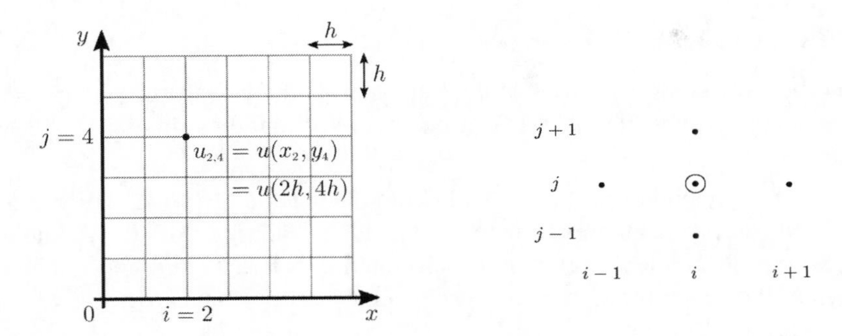

Figure 8.2. Left: Discretization of the square Ω: the grid points with $n = 6$. Right: The stencil for the scheme (8.13); the value at the center grid point equals the average on the neighboring grid points.

which after a little algebra reads

$$\boxed{u_{i,j} = \frac{u_{i-1,j} + u_{i+1,j} + u_{i,j-1} + u_{i,j+1}}{4}.} \tag{8.13}$$

But this is simply a discrete version of **the mean value property** (8.4); to find the value of the harmonic function u at a point on the grid, one averages the values over the neighboring grid points, the grid points which are exactly h away (see Figure 8.2, right). Thus, one again sees that the **mean value property** is the **essence of Laplace's equation**.

8.7. The Eigenfunctions and Eigenvalues of the Laplacian

Eigenfunctions and eigenvalues of the Laplacian are of great importance and indeed form the basis for **Fourier series and Fourier's method of separation of variables**. While we will address these topics in detail in Chapters 11 and 12, respectively, here we present some basic definitions and results.

Let Ω be a bounded domain in $\mathbb{R}^N$ ($N = 1, 2, 3$) with piecewise smooth boundary. Consider the following BVP with, for simplicity, Dirichlet boundary conditions. For $\lambda \in \mathbb{R}$, we seek to find a nontrivial solution $u \not\equiv 0$ to

$$\begin{cases} -\Delta u = \lambda u & \text{in } \Omega, \\ u = 0 & \text{on } \partial\Omega. \end{cases} \tag{8.14}$$

We call any nonzero solution to (8.14) an **eigenfunction** of (8.14) with corresponding **eigenvalue** λ, and we refer to the BVP (8.14) as an eigenvalue problem. The reader should immediately recall eigenvalues and eigenvectors of a matrix from **linear algebra**. These two notions are directly connected and (8.14) may be viewed as an "infinite-dimensional" analogue of the respective ideas in linear algebra. This connection is explored and discussed in Section 11.2.

A few comments are in order:

(i) The **minus sign** in front of the Laplacian is there for a reason: As we shall soon see, all the eigenvalues of (8.14) are positive. Without the minus sign, they would be negative.

(ii) Not every choice of $\lambda \in \mathbb{R}$ will be an eigenvalue of (8.14). For example, if $\lambda = 0$, (8.14) seeks a harmonic function in Ω with boundary values 0. We know from the Maximum Principle that the only such function is $u \equiv 0$, and hence 0 is **not** an eigenvalue. It turns out that there are also no eigenvalues which are negative.

(iii) If u is a solution to (8.14), so is cu for any constant c (note the analogy for the eigenvectors of a matrix).

We give four examples of eigenfunctions and eigenvalues of (8.14). The first two will seem familiar, while the latter two may involve some *special functions* which are new to the reader.

- Suppose $\Omega = (0,1)$ in $\mathbb{R}$ and note that in dimension 1, $-\Delta = -\frac{d^2}{dx^2}$. For any $n = 1, 2, 3, \ldots$, the function $v_n(x) = \sin n\pi x$ is an eigenfunction of (8.14) with corresponding eigenvalue $n^2\pi^2$. These eigenfunctions are the constituents of a Fourier series.
- Suppose Ω is the **unit cube** in $\mathbb{R}^3$, $(0,1) \times (0,1) \times (0,1)$. For any triple of positive integers (l, m, n), $v_{l,m,n}(x, y, z) := \sin l\pi x \, \sin m\pi y \, \sin n\pi z$ is an eigenfunction of (8.14) with corresponding eigenvalue $\lambda_{l,m,n} := \pi^2(l^2 + m^2 + n^2)$.
- Suppose Ω is the unit disc in $\mathbb{R}^2$. It can be seen (cf. Section 12.8.1) that in polar coordinates (r, θ), the eigenfunctions of (8.14) take the form $R_n(r)\Theta_n(\theta)$ where R_n are **Bessel functions** and the Θ_n are sines or cosines. The corresponding eigenvalues are in terms of the zeros of Bessel functions. Whereas we do not have a simple explicit formula for these zeros, they can be computed to any degree of precision.
- Suppose Ω is the unit ball in $\mathbb{R}^3$. It can be seen (cf. Section 12.9) that in spherical coordinates (r, θ, ψ), the eigenfunctions of (8.14) take the form $R(r)\Psi(\theta, \psi)$ where $R(r)$ are solutions to a **spherical Bessel equation** and the Ψ are known as **spherical harmonics** involving **Legendre polynomials**. Again, the corresponding eigenvalues are not explicit but can be computed to any degree of precision.

Before proceeding further, we need to define an inner product between two functions: a notion which generalizes the dot product of two vectors. For simplicity of notation, we present the definition for $N = 3$ (analogous definitions hold true in any space dimension). We work in the class of functions u defined on Ω which are square integrable on Ω; i.e., $\iiint_\Omega |u(\mathbf{x})|^2 \, d\mathbf{x} < \infty$. We call this class $L^2(\Omega)$. If $u, v \in L^2(\Omega)$, we define the L^2 **inner product** to be

$$(u, v) \; := \iiint_\Omega u(\mathbf{x}) \, v(\mathbf{x}) \, d\mathbf{x}. \tag{8.15}$$

We say u and v are **orthogonal** if $(u, v) = 0$. The L^2-norm of u is defined to be $\|u\| \; := (u, u)^{1/2} = \left(\iiint_\Omega |u(\mathbf{x})|^2 \, d\mathbf{x}\right)^{1/2}$.

Next, we state the following key results surrounding (8.14), a few of which we prove.

Properties of Eigenvalues and Eigenfunctions of (8.14).

Property 1: The eigenvalues of (8.14) are all **strictly positive**.
Property 2: The eigenvalues of (8.14) consist of a **countably** infinite set (i.e., a sequence) which we label as $0 < \lambda_1 \leq \lambda_2 \leq \lambda_3 \leq \cdots \leq \lambda_n \leq \cdots$, where $\lambda_n \to \infty$ as $n \to \infty$. Here, we include the equal signs as an eigenvalue could have more than one linearly independent[a] eigenfunction. Hence we repeat eigenvalues with multiplicity; in other words, each λ_n corresponds to (up to a constant) one eigenfunction.
Property 3: Eigenfunctions corresponding to distinct eigenvalues are **orthogonal** in the sense above.
Property 4: There exists a countable set consisting of the eigenfunctions (8.14) which constitutes a **"spanning set" for all reasonable functions**. This means that we can represent almost any function as a sum of appropriately chosen constants times these eigenfunctions. This property is the heart of Fourier series and Fourier's method for solving linear PDEs. This is indeed a bold and powerful statement which certainly needs to be made precise. We will do so (at least in dimension $N = 1$) in Chapter 11.
Property 5: Eigenvalues are directly related to minimizers of an **energy**.

[a] A finite set of functions is linealy independent if each function **cannot** be written as a linear combination of the other functions.

Properties 2 and 4 are deep; while we will present in Chapter 11 some partial results in dimension $N = 1$, full and complete proofs require the machinery of **functional analysis**. What we can readily do now is prove Properties 1 and 3, and then address Property 5.

To begin with, it is straightforward to show that all eigenvalues must be positive. Indeed, if Δ is an eigenvalue of (8.14) with corresponding eigenfunction u, then

$$\iiint_\Omega (-\Delta u)\, u \, d\mathbf{x} \;=\; \lambda \iiint_\Omega |u|^2 \, d\mathbf{x}.$$

On the other hand by applying Green's First Identity (cf. (A.16) in the Appendix), we find

$$\iiint_\Omega (-\Delta u)\, u \, d\mathbf{x} \;=\; \iiint_\Omega |\nabla u|^2 \, d\mathbf{x} \;>\; 0.$$

Thus $\lambda > 0$.

One can also readily show the orthogonality of any two eigenfunctions corresponding to **different** eigenvalues. Indeed, let v_i and v_j be eigenfunctions corresponding to eigenvalues λ_i and λ_j with $\lambda_i \neq \lambda_j$. Applying Green's Second Identity (cf. (A.17) in the Appendix) to v_i and v_j we find

$$\iiint_\Omega v_i\,(\Delta v_j) \;-\; v_j\,(\Delta v_i)\, d\mathbf{x} \;=\; 0,$$

since both functions vanish on the boundary. On the other hand, v_i and v_j are eigenfunctions and hence

$$\iiint_\Omega v_i\,(\Delta v_j) - v_j\,(\Delta v_i)\,d\mathbf{x} = (\lambda_j - \lambda_i)\iiint_\Omega v_i\, v_j\, d\mathbf{x}.$$

Since $\lambda_j - \lambda_i \neq 0$, we must have $(v_i, v_j) = \iiint_\Omega v_i\, v_j\, d\mathbf{x} = 0$.

In the next subsection, we address Property 5 and the relationship of the eigenvalues with the **Dirichlet energy** of Section 8.4.3.

8.7.1. Eigenvalues and Energy: The Rayleigh Quotient.

Given any w in

$$\mathcal{A} := \left\{ w \in C^2(\Omega) \cap C(\overline{\Omega}) \;\middle|\; w = 0 \text{ on } \partial\Omega,\ w \not\equiv 0 \right\},$$

we define its **Rayleigh quotient** to be

$$\mathcal{R}[w] := \frac{\|\nabla w\|^2}{\|w\|^2} = \frac{\iiint_\Omega |\nabla w|^2\, d\mathbf{x}}{\iiint_\Omega |w|^2\, d\mathbf{x}}.$$

We consider the variational problem:

$$\text{Minimize} \quad \mathcal{R}[w] \qquad \text{over } w \in \mathcal{A}. \tag{8.16}$$

We will assume this minimizer exists (proving this requires techniques beyond the scope of this text). First, let us note that this variational problem (8.16) is **equivalent to** (cf. Exercise **8.25**) the *constrained* Dirichlet energy problem:

$$\text{Minimize} \quad \iiint_\Omega |\nabla w|^2\, d\mathbf{x} \qquad \text{over } w \in \mathcal{A} \text{ with } \iiint_\Omega |w|^2\, d\mathbf{x} = 1. \tag{8.17}$$

Hence, minimizing the Rayleigh quotient is the same as minimizing the **Dirichlet energy** (8.8) with an integral constraint.

We have the following theorem which holds in any space dimension; for convenience, we state it in 3D.

Theorem 8.4. *Suppose $u \in \mathcal{A}$ is a minimizer of (8.16) with a minimizing value m. Then m is the **smallest** eigenvalue of (8.14) with corresponding eigenfunction u. In other words,*

$$\lambda_1 = m := \mathcal{R}[u] := \min_{w \in \mathcal{A}} \mathcal{R}[w] \qquad \textit{and} \qquad -\Delta u = \lambda_1 u \ \textit{ in } \Omega.$$

Proof. We follow the exact same reasoning as in one direction of the proof of Theorem 8.3. Fix for the moment $v \in \mathcal{A}$. For $\epsilon \in \mathbb{R}$, let $w = u + \epsilon v \in \mathcal{A}$ and

$$f(\epsilon) := \mathcal{R}[w] = \frac{\iiint_\Omega |\nabla(u+\epsilon v)|^2\, d\mathbf{x}}{\iiint_\Omega |u+\epsilon v|^2\, d\mathbf{x}} = \frac{\iiint_\Omega |\nabla u|^2 + 2\epsilon \nabla u \cdot \nabla v + \epsilon^2 |\nabla v|^2\, d\mathbf{x}}{\iiint_\Omega u^2 + 2\epsilon u v + \epsilon^2 v^2\, d\mathbf{x}}.$$

Since u is a minimizer, $f(\epsilon)$ must have a minimum at $\epsilon 0$; hence $f'(0) = 0$. Computing the derivative of f with respect to ϵ, we find

$$0 = f'(0) = \frac{\left(\iiint_\Omega |u|^2\, d\mathbf{x}\right)\left(2\iiint_\Omega \nabla u \cdot \nabla v\, d\mathbf{x}\right) - \left(\iiint_\Omega |\nabla u|^2\, d\mathbf{x}\right)\left(2\iiint_\Omega u v\, d\mathbf{x}\right)}{\left(\iiint_\Omega |u|^2\, d\mathbf{x}\right)^2}$$

and hence

$$\iiint_\Omega \nabla u \cdot \nabla v \, d\mathbf{x} = \frac{\iiint_\Omega |\nabla u|^2 \, d\mathbf{x}}{\iiint_\Omega |u|^2 \, d\mathbf{x}} \iiint_\Omega uv \, d\mathbf{x} = m \iiint_\Omega uv \, d\mathbf{x}.$$

Using Green's First Identity (cf. (A.16) in the Appendix) we arrive at

$$\iiint_\Omega -\Delta u v \, d\mathbf{x} = m \iiint_\Omega uv \, d\mathbf{x}.$$

Now we note that the above holds regardless of the choice of $v \in \mathcal{A}$. Hence the IPW Theorem, Theorem A.7 of the Appendix, yields that $-\Delta u = mu$ in Ω. Thus m is an eigenvalue corresponding to eigenfunction u.

We now show that m is the smallest eigenvalue. Let λ be **any** eigenvalue of (8.14) with corresponding eigenfunction v; i.e., $-\Delta v = \lambda v$. Our goal is to show that we must have $m \leq \lambda$. To see this, we apply Green's First Identity to find

$$m \leq \frac{\iiint_\Omega |\nabla v|^2 \, d\mathbf{x}}{\iiint_\Omega |v|^2 \, d\mathbf{x}} = \frac{\iiint_\Omega (-\Delta v)\, v \, d\mathbf{x}}{\iiint_\Omega |v|^2 \, d\mathbf{x}} = \frac{\iiint_\Omega \lambda v^2 \, d\mathbf{x}}{\iiint_\Omega |v|^2 \, d\mathbf{x}} = \lambda. \qquad \square$$

Example 8.7.1. We present a simple application of Theorem 8.4 in 1D. This is a question you can ask your fellow students and impress (or shock) them with the answer! What is the minimum value of $\int_0^1 |w'(x)|^2 \, dx$ over all $w \in C^1[0,1]$ with $w(0) = 0 = w(1)$ and $\int_0^1 |w(x)|^2 \, dx = 1$? The answer is π^2. Indeed, Theorem 8.4 and the equivalence of (8.16) and (8.17) implies that the answer is the first eigenvalue of $-\frac{d^2}{dx^2}$ on $[0,1]$ with Dirichlet boundary conditions. This is easily shown to be π^2 with corresponding minimizer (i.e., eigenfunction) $\sin \pi x$.

Note further that by computing the Rayleigh quotient for any particular choice of w, Theorem 8.4 immediately implies that we obtain an upper bound on λ_1. For example, since $w(x) = x(1-x) \in \mathcal{A}$, we can compute an upper bound on λ_1 via $\mathcal{R}[w]$ which is approximately 9.87. This value is slightly above π^2 but pretty close. However be careful here — we got lucky!

We remark that for general domains, it is impossible to find exact values for the eigenvalues. An important approximation method, known as **the Rayleigh-Ritz method**, is based upon the previous observation of using a particular test function w to approximate λ_1.

The eigenvalue λ_1 is often referred to as the **principal eigenvalue**; it gives the minimum energy. What about the other eigenvalues? They also satisfy a variational principle (i.e., they minimize something) **but** there are now additional **constraints**. To this end, let us assume Property 2 and denote the eigenvalues by $0 < \lambda_1 \leq \lambda_2 \leq \lambda_3 \leq \cdots \leq \lambda_n \leq \cdots$, repeating with multiplicity. Thus each eigenvalue λ_n has, up to a multiplicative constant, exactly one eigenfunction which we denote by v_n.

Theorem 8.5. *For each $n = 2, 3, \ldots$, we have*

$$\lambda_n = \min_{w \in \mathcal{A}_n} \mathcal{R}[w] \quad \textit{where} \quad \mathcal{A}_n := \left\{ w \in \mathcal{A} \,\middle|\, (w, v_i) = 0, i = 1, 2, \ldots, n-1 \right\},$$

where we assume that the minimizer exists.

Proof. Fix such an n, and define $m_n := \min_{w \in \mathcal{A}_n} \mathcal{R}[w]$. Let u denote the minimizer associated with m_n; i.e., $m_n = \mathcal{R}[u]$. By following the previous argument of considering a perturbation $w = u + \epsilon v$, we find

$$\iiint_\Omega (\Delta u + m_n u)\, v \, d\mathbf{x} = 0 \qquad \text{for every } v \in \mathcal{A}_n. \tag{8.18}$$

Note that we cannot yet use the IPW Theorem, Theorem A.7 of the Appendix, to conclude $\Delta u + m_n u = 0$ since the above holds only for $v \in \mathcal{A}_n$. We need to show that (8.18) holds for all $v \in \mathcal{A}$. However, what functions are in $\mathcal{A}$ which are not in $v \in \mathcal{A}_n$? These are precisely all linear combinations of the eigenfunctions $v_1, v_2, \ldots, v_{n-1}$.

Consider one of these eigenfunctions v_i, $i = 1, \ldots, n-1$, and apply Green's Second Identity (cf. (A.17) in the Appendix) to the minimizer u and v_i. This gives

$$\iiint_\Omega (\Delta u + m_n u)\, v_i \, d\mathbf{x} = \iiint_\Omega u(\Delta v_i + m_n v_i) \, d\mathbf{x} = (m_n - \lambda_i) \iiint_\Omega u v_i \, d\mathbf{x} = 0,$$

since $(u, v_i) = 0$. Thus (8.18) holds for all $v \in \mathcal{A}$ and the IPW Theorem implies that $-\Delta u = m_n u$ in Ω. Hence, m_n is also an eigenvalue.

Next we claim that if λ is another eigenvalue distinct from $\lambda_1, \ldots, \lambda_{n-1}$, then $m_n \le \lambda$. To this end, consider such a λ and let v^* denote the associated eigenfunction. Since λ is distinct from $\lambda_1, \ldots, \lambda_{n-1}$, v^* is orthogonal to the eigenfunctions v_i, $i = 1, \ldots, n-1$, and hence $v^* \in \mathcal{A}_n$. Thus

$$m_n \le \frac{\iiint_\Omega |\nabla v^*|^2 \, d\mathbf{x}}{\iiint_\Omega |v^*|^2 \, d\mathbf{x}} = \frac{\iiint_\Omega -\Delta v^* \, v^* \, d\mathbf{x}}{\iiint_\Omega |v^*|^2 \, d\mathbf{x}} = \frac{\iiint_\Omega \lambda v^* \, v^* \, d\mathbf{x}}{\iiint_\Omega |v^*|^2 \, d\mathbf{x}} = \lambda.$$

Lastly, since we have repeated eigenvalues with multiplicity, it may be possible for the value of m_n to be λ_i for certain $i = 1, \ldots, n-1$. However, we can argue by induction that the definition of m_n implies that $m_n \ge \lambda_i$ for all $1 \le i \le n-1$. Thus by our ordering of the eigenvalues, we must have $m_n = \lambda_n$. □

8.8. The Laplacian and Curvature

A problem in geometry is to quantify the "curvature" of the graph of a smooth function of two variables. This problem traditionally lies within the vast field of differential geometry and its study elucidates a profound distinction between **intrinsic and extrinsic** aspects of surfaces. In this section, we simply address the direct connection between curvature and the Laplacian. In particular, we do so with none of the formalism of differential geometry.

8.8.1. Principal Curvatures of a Surface.

When one first encounters the definition of the Laplacian operator, it may seem rather convoluted: $\Delta = \operatorname{div} \nabla$. The combination of the Divergence Theorem and the interpretation of the gradient guarantees its importance. However, given a function, say, $u(x, y)$, one could still ask exactly what Δu at a particular point tells us. An experimentally inclined mathematician might try plotting a particular u with its Laplacian superimposed (see Figure 8.3).

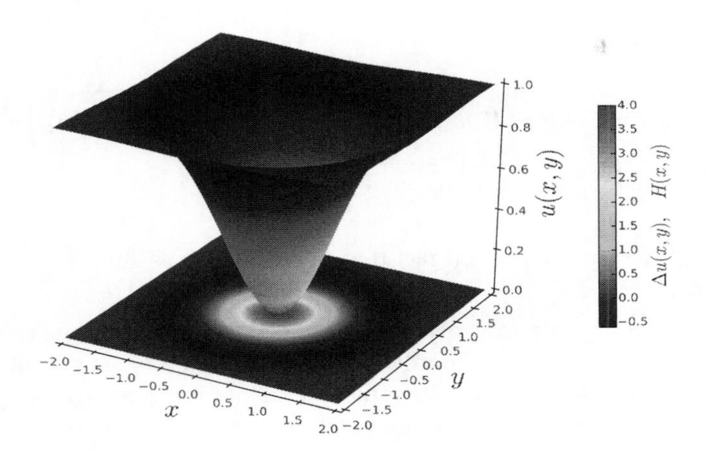

Figure 8.3. A plot of the surface $z = u(x, y)$ with a particular shading (the electronic version is in color): The contour plot in the xy-plane is for the Laplacian, while the surface is colored according to its the mean curvature H (defined in the text).

From Figure 8.3, it appears that minima of our scalar field $u(x, y)$ are peaks of u's Laplacian. Similarly, maxima of the scalar field are mapped to minima via the Laplacian. In other words, the values of (x, y) at which the u curve upwards or downwards are translated into peaks of u's Laplacian. This is certainly suggestive, but let us be more quantitative.

Suppose that $u(x, y)$ obtains a local minimum of 0 at the origin. Expanding about the origin by the **multivariable Taylor's Theorem**[6], we find

$$u(\mathbf{x}) = u(\mathbf{0}) + \nabla u^T(\mathbf{0})\mathbf{x} + \frac{1}{2}\mathbf{x}^T\mathbf{H}[u](\mathbf{0})\mathbf{x} + \mathcal{O}(|\mathbf{x}|^3)$$

where $\mathbf{H}[u](\mathbf{0})$ is the **Hessian matrix**[7] of second derivatives of u evaluated at the origin. Since $u(x, y)$ obtains a local minimum of 0 at the origin, we have $u(\mathbf{0}) = 0$ and $\nabla u(\mathbf{0}) = \mathbf{0}$. Thus, writing the above in terms of coordinates $\mathbf{x} = (x, y)$, we have

$$u(\mathbf{x}) = \frac{1}{2}\left(u_{xx}(0,0)x^2 + 2u_{xy}(0,0)xy + u_{yy}(0,0)y^2\right) + \mathcal{O}((x^2 + y^2)^{3/2}).$$

This translates as saying the behavior of the function u around $\mathbf{0}$ is dictated by the matrix $\mathbf{H}[u](\mathbf{0})$ and its eigenvalues. Given any 2×2 matrix $\mathbf{A}$, the eigenvalues $\lambda_\pm$ are determined by the formula

$$\lambda_\pm = \frac{1}{2}\left(\mathrm{Tr}\{\mathbf{A}\} \pm \sqrt{(\mathrm{Tr}\{\mathbf{A}\})^2 - 4\det\{\mathbf{A}\}}\right),$$

where Tr denotes the trace (sum of diagonal entries) of the matrix and det denotes its determinant. In terms of our 2×2 Hessian matrix $\mathbf{H}[u](\mathbf{0})$, this means the eigenvalues

[6]Since we will rarely use this important theorem in this text, it was omitted from the Appendix; however, the reader is encouraged to review it online.

[7]Recall that for a function u of two variables, the Hessian matrix is

$$\mathbf{H}[u] = \begin{bmatrix} u_{xx} & u_{xy} \\ u_{yx} & u_{yy} \end{bmatrix}.$$

are given by

$$\begin{aligned}\lambda_{\pm} &= \frac{1}{2}\left(\operatorname{Tr}\{\mathbf{H}[u](\mathbf{0})\} \pm \sqrt{(\operatorname{Tr}\{\mathbf{H}[u](\mathbf{0})\})^2 - 4\det\{\mathbf{H}[u](\mathbf{0})\}}\right)\\ &= \frac{1}{2}\left(\Delta u(\mathbf{0}) \pm \sqrt{(\Delta u(\mathbf{0}))^2 - 4\det\{\mathbf{H}[u](\mathbf{0})\}}\right).\end{aligned}$$

Here, we used the very important fact that the Laplacian is simply the trace of the Hessian matrix.

At a critical point, these eigenvalues of the Hessian define the **principal curvatures** κ_1 and κ_2 of the surface described by the equation $z = u(x, y)$ at the origin. Note that upon some thought, this aligns nicely with our intuition from single variable calculus, namely that the concavity of a graph is described by the second derivative. Hence we can generalize to infer that the curvature of a surface is described by the amplitudes of the principal quadratic approximation to its local minima. As we will soon see, curvature away from minima is a little more nuanced and requires further formalism.

8.8.2. Mean Curvature.

The **mean curvature**, H, is defined to be the average of the principal curvatures

$$H := \frac{1}{2}(\kappa_1 + \kappa_2). \tag{8.19}$$

As previously noted, at a critical point these principal curvatures are equal to the eigenvalues of the Hessian. Hence, the above analysis shows that at a critical point of $u(x, y)$, we have

$$\begin{aligned}H &= \frac{1}{2}(\lambda_+ + \lambda_-)\\ &= \frac{1}{4}\left(\left(\Delta u + \sqrt{(\Delta u)^2 - 4\det\mathbf{H}[u]}\right) + \left(\Delta u - \sqrt{(\Delta u)^2 - 4\det\mathbf{H}[u]}\right)\right)\\ &= \frac{1}{2}\Delta u.\end{aligned}$$

Thus as we observed, the mean curvature is indeed proportional to the Laplacian of our function at a critical point.

At a general point $\mathbf{x}$, it is not necessarily the case that $H = \frac{1}{2}\Delta u$. However, it is true in a coordinate system wherein the tangent plane to the surface is flat: parallel to the plane of the new coordinates. In general, there is a far more robust way of calculating mean curvature which can be applied at any point (not just critical points). Given any orientable surface in $\mathbb{R}^3$, there is a well-defined unit normal $\mathbf{n}$. Assuming the 3D divergence of this normal can be calculated, a very useful characterization is that

$$H = -\frac{1}{2}\operatorname{div}_{3D}\mathbf{n}. \tag{8.20}$$

Note that while this may at first sight seem surprising, it is intuitive: When the surface is flatter, the divergence of unit vectors will be smaller, and when the normal has wild deviations, the divergence will capture it. Our unit normal vector field to a graph of the level set $f(x, y, z) := z - u(x, y) = 0$ is given by its 3D gradient ∇f. Hence, we find

$$\mathbf{n} = \frac{(-\nabla u, 1)}{\sqrt{|\nabla u|^2 + 1}},$$

where the 2D gradient ∇ is with respect to x and y. Here we have taken the normal to the surface $z = u(x, y)$ which points upwards. Hence by (8.20), we have

$$\begin{aligned} H &= \frac{1}{2}\mathrm{div}_{2D}\left(\frac{\nabla u}{\sqrt{|\nabla u|^2 + 1}}\right) \\ &= \frac{1}{2}\left(\frac{1}{\sqrt{|\nabla u|^2 + 1}}\underbrace{\mathrm{div}_{2D}\,\nabla u}_{\Delta u} + \nabla\left(\frac{1}{\sqrt{|\nabla u|^2 + 1}}\right)\cdot \nabla u\right) \end{aligned} \tag{8.21}$$

where we have applied the product rule (cf. Exercise **A.13** in the Appendix A) for the divergence operator. We immediately see that if we are at a critical point wherein the gradient ∇u vanishes, the mean curvature reduces to our pure halved Laplacian, consistent with what we have previously found.

8.8.3. Curvature and Invariance. As was already alluded to, the concept of curvature is actually quite a nuanced one. Indeed, one may already ask why taking the mean of the principal curvatures is the correct/useful quantity to study. Indeed, why not study quantities like

$$\lambda_+\lambda_-, \qquad \left(\frac{1}{\lambda_+} + \frac{1}{\lambda_-}\right)^{-1}, \qquad \lambda_+^3 + \lambda_-^3?$$

In fact, the first quantity is what is known as the **Gaussian**[8] **curvature**, $K := \lambda_+\lambda_-$, which no longer has a direct connection with the Laplacian but is arguably an even more fundamental description of curvature than the mean. To understand the key differences between mean and Gaussian curvatures, one needs to understand how surfaces transform under different **embeddings**. The key difference between the two (loosely speaking) is that in a sense the Gaussian curvature is independent of how we visualize the surface. If we were to keep all distances between points on our surface locally identical but change the (embedding) space in which the surface resides, the Gaussian curvature will be invariant but the mean curvature will take on different values. Hence we differentiate the two by saying that Gaussian curvature is **intrinsic** to the surface while the mean curvature is **extrinsic.** In other words, if you were an ant living on the surface, you would only ever be able to measure the Gaussian curvature. The mean curvature is only available for computation to those living in the embedding space. A simple example would be a 2D **cylinder** which can be embedded in $\mathbb{R}^3$ as the zero level set of $f(x, y, z) = x^2 + y^2$. While the mean curvature is not zero, the Gaussian curvature is 0, indicating that this surface is intrinsically flat. You might say that it does not look flat when it is placed in three-dimensional space, but this is an extrinsic view of the surface.

[8]The German mathematician and physicist **Johann Carl Friedrich Gauss** (1777–1855) should not require a footnote — he was (and is) simply a giant of science.

A slightly more involved example is the **torus**, which can be defined as a Cartesian product of two rings. By a ring, we mean $\{\mathbf{x} \in \mathbb{R}^2 : |\mathbf{x}| = 1\}$. Indeed, imagine taking a large single ring and glueing several rings of a smaller radius to each point on the original ring. You will get the boundary of a donut! Note however that you can very naturally embed this set in $\mathbb{R}^4$ as $\{\mathbf{x} = (x_1, x_2, y_1, y_2) \in \mathbb{R}^4 : x_1^2 + x_2^2 = 1, y_1^2 + y_2^2 = 1\}$ instead of this "extrinsically curved" embedding in $\mathbb{R}^3$, and with more differential geometric machinery you can show that this intersection of cylinders in $\mathbb{R}^4$ is locally flat. Or alternatively, we can realize the torus as the quotient $\mathbb{R}^2/\mathbb{Z}^2$ identifying points on the plane differing by a standard unit vector: Try taking a piece of paper and identifying opposing edges with tape. The intrinsic flatness is thus inherited from the plane.

The intrinsic nature of the Gaussian curvature was a discovery of Gauss and launched the modern field of differential geometry[9]. On the other hand, the problem of finding a surface $z = u(x, y)$ with **specified** Gaussian curvature $K(x, y)$ entails (cf. Exercise **8.28**) solving an extremely difficult, fully nonlinear PDE called a **Monge-Ampère equation**:

$$\det \mathbf{H}[u] = K(x, y)(1 + |\nabla u|^2)^2.$$

8.9. Chapter Summary

- The **Laplacian** is a particular combination of derivatives stemming from the divergence of the gradient:

$$\Delta u = \nabla \cdot \nabla u = \operatorname{div} \nabla u = \sum_{i=1}^{n} u_{x_i x_i}.$$

- There are many important PDEs and boundary value problems (BVPs) involving the Laplacian. For example, **the Dirichlet problem** asks for a harmonic function inside a domain Ω, i.e., a function whose Laplacian vanishes in Ω, with specified boundary conditions. When we specify the value of the Laplacian throughout the domain Ω, we obtain **the Poisson equation**. These problems are ubiquitous in physics and are also associated with random walks in higher space dimensions.
- **Harmonic functions** are very special and represent concentration densities which are in **equilibrium**. They have several important properties:
  - **The Mean Value Property:** The value of any harmonic function at a point equals the average value over any sphere or ball centered at the point.
  - **The Maximum Principle:** The maximum value of any harmonic function on a bounded domain must be attained on the boundary of the domain.
  - **The Dirichlet Principle:** A harmonic function on a domain Ω with boundary values g is the function which minimizes $\int \cdots \int_\Omega |\nabla u|^2 \, d\mathbf{x}$, among all functions u with the same boundary values g.
  - **Smoothness:** Any harmonic function must be C^∞ on its domain of definition.

[9]For more details and a proof of the invariance theorem see the book *An Introduction to Differentiable Manifolds and Riemannian Geometry* by William M. Boothby, Academic Press, second edition, 2002.

- Given the rotational invariance of the Laplacian, we looked for all **radial solutions** of Laplace's equation in space dimensions $N \geq 2$. Except for the trivial case of constant solutions, all radial solutions were found to **blow up** at the origin, with the size/nature of the blow-up different in each dimension. In each dimension $N \geq 2$, we defined one such solution and called it the **fundamental solution of the Laplacian.** Its power will be unlocked in Chapter 10.
- If we **discretize** Laplace's equation on a **finite grid** and replace the second derivatives with central finite differences, we obtain a discrete version of the **mean value property**: The value at any grid point is the average of the values at neighboring grid points.
- The Laplacian is related to the notion of **mean curvature** of a surface. In particular, consider a surface described as the graph $z = u(x, y)$, and let $(x_0, y_0, u(x_0, y_0))$ be a point on the surface. Suppose further that the coordinate system (x and y) is chosen such that the tangent plane at the point is parallel to the xy-plane. Then the Laplacian of u at (x_0, y_0) gives the mean curvature of the surface at point $(x_0, y_0, u(x_0, y_0))$.

Exercises

8.1 Provide an example of a nonlinear function on $\mathbb{R}^2$ which is harmonic everywhere.

8.2 Let Ω be a connected domain in $\mathbb{R}^N$ and let $u : \Omega \to \mathbb{R}$ be harmonic such that u^2 is also harmonic in Ω. Prove that u is constant in Ω. Hint: First show that if u and v are harmonic, then the product uv is harmonic if and only if $\nabla u \cdot \nabla v = 0$.

8.3 (a) Show directly using the chain rule that the 2D Laplace equation in polar coordinates for a function $u(r, \theta)$ is $\left[\partial_{rr} + \frac{1}{r}\partial_r + \frac{1}{r^2}\partial_{\theta\theta}\right] u = 0$.
(b) Show directly using the chain rule that the 3D Laplace equation in spherical coordinates for a function $u(r, \phi, \theta)$ is

$$\left[\frac{1}{r^2}\partial_r\left(r^2\partial_r\right) + \frac{1}{r^2 \sin\phi}\partial_\phi\left[\sin\phi\, \partial_\phi\right] + \frac{1}{r^2 \sin^2\phi}\partial_{\theta\theta}\right] u = 0.$$

8.4 Show in 3D that the Laplacian is rotationally invariant. That is, suppose we rotated our $\mathbf{x} = (x_1, x_2, x_3)$ coordinate system by performing a rotation $\mathbf{R}$. Such an $\mathbf{R}$ is captured by a 3×3 orthogonal matrix and the new (rotated) coordinates are $\mathbf{x}' = \mathbf{R}\mathbf{x}$. Show that the Laplacian transforms in the new coordinate system $\mathbf{x}' = (x'_1, x'_2, x'_3)$ to

$$\Delta_{\mathbf{x}'} = u_{x'_1 x'_1} + u_{x'_2 x'_2} + u_{x'_3 x'_3}.$$

8.5 (a) Let $\mathbf{x} = (x_1, x_2, x_3)$ and consider the function defined for $\mathbf{x} \neq \mathbf{0}$, $u(\mathbf{x}) = \frac{1}{|\mathbf{x}|}$. Show by direct differentiation that for $\mathbf{x} \neq \mathbf{0}$, $\Delta u(\mathbf{x}) = 0$. Thus, u is a radially symmetric *harmonic* function away from the origin.
(b) Now consider the same function $\frac{1}{|\mathbf{x}|}$ in 2D. Show that for $\mathbf{x} \neq \mathbf{0}$, $\Delta u(\mathbf{x}) \neq 0$.

8.6 (**Harmonic Polynomials**) A polynomial in N variables which is harmonic on $\mathbb{R}^N$ is called a harmonic polynomial. A polynomial is homogeneous if all terms are of the same order. Let $N = 2$ and consider homogeneous harmonic polynomials in two variables of order m. Show that for each m, there are exactly two such linearly independent (i.e., not a scalar multiple of the other) homogeneous harmonic polynomials. In other words, the dimension of the vector space of all homogeneous harmonic polynomials in two variables of order m is two. Hint: Work in polar coordinates.

8.7 (a) Prove (8.5), the mean value property over solid balls in 3D.
(b) State the mean value property in 2D and prove it. The proof in 2D only requires a few minor modifications.

8.8 From the statement and proof of the mean value property, the reader may have noticed a similarity with **spherical means for the 3D wave equation**. Indeed, prove Theorem 8.1 by noting that a smooth 3D harmonic function is automatically a smooth steady-state (no time dependence) solution to the 3D wave equation. Then recall that the spherical means solve the Euler-Poisson-Darboux equation (4.19). Note that the key observation here is that the spherical means (which only depend on r) are a **smooth** solution to (4.19).

8.9 Suppose $u(\mathbf{x})$ is a smooth solution to $\Delta u = f$ in Ω (some bounded connected open subset of $\mathbb{R}^3$) and $\frac{\partial u}{\partial n} = g$ on $\partial\Omega$. Show that

$$\iiint_\Omega f \, d\mathbf{x} = \iint_{\partial\Omega} g \, dS.$$

8.10 (**Laplace's Equation Should Not Be Viewed as an Evolution Equation**) Suppose we consider Laplace's equation in dimension $N = 2$ and think of y as time t. Thus, we consider the initial value problem $u_{tt} + u_{xx} = 0$ with $u(x, 0) = f(x)$ and $u_t(x, 0) = g(x)$. Check directly (i.e., differentiate) that for any positive integer n, if $f(x) = 0$ and $g_n(x) = \frac{\sin nx}{n}$, then a solution is

$$u_n(x, t) = \frac{\sinh nt \, \sin nx}{n^2}.$$

As n tends to infinity, the initial data tends to zero (in fact uniformly). But what happens to the solution at any fixed later time t? Hence, is the problem *well-posed* in terms of stability with respect to small changes in the initial data?

8.11 Suppose that $u(r, \theta)$ is a harmonic function in the disk with center at the origin and radius 2 and that on the boundary $r = 2$ we have $u = 3\sin 2\theta + 1$. Without finding the solution, answer the following questions:
(a) What is the maximum value of u on the closed disk $\{(r, \theta) | r \leq 2\}$?
(b) What is the value of u at the origin?

8.12 (**Laplace's Equation in Fluid Dynamics**) This exercise requires the reader to be familiar with Section 2.8. Consider the spatial (Eulerian) description of the fluid $\mathbf{u}(\mathbf{x}, t)$. A fluid is called **irrotational** if for each t, curl $\mathbf{u} = \nabla \times \mathbf{u} = 0$. Show that in an irrotational fluid, at each time t there exists a function $\phi(\mathbf{x})$ (called a potential) such that $\mathbf{u} = -\nabla\phi$. Hence for irrotational flow, the incompressibility condition div $\mathbf{u} = 0$ is equivalent to $\Delta\phi = 0$. Hint: Look up the *Helmholtz decomposition*.

8.13 **(Liouville's Theorem)** Let $\Delta u(\mathbf{x}) = 0$ for $\mathbf{x}$ in the plane (i.e., u is harmonic everywhere). Suppose $|u|$ is bounded; i.e., suppose there exists a constant $M > 0$ such that $|u(\mathbf{x})| \leq M$ for all $\mathbf{x}$. Prove that u must be constant. Interpret this when u represents the temperature at a point $\mathbf{x}$ in the plane.

Hint: First note that if u is harmonic, then so is any derivative of u. Fix $\mathbf{x}_0$. For each derivative of u, apply the mean value property on $B(\mathbf{x}_0, r)$, together with the Divergence Theorem, to show that $|\nabla u(\mathbf{x}_0)| \leq \frac{CM}{r}$, for some constant C.

8.14 Show by example that the Maximum Principle and uniqueness for the Dirichlet problem are false if Ω is unbounded. Hint: Let $N = 2$ and let Ω be the top half-plane whose boundary is the x-axis. Look for a nonconstant harmonic function in $\mathbb{R}^2$ which vanishes on the x-axis.

8.15 **(Numerical Approximation to Laplace's Equation)**

(a) Consider the Dirichlet problem in 1D on the interval $\Omega = (0, 1)$: $u_{xx} = 0$ with $u(0) = -1$, $u(1) = 1$. Discretize the interval into 4 pieces of step size $h = \frac{1}{4}$. Label the interior vertices from 1, 2, 3 and let $\mathbf{v}$ be the vector denoting the values of u at the respective vertices. Write all the algebraic equations in the form $\mathbf{A}\,\mathbf{v} = \mathbf{b}$, for some 3×3 matrix $\mathbf{A}$ and 3 vector $\mathbf{b}$. Note that what you know (i.e., the boundary values) will go into $\mathbf{b}$. Solve this equation and plot the approximate solution. Compare with the exact solution.

(b) Consider the 2D BVP for the Poisson equation: $\Delta u = 1$ on Ω with $u = 0$ on $\partial\Omega$, where Ω is the interior of the 2D unit square $(0, 1) \times (0, 1)$. Divide the square into 9 interior grid points with step size $h = \frac{1}{4}$ and label these points $1, \ldots, 9$ as shown:

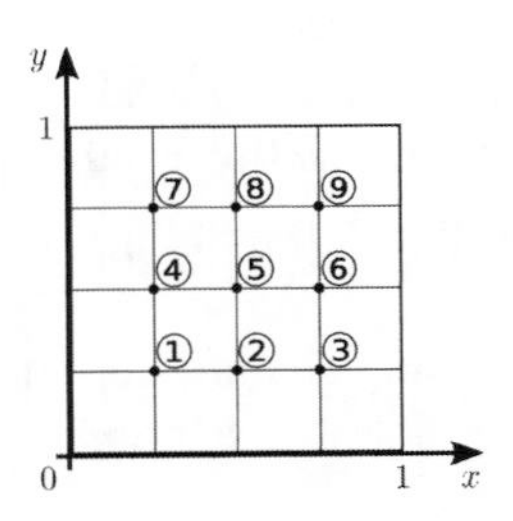

There will also be 16 boundary grid points at which you know the solution. Write the system of equations as $\mathbf{A}\,\mathbf{v} = \mathbf{b}$, for a 9×9 matrix $\mathbf{A}$ and 9 vector $\mathbf{b}$. Using any software (e.g., Matlab, Mathematica), invert the matrix, **solve** for $\mathbf{v}$, and **plot** the approximate solution.

8.16 **(Probabilistic Interpretation of Section 8.3)** Let Ω be the unit square whose boundary consists of two parts Γ_1 and Γ_2. Let $f(x, y)$ be a given function on space and let g be defined on $\partial\Omega$ by $g(x, y) = 1$ if $(x, y) \in \Gamma_1$ and $g(x, y) = 0$ if $(x, y) \in \Gamma_2$. Divide Ω into a rectangular grid, and at any grid point, perform a random walk with step size h ($h = 1/n$ for some integer n) and time step δt.

Suppose the following: (i) At each time step δt we receive a pay-off of $\delta t f(x, y)$ dollars, where (x, y) is the point at which we have arrived. (ii) When we reach the boundary we receive an additional dollar if we hit the Γ_1 part of the boundary; if we hit Γ_2, we receive nothing. For any grid point (x, y), let $u(x, y)$ be the expected pay-off of the game, assuming we start at grid point (x, y). Show that the expected pay-off $u(x, y)$ in the limit where h and δt both tend to 0 with σ^2 fixed solves

$$\begin{cases} u_{xx} + u_{yy} &= -4f(x, y)/\sigma^2 \qquad \text{on } \Omega, \\ \qquad u &= g \qquad \text{on } \partial\Omega. \end{cases}$$

8.17 (**Harmonic Functions in Complex Analysis**) This question is for students who have taken a course in complex variables. Consider a complex-valued function f of a complex variable z. Denoting a complex number by its real and imaginary parts, we can write $z = x + iy$ and $f(z) = u(x, y) + iv(x, y)$.
(a) Show that if f is differentiable with respect to z, then both $u(x, y)$ and $v(x, y)$ are harmonic functions for $(x, y) \in \mathbb{R}^2$.
(b) What are the analogous interpretations/results in complex variables of the mean value property and the Maximum Principle?

8.18 (**Harmonic Functions Are C^∞**) This question is for students interested in mathematical analysis. Prove that if u is a C^2 harmonic function on a bounded domain $\Omega \subset \mathbb{R}^3$, then $u \in C^\infty(\Omega)$. Hint: Consider $u_\epsilon := \phi_\epsilon * u$ which is defined on $\Omega_\epsilon := \{\mathbf{x} \in \Omega \,|\, \text{dist}(\mathbf{x}, \partial\Omega) \geq \epsilon\}$, where

$$\phi_\epsilon(\mathbf{x}) = C\phi_\epsilon(x_1)\phi_\epsilon(x_2)\phi_\epsilon(x_3) \qquad \text{with } \phi_\epsilon(x) := \begin{cases} e^{-\frac{1}{\epsilon^2 - x^2}} & \text{if } |x| < \epsilon, \\ 0 & \text{if } |x| \geq \epsilon, \end{cases}$$

and, for each $\epsilon > 0$, C is chosen so that $\iiint_{\mathbb{R}^3} \phi_\epsilon(\mathbf{x})\, d\mathbf{x} = 1$. Given that u satisfies the MVP, can u_ϵ be different than u?

8.19 (**Converse of the Mean Value Property**) Prove that if $u \in C^2(\Omega)$ and satisfies the mean value property (8.4) for all $B(\mathbf{x}_0, r) \subset \Omega$, then $\Delta u = 0$ in Ω. Hint: Consider the proof of Theorem 8.1 and reverse the arguments.

8.20 (**Subharmonic Functions**) Recall the definition of subharmonic functions from (8.7).
(a) Repeat the proof of the mean value property to show that if u is subharmonic, then for all $\mathbf{x}_0 \in \Omega$ and $r > 0$ such that $B(\mathbf{x}_0, r) \subset \Omega$, we have

$$u(\mathbf{x}_0) \leq \frac{1}{4\pi r^2} \iint_{\partial B(\mathbf{x}_0, r)} u(\mathbf{x})\, dS_{\mathbf{x}}.$$

(b) Show that the mean value inequality above is sufficient to deduce the maximum principle; that is, $\max_{\overline{\Omega}} u = \max_{\partial\Omega} u$.
(c) Define **superharmonic** functions. What are the analogous statements of (a) and (b) for superharmonic functions? Prove them.

The next two questions pertain to **comparison principles for the Laplacian** and require the Maximum Principle of the previous exercise. As in the previous exercise, it is convenient to work with $-\Delta$.

8.21 (a) Suppose Ω is a bounded domain and $u, v \in C^2(\Omega) \cap C(\overline{\Omega})$ satisfy

$$-\Delta u \leq 0, \; -\Delta v \geq 0 \quad \text{in } \Omega \quad \text{and} \quad u \leq v \text{ on } \partial\Omega.$$

Prove that $u \leq v$ in $\overline{\Omega}$.

(b) Note that the choice of a 0 right-hand side here is irrelevant: In other words, prove that the same result holds true if $-\Delta u \leq f(\mathbf{x})$ and $-\Delta v \geq f(\mathbf{x})$ for some function $f(\mathbf{x})$.

8.22 Let $f \in C(\overline{B(\mathbf{0},1)})$ and $g \in C(\partial B(\mathbf{0},1))$. Prove that there exists a constant A (depending only on the space dimension N) such that if $u \in C^2(B(\mathbf{0},1)) \cap C(\overline{B(\mathbf{0},1)})$ solves

$$\begin{cases} -\Delta u = f & \text{in} \quad B(\mathbf{0},1), \\ u = g & \text{on} \quad \partial B(\mathbf{0},1), \end{cases}$$

then

$$\max_{\overline{B(\mathbf{0},1)}} |u| \leq \max_{\partial B(\mathbf{0},1)} |g| + A \max_{\overline{B(\mathbf{0},1)}} |f|.$$

Hint: Use the comparison principle of the previous question and the fact that the function $\phi(\mathbf{x}) = \frac{1}{2N}\left(1 - |\mathbf{x}|^2\right)$ satisfies $-\Delta\phi = 1$.

8.23 (**Dirichlet's Principle for Poisson's Equation**). Consider the variant of the Dirichlet Principle wherein for some fixed $f \in C(\Omega)$ we minimize

$$E(w) := \iiint_\Omega \frac{1}{2}|\nabla w|^2 + wf \, d\mathbf{x}, \tag{8.22}$$

over $\mathcal{A}_h$. Prove that Theorem 8.3 holds true with (8.9) replaced with

$$\begin{cases} \Delta u = f & \text{in } \Omega, \\ u = h & \text{on } \partial\Omega. \end{cases}$$

8.24 (**Neumann and the Natural Boundary Conditions**) Consider a function $f \in C(\Omega)$ which integrates to 0. Suppose we minimize (8.22) over all $w \in C^2(\Omega) \cap C^1(\overline{\Omega})$ (no restriction on the boundary values). Show that the minimizer will solve the Neumann problem for Poisson's equation

$$\begin{cases} \Delta u = f & \text{in } \Omega, \\ \frac{\partial u}{\partial \mathbf{n}} = 0 & \text{on } \partial\Omega. \end{cases}$$

In other words, the Neumann boundary conditions *naturally pop up* when minimizing over an unconstrained class of functions. **Hint:** Follow the same reasoning as in the proof of Theorem 8.3 but use Green's First Identity.

8.25 Prove that the variational problems (8.16) and (8.17) involving the Rayleigh quotient are equivalent. That is, prove that the two minimum values agree and the respective minimizing functions are equal up to a multiplicative constant.

8.26 Let $\Omega \in \mathbb{R}^3$ be a bounded domain.
(a) Show that the BVP

$$\begin{cases} \Delta u = u & \text{in } \Omega, \\ u = 0 & \text{on } \partial\Omega \end{cases}$$

has only the trivial solution $u \equiv 0$. Hint: Multiply both sides of the PDE by u and integrate over Ω. Apply Green's First Identity (A.16).

(b) Prove that for any continuous function f (continuity is overkill here) defined on Ω, the BVP

$$\begin{cases} \Delta u = u - f & \text{in } \Omega, \\ u = 0 & \text{on } \partial\Omega \end{cases} \tag{8.23}$$

has a unique solution. Hint: Follow the same steps as in part (a) but now use the trivial inequality for $a, b \in \mathbb{R}$, $ab \leq \frac{a}{2} + \frac{b}{2}$, on the product uf.
(c) Use your work for part (b) to derive the following stability result for the BVP (8.23): Let $\epsilon > 0$ and suppose f_1 and f_2 are two functions on Ω such that

$$\iiint_\Omega |f_1(\mathbf{x}) - f_2(\mathbf{x})|^2 \, d\mathbf{x} < \epsilon.$$

Let u_i $(i = 1, 2)$ be a solution to the associated BVP (8.23) with $f = f_i$. Then

$$\iiint_\Omega |u_1(\mathbf{x}) - u_2(\mathbf{x})|^2 \, d\mathbf{x} < \epsilon \qquad \text{and} \qquad \iiint_\Omega |\nabla u_1(\mathbf{x}) - \nabla u_2(\mathbf{x})|^2 \, d\mathbf{x} < \epsilon.$$

8.27 (**The Laplacian and Curvature**)
(a) Use the chain rule to show from (8.21) that

$$\begin{aligned} H &= \frac{1}{2}\left(\frac{\Delta u}{\sqrt{|\nabla u|^2 + 1}} - \frac{1}{2}\left(\frac{1}{(|\nabla u|^2 + 1)^{3/2}}\right)(2|\nabla u|)\left(\frac{\nabla u}{|\nabla u|}\right) H[u] \cdot \nabla u\right) \\ &= \frac{1}{2\sqrt{1 + |\nabla u|^2}}\left(\Delta u - \frac{\nabla u^T H[u] \nabla u}{1 + |\nabla u|^2}\right). \end{aligned}$$

(b) Prove that the two notions of mean curvature (8.19) and (8.20) are equivalent at any general point of a graph. The principal curvatures away from the extrema are defined again by the Hessian matrix but upon rotating into a coordinate system where the tangent plane to the graph is parallel to an altered xy-plane. Namely, upon rotating back into the original coordinates show that the average of principal curvatures reduces to the formula from part (a).

8.28 (**Gaussian Curvature**) Show that for the graph of a smooth function $u(x, y)$, the Gaussian curvature K at (x, y) is given by

$$K(x, y) = \frac{\det \mathbf{H}[u]}{(1 + |\nabla u|^2)^2}.$$

8.29 Consider the discrete version of the Laplacian on a square $[0, 1]^2$. We discretize the inside and boundary of the square with points (x_i, y_j) where $i, j = 0, 1, 2, \ldots, n$. The relation between the values of u at the grid points is given by the scheme (8.13).
(a) Prove the **Discrete Maximum Principle**: If the max of u is attained at an interior grid point, then u is constant at all grid points.
(b) Now consider the **discrete Dirichlet problem** where we are given the values of u at the boundary grid points; i.e., the value of $u(x_i, y_j)$ is known if at least one index is 0 or n. We wish to use the scheme (8.13) to infer the value of u at the interior grid points. As before this will reduce to a system of coupled linear equations. Let $n = 4$, and let the boundary data be $u(x_i, y_j) = 1$ if either $i = 0$, $j = 1, 2, 3$ or $i = 4$, $j = 1, 2, 3$, and $u(x_i, y_j) = 0$ at all other boundary grid points. Numerically compute the solution at all other grid points.

8.30 (**A Simplified Version of Harnack's Inequality**) Let Ω be a domain in $\mathbb{R}^3$ and let B be a ball such that $\overline{B} \subset \Omega$. Then there exists a constant $C > 0$ such that

$$\max_{\mathbf{y}\in\overline{B}} u(\mathbf{y}) \leq C \min_{\mathbf{y}\in\overline{B}} u(\mathbf{y}) \tag{8.24}$$

for every nonnegative harmonic function u on Ω. The power of this statement is that the same constant C "works" for every nonnegative harmonic function.
(a) First let us try to get a sense of what this inequality says! Show that Harnack's inequality means that there exists a constant $C > 0$ such that for any nonnegative harmonic function u, we have

$$\frac{1}{C}u(\mathbf{y}) \leq u(\mathbf{x}) \leq Cu(\mathbf{y}) \qquad \text{for all } \mathbf{x} \text{ and } \mathbf{y} \text{ in } B.$$

Then argue that the inequality says that the values of **any** nonnegative harmonic function on B are *comparable* in the sense of very small values of u (u tending to 0) and very large values (u tending to $+\infty$). Interpret this statement in the context of equilibrium temperature distributions inside Ω.
(b) Use the mean value property to prove (8.24).

8.31 (**The Biharmonic Equation**) The biharmonic equation is a fourth-order linear PDE:

$$\Delta^2 u := \Delta\Delta u = 0.$$

Clearly any harmonic function trivially solves the biharmonic equation.
(a) Consider the radial function $u(\mathbf{x}) = \frac{1}{|\mathbf{x}|}$ defined on $\mathbb{R}^N\backslash\{\mathbf{0}\}$. Since in dimension $N = 3$ this function is harmonic on $\mathbb{R}^3\backslash\{\mathbf{0}\}$, it is also biharmonic. However, show that in dimension $N = 5$, u is biharmonic on $\mathbb{R}^5\backslash\{\mathbf{0}\}$ but not harmonic.
(b) From the point of view of applications, the biharmonic equation in $N = 2$ proves particularly important. Write out the equation in the x and y coordinates. Find all radial biharmonic solutions on $\mathbb{R}^2\backslash\{\mathbf{0}\}$.
(c) For those interested in continuum mechanics, research online to describe how this equation comes up in the linear elasticity of a beam.

8.32 (**The p-Laplacian**) For any $p > 1$, we define the p-Laplacian to the operator defined by

$$\Delta_p u := \nabla \cdot \left(|\nabla u|^{p-2}\nabla u\right).$$

Except for the case $p = 2$ (where we obtain the Laplacian), Δ_p is a **nonlinear** operator.
(a) Motivate this definition by proving (in analogy to the Dirichlet Prinicple of Theorem 8.3) that if $u \in \mathcal{A}_h$ minimizes

$$E_p(w) := \frac{1}{2}\iiint_\Omega |\nabla w|^p \, d\mathbf{x} \qquad \text{over all } w \in \mathcal{A}_h,$$

then $\Delta_p u = 0$ in Ω.
(b) Case $p = 1$: Describe the geometric interpretation of $\Delta_1 u$ in terms of curvature of a graph.
(c) For students with some exposure to real analysis, explore what happens as $p \to \infty$ and the resulting *infinity Laplacian*. By researching online, describe the relevance of the infinity Laplacian in mathematics and the sciences.

8.33 (**The Fractional Laplacian**) This exercise asks the reader to explore online a contemporary research topic of great interest in pure and applied mathematics.
(a) First off, write down the Laplacian in Fourier space (via the Fourier transform). Then for $0 < s < 1$, explore the notion of the fractional Laplacian $(-\Delta u)^s$ by first defining its Fourier transform. Write down the $(-\Delta u)^s$ as a *singular integral operator* and explain why we refer to the fractional Laplacian as **nonlocal** in contrast to the local Laplacian
(b) What properties and results of this chapter for the Laplacian and Laplace's equation extend to the fractional Laplacian?
(c) Briefly discuss the relevance of the fractional Laplacian in mathematics and the sciences.

8.34 (**The Laplacian on a Riemannian Manifold**) The theory behind the Laplacian and its eigenvalues beautifully extends to any Riemannian manifold. This exercise exposes you to a little of this fascinating subject. You can explore these notions online but we recommend consulting the excellent online undergraduate lecture notes of **Yaiza Canzani**[10].
(a) Define a Riemannian manifold $\mathcal{M}$ and the Laplacian on $\mathcal{M}$.
(b) Consider these notions for three simple examples of Riemannian manifolds: the circle S^1, the torus $T^2 = \mathbb{R}^2/\mathbb{Z}^2$, and the sphere S^2. What are the eigenvalues of the Laplacian in each case? We will address the case of the sphere in Section 12.9.3 of Chapter 12 on spherical harmonics.
(c) What are the analogous results of Section 8.7 for a general Riemannian manifold?

[10]Notes for Spectral Geometry, available at `https://canzani.web.unc.edu/wp-content/uploads/sites/12623/2018/07/Spectral-Geometry-Laval-minicourse.pdf`. Published in *Spectral Theory and Applications* (A. Girouard, ed.), Contemporary Mathematics, Vol. 720, Amer. Math. Soc., 2018.

Chapter 9

Distributions in Higher Dimensions and Partial Differentiation in the Sense of Distributions

In Chapter 5 we were concerned with distributions (including the delta function) *applied to* test functions of one variable. We now extend the theory of distributions to higher dimensions by considering **test functions of several variables**. What is fundamentally richer in this new territory is partial differentiation in the sense of distributions, in particular, partial differentiation of functions of several variables **but** in the sense of distributions. In Chapter 5 we saw that differentiating the Heaviside function, alternatively differentiating twice the absolute value function, in the sense of distributions gave rise to a delta function. The analogous statements in this chapter will pertain to the application of the divergence and the Laplacian in the sense of distributions to functions with blow-up discontinuities; in particular, we will discuss why in 3D

$$\operatorname{div}\left(\frac{\mathbf{x}}{|\mathbf{x}|^3}\right) = 4\pi\delta_{\mathbf{0}} \qquad \text{and} \qquad \Delta\left(\frac{1}{|\mathbf{x}|}\right) = -4\pi\delta_{\mathbf{0}}.$$

The second statement involving the Laplacian is one of the most far-reaching observations of this text. It will enable us to solve many linear (purely spatial) PDEs involving the Laplacian, and its proof and consequences will be the basis of Chapter 10.

9.1. • The Test Functions and the Definition of a Distribution

To extend distributions to higher dimensions, we consider test functions defined over $\mathbb{R}^N$; that is, we consider $\phi \in C_c^\infty(\mathbb{R}^N)$. The space $C_c^\infty(\mathbb{R}^N)$ is the space of all C^∞ functions which are identically zero *outside* some bounded set (cf. Figure 9.1). So, for

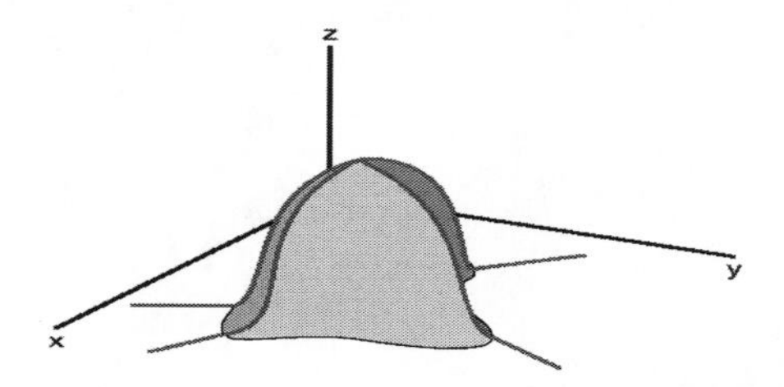

Figure 9.1. Graph of a function with compact support $z = \phi(x, y)$.

example, if $\phi_a(x) \in C_c^\infty(\mathbb{R})$ is defined by (5.13), then

$$\Phi_a(\mathbf{x}) := \phi_a(x_1) \cdots \phi_a(x_N) \in C_c^\infty(\mathbb{R}^N).$$

As in 1D, there are many functions in $C_c^\infty(\mathbb{R}^N)$; via a multivariable convolution with $\Phi_a(\mathbf{x})$, one can suitably approximate any function with compact support on $\mathbb{R}^N$ by a function in $C_c^\infty(\mathbb{R}^N)$.

With these test functions in hand, the definition of a distribution is exactly the same as before.

Definition of a Multivariable Distribution

Definition 9.1.1. A **distribution** F is a linear and continuous map from the space of test functions $C_c^\infty(\mathbb{R}^N)$ to the real numbers.

We adopt the the same notation for the action of a distribution F on a test function $\phi \in C_c^\infty(\mathbb{R}^N)$:

$$\langle F, \phi \rangle \in \mathbb{R}.$$

All the basic definitions for distributions in 1D carry over. In particular, two important types of distributions are the following:

The Two Main Types of Multivariable Distributions

- Any **locally integrable function** f on $\mathbb{R}^N$ can be thought of as a distribution F_f, where

$$\langle F_f, \phi \rangle = \int \cdots \int_{\mathbb{R}^N} f(\mathbf{x})\phi(\mathbf{x})\, d\mathbf{x}, \qquad \text{for any } \phi \in C_c^\infty(\mathbb{R}^N).$$

  As before, whenever we speak of a function $f(\mathbf{x})$ *in the sense of distributions* we mean F_f.
- The **multidimensional delta function** concentrated at the point $\mathbf{0} \in \mathbb{R}^N$ is the distribution $\delta_\mathbf{0}$, defined by

$$\langle \delta_\mathbf{0}, \phi \rangle = \phi(\mathbf{0}), \qquad \text{for any } \phi \in C_c^\infty(\mathbb{R}^N).$$

  More generally, for $\mathbf{y} \in \mathbb{R}^N$ the multidimensional delta function $\delta_\mathbf{y}$ concentrated at $\mathbf{y}$ is defined by

$$\langle \delta_\mathbf{y}, \phi \rangle = \phi(\mathbf{y}), \qquad \text{for any } \phi \in C_c^\infty(\mathbb{R}^N).$$

While the multidimensional delta function will play a major role in what is to come, we note that in higher dimensions there are many other possibilities for how a distribution can **concentrate**; in particular, they can concentrate on any lower-dimensional subset of the ambient space $\mathbb{R}^N$, for example, a one-dimensional curve in $\mathbb{R}^2$, and a one-dimensional curve or a two-dimensional surface in $\mathbb{R}^3$. We give two examples of such distributions which can be loosely viewed as a *delta functions uniformly distributed across a lower-dimensional set.*

- In $\mathbb{R}^2$ (viewed as the xy-plane), we can consider distributions which concentrate on the x-axis, for example, the distribution defined by

$$\langle F_{x-axis}, \phi \rangle := \int_{-\infty}^{\infty} \phi(x,0)\, dx \qquad \text{for any } \phi \in C_c^\infty(\mathbb{R}^2). \tag{9.1}$$

  One can think of the distribution as a continuum of one-dimensional delta functions concentrated along the x-axis. Note that it is tempting to informally write this distribution as $F_{x-axis} =$ "$\delta_0(y)$" with $\delta_0(y)$ being the usual 1D delta function with concentration at $y = 0$ but now viewed as an object ("function") on the xy-plane. The idea is that we are not considering the 2D delta function, but rather a 1D delta function on every line on which x equals a constant. Mathematicians call this distribution the *one-dimensional Hausdorff measure restricted to the x-axis.*

- In $\mathbb{R}^3$, we can consider a distribution which concentrates on the unit sphere $\partial B(\mathbf{0},1) = \{\mathbf{x} \in \mathbb{R}^3 \mid |\mathbf{x}| = 1\}$, for example, the distribution defined by

$$\langle F_{\text{sphere}}, \phi \rangle := \iint_{\partial B(\mathbf{0},1)} \phi(\mathbf{x})\, dS_{\mathbf{x}} \qquad \text{for any } \phi \in C_c^\infty(\mathbb{R}^3). \tag{9.2}$$

  Again, note that it is tempting to informally write this distribution as $F_{\text{sphere}} =$ "$\delta_0(|\mathbf{x}|^2 - 1)$" with $\delta_0(\cdot)$ being the usual 1D delta function with concentration at 0.

There is a notion of **the support of a distribution** which we address in Exercise **9.19**. The support gives us a precise way of stating that a distribution concentrates on a set. Indeed, the support of the distribution F_{x-axis} is the x-axis in $\mathbb{R}^2$, while the support of the distribution F_{sphere} is the unit sphere in $\mathbb{R}^3$

9.2. • Convergence in the Sense of Distributions

Analogous to 1D, we define convergence in the sense of distributions as follows:

Definition of Convergence in the Sense of Distributions

Definition 9.2.1. A sequence F_n of N-dimensional distributions converges to an N-dimensional distribution F if

$$\langle F_n, \phi \rangle \xrightarrow{n\to\infty} \langle F, \phi \rangle \quad \text{for all } \phi \in C_c^\infty(\mathbb{R}^N).$$

If this holds, we say $F_n \to F$ in the sense of distributions.

We will usually concern ourselves with a sequence of distributions generated by functions $f_n(\mathbf{x})$. Hence, given a sequence of functions $f_n(\mathbf{x})$ on $\mathbb{R}^N$, we say the sequence *convergences in the sense of distributions* to an N-dimensional distribution F if $\langle F_{f_n}, \phi\rangle \to \langle F, \phi\rangle$ for all $\phi \in C_c^\infty(\mathbb{R}^N)$, that is, if

$$\int \cdots \int_{\mathbb{R}^N} f_n(\mathbf{x})\,\phi(\mathbf{x})\,d\mathbf{x} \xrightarrow{n\to\infty} \langle F, \phi\rangle \quad \text{for all } \phi \in C_c^\infty(\mathbb{R}^N).$$

We denote this convergence by $f_n \to F$ in the sense of distributions.

It is straightforward to construct the multidimensional analogues of the sequence of spikes (5.23) and the sequence of Gaussians (5.25), both of which converge in the sense of distributions to $\delta_{\mathbf{0}}$. For example, the sequence of N-dimensional Gaussians is

$$f_n(\mathbf{x}) = \frac{1}{(4\pi\sigma_n)^{N/2}} \exp\left(-\frac{|\mathbf{x}|^2}{4\sigma_n}\right) \quad \text{with} \quad \sigma_n = \frac{1}{n}. \tag{9.3}$$

There are many other sequences of multidimensional functions which converge in the sense of distributions to $\delta_{\mathbf{0}}$. As before, the three properties of Section 5.5.3 must hold. For example, let f be any integrable function of $\mathbb{R}^3$ which is positive and integrates to one; i.e.,

$$\iiint_{\mathbb{R}^3} f(\mathbf{x})\,d\mathbf{x} = 1.$$

If we let

$$f_n(\mathbf{x}) := \frac{1}{\sigma_n^3} f\left(\frac{\mathbf{x}}{\sigma_n}\right) \qquad \text{with } \sigma_n = \frac{1}{n},$$

then

$$f_n \longrightarrow \delta_{\mathbf{0}}, \tag{9.4}$$

in the sense of distributions as $n \to \infty$; i.e., for all $\phi \in C_c^\infty(\mathbb{R}^3)$ we have

$$\iiint_{\mathbb{R}^3} f(\mathbf{x})\,\phi(\mathbf{x})\,d\mathbf{x} \longrightarrow \phi(\mathbf{0}).$$

Additionally, we see by the change of variables

$$\mathbf{y} = \frac{\mathbf{x}}{\sigma_n}, \qquad d\mathbf{y} = \frac{1}{\sigma_n^3}\,d\mathbf{x}$$

in the triple integral that we have

$$\begin{aligned}
\iiint_{\mathbb{R}^3} \frac{1}{\sigma_n^3} f\left(\frac{\mathbf{x}}{\sigma_n}\right) d\mathbf{x} &= \frac{1}{\sigma_n^3} \iiint_{\mathbb{R}^3} f\left(\frac{\mathbf{x}}{\sigma_n}\right) d\mathbf{x} \\
&= \frac{1}{\sigma_n^3} \iiint_{\mathbb{R}^3} f(\mathbf{y})\,\sigma_n^3\,d\mathbf{y} \\
&= \iiint_{\mathbb{R}^3} f(\mathbf{y})\,d\mathbf{y} \\
&= 1.
\end{aligned}$$

This was exactly why we chose the rescaling factor in the definition of f_n. Thus, the sequence $f_n(\mathbf{x})$ of positive functions retains the unit mass property. It also satisfies the critical third concentration property, which is that the functions concentrate more and more around $\mathbf{x} = 0$. This is straightforward to see if f had compact support; i.e., it is identically zero outside the interval $[-L, L]$ for some $L > 0$. Hence, for each n, $f_n(\mathbf{x})$ is identically zero outside the interval

$$[-\sigma_n L\,,\, \sigma_n L] \;=\; \left[-\frac{L}{n}, \frac{L}{n}\right],$$

and the size of this interval tends to 0 as $n \to \infty$. Thus, the functions concentrate their mass around 0. For general functions f, one must do some more work to prove (9.4). This is very similar to that of Theorem 5.3 and is left as an exercise (cf. Exercise **9.8**).

9.3. • Partial Differentiation in the Sense of Distributions

9.3.1. • The Notation and Definition. To talk about partial derivatives in the sense of distributions, we need to recall the **vector-index notation of Laurent Schwartz** first introduced in Section 6.11.1. Let F be a distribution over test functions $\phi \in C_c^\infty(\mathbb{R}^N)$ and let α denote a vector with N components whose entries are nonnegative integers. We agree that a nonnegative i-th entry α_i means α_i derivatives with respect to x_i, with the understanding that if $\alpha_i = 0$, no derivatives are taken. Thus, for any such index α we have an associated partial derivative ∂^α. So, for example, in dimension $N = 3$,

$$\partial^{(1,0,0)}\phi(x_1, x_2, x_3) \;=\; \frac{\partial}{\partial x_1}\phi(x_1, x_2, x_3), \qquad \partial^{(0,2,0)}\phi(x_1, x_2, x_3) \;=\; \frac{\partial^2}{\partial x_2^2}\phi(x_1, x_2, x_3),$$

$$\partial^{(2,3,2)}\phi(x_1, x_2, x_3) \;=\; \frac{\partial^7}{\partial x_1^2\,\partial x_2^3\,\partial x_3^2}\phi(x_1, x_2, x_3).$$

We denote by $|\alpha|$ the total number of derivatives taken, which by definition is simply the sum of the components α_i. So, in full generality, if $\alpha = (\alpha_1, \alpha_2, \dots, \alpha_N)$, then

$$\partial^\alpha \phi(x_1, x_2, \dots, x_N) \;=\; \frac{\partial^{|\alpha|}}{\partial x_1^{\alpha_1}\,\partial x_2^{\alpha_2}, \,\dots\,, \partial x_N^{\alpha_N}}\phi(x_1, x_2, \dots, x_N).$$

We can now define the partial derivative ∂^α of a distribution F in the sense of distributions.

Definition of the ∂^α Partial Derivative of a Distribution

Definition 9.3.1. The ∂^α derivative of a distribution F is the new distribution $\partial^\alpha F$ defined by

$$\langle \partial^\alpha F, \phi\rangle \;:=\; (-1)^{|\alpha|}\,\langle F, \partial^\alpha \phi\rangle.$$

So, for example, if F is a distribution on $\mathbb{R}^3$, then

$$\partial^{(1,2,0)}F \;=\; \frac{\partial^3}{\partial x_1\,\partial x_2^2}F$$

is the distribution G which assigns to each test function $\phi \in C_c^\infty(\mathbb{R}^3)$ the real number

$$\langle G, \phi\rangle \;=\; -\left\langle F, \frac{\partial^3}{\partial x_1\,\partial x_2^2}\phi\right\rangle.$$

Two comments are in order:

- **Order of differentiation:** Recall from calculus that the order in which one takes partial derivatives does not matter when the function is sufficiently smooth. Since a partial derivative ∂^α is always applied to a smooth test function, the order of the actual partial derivatives is irrelevant. For example, for any distribution F, we have
$$\frac{\partial F}{\partial x_1 \partial x_2} = \frac{\partial F}{\partial x_2 \partial x_1} \qquad \text{in the sense of distributions.}$$
- **Generalization of classical differentiation:** As with one variable, if a function u is sufficiently smooth, then differentiation in the sense of distributions sheds no new light. That is, if $u \in C^k$ (all partial derivatives up to and including order k exist and form continuous functions), then for any α with $|\alpha| \leq k$, the distributional partial derivative $\partial^\alpha u$ is simply the distribution generated by the classical pointwise partial derivative. The situation can change when **singularities are present**. We start with a simple example of a jump discontinuity in the next subsection.

9.3.2. • A 2D Jump Discontinuity Example.

This example can be considered as the two-dimensional analogue to the 1D Heaviside function. Consider the function

$$f(x,y) := \begin{cases} 1 & \text{if } y \geq 0, \\ 0 & \text{if } y < 0. \end{cases}$$

Let us first compute $f_y = \frac{\partial f}{\partial y}$ in the sense of distributions. Note that, for all (x,y) except when $y = 0$, the pointwise partial derivative of f with respect to y is 0. For any $\phi(x,y) \in C_c^\infty(\mathbb{R}^2)$, we compute

$$\begin{aligned} \langle f_y, \phi \rangle = -\langle f, \phi_y \rangle &= -\int_{-\infty}^{\infty}\int_{-\infty}^{\infty} f(x,y)\,\phi_y(x,y)\,dy\,dx \\ &= -\int_{-\infty}^{\infty}\int_{0}^{\infty} \phi_y(x,y)\,dy\,dx \\ &= \int_{-\infty}^{\infty} \phi(x,0)\,dx. \end{aligned} \tag{9.5}$$

Note that when passing to the last line, we used the compact support of ϕ to dispense with the values of $\phi(x,y)$ as $y \to \infty$. In words, equation (9.5) tells us that f_y is the distribution which assigns to any test function the integral of the values of ϕ along the x-axis. This distribution was previously defined as F_{x-axis} in (9.1) and **cannot** be captured (generated) by a function.

What about f_x? For this partial derivative y is fixed and, hence, the pointwise derivative is always 0, even when $y = 0$. In other words, we do not ever "see" the jump discontinuity when we differentiate pointwise with respect to x. One would expect that the distributional partial derivative should also be 0. Indeed, for any $\phi(x,y) \in C_c^\infty(\mathbb{R}^2)$

we compute

$$\langle f_x, \phi \rangle = -\langle f, \phi_x \rangle = -\int_{-\infty}^{\infty}\int_{-\infty}^{\infty} f(x,y)\,\phi_x(x,y)\,dy\,dx$$
$$= -\int_{0}^{\infty}\int_{-\infty}^{\infty} \phi_x(x,y)\,dx\,dy = 0. \tag{9.6}$$

As expected, the jump discontinuity is *not felt* in the sense of distributions by taking a derivative with respect to x. Note that we chose particular orders of integration in (9.5) and (9.6); however, in either case the order could be reversed, resulting in the same answer. In other words, in both cases it is legal to interchange the order of integration (cf. the Fubini-Tonelli Theorem, Theorem A.13, of the Appendix).

What about the mixed second partial derivatives f_{xy} and f_{yx} in the sense of distributions? As we have noted above, the order of distributional differentiation is irrelevant so we must have $f_{xy} = f_{yx}$ in the sense of distributions. You should check for yourself (Exercise **9.7**) that this mixed partial derivative in the sense of distributions is 0.

Conclusion: We saw that one of the distributional partial derivatives was also no longer generated by a function. It was not a delta function concentrated at a single point, but rather a distribution which was "concentrated" on the line of discontinuity of the function (the x-axis). One might wonder if one can ever find a distributional derivative of a function of several variables which is exactly $\delta_{\mathbf{0}}$. As we shall show in the next two sections, the answer is yes; however here the following is required:

- The function (or vector field) must have a blow-up singularity at the source point (rather than a jump discontinuity).
- One must consider a suitable combination of second-order partial derivatives, e.g., the divergence, curl, or Laplacian.

9.4. • The Divergence and Curl in the Sense of Distributions: Two Important Examples

9.4.1. • The Divergence of the Gravitational Vector Field.

Recall that if $\mathbf{F}(\mathbf{x}) = (f_1(\mathbf{x}), f_2(\mathbf{x}), f_3(\mathbf{x}))$ is a three-dimensional vector field on $\mathbb{R}^3$ with smooth component functions f_i, then

$$\operatorname{div}\mathbf{F} = \frac{\partial f_1}{\partial x_1} + \frac{\partial f_2}{\partial x_2} + \frac{\partial f_3}{\partial x_3}.$$

If, however, all we know about the component functions is that they are locally integrable functions, we can still interpret the divergence in the sense of distributions. To do this, we consider each derivative in the sense of distributions and then sum the results. Precisely, div **F** is the distribution defined for any $\phi \in C_c^\infty(\mathbb{R}^3)$ by

$$\begin{aligned}\langle \operatorname{div}\mathbf{F}, \phi\rangle &= \left\langle \frac{\partial f_1}{\partial x_1} + \frac{\partial f_2}{\partial x_2} + \frac{\partial f_3}{\partial x_3}, \phi \right\rangle \\ &= \left\langle \frac{\partial f_1}{\partial x_1}, \phi \right\rangle + \left\langle \frac{\partial f_2}{\partial x_2}, \phi \right\rangle + \left\langle \frac{\partial f_3}{\partial x_3}, \phi \right\rangle \\ &= -\iiint_{\mathbb{R}^3} f_1(\mathbf{x})\frac{\partial \phi}{\partial x_1}(\mathbf{x}) + f_2(\mathbf{x})\frac{\partial \phi}{\partial x_2}(\mathbf{x}) + f_3(\mathbf{x})\frac{\partial \phi}{\partial x_3}(\mathbf{x})\, d\mathbf{x} \\ &= -\iiint_{\mathbb{R}^3} \mathbf{F}(\mathbf{x}) \cdot \nabla\phi(\mathbf{x})\, d\mathbf{x}. \qquad (9.7)\end{aligned}$$

Now recall from your advanced calculus course the **gravitational vector field**[1]. If you have forgotten, open any multivariable calculus text where you will almost surely see a discussion about this vector field. Given a unit mass which lies at the origin, then the gravitational force on a particle of mass m, which is at point $\mathbf{x}$, is given by

$$\boxed{\mathbf{G}(\mathbf{x}) = -km\frac{\mathbf{x}}{|\mathbf{x}|^3} = \frac{-km}{|\mathbf{x}|^2}\left(\frac{\mathbf{x}}{|\mathbf{x}|}\right),}$$

where k is the gravitational constant. Recall two important related facts about the flux and divergence of **G**:

- The **gravitational flux out of** any sphere $\mathcal{S}$ is given by

$$\iint_{\mathcal{S}} \mathbf{G}\cdot\mathbf{n}\, dS = \begin{cases} 0 & \text{if } \mathbf{0} \text{ is not inside } \mathcal{S}, \\ -4\pi km & \text{if } \mathbf{0} \text{ is inside } \mathcal{S}. \end{cases} \qquad (9.8)$$

- The **divergence** of **G** is 0 except at the origin where it is undefine; i.e.,

$$\operatorname{div}\mathbf{G}(\mathbf{x}) = 0 \qquad \text{for all } \mathbf{x} \neq \mathbf{0}. \qquad (9.9)$$

How can we reconcile these two? It does not contradict the Divergence Theorem as the theorem applies to smooth vector fields and **G** is not smooth at the origin. They are reconciled by the fact that

$$\boxed{\operatorname{div}\mathbf{G} = -4\pi km\,\delta_{\mathbf{0}} \quad \text{in the sense of distributions.}} \qquad (9.10)$$

We now state and prove[2] this statement as a theorem. Note that, in order to interpret **G** as a distribution or, more precisely, a vector of distributions, we need each component to be locally integrable. This amounts to the fact that $|\mathbf{G}|$ is locally integrable. This is true and, in fact, we will need to show it in the course of proving (9.10). Since the constants k and m in **G** will play no role, we set them to 1.

Theorem 9.1. *Let* $\mathbf{x} \in \mathbb{R}^3$ *and* **G** *be defined by*

$$\mathbf{G}(\mathbf{x}) = -\frac{\mathbf{x}}{|\mathbf{x}|^3}.$$

Then

$$\operatorname{div}\mathbf{G} = -4\pi\delta_{\mathbf{0}} \quad \textit{in the sense of distributions.}$$

[1]This vector field is almost identical to the **electrostatic vector field** (cf. Section 10.6).

[2]A physics or engineering oriented reader might be put off by a so-called "rigorous proof" but then we strongly encourage them to try to convince themselves of what is going on and why the delta function pops up. Surely they will want some explanation/justification. In the end, we suspect they will begin to appreciate the arguments presented in the proof.

That is (cf. (9.7)), for all test functions $\phi \in C_c^\infty(\mathbb{R}^3)$ *we have*

$$-\iiint_{\mathbb{R}^3} \mathbf{G}(\mathbf{x}) \cdot \nabla\phi(\mathbf{x})\, d\mathbf{x} = -4\pi\phi(\mathbf{0}).$$

Before presenting the proof, let us sketch the path we will take. We start by recalling the steps involved in differentiating, in the sense of distributions, a piecewise C^1 function $f(x)$ with jump discontinuities: (1) We wrote down the definition of the derivative in the sense of distributions for some fixed test function ϕ. This gave rise to an integral over $\mathbb{R}$. (2) We split up the integral over subintervals between these discontinuities. (3) We integrated by parts (which was legal) on each subinterval where f was smooth. In doing so, we saw two possible "effects" on the test function ϕ: (i) integrating against the pointwise derivative and (ii) boundary terms which led to the evaluation of the test function at the end points. The latter could give rise to delta functions in the distributional derivative. Hence, the distributional derivative was, in general, comprised of both functions (which were the pointwise classical derivatives within each interval) and delta functions in the case of jump discontinuities.

The steps we now follow are similar in spirit but address two major differences: The space is three dimensional and the function $\frac{1}{|\mathbf{x}|}$ has a blow-up singularity at $\mathbf{0}$. The new steps are as follows:

- We write down the definition of the divergence in the sense of distributions for some fixed test function ϕ. This gives rise to an integral over $\mathbb{R}^3$.
- We first isolate the singularity by breaking up this integral into two pieces: a small ball $B(\mathbf{0}, \epsilon)$ centered at $\mathbf{0}$ and its complement, the "annular" region of space.
- In this annular region of space, all the fields (functions and vector fields) are smooth. Hence, in this region we can perform some "legitimate calculus" in the form of integration by parts to focus the effects on the test function ϕ. This produces boundary terms on $\partial B(\mathbf{0}, \epsilon)$.
- Since the previous steps will hold for all $\epsilon > 0$, we can then let $\epsilon \to 0^+$. The appearance of $\phi(\mathbf{0})$ will come from the limit of the boundary integral over $\partial B(\mathbf{0}, \epsilon)$. Here the exponent in the denominator, in relation to the space dimension, plays a pivotal role.

Proof of Theorem 9.1. Let $\phi \in C_c^\infty(\mathbb{R}^3)$ be a test function. Our goal is to prove that

$$-\iiint_{\mathbb{R}^3} \mathbf{G}(\mathbf{x}) \cdot \nabla\phi(\mathbf{x})\, d\mathbf{x} = -4\pi\phi(\mathbf{0}). \tag{9.11}$$

Let $\epsilon > 0$ be fixed for the moment, and break up the integral into two pieces:

$$\iiint_{\mathbb{R}^3} \mathbf{G}(\mathbf{x}) \cdot \nabla\phi(\mathbf{x})\, d\mathbf{x} = \iiint_{B(\mathbf{0},\epsilon)} \mathbf{G}(\mathbf{x}) \cdot \nabla\phi(\mathbf{x})\, d\mathbf{x} + \iiint_{\mathbb{R}^3\backslash B(\mathbf{0},\epsilon)} \mathbf{G}(\mathbf{x}) \cdot \nabla\phi(\mathbf{x})\, d\mathbf{x}, \tag{9.12}$$

where $\mathbb{R}^3\backslash B(\mathbf{0}, \epsilon)$ denotes the unbounded region outside the ball $B(\mathbf{0}, \epsilon)$.

Since ϕ has compact support, there exists a bounded domain Ω such that ϕ is identically zero on its complement (including $\partial\Omega$); in particular, $\phi = 0$ on $\partial\Omega$. We may assume $B(\mathbf{0}, \epsilon) \subset \Omega$. Hence,

$$\iiint_{\mathbb{R}^3 \backslash B(\mathbf{0},\epsilon)} \mathbf{G}(\mathbf{x}) \cdot \nabla\phi(\mathbf{x})\, d\mathbf{x} = \iiint_{\Omega \backslash B(\mathbf{0},\epsilon)} \mathbf{G}(\mathbf{x}) \cdot \nabla\phi(\mathbf{x})\, d\mathbf{x}.$$

On $\Omega \backslash B(\mathbf{0}, \epsilon)$, the vector field $\mathbf{G}$ is a smooth vector field with zero divergence. Hence, we can apply the vector field integration by parts formula (A.15) (from Section A.6.4) on the domain $\Omega \backslash B(\mathbf{0}, \epsilon)$ with $\mathbf{u} = \mathbf{G}$ and $v = \phi$ to find

$$\iiint_{\Omega \backslash B(\mathbf{0},\epsilon)} \mathbf{G}(\mathbf{x}) \cdot \nabla\phi(\mathbf{x})\, d\mathbf{x} = \iint_{\partial\left(\Omega \backslash B(\mathbf{0},\epsilon)\right)} (\phi\, \mathbf{G}) \cdot \mathbf{n}\, dS.$$

Now the boundary of $\Omega \backslash B(\mathbf{0}, \epsilon)$ consists of two pieces: the boundary of Ω and the boundary of the ball $B(\mathbf{0}, \epsilon)$. The integral is the sum over both pieces[3]. The surface integral over $\partial\Omega$ is zero since $\phi = 0$ on $\partial\Omega$. For the surface integral over $\partial B(\mathbf{0}, \epsilon)$, we note that the outer normal to $\Omega \backslash B(\mathbf{0}, \epsilon)$ on the spherical part of the boundary points **into** the ball. Hence, at any point $\mathbf{x} \in \partial B(\mathbf{0}, \epsilon)$ where $|\mathbf{x}| = \epsilon$, we have $\mathbf{n} = -\frac{\mathbf{x}}{\epsilon}$. Also, note that for $\mathbf{x} \in \partial B(\mathbf{0}, \epsilon)$, we have $\mathbf{x} \cdot \mathbf{x} = |\mathbf{x}|^2 = \epsilon^2$. So

$$\begin{aligned}
\iint_{\partial B(\mathbf{0},\epsilon)} (\phi\, \mathbf{G}) \cdot \mathbf{n}\, dS &= \iint_{\partial B(\mathbf{0},\epsilon)} \left(\phi(\mathbf{x})\left(-\frac{\mathbf{x}}{\epsilon^3}\right)\right) \cdot \left(-\frac{\mathbf{x}}{\epsilon}\right) dS \\
&= \iint_{\partial B(\mathbf{0},\epsilon)} \frac{\phi(\mathbf{x})}{\epsilon^2}\, dS \\
&= \frac{1}{\epsilon^2} \iint_{\partial B(\mathbf{0},\epsilon)} \phi(\mathbf{x})\, dS \\
&= 4\pi \left(\frac{1}{4\pi\epsilon^2} \iint_{\partial B(\mathbf{0},\epsilon)} \phi(\mathbf{x})\, dS \right). \qquad (9.13)
\end{aligned}$$

The expression in the parenthesis of the last line is simply **the average value of ϕ over the sphere** $\partial B(\mathbf{0}, \epsilon)$. But now the miracle happens. As ϵ gets smaller and smaller this average gets closer and closer to $\phi(\mathbf{0})$; i.e., by the Averaging Lemma (Lemma A.7.1), we have

$$\lim_{\epsilon \to 0} \frac{1}{4\pi\epsilon^2} \iint_{\partial B(\mathbf{0},\epsilon)} \phi(\mathbf{x})\, dS = \phi(\mathbf{0}).$$

Combining all this, we have shown that

$$\iiint_{\mathbb{R}^3} \mathbf{G}(\mathbf{x}) \cdot \nabla\phi(\mathbf{x})\, d\mathbf{x} = \iiint_{B(\mathbf{0},\epsilon)} \mathbf{G}(\mathbf{x}) \cdot \nabla\phi(\mathbf{x})\, d\mathbf{x} + 4\pi \left(\frac{1}{4\pi\epsilon^2} \iint_{\partial B(\mathbf{0},\epsilon)} \phi(\mathbf{x})\, dS \right).$$

Moreover, this holds for all ϵ (or to be pedantic, all ϵ sufficiently small with respect to the support of ϕ). Hence, we can let ϵ tend to 0 on both sides of the equality. There is no ϵ dependence on the left so that limit is easy. For the right-hand side, we have

[3] You might wonder if we should sum or subtract. What is important here is orientation through the outer normal $\mathbf{n}$. Hence, we do indeed add the terms but always consider the normal pointing **out** of $\Omega \backslash B(\mathbf{0}, \epsilon)$.

already seen that the second term approaches $4\pi\phi(\mathbf{0})$. Hence the only issue is to show that

$$\lim_{\epsilon\to 0} \iiint_{B(\mathbf{0},\epsilon)} \mathbf{G}(\mathbf{x}) \cdot \nabla\phi(\mathbf{x})\, d\mathbf{x} = 0. \tag{9.14}$$

This is not immediately obvious; yes, the size of the domain of integration is tending to 0 but the integrand is also blowing up. It is the trade-off we are interested in and it basically boils down to the fact that $|\mathbf{G}|$ is integrable over the origin (which, in fact, is necessary in order to interpret $\mathbf{G}$ as a distribution). To this end, we use some standard tricks to estimate (bound above) the integral:

$$\begin{aligned}
\left| \iiint_{B(\mathbf{0},\epsilon)} \mathbf{G}(\mathbf{x}) \cdot \nabla\phi(\mathbf{x})\, d\mathbf{x} \right| &\le \iiint_{B(\mathbf{0},\epsilon)} |\mathbf{G}(\mathbf{x}) \cdot \nabla\phi(\mathbf{x})|\, d\mathbf{x} \\
&\le \iiint_{B(\mathbf{0},\epsilon)} |\mathbf{G}(\mathbf{x})|\, |\nabla\phi(\mathbf{x})|\, d\mathbf{x} \\
&\le C \iiint_{B(\mathbf{0},\epsilon)} |\mathbf{G}(\mathbf{x})|\, d\mathbf{x} \\
&\le C \iiint_{B(\mathbf{0},\epsilon)} \frac{1}{|\mathbf{x}|^2}\, d\mathbf{x} \\
&\le C \int_0^{\epsilon} \frac{1}{r^2} 4\pi r^2\, dr \\
&= 4\pi C\epsilon.
\end{aligned} \tag{9.15}$$

Here we used a few simple facts: In the first line, the absolute value of an integral is less than or equal to the integral of the absolute value of the integrand; in the second line, the Cauchy-Schwarz inequality says that for two vectors $\mathbf{a}$ and $\mathbf{b}$, $|\mathbf{a} \cdot \mathbf{b}| \le |\mathbf{a}||\mathbf{b}|$; in the third line, since $|\nabla\phi|$ is continuous on $B(\mathbf{0}, \epsilon)$, it attains its maximum which we have denoted by C; in the fifth line, we converted to spherical coordinates. So this establishes (9.14), which implies (9.11). □

It is worth reflecting over this proof and why it worked out so nicely. It was based upon a few simple properties of the gravitational vector field:

- $\mathbf{G}$ was radial (this made the integral computations straightforward).
- $\mathbf{G}$ was divergence-free away from the origin.
- The length of $\mathbf{G}$,
$$|\mathbf{G}| = \frac{1}{|\mathbf{x}|^2},$$
had the right exponent of $|\mathbf{x}|$ in the denominator, i.e., the right power associated with the blow-up at $\mathbf{x} = \mathbf{0}$. The exponent was two[4]; consequently, this produced the average and facilitated the application of the Averaging Lemma for (9.13). This exponent also allowed us to conclude (9.14) via (9.15).

Note that in the latter two properties, the space dimension (here three) was crucial.

[4]Note that this is exactly what *Newton's Law of Gravitation* (or alternatively *Coulomb's Law in electrostatics*) dictates.

It is also worth reflecting on the role of the fundamental calculus identity (A.15). We have chosen the domain Ω to encompass the support of ϕ, and hence $\phi = 0$ on $\partial\Omega$. So why not just apply (A.15) with $\mathbf{u} = \mathbf{G}$ and $v = \phi$ on the **entire** domain Ω? Firstly, (A.15) is not valid unless $\mathbf{u}$ and v are sufficiently smooth (C^1 here would suffice), and $\mathbf{u} = \mathbf{G}$ blows up at $\mathbf{0} \in \Omega$. On the other hand, (A.15) is "valid" with the correct interpretation of the expression

$$\text{“}\iiint_{\Omega} \operatorname{div} \mathbf{u}\, v\, d\mathbf{x}\text{”}$$

on the right-hand side. Precisely, this is the action of the distribution $\operatorname{div}\mathbf{u}$ on the test function v. For the particular choice of $\mathbf{u} = \mathbf{G}$, this distribution is, up to a constant, a delta function. But we know this crucial point after completing the steps of the proof wherein we removed a small ball around the singularity at the origin and then took the limit as the size of the ball tended to 0.

Finally, let us address the Divergence Theorem with regard to the seemingly contradictory statements (9.8) and (9.9). Because of the singularity at the origin, the Divergence Theorem (as stated for functions) does not apply over any surface containing the origin. However, knowing now that $\operatorname{div}\mathbf{G} = -4\pi\delta_{\mathbf{0}}$ in the sense of distributions, we can now argue, albeit informally, that, yes,

$$\text{“}\iint_{\partial\mathcal{S}} \mathbf{G}\cdot\mathbf{n}\, dS \;=\; \int_{\mathcal{S}} \underbrace{\operatorname{div}\mathbf{G}}_{=-4\pi\delta_{\mathbf{0}}}\; d\mathbf{x} \;=\; -4\pi\text{”},$$

for any domain $\mathcal{S}$ containing the origin. As we shall see in Chapter 10, once we establish that a certain combination of derivatives of some function is, in fact, a delta function, we can be free to informally apply calculus identities like (A.15). After all, we will know that we can always furnish a proof by repeating the basic steps of isolating the singularity, working around it, and taking a limit.

9.4.2. The Curl of a Canonical Vector Field. Recall that the curl of a smooth vector field 3D $\mathbf{F}(x, y, z) = \big(F_1(x, y, z), F_2(x, y, z), F_2(x, y, z)\big)$ is defined as the 3D vector field

$$\begin{aligned} \operatorname{curl}\mathbf{F}(x, y, z) &= \left(\frac{\partial F_3}{\partial y} - \frac{\partial F_2}{\partial z}, \frac{\partial F_1}{\partial z} - \frac{\partial F_3}{\partial x}, \frac{\partial F_2}{\partial x} - \frac{\partial F_1}{\partial y}\right) \\ &= \left(\frac{\partial F_3}{\partial y} - \frac{\partial F_2}{\partial z}\right)\mathbf{i} + \left(\frac{\partial F_1}{\partial z} - \frac{\partial F_3}{\partial x}\right)\mathbf{j} + \left(\frac{\partial F_2}{\partial x} - \frac{\partial F_1}{\partial y}\right)\mathbf{k}. \end{aligned}$$

For a two-dimensional smooth vector field $\mathbf{F}(x, y) = \big(F_1(x, y), F_2(x, y)\big)$ one defines the curl as

$$\operatorname{curl}\mathbf{F}(x, y) \;:=\; \operatorname{curl}\big(F_1(x, y),\, F_2(x, y),\, 0\big) \;=\; \left(\frac{\partial F_2}{\partial x} - \frac{\partial F_1}{\partial y}\right)\mathbf{k}.$$

Hence, we may as well simply **associate** the curl of a 2D vector field with the **scalar**

$$\frac{\partial F_2}{\partial x} - \frac{\partial F_1}{\partial y}.$$

In almost all calculus texts you will find the following example of a 2D vector field:

$$\boxed{\mathbf{F}_c(x, y) \;=\; \left(\frac{-y}{x^2 + y^2}, \frac{x}{x^2 + y^2}\right).}$$

This example usually appears in sections dealing with circulation, curl, and/or Green's Theorem. What is remarkable about this example is the following:

- Away from the origin, the vector field is curl-free; i.e.,

$$\text{curl}\,\mathbf{F_c}(x,y) \;=\; \frac{\partial}{\partial x}\left(\frac{x}{x^2+y^2}\right) - \frac{\partial}{\partial y}\left(-\frac{y}{x^2+y^2}\right) = 0 \qquad \text{for all } (x,y)\neq 0.$$

- The circulation integral

$$\int_{\mathcal{C}} \mathbf{F_c}\cdot d\mathbf{r} = \int_{\mathcal{C}} \mathbf{F_c}\cdot \mathbf{T}\,ds = \int_{\mathcal{C}} \left(\frac{-y}{x^2+y^2}\right)dx + \left(\frac{x}{x^2+y^2}\right)dy = 2\pi,$$

for any closed curve containing the origin, oriented counterclockwise. Here, $\mathbf{T}$ denotes the unit tangent to the curve.

These two facts might seem slightly contradictory in regard to **Green's Theorem,** which states that for any 2D smooth vector field $\mathbf{F} = \langle F_1, F_2\rangle$ and any smooth closed curve $\mathcal{C}$, oriented counterclockwise and enclosing a region R in the plane, we have

$$\iint_R \left(\frac{\partial F_2}{\partial x} - \frac{\partial F_1}{\partial y}\right) dx\,dy = \int_{\mathcal{C}} \mathbf{F}\cdot d\mathbf{r} = \int_{\mathcal{C}} F_1\,dx + F_2\,dy.$$

The issue is that if the region R includes the origin, we cannot directly apply Green's Theorem to $\mathbf{F_c}$ as the vector field is not smooth in all of R.

We now show that we can completely describe what is happing by **interpreting the curl in the sense of distributions**. Note again that, since we are working in 2D, the curl is a scalar. If $\mathbf{F}$ is a 2D vector field whose components are locally integrable functions on $\mathbb{R}^2$, then curl $\mathbf{F}$ is the distribution whose assignment to each test function $\phi \in C_c^\infty(\mathbb{R}^2)$ is given by

$$\boxed{\langle \text{curl}\,\mathbf{F}, \phi\rangle = \iint_{\mathbb{R}^2} (-F_2\phi_x + F_1\phi_y)\,dx\,dy.}$$

Theorem 9.2. *We have* curl $\mathbf{F_C} = 2\pi\delta_0$ *in the sense of distributions; that is, for all* $\phi \in C_c^\infty(\mathbb{R}^2)$ *we have*

$$\iint_{\mathbb{R}^2} \left[\left(\frac{-x}{x^2+y^2}\right)\phi_x + \left(\frac{-y}{x^2+y^2}\right)\phi_y\right] dx\,dy = 2\pi\phi(0,0). \tag{9.16}$$

To prove the theorem we need our particular incarnation of integration by parts which will now follow directly from Green's Theorem.

The Necessary "Calculus Identity"

We first note the simple vector identity that for any **smooth** 2D vector field $\mathbf{F}$ and smooth function ϕ, we have

$$\operatorname{curl}(\mathbf{F}\phi) = \phi \operatorname{curl}\mathbf{F} + (F_2, -F_1) \cdot \nabla\phi.$$

Now let Ω be a bounded domain with boundary curve $\mathcal{C}$ given the positive orientation[a]. Integrating the above equation over Ω gives

$$\iint_\Omega \operatorname{curl}(\mathbf{F}\phi)\,dx\,dy = \iint_\Omega \phi \operatorname{curl}\mathbf{F}\,dx\,dy + \iint_\Omega (F_2, -F_1) \cdot \nabla\phi\,dx\,dy.$$

Green's Theorem, applied to the left-hand side, gives

$$\iint_\Omega \operatorname{curl}(\mathbf{F}\phi)\,dx\,dy = \int_\mathcal{C} \phi\mathbf{F} \cdot d\mathbf{r}.$$

Thus, combining these two, we find

$$\iint_\Omega (F_2, -F_1) \cdot \nabla\phi\,dx\,dy = -\iint_\Omega \phi \operatorname{curl}\mathbf{F}\,dx\,dy + \int_\mathcal{C} \phi\mathbf{F} \cdot d\mathbf{r}. \qquad (9.17)$$

[a] Recall that this means that as we traverse the boundary curve $\mathcal{C}$, the region Ω must be on our left.

Proof of Theorem 9.2. The proof will be very similar to that of Theorem 9.1 with the following steps: (i) We isolate the singularity at the origin by removing a small disk of radius ϵ; (ii) we apply our integration by parts formula (here (9.17)) away from the singularity; (iii) we consider the boundary term as ϵ tends to 0.

To this end, fix $\phi \in C_c^\infty(\mathbb{R}^2)$ and consider a domain Ω which contains the support of ϕ; i.e., $\phi \equiv 0$ on $\partial\Omega$ and outside. Without loss of generality, we can assume the origin lies in Ω (why?). Now consider $\epsilon > 0$ sufficiently small such that $B(\mathbf{0}, \epsilon) \subset \Omega$, and let $\Omega_\epsilon = \Omega \backslash B(\mathbf{0}, \epsilon) \subset \Omega$. Denoting F_1 and F_2 as the respective components of $\mathbf{F}_\mathbf{C}$, we have

$$\begin{aligned}\iint_{\mathbb{R}^2} (-F_2, F_1) \cdot \nabla\phi\,dx\,dy &= \iint_\Omega (-F_2, F_1) \cdot \nabla\phi\,dx\,dy \\ &= \iint_{\Omega_\epsilon} (-F_2, F_1) \cdot \nabla\phi\,dx\,dy + \iint_{B(\mathbf{0},\epsilon)} (-F_2, F_1) \cdot \nabla\phi\,dx\,dy.\end{aligned}$$

We will again show that as ϵ tends to zero the first term tends to $2\pi\phi(\mathbf{0})$ while the second tends to 0. To this end, let us start with the first term. On Ω_ϵ, the vector field $\mathbf{F}_\mathbf{C}$ is smooth (of course ϕ is always smooth). Moreover, since curl $\mathbf{F}_\mathbf{C} = 0$ on Ω_ϵ, we can apply (9.17) on Ω_ϵ to find

$$\iint_{\Omega_\epsilon} (-F_2, F_1) \cdot \nabla\phi\,dx\,dy = -\int_{\partial\Omega_\epsilon} \phi\mathbf{F}_\mathbf{C} \cdot d\mathbf{r}.$$

The boundary of Ω_ϵ has two pieces: Positive orientation dictates that the boundary of Ω is oriented counterclockwise while the circle $\partial B(\mathbf{0}, \epsilon)$ is oriented clockwise. Since ϕ is zero on $\partial\Omega$, we have

$$\iint_{\Omega_\epsilon} (-F_2, F_1) \cdot \nabla\phi\,dx\,dy = -\int_{\partial B(\mathbf{0},\epsilon)} \phi\mathbf{F}_\mathbf{C} \cdot d\mathbf{r} = -\int_{\partial B(\mathbf{0},\epsilon)} \phi\mathbf{F}_\mathbf{C} \cdot \mathbf{T}\,ds,$$

where $\mathbf{T}$ denotes the unit tangent (pointing clockwise) to $\partial B(\mathbf{0},\epsilon)$ and ds denotes integration with respect to arc length. On $\partial B(\mathbf{0},\epsilon)$, we have $\mathbf{F_C}\cdot\mathbf{T} = -\frac{1}{\epsilon}$ and hence

$$-\int_{\partial B(\mathbf{0},\epsilon)} \phi\,\mathbf{F_C}\cdot\mathbf{T}\,ds = -\left(-\frac{1}{\epsilon}\int_{\partial B(\mathbf{0},\epsilon)} \phi\,ds\right) = 2\pi\left(\frac{1}{2\pi\epsilon}\int_{\partial B(\mathbf{0},\epsilon)} \phi\,ds\right).$$

By the Averaging Lemma, the average in the parentheses tends to $2\pi\phi(\mathbf{0})$ as ϵ tends to 0.

It remains to show that

$$\iint_{B(\mathbf{0},\epsilon)} (-F_2, F_1)\cdot\nabla\phi\,dx\,dy \quad\longrightarrow\quad 0 \qquad \text{as } \epsilon\to 0^+.$$

To this end, note that

$$\begin{aligned}
\left|\iint_{B(\mathbf{0},\epsilon)} (-F_2, F_1)\cdot\nabla\phi\,dx\,dy\right| &\le \iint_{B(\mathbf{0},\epsilon)} |(-F_2, F_1)\cdot\nabla\phi|\;dx\,dy\\
&\le \iint_{B(\mathbf{0},\epsilon)} |\mathbf{F_C}|\cdot|\nabla\phi|\;dx\,dy\\
&\le C\iint_{B(\mathbf{0},\epsilon)} |\mathbf{F_C}|\;dx\,dy\\
&\le C\iint_{B(\mathbf{0},\epsilon)} \frac{1}{|\mathbf{x}|}\,dx\,dy\\
&\le C\int_0^{2\pi}\int_0^{\epsilon}\frac{1}{r}\,r\,dr\,d\theta\\
&= 2\pi C\epsilon,
\end{aligned}$$

which tends to 0 as $\epsilon\to 0^+$. Here, C denotes the maximum value of $|\nabla\phi|$. The proof is now complete. □

9.5. • The Laplacian in the Sense of Distributions and a Fundamental Example

In terms of the Schwartz notation, the Laplacian is simply

$$\sum_{i=1}^{n}\partial^{\alpha_i} \qquad \text{where for each } i \quad \alpha_i = (0,\dots,0,2,0,\dots,0),$$

with the 2 in the i-th position. We can thus apply the Laplacian to any distribution by summing the respective distributional partial derivatives. So, in three space dimensions, the PDE $\Delta u = f$, where u and f are locally integrable functions, can be interpreted in the sense of distributions to mean that

$$\langle\Delta u,\phi\rangle = \langle f,\phi\rangle \qquad \text{for all } \phi\in C_c^\infty(\mathbb{R}^3).$$

By definition of partial differentiation in the sense of distributions,

$$\langle\Delta u,\phi\rangle = \langle u,\Delta\phi\rangle = \iiint_{\mathbb{R}^3} u(\mathbf{x})\,\Delta\phi(\mathbf{x})\;d\mathbf{x}.$$

Notice the positive sign because of the fact that the Laplacian is comprised of second-order partial derivatives. Hence, $\Delta u = f$ in the sense of distributions means that

$$\iiint_{\mathbb{R}^3} u(\mathbf{x})\,\Delta\phi(\mathbf{x})\,d\mathbf{x} \;=\; \iiint_{\mathbb{R}^3} f(\mathbf{x})\,\phi(\mathbf{x})\,d\mathbf{x} \qquad \text{for all } \phi \in C_c^\infty(\mathbb{R}^3).$$

We can also consider the case where f is replaced with a distribution which cannot be represented by a function; as an example, for a locally integrable function u, the PDE

$$\Delta u = \delta_{\mathbf{0}} \qquad \text{in the sense of distributions}$$

means

$$\iiint u(\mathbf{x})\,\Delta\phi(\mathbf{x})\,d\mathbf{x} \;=\; \phi(\mathbf{0}) \qquad \text{for all } \phi \in C_c^\infty(\mathbb{R}^3).$$

In full generality, if F and G are distributions over test functions in $C_c^\infty(\mathbb{R}^3)$, then $\Delta F = G$ means that

$$\langle F, \Delta\phi\rangle \;=\; \langle G, \phi\rangle \qquad \text{for all } \phi \in C_c^\infty(\mathbb{R}^3).$$

A fundamental example: Consider the following function defined on $\mathbb{R}^3$:

$$\boxed{\Phi(\mathbf{x}) \;=\; \frac{1}{|\mathbf{x}|} \qquad \mathbf{x} \neq \mathbf{0}.}$$

Define Φ to be any number of your choosing at the origin $\mathbf{0}$; the choice will be irrelevant to the sequel. Recall from Section 8.5 that we derived this function by looking for radially symmetric harmonic functions in 3D. However, it is not harmonic at the origin; the derivatives in any direction blow up as we approach the origin. The space dimension associated with the function $\Phi(\mathbf{x}) = \frac{1}{|\mathbf{x}|}$ is critical in determining its character; it is not harmonic away from the origin in any other dimension.

So what do we make of this singularity at the origin? The answer lies in interpreting Φ as a distribution. First let us note that despite the singularity, the function is still locally integrable in 3D; by reverting to spherical coordinates, we have

$$\iiint_{B(\mathbf{0},1)} \Phi(\mathbf{x})\,d\mathbf{x} \;=\; \int_0^1 \frac{1}{r}\,4\pi r^2\,dr \;<\; \infty. \tag{9.18}$$

The function Φ is not integrable on $\mathbb{R}^3$ as a result of the slow decay in its tails. However, recall that in order to interpret a function as a distribution, we need to have a finite value of the integral of $\Phi\phi$ where ϕ has compact support. Hence, local integrability suffices.

Now we have the machinery (or in some sense "the vocabulary") to address the full Laplacian of Φ, capturing the effect of the singularity at $\mathbf{0}$. Indeed, we can state one of the most far-reaching observations in this text: In three space dimensions,

$$\boxed{\Delta\left(\frac{1}{|\mathbf{x}|}\right) = -4\pi\delta_{\mathbf{0}} \qquad \text{in the sense of distributions.}} \tag{9.19}$$

In other words, for any $\phi \in C_c^\infty(\mathbb{R}^3)$, we have

$$\iiint \frac{1}{|\mathbf{x}|}\,\Delta\phi(\mathbf{x})\,d\mathbf{x} \;=\; -4\pi\phi(\mathbf{0}).$$

This fact will be the cornerstone of Chapter 10 where it will be used to solve PDEs involving the Laplacian. A self-contained proof of (9.19) (i.e., an explanation of why it is true) will be presented in Section 10.1 (cf. Theorem 10.1).

But have we already proven (9.19)? It should look very similar to what was proved in Theorem 9.1; i.e.,

$$\operatorname{div}\left(-\frac{\mathbf{x}}{|\mathbf{x}|^3}\right) = -4\pi\delta_{\mathbf{0}} \quad \text{in the sense of distributions.}$$

Indeed, for all $\mathbf{x} \neq \mathbf{0}$, we have (pointwise)

$$\nabla\Phi(\mathbf{x}) = \nabla\frac{1}{|\mathbf{x}|} = -\frac{\mathbf{x}}{|\mathbf{x}|^3}. \tag{9.20}$$

Hence, can we conclude that

$$\Delta\Phi = \operatorname{div}\nabla\Phi(\mathbf{x}) = -4\pi\delta_{\mathbf{0}} \quad \text{in the sense of distributions?}$$

Not immediately; we would still need to prove that the pointwise gradient (9.20) of Φ is the same as its distributional gradient. This can of course be done, effectively proving (9.19) by tackling one order of distributional derivatives at a time. Because of its importance, we give a direct proof in Section 10.1 tackling both orders of distributional derivatives at the same time.

9.6. Distributions Defined on a Domain (with and without Boundary)

Pure domain (without boundary): We can also speak of distributions where the underlying space is not $\mathbb{R}^N$, but rather some domain $\Omega \subset \mathbb{R}^N$. Recall that a domain is simply an open subset of $\mathbb{R}^N$. In this case, the test functions ϕ must have compact support in Ω; we denote this space as $C_c^\infty(\Omega)$. But what exactly does compact support in Ω mean? A function ϕ is in $C_c^\infty(\Omega)$[5] if it is the **restriction** of a C^∞ function, defined on **all** of $\mathbb{R}^N$ (yes, not just on Ω but on all of $\mathbb{R}^N$), for which there exists a subset $K \subset \Omega$, bounded and closed in $\mathbb{R}^N$, such that

$$\phi(\mathbf{x}) = 0 \qquad \text{for all } \mathbf{x} \notin K.$$

Note that such a test function must vanish on the boundary $\partial\Omega$; i.e., $\phi(\mathbf{x}) = 0$ for $\mathbf{x} \in \partial\Omega$. Indeed, it must vanish on the complement of Ω in $\mathbb{R}^N$. However, more is true: As one approaches any boundary point from inside Ω, ϕ must be zero "before" reaching the boundary. Again, think of points on the boundary as points which are never "reached" or "touched" with our test functions. The main reason for this definition is that we want all the partial derivatives of ϕ to also vanish on $\partial\Omega$. So, for example, if $\Omega = B(\mathbf{0}, R)$, the ball in $\mathbb{R}^N$ of radius $R > 0$, and ϕ is taken to be

$$\phi(\mathbf{x}) = \phi_a(x_1)\cdots\phi_a(x_N),$$

where ϕ_a is the bump defined by (5.13), then $\phi \in C_c^\infty(\Omega)$ if $a < R$, but not if $a = R$.

[5] An equivalent definition is the following. We define the support of a function $\phi \in C^\infty(\mathbb{R}^N)$ to be

$$\mathcal{S} = \text{support}\,\phi := \{\mathbf{x} \in \Omega \,|\, \phi(\mathbf{x}) \neq 0\}.$$

Now, suppose we take the closure $\overline{\mathcal{S}}$ of $\mathcal{S}$ in $\mathbb{R}^N$ (not in Ω). Then if $\overline{\mathcal{S}} \subset \Omega$ and is bounded, we say ϕ has compact support in Ω.

With this class of test functions, we can define statements in the sense of distributions which are localized to Ω. For example, a locally integrable function u on a domain $\Omega \subset \mathbb{R}^3$ is a solution to $\Delta u = f$ on Ω in the sense of distributions if for all $\phi \in C_c^\infty(\Omega)$,

$$\iiint_\Omega u\,\Delta\phi\,d\mathbf{x} = \iiint_\Omega f\,\phi\,d\mathbf{x}.$$

Domain with boundary: It is often useful to incorporate boundary (or initial) conditions in a statement in the sense of distributions. This effectively means we want to incorporate the boundary/initial conditions in a more integral-based sense rather than a pointwise sense. The most common case will be when the domain is the upper half-plane $\Omega = \{(x,t)\,|\,t > 0\}$ and we want to incorporate the $t = 0$ axis and work on $\overline{\Omega} = \{(x,t)\,|\,t \geq 0\}$. This can be achieved by considering distributions defined on test functions with compact support in $\overline{\Omega}$ (a closed set). Now, the set where these test functions are nonzero **can include** parts of the $t = 0$ axis; hence, the distributions are "alive" on the $t = 0$ axis. We will discuss this further in Section 9.7.4.

9.7. Interpreting Many PDEs in the Sense of Distributions

By taking the sums of partial derivatives, we are now able to **interpret a wide subclass of linear and quasilinear PDEs in the sense of distributions**. In doing so, we can:

- address nonsmooth (discontinuous) functions as solutions to certain PDEs,
- incorporate objects like the delta function into a PDE (and/or initial conditions).

We provide three examples, in each case highlighting the additional insight that the distribution perspective gives. In each example, the distributional solutions are distributions generated by functions, albeit functions which can have singularities. In general, a distributional solution to a PDE might be a distribution which cannot be generated by a function; we present one such example at the end of Section 9.7.3.

9.7.1. Our First Example Revisited!

Recall what was essentially the first PDE of this text:

$$au_x + bu_y = 0. \tag{9.21}$$

We can interpret this PDE in the sense of distributions. To this end, $u(x,y)$, a locally integrable function on $\mathbb{R}^2$, is a solution to (9.21) on $\mathbb{R}^2$ if we have

$$\int_{-\infty}^{\infty}\int_{-\infty}^{\infty} \big(-au(x,y)\,\phi_x(x,y) - bu(x,y)\,\phi_y(x,y)\big)\,dx\,dy = 0 \qquad \text{for all } \phi \in C_c^\infty(\mathbb{R}^2),$$

or equivalently,

$$\int_{-\infty}^{\infty}\int_{-\infty}^{\infty} u(x,y)\big(a\,\phi_x(x,y) + b\,\phi_y(x,y)\big)\,dx\,dy = 0 \qquad \text{for all } \phi \in C_c^\infty(\mathbb{R}^2).$$

Ok, but does this weaker interpretation enlighten or enrich us in any way; for example, does it allow for more general solutions? We discovered early on that the general solution to (9.21) was

$$u(x,y) = f(bx - ay), \tag{9.22}$$

for any C^1 function of one variable. For a classical solution, we would need f to be C^1; otherwise we would not be able to compute u_x or u_y. However, the PDE only dictates that the solution be constant along the characteristics which are lines parallel to $\langle b, -a\rangle$. Hence, should it really matter if f was not C^1 or even continuous? For example, could we take

$$f(z) = \begin{cases} 1 & \text{if } z \geq 0, \\ -1 & \text{if } z < 0? \end{cases}$$

It would **not** matter for u, defined by (9.22), to be a solution in the sense of distributions. In fact, in Exercise **9.12** you are asked to show that u defined by (9.22) is a solution in the sense of distributions for any locally integrable function f. Herein lies the extra insight when we view (9.21) in the sense of distributions — we enrich the class of solutions. These enriched class of solutions are far from being just a mathematical curiosity but can represent physically relevant (observed) solutions, as discussed in the next example.

9.7.2. Burgers's Equation and the Rankine-Hugoniot Jump Conditions.

Recall from Section 2.3, the Burgers equation (without viscosity):

$$u_t + uu_x = 0.$$

We saw in Section 2.3 that for certain initial data, the solution develops a discontinuity at a certain time. What happens next can be inferred from interpreting the PDE in the sense of distributions.

The product of u with u_x in Burgers's equation might, at first, seem problematic in terms of an interpretation in the sense of distributions. However, it is a special product and, indeed, the PDE can be written in the equivalent form

$$u_t + \big(f(u)\big)_x = 0 \quad \text{where} \quad f(u) = \frac{1}{2}u^2. \tag{9.23}$$

This PDE can indeed be interpreted in the sense of distributions: We say $u(x,t)$, a locally integrable function on $\mathbb{R}\times(0,\infty)$, is a solution in the sense of distributions if for **all** $\phi(x,t) \in C_c^\infty(\mathbb{R}\times(0,\infty))$

$$\int_0^\infty \int_{-\infty}^\infty \big(u(x,t)\,\phi_t(x,t) + f(u(x,t))\,\phi_x(x,t)\big)\,dx\,dt = 0. \tag{9.24}$$

What insight does (9.24) yield? Firstly, it allows us to consider piecewise discontinuous solutions, i.e., solutions with jump discontinuities. Recall Example 2.3.1 from Section 2.3 where we considered the initial data

$$g(x) = \begin{cases} 1 & \text{if } x \leq 0, \\ 1-x & \text{if } 0 \leq x \leq 1, \\ 0 & \text{if } x \geq 1. \end{cases}$$

We plotted the profiles of u for times $t = 0, 0.5, 0.75$, and 1 in Figure 2.13 but stopped at $t = 1$, where a jump discontinuity formed. Recall that $u(x,1) = 1$ if $x \leq 1$ and $u(x,1) = 0$ otherwise. The time $t = 1$ was when the wave *broke*. What happens next (for $t > 1$)? Burgers's equation is a model for a physical problem wherein one can observe that this jump discontinuity prevails after $t > 1$ but with spacial point of the discontinuity propagating through space with a particular speed; as it turns out, the speed here would be 1/2. So for example, the solution at $t = 2$ would be $u(x, 2) = 1$ if $x \leq 1.5$ and 0 otherwise. A discontinuity in space which propagates through time is called a **shock wave**. This **physical condition** relating the speed of the jump discontinuity and the magnitude of the jump are old and are known as **the Rankine-Hugoniot jump conditions**. They are thus named in recognition of the 19th-century work done on gas dynamics by Scottish engineer and physicist William John Macquorn Rankine and French engineer Pierre-Henri Hugoniot. For the above example, they found that a discontinuity connecting the constant states 1 and 0 would propagate with speed 1/2. More generally, two constant states one on the left u_l and one on the right u_r could be pieced together with a discontinuity which propagated with speed

$$s = \frac{u_l + u_r}{2}. \tag{9.25}$$

Amazingly, this jump condition is **simply a direct consequence of considering the solution from the perspective of distributions**. To see this, let us consider initial data of the form

$$g(x) = \begin{cases} u_l & \text{if } x \leq 0, \\ u_r & \text{if } x > 0, \end{cases}$$

where u_l and u_r are two distinct constants. Suppose we purport that the discontinuity at $x = 0$ will propagate at a constant speed $s > 0$. This amounts to considering the piecewise constant solution

$$u(x, t) = \begin{cases} u_l & \text{if } x \leq st, \\ u_r & \text{if } x > st. \end{cases} \tag{9.26}$$

Note that $u(x, t)$ has a jump discontinuity precisely on the line $\mathcal{L} := \{(x, t) \,|\, x = st\}$.

But is (9.26) a (global) solution in the sense of distributions; i.e., does it satisfy (9.24)? It is instructive to recall differentiating the Heaviside function in the sense of distributions. There we saw that differentiating the jump in the sense of distributions gave rise to the delta function. A two-dimensional analogue was discussed in Section 9.3.2 wherein taking a partial derivative in the sense of distributions produced a continuum of "delta-like" functions concentrated on the x-axis. Similar issues will prevail here: Taking the partial derivatives u_t and u_x in the sense of distributions will produce "delta-like" distributions which are concentrated on the jump line $\mathcal{L}$ in space-time. What we are looking for are *conditions* which ensure that these "delta-like" distributions will cancel out. Let us write out a simple theorem and then present it proof.

Theorem 9.3. *The piecewise constant function defined by* (9.26) *satisfies* (9.24) *iff the Rankine-Hugoniot jump condition* (9.25) *holds true.*

In the following proof we will invoke the Divergence Theorem in the xt-plane to the left and right of the jump line $\mathcal{L}$. In doing so we will see the appearance of boundary terms in the form of line integrals over parts of the line $\mathcal{L}$.

Proof. Step 1: The definition of a solution in the sense of distributions, localization with an appropriate test function, and split-up of the integral. Suppose (9.24) holds true for all test functions ϕ. Fix such a test function ϕ whose support includes some part of the line $\mathcal{L}$. Note that by definition this means that $\phi \equiv 0$ **outside** of some closed disk $\mathcal{D}$ in the xt upper half-plane which lies strictly above the $t = 0$ axis (see Figure 9.2). Of course its exact support need not be $\mathcal{D}$ but some subset of it.

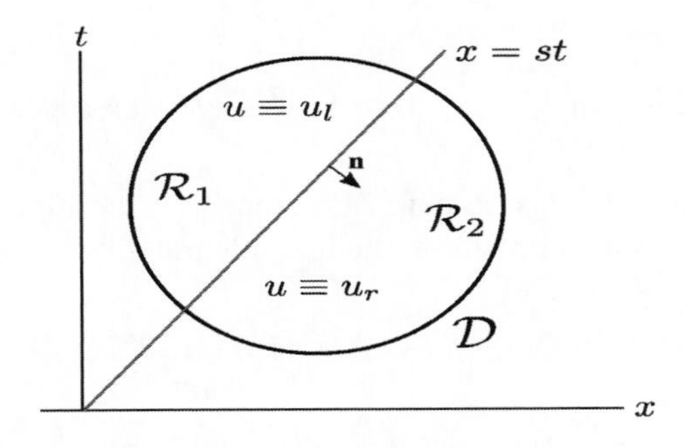

Figure 9.2

Divide the disk $\mathcal{D}$ into to two pieces, one to the left and the other to the right of the shock; i.e., consider

$$\mathcal{R}_1 = \mathcal{D} \cap \{(x,t)\,|\,x < st\} \quad \text{and} \quad \mathcal{R}_2 = \mathcal{D} \cap \{(x,t)\,|\,x > st\}.$$

Then

$$\begin{aligned} 0 &= \int_0^\infty \int_{-\infty}^\infty (u\phi_t + f(u)\phi_x)\,dxdt \\ &= \int_0^\infty \int_{-\infty}^{st} (u\phi_t + f(u)\phi_x)\,dxdt + \int_0^\infty \int_{st}^\infty (u\phi_t + f(u)\phi_x)\,dxdt \\ &= \iint_{\mathcal{R}_1} (u_l\phi_t + f(u_l)\phi_x)\,dxdt + \iint_{\mathcal{R}_2} (u_r\phi_t + f(u_r)\phi_x)\,dxdt. \end{aligned} \tag{9.27}$$

Step 2: Set-up for, and application of, the Divergence Theorem on $\mathcal{R}_1$ and $\mathcal{R}_2$. On either region $\mathcal{R}_1$ or $\mathcal{R}_2$, we note that u is a constant which is continuous up to its boundary, and hence so are the vector fields in the xt-plane

$$\langle f(u_l)\phi, u_l\phi\rangle \quad \text{and} \quad \langle f(u_r)\phi, u_r\phi\rangle. \tag{9.28}$$

Note that here we view the first component as the x component and the second component as the t component. Note that

$$\operatorname{div}\langle f(u_l)\phi, u_l\phi\rangle = u_l\phi_t + f(u_l)\phi_x \quad \text{and} \quad \operatorname{div}\langle f(u_r)\phi, u_r\phi\rangle = u_r\phi_t + f(u_r)\phi_x.$$

Thus (9.27) becomes

$$0 = \iint_{\mathcal{R}_1} \operatorname{div}\langle f(u_l)\phi, u_l\phi\rangle\,dx\,dt + \iint_{\mathcal{R}_2} \operatorname{div}\langle f(u_r)\phi, u_r\phi\rangle\,dx\,dt. \tag{9.29}$$

We now apply the 2D Divergence Theorem on each $\mathcal{R}_i$. Let $\mathbf{n}$ denote the unit normal to the line $\mathcal{L}$ which points to the right (i.e., into $\mathcal{R}_2$). Since the respective vector fields (9.28) are zero on all parts of $\partial\mathcal{R}_i$ except the part which lies along the line $\mathcal{L}$, we find

$$\iint_{\mathcal{R}_1} \operatorname{div} \langle f(u_l)\phi, u_l\phi\rangle \, dxdt = \int_{\mathcal{L}} \langle f(u_l)\phi, u_l\phi\rangle \cdot \mathbf{n} \, ds$$

and

$$\iint_{\mathcal{R}_2} \operatorname{div} \langle f(u_r)\phi, u_r\phi\rangle \, dxdt = \int_{\mathcal{L}} -\langle f(u_r)\phi, u_r\phi\rangle \cdot \mathbf{n} \, ds.$$

Note that the negative in the last term is there since the outer unit normal to $\mathcal{R}_2$ is $-\mathbf{n}$. Thus (9.29) becomes a statement about the line integral

$$\begin{aligned} \int_{\mathcal{L}} \Big(\langle f(u_l), u_l\rangle \cdot \mathbf{n} \; - \; \langle f(u_r), u_r\rangle \cdot \mathbf{n} \Big) \phi \, ds \\ = \Big(\langle f(u_l), u_l\rangle \cdot \mathbf{n} \; - \; \langle f(u_r), u_r\rangle \cdot \mathbf{n} \Big) \int_{\mathcal{L}} \phi \, ds \\ = 0. \end{aligned} \tag{9.30}$$

Step 3: Invoking the power of (9.24) **holding for all** ϕ. Since (9.30) holds for all ϕ with compact support, the constant prefactor must be zero; i.e.,

$$\langle f(u_l), u_l\rangle \cdot \mathbf{n} \; - \; \langle f(u_r), u_r\rangle \cdot \mathbf{n} \; = \; 0.$$

Denoting the components of $\mathbf{n}$ by $\langle n_x, n_t\rangle$, we have

$$(f(u_r) - f(u_l))n_x + (u_r - u_l)n_t = 0$$

or

$$\frac{f(u_r) - f(u_l)}{u_r - u_l} = -\frac{n_t}{n_x}.$$

Lastly, let us note that this ratio of the components of $\mathbf{n}$ is exactly s, the slope of the line. Hence, recalling that $f(u) = \frac{1}{2}u^2$, we arrive at (9.25). The reader can readily retrace the steps to verify the full if and only if. □

We make two remarks:

- Theorem 9.3 readily generalizes to a piecewise smooth solution, that is, a classical smooth solution on the left and on the right of some curve $x = \eta(t)$ which at any time t has a limit from the left and a limit from the right (cf. Exercise **9.24**).
- In a (hopefully) future volume on this subject, we will explore further such distributional solutions. In particular, we will observe that the IVP has many (in fact infinitely many) distributional solutions; all but one are nonphysical. An additional condition is required to weed out these unphysical solutions.

9.7.3. The Wave Equation with a Delta Function Source. An important problem involving the 1D wave equation is to find the solution to

$$u_{tt} - u_{xx} = \text{“}\delta_0(x)\,\delta_0(t)\text{”}.$$

Here, we use the parentheses around the product of delta functions; however, our intention is to solve

$$u_{tt} - u_{xx} = \delta_{\mathbf{0}} \qquad \text{in the sense of distributions,} \tag{9.31}$$

where $\delta_{\mathbf{0}}$ denotes the 2D delta function with concentration at the point $(x, t) = (0, 0)$. Now, both sides make precise sense as distributions and we can dispense with any parentheses. Before imposing any other conditions (i.e., initial conditions), let us interpret the PDE in the sense of distributions. In general, a distributional solution to a PDE might not be generated by a function, and this is indeed the case if we considered the analogue of this delta-driven wave equation in 3D. As we show now, in 1D the distributional solution is generated by a locally integrable function. Since the wave equation is time reversible, let us look for a distributional solution generated by a locally integrable function $u(x,t)$ on $\mathbb{R}^2$ $(x, t \in \mathbb{R})$. A locally integrable function $u(x,t)$ is a solution in the sense of distributions to (9.31) if for **all** $\phi(x,t) \in C_c^\infty(\mathbb{R}^2)$,

$$\int_{-\infty}^{\infty}\int_{-\infty}^{\infty} u(x,t)\left(\phi_{tt}(x,t) - \phi_{xx}(x,t)\right) dx\, dt = \phi(0,0). \tag{9.32}$$

This is what it means to be a solution in the sense of distributions, but what is an example of a solution? For that, we need some additional conditions (say, initial conditions given at $t = 0$), which, for simplicity, we take to be

$$u(x,0) = u_t(x,0) \equiv 0.$$

Note that the only nonzero information in the problem comes from the source which is enforced exactly and only at $t = 0$; hence, we can take the solution to be identically 0 for $t \leq 0$. For $t > 0$, there are two ways that one can directly derive a potential solution formula, a formula which is a good candidate for satisfying (9.32). The first is to look at the solution to the wave equation with a functional source term derived in Section 3.6. For a source function $f(x,t)$, $c = 1$, and zero initial conditions, the solution for $t > 0$ was

$$u(x,t) = \frac{1}{2}\iint_D f(y,s)\, dy\, ds,$$

where D was the domain of dependence on the initial data; that is,

$$D = \left\{(y,s)\,\middle|\, 0 \leq s \leq t, x - s < y < x + s\right\}.$$

Informally, taking "$f = \delta_{\mathbf{0}}$" and noting that the integral should be 1 if $\mathbf{0} = (0,0) \in D$ and 0 otherwise, yields

$$u(x,t) = \begin{cases} \frac{1}{2} & \text{if } |x| < t, t > 0, \\ 0 & \text{if } |x| \geq t, t > 0, \\ 0 & \text{if } t \leq 0. \end{cases} \tag{9.33}$$

This function can conveniently be written in terms of the Heaviside function H as

$$\boxed{u(x,t) = \frac{1}{2}H(t-|x|).}$$

One could also derive this formula via the Fourier transform in the sense of tempered distributions. As we will see in Section 13.2.3, this function is actually rather "fundamental" to the wave equation; in fact, it is known as the fundamental solution (or Green's function) for the 1D wave equation.

So we have derived our potential candidate solution (9.33). Let us now check that it satisfies (9.32). Recall that we proved Theorem 3.2 (the solution formula for the wave equation with source f) using Green's Theorem. Green's Theorem will also be used in this instance but only on the test function ϕ. To this end, let $\phi(x,t) \in C_c^\infty(\mathbb{R}^2)$. Since ϕ has compact support, there exists some value of time, $T > 0$, such that $\phi(x,t) \equiv 0$ for all x and $t > T$. Using the form of $u(x,t)$ given by (9.33) we have

$$\int_{-\infty}^{\infty}\int_{-\infty}^{\infty} u(x,t)\left(\phi_{tt}(x,t) - \phi_{xx}(x,t)\right)dx\,dt = \frac{1}{2}\int_0^T\int_{-t}^{t}\left(\phi_{tt}(x,t) - \phi_{xx}(x,t)\right)dx\,dt.$$

Following the proof of Theorem 3.2, we use Green's Theorem to find that

$$\frac{1}{2}\int_0^T\int_{-t}^{t}\left(\phi_{tt}(x,t) - \phi_{xx}(x,t)\right)dx\,dt = \frac{1}{2}\int_{\mathcal{C}} -\phi_t(x,t)\,dx - \phi_x(x,t)\,dt,$$

where $\mathcal{C}$ is the boundary of the upside down triangular region $0 < t < T, -t < x < t$ oriented counterclockwise. This curve consists of three pieces: $\mathcal{C}_1, \mathcal{C}_2, \mathcal{C}_3$, the top piece where $t = T$ and the two sides where $x = t$ and $x = -t$. On $\mathcal{C}_1$, ϕ and all its derivatives vanish; hence, there is no contribution to the line integral. On the other hand, on each of the sides, one finds (with the same calculation as in Theorem 3.2) that the line integral gives $\frac{1}{2}\phi(0,0)$. For example, $\mathcal{C}_2$, the segment of the line $x = t$, can be parametrized by $x = s, t = s$ for $0 \le s \le T$. Hence,

$$\begin{aligned}
\frac{1}{2}\int_{\mathcal{C}_2} -\phi_t(x,t)\,dx - \phi_x(x,t)\,dt &= -\frac{1}{2}\int_0^T \frac{d}{ds}\phi(s,s)\,ds \\
&= -\frac{1}{2}(\phi(T,T) - \phi(0,0)) = \frac{1}{2}\phi(0,0).
\end{aligned}$$

The integral over $\mathcal{C}_2$, the segment of the line $x = -t$, also gives $\frac{1}{2}\phi(0,0)$. In conclusion, (9.32) holds true.

As a good exercise (cf. Exercise **9.14**), make sense of

$$u_{tt} - u_{xx} = \text{“}\delta_0(x)\text{”}$$

in the sense of distributions, then solve it with the zero initial conditions.

Finally, be aware that the analogue of (9.31) in 3D,

$$u_{tt} - c^2\Delta u = \text{“}\delta_0(x)\delta_0(y)\delta_0(z)\delta_0(t)\text{”} \qquad \text{or} \qquad u_{tt} - c^2\Delta u = \delta_{\mathbf{0}}, \quad \mathbf{0} = (0,0,0,0),$$

can readily be interpreted in the sense of distributions **but** its distributional solution is **not** a distribution generated by a function (cf. Exercise **9.16**). It is the distribution which informally can be written as

$$\text{“}\frac{1}{4\pi c^2 t^2}\,\delta_0\,(ct - |\mathbf{x}|)\text{”}.$$

We will further address this distributional solution in Section 13.2.4.

9.7.4. Incorporating Initial Values into a Distributional Solution. It often proves convenient to bring the initial values into the integral-based statement of a solution in the sense of distributions. Let us suppose the solution sought (in the sense of a distribution) can be generated by a locally integrable function. For the sake of illustration, consider the initial value problem for the simple transport equation

$$\begin{cases} u_t + cu_x = 0, & x \in \mathbb{R}, t > 0, \\ u(x,0) = f(x). \end{cases} \tag{9.34}$$

The analogous steps and definition can be done for the IVP of other PDEs which can be interpreted in the sense of distributions. First off, we present the motivation for the definition. Suppose $u(x,t)$ is a smooth solution to (9.34) (you may also assume f is smooth). Let $\phi \in C_c^\infty((-\infty,\infty)\times[0,\infty))$. Recall from Section 9.6 that since the domain includes its boundary (the $t = 0$ axis), compact support on $(-\infty,\infty)\times[0,\infty)$ allows $\phi(x,0)$ to be nonzero for certain values of x. Since u is a smooth solution, we have

$$0 = \int_0^\infty \int_{-\infty}^\infty (u_t(x,t) + cu_x(x,t))\,\phi(x,t)\,dx\,dt.$$

We break up the integral on the right into its two pieces and integrate by parts on each. For the first integral, we use Fubini's Theorem, Theorem A.13 of the Appendix, and then integrate by parts with respect to t to find

$$\begin{aligned} &\int_0^\infty \int_{-\infty}^\infty u_t(x,t)\,\phi(x,t)\,dx\,dt \\ &\quad = \int_{-\infty}^\infty \int_0^\infty u_t(x,t)\,\phi(x,t)\,dt\,dx \\ &\quad = -\int_{-\infty}^\infty \int_0^\infty u(x,t)\,\phi_t(x,t)\,dt\,dx \;-\; \int_{-\infty}^\infty u(x,0)\,\phi(x,0)\,dx \\ &\quad = -\int_{-\infty}^\infty \int_0^\infty u(x,t)\,\phi_t(x,t)\,dt\,dx \;-\; \int_{-\infty}^\infty f(x)\,\phi(x,0)\,dx. \end{aligned}$$

For the second, we now integrate by parts with respect to x to find

$$\int_0^\infty \int_{-\infty}^\infty u_x(x,t)\,\phi(x,t)\,dx\,dt \;=\; -\int_{-\infty}^\infty \int_0^\infty u(x,t)\,\phi_x(x,t)\,dt\,dx.$$

Note that in performing the second integration by parts with respect to x, there was no nonzero boundary term. Thus we have that for all $\phi \in C_c^\infty((-\infty,\infty)\times[0,\infty))$,

$$0 = -\int_0^\infty \int_{-\infty}^\infty u(x,t)\,\left(\phi_t(x,t) + c\phi_x(x,t)\right)dx\,dt \;-\; \int_{-\infty}^\infty f(x)\,\phi(x,0)\,dx.$$

With our motivation in hand, we are ready for the following definition. A locally integrable function $u(x,t)$ is a solution in the sense of distributions to (9.34) if for all $\phi \in C_c^\infty((-\infty,\infty)\times[0,\infty))$, we have

$$\boxed{-\int_0^\infty \int_{-\infty}^\infty u(x,t)\,(\phi_t(x,t)+c\phi_x(x,t))\,dx\,dt \;-\; \int_{-\infty}^\infty f(x)\,\phi(x,0)\,dx \;=\; 0.} \tag{9.35}$$

With this definition, we now consider initial value problems where the initial data is either a function with singularities or a delta function.

Example 9.7.1 (Transport Equation with Heaviside Initial Data). Consider

$$\begin{cases} u_t + cu_x \;=\; 0, & x\in\mathbb{R}, t>0, \\ u(x,0) = H(x), \end{cases} \tag{9.36}$$

where $H(x)$ is the Heaviside function. Based upon our notion of characteristics, one would expect that a good candidate for the solution would be

$$u(x,t) \;=\; H(x-ct).$$

Indeed, we previously found that the general pointwise solution to the transport equation was $f(x-ct)$ for any function f. So why not replace f with $H(x)$, ignoring the fact that H has a singularity at 0! In Exercise **9.9** you are asked to verify that this function satisfies (9.35) and is hence a solution in the sense of distributions to the IVP.

Example 9.7.2 (Transport Equation with Delta Function Initial Data). Consider

$$\begin{cases} u_t + cu_x \;=\; 0, & x\in\mathbb{R}, t>0, \\ u(x,0) = \delta_0. \end{cases} \tag{9.37}$$

In this case, the solution in the sense of distributions is not generated by a locally integrable function. Hence, we must make sense of both the double integral and the single integral at $t=0$ which appear in (9.35) for a u which is a general distribution. We simply must generalize the statement (9.35) when u is a general distribution. To this end, a distribution F is a solution in the sense of distributions to (9.34) if we have

$$-\langle F\,,\,(\phi_t(x,t)+c\phi_x(x,t))\,\rangle \;-\; \phi(0,0) \;=\; 0 \qquad \text{for all } \phi\in C_c^\infty((-\infty,\infty)\times[0,\infty)).$$

Note that here we have replaced the integral on the $t=0$ axis

$$\int_{-\infty}^\infty f(x)\,\phi(x,0)\,dx \qquad \text{with} \qquad \langle\delta_0, \phi(x,0)\rangle \;=\; \phi(0,0).$$

We now claim that the solution to (9.37) in the sense of distributions is given by the distribution F which assigns to each $\phi\in C_c^\infty((-\infty,\infty)\times[0,\infty))$

$$\langle F\,,\,\phi\,\rangle \;=\; \int_0^\infty \phi(cs,s)\,ds.$$

To check this, note that

$$-\langle F\,,\,(\phi_t(x,t)+c\phi_x(x,t))\rangle \;=\; -\int_0^\infty \phi_t(cs,s)+c\phi_x(cs,s)\,ds$$
$$=\; -\int_0^\infty \frac{d}{ds}\phi(cs,s)\,ds \;=\; \phi(0,0).$$

Following the reasoning of Section 5.6, it is rather suggestive to write this distributional solution F informally as "$u(x,t)=\delta_0(x-ct)$". Indeed, the general solution of all functional solutions to the transport equation was $f(x-ct)$ for any function f, so why not "replace" f with δ! You can appreciate why this informal notation is quite descriptive and in a certain sense, the above sense, correct.

Finally let us give an example of the diffusion equation with a delta function initial condition. While we have catered the definition of the IVP distributional solution to the transport equation, the analogous definition holds for the diffusion equation. This example illustrates that even though the initial conditions entail an object which is **not** generated by a function, the distributional solution for $t>0$ **is** generated by a smooth function.

Example 9.7.3 (Diffusion Equation with Delta Function Initial Data). Consider the IVP

$$\begin{cases} u_t \;=\; \alpha u_{xx}, & x\in\mathbb{R}, t>0,\\ u(x,0)=\delta_0. \end{cases} \tag{9.38}$$

Similar to what was done for the transport equation in (9.35), we can incorporate the initial data into the statement of a distributional solution. What would we expect for the distributional solution? Recall that the fundamental solution,

$$\Phi(x,t) \;=\; \frac{1}{\sqrt{4\pi ct}}\, e^{-\frac{x^2}{4\alpha t}}, \tag{9.39}$$

was a smooth pointwise solution to the diffusion equation for any $t>0$ and

$$\Phi(\cdot,t)\longrightarrow \delta_0 \qquad \text{as } t\to 0^+ \text{ in the sense of distributions.}$$

In Exercise **9.18**, you are asked to formulate the distributional statement for (9.38) and verify that $u(x,t)=\Phi(x,t)$ is indeed such a solution.

A useful way to solve linear IVPs, like (9.38), is via the Fourier transform interpreted in the sense of tempered distributions. For (9.38), we take the distributional Fourier transform in the x (spatial) variable of the PDE and the initial conditions. This yields for each k an initial value ODE problem for $F(t)=\hat{u}(k,t)$:

$$F'(t)=-\alpha k^2F(t) \quad \text{with} \quad F(0)=1. \tag{9.40}$$

This is similar to what we did in Section 6.8.2 but the new feature here is that we take the Fourier transform of the delta function initial condition. One can readily check that one obtains the fundamental solution (9.39) by solving (9.40) and then taking the inverse Fourier transform.

9.7.5. Not All PDEs Can Be Interpreted in the Sense of Distributions. By our definition of differentiation in the sense of distributions, we have imposed certain restrictions on the structure of a PDE in order to interpret it in the sense of distributions. Distributions have a **linear structure** embedded in them; note that we can add two distributions to generate another but we cannot multiply them. We can certainly interpret **linear PDEs with constant coefficients** in the sense of distributions. However, we cannot interpret in the sense of distributions the following linear transport, wave, and diffusion equations with a variable coefficient:

$$u_t = c(x)u_x, \qquad u_{tt} = c^2(x)u_{xx}, \qquad u_t = c(x)u_{xx}.$$

Quasilinear equations like Burgers's equation, which have a certain form, can also be interpreted in the sense of distributions. Yet with fully nonlinear equations we have a problem. For example, we cannot interpret[6] the Hamilton-Jacobi equation

$$u_t + (u_x)^2 = 0$$

in the sense of distributions.

9.8. A View Towards Sobolev Spaces

For our purposes, it sufficed to only consider distributional partial derivatives of functions which were smooth except at certain points (or on certain sets). On the other hand, recall that the theory of distributions allows us to differentiate functions which may not be differentiable at any point (in the pointwise sense). There exists a wide class of functions of several variables with complex singularities but whose partial derivatives in the sense of distributions are **still generated by locally integrable functions**. In other words, their partial derivatives in the sense of distributions are still functions, but functions which are interpreted as distributions. Surprisingly, this class includes functions whose classical pointwise partial derivatives fail to exist at all points; in fact, one can have functions which are nowhere continuous (discontinuous at every point) but still have a distributional derivative which is generated by a locally integrable function!

It turns out to be very useful to consider the integrability properties of **powers** of distributional partial derivative functions, i.e., whether or not the function generating the distributional partial derivative, raised to a certain power $p \geq 1$, is integrable on the domain. In doing so, this gives rise to very important classes (or spaces) of functions called **Sobolev spaces**. While we plan to address a certain Sobolev space in a future volume on this subject, let me provide a few details and hints into the realm of Sobolev spaces. We present the definition and one example.

[6]There is a generalization of solutions to Hamilton-Jacobi equations which in many ways resembles the spirit of distributions. This weaker form of a solution is called a **viscosity solution**.

Definition of the Sobolev Space $W^{k,p}(\mathbb{R}^N)$

Let $k \in \mathbb{N}$, $p \geq 1$, and let $u : \mathbb{R}^N \to \mathbb{R}$ be a locally integrable function. Suppose the following:

- For any multi-index α of order less than or equal to k, $\partial^\alpha u$ in the sense of distributions is generated by a locally integrable function v_α; that is, for all $\phi \in C_c^\infty(\mathbb{R}^N)$, we have

$$\begin{aligned} \langle \partial^\alpha u, \phi \rangle = (-1)^{|\alpha|} \langle u, \partial^\alpha \phi \rangle &= (-1)^{|\alpha|} \int \dots \int_{\mathbb{R}^N} u(\mathbf{x})\, \partial^\alpha \phi(\mathbf{x})\, d\mathbf{x} \\ &= \int \dots \int_{\mathbb{R}^N} v_\alpha(\mathbf{x})\, \phi(\mathbf{x})\, d\mathbf{x}. \end{aligned} \tag{9.41}$$

- The function v_α satisfies

$$\int \dots \int_{\mathbb{R}^N} |v_\alpha(\mathbf{x})|^p\, d\mathbf{x} < \infty.$$

Then we say u belongs to the Sobolev space $W^{k,p}(\mathbb{R}^N)$.
If $\Omega \subset \mathbb{R}^N$ is any domain, we can analogously define the Sobolev space $W^{k,p}(\Omega)$ by focusing all integrals to be over Ω and considering test functions with compact support in Ω (cf. Section 9.6).

The standard example will be, not surprisingly, functions which blow up at the origin. To this end, let $\Omega = B(\mathbf{0}, 1)$ and

$$u(\mathbf{x}) = \frac{1}{|\mathbf{x}|^\beta} \qquad \text{for } \beta > 0.$$

A good exercise (cf. Exercise **9.23**) is to verify that

$$u \in W^{1,p}(B(\mathbf{0}, 1)) \qquad \text{if and only if} \qquad \beta < \frac{N-p}{p}.$$

To this end, write out (9.41), with $\mathbb{R}^N$ replaced by $B(\mathbf{0}, 1)$, for any partial derivative, isolate the origin with a ball of radius ϵ, and perform the necessary calculus to infer the limit as $\epsilon \to 0$.

In this example the function was classically differentiable at all points except the origin (where the function experienced a blow-up discontinuity). **Warning:** There are examples of functions in $W^{1,p}$ which are nowhere continuous (hence nowhere differentiable)!

What is particularly remarkable in the study of Sobolev spaces is the fact that degree (size) of integrability p controls regularity (smoothness) properties of the function. For example, if a function is in $W^{1,p}$ for $p > N$, it must be (up to redefinition on a negligible set) continuous! In other words, all discontinuities are "removable" (in the standard calculus sense). If a function is in $W^{2,p}$ for all $p > N$, then it is (up to redefinition on a negligible set) C^1. For more on Sobolev spaces see Evans [**10**] or the classic monograph of Brezis [**19**].

9.9. Fourier Transform of an N-dimensional Tempered Distribution

In Section 6.9 we defined the Fourier transform of a one-dimensional tempered distribution. One can similarly define the Fourier transform of an N-dimensional tempered distribution. An N-dimensional tempered distribution acts on functions of N variables which lie in the N-dimensional Schwartz class $\mathcal{S}(\mathbb{R}^N)$. These are complex-valued C^∞ functions on $\mathbb{R}^N$ for which

$$\lim_{|\mathbf{x}|\to\infty} |\mathbf{x}^\beta|\ |\partial^\alpha \phi| \;=\; 0,$$

for all multi-indices β and α. In other words, the function and all its partial derivatives have rapidly decreasing tails.

Then, as before, if F is an N-dimensional tempered distribution, the Fourier transform of F is the tempered distribution $\widehat{F}$ defined by

$$\langle \widehat{F}, \phi\rangle \;:=\; \langle F, \widehat{\phi}\rangle \qquad \text{for any } \phi \in \mathcal{S}(\mathbb{R}^N).$$

Similarly, we define the inverse Fourier transform $\check{F}$ by

$$\langle \check{F}, \phi\rangle \;:=\; \langle F, \check{\phi}\rangle \qquad \text{for any } \phi \in \mathcal{S}(\mathbb{R}^N).$$

The following presents a few useful examples.

(i) (**Examples with the N-dimensional delta function**)
We have

$$\boxed{\hat{1} \;=\; (2\pi)^N\, \delta_{\mathbf{0}} \quad \text{in the sense of tempered distributions.}} \tag{9.42}$$

On the other hand,

$$\boxed{\widehat{\delta_{\mathbf{0}}} \;=\; 1 \quad \text{in the sense of tempered distributions.}}$$

More generally, if $\mathbf{a} \in \mathbb{R}^N$,

$$\boxed{\widehat{e^{i\mathbf{a}\cdot\mathbf{x}}} \;=\; (2\pi)^N\, \delta_{\mathbf{a}} \quad \text{and} \quad \widehat{\delta_{\mathbf{a}}} \;=\; e^{i\mathbf{a}\cdot\mathbf{x}} \quad \text{in the sense of tempered distributions.}}$$

(ii) (**Some radial rational functions**)
An important class of functions on $\mathbb{R}^N$ takes the form of

$$\frac{1}{|\mathbf{x}|^p},$$

for $p > 0$. The relationship between p and the dimension N is critical for determining the *character* of the function. Let us consider three dimensions ($N = 3$) and $p = 1$ and 2. Note that, in either case, the function is not integrable on $\mathbb{R}^3$. However, one can consider these functions as tempered distributions, enabling the computation of their Fourier transforms. By considering limits as $a \to 0$ in (6.66) one would find

$$\boxed{\widehat{\frac{1}{|\mathbf{x}|}} \;=\; \frac{4\pi}{|\mathbf{k}|^2} \quad \text{in the sense of tempered distributions on } \mathbb{R}^3.} \tag{9.43}$$

One can also check that

$$\boxed{\widehat{\frac{1}{|\mathbf{x}|^2}} \;=\; \frac{2\pi^2}{|\mathbf{k}|} \quad \text{in the sense of tempered distributions on } \mathbb{R}^3.} \tag{9.44}$$

9.10. Using the Fourier Transform to Solve Linear PDEs, III: Helmholtz and Poisson Equations in Three Space

We now use the multidimensional Fourier transform to solve some purely spatial PDEs involving the Laplacian. While these PDEs will be the focus of Chapter 10, here we assume we have a solution and then use the Fourier transform to derive solution formulas. These formulas will be studied in detail in Chapter 10.

Consider first the **Helmholtz equation** on $\mathbb{R}^3$ which, for a fixed $a \in \mathbb{R}$, $a \neq 0$, and f an integrable (or square-integrable) function on $\mathbb{R}^3$, is given by

$$\boxed{-\Delta u(\mathbf{x}) + a^2 u(\mathbf{x}) = f(\mathbf{x}) \qquad \text{for } \mathbf{x} \in \mathbb{R}^3.} \tag{9.45}$$

You will shortly see why we included a minus sign in front of the Laplacian. Let us now assume that $u(\mathbf{x})$ is an integrable function which solves the PDE and has a Fourier transform in the classical sense. Then taking the 3D Fourier transform of both sides of (9.45), we find

$$|\mathbf{k}|^2 \hat{u}(\mathbf{k}) + a^2 \hat{u}(\mathbf{k}) = \hat{f}(\mathbf{k}).$$

Note that the minus sign canceled the minus sign resulting from Fourier transforming the Laplacian. Hence, solving for $\hat{u}(\mathbf{k})$ we find

$$\hat{u}(\mathbf{k}) = \frac{1}{|\mathbf{k}|^2 + a^2} \hat{f}(\mathbf{k}).$$

Now we take the inverse Fourier transform and note the following:

- Since

$$\widehat{(\phi * \psi)}(\mathbf{k}) = \hat{\phi}(\mathbf{k}) \hat{\psi}(\mathbf{k}),$$

  we have with

$$\phi = \mathcal{F}^{-1}\left(\frac{1}{|\mathbf{k}|^2 + a^2}\right) \qquad \text{and} \qquad \psi = f$$

  that

$$u(\mathbf{x}) = (\phi * \psi)(\mathbf{x}).$$

- Computing the inverse Fourier transform gives

$$\mathcal{F}^{-1}\left(\frac{1}{|\mathbf{k}|^2 + a^2}\right) = \frac{e^{-a|\mathbf{x}|^2}}{4\pi|\mathbf{x}|}. \tag{9.46}$$

Thus, we find

$$\boxed{u(\mathbf{x}) = \iiint_{\mathbb{R}^3} \frac{e^{-a|\mathbf{x}-\mathbf{y}|^2}}{4\pi|\mathbf{x}-\mathbf{y}|} f(\mathbf{y})\, d\mathbf{y}.} \tag{9.47}$$

The reader should note that in this calculation, we did not need the theory of distributions; the inverse Fourier transform (9.46) was in the classical functional sense. In particular, the reader might wonder as to the placement of this section in this chapter. The answer comes from asking what happens if we take $a = 0$? The corresponding PDE is the Poisson equation on all of space:

$$-\Delta u(\mathbf{x}) = f(\mathbf{x}) \qquad \text{for } \mathbf{x} \in \mathbb{R}^3. \tag{9.48}$$

Naively, one would expect that we can simply let $a = 0$ in (9.47), note that $e^{-a|\mathbf{x}-\mathbf{y}|^2}$ is identically 1, and derive the solution formula. This will in fact work.

Of course things change with respect to the Fourier transform when $a = 0$. However considering the equation and the transforms in the sense of tempered distributions justifies the necessary steps. Recall from the previous section that

$$\mathcal{F}^{-1}\left(\frac{1}{|\mathbf{k}|^2}\right) = \frac{1}{4\pi|\mathbf{x}|} \qquad \text{in the sense of tempered distributions.}$$

Note that one way to arrive at the above is to take the limit as $a \to 0$ in (9.46). Hence, either way, we are effectively making a calculation for $a > 0$ and then letting a tend to 0. We arrive at

$$\boxed{u(\mathbf{x}) = \iiint_{\mathbb{R}^3} \frac{1}{4\pi|\mathbf{x}-\mathbf{y}|} f(\mathbf{y})\, d\mathbf{y}} \tag{9.49}$$

as the solution formula for (9.48). We will show in the next chapter that it is indeed the solution.

We end with a short discussion surrounding the appearance of the function $\frac{1}{4\pi|\mathbf{x}|}$, the same function presented earlier as a "fundamental" example in Section 9.5. Consider the Poisson equation (9.48) with f replaced by the delta function; that is, consider the PDE $-\Delta u = \delta_{\mathbf{0}}$ in the sense of tempered distributions. Then, taking the Fourier transform of both sides as tempered distributions and noting that $\widehat{\delta_{\mathbf{0}}} = 1$, we arrive at

$$\hat{u} = \frac{1}{|\mathbf{k}|^2}.$$

Hence, $u(\mathbf{x})$ is the tempered distribution generated by a function, specifically

$$u(\mathbf{x}) = \frac{1}{4\pi|\mathbf{x}|}.$$

We have arrived at the fundamental fact announced[7] in Section 9.5:

$$-\Delta\left(\frac{1}{4\pi|\mathbf{x}|}\right) = \delta_{\mathbf{0}} \qquad \text{in the sense of tempered distributions.} \tag{9.50}$$

Note that we have not actually proven (9.50); the underlying premise behind any Fourier transform calculation is that a solution exists. Regardless of how convincing this calculation might seem, (9.50) is so important that we will devote an entire chapter to it and its consequences.

9.11. Chapter Summary

- Distributions can be extended to higher space dimensions via test functions of several variables which are smooth and have compact support.
- As in 1D, **the theory of distributions** allows us to both
  - interpret and analyze (i.e., do calculus with) classical functions of several variables, not in a pointwise — input/output — sense, but from the point of view of averaging and integration;

[7] To be precise, we announced this in the sense of distributions, which is a slightly weaker statement.

- provide precise meaning to, and find occurrences of, objects like the multidimensional delta function $\delta_{\mathbf{0}}$, which cannot be captured by a pointwise-defined classical function.

- In general, there are many types of multidimensional distributions. We were primarily concerned with two important classes:
  (1) Distributions **generated by locally integrable functions**.
  (2) Distributions which **concentrate at points**; for example, the multidimensional **delta function** $\delta_{\mathbf{0}}$, which is defined by $\langle \delta_{\mathbf{0}}, \phi \rangle = \phi(\mathbf{0})$, for any test function ϕ.

  We also saw examples of distributions which concentrated on lower-dimensional sets, for example the x-axis in $\mathbb{R}^2$, the unit sphere in $\mathbb{R}^3$, the wave cone in $\mathbb{R}^4$.

- Analogous to 1D, **convergence in the sense of distributions** allows us to capture, and calculate with, the limiting behavior of functions which concentrate, converging in the sense of distributions to the multidimensional delta function $\delta_{\mathbf{0}}$.

- We can take the **partial derivative** of any distribution by considering the action of the respective derivative on the test function. Here, it is convenient to adopt the ∂^α notation of Schwartz. Any such partial derivative is also a distribution over the same class of test functions. For our purposes, we focused on distributional partial derivatives of **distributions generated by functions**, in particular, partial derivatives of functions with singularities in the sense of distributions.

- By taking **linear combinations** of partial derivatives, we can interpret **the divergence and Laplacian** in the sense of distributions. We focused on their application to functions that were smooth with the exception of having a blow-up singularity at the origin. In particular, we considered the following:
  - the distributional divergence of a vector field of several variables with a blow-up discontinuity;
  - the distributional Laplacian of a function of several variables with a blow-up discontinuity.

  Important examples in 3D where such distributional combination of partial derivatives yield the delta function are

$$\operatorname{div}\left(\frac{\mathbf{x}}{|\mathbf{x}|^3}\right) = 4\pi\delta_{\mathbf{0}} \qquad \text{and} \qquad \Delta\left(\frac{1}{|\mathbf{x}|}\right) = -4\pi\delta_{\mathbf{0}},$$

  in the sense of distributions. We motivated and proved the first statement. The second, involving the Laplacian, will form the basis of Section 10.1. In these examples there is no nonzero functional component to the distributional derivatives, but rather, only the delta function.

- In finding partial derivatives and the divergence in the sense of distributions of functions with singularities, the steps were as follows:
  (i) We fixed a test function and wrote down the integral of the function times the derivative (partial derivative) of the test function.
  (ii) We **isolated** the singularity (singularities). For the blow-up singularity of the gravitational vector field, this amounted to breaking up the domain into a small ball of radius ϵ about the singularity and its complement.

(iii) On regions of integration where the original function is smooth, we performed some version of **integration by parts** to place the derivatives back on the original function, at the expense of additional boundary terms. A delta function is **detected** from boundary terms or, as in the case of the blow-up singularity, from limits of boundary terms.

- We can consider solutions in the sense of distributions to certain linear and quasi-linear PDEs, thereby providing insight into the potential discontinuous structure of solutions.
- By considering **tempered distributions** over the multidimensional Schwartz space, we can find the **Fourier transform** of several functions which are not integrable.

Exercises

9.1 Let $g \in C^\infty(\mathbb{R}^3)$. If F is any distribution acting on test functions in the class $C_c^\infty(\mathbb{R}^3)$, we can define the product of g with F as a distribution; for any $\phi \in C^\infty(\mathbb{R}^3)$, $\langle gF, \phi\rangle = \langle F, g\phi\rangle$.
(a) Why was this a reasonable definition?
(b) Prove that $|\mathbf{x}|^2\,\Delta\delta_0 = 6\,\delta_0$ in the sense of distributions where $\mathbf{x} \in \mathbb{R}^3$ and Δ is the 3D Laplacian.

9.2 (**Direct Product of Distributions**) Suppose F and G are 1D distributions over independent variables x and y, respectively. We define the direct product $F \otimes G$ to be a 2D distribution defined for all $\phi(x, y) \in C_c^\infty(\mathbb{R}^2)$ by

$$\langle F \otimes G, \phi(x,y)\rangle \;:=\; \langle\, F\,,\, \langle G, \phi(x,y)\rangle\,\rangle\,.$$

(a) First argue that this definition makes sense with respect to the test functions; for example, we may view $\langle G, \phi(x,y)\rangle$ as a test function in x.
(b) Show that the definition is consistent with the classical case when F and G are generated by smooth functions.
(c) Note that we can easily extend the definition to direct products over N variables. Show that if δ_{a_i} is the 1D delta function with respect to independent variable x_i concentrated at $x_i = a_i$, then $\delta_{a_1} \otimes \delta_{a_2} \otimes \cdots \otimes \delta_{a_N} = \delta_{\mathbf{a}}$ where $\delta_{\mathbf{a}}$ is the N-dimensional delta function with concentration at $\mathbf{a} = (a_1, \ldots, a_N)$.

9.3 Consider the function defined for $\mathbf{x} = (x_1, x_2, x_3) \neq \mathbf{0}$ in $\mathbb{R}^3$, $u(\mathbf{x}) = \frac{1}{|\mathbf{x}|^2}$. Why can we consider this function as a distribution? What is Δu in the sense of distributions? Start by considering two cases for the test functions ϕ: Case (i): $\mathbf{0} \notin \operatorname{supp} \phi$. Case (ii): $\mathbf{0} \in \operatorname{supp} \phi$. What is the conclusion?

9.4 Consider the PDE $\operatorname{div} \mathbf{u} = 0$ where $\mathbf{u}(x, y) = (u_1(x, y), u_2(x, y))$ is a 2D vector field. Write down what it means for a 2D vector field $\mathbf{u}$ to be a solution in the sense of distributions.

The simplest divergence-free vector field is a constant vector field. Now consider a piecewise constant vector field which takes on two constant vector fields, one to the left and one to the right of some curve

$$\Gamma := \{(x,y) \mid y = f(x)\}.$$

Here, f is some smooth function. More precisely, let

$$u(x,y) = \begin{cases} \mathbf{a} & \text{if } y \geq f(x), \\ \mathbf{b} & \text{if } y < f(x), \end{cases}$$

where $\mathbf{a} \neq \mathbf{b}$. Show that, in general, $\operatorname{div} \mathbf{u} \neq 0$ in the sense of distributions. Find a condition on $\mathbf{a}, \mathbf{b}$ and the normal to the curve Γ, which ensures that $\operatorname{div} \mathbf{u} = 0$ in the sense of distributions. These are called jump conditions.

9.5 Prove that in dimension $N = 2$, $\Delta(\log |\mathbf{x}|) = 2\pi\, \delta_{\mathbf{0}}$ in the sense of distributions.

9.6 (a) Let $\mathbf{a} \in \mathbb{R}^N$ and let $\delta_{\mathbf{0}}$ be the N-dimensional delta function. Prove that

$$\operatorname{div}(\mathbf{a}\, \delta_{\mathbf{0}}) = \mathbf{a} \cdot \nabla \delta_{\mathbf{0}} \quad \text{in the sense of distributions.}$$

(b) This is the vector analogue of Exercise **5.21**. As usual, let $\mathbf{x}$ be the underlying independent variable of the distribution $\delta_{\mathbf{y}}$ (for some $\mathbf{y} \in \mathbb{R}^N$). Now we wish to take the gradient of $\delta_{\mathbf{y}}$ **not** with respect to the underlying variable $\mathbf{x}$, but with respect to the source parameter $\mathbf{y}$. Rather than make a precise distributional definition for this derivative, simply justify the informal equation

$$\text{“}\mathbf{a} \cdot \nabla_{\mathbf{y}} \delta_{\mathbf{y}} = -\mathbf{a} \cdot \nabla_{\mathbf{x}} \delta_{\mathbf{y}}\text{”}.$$

9.7 (**Changing the Order of Differentiation**) In multivariable calculus you are told that for a function of two variables $f(x,y)$, it is not always the case that $f_{xy}(x,y) = f_{yx}(x,y)$, but if the function is nice (say, C^2), then equality does hold true.

First, show that in the sense of distributions, it is always the case that $f_{xy} = f_{yx}$.

Consider the example

$$f(x,y) = \begin{cases} 1 & \text{if } y \geq x^3, \\ 0 & \text{if } y < x^3. \end{cases}$$

What is f_{xy} in the sense of distributions? How does this compare with the classical pointwise derivatives f_{xy} and f_{yx} which are defined at all points (x,y) except when $y = x^3$.

9.8 Prove (9.4) by following the steps of the proof of Theorem 5.2.

9.9 Prove that $u(x,t) = H(x-ct)$ is a solution in the sense of distributions on $\mathbb{R}\times[0,\infty)$ to (9.36). Note here the incorporation of the initial values into the statement of a distributional solution.

9.10 Let $\delta_{\mathbf{0}}$ be the delta function in 3D and let $\mathbf{A}$ be a 3×3 matrix such that $\det \mathbf{A} \neq 0$. Justify the informal equation

$$\text{“}\delta_{\mathbf{0}}(\mathbf{A}\mathbf{x}) = \frac{1}{|\det \mathbf{A}|}\, \delta_{\mathbf{0}}(\mathbf{x})\text{”} = \frac{1}{|\det \mathbf{A}|}\, \delta_{\mathbf{0}}.$$

9.11 (a) (**The 3D Delta Function $\delta_{\mathbf{x}_0}$ in Cylindrical Coordinates**) Justify the expression that

$$\text{“}\,\delta_{\mathbf{0}}(\mathbf{x}-\mathbf{x}_0) = \frac{1}{r}\delta_0(r-r_0)\delta_0(\theta-\theta_0)\delta_0(z-z_0)\text{”},$$

where (r, θ, z) are the cylindrical coordinates of $\mathbf{x}$ and $\mathbf{x}_0 \neq \mathbf{0}$. What would the above equality look like when $\mathbf{x}_0 = \mathbf{0}$, i.e., an expression for the Delta function with concentration at $\mathbf{0}$?

(b) (**The 3D Delta Function $\delta_{\mathbf{x}_0}$ in Spherical Coordinates**) Find the analogous equations in terms of spherical coordinates.

(c) (**The 2D Delta Function $\delta_{\mathbf{0}}$ in Polar Coordinates**) Consider $\delta_{\mathbf{0}}$, the 2D delta function which is informally written as "$\delta_{\mathbf{0}}(x, y)$". Let "$\delta_0(r)$" informally denote the 1D delta function in variable r. Suppose we work in polar coordinates (r, θ). Justify the expression

$$\text{“}\delta_{\mathbf{0}}(x, y) = \frac{1}{2\pi}\frac{\delta_0(r)}{r}\text{”}.$$

One can think of the right-hand side as an expression of the delta function in polar coordinates.

9.12 Let f be a locally integrable function of one variable. Prove that u, defined by $u(x, y) = f(bx - ay)$, is a solution in the sense of distributions to $au_x + bu_y = 0$.

9.13 Consider the IVP for the transport equation with a delta function source

$$u_t + cu_x = \delta_{\mathbf{0}}, \qquad u(x, 0) = g(x),$$

where $\delta_{\mathbf{0}}$ is the 2D delta function with concentration at $(0, 0)$. What does it mean to have a solution to this PDE in the sense of distributions? Find this solution and prove that it is indeed a solution in the sense of distributions. Hint: To find the solution you can proceed by solving $u_t + cu_x = f(x, t)$ and then guessing the solution formula for the above by informally "putting" $\delta_{\mathbf{0}}$ in for $f(x, t)$.

9.14 Consider the following informal IVP involving the wave equation with source

$$u_{tt} - u_{xx} = \text{“}\delta(x)\text{”}, \quad \text{with} \quad u(x, 0) \equiv 0.$$

Make sense of this PDE in the sense of distributions. Find a solution formula and show that it is indeed a solution in the sense of distributions. To find the candidate for the solution formula proceed as follows: Fix t and consider both sides of the PDE as tempered distributions of one variable (the underlying variable x). Take the Fourier transform in x and proceed from there.

9.15 Repeat Exercise **9.14** for the wave equation with source

$$u_{tt} - u_{xx} = \text{“}\delta(x)\text{”}, \quad \text{with} \quad u(x, 0) = u_t(x, 0) \equiv 0.$$

9.16 Consider the IVP for the 3D wave equation with delta function source:

$$u_{tt} - c^2\Delta u = \delta_{\mathbf{0}}, \qquad u(\mathbf{x}, 0) = u_t(\mathbf{x}, 0) = 0,$$

where $\delta_{\mathbf{0}}$ is the 4D delta function with concentration at $(\mathbf{x}, t) = (0, 0, 0, 0)$; i.e., informally

$$\text{“}\delta_{\mathbf{0}} = \delta_0(x_1)\delta_0(x_2)\delta_0(x_3)\delta_0(t).\text{”}$$

What does it mean for a distribution F to be a solution in the sense of distributions? Find this distribution and show that it is indeed a solution in the sense of distributions.

To find the candidate distributional solution first do Exercise **4.11**, where you solve the 3D equation with a functional source term $f(\mathbf{x}, t)$. Informally place $\delta_{\mathbf{0}}$ in for f and decide on a description for the candidate distributional solution. Alternatively, you can come up with the same candidate via the Fourier transform. Note that your candidate solution will be a distribution which is **not** generated by a function.

9.17 Consider the following informal IVP involving the 3D wave equation with source

$$u_{tt} - c^2 \Delta u = \text{“}\delta(x_1)\delta(x_2)\delta(x_3)\text{”}, \qquad u(\mathbf{x}, 0) = u_t(\mathbf{x}, 0) \equiv 0.$$

Note that the product on the right represents the 3D delta function in the spatial variables. Make sense of this PDE in the sense of distributions. Find a solution formula and show that it is indeed a solution in the sense of distributions. Is your distributional solution generated by a locally integrable function? To find the distributional solution candidate proceed as follows: Fix t and consider both sides of the PDE as tempered distributions of three variables, i.e., the underlying spatial variable $\mathbf{x}$. Take the Fourier transform in $\mathbf{x}$ and proceed from there.

9.18 (a) Formulate the distributional statement for the IVP (9.38) and verify that the fundamental solution (9.39) satisfies the statement.
(b) Directly solve (9.38) using the Fourier transform.

9.19 (**The Support of a Distribution**) This exercise is for students interested in analysis and allows us to make precise sense when we say a distribution is concentrated on a set. Let $\Omega \subset \mathbb{R}^N$ be an open set (you can think of N as being 1, 2, or 3). We say a distribution F defined on $C_c^\infty(\mathbb{R}^N)$ **vanishes** in Ω if

$$\langle F, \phi \rangle = 0 \qquad \text{for all } \phi \in C_c^\infty(\mathbb{R}^N) \text{ for which } \phi(\mathbf{x}) = 0 \text{ if } \mathbf{x} \in \Omega^c.$$

We define **the support of the distribution** F to be the complement of the largest open set on which F vanishes. This means that the support equals the complement of an open set Ω_0 with the property that if F vanishes on an open set Ω, then $\Omega \subset \Omega_0$. You may take it at face value that the support exists (this requires some tools in analysis). Note that it is a closed set.
(a) Show that if f is a locally integrable function whose support, as defined by (A.1), is the set S, then the support of f in the sense of distributions (i.e., the support of the distribution F_f) is also S.
(b) Show that the support of $\delta_{\mathbf{0}}$ is the singleton set $\{\mathbf{0}\}$.
(c) Show that the support of the distribution $f_y = \frac{\partial f}{\partial y}$, where

$$f(x, y) = \begin{cases} 1 & \text{if } y \geq 0, \\ 0 & \text{if } y < 0 \end{cases}$$

is the x-axis.

9.20 (An Alternate Approach to Kirchhoff's Formula for the 3D Wave Equation) Recall the IVP (4.9) for the 3D wave equation where ϕ and ψ are in $C_c^\infty(\mathbb{R}^3)$. Let us now use the methods of this chapter to derive Kirchhoff's formula for the solution to this IVP at a fixed point $\mathbf{x}_0, t_0$. To this end, for any $\mathbf{x} \in \mathbb{R}^3$ define

$$\mathbf{V}(\mathbf{x}) := \left(\frac{\nabla u(x,t)}{|\mathbf{x}-\mathbf{x}_0|} + \frac{\mathbf{x}-\mathbf{x}_0}{|\mathbf{x}-\mathbf{x}_0|^3}\, u(x,t) + \frac{\mathbf{x}-\mathbf{x}_0}{|\mathbf{x}-\mathbf{x}_0|^2}\, u_t(x,t) \right)\Bigg|_{t=t_0-|\mathbf{x}-\mathbf{x}_0|}.$$

Note that while we will be interested in $\mathbf{V}$ for $\mathbf{x} \in B(\mathbf{x}_0, t_0)$, by considering the solution for $t < 0$ as well, $\mathbf{V}$ is defined for all $\mathbf{x} \in \mathbb{R}^3$.

(a) Show that $\operatorname{div} \mathbf{V} = 0$ for all $\mathbf{x} \neq \mathbf{x}_0$.

(b) As we have done many times in this chapter, fix $\epsilon > 0$ and integrate $\operatorname{div} \mathbf{V}$ over

$$B(\mathbf{x}_0, t_0) \backslash B(\mathbf{x}_0, \epsilon),$$

apply the Divergence Theorem, and then let $\epsilon \to 0^+$. Kirchhoff's formula should pop out!

9.21 (The Gradient in the Sense of Distributions and the Normal to a Set) Let $u : \mathbb{R}^2 \to \mathbb{R}$ be a locally integrable function; then we can view its gradient in the sense of distributions as a 2 vector whose components are the respective distributions u_x and u_y. We can capture these distributions via a vector-valued function ϕ with compact support; i.e., $\phi = (\phi_1, \phi_2)$ with $\phi_i \in C_c^\infty(\mathbb{R}^2)$. Thus

$$\langle \nabla u, \phi \rangle := \Big(\langle u_x, \phi_1 \rangle, \langle u_y, \phi_2 \rangle \Big) = \left(\iint_{\mathbb{R}^2} u\, \phi_x \, dx\, dy, \iint_{\mathbb{R}^2} u\, \phi_y \, dx\, dy \right).$$

Now let Ω be a subset of $\mathbb{R}^2$ and consider u to be the characteristic function of Ω; that is,

$$u(x,y) = \chi_\Omega(x,y) := \begin{cases} 1 & \text{if } (x,y) \in \Omega, \\ 0 & \text{otherwise.} \end{cases}$$

We want to investigate why ∇u in the sense of distributions gives us a generalized notion of a normal vector.

(a) To do this, let $\Omega = B(\mathbf{0}, 1)$. Find ∇u in the sense of distributions and in doing so observe the following two facts: (i) Both components are distributions which concentrate on the circular boundary $\partial\Omega$. This can be made precise by showing that the support (cf. Exercise **9.19**) of the components is $\partial\Omega$; (ii) these distributional components have scalar multiples (functions) and the vector of these scalars is the outer norm to $\partial\Omega$.

(b) Repeat this exercise for u being the characteristic function of the upper half-plane and the characteristic function of the unit square $[0,1]^2$.

9.22 Let $\mathbf{x}(t)$ be an infinitely smooth path of a particle in $\mathbb{R}^3$ whose velocity vector at time t is given by $\dot{\mathbf{x}}(t)$. You can also think of $\mathbf{x}(t)$ as a smooth parametric curve in $\mathbb{R}^3$. At any fixed t, we can consider $\delta_{\mathbf{x}(t)}$, a three-dimensional delta function with concentration at $\mathbf{x} = \mathbf{x}(t)$. Now consider time as a variable and $\delta_{\mathbf{x}(t)}$ as a "continuum of delta functions" concentrated on the curve $\mathbf{x}(t)$.

(a) First, make proper sense of $\delta_{\mathbf{x}(t)}$ as a distribution on $\mathbb{R}^4$, that is, with respect to test functions $\phi(\mathbf{x}, t) \in C_c^\infty(\mathbb{R}^4)$.

(b) Show that

$$\partial_t \, \delta_{\mathbf{x}(t)} \;=\; -\mathrm{div}\left(\dot{\mathbf{x}}(t)\, \delta_{\mathbf{x}(t)}\right) \qquad \text{in the sense of distributions,}$$

where div is the spatial (3D) divergence.

9.23 (**Examples of Functions with Blow-Up in a Sobolev Space**) Let $B(\mathbf{0}, 1)$ be the unit ball in $\mathbb{R}^N$, $p > 0$ and

$$u(\mathbf{x}) \;=\; \frac{1}{|\mathbf{x}|^{\beta}} \qquad \text{for } \beta > 0.$$

Prove that

$$u \in W^{1,p}(B(\mathbf{0}, 1)) \qquad \text{if and only if} \qquad \beta < \frac{N-p}{p}.$$

Proceed via the following steps:
(i) For any $i = 1, \ldots, N$, take α to be the multi-index of order 1 which is exactly 1 in the i-th position and 0 otherwise.
(ii) Fix a test function $\phi \in C_c^\infty(B(\mathbf{0}, 1))$ and write out the integral in (9.41) in two pieces: $B(\mathbf{0}, \epsilon)$ and $B(\mathbf{0}, 1)\backslash B(\mathbf{0}, \epsilon)$.
(iii) Find conditions on β and N so that the integral over $B(\mathbf{0}, \epsilon)$ vanishes as $\epsilon \to 0^+$ and use integration by parts to place the derivative back on u in the integral over $B(\mathbf{0}, 1)\backslash B(\mathbf{0}, \epsilon)$. Conclude that this pointwise derivative of u (away from the origin) is the distributional derivative of u.
(iv) Find conditions on p such that this distributional derivative function to the p-th power is integrable.

9.24 (**The Full Rankine-Hugoniot Jump Conditions**) Prove the following: Let $x = \eta(t)$ be a differentiable curve in the xt upper half-pane. Suppose $u(x, t)$ is a C^1 function at (x, t) if either $x < \eta(t)$ or $x > \eta(t)$ which solves (9.23) at all points (x, t) to the left and to the right of $x = \eta(t)$. Assume the limits

$$u^- \;=\; u^-(\eta(t), t) \;:=\; \lim_{x \to \eta(t)^-} u(x, t),$$

$$u^+ \;=\; u^+(\eta(t), t) \;=\; \lim_{x \to \eta(t)^+} u(x, t)$$

exist. Then u is a solution in the sense of distributions if and only if at all points $(\eta(t), t)$ we have

$$\eta'(t) \;=\; \frac{f(u^+) - f(u^-)}{u^+ - u^-}.$$

Hint: The proof is almost identical to the proof of Theorem 9.3.

9.25 Consider the following extension of the fundamental solution $\Phi(x, t)$ of the 1D diffusion equation to all $t \in \mathbb{R}$:

$$\Phi(x, t) \;=\; \begin{cases} \frac{1}{\sqrt{4\pi\alpha t}}\, e^{-\frac{x^2}{4\alpha t}} & \text{if } t > 0, \\ 0 & \text{if } t \le 0. \end{cases} \tag{9.51}$$

Prove that this extended $\Phi(x, t)$ solves $u_t - \alpha u_{xx} \;=\; \delta_{\mathbf{0}}$ in the sense of distributions on $\mathbb{R}^2$. Note that here $\delta_{\mathbf{0}}$ is the 2D delta function which we can informally write as "$\delta_0(x)\delta_0(t)$".

Chapter 10

The Fundamental Solution and Green's Functions for the Laplacian

Chapter 8 was focused on harmonic functions and their nice "smooth" properties. It may thus come as a surprise that the crucial tool in finding designer harmonic functions with prescribed boundary values is a particular function with a blow-up singularity, namely **the fundamental solution** introduced in Definition 8.5.1 of Section 8.5. Recall that in 3D, the fundamental solution was $\Phi(\mathbf{x}) = -\frac{1}{4\pi|\mathbf{x}|}$. In this chapter we will first prove that $\Delta\Phi(\mathbf{x}) = \delta_0$ in the sense of distributions and then explore the consequences. We will also introduce the notion of **Green's functions**, which are **modified fundamental solutions with boundary conditions**. Most of what we present in this chapter will be for Green's functions with the Dirichlet boundary condition.

This property of **focusing** (or **concentrating**) the Laplacian at a singleton (a point) gives Φ and Green's functions their defining character and enables us to use them to solve several PDEs and BVPs involving the Laplacian. While we will first present the mathematical details, all these functions, methods, and ideas are well illustrated via the electric potential in electrostatics. We discuss this in the final section of this chapter, but students who have taken a first-year university course in physics are encouraged to read Section 10.6 in conjunction with the subsequent purely mathematical sections.

10.1. • The Proof for the Distributional Laplacian of $\frac{1}{|\mathbf{x}|}$

In one independent variable, we saw that when a function has singularities, its derivative in the sense of distributions can be

- simply the pointwise derivative (taken away from the singularity) considered as a distribution,
- a distribution with components not captured by a function, for example, a delta function.

The same is true for the Laplacian of a function of several variables, specifically a function with a blow-up discontinuity at the origin. As we shall see, the rate of the blow-up in comparison to the space dimension is crucial to determining the character of its Laplacian. We start with perhaps the most important calculation presented in this text! Recall the definition of the fundamental solution in 3D

$$\Phi(\mathbf{x}) = -\frac{1}{4\pi|\mathbf{x}|}, \qquad \mathbf{x} \in \mathbb{R}^3, \mathbf{x} \neq \mathbf{0}. \tag{10.1}$$

Define Φ to be any number of your choosing at $\mathbf{x} = \mathbf{0}$. Recall[1] the important fact about the pointwise Laplacian of this radial function u: $\Delta\Phi(\mathbf{x}) = 0$ for all $\mathbf{x} \neq \mathbf{0}$. But what about the singularity at $\mathbf{x} = \mathbf{0}$? Does it affect the derivative? To answer this we must look at u as a distribution. Recall from (9.18) that u defined by (10.1) is locally integrable on $\mathbb{R}^3$, and hence we can consider u as a distribution and find the **distributional Laplacian**, i.e., the Laplacian in the sense of distributions.

Distributional Laplacian of $\frac{1}{|\mathbf{x}|}$ in 3D

Theorem 10.1. *Let* $N = 3$. *Then*

$$\Delta\left(\frac{1}{|\mathbf{x}|}\right) = -4\pi\delta_{\mathbf{0}} \qquad \textit{in the sense of distributions.}$$

That is, for any $\phi \in C_c^\infty(\mathbb{R}^3)$,

$$\iiint_{\mathbb{R}^3} \frac{1}{|\mathbf{x}|}\Delta\phi(\mathbf{x})\,d\mathbf{x} = -4\pi\phi(\mathbf{0}). \tag{10.2}$$

Hence, we have $\Delta\Phi = \delta_{\mathbf{0}}$ *in the sense of distributions.*

To prove (10.2) we will follow closely the proof of Theorem 9.1 for the divergence of the gravitational vector field. The analogous steps will be as follows:

- We isolate the singularity by removing a small ball centered at $\mathbf{0}$, $B(\mathbf{0}, \epsilon)$, and focus on the annular region of space.
- We integrate by parts on this annular region (this amounts to applying Green's Second Identity (A.17)) to focus the effects on the test function ϕ. This will produce boundary terms on $\partial B(\mathbf{0}, \epsilon)$.
- Since the previous steps will hold for all $\epsilon > 0$, we let $\epsilon \to 0^+$. The appearance of $\phi(\mathbf{0})$ will come from the limit of one of the boundary integrals over $\partial B(\mathbf{0}, \epsilon)$.

Proof of Theorem 10.1. Let $0 < \epsilon < 1$ and $\phi \in C_c^\infty(\mathbb{R}^3)$, and write

$$\iiint_{\mathbb{R}^3} \frac{1}{|\mathbf{x}|}\Delta\phi(\mathbf{x})\,d\mathbf{x} = \iiint_{\mathbb{R}^3\setminus B(\mathbf{0},\epsilon)} \frac{1}{|\mathbf{x}|}\Delta\phi(\mathbf{x})\,d\mathbf{x} + \iiint_{B(\mathbf{0},\epsilon)} \frac{1}{|\mathbf{x}|}\Delta\phi(\mathbf{x})\,d\mathbf{x}.$$

Note that the above is true for all ϵ. Hence,

$$\iiint_{\mathbb{R}^3} \frac{1}{|\mathbf{x}|}\Delta\phi(\mathbf{x})\,d\mathbf{x} = \lim_{\epsilon\to 0^+}\left[\iiint_{\mathbb{R}^3\setminus B(\mathbf{0},\epsilon)} \frac{1}{|\mathbf{x}|}\Delta\phi(\mathbf{x})\,d\mathbf{x} + \iiint_{B(\mathbf{0},\epsilon)} \frac{1}{|\mathbf{x}|}\Delta\phi(\mathbf{x})\,d\mathbf{x}\right]. \tag{10.3}$$

[1] If you don't recall doing this calculation, do it now.

Step (i): We first show that

$$\lim_{\epsilon \to 0^+} \iiint_{B(\mathbf{0},\epsilon)} \frac{1}{|\mathbf{x}|} \Delta\phi(\mathbf{x})\, d\mathbf{x} = 0.$$

This is essentially a consequence of the fact that $\frac{1}{|\mathbf{x}|}$ is locally integrable. For completeness, we present the short calculation. For any $0 < \epsilon < 1$,

$$\left| \iiint_{B(\mathbf{0},\epsilon)} \frac{1}{|\mathbf{x}|} \Delta\phi(\mathbf{x})\, d\mathbf{x} \right| \le \iiint_{B(\mathbf{0},\epsilon)} \frac{1}{|\mathbf{x}|} |\Delta\phi(\mathbf{x})|\, d\mathbf{x} \le C \iiint_{B(\mathbf{0},\epsilon)} \frac{1}{|\mathbf{x}|}\, d\mathbf{x},$$

where the last inequality follows from boundedness of $|\Delta\phi(\mathbf{x})|$ on $B(\mathbf{0},1)$; i.e., there is a constant C, depending on ϕ, such that

$$C = \max_{B(\mathbf{0},1)} |\Delta\phi(\mathbf{x})| < \infty.$$

Switching to spherical coordinates (using r in place of ρ), we have

$$\left| \iiint_{B(\mathbf{0},\epsilon)} \frac{1}{|\mathbf{x}|} \Delta\phi(\mathbf{x})\, d\mathbf{x} \right| \le C \int_0^\epsilon \frac{1}{r}(4\pi r^2)\, dr = 4\pi C \int_0^\epsilon r\, dr \xrightarrow{\epsilon \to 0^+} 0. \tag{10.4}$$

Step (ii): Since ϕ has compact support, there exists a bounded domain Ω such that the following hold:

- $B(\mathbf{0},1) \subset \Omega$.
- ϕ and all its derivatives are identically 0 on Ω^c (the complement of Ω), in particular on $\partial\Omega$.

Note that $0 < \epsilon < 1$ implies that Ω contains $B(\mathbf{0},\epsilon)$. Therefore,

$$\iiint_{\mathbb{R}^3 \backslash B(\mathbf{0},\epsilon)} \frac{1}{|\mathbf{x}|} \Delta\phi(\mathbf{x})\, d\mathbf{x} = \iiint_{\Omega \backslash B(\mathbf{0},\epsilon)} \frac{1}{|\mathbf{x}|} \Delta\phi(\mathbf{x})\, d\mathbf{x}.$$

We now apply Green's Second Identity (A.17) with $v = 1/|\mathbf{x}|$ and $u = \phi$ on the domain $\Omega \backslash B(\mathbf{0},\epsilon)$. Since the boundary of this domain consists of two pieces, $\partial\Omega$ and $\partial B(\mathbf{0},\epsilon)$, this gives

$$\iiint_{\Omega \backslash B(\mathbf{0},\epsilon)} \frac{1}{|\mathbf{x}|} \Delta\phi(\mathbf{x}) - \phi(\mathbf{x}) \Delta \frac{1}{|\mathbf{x}|}\, d\mathbf{x}$$

$$= \iint_{\partial B(\mathbf{0},\epsilon)} \frac{1}{|\mathbf{x}|} \frac{\partial \phi}{\partial \mathbf{n}} - \phi(\mathbf{x}) \frac{\partial}{\partial \mathbf{n}} \frac{1}{|\mathbf{x}|}\, dS + \iint_{\partial\Omega} \frac{1}{|\mathbf{x}|} \frac{\partial \phi}{\partial \mathbf{n}} - \phi(\mathbf{x}) \frac{\partial}{\partial \mathbf{n}} \frac{1}{|\mathbf{x}|}\, dS.$$

Note that on $\Omega \backslash B(\mathbf{0},\epsilon)$, $\frac{1}{|\mathbf{x}|}$ is a harmonic function and, by choice of Ω, ϕ and all its derivatives vanish on $\partial\Omega$. Thus, the boundary term over $\partial\Omega$ is zero and we have

$$\iiint_{\Omega \backslash B(\mathbf{0},\epsilon)} \frac{1}{|\mathbf{x}|} \Delta\phi(\mathbf{x})\, d\mathbf{x} = \iint_{\partial B(\mathbf{0},\epsilon)} \frac{1}{|\mathbf{x}|} \frac{\partial \phi}{\partial \mathbf{n}} - \phi(\mathbf{x}) \frac{\partial}{\partial \mathbf{n}} \frac{1}{|\mathbf{x}|}\, dS. \tag{10.5}$$

Step (iii): Let us show that the first integral above on the right-hand side vanishes. We first note that since $\left|\frac{\partial \phi}{\partial \mathbf{n}}\right| = |\nabla\phi(\mathbf{x}) \cdot \mathbf{n}| \leq |\nabla\phi(\mathbf{x})|$, $\left|\frac{\partial \phi}{\partial \mathbf{n}}\right|$ is bounded by a constant C' on $\partial B(\mathbf{0}, \epsilon)$ where

$$C' := \max_{B(\mathbf{0},1)} |\nabla\phi(\mathbf{x})| < \infty.$$

Hence, we have

$$\begin{aligned}
\left|\iint_{\partial B(\mathbf{0},\epsilon)} \frac{1}{|\mathbf{x}|}\frac{\partial \phi}{\partial \mathbf{n}}\, dS\right| \leq \iint_{\partial B(\mathbf{0},\epsilon)} \frac{1}{|\mathbf{x}|}\left|\frac{\partial \phi}{\partial \mathbf{n}}\right| dS &\leq \frac{1}{\epsilon}\iint_{\partial B(\mathbf{0},\epsilon)} |\nabla\phi|\, dS \\
&\leq \frac{1}{\epsilon}\iint_{\partial B(\mathbf{0},\epsilon)} C'\, dS \\
&\leq \frac{1}{\epsilon}(4C'\pi\epsilon^2) = 4C'\pi\epsilon.
\end{aligned}$$

The last line tends to 0 as $\epsilon \to 0^+$, and hence

$$\lim_{\epsilon\to 0^+} \iint_{\partial B(\mathbf{0},\epsilon)} \frac{1}{|\mathbf{x}|}\frac{\partial \phi}{\partial \mathbf{n}}\, dS = 0. \tag{10.6}$$

Step (iv): Finally we consider the last term on the right-hand side of (10.5). On $\partial B(\mathbf{0}, \epsilon)$ with the normal pointing into $B(\mathbf{0}, \epsilon)$, $\frac{\partial}{\partial \mathbf{n}} = -\frac{\partial}{\partial r}$. Hence,

$$\frac{\partial}{\partial \mathbf{n}}\frac{1}{|\mathbf{x}|} = -\frac{\partial}{\partial r}\frac{1}{r} = \frac{1}{r^2} = \frac{1}{\epsilon^2},$$

since on $\partial B(\mathbf{0}, \epsilon)$ the radius r is always equal to ϵ. Therefore,

$$\begin{aligned}
\iint_{\partial B(\mathbf{0},\epsilon)} -\phi(\mathbf{x})\frac{\partial}{\partial \mathbf{n}}\frac{1}{|\mathbf{x}|}\, dS &= \iint_{\partial B(\mathbf{0},\epsilon)} -\frac{\phi(\mathbf{x})}{\epsilon^2}\, dS \\
&= -4\pi\left(\frac{1}{4\pi\epsilon^2}\iint_{\partial B(\mathbf{0},\epsilon)} \phi(\mathbf{x})\, dS\right).
\end{aligned} \tag{10.7}$$

Combining Steps (i)–(iv): By (10.3), (10.4), (10.5), (10.6), and (10.7), we have

$$\iiint_{\mathbb{R}^3} \frac{1}{|\mathbf{x}|}\Delta\phi(\mathbf{x})\, d\mathbf{x} = \lim_{\epsilon\to 0^+} -4\pi\left(\frac{1}{4\pi\epsilon^2}\iint_{\partial B(\mathbf{0},\epsilon)} \phi(\mathbf{x})\, dS\right).$$

By the Averaging Lemma, Lemma A.7.1, the limit on the right is simply $-4\pi\phi(\mathbf{0})$.
□

The analogous steps prove that for any fixed $\mathbf{x}_0$, we have

$$\Delta_{\mathbf{x}}\left(-\frac{1}{4\pi|\mathbf{x}-\mathbf{x}_0|}\right) = \delta_{\mathbf{x}_0} \quad \text{in the sense of distributions.}$$

10.2. • Unlocking the Power of the Fundamental Solution for the Laplacian

Let us begin by recalling Section 8.5:

- After noting the symmetry and rotational invariance of Laplace's equation, we looked for radial solutions.

- In each space dimension $N \geq 2$, we found exactly one such solution (up to a multiplicative constant). However, this solution was singular at the origin.
- Motived by these findings, we defined the fundamental solution for[2] Δ to be

$$\Phi(\mathbf{x}) := \begin{cases} \frac{1}{2\pi} \log |\mathbf{x}| & \text{in dimension } N = 2, \\ -\frac{1}{4\pi|\mathbf{x}|} & \text{in dimension } N = 3, \\ -\frac{1}{N(N-2)\omega_N |\mathbf{x}|^{N-2}} & \text{in dimension } N > 3, \end{cases}$$

where ω_N is the (hyper)volume of the unit ball in N-dimensional space.

So in any space dimension $N \geq 2$, the fundamental solution is not defined at the origin; it can de defined to be any number at the origin as this single value will be irrelevant to its character and use. However, the blow-up has a profound effect on the distributional Laplacian as, indeed, we just proved in 3D that $\Delta\Phi = \delta_0$ in the sense of distributions. The proof in other dimensions is similar (cf. Exercise **10.2**). We will now show that this key property of concentrating the Laplacian at one point enables us to use the fundamental solution Φ to (i) solve the Poisson equation and (ii) derive a formula for a harmonic function u inside a domain Ω solely in terms of boundary information of u and its derivatives on $\partial\Omega$.

10.2.1. • The Fundamental Solution Is Key to Solving Poisson's Equation.

For simplicity of expression, we work in this and the next subsection with dimension $N = 3$. However, all the results hold verbatim in any space dimension. Consider Poisson's equation on $\mathbb{R}^3$,

$$\Delta u = f, \tag{10.8}$$

where f is a given function in $C_c^\infty(\mathbb{R}^3)$. The following theorem shows that the solution is simply given by **convolving** f with the fundamental solution Φ (recall the convolution of two functions from Section 6.3).

Theorem 10.2 (Solution to Poisson's Equation). *A solution to* (10.8) *is the* C^∞ *function*

$$\boxed{u(\mathbf{x}) = \iiint_{\mathbb{R}^3} \Phi(\mathbf{x}-\mathbf{y})\, f(\mathbf{y})\, d\mathbf{y}.} \tag{10.9}$$

Before proving the theorem, we will demonstrate why it is intuitively true. We want to compute the Laplacian in $\mathbf{x}$ of the right-hand side of (10.9) at a point $\mathbf{x}$ and show that it is $f(\mathbf{x})$. If we were able to bring the $\mathbf{x}$ Laplacian into the $\mathbf{y}$ integral, we would find

$$\Delta_{\mathbf{x}} u(\mathbf{x}) = \text{“} \iiint_{\mathbb{R}^3} \Delta_{\mathbf{x}}\Phi(\mathbf{x}-\mathbf{y}) f(\mathbf{y})\, d\mathbf{y} = \iiint_{\mathbb{R}^3} \underbrace{\Delta_{\mathbf{y}}\Phi(\mathbf{x}-\mathbf{y})}_{\delta_{\mathbf{x}}}\, f(\mathbf{y})\, d\mathbf{y} = f(\mathbf{x}).\text{”}$$

Here we use the fact that, whatever these objects are, the expectation is that

$$\text{“}\Delta_{\mathbf{x}}\Phi(\mathbf{x}-\mathbf{y}) = \Delta_{\mathbf{y}}\Phi(\mathbf{x}-\mathbf{y})\text{”}.$$

[2] Recall the warning in Section 8.5 that most mathematical texts address the fundamental solution for $-\Delta$ instead of Δ.

The pointwise Laplacian of $\Phi(\mathbf{x}-\mathbf{y})$ with respect to either $\mathbf{x}$ or $\mathbf{y}$ is zero except at $\mathbf{y} = \mathbf{x}$, where it is undefined. Within the realm of pointwise functions, bringing the Laplacian into the integral where it hits the singular Φ is illegal. To make this informal argument precise (via the distributional viewpoint), we change variables in the convolution so that the $\mathbf{x}$ dependence is solely in f and not in Φ; then it is legal to differentiate under the integral sign in the usual pointwise sense. As we show in the proof below, the Laplacian now only "hits" f, which can be viewed as a test function — the essence of differentiation in the sense of distributions.

Proof of Theorem 10.2. First note that by changing variables[3], we have

$$u(\mathbf{x}) = \iiint_{\mathbb{R}^3} \Phi(\mathbf{x}-\mathbf{y}) f(\mathbf{y})\, d\mathbf{y} = \iiint_{\mathbb{R}^3} \Phi(\mathbf{y}) f(\mathbf{x}-\mathbf{y})\, d\mathbf{y}.$$

This is actually just the commutative property of convolution. Since f is $C_c^\infty(\mathbb{R}^3)$ and $\Phi(\cdot)$ is locally integrable, it is perfectly legitimate to differentiate inside the integral and find

$$\Delta u(\mathbf{x}) = \iiint_{\mathbb{R}^3} \Phi(\mathbf{y})\, \Delta_{\mathbf{x}} f(\mathbf{x}-\mathbf{y})\, d\mathbf{y}.$$

Now note that as (smooth) functions of $\mathbf{y}$ and $\mathbf{x}$, it is certainly true that

$$\Delta_{\mathbf{x}} f(\mathbf{x}-\mathbf{y}) = \Delta_{\mathbf{y}} f(\mathbf{x}-\mathbf{y}) \quad \text{for all } \mathbf{x}, \mathbf{y}.$$

Hence,

$$\Delta u(\mathbf{x}) = \iiint_{\mathbb{R}^3} \Phi(\mathbf{y})\, \Delta_{\mathbf{y}} f(\mathbf{x}-\mathbf{y})\, d\mathbf{y}.$$

Noting that

$$\Delta_{\mathbf{y}} \Phi(\mathbf{y}) = \delta_{\mathbf{0}} \quad \text{in the sense of distributions,}$$

we may view $f(\mathbf{x}-\mathbf{y})$ as a test function in the $\mathbf{y}$ variable. Hence,

$$\iiint_{\mathbb{R}^3} \Phi(\mathbf{y})\, \Delta_{\mathbf{y}} f(\mathbf{x}-\mathbf{y})\, d\mathbf{y} = f(\mathbf{x}-\mathbf{y})|_{\mathbf{y}=\mathbf{0}} = f(\mathbf{x}).$$

We have shown that $\Delta u(\mathbf{x}) = f(\mathbf{x})$ (in the regular pointwise sense!). □

We conclude this subsection with a few remarks.

(i) **The role of concentration.** It is instructive to reflect on Theorem 10.2 and its proof. All the ingredients in the statement of the theorem are smooth functions. However, crucial to the solution, i.e., the solution formula (10.9) and its proof, is **concentration of the Laplacian** via the delta function as a distribution. Section 10.6 provides the physical interpretation of this concentration phenomenon and its role in solving Poisson's equation, the differential form of Gauss's Law (cf. Section 10.6.1.1).

[3] Let $\mathbf{z} = \mathbf{x}-\mathbf{y}$ and rewrite the $\mathbf{y}$ integral as an iterated integral with respect to dz_1, dz_2, dz_3, noting that the y_i limits from $-\infty$ to ∞ also reverse with the change of variable to z_i.

(ii) **Assumptions on f.** While the assumption that $f \in C_c^\infty$ was convenient for the above proof, two comments are in order.

- The same proof works with $f \in C_c^2$ as we only need two derivatives to place on f. However, we would then only have that u, as defined by (10.9), is C^2 in Ω (not necessarily C^∞): To interpret the Laplacian in the sense of distributions we really only need two derivatives on the test functions.
- Via more sophisticated mathematical arguments, one can solve the Poisson equation with a similar formula for very general f (for example, f need not be differentiable anywhere). This is not surprising since we previously found a distributional solution when f is a delta function, namely the fundamental solution.

(iii) **Uniqueness of the solution.** One might ask if the solution (10.9) is the unique solution to (10.8). The answer is certainly no, as one could add any harmonic function to (10.9) and still have a solution to (10.8). However, it is true that up to adding a constant, (10.9) is the only solution to (10.8) which remains bounded as $|\mathbf{x}| \to \infty$. Why?

10.2.2. • The Fundamental Solution Gives a Representation Formula for Any Harmonic Function in Terms of Boundary Data. In the subsequent sections, we will **shift notation** from using $\mathbf{x}, \mathbf{y}$ to $\mathbf{x}, \mathbf{x}_0$. Basically, $\mathbf{x}_0$ will be the point at which we wish to find the solution u and $\mathbf{x}$ will be used as a variable of integration for data.

Recall that the Dirichlet problem on a domain Ω involves finding a harmonic function with prescribed boundary values on $\partial\Omega$. In the remaining sections, we shall see how the fundamental solution is key to solving the Dirichlet problem. This should at first sight seem strange; while the solution to the Dirichlet problem is smooth and harmonic everywhere inside Ω, the fundamental solution is not harmonic everywhere and, in fact, blows up at its singularity. But the essence of the fundamental solution is that its Laplacian **concentrates at one point**, and the *key* to unlocking the power of this property is the following **representation formula**. It shows that the fundamental solution can be used to write the value of **any** harmonic function at a point in terms of a **weighted average of its boundary data**. In particular, if we know information about the values of a harmonic function and its derivatives on $\partial\Omega$, we know the values inside Ω as well. The formula holds in any space dimension but, for simplicity of expression, we state and prove it in dimension 3.

The Representation Formula

Theorem 10.3. *Let $u \in C^2(\overline{\Omega})$ and suppose that $\Delta u = 0$ in Ω. Let $\mathbf{x}_0 \in \Omega$; then*

$$u(\mathbf{x}_0) = \iint_{\partial\Omega} \left[u(\mathbf{x}) \frac{\partial\, \Phi(\mathbf{x}-\mathbf{x}_0)}{\partial \mathbf{n}} - \Phi(\mathbf{x}-\mathbf{x}_0) \frac{\partial u(\mathbf{x})}{\partial \mathbf{n}} \right] dS_{\mathbf{x}}. \tag{10.10}$$

The reason why (10.10) holds true is simply that $\Delta\Phi(\mathbf{x}) = \delta_0$ in the sense of distributions (Theorem 10.1). Using this fact, we first give an intuitive argument for the representation formula. Then we show that the actual proof is a simple repetition of the steps used in the proof of Theorem 10.1. To this end, note that we stated and proved Green's Second Identity for smooth functions $C^2(\overline{\Omega})$, and hence we **cannot** directly apply it to $u(\mathbf{x})$ and

$$v(\mathbf{x}) = \Phi(\mathbf{x} - \mathbf{x}_0) = -\frac{1}{4\pi|\mathbf{x} - \mathbf{x}_0|}$$

since $v(\mathbf{x})$ is not smooth at $\mathbf{x} = \mathbf{x}_0$. However, if we think *outside the box* viewing Δv as the delta function $\delta_{\mathbf{x}_0}$, we find

$$\text{“}\underbrace{\iiint_\Omega (u \underbrace{\Delta v}_{=\delta_{\mathbf{x}_0}} - v \underbrace{\Delta u}_{=0})\, d\mathbf{x}}_{u(\mathbf{x}_0)} = \iint_{\partial\Omega} \left(u \frac{\partial v}{\partial \mathbf{n}} - v \frac{\partial u}{\partial \mathbf{n}} \right) dS.\text{”}$$

As usual, note the use of quotation marks. The right-hand side matches the right-hand side of (10.10). What about the left-hand side? Since $\Delta u = 0$ everywhere in Ω the second term vanishes. For the first term involving Δv, we should **not** make sense of it pointwise, but we can in the sense of distributions where we can *think of u* here[4] as the test function. In this case, we get $u(\mathbf{x}_0)$ and we are done.

We can make all these informal calculations precise, circumventing the direct application of Green's Second Identity (A.17) over Ω and any explicit mention of a delta function, by removing a small ball of radius ϵ around $\mathbf{x}_0$ and examining what happens as $\epsilon \to 0^+$. We have already seen the spirit of this argument a few times. For completeness and because of its importance, we present it one more time!

Proof of Theorem 10.3. Fix some $\epsilon_0 > 0$ such that $B(\mathbf{x}_0, \epsilon_0) \subset \Omega$. Let $0 < \epsilon < \epsilon_0$ and consider the annular region $\Omega_\epsilon = \Omega \backslash B(\mathbf{x}_0, \epsilon)$ (illustrated in Figure 10.1) obtained by removing $B(\mathbf{x}_0, \epsilon)$ from Ω.

We can apply Green's Second Identity (A.17) on Ω_ϵ to u and $v = \Phi(\mathbf{x} - \mathbf{x}_0) = -1/(4\pi|\mathbf{x} - \mathbf{x}_0|)$. This is perfectly legitimate since both of these functions are smooth on Ω_ϵ, for any $\epsilon > 0$. We obtain

$$\iiint_{\Omega_\epsilon} \left(u(\mathbf{x})\Delta\Phi(\mathbf{x} - \mathbf{x}_0) - \Phi(\mathbf{x} - \mathbf{x}_0)\Delta u(\mathbf{x})\right) d\mathbf{x}$$
$$= \iint_{\partial\Omega_\epsilon} \left(u(\mathbf{x}) \frac{\partial \Phi(\mathbf{x} - \mathbf{x}_0)}{\partial \mathbf{n}} - \Phi(\mathbf{x} - \mathbf{x}_0) \frac{\partial u}{\partial \mathbf{n}} \right) dS_{\mathbf{x}},$$

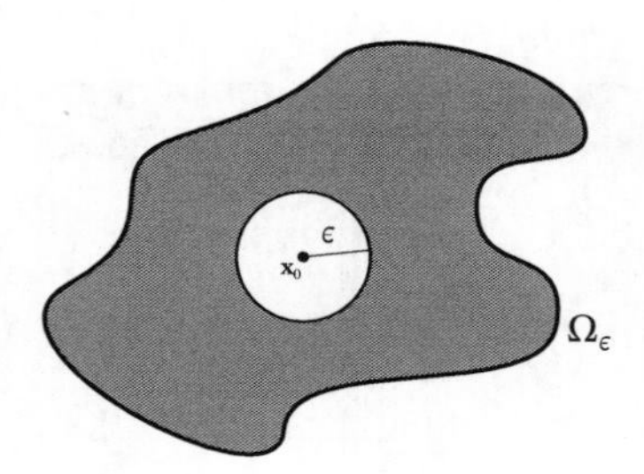

Figure 10.1. Cartoon associated with proving Theorem 10.3.

[4]In general, we note that u does not have compact support. However, we do not need u to have compact support.

where all derivatives and integrals are with respect to the $\mathbf{x}$ variable (recall $\mathbf{x}_0$ is fixed in this proof). However, now the left-hand side, i.e., the triple integral, is zero since both u and $\Phi(\mathbf{x}-\mathbf{x}_0)$ are harmonic in Ω_ϵ. On the other hand, the boundary integral is now defined over two pieces, $\partial\Omega$ and $\partial B(\mathbf{x}_0,\epsilon)$. Thus, we have

$$\begin{aligned} 0 \;=\; & \iint_{\partial\Omega} \left(u(\mathbf{x}) \frac{\partial \Phi(\mathbf{x}-\mathbf{x}_0)}{\partial \mathbf{n}} - \Phi(\mathbf{x}-\mathbf{x}_0)\frac{\partial u}{\partial \mathbf{n}} \right) dS_{\mathbf{x}} \\ & + \iint_{\partial B(\mathbf{x}_0,\epsilon)} \left(u(\mathbf{x}) \frac{\partial \Phi(\mathbf{x}-\mathbf{x}_0)}{\partial \mathbf{n}} - \Phi(\mathbf{x}-\mathbf{x}_0)\frac{\partial u}{\partial \mathbf{n}} \right) dS_{\mathbf{x}}. \end{aligned} \tag{10.11}$$

The above expression holds for all $0 < \epsilon < \epsilon_0$. Our goal is now to show that the boundary integral over $\partial B(\mathbf{x}_0,\epsilon)$ will tend to $-u(\mathbf{x}_0)$ as $\epsilon \to 0^+$. This is just a repetition of Steps (iii) and (iv) in the proof of Theorem 10.1. Because of the importance of these steps, let us repeat them one more time. Consider the second boundary integral in (10.11) over $\partial B(\mathbf{x}_0,\epsilon)$. We first show that the second term tends to 0 as $\epsilon \to 0^+$. To this end,

$$\begin{aligned} \left| \iint_{\partial B(\mathbf{x}_0,\epsilon)} \frac{-1}{4\pi|\mathbf{x}-\mathbf{x}_0|}\frac{\partial u}{\partial \mathbf{n}}\, dS_{\mathbf{x}} \right| & \le \iint_{\partial B(\mathbf{x}_0,\epsilon)} \frac{1}{4\pi|\mathbf{x}-\mathbf{x}_0|}\left|\frac{\partial u}{\partial \mathbf{n}}\right| dS_{\mathbf{x}} \\ & = \frac{1}{4\pi\epsilon}\iint_{\partial B(\mathbf{x}_0,\epsilon)} \left|\frac{\partial u}{\partial \mathbf{n}}\right| dS_{\mathbf{x}} \\ & = \epsilon\left(\frac{1}{4\pi\epsilon^2}\iint_{\partial B(\mathbf{x}_0,\epsilon)} \left|\frac{\partial u}{\partial \mathbf{n}}\right| dS_{\mathbf{x}} \right). \end{aligned}$$

The last term in the parentheses is the average value of $|\frac{\partial u}{\partial \mathbf{n}}|$ on $\partial B(\mathbf{x}_0,\epsilon)$. Since $|\frac{\partial u}{\partial \mathbf{n}}|$ is a continuous function, Lemma A.7.1 implies that this average tends to a constant as ϵ tends to 0. Thus, the last line tends to 0 as $\epsilon \to 0^+$. Hence

$$\lim_{\epsilon\to 0^+} \iint_{\partial B(\mathbf{x}_0,\epsilon)} \frac{-1}{4\pi|\mathbf{x}-\mathbf{x}_0|}\frac{\partial u}{\partial \mathbf{n}}\, dS_{\mathbf{x}} \;=\; 0. \tag{10.12}$$

We have already addressed the limit of the first term, but let us repeat the details here. If we let $r = |\mathbf{x}-\mathbf{x}_0|$, then we have $\frac{\partial}{\partial \mathbf{n}} = -\frac{\partial}{\partial r}$. The minus sign appears because the outer normal to $\partial\Omega_\epsilon$ points into $B(\mathbf{x}_0,\epsilon)$. Hence,

$$\frac{\partial}{\partial \mathbf{n}}\left(\frac{-1}{4\pi|\mathbf{x}-\mathbf{x}_0|}\right) = \frac{\partial}{\partial r}\frac{1}{4\pi r} = -\frac{1}{4\pi r^2}.$$

Note that on $\partial B(\mathbf{x}_0,\epsilon)$, $r = \epsilon$. Therefore,

$$\begin{aligned} \iint_{\partial B(\mathbf{x}_0,\epsilon)} u(\mathbf{x})\frac{\partial}{\partial \mathbf{n}}\left(\frac{-1}{4\pi|\mathbf{x}-\mathbf{x}_0|}\right) dS_{\mathbf{x}} & = \iint_{\partial B(\mathbf{x}_0,\epsilon)} -\frac{u(\mathbf{x})}{4\pi r^2}\, dS_{\mathbf{x}} \\ & = \iint_{\partial B(\mathbf{x}_0,\epsilon)} -\frac{u(\mathbf{x})}{4\pi\epsilon^2}\, dS_{\mathbf{x}} \\ & = -\left(\frac{1}{4\pi\epsilon^2}\iint_{\partial B(\mathbf{x}_0,\epsilon)} u(\mathbf{x})\, dS_{\mathbf{x}}\right). \end{aligned} \tag{10.13}$$

Lemma A.7.1 implies that the last line tends to $-u(\mathbf{x}_0)$ as $\epsilon \to 0^+$. Thus, letting $\epsilon \to 0^+$ in (10.11) gives (10.10). □

10.3. • Green's Functions for the Laplacian with Dirichlet Boundary Conditions

Green's functions are **modified fundamental solutions with boundary conditions.** In this section we will focus on the **Dirichlet boundary condition**. However, other boundary conditions are also important, and we discuss the Neumann boundary condition in Section 10.5.

Let Ω be a domain in $\mathbb{R}^N$. Green's functions with Dirichlet boundary conditions are **tailor made** to solve the Dirichlet problem:

$$\begin{cases} \Delta u = 0 & \text{on } \Omega, \\ u = g & \text{on } \partial\Omega. \end{cases} \tag{10.14}$$

We will follow our usual approach of assuming we have a solution and then manipulating/exploring its structure in such a way that we obtain an explicit formula for the solution. We then check that this formula does indeed do the trick. To this end, suppose u solves (10.14). The representation theorem (10.10) gives a formula for u in terms of its boundary values (this is g) **and** the values of the normal derivative on the boundary. The problem is that we were not given the values of $\frac{\partial u}{\partial \mathbf{n}}$ on the boundary. The idea is to produce a modified fundamental solution tailor-made for the domain Ω so that it vanishes on its boundary.

10.3.1. • The Definition of the Green's Function with Dirichlet Boundary Conditions.

The Green's function will only (and need only) be defined on $\overline{\Omega}$. Precisely, for any $\mathbf{x}_0 \in \Omega$, we want a new function, $G(\mathbf{x}, \mathbf{x}_0)$ of $\mathbf{x} \in \overline{\Omega}$, which has the following two properties:

(i) $\Delta_{\mathbf{x}}\, G(\mathbf{x}, \mathbf{x}_0) = \delta_{\mathbf{x}_0}$ in the sense of distributions on Ω.

(ii) For all $\mathbf{x} \in \partial\Omega$, $G(\mathbf{x}, \mathbf{x}_0) = 0$.

Note that $G(\mathbf{x}, \mathbf{x}_0)$ is defined for $\mathbf{x} \in \overline{\Omega}$, and hence in order to consider it as a distribution (in the variable $\mathbf{x}$), we need test functions which are compactly supported in Ω (see Section 9.6).

It will be convenient and instructive to rephrase the first property (i) in a way that we do not have to explicitly mention distributions or test functions. To this end, the key idea is to note that the fundamental solution $\Phi(\mathbf{x}-\mathbf{x}_0)$ has exactly property (i) and, moreover, if we add to it a harmonic function, we preserve this property. Hence, we can rephrase property (i) by saying that

$$G(\mathbf{x}, \mathbf{x}_0) \;=\; \Phi(\mathbf{x} - \mathbf{x}_0) \;+\; H_{\mathbf{x}_0}(\mathbf{x}) \qquad \text{for some harmonic function } H_{\mathbf{x}_0}(\mathbf{x}).$$

The function $H_{\mathbf{x}_0}$ is often referred to as the **corrector**. Since the choice of function changes with $\mathbf{x}_0$, we label with the subscript. We use this fact in the following equivalent definition, stated in dimension 3.

Definition of the Green's Function for Δ on $\Omega \subset \mathbb{R}^N$ with Dirichlet Boundary Conditions

Definition 10.3.1. The **Green's function** for Δ on $\Omega \subset \mathbb{R}^N$ with source point $\mathbf{x}_0 \in \Omega$, written $G(\mathbf{x}, \mathbf{x}_0)$, is a function defined for all $\mathbf{x} \in \overline{\Omega}$, except at $\mathbf{x} = \mathbf{x}_0$, such that the following hold:

(i) For each $\mathbf{x}_0 \in \Omega$, there exists a function $H_{\mathbf{x}_0}(\mathbf{x}) \in C(\overline{\Omega})$ which is smooth and harmonic in Ω such that

$$G(\mathbf{x}, \mathbf{x}_0) = \Phi(\mathbf{x} - \mathbf{x}_0) + H_{\mathbf{x}_0}(\mathbf{x}) \qquad \text{for all } \mathbf{x} \neq \mathbf{x}_0.$$

(ii) $G(\mathbf{x}, \mathbf{x}_0) = 0$ for any $\mathbf{x} \in \partial\Omega$.

A few remarks on this definition are in order:

- The first property implies that $G(\mathbf{x}, \mathbf{x}_0)$ away from $\mathbf{x} = \mathbf{x}_0$ is a smooth harmonic function **and** has the *exact same blow-up behavior* as the fundamental solution around the **source point** $\mathbf{x} = \mathbf{x}_0$.
- The fundamental solution $\Phi(\mathbf{x} - \mathbf{x}_0)$ can be viewed as the Green's function for Δ associated with $\Omega = \mathbb{R}^3$ which has no boundary.
- Finding the Green's function for a domain Ω amounts to finding a harmonic function $H_{\mathbf{x}_0}$ in Ω whose values on $\partial\Omega$ agree with those of the fundamental solution $\Phi(\mathbf{x} - \mathbf{x}_0)$. In other words, we want to solve a particular Dirichlet problem

$$\begin{cases} \Delta_{\mathbf{x}} H_{\mathbf{x}_0}(\mathbf{x}) = 0 & \text{on } \Omega, \\ H_{\mathbf{x}_0}(\mathbf{x}) = -\Phi(\mathbf{x} - \mathbf{x}_0) & \text{for } \mathbf{x} \in \partial\Omega. \end{cases}$$

10.3.2. Using the Green's Function to Solve the Dirichlet Problem for Laplace's Equation. Let Ω be a **bounded** domain. We show that **if** we can find a Green's function G for a particular bounded domain Ω and any source point $\mathbf{x}_0 \in \Omega$, **then** we can use it to explicitly solve the Dirichlet problem (10.14) for **any** data function g. For simplicity of expression, in this subsection and the next we work in dimension $N = 3$. All the results hold in any space dimension.

The Green's Function and the Dirichlet Problem for Laplace's Equation

Theorem 10.4. *Let Ω be a bounded domain in $\mathbb{R}^3$. For $\mathbf{x}_0 \in \Omega$, let $G(\mathbf{x}, \mathbf{x}_0)$ be the associated Green's function on Ω, i.e., the function which satisfies all the conditions of* DEFINITION 10.3.1. *Let u be a function in $C(\overline{\Omega}) \cap C^2(\Omega)$ that solves* (10.14). *Then*

$$u(\mathbf{x}_0) = \iint_{\partial\Omega} g(\mathbf{x}) \frac{\partial}{\partial \mathbf{n}} G(\mathbf{x}, \mathbf{x}_0)\, dS_{\mathbf{x}} \qquad \textit{for all } \mathbf{x}_0 \in \Omega. \tag{10.15}$$

Proof. By the representation formula (10.10), we know that

$$u(\mathbf{x}_0) = \iint_{\partial\Omega} \left(u(\mathbf{x}) \frac{\partial v(\mathbf{x})}{\partial \mathbf{n}} - \frac{\partial u(\mathbf{x})}{\partial \mathbf{n}} v(\mathbf{x}) \right) dS_{\mathbf{x}}, \tag{10.16}$$

where

$$v(\mathbf{x}) = -\frac{1}{4\pi|\mathbf{x}-\mathbf{x}_0|}.$$

By definition, $G(\mathbf{x}, \mathbf{x}_0) = v(\mathbf{x}) + H_{\mathbf{x}_0}(\mathbf{x})$, where $H_{\mathbf{x}_0}$ is harmonic. If we apply Green's Second Identity (A.17) to $u(\mathbf{x})$ and $H_{\mathbf{x}_0}(\mathbf{x})$, we find

$$0 = \iint_{\partial\Omega} \left(u(\mathbf{x})\frac{\partial H_{\mathbf{x}_0}(\mathbf{x})}{\partial \mathbf{n}} - \frac{\partial u(\mathbf{x})}{\partial \mathbf{n}} H_{\mathbf{x}_0}(\mathbf{x}) \right) dS_{\mathbf{x}}. \tag{10.17}$$

Adding (10.16) and (10.17) and noting that if $\mathbf{x} \in \partial\Omega$, then $G(\mathbf{x}, \mathbf{x}_0) = 0$ and $u(\mathbf{x}) = g(\mathbf{x})$, gives (10.15). □

One can also prove the converse; that is, if u is defined by (10.15), then u is a continuous function on $\overline{\Omega}$, C^2 in Ω, and satisfies (10.14). We will prove this for the case when Ω is the ball (Theorem 10.9). As we have seen before with the diffusion equation, the proof of boundary value attainment is subtle.

The Green's function G can be used to solve the **Dirichlet problem for Poisson's equation**: For any $f \in C(\overline{\Omega})$ and g continuous on $\partial\Omega$,

$$\begin{cases} \Delta u = f & \text{on } \Omega, \\ u = g & \text{on } \partial\Omega. \end{cases} \tag{10.18}$$

The proof of the following theorem follows from a similar representation formula to that of (10.10) and the Definition 10.3.1. It is left as an exercise (Exercise **10.6**).

The Green's Function and the Dirichlet Problem for Poisson's Equation

Theorem 10.5. *Let Ω be a bounded domain in $\mathbb{R}^3$. For $\mathbf{x}_0 \in \Omega$, let $G(\mathbf{x}, \mathbf{x}_0)$ be the associated Green's function on Ω, i.e., the function which satisfies all the conditions of* DEFINITION 10.3.1. *Let u be a function in $C(\overline{\Omega}) \cap C^2(\Omega)$ that solves* (10.18). *Then*

$$u(\mathbf{x}_0) = \iint_{\partial\Omega} g(\mathbf{x})\frac{\partial}{\partial \mathbf{n}} G(\mathbf{x}, \mathbf{x}_0)\, dS_{\mathbf{x}} + \iiint_{\Omega} G(\mathbf{x}, \mathbf{x}_0) f(\mathbf{x})\, d\mathbf{x} \qquad \textit{for all } \mathbf{x}_0 \in \Omega. \tag{10.19}$$

10.3.3. • Uniqueness and Symmetry of the Green's Function. Given a domain Ω, a natural question is: Does a Green's function exist, and if so, how can we "find it"? We will address this (in part) shortly but let us note here that, by the Maximum Principle, there can exist **at most one** Green's function for any domain Ω. Indeed, we have the following uniqueness theorem.

Theorem 10.6 (Uniqueness of the Green's Function). *Let Ω be a bounded domain and fix $\mathbf{x}_0 \in \Omega$. Let $G_1(\mathbf{x}, \mathbf{x}_0)$ and $G_2(\mathbf{x}, \mathbf{x}_0)$ be two Green's functions for the Laplacian on Ω, i.e., two functions which both satisfy Definition* 10.3.1. *Then for all $\mathbf{x} \in \overline{\Omega}$, we have $G_1(\mathbf{x}, \mathbf{x}_0) = G_2(\mathbf{x}, \mathbf{x}_0)$.*

Proof. By definition, the function $F(\mathbf{x}) = G_1(\mathbf{x}, \mathbf{x}_0) - G_2(\mathbf{x}, \mathbf{x}_0)$ is smooth and harmonic in Ω. Moreover, $F(\mathbf{x})$ continuously attains boundary values which are identically 0. By the Maximum Principle (Theorem 8.2), $F(\mathbf{x}) \equiv 0$ in Ω. □

Let us also note the following useful, yet initially surprising, symmetry property of the Green's function. In formula (10.15) there is a clear separation between the roles of $\mathbf{x}_0$ and $\mathbf{x}$; while the **source point** $\mathbf{x}_0$ is a fixed point in the domain Ω, $\mathbf{x}$ lies on $\partial\Omega$ and is a variable over which we integrate. However, a remarkable property of Green's functions is that they are completely **interchangeable** in the following sense.

Theorem 10.7 (Symmetry of the Green's Function). *For any $\mathbf{x}, \mathbf{x}_0 \in \Omega$ with $\mathbf{x} \neq \mathbf{x}_0$, we have*

$$G(\mathbf{x}, \mathbf{x}_0) = G(\mathbf{x}_0, \mathbf{x}).$$

As we did for the representation formula, we first provide the intuitive explanation, treating the delta function as a function-like object which can appear in an integrand. We then outline the steps needed to use the arguments of Theorem 10.1 to generate a proof. To proceed, fix $\mathbf{a} \neq \mathbf{b} \in \Omega$ and let

$$u(\mathbf{x}) := G(\mathbf{x}, \mathbf{a}) \qquad \text{and} \qquad v(\mathbf{x}) := G(\mathbf{x}, \mathbf{b}).$$

Our goal here is to show that $u(\mathbf{b}) = v(\mathbf{a})$. Suppose we illegally applied Green's Second Identity (A.17) to u, v. Then we would find

$$\text{“}\iiint_\Omega (u\Delta v - v\Delta u)\, d\mathbf{x} = \iint_{\partial\Omega} \left(u\frac{\partial v}{\partial \mathbf{n}} - v\frac{\partial u}{\partial \mathbf{n}}\right) dS = 0.\text{”}$$

Since the right-hand side is 0 (u and v vanish on the boundary), this means

$$\text{“}\iiint_\Omega (u \underbrace{\Delta v}_{=\delta_{\mathbf{b}}} - v \underbrace{\Delta u}_{=\delta_{\mathbf{a}}})\, d\mathbf{x} = 0.\text{”}$$

Taking u as the test function in the first integral and v in the second would give

$$u(\mathbf{b}) - v(\mathbf{a}) = 0 \qquad \text{or} \qquad u(\mathbf{b}) = v(\mathbf{a}),$$

and we are (seemingly) done.

The reader might correctly object to the interpretation of u and v as test functions; $u(\mathbf{x})$, for example, is singular at $\mathbf{x} = \mathbf{a}$. However, since Δv will concentrate at $\mathbf{x} = \mathbf{b}$, in principle we only require u to be continuous around $\mathbf{b}$. To actually prove Theorem 10.7, we can circumvent this "illegal" application of Green's Second Identity (A.17) to the entire region by arguing as in the proof of the representation formula. To set this up, we let $\epsilon > 0$ and consider the region obtained by removing two small balls, $B(\mathbf{a}, \epsilon)$ and $B(\mathbf{b}, \epsilon)$, from Ω. Call this annular region Ω_ϵ (see Figure 10.2). We apply Green's

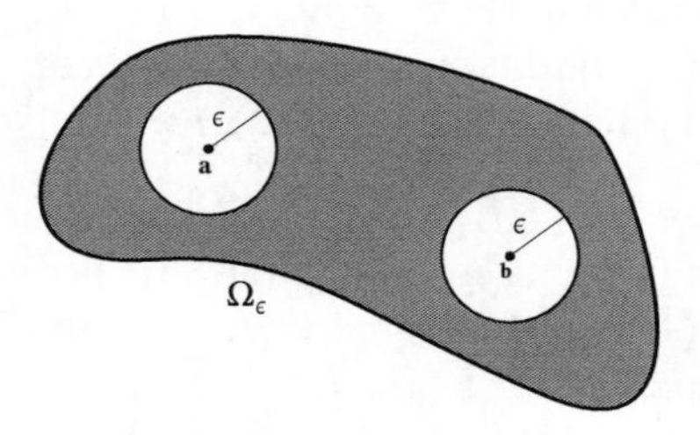

Figure 10.2. Cartoon associated with proving Theorem 10.7.

Second Identity (A.17) on Ω_ϵ to both u and v, noting that the only nonzero terms are boundary terms on $\partial B(\mathbf{a}, \epsilon)$ and $\partial B(\mathbf{b}, \epsilon)$. By writing u and v as

$$u(\mathbf{x}) = -\frac{1}{4\pi|\mathbf{x}-\mathbf{a}|} + H_{\mathbf{a}}(\mathbf{x}) \qquad \text{and} \qquad v(\mathbf{x}) = -\frac{1}{4\pi|\mathbf{x}-\mathbf{b}|} + H_{\mathbf{b}}(\mathbf{x}),$$

for respective harmonic (smooth) functions $H_{\mathbf{a}}$ and $H_{\mathbf{b}}$, we can then examine what happens as ϵ tends to 0. These steps are almost identical to that performed in the proof of Theorem 10.1. It is a good exercise to fill in and complete these steps (cf. Exercise **10.5**).

In the next section, we embark on finding Green's functions for certain domains in $\Omega \subset \mathbb{R}^3$. Recall that this boils down to solving, for each $\mathbf{x}_0 \in \Omega$, the **specific** Dirichlet problem for $H_{\mathbf{x}_0}(\mathbf{x})$:

$$\begin{cases} \Delta H_{\mathbf{x}_0} = 0 & \text{on } \Omega, \\ H_{\mathbf{x}_0} = \frac{1}{4\pi|\mathbf{x}-\mathbf{x}_0|} & \text{on } \partial\Omega. \end{cases}$$

Then we have our Green's function, $G(\mathbf{x}, \mathbf{x}_0)$, by taking the sum of $H_{\mathbf{x}_0}(\mathbf{x})$ with the fundamental solution $\Phi(\mathbf{x}-\mathbf{x}_0)$. The reader should note the role of the source point $\mathbf{x}_0$ in this process. For general domains, it is impossible to find explicit formulas for these Green's functions. However, in cases where Ω exhibits some **symmetry**, one can exploit this fact to determine $H_{\mathbf{x}_0}(\mathbf{x})$ explicitly. We will show this for the half-space and the ball in dimension 3, leaving the respective problems in dimension 2 as exercises. However, we first pause to address Green's functions in one space dimension.

10.3.4. • The Fundamental Solution and Green's Functions in One Space Dimension.

The fundamental solution for the **1D Laplacian** $\frac{d^2}{dx^2}$ is

$$\Phi(x - x_0) = \frac{1}{2}|x - x_0|. \tag{10.20}$$

Note that the fundamental solution is only unique up to the addition of a harmonic function (in 1D this means a linear function). Hence, you can add any linear function to this one and generate an equivalent fundamental solution. The fundamental solution has the property that

$$\frac{d^2}{dx^2}\Phi(x - x_0) = \delta_{x_0} \qquad \text{in the sense of distributions.} \tag{10.21}$$

Note that, analogous to Theorem 10.2, we can use this fundamental solution to solve the **Poisson problem in 1D**: For any function $f \in C_c^\infty(\mathbb{R})$ find u such that

$$\frac{d^2u}{dx^2} = f.$$

Indeed, we have

$$u(x) = \int_{-\infty}^{\infty} \Phi(x-y)\, f(y)\, dy = \int_{-\infty}^{\infty} \Phi(y)\, f(x-y)\, dy. \tag{10.22}$$

As with Theorem 10.2, we can prove this by focusing on the form on the right and bringing the second derivative with respect to x into the integral to find

$$\frac{d^2u}{dx^2} = \int_{-\infty}^{\infty} \Phi(y)\, \frac{\partial^2}{\partial x^2} f(x-y)\, dy.$$

As functions of x and y, we have

$$\frac{\partial^2}{\partial x^2} f(x-y) = \frac{\partial^2}{\partial y^2} f(x-y).$$

Hence, we have

$$\frac{d^2u}{dx^2} = \int_{-\infty}^{\infty} \Phi(y)\, \frac{\partial^2}{\partial y^2} f(x-y)\, dy.$$

But since $\frac{d^2}{dx^2}\Phi(x) = \delta_0$ in the sense of distributions, the above integral on the right is $f(x)$.

On the other hand, we could prove (10.21) and (10.22) together by including the precise form of Φ (i.e., (10.20)) into the integral on the right of (10.22), splitting this integral into two pieces for positive and negative y, and integrating by parts. We have done this before but (why not?) let us repeat it — it's easy!

$$\begin{aligned}
\frac{d^2u}{dx^2} &= \int_{-\infty}^{\infty} \Phi(y)\, \frac{\partial^2}{\partial y^2} f(x-y)\, dy \\
&= \frac{1}{2}\int_{-\infty}^{\infty} |y|\, \frac{\partial^2}{\partial y^2} f(x-y)\, dy \\
&= \frac{1}{2}\int_{-\infty}^{0} -y\, \frac{\partial^2}{\partial y^2} f(x-y)\, dy + \frac{1}{2}\int_{0}^{\infty} y\, \frac{\partial^2}{\partial y^2} f(x-y)\, dy \\
&= \frac{1}{2}\int_{-\infty}^{0} \frac{\partial}{\partial y} f(x-y)\, dy - \frac{1}{2}\int_{0}^{\infty} \frac{\partial}{\partial y} f(x-y)\, dy \\
&= \frac{1}{2} f(x) - \left(-\frac{1}{2} f(x)\right) = f(x).
\end{aligned}$$

Note that the boundary term in the second to last line (from integration by parts) vanished since f had compact support.

Now let us find the **Green's function for the 1D Laplacian** on the domain $\Omega = (-1, 1)$, i.e., the unit ball in 1D centered at the origin. For any $x_0 \in (-1, 1)$, we want to add to $\Phi(x - x_0)$ a harmonic function $H_{x_0}(x)$ (in 1D this means linear) which cancels out its boundary values. Note that at $x = 1$, $\Phi = \frac{1-x_0}{2}$ and at $x = -1$, $\Phi = \frac{1+x_0}{2}$. Thus, we choose

$$H_{x_0}(x) = -\frac{1}{2}(1 - x_0 x),$$

and we arrive at our Green's function

$$G(x, x_0) = \Phi(x - x_0) + H_{x_0}(x) = \frac{1}{2}|x - x_0| - \frac{1}{2}(1 - x_0 x).$$

10.4. • Green's Functions for the Half-Space and Ball in 3D

In this section we explicitly find the Green's functions for two rather particular 3D domains. We then use these Green's functions to solve the respective Dirichlet problems. In each case, we follow three steps:

Step 1. We use the symmetry of the domain and the "**method of images**" to easily find the Green's function.

Step 2. We then use (10.15) to write down a candidate solution to the Dirichlet problem. This involves computing the normal derivative of the Green's function.

Step 3. The basis of formula (10.15) was Theorem 10.4 which was proved for a bounded domain (note the half-space is unbounded). Moreover, the underlying premise of the theorem was that **if** a smooth solution exists, **then** it is given by (10.15). Thus, as usual, we now must check that (10.15) does indeed solve the Dirichlet problem.

10.4.1. • Green's Function for the Half-Space.

We will use (x, y, z) to be the components of $\mathbf{x} \in \mathbb{R}^3$, and we wish to find Green's function on the half-space

$$\mathcal{H} := \{\mathbf{x} = (x, y, z) \in \mathbb{R}^3 \mid z > 0\}.$$

This is an **unbounded** domain with boundary being the xy-plane; i.e.,

$$\partial\mathcal{H} = \{\mathbf{x} = (x, y, z) \mid z = 0\} \quad \text{and} \quad \overline{\mathcal{H}} = \{\mathbf{x} = (x, y, z) \in \mathbb{R}^3 \mid z \geq 0\}.$$

Hence the results of Section 10.3.2 do not apply. However, let us proceed by explicitly finding the Green's function and then verifying that the potentially illegal formula (10.15) for the solution to

$$\begin{cases} \Delta u = 0 & \text{in } \mathcal{H}, \\ u = g & \text{on } \partial\mathcal{H} \end{cases} \tag{10.23}$$

does indeed hold true. We assume here that g is a continuous function on $\partial\mathcal{H}$ (the xy-plane).

Step 1 (Finding the Green's Function via the Method of Images). Fix a point $\mathbf{x}_0 = (x_0, y_0, z_0) \in \mathcal{H}$. We need to find a function $H_{\mathbf{x}_0}$ defined on $\overline{\mathcal{H}}$ such that the following hold:

- $H_{\mathbf{x}_0}(\mathbf{x})$ is harmonic in $\mathcal{H}$.
- $-\frac{1}{4\pi|\mathbf{x}-\mathbf{x}_0|} + H_{\mathbf{x}_0}(\mathbf{x}) = 0$ for $\mathbf{x} \in \partial\mathcal{H}$.

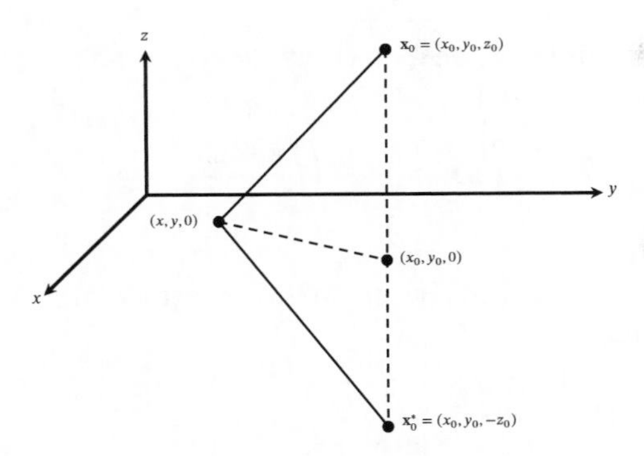

Figure 10.3. Depiction of $\mathbf{x}_0$ and $\mathbf{x}_0^*$ in the upper and lower half-spaces respectively.

In other words, we need $H_{\mathbf{x}_0}$ to be harmonic on $\mathcal{H}$ and to cancel out the values of the fundamental solution on the boundary. We will then take our Green's function to be

$$G(\mathbf{x}, \mathbf{x}_0) = -\frac{1}{4\pi|\mathbf{x} - \mathbf{x}_0|} + H_{\mathbf{x}_0}(\mathbf{x}). \tag{10.24}$$

Here it is easy to *find* the $H_{\mathbf{x}_0}$ that works. Given the point $\mathbf{x}_0 = (x_0, y_0, z_0) \in \mathcal{H}$, let $\mathbf{x}_0^* = (x_0, y_0, -z_0)$ denote its **reflection by the xy-plane**. Now define

$$H_{\mathbf{x}_0}(\mathbf{x}) = \frac{1}{4\pi|\mathbf{x} - \mathbf{x}_0^*|}.$$

Note that $H_{\mathbf{x}_0}$ is harmonic on $\mathcal{H}$ because the singularity $\mathbf{x}_0^*$ of $H_{\mathbf{x}_0}$ is **not** in the upper half-plane. Moreover, if $\mathbf{x} \in \partial\mathcal{H}$, then $|\mathbf{x} - \mathbf{x}_0| = |\mathbf{x} - \mathbf{x}_0^*|$ and, hence, $G(\mathbf{x}, \mathbf{x}_0) = 0$. So we arrive at our Green's function:

$$\boxed{G(\mathbf{x}, \mathbf{x}_0) = -\frac{1}{4\pi|\mathbf{x} - \mathbf{x}_0|} + \frac{1}{4\pi|\mathbf{x} - \mathbf{x}_0^*|}.} \tag{10.25}$$

This idea of using the reflected source (the mirrored image) to cancel the boundary values is known as **the method of images**.

Step 2 (Writing Down Our Candidate Solution for the Dirichlet Problem (10.23)). The natural parametrization of $\partial\mathcal{H}$ (the xy-plane) is with x and y each spanning the entire real line. Hence, the boundary data can be written as $g(x, y)$.[5] Moreover, on this boundary, the surface area element dS is simply $dxdy$. Note that the half-space is **unbounded** so technically our previous work, i.e., Theorem 10.4, does not apply. However, in the end, we can check that our solution formula will solve (10.23); hence, let us follow the basic reasoning of Theorem 10.4 to *propose* that the solution is given by

$$u(\mathbf{x}_0) = \int_{-\infty}^{\infty}\int_{-\infty}^{\infty} g(x, y)\,\frac{\partial G(\mathbf{x}, \mathbf{x}_0)}{\partial \mathbf{n}}\,dx\,dy \qquad \text{where } \mathbf{x} = (x, y, 0).$$

This is in fact the case!

[5] Strictly speaking we should denote points on the xy-plane ($\partial\mathcal{H}$) by $(x, y, 0)$. However, by abusing notation slightly, it is convenient to omit the 0.

To unravel this formula, we need to compute $\partial G/\partial \mathbf{n}$ on the xy-plane. To this end, we first compute that

$$\frac{\partial G}{\partial \mathbf{n}} = -\frac{\partial G}{\partial z}(\mathbf{x}, \mathbf{x}_0) = \frac{1}{4\pi}\left(\frac{z+z_0}{|\mathbf{x}-\mathbf{x}_0^*|^3} - \frac{z-z_0}{|\mathbf{x}-\mathbf{x}_0|^3}\right),$$

but where do we want to evaluate $\partial G/\partial z$? The answer is that we want to evaluate it on the xy-plane, i.e., when $z = 0$. By construction (see Figure 10.3), if $z = 0$, we have $|\mathbf{x} - \mathbf{x}_0^*| = |\mathbf{x} - \mathbf{x}_0|$ and so

$$-\left.\frac{\partial G}{\partial z}\right|_{z=0} = \frac{1}{2\pi}\frac{z_0}{|\mathbf{x}-\mathbf{x}_0|^3}.$$

Therefore, the solution u is given for any $\mathbf{x}_0 \in \mathcal{H}$ by

$$\boxed{u(\mathbf{x}_0) = \frac{z_0}{2\pi}\int_{-\infty}^{\infty}\int_{-\infty}^{\infty}\frac{g(x,y)}{|\mathbf{x}-\mathbf{x}_0|^3}\,dx\,dy.} \tag{10.26}$$

In terms of components, we can write (10.26) as

$$u(x_0, y_0, z_0) = \frac{z_0}{2\pi}\int_{-\infty}^{\infty}\int_{-\infty}^{\infty}\frac{g(x,y)}{\left[(x-x_0)^2+(y-y_0)^2+z_0^2\right]^{3/2}}\,dx\,dy.$$

If for $\mathbf{x}_0 = (x_0, y_0, z_0) \in \mathcal{H}$ and $\mathbf{x} = (x, y, 0) \in \partial\mathcal{H}$, we define the kernel

$$k(\mathbf{x}, \mathbf{x}_0) = \frac{z_0}{2\pi\left[(x-x_0)^2+(y-y_0)^2+z_0^2\right]^{3/2}};$$

then we can rewrite the solution formula (10.26) as

$$u(\mathbf{x}_0) = \int_{-\infty}^{\infty}\int_{-\infty}^{\infty} k(\mathbf{x}, \mathbf{x}_0)\, g(x,y)\, dxdy. \tag{10.27}$$

It is important when you see complicated formulas such as (10.26) or (10.27) to not let the notation get the better hand. Decide which variables are fixed and which are integrated over. Here, $\mathbf{x}_0 = (x_0, y_0, z_0)$ is a fixed point in $\mathcal{H}$. The variable $\mathbf{x}$ is used as a dummy variable of integration and it traverses the xy-plane. Hence, in terms of components $\mathbf{x} = (x, y, 0)$, g is the data function defined for (x, y) on the xy-plane.

Step 3: Verifying the Candidate Solution. Finally, we need to check that the function defined by (10.27) does indeed solve the Dirichlet problem on the half-space (10.23). First off, it does define a harmonic function in $\mathcal{H}$. This is simply a consequence of four things:

- One can differentiate under the integral sign.
- The Green's function is harmonic away from its singularity.
- The Green's function is symmetric.
- Any partial derivative (or linear combination of derivatives) of a harmonic function is itself harmonic.

The details of these steps are presented for the ball in Section 10.4.3: The argument for the half-space is identical.

But what about the boundary values? Looking at the formula (10.26), we see that it is defined for every $\mathbf{x}_0 \in \mathcal{H}$ but if $\mathbf{x}_0 \in \partial\mathcal{H}$, then, by definition, $z_0 = 0$. Does this mean the formula gives 0? If so, something has gone terribly wrong. First note that it is not 0 since, with $z_0 = 0$, the double integral diverges. But then what is 0 times ∞? Even though the formula is not strictly defined on the boundary, what matters to us is that this function u, defined on $\mathcal{H}$, has a **continuous limit** as we approach the boundary and that this limit is g. If this is the case, then one can *naturally* extend formula (10.26) to all of $\overline{\mathcal{H}}$ by defining u to be g on $\partial\mathcal{H}$. In doing so, we find the extension is continuous on $\overline{\mathcal{H}}$ and solves the Dirichlet problem (10.23).

Let us pursue this a bit further. One can directly check (though it is not so easy) that for any $\mathbf{x}_0 \in \Omega$,

$$\int_{-\infty}^{\infty}\int_{-\infty}^{\infty} k(\mathbf{x}, \mathbf{x}_0)\, dx\, dy = 1. \tag{10.28}$$

Now let $(x^*, y^*, 0) \in \partial\mathcal{H}$. The reason why we get the right boundary values is that as $(x_0, y_0, z_0) \to (x^*, y^*, 0)$, the associated kernel $k(\mathbf{x}, \mathbf{x}_0)$ as a function of x, y approaches (in the sense of distributions) a delta function with concentration at $(x, y) = (x^*, y^*)$. Thus,

$$\lim_{(x_0,y_0,z_0)\to(x^*,y^*,0)} u(\mathbf{x}_0) = g(x^*, y^*). \tag{10.29}$$

We now provide a precise statement. The proof is left as an exercise (cf. Exercise **10.7**) but note that the proof is nearly identical to the proof for the ball (Theorem 10.9) which is presented in full detail in Section 10.4.3.

Theorem 10.8. *Let g be a continuous and integrable function on $\mathbb{R}^2$ and let u be defined on $\mathcal{H}$ by* (10.26)*, or equivalently* (10.27)*. Then:*

(i) *u is a C^∞ function in $\mathcal{H}$.*

(ii) *u is harmonic in $\mathcal{H}$.*

(iii) *For any $(x^*, y^*) \in \mathbb{R}^2$,* (10.29) *holds true.*

Thus u naturally extends to a continuous function on $\overline{\mathcal{H}}$ which satisfies (10.23).

10.4.2. • Green's Function for the Ball.

Consider $\Omega = B(\mathbf{0}, a)$, the ball centered at the origin of radius a. Note that $\partial\Omega = \{\mathbf{x} : |\mathbf{x}| = a\}$, the sphere of radius a. We wish to construct the Green's function and use it to solve

$$\begin{cases} \Delta u = 0 & \text{in } B(\mathbf{0}, a), \\ u = g & \text{on } \partial B(\mathbf{0}, a), \end{cases} \tag{10.30}$$

where g is any continuous function defined on the sphere of radius a.

Step 1 (Finding the Green's Function via the Method of Images). Fix $\mathbf{x}_0 \neq 0$. Constructing the Green's function is now slightly more involved than it was for the half-space because the nature of the reflection (mirrored image) is now more complicated (see Figure 10.4). What will work is a notion of reflection by the sphere: We choose $\mathbf{x}_0^*$ to be an "extension" of the vector $\mathbf{x}_0$ such that $|\mathbf{x}_0|\,|\mathbf{x}_0^*| = a^2$. This means we choose

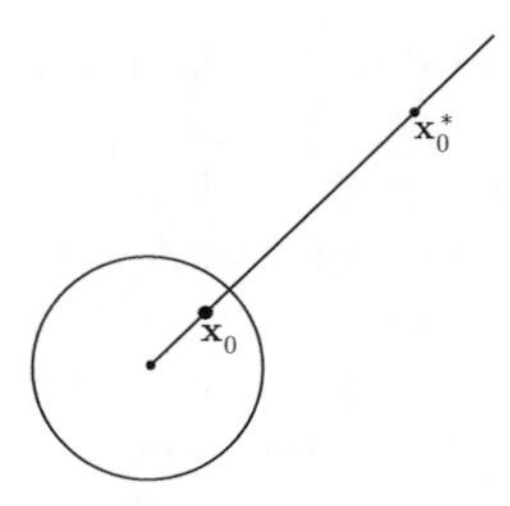

Figure 10.4. 2D depiction of reflection by the sphere.

$\mathbf{x}_0^* = c\frac{\mathbf{x}_0}{|\mathbf{x}_0|}$ where $c = |\mathbf{x}_0^*| = a^2/|\mathbf{x}_0|$; i.e.,

$$\mathbf{x}_0^* = \frac{a^2\mathbf{x}_0}{|\mathbf{x}_0|^2}.$$

With this reflected point, we claim that the corrector is

$$H_{\mathbf{x}_0}(\mathbf{x}) := \frac{a}{|\mathbf{x}_0|}\frac{1}{4\pi|\mathbf{x}-\mathbf{x}_0^*|}.$$

That is, the following choice for the Green's function will work:

$$\boxed{G(\mathbf{x},\mathbf{x}_0) = -\frac{1}{4\pi|\mathbf{x}-\mathbf{x}_0|} + \frac{a}{|\mathbf{x}_0|}\frac{1}{4\pi|\mathbf{x}-\mathbf{x}_0^*|}.} \tag{10.31}$$

Note here the additional factor of $\frac{a}{|\mathbf{x}_0|}$ in the second term. One could certainly complain here that this definition is unmotivated. Yes, it is true that the second term is harmonic in $B(\mathbf{0}, a)$ since the source point $\mathbf{x}_0^*$ lies outside the ball, but the particular choice of $\mathbf{x}_0^*$ and the scaling factor is only justified after the fact that if $\mathbf{x} \in \partial\Omega$ (i.e., $|\mathbf{x}| = a$), then $G(\mathbf{x},\mathbf{x}_0) = 0$. Let us verify this now. It suffices to show that if $|\mathbf{x}| = a$, then

$$\frac{|\mathbf{x}_0|^2}{a^2}|\mathbf{x}-\mathbf{x}_0^*|^2 = |\mathbf{x}-\mathbf{x}_0|^2. \tag{10.32}$$

To this end, we note that

$$\frac{|\mathbf{x}_0|^2}{a^2}|\mathbf{x}-\mathbf{x}_0^*|^2 = \frac{|\mathbf{x}_0|^2}{a^2}(\mathbf{x}-\mathbf{x}_0^*)\cdot(\mathbf{x}-\mathbf{x}_0^*) = \frac{|\mathbf{x}_0|^2}{a^2}\left(a^2 - 2\mathbf{x}\cdot\mathbf{x}_0^* + |\mathbf{x}_0^*|^2\right)$$

and we notice that since

$$\mathbf{x}_0^* = \frac{a^2}{|\mathbf{x}_0|^2}\mathbf{x}_0 \qquad \text{and} \qquad |\mathbf{x}| = a,$$

we have

$$\begin{aligned}\frac{|\mathbf{x}_0|^2}{a^2}|\mathbf{x}-\mathbf{x}_0^*|^2 &= \frac{|\mathbf{x}_0|^2}{a^2}\left(a^2 - 2\frac{\mathbf{x}\cdot\mathbf{x}_0 a^2}{|\mathbf{x}_0|^2} + \frac{a^4}{|\mathbf{x}_0|^2}\right)\\ &= |\mathbf{x}_0|^2 - 2\mathbf{x}\cdot\mathbf{x}_0 + a^2 = (\mathbf{x}_0-\mathbf{x})\cdot(\mathbf{x}_0-\mathbf{x}) = |\mathbf{x}-\mathbf{x}_0|^2.\end{aligned}$$

Step 2 (Writing Down Our Candidate Solution for the Dirichlet Problem). We must compute the normal derivative of G. To this end, it is convenient to define

$$\rho = |\mathbf{x}-\mathbf{x}_0| \qquad \text{and} \qquad \rho^* = |\mathbf{x}-\mathbf{x}_0^*|$$

and write the Green's function in terms of ρ, ρ^* to arrive at

$$G(\mathbf{x}, \mathbf{x}_0) = -\frac{1}{4\pi\rho} + \frac{a}{|\mathbf{x}_0|}\frac{1}{4\pi\rho^*}.$$

Since

$$\nabla\rho = \frac{(\mathbf{x}-\mathbf{x}_0)}{\rho}, \qquad \nabla\rho^* = \frac{(\mathbf{x}-\mathbf{x}_0^*)}{\rho^*},$$

we have

$$\nabla G(\mathbf{x}, \mathbf{x}_0) = \frac{\mathbf{x}-\mathbf{x}_0}{4\pi\rho^3} - \frac{a}{|\mathbf{x}_0|}\frac{\mathbf{x}-\mathbf{x}_0^*}{4\pi(\rho^*)^3}.$$

Note that by (10.32), $\rho^* = \frac{a\rho}{|\mathbf{x}_0|}$. Hence by definition of $\mathbf{x}_0^*$ we find, after a little algebra, that for $\mathbf{x} \in \partial\Omega$ (i.e., when $|\mathbf{x}| = a$),

$$\nabla G(\mathbf{x}, \mathbf{x}_0) = \frac{\mathbf{x}}{4\pi\rho^3}\left(1 - \frac{|\mathbf{x}_0|^2}{a^2}\right),$$

and the normal derivative with $\mathbf{n} = \mathbf{x}/a$ is

$$\frac{\partial G}{\partial \mathbf{n}}(\mathbf{x}, \mathbf{x}_0) = \nabla G \cdot \frac{\mathbf{x}}{a} = \frac{a^2 - |\mathbf{x}_0|^2}{4\pi a\rho^3}.$$

Based upon Theorem 10.4, our candidate solution to (10.30) is

$$\boxed{u(\mathbf{x}_0) = \frac{a^2 - |\mathbf{x}_0|^2}{4\pi a}\iint_{\partial B(\mathbf{0},a)} \frac{g(\mathbf{x})}{|\mathbf{x}-\mathbf{x}_0|^3}\, dS_{\mathbf{x}}, \quad \text{for } \mathbf{x}_0 \neq \mathbf{0}.} \tag{10.33}$$

Note that, for this reflection argument to work, we needed $\mathbf{x}_0 \neq \mathbf{0}$. However, at the origin, we know the value of u by the mean value property:

$$u(\mathbf{0}) = \frac{1}{4\pi a^2}\iint_{\partial B(\mathbf{0},a)} g(\mathbf{x})\, dS_{\mathbf{x}}.$$

On the other hand, note that this is exactly what we get by plugging $\mathbf{x}_0 = \mathbf{0}$ into (10.33), since if $\mathbf{x}_0 = \mathbf{0}$, $|\mathbf{x}-\mathbf{x}_0| = a$. Thus, we can conclude that for all $\mathbf{x}_0 \in B(\mathbf{0}, a)$,

$$\boxed{u(\mathbf{x}_0) = \iint_{\partial B(\mathbf{0},a)} k(\mathbf{x}, \mathbf{x}_0)\, g(\mathbf{x})\, dS_{\mathbf{x}}, \quad \text{where} \quad k(\mathbf{x}, \mathbf{x}_0) = \frac{a^2 - |\mathbf{x}_0|^2}{4\pi a|\mathbf{x}-\mathbf{x}_0|^3}.}$$

The function k is called the **Poisson kernel** and (10.33) is often called the (3D) **Poisson formula** for the solution to the Dirichlet problem on the ball.

Step 3: Verifying the Candidate Solution. The Poisson kernel possesses the following integrability condition.

Lemma 10.4.1. *For any* $\mathbf{x}_0 \in B(\mathbf{0}, a)$, *we have*

$$\iint_{\partial B(\mathbf{0},a)} k(\mathbf{x}, \mathbf{x}_0)\, dS_{\mathbf{x}} = 1.$$

Proof. One can prove this directly by brute force integration. On the other hand, one can simply argue as follows: The function $u(\mathbf{x}) \equiv 1$ solves the Dirichlet problem with

$g \equiv 1$. We have found the Green's function G for the ball, i.e., the function which satisfies the conditions of Definition 10.3.1. Hence, by Theorem 10.4, we have

$$1 = \frac{1}{4\pi a} \iint_{\partial B(\mathbf{0},a)} \frac{a^2 - |\mathbf{x}|^2}{|\mathbf{x}-\mathbf{y}|^3}\, dS_{\mathbf{x}}. \qquad \square$$

Now we need to check that if u is given by (10.33), then u indeed solves the Dirichlet problem (10.30). As with the half-space, if we plug in a value of $\mathbf{x}_0$ which is on $\partial B(\mathbf{0}, a)$, we get 0 times an improper integral which blows up. However, we will now prove that if $\mathbf{x}_0^* \in \partial B(\mathbf{0}, a)$, then

$$\lim_{\mathbf{x}_0 \to \mathbf{x}_0^*} u(\mathbf{x}_0) = g(\mathbf{x}_0^*).$$

In other words, we will again have a case of convergence in the sense of distributions to the delta function. Here the culprit is the function $k(\mathbf{x}, \mathbf{x}_0)$ viewed as a function of $\mathbf{x}$ on $\partial B(\mathbf{0}, a)$ with $\mathbf{x}_0$ being the (continuum) index of the continuum sequence. Essentially, we require that

$$k(\mathbf{x}, \mathbf{x}_0) \longrightarrow \delta_{\mathbf{x}_0^*} \qquad \text{as } \mathbf{x}_0 \to \mathbf{x}_0^* \qquad \text{in the sense of distributions on } \partial B(\mathbf{0}, a),$$

that is, for the test functions defined on $\partial B(\mathbf{0}, a)$. We make a precise statement in the following theorem whose proof is presented in the next subsection. Note that, in addition to the boundary values, it is not so clear that u defined by (10.33) is harmonic inside the ball.

In Theorem 10.9 below, we will change the notation as follows: $\mathbf{x}$, instead of $\mathbf{x}_0$, for the source point; $\mathbf{y}$, instead of $\mathbf{x}$, as the dummy variable for the surface integral over the boundary (sphere); $\mathbf{x}_0$ for a fixed point on the boundary.

Theorem 10.9. *Let g be continuous on $\partial B(\mathbf{0}, a)$ and let u be defined on $B(\mathbf{0}, a)$ by*

$$u(\mathbf{x}) = \iint_{\partial B(\mathbf{0},a)} k(\mathbf{y}, \mathbf{x})\, g(\mathbf{y})\, dS_{\mathbf{y}} = \frac{a^2 - |\mathbf{x}|^2}{4\pi a} \iint_{\partial B(\mathbf{0},a)} \frac{g(\mathbf{y})}{|\mathbf{y}-\mathbf{x}|^3}\, dS_{\mathbf{y}}. \tag{10.34}$$

Then:

(i) *u is a C^∞ function in $B(\mathbf{0}, a)$.*

(ii) *u is harmonic in $B(\mathbf{0}, a)$.*

(iii) *For any $\mathbf{x}_0 \in \partial B(\mathbf{0}, a)$,*

$$\lim_{\mathbf{x} \to \mathbf{x}_0} u(\mathbf{x}) = g(\mathbf{x}_0). \tag{10.35}$$

Thus, u naturally extends to a continuous function on $\overline{B}(\mathbf{0}, a)$ and satisfies (10.30).

10.4.3. The Proof for Theorem 10.9.

Proof. We follow the proof in Evans [**10**].

(i) First we note that if $\mathbf{x} \in B(\mathbf{0}, a)$, for any $\mathbf{y} \in \partial B(\mathbf{0}, a)$ the integrand of (10.34) is a smooth function, in fact C^∞. Hence, when computing derivatives of u (with respect to the components of $\mathbf{x}$) it is legal to bring these derivatives inside the integral (cf. Section A.9). These observations imply that u is a C^∞ function in $B(\mathbf{0}, a)$ and

$$\Delta_{\mathbf{x}} u(\mathbf{x}) = \iint_{\partial B(\mathbf{0},a)} \Delta_{\mathbf{x}} k(\mathbf{y}, \mathbf{x})\, g(\mathbf{y})\, dS_{\mathbf{y}}.$$

Hence, u is harmonic inside $B(\mathbf{0}, a)$ if $k(\mathbf{y}, \mathbf{x})$ (as a function of $\mathbf{x}$) is harmonic inside $B(\mathbf{0}, a)$.

(ii) Next we address why u is harmonic inside $B(\mathbf{0}, a)$. We could check this directly by verifying that $\Delta_{\mathbf{x}} k(\mathbf{y}, \mathbf{x}) = 0$, but we actually do not need to do any tedious calculations of derivatives. Let us recall where k came from; it came from the normal derivative of the Green's function, specifically

$$k(\mathbf{y}, \mathbf{x}) = \frac{\partial G(\mathbf{y}, \mathbf{x})}{\partial \mathbf{n}},$$

where the normal derivative is with respect to the $\mathbf{y}$ variable. But note the following:

For fixed $\mathbf{x}$, $G(\mathbf{y}, \mathbf{x})$ as a function of $\mathbf{y}$ is harmonic except at $\mathbf{y} = \mathbf{x}$.

By symmetry of the Green's function, this then means the following:

For fixed $\mathbf{y}$, $G(\mathbf{y}, \mathbf{x})$ as a function of $\mathbf{x}$ is harmonic except at $\mathbf{x} = \mathbf{y}$.

Moreover, since G is infinitely smooth away from its singularity, any partial derivative (or linear combination thereof) of G is also harmonic away from the singularity. Thus,

$$\text{for fixed } \mathbf{y}, \quad \frac{\partial G(\mathbf{y}, \mathbf{x})}{\partial \mathbf{n}} \quad \text{as a function of } \mathbf{x} \text{ is harmonic except at } \mathbf{x} = \mathbf{y}.$$

Since $\mathbf{y}$ lies on the boundary sphere whereas $\mathbf{x}$ lies in the interior ball, this gives that for fixed $\mathbf{y}$, $\Delta_{\mathbf{x}}\, k(\mathbf{y}, \mathbf{x}) = 0$. Thus, u defined by (10.34) is a harmonic function for $\mathbf{x} \in B(\mathbf{0}, a)$.

(iii) Lastly we prove (10.35). This means that for every $\epsilon > 0$, we must prove there exists a $\delta > 0$ such that

$$\text{if} \quad |\mathbf{x} - \mathbf{x_0}| < \delta, \qquad \text{then} \qquad |u(\mathbf{x}) - g(\mathbf{x_0})| < \epsilon. \tag{10.36}$$

This is, so far, the most involved of our *delta function convergence proofs*. But please bear with me — it is not so bad! Let $\epsilon > 0$ and first note that by Lemma 10.4.1,

$$1 = \frac{1}{4\pi a} \iint_{\partial B(\mathbf{0},a)} \frac{a^2 - |\mathbf{x}|^2}{|\mathbf{x} - \mathbf{y}|^3}\, dS_{\mathbf{y}}. \tag{10.37}$$

Let $\mathbf{x_0}$ be a fixed point on the boundary, i.e., $|\mathbf{x_0}| = a$, and let $\mathbf{x}$ be such that $|\mathbf{x}| < a$. The variable $\mathbf{y}$ is the integration variable in the Poisson formula which traverses the circle of radius a (cf. Figure 10.5).

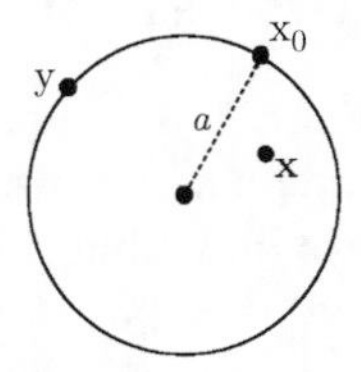

Figure 10.5

By (10.34) and (10.37), we have

$$\begin{aligned} |u(\mathbf{x}) - g(\mathbf{x_0})| &\leq \frac{1}{4\pi a} \iint_{\partial B(\mathbf{0},a)} \frac{a^2 - |\mathbf{x}|^2}{|\mathbf{x}-\mathbf{y}|^3} |g(\mathbf{y}) - g(\mathbf{x_0})| \, dS_{\mathbf{y}} \\ &= I_1 + I_2, \end{aligned}$$

where

$$I_1 := \frac{1}{4\pi a} \iint_{\{\mathbf{y}\,|\,|\mathbf{y}|=a,\ |\mathbf{y}-\mathbf{x_0}|<\delta_0\}} \frac{a^2 - |\mathbf{x}|^2}{|\mathbf{x}-\mathbf{y}|^3} |g(\mathbf{y}) - g(\mathbf{x_0})| \, dS_{\mathbf{y}},$$

$$I_2 := \frac{1}{4\pi a} \iint_{\{\mathbf{y}\,|\,|\mathbf{y}|=a,\ |\mathbf{y}-\mathbf{x_0}|\geq\delta_0\}} \frac{a^2 - |\mathbf{x}|^2}{|\mathbf{x}-\mathbf{y}|^3} |g(\mathbf{y}) - g(\mathbf{x_0})| \, dS_{\mathbf{y}},$$

for any $\delta_0 > 0$. We will choose δ_0 shortly, which will in turn allow us to choose our δ for (10.36).
Since g is continuous, we can find $\delta_0 > 0$ (depending on ϵ and g) such that

$$|g(\mathbf{y}) - g(\mathbf{x_0})| < \frac{\epsilon}{2} \qquad \text{if } |\mathbf{y}-\mathbf{x_0}| < \delta_0 \text{ with } |\mathbf{y}| = a.$$

Hence, by (10.37), $I_1 < \frac{\epsilon}{2}$. Turning to I_2, we will need to refine the choice of δ even further (hence, the use of δ_0 above). We will quantify the fact that the integrand of I_2 is well-behaved (i.e., is bounded above) if $\mathbf{x}$ is sufficiently close to $\mathbf{x_0}$. To this end, first note that if we denote M to be the maximum of g on $\partial B(\mathbf{0}, a)$, then

$$I_2 \leq 2M \frac{1}{4\pi a} \iint_{\{\mathbf{y}\,|\,|\mathbf{y}|=a,\ |\mathbf{y}-\mathbf{x_0}|\geq\delta_0\}} \frac{a^2 - |\mathbf{x}|^2}{|\mathbf{x}-\mathbf{y}|^3} \, dS_{\mathbf{y}}.$$

Next, note that if $|\mathbf{x}-\mathbf{x_0}| < \frac{\delta_0}{2}$, then on the region of integration wherein $|\mathbf{y}-\mathbf{x_0}| > \delta_0$, we have

$$\delta_0 < |\mathbf{y}-\mathbf{x_0}| \leq |\mathbf{y}-\mathbf{x}| + |\mathbf{x}-\mathbf{x_0}| \leq |\mathbf{y}-\mathbf{x}| + \frac{\delta_0}{2}.$$

Thus, for I_2, we have that if $|\mathbf{x}-\mathbf{x_0}| < \frac{\delta_0}{2}$, then $|\mathbf{x}-\mathbf{y}| \geq \frac{\delta_0}{2}$, and hence

$$\frac{a^2 - |\mathbf{x}|^2}{|\mathbf{x}-\mathbf{y}|^3} \leq \frac{8(a^2 - |\mathbf{x}|^2)}{\delta_0^3}.$$

Our choice of δ_0, which depended on ϵ, will not suffice for (10.36). However, given this choice of δ_0 we can choose a smaller δ with $\delta < \frac{\delta_0}{2}$ and

$$\frac{32Ma^2\delta}{\delta_0^3} < \frac{\epsilon}{2}.$$

Note that this choice of δ will depend on ϵ **and** the previous δ_0 (which depended only on ϵ and the function g). Let us now show that this choice of δ gives (10.36). To this end, for $\mathbf{x} \in B(\mathbf{0}, a)$ and $\mathbf{x}_0 \in \partial B(\mathbf{0}, a)$ (i.e., $|\mathbf{x_0}| = a$), if $|\mathbf{x}-\mathbf{x_0}| < \delta$, then $a - |\mathbf{x}| < \delta$ and hence,

$$a^2 - |\mathbf{x}|^2 = (a - |\mathbf{x}|)(a + |\mathbf{x}|) < 2a\delta.$$

Thus, for $|\mathbf{x} - \mathbf{x_0}| < \delta$, we have

$$\begin{aligned}
|u(\mathbf{x}) - g(\mathbf{x_0})| &\leq I_1 + I_2 \\
&\leq \frac{\epsilon}{2} + I_2 \\
&\leq \frac{\epsilon}{2} + 2M\frac{1}{4\pi a}\iint_{\{\mathbf{y}\,|\,|\mathbf{y}|=a,\ |\mathbf{y}-\mathbf{x_0}|\geq\delta_0\}} \frac{a^2 - |\mathbf{x}|^2}{|\mathbf{x}-\mathbf{y}|^3}\, dS_{\mathbf{y}} \\
&\leq \frac{\epsilon}{2} + 2M\frac{1}{4\pi a}\iint_{\{\mathbf{y}\,|\,|\mathbf{y}|=a,\ |\mathbf{y}-\mathbf{x_0}|\geq\delta_0\}} \frac{16a\delta}{\delta_0^3}\, dS_{\mathbf{y}} \\
&< \frac{\epsilon}{2} + 32Ma\frac{\delta}{\delta_0^3}\frac{1}{4\pi a}(4\pi a^2) \\
&= \frac{\epsilon}{2} + \frac{32Ma^2\delta}{\delta_0^3} \\
&< \frac{\epsilon}{2} + \frac{\epsilon}{2} = \epsilon.
\end{aligned}$$

□

10.4.4. Green's Functions for Other Domains and Differential Operators.

In this subsection, we make two remarks:

(1) There are a **few** other domains where we can use similar reflections to produce the Green's function for the Laplacian with Dirichlet boundary conditions, e.g., the octant in 3D $\{(x, y, z) \,|\, x > 0, y > 0, z > 0\}$, or the quadrant in 2D. For bounded domains with a certain simple symmetric structure, like cubes and rectangular boxes, one can adopt the method of separation of variables (cf. Chapter 12): This will yield the Green's function as an infinite series of special functions. However, for a generic domain, even one with a smooth boundary, there is no method to obtain an explicit formula. There are two (related) ways one may proceed.

- Adapting techniques in **functional analysis** to prove the existence and properties of Green's functions. In a future book on the subject, we will present a brief introduction to these methods and the approaches taken, which are directly related to considering a form of a *weak solution* in the spirit of distributions.
- Invoking **numerical methods** to simulate (approximate) the Green's function. One of the most widely used methods is the **finite element method** which is directly related to the previous functional analytic approach.

(2) The notions of the fundamental solution and Green's functions extend to **any linear** differential operator with constant coefficients and indeed many other linear differential operators with a certain structure. We will discuss this further in Section 13.2 but let me provide one example here. Consider the operator $\Delta + c$, where c is any constant. The action of this operator on any function u is simply $\Delta u + cu$. Then the fundamental solution to this operator $\Phi(x)$ is a function such that

$$\Delta\Phi + c\Phi = \delta_0 \qquad \text{in the sense of distributions.}$$

You are asked to find this function in Exercise **10.14**. In doing so, one can find an explicit solution to

$$\Delta u + cu = f,$$

for any $f \in C_c^\infty(\mathbb{R}^N)$. If we consider the related BVP on a domain Ω, then we deal with the Green's function for the operator $\Delta + c$ on Ω.

10.5. Green's Functions for the Laplacian with Neumann Boundary Conditions

The notion of Green's functions extends to other boundary conditions, for example Neumann and periodic boundary conditions. In this section, we address Green's functions for the Laplacian with Neumann boundary conditions. These functions are sometimes referred to as **Neumann functions** or, occasionally, **Green's functions of the second kind**.

For simplicity, let us work in dimension $N = 3$. Let $\Omega \subset \mathbb{R}^3$ be a **bounded** domain with smooth boundary. Our goal is to write the solution to the Poisson equation with Neuman boundary conditions

$$\begin{cases} \Delta u(\mathbf{x}) = f(\mathbf{x}) & \text{in } \Omega, \\ \dfrac{\partial u}{\partial \mathbf{n}}(\mathbf{x}) = g(\mathbf{x}) & \text{on } \partial\Omega \end{cases} \tag{10.38}$$

in terms of a new Neumann Green's function, which we label as $N(\mathbf{x}, \mathbf{x}_0)$.

We start by pointing out two particularities of the Neuman problem (10.38):

(1) If u is a solution to (10.38), then so is $u + C$ for any constant C.

(2) We cannot hope to find a solution for any given f and g. More precisely, in order to have an equation that is not overdetermined we will require a **compatibility condition** between f and g. To this end, by the Divergence Theorem, we have

$$\iiint_\Omega f(x)d\mathbf{x} = \iiint_\Omega \Delta u(\mathbf{x})\, d\mathbf{x} = \iint_{\partial\Omega} \frac{\partial u(x)}{\partial \mathbf{n}}\, dS = \iint_{\partial\Omega} g(\mathbf{x})\, dS.$$

Hence (10.38) can only be solved if

$$\iiint_\Omega f(x)d\mathbf{x} = \iint_{\partial\Omega} g(\mathbf{x})\, dS. \tag{10.39}$$

Following our reasoning in Section 10.3.1, it would be tempting to define the (Neumann) Green's function $N(\mathbf{x}, \mathbf{x}_0)$ to be the functions with the following two properties:

$$\begin{cases} \Delta_{\mathbf{x}} N(\mathbf{x}, \mathbf{x}_0) = \delta_{\mathbf{x}_0} & \text{in the sense of distributions on } \Omega, \\ \frac{\partial N(\mathbf{x}, \mathbf{x}_0)}{\partial \mathbf{n}} = 0 & \text{for all } \mathbf{x} \in \partial\Omega, \end{cases} \tag{10.40}$$

where the normal derivative is with respect to the $\mathbf{x}$ variable. However, it is easy to see that no such function exists. While we cannot officially place $f = \delta_{\mathbf{x}_0}, g \equiv 0$ into (10.39), it is clear what would happen if we did. We would find that

$$\text{“}\underbrace{\iiint_\Omega \delta_{\mathbf{x}_0}\, d\mathbf{x}}_{=1}\text{”} = \iint_{\partial\Omega} 0\, dS = 0. \tag{10.41}$$

This can certainly be made precise but regardless, we have an issue. There are two ways to overcome this issue by modifying (10.40): **Either** subtract an appropriate constant, namely the reciprocal of the volume of Ω, from the delta function in the first equation **or** replace the 0 in the Neumann boundary condition with an appropriate constant, namely the reciprocal of the area of $\partial\Omega$. Let us choose the first option, and define the (Neumann) Green's function $N(\mathbf{x},\mathbf{x}_0)$ to be the function with the following two properties:

$$\begin{cases} \Delta_{\mathbf{x}} N(\mathbf{x},\mathbf{x}_0) = \delta_{\mathbf{x}_0} - \frac{1}{\mathrm{vol}(\Omega)} & \text{in the sense of distributions on } \Omega, \\ \frac{\partial N(\mathbf{x},\mathbf{x}_0)}{\partial \mathbf{n}} = 0 & \text{for all } \mathbf{x} \in \partial\Omega. \end{cases} \tag{10.42}$$

Notice that now the issue surrounding (10.41) is no longer present.

We can rephrase (10.42) in terms of a corrector $H_{\mathbf{x}_0}(\mathbf{x})$ which is now a solution to

$$\begin{cases} \Delta_{\mathbf{x}} H_{\mathbf{x}_0}(\mathbf{x}) = -\frac{1}{\mathrm{vol}(\Omega)} & \text{on } \Omega, \\ \frac{\partial H_{\mathbf{x}_0}(\mathbf{x})}{\partial \mathbf{n}}(\mathbf{x}) = -\frac{\partial \Phi(\mathbf{x}-\mathbf{x}_0)}{\partial \mathbf{n}} & \text{for } \mathbf{x} \in \partial\Omega. \end{cases}$$

Hence the Green's function defined by (10.42) is simply

$$\boxed{N(\mathbf{x},\mathbf{x}_0) := \Phi(\mathbf{x}-\mathbf{x}_0) + H_{\mathbf{x}_0}(\mathbf{x}).}$$

With the definition (10.42) in hand, we turn to the Representation Formula Theorem, Theorem 10.3. First note that while this formula was proved for a harmonic function u, the identical steps apply for a function with nonzero Laplacian: Indeed, one has that if u is $C^2(\overline{\Omega})$, then for all $\mathbf{x}_0 \in \Omega$, we have

$$u(\mathbf{x}_0) = \iint_{\partial\Omega} \left[\Phi(\mathbf{x}-\mathbf{x}_0)\frac{\partial u(\mathbf{x})}{\partial \mathbf{n}} - u(\mathbf{x})\frac{\partial \Phi}{\partial \mathbf{n}}(\mathbf{x}-\mathbf{x}_0) \right] dS_{\mathbf{x}} + \iiint_{\Omega} \Phi(\mathbf{x}-\mathbf{x}_0)\Delta u(\mathbf{x})\, d\mathbf{x}. \tag{10.43}$$

We now apply the same steps performed in Section 10.3.2 for Dirichlet boundary conditions to arrive at the following: If u is a $C^1(\overline{\Omega}) \cap C^2(\Omega)$ function which solves (10.38) and $N(\mathbf{x},\mathbf{x}_0)$ solves (10.42), then for any $\mathbf{x}_0 \in \Omega$ we have

$$u(\mathbf{x}_0) = \iint_{\partial\Omega} N(\mathbf{x},\mathbf{x}_0)\, g(\mathbf{x})\, dS_{\mathbf{x}} + \iiint_{\Omega} \left[N(\mathbf{x},\mathbf{x}_0)\, f(\mathbf{x}) + u(\mathbf{x})\frac{1}{\mathrm{vol}(\Omega)} \right] d\mathbf{x}.$$

The additional term $\iiint_{\Omega} u(\mathbf{x})\frac{1}{\mathrm{vol}(\Omega)}\,\mathbf{x}$, which is simply the average of u over Ω, is just a constant. As we first mentioned the solution to (10.38) can only be unique up to an additive constant. We arrive at the analogue of Theorem 10.5 for Neumann boundary conditions:

The Green's Function with Neumann Boundary Conditions and the Neumann Problem for Poisson's Equation

Theorem 10.10. *Let Ω be a bounded domain in $\mathbb{R}^3$. For $\mathbf{x}_0 \in \Omega$, let $N(\mathbf{x}, \mathbf{x}_0)$ be the Green's function on Ω defined by* (10.42)*. Let u be a function in $C^1(\overline{\Omega}) \cap C^2(\Omega)$ that solves* (10.38)*. Then, up to an additive constant,*

$$u(\mathbf{x}_0) = \iint_{\partial\Omega} N(\mathbf{x}, \mathbf{x}_0)\, g(\mathbf{x})\, dS_{\mathbf{x}} + \iiint_{\Omega} N(\mathbf{x}, \mathbf{x}_0)\, f(\mathbf{x})\, d\mathbf{x} \qquad \textit{for all } \mathbf{x}_0 \in \Omega.$$

Following the presentation for the Green's function with Dirichlet boundary conditions, it would seem most natural to now derive the Neumann Green's function for the half-space and the ball. The half-space is very easy; the ball is another matter.

10.5.1. Finding the Neumann Green's Function for the Half-Space in $\mathbb{R}^3$.

The half-space $\mathcal{H}$ has infinite volume and its boundary, the xy-plane, has infinite area and this makes finding the Neumann Green's function easy. Our definition (10.42) was for a bounded domain and entailed subtracting the reciprocal of the volume of Ω from the delta function. As Ω becomes unbounded, it would appear this correction term vanishes, and we could simply look for $N(\mathbf{x}, \mathbf{x}_0)$ which satisfies

$$\begin{cases} \Delta_{\mathbf{x}} N(\mathbf{x}, \mathbf{x}_0) = \delta_{\mathbf{x}_0} & \text{in the sense of distributions on } \mathcal{H}, \\ \frac{\partial N(\mathbf{x}, \mathbf{x}_0)}{\partial \mathbf{n}} = 0 & \text{for all } \mathbf{x} \in \partial\mathcal{H}. \end{cases} \tag{10.44}$$

Note that for this unbounded domain, we cannot directly apply the Divergence Theorem as in (10.39) to contradict the existence of such an N.

Finding the $N(\mathbf{x}, \mathbf{x}_0)$ which satisfies (10.44) is easy and requires a minor modification to (10.25), the Dirichlet Green's function for the half-space. Indeed, we claim for $\mathbf{x} = (x, y, z)$, $\mathbf{x}_0 = (x_0, y_0, z_0)$, and the same reflected source $\mathbf{x}_0^* = (x_0, y_0, -z_0)$,

$$N(\mathbf{x}, \mathbf{x}_0) = -\frac{1}{4\pi|\mathbf{x} - \mathbf{x}_0|} - \frac{1}{4\pi|\mathbf{x} - \mathbf{x}_0^*|}$$

does the trick — just a reversal of sign in the second term! To this end, we compute

$$\nabla_{\mathbf{x}} N(\mathbf{x}, \mathbf{x}_0) = \frac{\mathbf{x} - \mathbf{x}_0}{4\pi|\mathbf{x} - \mathbf{x}_0|^3} + \frac{\mathbf{x} - \mathbf{x}_0^*}{4\pi|\mathbf{x} - \mathbf{x}_0^*|^3}.$$

Hence for $\mathbf{x} \in \partial\mathcal{H}$, where $z = 0$, and $\mathbf{n} = \langle 0, 0, -1 \rangle$, we have $|\mathbf{x} - \mathbf{x}_0| = |\mathbf{x} - \mathbf{x}_0^*|$ and

$$\frac{\partial N(\mathbf{x}, \mathbf{x}_0)}{\partial \mathbf{n}} = \nabla_{\mathbf{x}} N(\mathbf{x}, \mathbf{x}_0) \cdot \mathbf{n} = \frac{z - z_0 + z + z_0}{4\pi|\mathbf{x} - \mathbf{x}_0|^3} = 0.$$

10.5.2. Finding the Neumann Green's Function for the Ball in $\mathbb{R}^3$.

The obvious and important task of finding the Neumann Green's function for the ball is surprisingly lacking in the literature. Almost no undergraduate nor graduate PDE texts include it. In the rare books which do include it, the exposition is often terse and unmotivated. Recently, this issue has been rectified by a very nice and short article[6] by Benedikt Wirth appearing in the *American Mathematical Monthly*. Let us present Wirth's derivation which only requires the notion of, and manipulations with, the delta function — so it's perfect for this text! The derivation will work in any space dimension; for simplicity we present it here in dimension three and for the unit ball. We will use slightly different notation than in Wirth's paper.

Since we will need to differentiate with respect to the source point $\mathbf{x}_0$, it is convenient to switch the notation from $\mathbf{x}_0$ to $\mathbf{y}$. We are thus looking for the solution $N(\mathbf{x}, \mathbf{y})$ to the following:

$$\begin{cases} \Delta_{\mathbf{x}} N(\mathbf{x}, \mathbf{y}) = \delta_{\mathbf{y}} - \frac{3}{4\pi} & \text{in the sense of distributions on } B(\mathbf{0}, 1), \\ \frac{\partial N(\mathbf{x}, \mathbf{y})}{\partial \mathbf{n}} = 0 & \text{for all } \mathbf{x} \in \partial B(\mathbf{0}, 1). \end{cases} \tag{10.45}$$

Unlike for the half-space, the presence of $-\frac{3}{4\pi}$ is essential for compatibility with the Neumann boundary conditions, and this presence makes a direct application of the method of images impossible. Our approach, effectively, will be to take a **directional derivative** of both sides of the PDE in (10.45). This will result in a new BVP (cf. (10.48) below) which can be easily solved by applying *the method of images* on the directional derivative of the fundamental solution. We will then integrate back to arrive at N. This new BVP involves a Poisson equation with a *dipole moment* right-hand side. While we will not discuss this physical interpretation here, the reader is encouraged to consider Exercise **10.29**, after reading Section 10.6.1 on the electrostatics interpretation of the present chapter. Let us now carefully present the details.

Notation: In the following, a 3D unit vector $\mathbf{e}$ will play a **double role**. It will be used as a direction (for example, upon which we take directional derivatives), and it will also be used in the source point $\mathbf{y}$. Henceforth, let us write the source point as

$$\mathbf{y} = c\mathbf{e}, \qquad \text{for some } 0 \le c < 1.$$

In principle, the direction of derivatives and the source point are unrelated; however, in this derivation they are coupled.

Step 1: Constructing the new BVP with dipole moment right-hand side. Let us take a distributional directional derivative in the $\mathbf{e}$ direction on both sides of the equation in (10.45). However, it is important that we take this directional derivative with respect to $\mathbf{y}$ rather than $\mathbf{x}$. This informally yields

$$\begin{cases} \text{“}\Delta_{\mathbf{x}}(\mathbf{e} \cdot \nabla_{\mathbf{y}} N(\mathbf{x}, \mathbf{y})) = \operatorname{div}_{\mathbf{y}} (\mathbf{e}\delta_{\mathbf{y}})\text{”} & \text{on } B(\mathbf{0}, 1), \\ \frac{\partial (\mathbf{e} \cdot \nabla_{\mathbf{y}} N(\mathbf{x}, \mathbf{y}))}{\partial \mathbf{n}} = 0 & \text{for all } \mathbf{x} \in \partial B(\mathbf{0}, 1). \end{cases} \tag{10.46}$$

Note the use of quotation marks for the top equation as we have not made precise the distributional definition of the $\mathbf{y}$ derivative of $\delta_{\mathbf{y}}$ on the right-hand side (differentiating the delta function with respect to the source). But as in Exercise **9.6**(b), we can informally make sense of $\operatorname{div}_{\mathbf{y}} (\mathbf{e}\delta_{\mathbf{y}})$ and argue that

$$\text{“}\operatorname{div}_{\mathbf{y}} (\mathbf{e}\delta_{\mathbf{y}}) = \mathbf{e} \cdot \nabla_{\mathbf{y}} \delta_{\mathbf{y}} = -\mathbf{e} \cdot \nabla_{\mathbf{x}} \delta_{\mathbf{y}} = -\operatorname{div}_{\mathbf{x}} (\mathbf{e}\delta_{\mathbf{y}})\text{”}. \tag{10.47}$$

[6] **Benedikt Wirth**, Green's Function for the Neumann-Poisson Problem on n-Dimensional Balls, *The American Mathematical Monthly*, 127:8, 737–743, 2020.

Note that the last two expressions above are well-defined distributions, and the last equality is readily verified in the sense of distributions (cf. Exercise **9.6**(a)). Thus we may rewrite (10.46) as the well-defined BVP (no need for quotation marks) for a function $F_{\mathbf{e}}(\mathbf{x}, \mathbf{y})$:

$$\begin{cases} \Delta_{\mathbf{x}} F_{\mathbf{e}}(\mathbf{x}, \mathbf{y}) = \operatorname{div}_{\mathbf{x}} (\mathbf{e}\delta_{\mathbf{y}}) & \text{in the sense of distributions on } B(\mathbf{0}, 1), \\ \frac{\partial F_{\mathbf{e}}(\mathbf{x}, \mathbf{y})}{\partial \mathbf{n}} = 0 & \text{for all } \mathbf{x} \in \partial B(\mathbf{0}, 1). \end{cases} \tag{10.48}$$

As usual, think of the source $\mathbf{y}$ as a parameter in this BVP. Because of the sign change in (10.47), note that $-\mathbf{e} \cdot \nabla_{\mathbf{y}} N(\mathbf{x}, \mathbf{y})$ solves (10.48).

Step 2: Finding a solution $F_{\mathbf{e}}(\mathbf{x}, \mathbf{y})$ for the BVP (10.48). Define

$$\Psi(\mathbf{x}) := \nabla \Phi(\mathbf{x}) = \nabla \left(-\frac{1}{4\pi |\mathbf{x}|} \right) = \frac{\mathbf{x}}{4\pi |\mathbf{x}|^3},$$

where as usual $\Phi(\mathbf{x})$ is the fundamental solution for the Laplacian in 3D. We consider now the directional derivative $\mathbf{e} \cdot \Psi$ and wish to compute its distributional Laplacian. Since (cf. Exercise **9.6**(a)) for any $\mathbf{y} \in \mathbb{R}^3$, $\mathbf{e} \cdot \nabla_{\mathbf{x}} \delta_{\mathbf{y}} = \operatorname{div}_{\mathbf{x}} (\mathbf{e}\delta_{\mathbf{y}})$ in the sense of distributions, it follows that, in the sense of distributions

$$\begin{aligned} \Delta_{\mathbf{x}}(\mathbf{e} \cdot \Psi(\mathbf{x} - \mathbf{y})) = \Delta_{\mathbf{x}}(\mathbf{e} \cdot \nabla_{\mathbf{x}} \Phi(\mathbf{x} - \mathbf{y})) &= \mathbf{e} \cdot \nabla_{\mathbf{x}}(\Delta_{\mathbf{x}} \Phi(\mathbf{x} - \mathbf{y})) \\ &= \mathbf{e} \cdot \nabla_{\mathbf{x}} \delta_{\mathbf{y}} \\ &= \operatorname{div}_{\mathbf{x}} (\mathbf{e}\delta_{\mathbf{y}}). \end{aligned} \tag{10.49}$$

Above we used that $\Delta_{\mathbf{x}} \Phi(\mathbf{x} - \mathbf{y}) = \delta_{\mathbf{y}}$ and the fact that the distributional derivatives commute.

Now let us restrict our attention to the ball $B(\mathbf{0}, 1)$. We use the method of images to modify $\mathbf{e} \cdot \Psi(\mathbf{x} - \mathbf{y})$, as a function of $\mathbf{x}$, to have homogenous Neumann boundary conditions on $\partial B(\mathbf{0}, 1)$, while preserving its distributional Laplacian. To this end, as in Section 10.4.2, for $\mathbf{y} \neq 0$ in $B(\mathbf{0}, 1)$ we define the reflected source $\mathbf{y}^* = \frac{\mathbf{y}}{|\mathbf{y}|^2}$. We now define

$$\begin{aligned} F_{\mathbf{e}}(\mathbf{x}, \mathbf{y}) &:= \mathbf{e} \cdot \Psi(\mathbf{x} - \mathbf{y}) - \mathbf{e} \cdot \frac{1}{|\mathbf{y}|^3} \Psi(\mathbf{x} - \mathbf{y}^*) \\ &= \frac{\mathbf{e}}{4\pi} \cdot \left(\frac{\mathbf{x} - \mathbf{y}}{|\mathbf{x} - \mathbf{y}|^3} - \frac{\mathbf{x} - \mathbf{y}^*}{|\mathbf{y}|^3 |\mathbf{x} - \mathbf{y}^*|^3} \right). \end{aligned} \tag{10.50}$$

Note here the similarity with the definition for the Dirichlet Green's function on the ball in (10.31); however, since we are now working at the level of gradients, the power of $|\mathbf{y}|$ in the denominator of the second term is different. Indeed, let us verify that this definition does enforce the right boundary conditions; that is, for all $\mathbf{x} \in \partial B(\mathbf{0}, 1)$ and $\mathbf{n} = \mathbf{x}$ (unit outer normal to $B(\mathbf{0}, 1)$), we have

$$\frac{\partial F_{\mathbf{e}}(\mathbf{x}, \mathbf{y})}{\partial \mathbf{n}} = \nabla_{\mathbf{x}} F_{\mathbf{e}}(\mathbf{x}, \mathbf{y}) \cdot \mathbf{n} = 0.$$

To this end, we find

$$\begin{aligned}
\nabla_{\mathbf{x}} F_{\mathbf{e}}(\mathbf{x},\mathbf{y}) \cdot \mathbf{x} &= \frac{\mathbf{e}}{4\pi}\left(\frac{1}{|\mathbf{x}-\mathbf{y}|^3} - \frac{1}{|\mathbf{y}|^3|\mathbf{x}-\mathbf{y}^*|^3}\right)\cdot \mathbf{x} \\
&\quad - \frac{3}{4\pi}\left(\frac{\mathbf{e}\cdot(\mathbf{x}-\mathbf{y})}{|\mathbf{x}-\mathbf{y}|^5}(\mathbf{x}-\mathbf{y}) - \frac{\mathbf{e}\cdot(\mathbf{x}-\mathbf{y}^*)}{|\mathbf{y}|^3|\mathbf{x}-\mathbf{y}^*|^5}(\mathbf{x}-\mathbf{y}^*)\right)\cdot \mathbf{x} \\
&= -\frac{3}{4\pi}\left(\frac{\mathbf{e}\cdot(\mathbf{x}-\mathbf{y})}{|\mathbf{x}-\mathbf{y}|^5}(|\mathbf{x}|^2 - \mathbf{x}\cdot\mathbf{y}) - \frac{|\mathbf{y}|^2\,\mathbf{e}\cdot(\mathbf{x}-\mathbf{y}^*)}{|\mathbf{x}-\mathbf{y}|^5}(|\mathbf{x}|^2 - \mathbf{x}\cdot\mathbf{y}^*)\right),
\end{aligned}$$

where to pass to the second line we used twice the fact that (as in Section 10.4.2) for any $\mathbf{x} \in \partial B(\mathbf{0},1)$, $\mathbf{y} \in B(\mathbf{0},1)$, we have $|\mathbf{x}-\mathbf{y}| = |\mathbf{y}||\mathbf{x}-\mathbf{y}^*|$. Writing $\mathbf{y}$ in terms of $\mathbf{e}$ and $c \in [0,1)$, we find

$$\nabla_{\mathbf{x}} F_{\mathbf{e}}(\mathbf{x}, c\mathbf{e}) \cdot \mathbf{x} = 3\frac{(\mathbf{e}\cdot\mathbf{x} - c)(1 - c\mathbf{e}\cdot\mathbf{x}) - (c^2\mathbf{e}\cdot\mathbf{x} - c)(1 - \frac{1}{c}\mathbf{e}\cdot\mathbf{x})}{4\pi|\mathbf{x}-\mathbf{y}|^5} = 0.$$

On the other hand by definition of $F_{\mathbf{e}}(\mathbf{x},\mathbf{y})$ and (10.49), we have

$$\Delta_{\mathbf{x}} F_{\mathbf{e}}(\mathbf{x},\mathbf{y}) = \operatorname{div}_{\mathbf{x}}(\mathbf{e}\delta_{\mathbf{y}}) \quad \text{in the sense of distributions on } B(\mathbf{0},1).$$

We have shown that for each unit vector $\mathbf{e}$, $F_{\mathbf{e}}(\mathbf{x},\mathbf{y})$ defined by (10.50) solves (10.48).

Step 3: Relating $F_{\mathbf{e}}(\mathbf{x},\mathbf{y})$ to $N(\mathbf{x},\mathbf{y})$. The natural questions now are: How is $F_{\mathbf{e}}(\mathbf{x},\mathbf{y})$ related to $N(\mathbf{x},\mathbf{y})$, and how do we find N? Since both the function $-\mathbf{e}\cdot\nabla_{\mathbf{y}}N(\mathbf{x},\mathbf{y})$ and $F_{\mathbf{e}}(\mathbf{x},\mathbf{y})$ solve the BVP (10.48), the two functions (as functions of $\mathbf{x}$) must agree up to an additive constant which may depend of $\mathbf{y}$ (why?). Denoting this constant by $g(\mathbf{y})$, we have

$$F_{\mathbf{e}}(\mathbf{x},\mathbf{y}) = -\mathbf{e}\cdot\nabla_{\mathbf{y}}N(\mathbf{x},\mathbf{y}) + g(\mathbf{y}).$$

Writing $\mathbf{y}$ in terms of $\mathbf{e}$ and $c \in [0,1)$, we have

$$F_{\mathbf{e}}(\mathbf{x}, c\mathbf{e}) = -\mathbf{e}\cdot\nabla_{\mathbf{y}}N(\mathbf{x}, c\mathbf{e}) + g(c) = -\frac{\partial}{\partial c}N(\mathbf{x}, c\mathbf{e}) + g(c), \tag{10.51}$$

where we abuse notation in writing $g(c)$ for $g(c\mathbf{e})$ with $\mathbf{e}$ fixed. Finally, we recall from (10.50) the explicit expression for $F_{\mathbf{e}}(\mathbf{x}, c\mathbf{e})$:

$$F_{\mathbf{e}}(\mathbf{x}, c\mathbf{e}) = \frac{1}{4\pi}\left(\frac{\mathbf{x}\cdot\mathbf{e} - c}{|\mathbf{x}-c\mathbf{e}|^3} - \frac{\mathbf{x}\cdot\mathbf{e} - \frac{1}{c}}{c^3|\mathbf{x} - \frac{\mathbf{e}}{c}|^3}\right). \tag{10.52}$$

Step 4: Integrating to recover $N(\mathbf{x},\mathbf{y})$. In view of (10.51), it would seem natural to recover $N(\mathbf{x},\mathbf{y})$ by finding an antiderivative of $F_{\mathbf{e}}(\mathbf{x}, c\mathbf{e})$ in the c variable. First off, the expression (10.52) reveals that close to $c = 0$, $F_{\mathbf{e}}$ fails to be integrable with respect to c. Specifically, $F_{\mathbf{e}}(\mathbf{x}, s\mathbf{e})$ behaves like $\frac{1}{4\pi s}$ close to $s = 0$. But as remarked in the previous step, the crucial properties of $F_{\mathbf{e}}(\mathbf{x}, s\mathbf{e})$, namely that it solves (10.48), remain with the addition of any function of s (this is the function g in (10.51)). Hence we may choose this function g to cancel off the singularity at 0 and define

$$E_c(\mathbf{x}) := -\int_0^c \left(F_{\mathbf{e}}(\mathbf{x}, s\mathbf{e}) - \frac{1}{4\pi s}\right) ds.$$

By (10.51), one would expect that $N(\mathbf{x}, c\mathbf{e}) = N(\mathbf{x}, 0) + E_c(\mathbf{x})$. This is indeed the case. To this end, let us look at what BVP $E_c(\mathbf{x})$ solves. Here, we will informally manipulate the delta function treating it as a true function. Hence, we will use quotation marks but as usual, this informal argument will lead us to the correct answer (which one can a posteriori confirm). Taking the Laplacian in $\mathbf{x}$ of $E_c(\mathbf{x})$, we find

$$
\begin{aligned}
\text{“}\,\Delta_{\mathbf{x}} E_c(\mathbf{x}) = -\int_0^c \Delta_{\mathbf{x}} F_{\mathbf{e}}(\mathbf{x}, s\mathbf{e})\, ds \quad &\overset{(10.48)}{=} \quad -\int_0^c \operatorname{div}_{\mathbf{x}}(\mathbf{e}\delta_{s\mathbf{e}})\, ds \\
&= \quad -\int_0^c \mathbf{e}\cdot\nabla_{\mathbf{x}}\delta_{s\mathbf{e}}\, ds \\
&\overset{(10.47)}{=} \quad \int_0^c \mathbf{e}\cdot\nabla_{\mathbf{y}}\delta_{s\mathbf{e}}\, ds \\
&= \quad \int_0^c \frac{\partial}{\partial s}\delta_{s\mathbf{e}}\, ds \\
&= \quad \delta_{c\mathbf{e}} - \delta_{\mathbf{0}}.\text{”}
\end{aligned}
$$

On the other hand, if $\mathbf{x} \in \partial B(\mathbf{0}, 1)$, then

$$\nabla_{\mathbf{x}} E_c(\mathbf{x}) \cdot \mathbf{n} = -\int_0^c \nabla_{\mathbf{x}} F_{\mathbf{e}}(\mathbf{x}, s\mathbf{e}) \cdot \mathbf{n}\, ds = 0.$$

Thus to find $N(\mathbf{x}, c\mathbf{e})$, which solves (10.45), we add to $E_c(\mathbf{x})$ the term $N(\mathbf{x}, \mathbf{0})$, a function with Neumann boundary conditions whose Laplacian is $\delta_{\mathbf{0}} - \frac{3}{4\pi}$. This term is readily found to be

$$N(\mathbf{x}, \mathbf{0}) = \Phi(\mathbf{x}) - \frac{1}{8\pi}|\mathbf{x}|^2.$$

Thus we have

$$
\begin{aligned}
N(\mathbf{x}, c\mathbf{e}) &= E_c(\mathbf{x}) + N(\mathbf{x}, \mathbf{0}) \\
&= -\frac{1}{4\pi}\int_0^c \left(\frac{\mathbf{x}\cdot\mathbf{e} - s}{|\mathbf{x} - s\mathbf{e}|^3} - \frac{\mathbf{x}\cdot\mathbf{e} - \frac{1}{s}}{s^3|\mathbf{x} - \frac{\mathbf{e}}{s}|^3} - \frac{1}{s} \right) ds + \Phi(\mathbf{x}) - \frac{1}{8\pi}|\mathbf{x}|^2.
\end{aligned}
$$

Finally, one directly computes that

$$-\frac{\partial}{\partial s}\Phi(\mathbf{x} - s\mathbf{e}) = \frac{\partial}{\partial s}\left(\frac{1}{4\pi|\mathbf{x} - s\mathbf{e}|}\right) = \frac{1}{4\pi}\frac{\mathbf{x}\cdot\mathbf{e} - s}{|\mathbf{x} - s\mathbf{e}|^3},$$

to find that

$$N(\mathbf{x}, c\mathbf{e}) = \Phi(\mathbf{x} - c\mathbf{e}) + \frac{1}{4\pi}\int_0^c \left(\frac{\mathbf{x}\cdot\mathbf{e} - \frac{1}{s}}{s^3|\mathbf{x} - \frac{\mathbf{e}}{s}|^3} + \frac{1}{s} \right) ds - \frac{1}{8\pi}|\mathbf{x}|^2.$$

Reverting back to $\mathbf{y} = c\mathbf{e}$ (note $|\mathbf{y}| = c$), we have arrived at our formula for $N(\mathbf{x}, \mathbf{y})$:

$$\boxed{N(\mathbf{x}, \mathbf{y}) = \Phi(\mathbf{x} - \mathbf{y}) + \frac{1}{4\pi}\int_0^{|\mathbf{y}|} \left(\frac{\mathbf{x}\cdot\frac{\mathbf{y}}{|\mathbf{y}|} - \frac{1}{s}}{\left|s\mathbf{x} - \frac{\mathbf{y}}{|\mathbf{y}|}\right|^3} + \frac{1}{s} \right) ds - \frac{1}{8\pi}|\mathbf{x}|^2.} \tag{10.53}$$

Step 5: Evaluating the integral. Our last task is to explicitly compute the integral in (10.53). The details are left as an exercise (Exercise **10.23**) but note that they are also explicitly outlined in the appendix of Wirth's article. One arrives at the explicit formula for the 3D Neumann Green's function for the unit ball:

$$N(\mathbf{x},\mathbf{y}) = \Phi(\mathbf{x}-\mathbf{y}) + \frac{1}{4\pi}\left[1 - \frac{1}{|\mathbf{y}||\mathbf{x}-\frac{\mathbf{y}}{|\mathbf{y}|^2}|} + \log\left(\frac{|\mathbf{y}|}{2}\sqrt{|\mathbf{x}|^2 - \left(\frac{\mathbf{x}\cdot\mathbf{y}}{|\mathbf{y}|}\right)^2}\right)\right.$$
$$\left. - \operatorname{arctanh}\left(\frac{\mathbf{y}\cdot\mathbf{x}-1}{|\mathbf{y}||\mathbf{x}-\frac{\mathbf{y}}{|\mathbf{y}|^2}|}\right)\right] - \frac{1}{8\pi}|\mathbf{x}|^2. \tag{10.54}$$

10.6. A Physical Illustration in Electrostatics: Coulomb's Law, Gauss's Law, the Electric Field, and Electrostatic Potential

The fact that the Laplacian and harmonic functions have many physical applications should be clear. However, while the previous sections on the fundamental solution and Green's functions are in principle self-contained, these functions and the method of images used to find certain Green's functions may still seem slightly abstract and daunting. They can readily be made concrete by illustrating their presence in the central physical phenomenon of **electrostatics** (the electric field and electric potential due to stationary charge distributions). In fact, the analysis and concepts of this chapter are completely intertwined with electrostatics. From one perspective, this is how the mathematics (PDEs, solutions, and concepts) were first discovered. On the other hand, the subject of electrostatics is pretty much the study of Laplace's equation.

In this section, we will briefly revisit all the previous notions of this chapter from the point of view of electrostatics; in doing so it is instructive to pay attention to the physical dimensions and units of the quantities in question (see Section A.11 if these terms seem foreign). Besides the usual suspects (length, time, and mass), we need the notion (dimension) of charge. The SI unit for charge is a **coulomb**. With this in hand, we have the dimension of **energy per unit charge** which in the SI system is denoted by a **volt** where

$$Volt = \frac{joule}{coulomb} = \frac{newton\ meter}{coulomb}$$

and the dimension of **electrical capacitance** which in the SI system is a **farad** where

$$farad = \frac{coulomb}{volt}.$$

10.6.1. Coulomb's Law and the Electrostatic Force.

Consider two stationary electrical charges at positions $\mathbf{x}_1$ and $\mathbf{x}_2$ in $\mathbb{R}^3$. Coulomb's Law states that *the two charges will repel or attract each other with a force proportional to the square of the distance between them.* Precisely, if q_1 and q_2 represent scalars associated with the magnitude and sign of the respective charges, then the electrostatic force due to charge 1 on charge 2 is given by[7]

$$\mathbf{F}_{1,2} = k_e \frac{q_1 q_2}{|\mathbf{x}_2 - \mathbf{x}_1|^2}\, \mathbf{r}_{1,2},$$

[7]The reader should note the similarity with Newton's Gravitational Law between two bodies. The law is essentially the same and one can rephrase much of this material in terms of gravitational fields.

where $\mathbf{r}_{1,2}$ denotes the unit vector which points from $\mathbf{x}_1$ to $\mathbf{x}_2$, i.e.,

$$\mathbf{r}_{1,2} = \frac{\mathbf{x}_2 - \mathbf{x}_1}{|\mathbf{x}_2 - \mathbf{x}_1|},$$

and k_e is **Coulomb's constant** which, in terms of the **electric constant** ϵ_0, is given by

$$k_e = \frac{1}{4\pi\epsilon_0}. \tag{10.55}$$

The electric constant is the fundamental physical constant of this section; also known as the vacuum permittivity or the permittivity of free space, it is approximately equal to (in SI units) $\epsilon_0 = 8.8541 \times 10^{-12}$ farads per meter. The magnitude of the Coulombic force is

$$|\mathbf{F}_{1,2}| = \frac{1}{4\pi\epsilon_0} \frac{q_1 q_2}{|\mathbf{x}_1 - \mathbf{x}_2|^2}.$$

Now consider a collection of n charges with charge q_i at fixed positions $\mathbf{x}_i$. Next, we place a charge q_0 at an arbitrary location $\mathbf{x} \in \mathbb{R}^3$. The total (net) force on q_0 due to the n charges q_i is

$$\mathbf{F}(\mathbf{x}) = \frac{1}{4\pi\epsilon_0} \sum_{i=1}^{n} \frac{q_i q_0}{|\mathbf{x} - \mathbf{x}_i|^2} \mathbf{r}_i = \frac{1}{4\pi\epsilon_0} \sum_{i=1}^{n} \frac{q_i q_0}{|\mathbf{x} - \mathbf{x}_i|^2} \frac{\mathbf{x} - \mathbf{x}_i}{|\mathbf{x} - \mathbf{x}_i|}.$$

If, instead of point charges, we had a **continuum** of charges based upon a charge density $\rho(\mathbf{x})$, then the resulting force on a charge q_0 at position $\mathbf{x}$ is

$$\mathbf{F}(\mathbf{x}) = \frac{1}{4\pi\epsilon_0} \iiint_{\mathbb{R}^3} \frac{q_0 \rho(\mathbf{y})}{|\mathbf{x} - \mathbf{y}|^2} \mathbf{r}_{\mathbf{y}} \, d\mathbf{y},$$

where $\mathbf{r}_{\mathbf{y}}$ denotes the unit vector from $\mathbf{y}$ to $\mathbf{x}$; i.e.,

$$\mathbf{r}_{\mathbf{y}} = \frac{\mathbf{x} - \mathbf{y}}{|\mathbf{x} - \mathbf{y}|}. \tag{10.56}$$

Note that the integral above is vector valued; each component is obtained by integrating the respective component of the integrand.

10.6.1.1. *The Electric Field and Gauss's Law.* Given n point charges q_i at positions $\mathbf{x}_i$, one defines the electric field generated by n point charges as **the force per unit charge**. Thus, the electric field is a vector field $\mathbf{E}$ where, for any $\mathbf{x} \in \mathbb{R}^3$, $\mathbf{E}(\mathbf{x})$ is the resulting net force per unit charge that a particle would feel at $\mathbf{x}$ as a result of the n point charges. While $\mathbf{E}$ is a three-dimensional vector, the dimensions of any component are force per unit charge. You can think of $\mathbf{E}(\mathbf{x})$ as the resulting net force on a unit charge at $\mathbf{x}$. Specifically, we have

$$\mathbf{E}(\mathbf{x}) = \frac{1}{4\pi\epsilon_0} \sum_{i=1}^{n} \frac{q_i}{|\mathbf{x} - \mathbf{x}_i|^2} \mathbf{r}_{\mathbf{x}_i}.$$

Now consider a continuum of charges described via a charge density $\rho(\mathbf{y})$. To yield the total force on a unit-charged test particle at position $\mathbf{x}$, we consider the force generated by a charge "$\rho(\mathbf{y})\, d\mathbf{y}$" and integrate:

$$\mathbf{E}(\mathbf{x}) = \iiint_{\mathbb{R}^3} \frac{1}{4\pi\epsilon_0} \frac{\rho(\mathbf{y})}{|\mathbf{x}-\mathbf{y}|^2}\, \mathbf{r}_{\mathbf{y}}\, d\mathbf{y},$$

where $\mathbf{r}_{\mathbf{y}}$ is defined by (10.56). Note that the above is a vector-valued integral.

Given **any** electric field $\mathbf{E}$, we can consider the electric flux out of any closed surface $\mathcal{S}$:

$$\iint_{\mathcal{S}} \mathbf{E} \cdot \mathbf{n}\, dS,$$

where $\mathbf{n}$ denotes the unit outer normal to $\mathcal{S}$. One of the fundamental laws of electrostatics is

Gauss's Law

The net electric flux through any closed surface $\mathcal{S}$ is equal to $\frac{1}{\epsilon_0}$ times the net electric charge within $\mathcal{S}$.

10.6.2. The Electrostatic Potential: The Fundamental Solution and Poisson's Equation. Suppose we are under the presence of an electric field $\mathbf{E}$ in $\mathbb{R}^3$ (generated by either a discrete collection or continuum of charges). What is the **work done by the field** in moving a unit charge from point $\mathbf{a}$ to point $\mathbf{b}$? The work done is the line integral

$$\int_{\mathcal{C}} \mathbf{E} \cdot d\mathbf{s},$$

where $\mathcal{C}$ denotes the path curve which goes from $\mathbf{a}$ to $\mathbf{b}$. But will this line integral depend on the particular choice of path? Recall the important result in advanced calculus where a vector field is independent of path if and only if it is a **gradient vector field**; that is, it is the gradient of some scalar function Φ. We refer to such vector fields as **conservative**. In the case of a conservative vector field $\mathbf{E}$, there exists a function Φ such that $\mathbf{E} = -\nabla\Phi$ and for any path (curve) $\mathcal{C}$ from $\mathbf{a}$ to $\mathbf{b}$,

$$\int_{\mathcal{C}} \mathbf{E} \cdot d\mathbf{s} = \Phi(\mathbf{a}) - \Phi(\mathbf{b}).$$

We call the function Φ the **potential** associated with the vector field. Note that the potential is only unique up to an additive constant. This notion of path independence greatly simplifies the quest to solve for a general electric field which entails three component functions: Now it only depends on finding one function, the potential, which induces the electric field through its gradient.

The electric field is a conservative field, and the associated potential is known either as the **electric potential or electrostatic potential**. For example, consider the case of an electric field generated by one charge q_0 at position $\mathbf{x}_0$,

$$\mathbf{E}(\mathbf{x}) = \frac{1}{4\pi\epsilon_0} \frac{q_0}{|\mathbf{x}-\mathbf{x}_0|^2} \frac{\mathbf{x}-\mathbf{x}_0}{|\mathbf{x}-\mathbf{x}_0|}.$$

A quick calculation (check this) shows that $\mathbf{E} = -\nabla\Phi_{\mathbf{x}_0}$ where

$$\boxed{\Phi_{\mathbf{x}_0}(\mathbf{x}) = \frac{1}{4\pi\epsilon_0}\frac{q_0}{|\mathbf{x}-\mathbf{x}_0|}.} \tag{10.57}$$

This should look familiar! Modulo the sign and the factor $\frac{q_0}{\epsilon_0}$, it is **the fundamental solution of the Laplacian** with singularity at $\mathbf{x}_0$. The reader should check that the physical dimension of the electrostatic potential $\Phi_{\mathbf{x}_0}$ is energy per unit charge (a **volt** in the SI system). Consequently, the difference between the electrostatic potential at two points is called **voltage**. At any given point $\mathbf{x} \in \mathbb{R}^3$, $\Phi_{\mathbf{x}_0}(\mathbf{x})$ is the work done against the electric field $\mathbf{E}$ in moving a unit charge from "infinitely far off" to the point $\mathbf{x}$. This is a good way to envision the fundamental solution.

For the electric field generated by a continuum of charges $\rho(\mathbf{y})$, which gives the density of charge at every point $\mathbf{y}$, we must *sum* all the corresponding electric potentials generated by the amount of charge at $\mathbf{y}$. In this continuum setting, this *sum* is an integral, and one finds that the total electric field $\mathbf{E} = -\nabla\Phi_\rho$ where

$$\boxed{\Phi_\rho(\mathbf{x}) = \frac{1}{4\pi\epsilon_0}\iiint_{\mathbb{R}^3}\frac{\rho(\mathbf{y})}{|\mathbf{x}-\mathbf{y}|}\,d\mathbf{y}.} \tag{10.58}$$

Next, we see that the connection between the distributional Laplacian of the fundamental solution and the delta function, as well as the solution to Poisson's equation, are simple consequences of Gauss's Law. Indeed, in the context of a continuous charge density ρ defined on $\mathbb{R}^3$, Gauss's Law asserts that for any bounded domain $\Omega \subset \mathbb{R}^3$,

$$\iint_{\partial\Omega}\mathbf{E}\cdot\mathbf{n}\,dS = \frac{1}{\epsilon_0}\iiint_\Omega \rho(\mathbf{y})\,d\mathbf{y}.$$

We can apply the Divergence Theorem to conclude that

$$\iint_{\partial\Omega}\mathbf{E}\cdot\mathbf{n}\,dS = \iiint_\Omega \operatorname{div}\mathbf{E}\,d\mathbf{y} = \frac{1}{\epsilon_0}\iiint_\Omega \rho(\mathbf{y})\,d\mathbf{y}.$$

Noting that $\mathbf{E} = -\nabla\Phi_\rho$ with Φ_ρ given by (10.58), we find

$$\iiint_\Omega \Delta\Phi_\rho\,d\mathbf{y} = -\frac{1}{\epsilon_0}\iiint_\Omega \rho(\mathbf{y})\,d\mathbf{y}$$

for all domains $\Omega \subset \mathbb{R}^3$. By the IPW Theorem, this implies

$$\Delta\Phi_\rho = -\frac{1}{\epsilon_0}\rho,$$

which is Poisson's equation. The reader should recall from Section 10.2.1 that the derivation of its solution yielded exactly (10.58). We can rewrite this equation in terms of $\mathbf{E} = -\nabla\Phi_\rho$ to find

$$\operatorname{div}\mathbf{E} = \frac{1}{\epsilon_0}\rho,$$

which is the differential version of Gauss's Law and which constitutes one of the components of **Maxwell's equations** (cf. Section 4.1.1).

For the case of a point charge of magnitude q_0 at position $\mathbf{x}_0$, where the electric potential $\Phi_{\mathbf{x}_0}$ given by (10.57) is essentially the fundamental solution, we can interpret Gauss's Law as

$$\text{“}\iiint_\Omega \Delta\Phi_{\mathbf{x}_0}\, d\mathbf{y}\text{”} = \begin{cases} -\frac{q_0}{\epsilon_0} & \text{if } \mathbf{x}_0 \in \Omega, \\ 0 & \text{if } \mathbf{x}_0 \notin \Omega. \end{cases}$$

This translates as

$$\Delta\Phi_{\mathbf{x}_0} = -\frac{q_0}{\epsilon_0}\delta_{\mathbf{x}_0} \qquad \text{in the sense of distributions,}$$

or, in terms of the electric field $\mathbf{E} = -\nabla\Phi_{\mathbf{x}_0}$,

$$\operatorname{div}\mathbf{E} = \frac{q_0}{\epsilon_0}\delta_{\mathbf{x}_0} \qquad \text{in the sense of distributions.}$$

10.6.3. Green's Functions: Grounded Conducting Plates, Induced Charge Densities, and the Method of Images. We now turn to the notions surrounding Green's functions with Dirichlet boundary conditions. For illustration, we focus on the half-space of Section 10.4.1. Consider an **infinite, grounded, thin conducting plate** which lies on the xy-plane. Suppose we have a point charge at position $\mathbf{x}_0$ above the plate. We wish to find the resulting electrostatic potential at all points $\mathbf{x} = (x, y, z)$ above this conducting plate. The term **grounded** refers to the fact that the potential vanishes on the plate. Because of the effects of the grounded plate, the potential is not simply given by (10.57); rather, it is given by the solution to

$$\begin{cases} -\Delta\Phi_{\text{plate}}(x,y,z) = \frac{q}{\epsilon_0}\delta_{\mathbf{x}_0} & \text{in the sense of distributions on } \{(x,y,z)\,|\,z>0\}, \\ \Phi_{\text{plate}}(x,y,0) = 0. & \end{cases}$$

We have solved this problem before! The solution is

$$\boxed{\Phi_{\text{plate}}(x,y,z) = -\frac{q}{\epsilon_0}G(\mathbf{x},\mathbf{x}_0),} \tag{10.59}$$

where G is the Green's function for the half-space derived in Section 10.4.1. Note that G has dimensions of length^{-1}.

Substituting the form of this solution yields

$$\begin{aligned} \Phi_{\text{plate}}(x,y,z) = \frac{1}{4\pi\epsilon_0}\Bigg(&\frac{q}{\sqrt{(x-x_0)^2+(y-y_0)^2+(z-z_0)^2}} \\ &+ \frac{-q}{\sqrt{(x-x_0)^2+(y-y_0)^2+(z+z_0)^2}}\Bigg), \end{aligned} \tag{10.60}$$

informing us that the grounding of the xy-plane is **physically equivalent** to the presence of a second opposing charge reflected across the plane (with nothing being explicitly grounded at all!). This second charge is often called a **ghost** or **image charge** (cf. Figure 10.6).

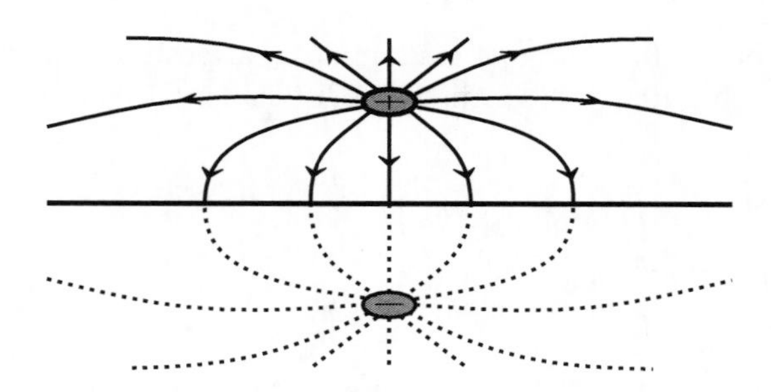

Figure 10.6. Using a ghost charge (−) to simulate the effect of a grounded plate on a charge (+).

Likewise, for the problem of a grounded ball, the grounding is physically equivalent to the existence of an image charge with an altered charge and location as prescribed by the symmetry of the problem (cf. Section 10.4.2). The exploitation of this equivalence is often referred to as **the method of images**.

This physical equivalence of systems can actually yield quantitatively new information about our solution. For example, in a scenario of two dual charges mirroring each other across the xy-plane, one may note that the total charge of the configuration is zero. How does this charge manifest itself in the physically equivalent situation of a grounded plane? There must be a charge build-up somewhere to cancel the point charge about the plane. But how can this be? Poisson's equation tells us that the only charge density is that of the delta function at $\mathbf{x}_0$. The key lies in the fact that Poisson's equation pertains only to charge density **per unit volume**. Charge will build up on the infinitely thin xy-plane with a charge density **per unit area** which one can show is given by

$$\sigma(x,y) := \epsilon_0 \frac{\partial \Phi_{\text{plate}}}{\partial \mathbf{n}} = -\epsilon_0 \frac{\partial \Phi_{\text{plate}}}{\partial z}(x,y,0) \tag{10.61}$$

where $\mathbf{n}$ is the outward pointing normal.[8] We call σ the **induced charge density**.

Now comes the amazing part. If there is indeed a true physical correspondence between the grounded plane and the dual charges, the total induced surface charge, Q, on our plate should exactly cancel the source charge q at position $\mathbf{x}_0$. This proves to be exactly the case since

$$\begin{aligned}\iint_{\partial\mathcal{H}} \sigma(x,y)\,dS &= \iint_{\partial\mathcal{H}} \epsilon_0 \frac{\partial \Phi_{plate}}{\partial \mathbf{n}}\,dS = \iint_{\partial\mathcal{H}} \epsilon_0 \nabla \Phi_{plate} \cdot \mathbf{n}\,dS \\ &= -q \iint_{\partial\mathcal{H}} \nabla_{\mathbf{x}} G(\mathbf{x},\mathbf{x}_0)\cdot \mathbf{n}\,dS \\ &= \text{``} -q \iiint_{\mathcal{H}} \underbrace{\Delta_{\mathbf{x}} G(\mathbf{x},\mathbf{x}_0)}_{\delta_{\mathbf{x}_0}}\,d\mathbf{x}\text{''} = -q. \end{aligned} \tag{10.62}$$

Note that the use of the Divergence Theorem in the last line is in quotation marks but, as usual, it can be justified by reverting to the sense of distributions. Thus, the fact that

[8]See any text on undergraduate electromagnetism for a derivation of this from Gauss's Law, for example, **David J. Griffiths**, Introduction to Electrodynamics, fourth edition, Pearson.

the normal derivative of the Green's function always integrates to unity is a manifestation of the global conservation of charge. This property extends to any domain Ω that defines a Green's function; that is, $\iint_{\partial\Omega} \frac{\partial G}{\partial \mathbf{n}}\, dS = 1$. Hence, even if the domain is not sufficiently symmetric to afford a simple analytical image charge configuration, one always exists to cancel the source.

10.6.4. Interpreting the Solution Formula for the Dirichlet Problem. For a domain $\Omega \subset \mathbb{R}^3$, recall the Dirichlet problem

$$\begin{cases} \Delta\Phi_D = 0 & \text{on } \Omega, \\ \Phi_D = g & \text{on } \partial\Omega. \end{cases}$$

Note the change in notation: We now use Φ_D (D for Dirichlet) in place of u. In our present context, the problem amounts to finding the electrostatic potential in Ω in a scenario where the following hold: (i) There is no charge inside Ω, but (ii) the potential on the boundary is held fixed at g. Recall from (10.15) that the solution is given for any $\mathbf{x}_0 \in \Omega$ by

$$\Phi_D(\mathbf{x}_0) = \iint_{\partial\Omega} \frac{\partial G}{\partial \mathbf{n}}(\mathbf{y}, \mathbf{x}_0)\, g(\mathbf{y}) dS_{\mathbf{y}}, \tag{10.63}$$

where $G(\mathbf{y}, \mathbf{x}_0)$ is the Green's function for Ω associated with a **unit** charge ($q = 1$) at $\mathbf{x}_0$; that is,

$$G(\mathbf{y}, \mathbf{x}_0) = -\frac{\epsilon_0}{q}\Phi_{\mathbf{x}_0}(\mathbf{y}) = -\epsilon_0\Phi_{\mathbf{x}_0}(\mathbf{y})$$

where $\Phi(\mathbf{y})$ solves

$$\begin{cases} -\Delta\Phi_{\mathbf{x}_0}(\mathbf{y}) = \frac{q}{\epsilon_0}\delta_{\mathbf{x}_0} = \frac{1}{\epsilon_0}\delta_{\mathbf{x}_0} & \text{in the sense of distributions on } \Omega, \\ \Phi_{\mathbf{x}_0}(\mathbf{y}) = 0 & \text{for } \mathbf{y} \in \partial\Omega. \end{cases}$$

Let us now use the previous discussion to interpret formula (10.63). First note that

$$\frac{\partial G}{\partial \mathbf{n}}(\mathbf{y}, \mathbf{x}_0)\, dS_{\mathbf{y}}$$

is dimensionless. In the previous subsection, we noted that if a unit charge is placed inside a region Ω with grounded boundary, a surface charge accumulates along $\partial\Omega$ with charge density given by the normal derivative of the Green's function evaluated along this surface. Equation (10.63) gives the potential at a point $\mathbf{x}_0 \in \Omega$ as a **weighted average** of the boundary's potential g; the weight is determined by the charge density which a unit source charge at $\mathbf{x}_0$ would induce along the boundary. As you bring the point $\mathbf{x}_0$ closer and closer to the boundary, the induced charge density will concentrate at that point! In other words, for any $\mathbf{x}_0 \in \partial\Omega$, we have

$$\lim_{\mathbf{x} \to \mathbf{x}_0} \frac{\partial G}{\partial \mathbf{n}}(\mathbf{y}, \mathbf{x}) = \delta_{\mathbf{x}_0}$$

in the sense of distributions on $\partial\Omega$, i.e., with respect to test functions defined on $\mathbf{y} \in \partial\Omega$.

It is a simple, yet useful, exercise (cf. Exercise **10.17**) to perform a **dimensional accounting** of this section. This amounts to the following: (i) finding the physical dimensions (and SI units) of all the dependent and independent variables and (ii) for each equality, ensuring that the dimensions of the left- and right-hand sides are in agreement.

10.7. Chapter Summary

- All solutions to PDEs and BVPs involving the Laplacian involve a certain **averaging of data**. Key to this averaging is the **fundamental solution**, which in 3D is $\Phi(\mathbf{x}) = -\frac{1}{4\pi|\mathbf{x}|}$.

- The fundamental solution is different for each space dimension and has the property that its pointwise Laplacian is 0, except at one point (the origin); however its singularity at the origin produces a delta function when we consider the Laplacian in the sense of distributions. Precisely, we have $\Delta\Phi(\mathbf{x}) = \delta_{\mathbf{0}}$ in the sense of distributions. In this manner, we say its distributional Laplacian is **concentrated at a point**. It is convenient to translate the fundamental singularity from **0** to a **source point** $\mathbf{x}_0$ and consider in 3D $\Phi(\mathbf{x}-\mathbf{x}_0) = -\frac{1}{4\pi|\mathbf{x}-\mathbf{x}_0|}$. In this case, we have $\Delta\Phi(\mathbf{x}-\mathbf{x}_0) = \delta_{\mathbf{x}_0}$ in the sense of distributions.

- The fundamental solution allows us to solve many problems involving the Laplacian. We focused on two of these problems.
(i) Poisson's equation $\Delta u = f$ is easily solved in all of space by the convolution of the fundamental solution with the function f.
(ii) Boundary value problems involving harmonic functions (the Dirichlet and Neumann Problems) can be solved via modified fundamental solutions with boundary conditions (Green's functions). Key to this is the representation formula (Theorem 10.3), which tells us that the values of any harmonic function u at a point $\mathbf{x}_0$ inside the domain Ω can be found by the **weighted averages** of the values and derivatives of u, on the boundary $\partial\Omega$. The precise weights are given by the fundamental solution (with singularity at $\mathbf{x}_0$) and its derivatives.

- We mainly focused on **Green's functions** with the Dirichlet boundary condition on a domain Ω. Precisely, the Green's function with source point $\mathbf{x}_0$ is the function of $\mathbf{x}$, written $G(\mathbf{x},\mathbf{x}_0)$, which satisfies the following: **(i)** $\Delta_{\mathbf{x}} G(\mathbf{x},\mathbf{x}_0) = \delta_{\mathbf{x}_0}$ in the sense of distributions on Ω and **(ii)** $G(\mathbf{x},\mathbf{x}_0) = 0$ for all $\mathbf{x} \in \partial\Omega$. Such functions can only be found explicitly in a few cases of highly symmetric domains Ω. Two standard cases which we visited are the half-space (or half-plane in 2D) and the ball (or disc in 2D). In these instances, the Green's function is found by adding to the fundamental solution another fundamental solution whose singularity lies outside Ω. This outside singularity is a certain type of reflection of $\mathbf{x}_0$ and is chosen to "cancel out" the value of the fundamental solution for $\mathbf{x} \in \partial\Omega$. Note that in all cases the Green's function is indeed a function (you can define it to be any value at the singularity). Its blow-up at the singularity is, however, key to its character and resembles the blow-up of the fundamental solution.
  The notion of a Green's function extends to many other boundary conditions, for example the Neumann and periodic boundary conditions. We briefly discussed the Neumann boundary condition.

- Via the electrostatic potential, **electrostatics** provides an **ideal physical paradigm** in which to illustrate this material. In particular, we have the following table:

Mathematical object associated with the Laplacian	**Corresponding quantity in electrostatics (up to physical constants and sign)**
The **fundamental solution** $\frac{-1}{4\pi\|\mathbf{x}-\mathbf{x}_0\|}$ in $\mathbb{R}^3$	The **electrostatic potential** at a point $\mathbf{x}$ associated with a **unit charge** at $\mathbf{x}_0$.
The solution to **Poisson's equation** $\Delta u = f$ on $\mathbb{R}^3$: $u(\mathbf{x}) = \iiint_{\mathbb{R}^3} -\frac{f(\mathbf{y})}{4\pi\|\mathbf{x}-\mathbf{y}\|}\, d\mathbf{y}$	The **electrostatic potential** in $\mathbb{R}^3$ associated with a **continuous charge distribution** given by f. We integrate (add up) the potentials associated with charge $f(\mathbf{y})$ at position $\mathbf{y}$.
The **Green's function** $G(\mathbf{x}, \mathbf{x}_0)$ associated with domain Ω	The **electrostatic potential** at point $\mathbf{x} \in \Omega$ associated with a **point charge** at $\mathbf{x}_0$ inside Ω with a **grounded, conducting** boundary $\partial\Omega$.
The solution to the **Dirichlet problem** $\begin{cases} \Delta u = 0 & \text{in } \Omega, \\ u = g & \text{on } \partial\Omega \end{cases}$	The **electrostatic potential** in a domain Ω which is free of any charges but whose boundary is held at **a fixed potential** g.

Exercises

10.1 **(a)** Let $\mathbf{x} = (x_1, x_2, x_3)$ and consider the function defined for $\mathbf{x} \neq \mathbf{0}$, $u(\mathbf{x}) = \frac{1}{|\mathbf{x}|}$. Show by direct differentiation that for $\mathbf{x} \neq \mathbf{0}$, $\Delta u(\mathbf{x}) = 0$. Thus, u is a radially symmetric *harmonic* function away from the origin.
(b) Now consider the same function $\frac{1}{|\mathbf{x}|}$ in two dimensions. Show that for $\mathbf{x} \neq \mathbf{0}$, $\Delta u(\mathbf{x}) \neq 0$.

10.2 **(a)** Analogous to the 3D theorem, Theorem 10.1, prove that $\Delta\Phi(\mathbf{x}) = \delta_0$ in the sense of distributions in 2D, where $\Phi(\mathbf{x})$ is the fundamental solution in 2D defined in Definition 8.5.1.
(b) Prove the analogous statement in dimensions $N > 3$.

10.3 Consider the Green's function for the 3D Δ on a domain Ω. What can you say about the sign of G; i.e., is it always positive, always negative, or does the answer depend on the source point $\mathbf{x}_0$? Prove your answer. **Hint:** Use the Maximum Principle for harmonic functions in appropriate regions of space.

10.4 Write down the analogous representation formula in Theorem 10.3 in 2D and prove it.

10.5 Prove the symmetry of the Green's function (Theorem 10.7) by using Green's Second Identity (A.17) in an appropriate region of space.

10.6 Prove Theorem 10.5. Hint: Apply Green's Second Identity (A.17) to $H(\mathbf{x})$ and u in Ω and "add" the result to the representation formula.

10.7 Prove Theorem 10.8; the basic steps are very similar to the proof for the ball. You may assume (10.28) holds true.

10.8 Starting with the fundamental solution for Δ in dimension $N = 2$ (cf. Definition 8.5.1), find the Green's function for the open disc of radius a: $D = \{(x, y) | x^2 + y^2 < a^2\}$. Use the Green's function to show that the solution to the Dirichlet problem in D, $\Delta u = 0$ in D, $u = g$ on ∂D, is given by

$$u(\mathbf{x}_0) = \frac{1}{2\pi a} \int_{|\mathbf{x}|=a} \frac{a^2 - |\mathbf{x}_0|^2}{|\mathbf{x} - \mathbf{x}_0|^2} g(\mathbf{x})\, dS_{\mathbf{x}}.$$

This is known as the Poisson formula in 2D. Here g is a continuous function on the boundary circle.

10.9 Let $\mathbf{x} \in \mathbb{R}^N$ and consider the function defined for $\mathbf{x} \neq \mathbf{0}$, $u(\mathbf{x}) = \frac{1}{|\mathbf{x}|}$.
(a) For which dimensions N is u locally integrable? That is, for which dimensions N is $\int \cdots \int_{B(\mathbf{0},1)} u(\mathbf{x})\, d\mathbf{x} < \infty$?
(b) For these values of N, we can interpret u in the sense of distributions. For $N = 3$, we showed that $\Delta u = -4\pi\delta_{\mathbf{0}}$ in the sense of distributions. For the other values of N in part (a), describe (compute) Δu in the sense of distributions.

10.10 Let G_1 be the Green's function for the 3D Laplacian Δ on $B(\mathbf{0}, 1) \subset \mathbb{R}^3$ and let G_2 be the Green's function for the 3D Laplacian Δ on $B(\mathbf{0}, 2) \subset \mathbb{R}^3$. Let $\mathbf{x}_0 \in B(\mathbf{0}, 1)$. Prove that for all $\mathbf{x} \in B(\mathbf{0}, 1)$ such that $\mathbf{x} \neq \mathbf{x}_0$, $G_1(\mathbf{x}, \mathbf{x}_0) > G_2(\mathbf{x}, \mathbf{x}_0)$. **Hint:** Use Exercise **10.3**.

10.11 (a) Find the Green's function for the 1D Laplacian $\frac{d^2}{dx^2}$ on $(-2, 3)$.
(b) Does there exist a Neumann function for $\frac{d^2}{dx^2}$ on $(-1, 1)$?

10.12 (a) Find the fundamental solution of the operator $\frac{d}{dx} + a$, for any $a \neq 0$. That is, find a function $\Phi(x)$ such that $\Phi' + a\Phi = \delta_0$ in the sense of distributions.
(b) Use the fundamental solution to write down the solution to the ODE $u'(x) + au(x) = f(x)$, where $f \in C_c^1(\mathbb{R})$.

10.13 Prove that (10.9) is the only solution to (10.8) which remains bounded as $|\mathbf{x}| \to \infty$.

10.14 Let $a > 0$ and $f \in C_c^2(\mathbb{R}^3)$. Derive an explicit formula for the solution to $\Delta u + au = f$ in $\mathbb{R}^3$. Hint: First find the fundamental solution $\Phi(\mathbf{x})$, where $\Delta\Phi + a\Phi = \delta_{\mathbf{0}}$ in the sense of distributions, by looking for radially symmetric solutions of $\Delta u + au = 0$.

10.15 **(The Solution to Poisson's Equation from the Solution to the Diffusion Equation)** Let g be a continuous function with compact support in $\mathbb{R}^3$ and let $u(\mathbf{x}, t)$ be a solution to the 3D diffusion equation $u_t = \Delta u$ on $\mathbb{R}^3 \times [0, \infty)$ with $u(\mathbf{x}, 0) = g(\mathbf{x})$.

(a) Recall the explicit formula for u from Section 7.9 and use it to show that for all $\mathbf{x}$, $\int_0^\infty |u(\mathbf{x},t)|\,dt < \infty$. Hence, v defined by $v(\mathbf{x}) := \int_0^\infty u(\mathbf{x},t)\,dt$ is well-defined. Moreover, use your calculation to conclude that v is bounded; i.e., there exists a constant C such that $|v(\mathbf{x})| < C$ for all $\mathbf{x} \in \mathbb{R}^3$.
(b) Show that v solves the Poisson equation $\Delta v = -g(\mathbf{x})$ on $\mathbb{R}^3$.
(c) Use the explicit formula for u (hence v) to recover the usual solution formula for Poisson's equation in terms of the 3D fundamental solution for Δ.

10.16 (**Dirichlet to Neumann Map**) Given a smooth function g on $\partial\Omega$, let u_g denote the unique solution to the Dirichlet problem (10.14). Now consider the mapping which takes

$$g \longrightarrow \frac{\partial u_g}{\partial \mathbf{n}} \quad \text{on } \partial\Omega.$$

This map, which takes a function defined on $\partial\Omega$ to another function defined on $\partial\Omega$, is known as the **Dirichlet to Neumann map**. It is an example of a Poincaré-Steklov operator and turns out to be very important in many inverse problems.
(a) Find the physical significance (meaning) of this map in both the contexts of heat (temperature) flow and electrostatics. In the latter case, address why the Dirichlet to Neumann map is referred to as the *voltage to current density* map.
(b) Let $\Omega = B(\mathbf{0},1) \subset \mathbb{R}^3$. Using the Poisson formula for the solution u_g write down an explicit form of the Dirichlet to Neumann map.

10.17 Perform a **dimensional accounting** of Section 10.6; that is, in SI units find the physical dimensions of **all** dependent (and independent) variables, and for each equality, make sure the dimensions agree.

10.18 Let Ω be the complement of $\overline{B(\mathbf{0},1)} \subset \mathbb{R}^3$; that is, $\Omega = \{\mathbf{x} \in \mathbb{R}^3 \,|\, |\mathbf{x}| > 1\}$. Find the Green's function for the Laplacian with Dirichlet boundary conditions on Ω.

10.19 Let Ω^+ be the rotated half-space; i.e., for any fixed $a, b, c \in \mathbb{R}$ (at least one of them nonzero) the domain $\Omega^+ := \{(x,y,z) \in \mathbb{R}^3 \,|\, ax + by + cz > 0\}$ with boundary $\partial\Omega^+ := \{(x,y,z) \in \mathbb{R}^3 \,|\, ax + by + cz = 0\}$. Find the Green's function for the Laplacian with Dirichlet boundary conditions on Ω^+.

10.20 Consider the half-ball $B^+ = \{(x,y,z) \in \mathbb{R}^3 \,|x^2 + y^2 + z^2 < a, z > 0\}$. Note ∂B^+ consists of a hemisphere and a disc lying in the xy-plane. Find the Green's function for the Laplacian with Dirichlet boundary conditions on B^+.

10.21 Find the Green's function with Dirichlet boundary conditions for the operator $\Delta + c$ on the 3D half-space. Use it to write down the solution to $\Delta u + cu = 0$ in $\mathcal{H}$ with $u = g$ on $\partial\mathcal{H}$.

10.22 This exercise illustrates the usefulness of Dirac's intuitive approach (cf. Section 5.6). Let $a_i > 0$ and consider the modified Laplacian in $\mathbb{R}^3$:

$$Lu := a_1^2 u_{x_1x_1} + a_2^2 u_{x_2x_2} + a_3^2 u_{x_3x_3}.$$

We want you to find the fundamental solution of L, i.e., a function $\Phi(\mathbf{x})$ such that $L\Phi = \delta_0$ in the sense of distributions on $\mathbb{R}^3$. As with the Laplacian, this fundamental solution will not be unique unless we impose some other condition; hence, impose the decay condition $\Phi(\mathbf{x}) \to 0$ as $\mathbf{x} \to \infty$.

(a) Let us first proceed informally as Dirac would. First recall Exercise **9.10** where you were asked to justify that if $\mathbf{A}$ is a 3×3 matrix such that $\det \mathbf{A} \neq 0$, then "$\delta_{\mathbf{0}}(\mathbf{A}\mathbf{x}) = \frac{1}{|\det \mathbf{A}|}\,\delta_{\mathbf{0}}(\mathbf{x})$". Assume this is the case and change variables with $y_i = x_i/a_i$ in the following informal equation:

$$\text{“}a_1^2\Phi_{x_1x_1} + a_2^2\Phi_{x_2x_2} + a_3^2\Phi_{x_3x_3} = \delta_0(x_1)\delta_0(x_2)\delta_0(x_3).\text{”}$$

By reducing to the fundamental solution for the usual Laplacian, find $\Phi(\mathbf{x})$.

(b) With Φ in hand, verify that it is indeed a distributional solution to $L\Phi = \delta_{\mathbf{0}}$.

10.23 By explicitly computing the integral in (10.52) arrive at formula (10.54). Hint: Look at the appendix of the article of Wirth.

10.24 Find an explicit formula for the Neumann function of the unit disk in 2D.

10.25 Consider the BVP for Δ on a bounded domain Ω with Robin boundary conditions. Define the Green's function associated with this BVP.

10.26 Suppose $\mathcal{S}$ is a closed surface containing one point charge of magnitude q. State precisely and then prove **Gauss's Law** for this scenario.

10.27 **(Green's Reciprocity Theorem)** In each of the following scenarios, consider two entirely **separate** universes.

(a) Suppose that in the first universe a charge density $\rho_1(\mathbf{x})$ produces a potential $\Phi_{\rho_1}(\mathbf{x})$ for $\mathbf{x} \in \mathbb{R}^3$, while in the second universe a charge density $\rho_2(\mathbf{x})$ produces a potential $\Phi_{\rho_2}(\mathbf{x})$. Prove Green's Reciprocity Theorem:

$$\iiint_{\mathbb{R}^3} \rho_1(\mathbf{x})\,\Phi_{\rho_2}(\mathbf{x})\,d\mathbf{x} = \iiint_{\mathbb{R}^3} \rho_2(\mathbf{x})\,\Phi_{\rho_1}(\mathbf{x})\,d\mathbf{x}.$$

(b) Now consider in each universe an electrical conductor in the shape of $\Omega \subset \mathbb{R}^3$ with conducting boundary surface $\partial\Omega = \mathcal{S}$. In the first universe a continuous charge density $\rho_1(\mathbf{x})$ in Ω produces a potential $\Phi_{\rho_1}(\mathbf{x})$ for $\mathbf{x} \in \Omega$ and a resulting surface charge density $\sigma_1(\mathbf{x})$ on the boundary $\mathcal{S}$. In the second universe, $\rho_2(\mathbf{x})$ produces a potential $\Phi_{\rho_2}(\mathbf{x})$ for $\mathbf{x} \in \Omega$ and a resulting surface charge density $\sigma_2(\mathbf{x})$ on $\mathcal{S}$. Prove Green's Reciprocity Theorem:

$$\iiint_{\Omega} \rho_1(\mathbf{x})\,\Phi_{\rho_2}(\mathbf{x})\,d\mathbf{x} + \iint_{\mathcal{S}} \sigma_1(\mathbf{x})\,\Phi_{\rho_2}(\mathbf{x})\,dS$$
$$= \iiint_{\Omega} \rho_2(\mathbf{x})\,\Phi_{\rho_1}(\mathbf{x})\,d\mathbf{x} + \iint_{\mathcal{S}} \sigma_2(\mathbf{x})\,\Phi_{\rho_1}(\mathbf{x})\,dS.$$

10.28 Following the electrostatics interpretation of Section 10.6, give a physical interpretation of the Neumann Green's function for the half-space and the method of images used to find it.

10.29 Here we explore the physical interpretation of (10.46) in the context of electricity. Explain why we refer to $\operatorname{div}_{\mathbf{y}}(\mathbf{e} \cdot \delta_{\mathbf{y}})$ as a **dipole moment**. Interpret the solution to the Poisson equation with dipole moment right-hand side. Interpret this in the context of (10.46) with vector $\mathbf{e}$ and function $N(\mathbf{x}, \mathbf{y})$. By researching online, connect this problem with **electroencephalography** in medical imaging.

Chapter 11

Fourier Series

> "La chaleur pénètre, comme la gravité, toutes les substances de l'univers, ses rayons occupent toutes les parties de l'espace. Le but de notre ouvrage est d'exposer les lois mathématiques que suit cet element. Cette théorie formera désormais une des branches importantes de la physique générale.."
>
> - **Jean Baptiste Joseph Fourier**[a]
>
> Rough translation: "Heat penetrates, like gravity, all the substances of the universe; its rays occupy all parts of the space. The goal of our work is to expose the mathematical laws that this element follows. This theory will now form one of the important branches of general physics."
>
> [a]From the second paragraph of *Théorie Analytique de la Chaleur*, 1822.

In 1807, Jean Baptiste Joseph Fourier wrote a paper[1] entitled *Mémoire sur la propagation de la chaleur dans les corps solides* which translates as *Memoir on the propagation of heat in solid bodies.* In this article, he expanded on the previous work of, among others, Euler, D'Alembert, and Bernoulli on trigonometric series (sums of sines and cosines) and introduced us to Fourier series. Roughly speaking, his novel approach was to write (almost) any function as an infinite sum of sines or cosines with increasing frequency. Fourier's ideas were met with great skepticism and it was not until 1822 that he published his full treatise in a book entitled *Théorie Analytique de la Chaleur* (*The Analytic Theory of Heat*). There is absolutely no question that Fourier series and

[1]The paper was not well received, and the history surrounding Fourier series and also Fourier's life is very interesting and well worth looking up online.

its extensions, e.g., the Fourier and other transforms, have revolutionized all of science, engineering, and medicine. The above referenced quotation demonstrates that Fourier, in his great wisdom (coupled with a touch of arrogance), was fully aware of this fact!

While it is true that we have already devoted an entire, rather long, chapter to what is in fact a generalization of Fourier series (Chapter 6 on the Fourier transform), we will assume **no knowledge of the Fourier transform** in this chapter. This means that from the point of view of Fourier's vision, we are starting again from scratch with rather different perspectives and motivations from Chapter 6. However, we will end this chapter by reconnecting with the Fourier transform.

The guiding idea that Fourier introduced was to break up (or decompose) a function (a signal in engineering language) into simpler functions. This decomposition into simpler functions works both ways in the sense that one can reconstruct the function from its simpler components. While there are, in general, many classes of simpler functions for the decomposition, the most famous are the trigonometric functions of sine and cosine with certain frequencies. It is tremendously useful and insightful to view Fourier series and its generalizations in the context of what we will call "**infinite-dimensional linear algebra**" and, additionally, the **eigenvalues and eigenfunctions** associated with one-dimensional boundary value problems involving the second derivative. This interpretation will naturally lead into the next chapter on the method of separation of variables.

11.1. • Prelude: The Classical Fourier Series — the Fourier Sine Series, the Fourier Cosine Series, and the Full Fourier Series

If the readers are already familiar with the basic notion of the Fourier sine and cosine series, they may skip this introductory section and proceed directly to Section 11.2.

11.1.1. • The Fourier Sine Series.

We begin by considering a function $\phi(x)$ defined on $(0, l)$ and **attempt** to find coefficients $b_n, n = 1, 2 \dots,$ such that

$$\phi(x) = \sum_{n=1}^{\infty} b_n \sin\left(\frac{n\pi x}{l}\right). \tag{11.1}$$

First note that the π/l factor ensures that all the functions $\sin\left(\frac{n\pi x}{l}\right)$ have nodes (in this case, zeros) at the end points 0 and l. These functions have increasing frequencies (alternatively, their period decreases): It is useful to plot a few instances of particular values of n. Second, let us note what equality means in (11.1). It constitutes a certain form of **convergence**; that is, if we let

$$S_N(x) := \sum_{n=1}^{N} b_n \sin\left(\frac{n\pi x}{l}\right)$$

denote a **partial sum**, then for any $x \in (0, l)$ the sequence of real numbers $\{S_N(x)\}_{N=1}^{\infty}$ converges to the number $\phi(x)$ as $N \to \infty$. Of course, the only way this can possibly happen is if the coefficients b_n are selected based upon the particular function ϕ; indeed, these coefficients must be tailor-made for ϕ.

Right away, we are confronted with the following questions:

(1) Does this really work? That is, by taking more and more suitable linear combinations of the functions $\sin\left(\frac{n\pi x}{l}\right)$, do we obtain a better and better approximation of the function ϕ? In the limit of infinitely many linear combinations, do we get exact accuracy; i.e., does the Fourier sine series at a value $x \in (0, l)$ **converge** to $\phi(x)$?

(2) If so, what are the values of b_n that allow this to work?

(3) What is special about sines and cosines with increasing frequencies which makes this possible?

Questions (1) and (3) require more thought and we will address them shortly. What is straightforward to address now is (2). This is because of a certain property, called orthogonality, that these sines and cosines possess (this provides a partial explanation for question (3)). To this end, first note that

$$\int_0^l \sin\left(\frac{n\pi x}{l}\right)\sin\left(\frac{m\pi x}{l}\right)\,dx = 0, \qquad n \neq m, \tag{11.2}$$

and

$$\int_0^l \left[\sin\left(\frac{n\pi x}{l}\right)\right]^2\,dx = \frac{l}{2}. \tag{11.3}$$

These equalities can be directly checked by performing the integration and the reader is encouraged to verify them; however, as we shall see in Section 11.2.3, they easily follow by a clever application of Green's Second Identity (A.17) in 1D. The equations (11.2) are called **orthogonality relations** and we will shortly motivate why we use the term "orthogonality".

Let us assume (11.1) holds true in an appropriate sense of convergence. Next, we compute as follows:[2]

$$\begin{aligned}
\int_0^l \phi(x)\sin\left(\frac{n\pi x}{l}\right)\,dx &= \int_0^l \sum_{m=1}^{\infty} b_m \sin\left(\frac{n\pi x}{l}\right)\sin\left(\frac{m\pi x}{l}\right)\,dx \\
&= \sum_{m=1}^{\infty} b_m \int_0^l \sin\left(\frac{n\pi x}{l}\right)\sin\left(\frac{m\pi x}{l}\right)\,dx \\
&\overset{(11.2)}{=} b_n \int_0^l \left[\sin\left(\frac{n\pi x}{l}\right)\right]^2\,dx \\
&\overset{(11.3)}{=} \frac{l}{2} b_n.
\end{aligned}$$

Thus,

$$b_n = \frac{2}{l}\int_0^l \phi(x)\sin\left(\frac{n\pi x}{l}\right)\,dx. \tag{11.4}$$

[2]Note that we performed a rather dangerous operation in passing to the second line: We interchanged the order of the **infinite** sum and the integral. This amounts to changing the order of two limits, and it is not always the case that the order does not matter. However, for the Fourier series we will encounter, such an interchange is valid (see Section 11.6.1 for more details).

Definition of the Fourier Sine Series

Definition 11.1.1. Define b_n as in (11.4); then

$$\sum_{n=1}^{\infty} b_n \sin\left(\frac{n\pi x}{l}\right)$$

is called the **Fourier sine series** of the function $\phi(x)$ on $(0, l)$.

11.1.2. • The Fourier Cosine Series. It is natural to ask if we can do the same thing with cosines. The answer is yes! The analogous statements are

$$\phi(x) = \frac{1}{2}a_0 + \sum_{n=1}^{\infty} a_n \cos\left(\frac{n\pi x}{l}\right) \tag{11.5}$$

with the orthogonality relations for cosines given by

$$\int_0^l \cos\left(\frac{n\pi x}{l}\right)\cos\left(\frac{m\pi x}{l}\right) dx = \begin{cases} 0 & n \neq m, \\ \frac{l}{2} & n = m. \end{cases} \tag{11.6}$$

Note that the above also holds true when exactly one of the indices n, m is 0. Hence, the previous arguments yield

$$\int_0^l \phi(x)\cos\left(\frac{n\pi x}{l}\right) dx = \frac{l}{2}a_n, \qquad n = 1, 2, \ldots,$$

while if $n = 0$, we find

$$\int_0^l \phi(x)\, dx = \frac{1}{2}a_0 \int_0^l 1^2\, dx = \frac{l}{2}a_0.$$

Thus we have

$$a_n = \frac{2}{l}\int_0^l \phi(x)\cos\left(\frac{n\pi x}{l}\right) dx, \quad n = 0, 1, \ldots. \tag{11.7}$$

Please note that we included the factor of one half for a_0 in (11.5) in order to have the same formula for a_n hold true for $n = 0$ and for $n \geq 1$: This issue with the $1/2$ can cause confusion so the reader should digest its presence now. Analogous to the Fourier sine series, we may define the **Fourier cosine series**:

Definition of the Fourier Cosine Series

Definition 11.1.2. Define a_n as in (11.7); then

$$\frac{1}{2}a_0 + \sum_{n=1}^{\infty} a_n \cos\left(\frac{n\pi x}{l}\right)$$

is called the **Fourier cosine series** of the function $\phi(x)$ on $(0, l)$.

11.1.3. • The Full Fourier Series. Here we are given ϕ defined on $(-l, l)$ and consider writing

$$\phi(x) = \frac{1}{2}a_0 + \sum_{n=1}^{\infty}\left[a_n \cos\left(\frac{n\pi x}{l}\right) + b_n \sin\left(\frac{n\pi x}{l}\right)\right].$$

First note, if we pick any two functions from

$$\left\{1, \cos\left(\frac{n\pi x}{l}\right), \sin\left(\frac{n\pi x}{l}\right) \;:\; n = 1, 2, \dots\right\}$$

and multiply them and integrate over $(-l, l)$, we will arrive at 0. On the other hand, we have

$$\int_{-l}^{l} \cos^2\left(\frac{n\pi x}{l}\right) dx = l = \int_{-l}^{l} \sin^2\left(\frac{n\pi x}{l}\right) dx \qquad \text{and} \qquad \int_{-l}^{l} 1^2\, dx = 2l.$$

Therefore, the exact same argument as before leads to the following conclusions:

$$a_n = \frac{1}{l}\int_{-l}^{l} \phi(x)\cos\left(\frac{n\pi x}{l}\right) dx, \quad n = 0, 1, \dots, \tag{11.8}$$

and

$$b_n = \frac{1}{l}\int_{-l}^{l} \phi(x)\sin\left(\frac{n\pi x}{l}\right) dx, \quad n = 1, 2, \dots. \tag{11.9}$$

This brings us to what is known as the full Fourier series:

Definition of the Full Fourier Series

Definition 11.1.3. Define a_n and b_n by (11.8) and (11.9); then

$$\frac{1}{2}a_0 + \sum_{n=1}^{\infty}\left[a_n \cos\left(\frac{n\pi x}{l}\right) + b_n \sin\left(\frac{n\pi x}{l}\right)\right]$$

is called the **full Fourier series** of the function $\phi(x)$ on $(-l, l)$.

11.1.4. • Three Examples. We now visit three empirical examples, which at the very least provide some empirical evidence for question (1) of Section 11.1.1; that is, one can indeed approximate any reasonable function to any degree of accuracy by a combination of sines and/or cosines. If you have never seen Fourier series before, surely these three examples should spark your curiosity as to why this works.

Example 11.1.1. We can construct the Fourier sine series for the function $\phi \equiv 1$ on $(0, l)$. The coefficients b_n are given by

$$b_n = \frac{2}{l}\int_{0}^{l} \sin\left(\frac{n\pi x}{l}\right) dx = \frac{2}{n\pi}[1 - (-1)^n] = \begin{cases} \frac{4}{n\pi} & \text{if } n \text{ odd}, \\ 0 & \text{if } n \text{ even}, \end{cases}$$

resulting in the Fourier sine series

$$\sum_{n=1}^{\infty} b_n \sin\left(\frac{n\pi x}{l}\right) = \frac{4}{\pi}\sum_{m=0}^{\infty} \frac{1}{2m+1}\sin\left(\frac{(2m+1)\pi x}{l}\right).$$

On the other hand, for the Fourier cosine series for $\phi(x) = 1$ on $(0, l)$ we find $a_0 = 2$ and $a_n = 0$ for $n \geq 1$. Thus,

$$\frac{1}{2}a_0 + \sum_{n=1}^{\infty} a_n \cos\left(\frac{n\pi x}{l}\right) = 1 + 0 + \cdots + 0 = 1.$$

We obtain here a trivial cosine series precisely because the function 1 is $\cos\left(\frac{n\pi x}{l}\right)$ for $n = 0$. One could also construct the full Fourier series of ϕ defined on $(-l, l)$. However, since the chosen function ϕ is even, we would simply recover the trivial Fourier cosine series: $a_0 = 1$ and $a_n = b_n = 0$ for $n \geq 1$.

Figure 11.1 plots an approximation of the Fourier sine series (with $l = 2$) by considering the sums of the first $N = 2, 5, 10$, and 50 terms in the series (including the zero terms). More precisely, we plot the **partial sums**

$$S_N(x) := \sum_{n=1}^{N} b_n \sin\left(\frac{n\pi x}{l}\right) \quad \text{for } N = 2, 5, 10, \text{ and } 50.$$

What do we observe? On the sampled interval $(0, 2)$, we appear to have convergence. However, over the larger interval $(-4, 4)$, we appear to be converging to the odd extension of 1 on $(-l, 0)$ which is then repeated periodically with period $2l = 4$.

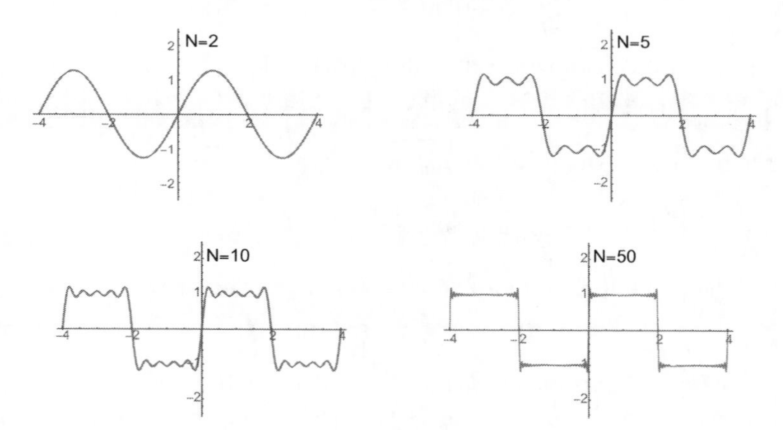

Figure 11.1. The partial sums of the Fourier sine series for $\phi(x) = 1$ on interval $(0, 2)$.

Example 11.1.2. Now we consider the function $\phi(x) = x$ on $(0, l)$. Unlike the previous example where 1 was a cosine with $n = 0$, we will now have a nontrivial Fourier sine and cosine series. The Fourier sine coefficients are

$$b_n = \frac{2}{l}\int_0^l x \sin\left(\frac{n\pi x}{l}\right) dx = \frac{2l(-1)^{n+1}}{n\pi},$$

and the Fourier sine series for x on $(0, l)$ is

$$\sum_{n=1}^{\infty} b_n \sin\left(\frac{n\pi x}{l}\right) = \frac{2l}{\pi}\sum_{n=1}^{\infty} \frac{(-1)^{n+1}}{n} \sin\left(\frac{n\pi x}{l}\right) \quad n = 1, 2, \ldots.$$

The Fourier cosine coefficients are given by

$$a_0 = l \quad \text{and} \quad a_n = \frac{2}{l}\int_0^l x\cos\Big(\frac{n\pi x}{l}\Big)\,dx = \begin{cases} -\frac{4l}{n^2\pi^2} & \text{if } n \text{ odd}, \\ 0 & \text{if } n \text{ even}, \end{cases} \qquad n = 1, 2, \dots,$$

and the Fourier cosine series for x on $(0, l)$ is

$$\frac{1}{2}a_0 + \sum_{n=1}^{\infty} a_n \cos\Big(\frac{n\pi x}{l}\Big) = \frac{l}{2} - \sum_{m=0}^{\infty} \frac{4l}{(2m+1)^2\pi^2}\cos\Big(\frac{(2m+1)\pi x}{l}\Big).$$

Since $\phi(x) = x$ is an odd function, the full Fourier series of $\phi(x) = x$ on $(-l, l)$ is identical to the Fourier sine series of x on $(0, l)$. If this is not clear at this stage, do not worry; it will become clear very shortly.

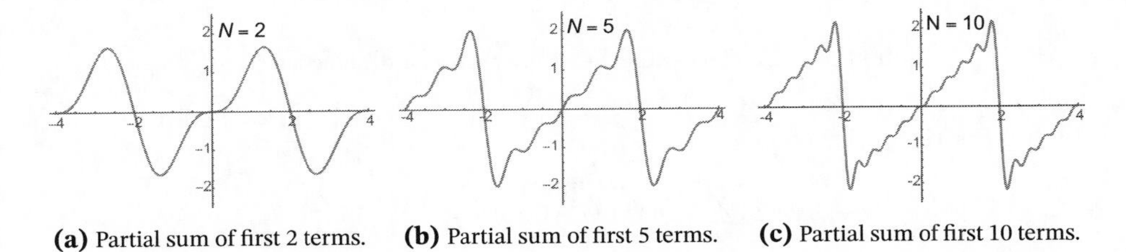

(a) Partial sum of first 2 terms. **(b)** Partial sum of first 5 terms. **(c)** Partial sum of first 10 terms.

Figure 11.2. The partial sums of the Fourier sine series for $\phi(x) = x$ on interval $(0, 2)$.

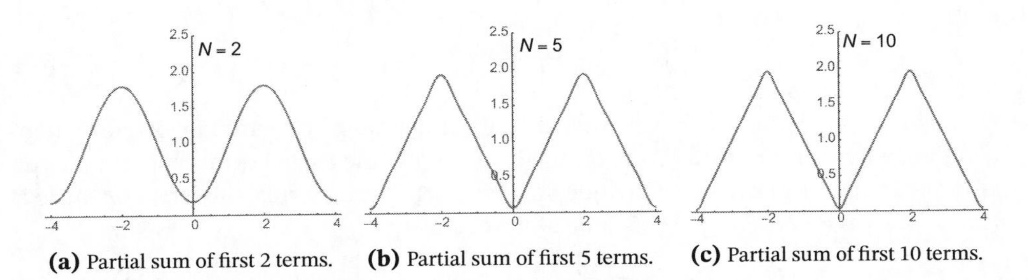

(a) Partial sum of first 2 terms. **(b)** Partial sum of first 5 terms. **(c)** Partial sum of first 10 terms.

Figure 11.3. The figure illustrates the partial sums of the Fourier cosine series for $\phi(x) = x$ on interval $(0, 2)$.

In Figures 11.2 and 11.3, respectively, we plot the partial sums of the Fourier sine and cosine series for $\phi(x) = x$ on $(0, 2)$. Here, we include in our numbering the first (constant) term $\frac{a_0}{2}$. What do we observe? On the sampled interval $(0, 2)$, we appear to have convergence. However, over the larger interval $(-4, 4)$, what seems to emerge is an odd extension of x which is then repeated periodically with period $2l = 4$.

Example 11.1.3. In our last example, we plot several of the partial sums of the full Fourier series for the following function ϕ defined on $(-2, 2)$:

$$\phi(x) = \begin{cases} 0 & \text{if } x \in (-2, 0], \\ x & \text{if } x \in (0, 2). \end{cases} \tag{11.10}$$

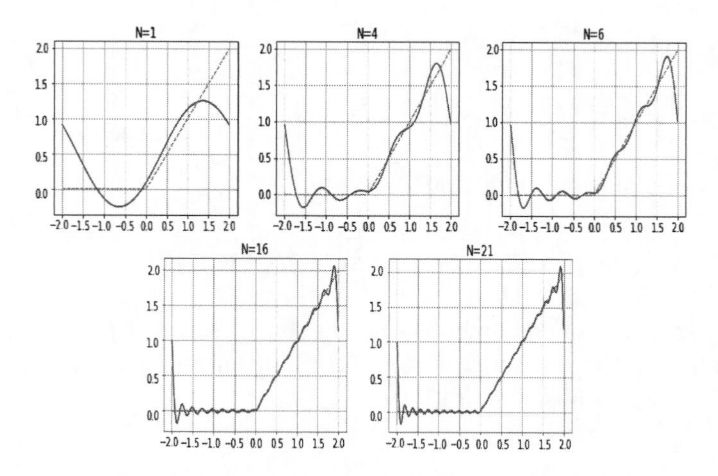

Figure 11.4. Plots of the partial full Fourier sums for the function (11.10).

The full Fourier series is

$$\frac{1}{2}a_0 + \sum_{n=1}^{\infty}\left[a_n \cos\left(\frac{n\pi x}{2}\right) + b_n \sin\left(\frac{n\pi x}{2}\right)\right]$$

where

$$a_n = \frac{1}{2}\int_{-2}^{2} \phi(x)\cos\left(\frac{n\pi x}{2}\right)\,dx, \quad n = 0, 1, \ldots,$$

and

$$b_n = \frac{1}{2}\int_{-2}^{2} \phi(x)\sin\left(\frac{n\pi x}{2}\right)\,dx, \quad n = 1, 2, \ldots.$$

One can readily compute these either directly (integration by parts) or by computer (symbolically or numerically). In Figure 11.4 we plot some partial sums (superimposed with the function ϕ) where we include, in addition to the a_0 terms, the terms a_n, b_n for $n = 1, \ldots, N$ with $N = 1, 5, 10, 15$, and 20, respectively.

11.1.5. • Viewing the Three Fourier Series as Functions over $\mathbb{R}$. Note that the interval on which we sample the function ϕ is $(-l, l)$ for the full Fourier series **but** $(0, l)$ for the sine and the cosine series. While, in principle, we can define these series for a function given on any finite interval, there is a reason for this difference. This is highlighted by the following important remarks concerning the three Fourier series viewed as functions on $\mathbb{R}$.

(i) The Fourier sine series for $\phi(x)$ on $(0, l)$. While ϕ is defined on $(0, l)$, the Fourier sine series is defined for any $x \in \mathbb{R}$. Since it is comprised entirely of odd functions which are periodic with period $2l$, the Fourier sine series is an **odd** function that is **periodic with period** $2l$.

(ii) The Fourier cosine series for $\phi(x)$ on $(0, l)$. While ϕ is defined on $(0, l)$, the Fourier cosine series is defined for any $x \in \mathbb{R}$. Since it is comprised entirely of even functions which are periodic with period $2l$, the Fourier cosine series is an **even** function that is **periodic with period** $2l$.

(iii) The full Fourier series for $\phi(x)$ on $(-l, l)$. While ϕ is defined on $(-l, l)$, the series is defined for any $x \in \mathbb{R}$. Since it is comprised entirely of functions which are periodic with period $2l$, the full Fourier series is a **periodic function with period** $2l$.

11.1.6. • Convergence, Boundary Values, Piecewise Continuity, and Periodic Extensions. Let us accept (assume) for the moment that for reasonable functions ϕ, any of these three Fourier series will converge pointwise to ϕ on the **sampled interval**, that is, the interval upon which the Fourier coefficients are based. It is important to note that in finding a Fourier series of ϕ on any interval, the values of the function (if they exist) at the end points are **irrelevant**. This is because all the information about ϕ which goes into the Fourier series comes from integrals over the interval. Recall that to find the value of an integral of a function f from a to b we do not really need to know the values of f at a and b. This is not to say that boundary values are not important but, as we saw in Example 11.1.1, regardless of what the original function was at the boundary points, the Fourier series will do what its constituent functions dictate. Indeed, the partial sums for the Fourier sine series of $\phi(x) \equiv 1$ suggest that the value at $x = 0$ or l ($l = 2$ in Figure 11.1) is 0. This is a consequence of the fact that there is an inherent discontinuity building up the function $\phi(x) \equiv 1$ on $(0, l)$ with sines (which are odd functions). Hence, the series will have values closer to -1 to the left of $x = 0$. Boundary convergence will be discussed more carefully when we deal with pointwise convergence in Section 11.5.

Since we will often refer to **piecewise continuous** functions in this chapter, let us present a precise definition.

Definition of a Piecewise Continuous Function

Definition 11.1.4. A function ϕ on an interval (a, b) is piecewise continuous if the following hold true:

(i) It is continuous at all points in (a, b) except at perhaps a finite number of points.

(ii) At a discontinuity point x, it has a jump discontinuity, meaning that the limits from the right and left, $\phi(x+)$ and $\phi(x-)$, where

$$\phi(x+) = \lim_{h\to 0^+} \phi(x+h) \quad \text{and} \quad \phi(x-) = \lim_{h\to 0^-} \phi(x+h) \tag{11.11}$$

both exist and are finite.

(iii) The limits $\phi(a+)$ and $\phi(b-)$ exist and are finite.

A function defined on all of $\mathbb{R}$ is piecewise continuous if it is continuous except possibly at a sequence of values wherein it has a jump discontinuity. Moreover, the distance between any of these discontinuities is bounded below by some positive number.

Given a continuous function ϕ defined on an interval (a, b) such that $\phi(a+)$ and $\phi(b-)$ both exist and are equal, we may define $\phi(a) = \phi(b) = \phi(a+) = \phi(b-)$; then **extend** ϕ to all of $\mathbb{R}$ by periodicity to construct a continuous function on $\mathbb{R}$ which is periodic with period $b - a$. In other words, for $x \in (b, b + (b - a))$ we repeat the function ϕ as on (a, b) and so forth. At all the "joining points", we define the extension to be $\phi(a) = \phi(b)$. We call this extension to all of $\mathbb{R}$ **the periodic extension of** ϕ. Because of the fact that $\phi(a) = \phi(b)$, there is no discontinuity at the "joining points". The best way to visualize this is to draw a picture of some function on some interval and perform this extension for yourself.

If either ϕ has a finite number of jump discontinuities in (a, b) or $\phi(a+) \neq \phi(b-)$, then we can still consider the periodic extension to all of $\mathbb{R}$ except we need to define the extension at the "joining points" (for example at $x = a$ and b). For consistency, we could always define it to be the limit from the right. The extension then becomes a well-defined piecewise continuous function on all of $\mathbb{R}$.

Under certain smoothness assumptions on ϕ, the following convergence results hold true (cf. Section 11.5).

Convergence of the Three Fourier Series

(i) For any fixed x **inside** the sampled interval at which ϕ is continuous, each of the Fourier series converges to $\phi(x)$.

(ii) The Fourier sine series for $\phi(x)$ on $(0, l)$ is defined for all $x \in \mathbb{R}$ and converges to the **odd-periodic extension of** ϕ. This extension is obtained by considering the odd reflection on $(-l, 0)$ and extending the resulting function by periodicity (period $2l$) to all of $\mathbb{R}$. The value of the series at integer multiplies of l will depend on the properties of ϕ defined on $(0, l)$.

(iii) The Fourier cosine series for $\phi(x)$ on $(0, l)$ is defined for all $x \in \mathbb{R}$ and converges to the **even-periodic extension of** ϕ. This extension is obtained by considering the even reflection on $(-l, 0)$ and extending the resulting function by periodicity (period $2l$) to all of $\mathbb{R}$. The value of the series at integer multiples of l will depend on the properties of ϕ defined on $(0, l)$.

(iv) The full Fourier series for $\phi(x)$ on $(-l, l)$ is defined for all $x \in \mathbb{R}$ and converges to the **periodic extension of** ϕ. The values of the series at $(2k+1)l$, $k \in \mathbb{Z}$, will depend on the properties of the original ϕ defined on $(-l, l)$.

(v) At any point where the odd/even-periodic extension has a **jump discontinuity**, the Fourier series will converge to the **average** of the limit of ϕ from the right and left. Such jump discontinuities can occur at either of the following:
- points inside the sampled interval (and their odd/even-periodic extensions) where the function ϕ has a jump discontinuity,
- end points of the sampled interval (and their odd/even-periodic extensions).

As an illustration of these remarks, consider the following example involving the mother function $\phi(x) = e^x$. Let

$$f_1(x) = \sum_{n=1}^{\infty} b_n \sin n\pi x, \qquad f_2(x) = \frac{a_0}{2} + \sum_{n=1}^{\infty} a_n \cos n\pi x,$$

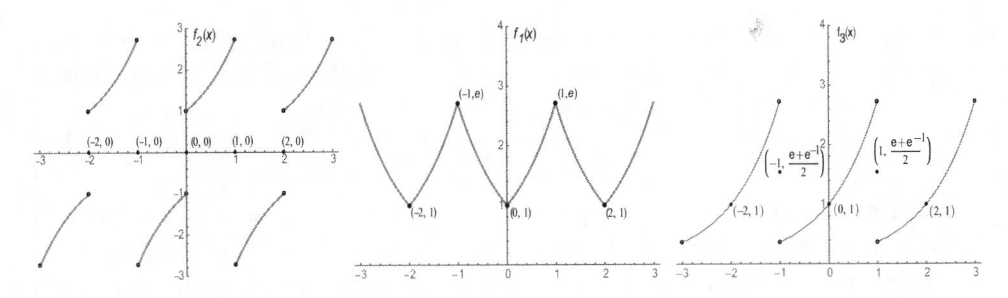

Figure 11.5. Plots of $f_1(x), f_2(x), f_3(x)$, Fourier series stemming from e^x.

and

$$f_3(x) = \frac{c_0}{2} + \sum_{n=1}^{\infty} (c_n \cos n\pi x + d_n \sin n\pi x),$$

where

$$a_n = 2\int_0^1 e^x \cos n\pi x \, dx, \quad n = 0, 1, \ldots, \qquad b_n = 2\int_0^1 e^x \sin n\pi x \, dx, \quad n = 1, 2, \ldots,$$

and

$$c_n = \int_{-1}^1 e^x \cos n\pi x \, dx, \quad n = 0, 1, \ldots, \qquad d_n = \int_{-1}^1 e^x \sin n\pi x \, dx, \quad n = 1, 2, \ldots.$$

First note that these three functions are, respectively, the Fourier sine and cosine series of e^x on $(0, 1)$ and the full Fourier series on $(-1, 1)$. What do these three functions look like over $\mathbb{R}$? That is, what is the graph of the pointwise limits of the respective Fourier series? We plot these for $x \in (-3, 3)$ in Figure 11.5. These plots come directly from the following: (a) the fact, taken here at face value, that we have convergence to e^x on the sampled interval, i.e., $(0, 1)$ for f_1, f_2 and $(-1, 1)$ for f_3; and (b) the basic structure of the component functions of the respective Fourier series.

If we had computed the coefficients (either by hand or with a computer) and plotted some partial sums, we would obtain graphs close to those of Figure 11.5 except at the break points (points of discontinuities). Indeed, the partial sums will always be continuous (in fact C^∞) and we only obtain a discontinuous function in the limit[3] of infinitely many terms.

11.1.7. Complex Version of the Full Fourier Series.

We end this subsection by noting a convenient way to write the full Fourier series using complex numbers. From Euler's formula, $e^{in\theta} = \cos n\theta + i \sin n\theta$, we see that the cosines and sines of the full Fourier series are the real and imaginary parts of one complex-valued function. In particular, we can invert the relation to find

$$\sin n\theta = \frac{e^{in\theta} - e^{-in\theta}}{2i}, \qquad \cos n\theta = \frac{e^{in\theta} + e^{-in\theta}}{2}.$$

[3]This should not sound so strange; one can easily concoct a sequence of smooth functions whose pointwise limit is discontinuous.

Hence, instead of writing the Fourier series in terms of infinite linear combinations of $\cos\frac{n\pi x}{l}$ and $\sin\frac{n\pi x}{l}$ with real coefficients, we can equivalently write the series in terms of infinite linear combinations of

$$\left\{e^{\frac{in\pi x}{l}} \;:\; n \in \mathbb{Z}\right\}, \tag{11.12}$$

with complex coefficients. Note that in order to exploit Euler's formula, we will need the frequency index n in (11.12) to take on negative integers as well as positive.

To provide some details, one can check directly that the complex-valued functions in (11.12) form an orthogonal set, and hence we can write the full Fourier series for a real-valued function ϕ as

$$\boxed{\phi(x) = \sum_{n=-\infty}^{\infty} c_n e^{in\pi x/l} \quad \text{where} \quad c_n = \frac{1}{2l}\int_{-l}^{l} \phi(x)\, e^{-in\pi x/l}\, dx.} \tag{11.13}$$

Note here that, while both the c_n and the functions $e^{\frac{in\pi x}{l}}$ are complex valued, the infinite sum is real valued and moreover is exactly the full Fourier series of $\phi(x)$. To see this, for $n = 1, 2 \ldots$, we couple the terms corresponding to n and $-n$; that is,

$$\sum_{n=-\infty}^{\infty} c_n e^{in\pi x/l} = c_0 + \sum_{n=1}^{\infty}\left[c_n e^{in\pi x/l} + c_{-n} e^{-in\pi x/l}\right].$$

On the other hand, note that by definition of the c_n we have

$$c_0 = \frac{1}{2}a_0, \qquad c_n = \frac{a_n - ib_n}{2}, \qquad c_{-n} = \frac{a_n + ib_n}{2},$$

where

$$a_n = \frac{1}{l}\int_{-l}^{l} \phi(x)\cos\Big(\frac{n\pi x}{l}\Big)\, dx, \quad n = 0, 1, \ldots,$$

$$b_n = \frac{1}{l}\int_{-l}^{l} \phi(x)\sin\Big(\frac{n\pi x}{l}\Big)\, dx, \quad n = 1, 2, \ldots.$$

Hence, we have

$$\begin{aligned} c_n e^{in\pi x/l} + c_{-n} e^{-in\pi x/l} &= \left(\frac{a_n - ib_n}{2}\right) e^{in\pi x/l} + \left(\frac{a_n + ib_n}{2}\right) e^{-in\pi x/l} \\ &= a_n\left(\frac{e^{in\pi x/l} + e^{-in\pi x/l}}{2}\right) + b_n\left(\frac{e^{in\pi x/l} - e^{-in\pi x/l}}{2i}\right) \\ &= a_n \cos\Big(\frac{n\pi x}{l}\Big) + b_n \sin\Big(\frac{n\pi x}{l}\Big) \end{aligned}$$

and

$$\sum_{n=-\infty}^{\infty} c_n e^{in\pi x/l} = \frac{1}{2}a_0 + \sum_{n=1}^{\infty}\left[a_n \cos\Big(\frac{n\pi x}{l}\Big) + b_n \sin\Big(\frac{n\pi x}{l}\Big)\right].$$

For most of the sequel, we will stick with sines and cosines, unless the complex notation provides additional insight. This will be the case, for example, when relating the Fourier transform to the Fourier series (cf. Section 11.8).

11.2. • Why Cosines and Sines? Eigenfunctions, Eigenvalues, and Orthogonality

Much of the theory behind Fourier series is a direct generalization of a fundamental result (Theorem 11.1 below) from **linear algebra**. To this end, let us first recall some basic concepts and structures from linear algebra. Then we will see that all these notions will have direct generalizations in the context of functions (which as we will argue are like infinite-dimensional vectors) and Fourier series.

11.2.1. • Finite Dimensions — the Linear Algebra of Vectors.

We first recall the **dot product** of two vectors in $\mathbb{R}^N$,

$$\mathbf{x} \cdot \mathbf{y} = \sum_{i=1}^{N} x_i y_i.$$

Note that $|\mathbf{x}|^2 = \mathbf{x} \cdot \mathbf{x}$ and two vectors $\mathbf{x}, \mathbf{y} \in \mathbb{R}^N$ are perpendicular (we will use the word **orthogonal**) when $\mathbf{x} \cdot \mathbf{y} = 0$. In fact, the dot product is directly related to angles. Any two vectors $\mathbf{x}$ and $\mathbf{y}$ in any dimension $N \geq 2$ lie in some two-dimensional plane and have a well-defined **angle** θ between them. The dot product tells us that

$$\cos\theta = \frac{\mathbf{x} \cdot \mathbf{y}}{|\mathbf{x}||\mathbf{y}|}. \tag{11.14}$$

In particular, given two vectors $\mathbf{x}$ and $\mathbf{y}$, the **vector projection** of $\mathbf{x}$ **onto** $\mathbf{y}$ is given by the vector

$$\mathrm{Proj}_{\mathbf{y}}\, \mathbf{x} := (\cos\theta\, |\mathbf{x}|)\, \frac{\mathbf{y}}{|\mathbf{y}|} = \left(\frac{\mathbf{x} \cdot \mathbf{y}}{|\mathbf{x}||\mathbf{y}|}\, |\mathbf{x}|\right) \frac{\mathbf{y}}{|\mathbf{y}|} = \mathbf{x} \cdot \mathbf{y}\, \frac{\mathbf{y}}{|\mathbf{y}|^2}. \tag{11.15}$$

A set $\{\mathbf{x}_i\}_{i=1}^N$ of N vectors in $\mathbb{R}^N$ is called a **basis** if the following hold:

- It is linearly independent; i.e., if for some choice of real numbers $c_1, \ldots, c_N$, we have

$$c_1\mathbf{x}_1 + \cdots + c_N\mathbf{x}_N = \mathbf{0}, \qquad \text{then} \quad c_1 = c_2 = \cdots = c_N = 0.$$

- It spans all of $\mathbb{R}^N$; i.e., for every $\mathbf{x} \in \mathbb{R}^N$ there exist real numbers $c_1, \ldots, c_N$ such that

$$\mathbf{x} = \sum_{i=1}^{N} c_i \mathbf{x}_i.$$

A basis $\{\mathbf{x}_i\}_{i=1}^N$ is called an **orthogonal basis** if for each $i \neq j$, we have $\mathbf{x}_i \cdot \mathbf{x}_j = 0$. We call it an **orthonormal** basis if, additionally, $|\mathbf{x}_i| = 1$ for all i. Based upon our Cartesian coordinate system, we have the **standard basis** $\{\mathbf{e}_i\}_{i=1}^N$ of unit vectors in the respective coordinate directions. So $\mathbf{e}_i \in \mathbb{R}^N$ is the vector whose components are all 0 except for its i-th component which is 1.

A fundamental object in linear algebra is a **linear transformation** from $\mathbb{R}^N$ to $\mathbb{R}^N$. With a set basis for $\mathbb{R}^N$, such a linear transformation can be described by an $N \times N$ array of numbers, commonly known as a matrix, $\mathbf{A} = (a_{ij})_{i,j=1}^N$. If no other information is given, we assume the components a_{ij} are with respect to the standard basis.

We say $\mathbf{A}$ is a symmetric $N \times N$ matrix if, in the terms of the matrix components a_{ij} of $\mathbf{A}$, we have $a_{ij} = a_{ji}$ for all $i, j \in \{1, \dots, N\}$. Alternatively, this is equivalent to saying that $\mathbf{A}$ is a symmetric linear transformation if

$$\mathbf{Ax} \cdot \mathbf{y} = \mathbf{x} \cdot \mathbf{Ay} \qquad \text{for any } \mathbf{x}, \mathbf{y} \in \mathbb{R}^N. \tag{11.16}$$

A nonzero vector $\mathbf{v}$ in $\mathbb{R}^N$ is a (real) **eigenvector** of an $N \times N$ matrix $\mathbf{A}$ with corresponding (real) **eigenvalue** λ if $\mathbf{Av} = \lambda\mathbf{v}$.

One of the fundamental results in linear algebra is the following:

Spectral Theorem from Linear Algebra

Theorem 11.1. *Let $\mathbf{A}$ be a symmetric $N \times N$ matrix. Then all eigenvalues of $\mathbf{A}$ are real and there exists an **orthogonal basis** of $\mathbb{R}^N$ consisting of **eigenvectors** of $\mathbf{A}$.*

The Spectral Theorem says the following for a symmetric matrix $\mathbf{A}$:

- There is a set of N (real) eigenvectors $\mathbf{v}_1, \dots, \mathbf{v}_N$ of $\mathbf{A}$ (with corresponding real eigenvalues) which are mutually orthogonal; i.e.,
$$\mathbf{v}_i \cdot \mathbf{v}_j = 0, \qquad i \neq j.$$
- The vectors $\mathbf{v}_i$ span all of $\mathbb{R}^N$; i.e., for every $\mathbf{x} \in \mathbb{R}^N$, there are real numbers c_i such that
$$\mathbf{x} = \sum_{i=1}^{N} c_i \mathbf{v}_i. \tag{11.17}$$

Because of the orthogonality property of the $\mathbf{v}_i$, we can fix an index j and take the dot product of (11.17) with $\mathbf{v}_j$ to find

$$\begin{aligned}
\mathbf{x} \cdot \mathbf{v}_j &= \left(\sum_{i=1}^{N} c_i \mathbf{v}_i \right) \cdot \mathbf{v}_j \\
&= \sum_{i=1}^{N} (c_i \mathbf{v}_i \cdot \mathbf{v}_j) \\
&= c_j |\mathbf{v}_j|^2.
\end{aligned}$$

Hence for any $i = 1, \dots, N$, we have

$$c_i = \frac{\mathbf{x} \cdot \mathbf{v}_i}{|\mathbf{v}_i|^2}. \tag{11.18}$$

In terms of vector projections as written in (11.15), the equation (11.17) is simply the statement that any vector can be written as the sum of its vector projections onto the eigenvectors; i.e.,

$$\mathbf{x} = \sum_{i=1}^{N} \operatorname{Proj}_{\mathbf{v}_i} \mathbf{x} = \sum_{i=1}^{N} \frac{\mathbf{x} \cdot \mathbf{v}_i}{|\mathbf{v}_i|^2} \mathbf{v}_i.$$

The reader will soon begin to appreciate that the above sum can be thought of as a "*finite-dimensional Fourier series associated with the linear transformation* $\mathbf{A}$".

11.2.2. • Infinite Dimensions — the Linear Algebra of Functions. We now want to extend these notions — vectors, dot product, linear transformations, eigenvectors — to infinite dimensions. To do this, we will need a link between **vectors** and **functions.**

An N-dimensional **vector** $\mathbf{x} = (x_1, \ldots, x_N)$ is an object with N components. A **sequence** $\{a_n\}_{n=0}^{\infty}$ can be viewed as an "infinite-dimensional vector" with a **countable** number of components; the value of a_n is the n-th component of the (generalized) vector. Going one step further, one can think of a **function** $f(x)$ defined on an interval (a, b) as an infinite-dimensional vector with a **continuum** of components; for any $x \in (a, b)$, we can *think of* the output value $f(x)$ as the the component of f corresponding to the "index" x. We could even call the value of $f(x)$, the "x-th *component*": Naturally, there are now a lot more components, indeed a continuum of components!

The inner product for functions: Now suppose we have two **real-valued functions**[4] f, g defined on (a, b). What is the natural generalization of the **dot product** between vectors? For the dot product of two vectors, we multiplied *respective* components and added. With two functions f and g, we define the generalized dot product, **now called the inner product** and written (f, g), by multiplying the *respective* components $f(x)$ and $g(x)$ and then integrating[5]; that is,

$$\boxed{(f, g) := \int_a^b f(x)g(x)\, dx.}$$

This is an important definition with huge ramifications, and as we will explain in Section 11.4, it is known as the L^2 inner product. Note that the Fourier coefficients of the previous section, for example (11.4) for the Fourier sine series on $(0, l)$, are simply (up to a normalization factor) the inner product between ϕ and $\sin \frac{n\pi x}{l}$.

Length of functions: With the L^2 inner product in hand, we can now define a generalized notion of length (called the L^2 **norm**) of a function f as

$$\boxed{\|f\|^2 := (f, f) = \int_a^b f^2(x)\, dx.}$$

Since we are now talking about functions instead of vectors, we use double lines for the "length" or "size". Note that the normalizing term which we divide by in, for example, the Fourier sine coefficients (11.4) is exactly $\| \sin \frac{n\pi x}{l} \|^2$.

Orthogonality for functions: Recall that two vectors are perpendicular (orthogonal) if and only if their dot product is zero. We now say that two functions f, g defined on (a, b) are **orthogonal** if and only if

$$(f, g) = 0.$$

---

[4]For **complex-valued** functions f, g, the inner product is defined by

$$(f, g) := \int_a^b f(x)\overline{g}(x)\, dx. \tag{11.19}$$

Note that the full Fourier series coefficients of a complex-valued function ϕ on $(-l, l)$ as given by (11.13) are simply the normalized complex inner product of ϕ with $e^{in\pi x/l}$.

[5]We will assume for the rest of this section that we are dealing with functions for which such integrals exist and are finite.

Note that the only way that a function f can be orthogonal to all functions is for it to be the zero function. For continuous functions, this follows from an IPW theorem (cf. Section A.7.1). However, these functions need not be continuous, and hence the correct terminology is to say that the function will be identically zero except on a **negligible set**[6], here negligible in the sense that it has no effect on integration; if you change the value of a function on a negligible set, its integral remains the same.

Projections and angles for functions: The inner product gives us more than just a notion of orthogonality; it also gives us a notion of "angle" between two functions and, specifically, a notion of projection of one function onto another. Following (11.15), if f and g are two functions, we define the projection of f onto g as

$$\text{Proj}_g f \;:=\; \frac{(f,g)}{||g||^2}\, g. \tag{11.20}$$

Linear transformations on functions: So far, we have a generalization to functions of the previous notions of the dot product, length, projections, and angles. The most important object in linear algebra is a **linear transformation (operator)** on a space of vectors. A linear transformation on $\mathbb{R}^N$ is completely characterized by an $N \times N$ matrix $\mathbf{A}$. What is an example of a linear transformation on a space of functions? Well, if the functions are also assumed to be smooth, then how about **differentiation**!

11.2.3. • The Linear Operator $\mathcal{A} = -\frac{d^2}{dx^2}$ and Symmetric Boundary Conditions. Consider the class of smooth (say, C^2) functions defined on $[a, b]$ which satisfy a certain **boundary condition** at $x = a$ and $x = b$. On this class, we define the differential operator

$$\boxed{\mathcal{A} := -\frac{d^2}{dx^2}.} \tag{11.21}$$

Why do we call this an operator? Let $X(x)$ be a function in this class. We can then **apply** $\mathcal{A}$ to X and obtain another function; that is, $\mathcal{A}(X) = \mathcal{A}X = -X''$. Thus $\mathcal{A}$ **operates** on a function and **generates** another function, and as such, we say $\mathcal{A}$ is an **operator** on this space of functions. Note that it is a **linear** operator in the sense that

$$\mathcal{A}(c_1X_1 \,+\, c_2X_2) \;=\; c_1\mathcal{A}(X_1) \,+\, c_2\mathcal{A}(X_2),$$

for functions X_1 and X_2 and scalars c_1 and c_2. This is simply a consequence of the fact that differentiation is linear; i.e., the derivative of a sum of functions is the sum of the derivatives.

Why the minus sign? The reason for the minus sign in $\mathcal{A}$ is partly convention and partly motivated by integration by parts. It is analogous to why mathematicians prefer to work with $-\Delta$ rather than Δ. As you read on, it will hopefully become clearer why we include the minus sign.

We now have our space of infinite vectors (functions) and a linear transformation or operator on this space. What about generalizing the idea of **eigenvectors and eigenvalues**? To this end, we would want to find $\lambda \in \mathbb{R}$ and a nonzero function X satisfying the boundary conditions, such that $\mathcal{A}X = \lambda X$. A function X that satisfies this

[6]There is a very rich mathematical theory (called **measure theory**) by which such a negligible set of **measure zero** can be made precise.

is called an **eigenfunction** and λ is called the **associated eigenvalue**. Start thinking about this right now: Are there functions with the property that their second derivatives are simply scalar multiples of themselves? As we shall see, the choice of boundary conditions determines the class of eigenfunctions. Following are three widely used boundary conditions for a continuous function on the closed interval $[a, b]$:

Three Widely Used Boundary Conditions

- **Dirichlet** boundary conditions for a function $X(x)$ on $[a, b]$ are the conditions
$$X(a) = 0 = X(b).$$
- **Neumann** boundary conditions for a function $X(x)$ on $[a, b]$ are the conditions
$$X'(a) = 0 = X'(b).$$
- **Periodic** boundary conditions for a function $X(x)$ on $[a, b]$ are the conditions
$$X(a) = X(b) \quad \text{and} \quad X'(a) = X'(b).$$

The key observation is the following: For a wide class of boundary conditions (including the three above), the **eigenfunctions corresponding to different eigenvalues are orthogonal**.

To see this we use a calculus identity which is simply Green's Second Identity (cf. (A.17)) in 1D! We can rederive it by noting that

$$-X_1''X_2 + X_1X_2'' = \frac{d}{dx}\Big(- X_1'X_2 + X_1X_2'\Big)$$

and then integrating from a to b to obtain

$$\int_a^b -X_1''X_2 + X_1X_2''\, dx = -X_1'(b)X_2(b) + X_1(b)X_2'(b) \\ - \big(- X_1'(a)X_2(a) + X_1(a)X_2'(a)\big). \tag{11.22}$$

The identity (11.22) is true for all smooth functions X_1 and X_2 on $[a, b]$. For a wide class of boundary conditions, the right-hand side vanishes. You should immediately check that this is the case for the three standard boundary conditions (Dirichlet, Neumann, and periodic). To this end, we make the following important definition:

Definition of a Symmetric Boundary Condition

Definition 11.2.1. A symmetric boundary condition on $[a, b]$ is any boundary condition for which the following is true: If $X_1(x)$ and $X_2(x)$ both satisfy the boundary condition, then

$$X_1(b)X_2'(b) - X_1'(b)X_2(b) = X_1(a)X_2'(a) - X_1'(a)X_2(a). \tag{11.23}$$

Thus if X_1, X_2 satisfy (11.23), the identity (11.22) becomes

$$\int_a^b -X_1''X_2 + X_1X_2''\,dx = 0 \quad \text{or} \quad \int_a^b (-X_1'')X_2\,dx = \int_a^b X_1(-X_2'')\,dx. \tag{11.24}$$

Hence, the operator $\mathcal{A} = -\frac{d^2}{dx^2}$ satisfies

$$\boxed{(\mathcal{A}X_1, X_2) = (X_1, \mathcal{A}X_2).} \tag{11.25}$$

This is exactly the generalization of symmetry condition (11.16) for matrices in the finite-dimensional setting of linear algebra. Note that the symmetry is not entirely due to the second derivative but crucially depends on the boundary condition: This motivates the use of the adjective *symmetric* for the boundary condition. Hence, we say $\mathcal{A}$ is a **symmetric operator** on the space of functions satisfying a symmetric boundary condition.

Application of (11.24) **to eigenfunctions:** Now suppose, in addition, X_1 and X_2 are eigenfunctions corresponding to different eigenvalues λ_1 and λ_2. This means that

$$-X_1'' = \lambda_1 X_1 \quad \text{and} \quad -X_2'' = \lambda_2 X_2.$$

By (11.24), we have

$$0 = \int_a^b -X_1''X_2 + X_1X_2''\,dx = (\lambda_1 - \lambda_2)\int_a^b X_1X_2\,dx = (\lambda_1 - \lambda_2)(X_1, X_2).$$

Hence if $\lambda_1 \neq \lambda_2$, then X_1 and X_2 must be orthogonal.

11.3. • Fourier Series in Terms of Eigenfunctions of $\mathcal{A}$ with a Symmetric Boundary Condition

Now let us pursue the idea of writing any function ϕ defined on an interval (a, b) as a linear combination of the eigenfunctions of $\mathcal{A}$, where $\mathcal{A}$ is defined by (11.21). We rewrite the eigenfunction/eigenvalue relation as

$$\boxed{X'' + \lambda X = 0} \tag{11.26}$$

with X satisfying some symmetric boundary conditions. We call such a problem **an eigenvalue problem**.

Now, suppose the following:

- There are a (countably many) infinite number of eigenfunctions and eigenvalues which we label (index), respectively, as

$$\{X_n(x)\}_{n=1}^{\infty} \quad \text{and} \quad \{\lambda_n\}_{n=1}^{\infty}, \tag{11.27}$$

  where each X_n corresponds to a distinct eigenvalue λ_n.

- It is possible to write

$$\phi(x) = \sum_{n=1}^{\infty} A_n X_n(x). \tag{11.28}$$

In the previous section, we showed that the $X_n(x)$ are mutually orthogonal. The mutual orthogonality implies that

$$A_n = \frac{(\phi, X_n)}{\|X_n\|^2} = \frac{\int_a^b \phi(x)\, X_n(x)\, dx}{\int_a^b X_n^2(x)\, dx}. \tag{11.29}$$

As in Section 11.1, we obtain (11.29) by noting that

$$\begin{aligned}
(\phi, X_n) = \left(\sum_{i=1}^{\infty} A_i X_i\,,\, X_n\right) &= \int_a^b \left(\sum_{i=1}^{\infty} A_i X_i\right) X_n\, dx \\
&= \int_a^b \sum_{i=1}^{\infty} [A_i X_i X_n]\, dx \\
&= \sum_{i=1}^{\infty} \int_a^b A_i X_i X_n\, dx \\
&= \sum_{i=1}^{\infty} (A_i X_i, X_n) \\
&= A_n\, (X_n, X_n),
\end{aligned}$$

where we used the orthogonality property of the X_i in the last line. Note again that we have assumed that *the integral of an infinite sum is the infinite sum of the integrals*. This is not always true but does hold under certain assumptions on ϕ — see Section 11.6.1 for more details.

The above calculation motivates the following definition.

Definition of a General Fourier Series

Definition 11.3.1. Consider a symmetric boundary condition and assume that there exists a countably infinite number of associated eigenfunctions and eigenvalues as labeled in (11.27). Then the infinite series

$$\sum_{n=1}^{\infty} A_n X_n(x) \quad \text{where} \quad A_n := \frac{(\phi, X_n)}{\|X_n\|^2} = \frac{\int_a^b \phi(x)\, X_n(x)\, dx}{\int_a^b X_n^2(x)\, dx}$$

is called a **general Fourier Series of the function** ϕ.

For the cases of Dirichlet, Neumann, and periodic boundary conditions, we call the resulting series a **classical Fourier series of the function** ϕ. In the case of classical Fourier series, we will soon see that the X_n will consist of sines and cosines with appropriately increasing frequencies.

Using our notion of projection as defined in (11.20), a Fourier series of a function ϕ based upon eigenfunctions $\{X_n\}$ is simply an infinite sum of all the projections of ϕ onto the X_n; that is,

$$\sum_{n=1}^{\infty} \text{Proj}_{X_n} \phi.$$

Two notational warnings are in order:

(i) Note that we have chosen to generically index the eigenfunctions from $n = 1$ to ∞. This is a choice we made — we could just as well have taken $n = 0$ to ∞. The choice of indexing is irrespective of whether or not 0 is an eigenvalue. Of course, when 0 is an eigenvalue, it is more suggestive to index from 0 to ∞.

(ii) It is actually not true that for all symmetric boundary conditions there will be only one linearly independent eigenfunction corresponding to each eigenvalue (cf. assumption (11.27)). This will be, for example, the case for periodic boundary conditions where there will be precisely two linearly independent and mutually orthogonal eigenfunctions corresponding to all nonzero eigenvalues. As you will see in the next subsection, we will include each of them as side-by-side terms in the general Fourier series. These eigenvalues are called **degenerate**. You should recall from linear algebra this notion of a degenerate eigenvalue and the corresponding linearly independent eigenvectors.

The above definition of a general Fourier series was directly inspired/motivated by the two assumptions associated with (11.27) and (11.28). We tackle the first assumption by considering separately the four basic symmetric boundary conditions in the next subsection.

We end this subsection by proving a very simple result which will dispense with the possibility of complex eigenvalues for our eigenvalue problem (11.26). Recall from linear algebra that it is possible to find eigenvalues of a real matrix A which are complex valued. The associated eigenvectors would also have complex-valued entries. Moreover, often these complex eigenvalues proved to be important for understanding the linear transformation A. However, the Spectral Theorem (Theorem 11.1) stated that all eigenvalues of a **symmetric** matrix are real; i.e., no complex eigenvalues exist. The **symmetry** of the operator $\mathcal{A} = -\frac{d^2}{dx^2}$, as defined in (11.25), also yields that all its eigenvalues must be real. That is:

Proposition 11.3.1. *All eigenvalues of the eigenvalue problem $X'' + \lambda X = 0$ over functions $X(x)$, $x \in [a, b]$, satisfying symmetric boundary conditions are real.*

Proof. We consider the possibility of a complex eigenvalue λ. Hence there would be complex-valued function $X(x)$ on $[a, b]$ satisfying the symmetric boundary condition such that $\mathcal{A}X = \lambda X$. Our goal is to show that λ is real, or equivalently, $\lambda = \overline{\lambda}$, where as usual the overline denotes complex conjugation. By (11.25), the symmetry of $\mathcal{A}$, we have

$$(\overline{X}, \mathcal{A}X) - (\mathcal{A}\overline{X}, X) = 0.$$

Since $\mathcal{A}X = \lambda X$, we have $\mathcal{A}\overline{X} = \overline{\lambda}\,\overline{X}$. Thus

$$(\overline{X}, \lambda X) - (\overline{\lambda}\,\overline{X}, X) = 0;$$

that is,

$$\int_a^b \overline{X}\lambda X - \overline{\lambda}\overline{X}X\,dx = (\lambda - \overline{\lambda})\int_a^b X\overline{X}\,dx = 0.$$

Since $X\overline{X} = |X|^2$ and with the fact that X is not the zero function, we must have $\lambda = \overline{\lambda}$. □

11.3.1. • The Eigenfunctions and Respective Fourier Series Associated with the Four Standard Symmetric Boundary Conditions. Consider the case of **Dirichlet boundary conditions** and let us find all the eigenvalues and eigenfunctions. By Proposition 11.3.1, we need only look for real eigenvalues. It is convenient to translate the interval $[a, b]$ so that it has the form $[0, l]$, for some $l > 0$. This will allow us to write the eigenfunctions in a simple form. We consider the eigenvalue problem for $X(x)$ on $[0, l]$:

$$X'' + \lambda X = 0, \qquad X(0) = 0 = X(l).$$

What are all the solutions (eigenvalues and eigenfunctions)? Since λ is real, there are three cases to consider.

Case 1: Suppose $\lambda < 0$. Then $X'' = -\lambda X$ where $-\lambda > 0$. The general solution to this ODE is

$$X(x) = Ae^{\sqrt{-\lambda}x} + Be^{-\sqrt{-\lambda}x},$$

for any constants A and B. However, if we impose the boundary condition $X(0) = 0 = X(l)$, we find

$$A + B = 0 \qquad \text{and} \qquad Ae^{\sqrt{-\lambda}l} + Be^{-\sqrt{-\lambda}l} = 0.$$

Thus $B = -A$ and

$$A\left(e^{\sqrt{-\lambda}l} - e^{-\sqrt{-\lambda}l}\right) = 0.$$

Since e^x is 1-1 on $\mathbb{R}$, the only way the above can hold is for A to be 0. But in this case, $A = B = 0$, yielding the trivial solution. Thus, we find no nontrivial solutions and hence no negative eigenvalues.

Case 2: Suppose $\lambda = 0$. Then $X'' = 0$ and the general solution is $X(x) = A + Bx$. But again the boundary values imply $A = B = 0$. Consequently, $\lambda = 0$ is not an eigenvalue.

Case 3: Suppose $\lambda > 0$. In this case, the general solution is

$$X(x) = A\cos\sqrt{\lambda}x + B\sin\sqrt{\lambda}x,$$

for any constants A and B. The condition $X(0) = 0$ implies $A = 0$. On the other hand, the condition $X(l) = 0$ implies either $B = 0$ (and again, we only have the trivial solutions) or $\sin\sqrt{\lambda}l = 0$. The latter holds if $\sqrt{\lambda}l$ is an integer multiple of π, that is, if

$$\lambda = \frac{n^2\pi^2}{l^2} \qquad \text{for any } n = 1, 2, \ldots.$$

The corresponding eigenfunctions are

$$\sin\frac{n\pi x}{l}, \qquad n = 1, 2, \ldots.$$

Thus, we have a countable number of eigenvalues λ_n and corresponding eigenfunctions X_n where

$$\lambda_n = \frac{n^2\pi^2}{l^2}, \qquad X_n(x) = \sin\frac{n\pi x}{l}.$$

So the "Fourier series" defined by Definition 11.3.1 is simply the Fourier sine series of the previous section! Moreover, by (11.29), the coefficients are given by

$$A_n = \frac{(\phi, X_n)}{||X_n||^2} = \frac{2}{l}\int_0^l \phi(x)\sin\left(\frac{n\pi x}{l}\right)dx,$$

just as before.

For the other typical boundary conditions, one can analogously find (cf. Exercise **11.10**) the corresponding countable number of eigenvalues and eigenfunctions and, thereby, generate the associated general Fourier series. As previously noted, the case of periodic boundary conditions is slightly different from the others in that there are two linearly independent and mutually orthogonal eigenfunctions corresponding to all nonzero eigenvalues. To this end, let us provide some details. For periodic boundary conditions, it is convenient to translate the interval $[a, b]$ to $[-l, l]$ in order to have a simple and consistent form for the eigenvalues. In this case, the boundary conditions

$$X(-l) = X(l) \quad \text{and} \quad X'(-l) = X'(l)$$

give rise to eigenvalues

$$\lambda_n = \frac{n^2\pi^2}{l^2}, \qquad n = 0, 1, \ldots.$$

However except for $\lambda_0 = 0$, there are now two corresponding eigenfunctions

$$Y_n(x) = \cos\left(\frac{n\pi x}{l}\right) \quad \text{and} \quad Z_n(x) = \sin\left(\frac{n\pi x}{l}\right), \qquad n = 1, 2, \ldots,$$

which are not scalar multiples of each other. As in linear algebra, we use the phrase two **linearly independent** eigenfunctions. Note that they are orthogonal to each other on $(-l, l)$; i.e.,

$$\int_{-l}^{l} \cos\left(\frac{n\pi x}{l}\right)\sin\left(\frac{n\pi x}{l}\right)dx = 0 \qquad \text{for all } n = 1, 2, \ldots.$$

Hence, the linearly independent eigenfunctions

$$\left\{1, \{Y_n(x), Z_n(x)\}_{n=1}^{\infty}\right\}$$

constitute an orthogonal set of functions. The "generic set" $\{X_n\}_{n=1}^{\infty}$ must be an ordering of all of them. In general there are many ways to order (label) them. We order them via their eigenvalues; that is, the corresponding Fourier series is

$$\frac{1}{2}A_0 + \sum_{n=1}^{\infty}\left[A_n\cos\left(\frac{n\pi x}{l}\right) + B_n\sin\left(\frac{n\pi x}{l}\right)\right],$$

where

$$A_n = \frac{1}{l}\int_{-l}^{l} \phi(x)\cos\Big(\frac{n\pi x}{l}\Big)\,dx, \quad n = 0, 1, \dots,$$

$$B_n = \frac{1}{l}\int_{-l}^{l} \phi(x)\sin\Big(\frac{n\pi x}{l}\Big)\,dx, \quad n = 1, 2, \dots.$$

Let us conclude by formally documenting the general Fourier series for the four most common boundary conditions.

(i) The Fourier series corresponding to Dirichlet boundary conditions. Dirichlet boundary conditions on $[0, l]$, $X(0) = 0 = X(l)$, give rise to eigenfunctions and eigenvalues

$$X_n(x) = \sin\frac{n\pi x}{l} \quad \text{with} \quad \lambda_n = \frac{n^2\pi^2}{l^2}, \; n = 1, 2, \dots.$$

The corresponding Fourier series is

$$\sum_{n=1}^{\infty} A_n \sin\frac{n\pi x}{l}$$

where

$$A_n = \frac{(\phi, X_n)}{\|X_n\|^2} = \frac{2}{l}\int_0^l \phi(x)\sin\Big(\frac{n\pi x}{l}\Big)\,dx, \quad n = 1, 2, \dots.$$

This is what we previously called the **Fourier sine series** of the function ϕ.

(ii) The Fourier series corresponding to Neumann boundary conditions. Neumann boundary conditions on $[0, l]$, $X'(0) = 0 = X'(l)$, give rise to eigenfunctions and eigenvalues

$$X_n(x) = \cos\frac{n\pi x}{l} \quad \text{with} \quad \lambda_n = \frac{n^2\pi^2}{l^2}, \; n = 0, 1, \dots.$$

The corresponding Fourier series for a function ϕ is

$$\frac{A_0}{2} + \sum_{n=1}^{\infty} A_n \cos\frac{n\pi x}{l}$$

where

$$A_n = \frac{(\phi, X_n)}{\|X_n\|^2} = \frac{2}{l}\int_0^l \phi(x)\cos\Big(\frac{n\pi x}{l}\Big)\,dx, \quad n = 0, 1, \dots.$$

This is what we previously called the **Fourier cosine series** of the function ϕ.

(iii) The Fourier series corresponding to periodic boundary conditions. Periodic boundary conditions on $[-l, l]$,

$$X(-l) = X(l) \quad \text{and} \quad X'(-l) = X'(l),$$

give rise to eigenfunctions

$$Y_n(x) = \cos\Big(\frac{n\pi x}{l}\Big), \; n = 0, 1, \dots, \qquad \text{and} \qquad Z_n(x) = \sin\Big(\frac{n\pi x}{l}\Big), \; n = 1, 2, \dots,$$

corresponding to eigenvalues $\lambda_n = \frac{n^2\pi^2}{l^2}$, $n = 0, 1, \ldots$. The corresponding Fourier series is

$$\frac{1}{2}A_0 + \sum_{n=1}^{\infty}\left[A_n \cos\left(\frac{n\pi x}{l}\right) + B_n \sin\left(\frac{n\pi x}{l}\right)\right],$$

where

$$A_n = \frac{(\phi, Y_n)}{||Y_n||^2} = \frac{1}{l}\int_{-l}^{l} \phi(x) \cos\left(\frac{n\pi x}{l}\right) dx, \quad n = 0, 1, \ldots,$$

and

$$B_n = \frac{(\phi, Z_n)}{||Z_n||^2} = \frac{1}{l}\int_{-l}^{l} \phi(x) \sin\left(\frac{n\pi x}{l}\right) dx, \quad n = 1, 2, \ldots.$$

This is what we previously called the **full Fourier series** of the function ϕ.

(iv) The Fourier series corresponding to a mixed boundary condition. The mixed boundary condition on $[0, l]$,

$$X(0) = 0 = X'(l),$$

gives rise to eigenfunctions and eigenvalues

$$X_n(x) = \sin\frac{(n+1/2)\pi x}{l} \quad \text{with} \quad \lambda_n = \frac{(n+1/2)^2\pi^2}{l^2}, \quad n = 0, 1, \ldots.$$

One can check by direct integration that for all $n = 0, 1, 2, \ldots$ we have

$$||X_n||^2 = \int_0^l \sin^2\frac{(n+1/2)\pi x}{l}\, dx = \frac{l}{2}.$$

The corresponding Fourier series is

$$\sum_{n=0}^{\infty} A_n \sin\frac{(n+1/2)\pi x}{l} \quad \text{where} \quad A_n = \frac{(\phi, X_n)}{||X_n||^2} = \frac{2}{l}\int_0^l \phi(x) \sin\frac{(n+1/2)\pi x}{l}\, dx.$$

11.3.2. • The Miracle: These Sets of Eigenfunctions Span the Space of All Reasonable Functions. Thus far we have not considered whether or not it is indeed possible to write ϕ as an infinite linear combination of the X_n; that is, do any particular set of eigenfunctions $\{X_n\}_{n=1}^{\infty}$ (corresponding to some symmetric boundary condition) **span** all "reasonable" functions? Drawing on our analogy with linear algebra (vectors in $\mathbb{R}^N$), an **orthogonal basis** of vectors also has the property that all vectors can be written as a linear combination of projections onto the basis vectors. If one were to remove one of these basis vectors, then the sum of the projections of an arbitrary vector x onto the remaining basis vectors would not in general yield x. In other words, the set of basis vectors minus one of them would not be "complete". When jumping from vectors to functions, the spanning property for eigenfunctions is often referred to as **completeness**; that is, the $\{X_n\}_{n=1}^{\infty}$ consists of a complete set.

So, whereas the motivation for defining the coefficients in all our Fourier series of a function ϕ was the equality (11.28), i.e.,

$$\phi(x) = \sum_{n=1}^{\infty} A_n X_n(x),$$

we still need to determine whether or not ϕ at any x is indeed equal to the infinite Fourier series at that x. The basic issue surrounding the spanning property (completeness), i.e., the equality above, is **convergence**. Convergence means convergence of the partial sums; that is, if we define the partial sums by

$$S_N(x) := \sum_{n=1}^{N} A_n X_n(x) \quad \text{where } A_n \text{ are given by (11.29)}, \tag{11.30}$$

then the goal is to show that **the functions $S_N(x)$ converge to the function $\phi(x)$ as** $N \to \infty$. Under certain conditions on the function ϕ, this is indeed the case. However, there are many ways in which a function can converge to another function. In the next few sections, we will discuss three ways that a Fourier series can converge to its associated function ϕ:

- L^2 (or mean-square) convergence,
- pointwise convergence,
- uniform convergence.

In each case, we will give **sufficient** conditions on ϕ which guarantee convergence. The proof that we will provide for pointwise convergence of the classical full Fourier series is particularly illuminating in that it highlights the role of the **delta function**.

In order to give a complete and fully satisfactory treatment of the spanning property (completeness) of these eigenfunctions, one needs machinery from **functional analysis**. In particular, the key result is what is known as the **Spectral Theorem for a compact, symmetric operator on a Hilbert space**. This result beautifully generalizes the Spectral Theorem from Linear Algebra, Theorem 11.1, to infinite dimensions. In fact, the reader should have already made the connection: A Fourier series is simply a way of writing an arbitrary function as the sum of its projections onto an orthogonal basis of eigenfunctions of $\mathcal{A}$ with a symmetric boundary condition.

11.4. • Convergence, I: The L^2 Theory, Bessel's Inequality, and Parseval's Equality

We first introduce a notion of convergence of functions called L^2 convergence. It is also known as **mean square convergence**.

11.4.1. • L^2 Convergence of a Sequence of Functions.

Definition of L^2 or Mean Square Convergence of Functions

Definition 11.4.1. We say a sequence of functions f_N defined on an interval (a, b) converges in L^2 to f if and only if

$$\int_a^b |f_N(x) - f(x)|^2 \, dx \xrightarrow{N\to\infty} 0.$$

Note that the actual values on the boundary points $x = a$ or $x = b$ are not important for L^2 convergence.

Loosely speaking, **the function space** L^2 (or more precisely $L^2((a,b))$) is the set of **all** real-valued functions f defined on (a,b) such that

$$\int_a^b |f(x)|^2\,dx < \infty. \tag{11.31}$$

If a function f on (a,b) satisfies (11.31), then we write $f \in L^2((a,b))$. In this definition, we make no requirement on the smoothness properties of f: They are just functions whose square can be integrated to yield a finite number. Clearly any piecewise continuous function on (a,b) is square integrable. As in Section 11.2.2, we define an **inner product** of two functions f, g in $L^2((a,b))$ and **norm** on f in $L^2((a,b))$ by, respectively,

$$(f,g) := \int_a^b f(x)\,g(x)\,dx \qquad \text{and} \qquad \|f\| := (f,f)^{1/2} = \left(\int_a^b |f(x)|^2\,dx\right)^{1/2}.$$

Note that f_N converges to f in L^2 as $N \to \infty$ if and only if $\|f_N - f\| \xrightarrow{N\to\infty} 0$.

11.4.2. • L^2 Convergence of Fourier Series.

Let $\phi \in L^2((a,b))$ and consider a general Fourier series in the form

$$\sum_{n=1}^{\infty} A_n X_n(x) \qquad \text{where} \quad A_n = \frac{(\phi, X_n)}{\|X_n\|^2} = \frac{\int_a^b \phi(x)\,X_n(x)\,dx}{\int_a^b X_n^2(x)\,dx}. \tag{11.32}$$

If we define the partial sums of the general Fourier series as

$$S_N(x) = \sum_{n=1}^{N} A_n X_n(x), \tag{11.33}$$

then L^2 convergence of the general Fourier series means $S_N \to \phi$ in L^2; that is

$$\|S_N(x) - \phi(x)\| \to 0, \qquad \text{as } N \to \infty.$$

The following L^2 convergence theorem is simple and elegant to state:

L^2 Convergence Theorem for General Fourier Series

Theorem 11.2. *Suppose $\phi \in L^2((a,b))$; then its general Fourier series associated with any symmetric boundary conditions converges to ϕ in the L^2 sense (mean square) on the interval (a,b). In this way, we say the eigenfunctions associated with a symmetric boundary condition are* ***complete*** *in $L^2((a,b))$.*

We will not be able to provide a full proof; however, we can provide a partial proof and, in particular, one which illustrates a fundamental inequality called **Bessel's inequality**. In doing so, we **reduce** the proof of Theorem 11.2 to proving **equality** in Bessel's inequality; this equality is known as **Parseval's equality**.

11.4.3. • Bessel's Inequality and Reducing the L^2 Convergence Theorem to Parseval's Equality. We begin by asking a general question. Let $\{X_n, n \geq 1\}$ be **any set** of orthogonal functions on an interval (a, b), and let ϕ be a function defined on (a, b) such that $||\phi|| < \infty$. Fix N and look at

$$E_N \ := \ \left|\left|\phi - \sum_{n=1}^{N} c_n X_n\right|\right|^2$$

where $c_1, \ldots, c_N \in \mathbb{R}$. We ask the following:

Among all combinations of c_i, which have the smallest value of E_N?

This question is often referred to as a problem of **least squares**. The answer is simple: The optimal choice is given by taking the c_n to be A_n where

$$A_n \ := \ \frac{(\phi, X_n)}{||X_n||^2}.$$

To see this, we write

$$\begin{aligned} E_N &= \int_a^b \left|\phi(x) - \sum_{n=1}^{N} c_n X_n(x)\right|^2 dx \\ &= \int_a^b |\phi(x)|^2\, dx - 2\sum_{n=1}^{N} c_n \int_a^b \phi(x) X_n(x)\, dx + \sum_{m=1}^{N}\sum_{n=1}^{N} c_n c_m \int_a^b X_n(x) X_m(x)\, dx \\ &= ||\phi||^2 - 2\sum_{n=1}^{N} c_n(\phi, X_n) + \sum_{n=1}^{N} c_n^2 ||X_n||^2 \end{aligned}$$

by orthogonality of the X_n. Viewing the above as a quadratic equation in each c_n, we complete the square to find

$$E_N \ = \ \sum_{n=1}^{N} ||X_n||^2 \left(c_n - \frac{(\phi, X_n)}{||X_n||^2}\right)^2 + ||\phi||^2 - \sum_{n=1}^{N} \frac{(\phi, X_n)^2}{||X_n||^2}. \tag{11.34}$$

Clearly, we can see that (11.34) is smallest when

$$c_n \ = \ \frac{(\phi, X_n)}{||X_n||^2}.$$

For notational consistency, let us call these optimal choices A_n. Then with $c_n = A_n$, we see from (11.34) that

$$E_N \ = \ ||\phi||^2 - \sum_{n=1}^{N} A_n^2 ||X_n||^2.$$

Since by definition $E_N \geq 0$, we have for all N,

$$\sum_{n=1}^{N} A_n^2 \int_a^b |X_n(x)|^2\, dx \leq \int_a^b |\phi(x)|^2\, dx.$$

Letting $N \to \infty$, we obtain

$$\sum_{n=1}^{\infty} A_n^2 \int_a^b |X_n(x)|^2\, dx \ \leq \ \int_a^b |\phi(x)|^2\, dx. \tag{11.35}$$

This is called **Bessel's inequality** and it holds for **any orthogonal set of functions**. Because of its importance, let us summarize it in the following theorem (which we have just proved).

Bessel's Inequality

Theorem 11.3. *Let $\phi \in L^2((a,b))$ and suppose $\{X_n(x)\}$ is an orthogonal family of functions on (a,b). Define A_n by*

$$A_n := \frac{(\phi, X_n)}{\|X_n\|^2} = \frac{\int_a^b \phi(x)\, X_n(x)\, dx}{\int_a^b |X_n(x)|^2\, dx}.$$

Then

$$\sum_{n=1}^{\infty} A_n^2 \int_a^b |X_n(x)|^2\, dx \le \int_a^b |\phi(x)|^2\, dx.$$

While Bessel's inequality applies to any orthogonal set of functions, let us now consider the case where the $\{X_n\}$ **are eigenfunctions** of $\mathcal{A}$ subject to a symmetric boundary condition. The "optimal" A_n above are simply the Fourier coefficients of the general Fourier series. Suppose we apply Bessel's inequality (Theorem 11.3) to this general Fourier series. By definition of L^2 convergence, **the general Fourier series converges in L^2 if and only if the error $E_N \to 0$ as $N \to \infty$.** On the other hand, $E_N \to 0$ if and only if (11.35) is an equality, in other words, if and only if Bessel's inequality is an equality. When equality holds, we call it **Parseval's equality**.

Alternately, we may say that the orthogonal set of functions X_n is a **basis** for (or **complete in**) the space of all functions in L^2 if there is **no gap** in Bessel's inequality; i.e.,

$$\sum_{n=1}^{\infty} A_n^2 \int_a^b |X_n(x)|^2\, dx = \int_a^b |\phi(x)|^2\, dx.$$

So in order to prove Theorem 11.2, the L^2 convergence theorem, we need to prove the following:

Parseval's Equality

Theorem 11.4. *Let $\phi \in L^2((a,b))$ and let $X_n(x)$ and A_n be, respectively, the eigenfunctions and Fourier coefficients of any general Fourier series for ϕ. Then*

$$\sum_{n=1}^{\infty} A_n^2 \int_a^b |X_n(x)|^2\, dx = \int_a^b |\phi(x)|^2\, dx.$$

Unfortunately, the proof of Theorem 11.4 requires some mathematical analysis beyond the scope of this book.

Notation Warning: In Bessel's inequality and Parseval's equality, we have chosen to use our convention of numbering (labeling) the eigenfunctions associated with the general Fourier series starting with $n = 1$. If, for example, you were to apply this to the Fourier cosine series on $(0,l)$, you would find that 0 is indeed an eigenvalue corresponding to eigenfunction 1. Consequently, we usually number (label) the eigenvalues and eigenfunctions starting with $n = 0$. This only serves as notation for the labeling. We could certainly apply Bessel's inequality as stated above here; however, in the form (notation) that we adopt above, the first eigenfunction, i.e., 1 corresponding to eigenvalue 0, is indexed by $n = 1$.

The point is that regardless of numbering conventions, view Bessel's inequality and Parseval's equality as statements incorporating **all** the eigenfunctions. This point will be made clearer in the next two examples.

11.4.4. The Riemann-Lebesgue Lemma and an Application of Parseval's Equality. We note the following corollary of Bessel's inequality (Theorem 11.3) applied to the classical Fourier series. For the classical Fourier series (i.e., those corresponding to the eigenvalue problem with either Dirichlet, Neumann, or periodic boundary conditions), we can show that for each n,

$$\int_a^b |X_n(x)|^2\,dx$$

is some fixed number independent of n (check this by evaluating some simple integrals with cosines or sines). Then Bessel's inequality implies that the infinite sum of the squares of the classical Fourier coefficients must converge. In particular, we must have the following:

Theorem 11.5 (The Riemann-Lebesgue Lemma). *If $\phi \in L^2((a,b))$, then the classical Fourier coefficients of ϕ tend to 0 as $n \to \infty$.*

Applying this to the case of Dirichlet boundary conditions on $[0,l]$, Theorem 11.5 implies that for every ϕ defined on $(0,l)$ such that $\|\phi\|^2 < \infty$,

$$\int_0^l \phi(x)\sin\frac{n\pi x}{l}\,dx \longrightarrow 0, \qquad \text{as } n \to \infty. \tag{11.36}$$

Since C^∞ functions with compact support in $(0,l)$ are certainly square integrable, this implies that $\sin\frac{n\pi x}{l}$ converges to 0 in the sense of distributions. We already witnessed a direct proof of this fact in Section 5.5.5. However, the above holds for any square-integrable functions (even ones which are not continuous or nowhere differentiable). In other words, Theorem 11.5 implies (11.36) holds for a much wider class of test functions than the usual "safe" choice of C_c^∞.

A Surprising Application of Parseval's Equality to Certain Infinite Series: We show, by example, how Parseval's equality can be used to compute the values of certain infinite series. Let $\phi \equiv 1$ on $(0,\pi)$, which is clearly square integrable. Using the classical Fourier sine series we have $X_n(x) = \sin(nx)$ and $A_n = 4/n\pi$ for n odd, $A_n = 0$ for n even. By Parseval's equality we have

$$\sum_{n\text{ odd}}\left(\frac{4}{n\pi}\right)^2\frac{\pi}{2} = \int_0^\pi 1\,dx = \pi, \qquad \text{and hence, } \sum_{n\text{ odd}}\frac{1}{n^2} = \frac{\pi^2}{8}.$$

11.5. • Convergence, II: The Dirichlet Kernel and Pointwise Convergence of the Full Fourier Series

In this section, we will focus on the full Fourier series on $(-\pi, \pi)$, that is, the general Fourier series on $(-\pi, \pi)$ corresponding to periodic boundary conditions. One can prove analogous results for other general Fourier series.

11.5.1. • Pointwise Convergence of a Sequence of Functions. The most obvious way that a sequence of functions can converge is pointwise; that is, for any fixed input, the sequence of outputs converge. We have already referred to this type of convergence on many occasions, but for completeness let us give a precise definition.

Definition of Pointwise Convergence of Functions

Definition 11.5.1. We say a sequence of functions f_N defined on an interval $[a, b]$ (or (a, b)) converges pointwise to f if and only if for every $x \in [a, b]$ $(x \in (a, b))$,

$$f_N(x) \to f(x) \text{ as } N \to \infty.$$

As a simple example, consider the sequence of functions

$$f_N(x) := \frac{x^2}{N}.$$

Then f_N converges pointwise to $f(x) \equiv 0$ on $\mathbb{R}$. This is simply a consequence of the fact that for any fixed number $x = a$, the sequence $\frac{a^2}{N}$ tends to 0 as $N \to \infty$.

In terms of a general Fourier series of the form (11.32), we say the general Fourier series converges pointwise to ϕ on (a, b) if and only if for any fixed $x \in (a, b)$, $S_N(x) \xrightarrow{N\to\infty} \phi(x)$, where the S_N are the partial sums defined by (11.33). In other words, for any $x \in (a, b)$, we have

$$|S_N(x) - \phi(x)| \to 0 \text{ as } N \to \infty.$$

11.5.2. • Pointwise Convergence of the Full Fourier Series: The Dirichlet Kernel and the Delta Function. Before stating and proving a theorem about pointwise convergence, we do a little exploration of the basic structure of the classical Fourier series to unlock some of its mystery. For simplicity let $l = \pi$ and consider a nice smooth function $\phi(x)$ on $(-\pi, \pi)$. The goal here is to uncover *how on earth* for any $x \in (-\pi, \pi)$,

$$\phi(x) = \frac{1}{2}a_0 + \sum_{n=1}^{\infty} \left[a_n \cos(nx) + b_n \sin(nx)\right],$$

where

$$a_n = \frac{1}{\pi}\int_{-\pi}^{\pi} \phi(y)\cos(ny)\,dy, \qquad b_n = \frac{1}{\pi}\int_{-\pi}^{\pi} \phi(y)\sin(ny)\,dy. \tag{11.37}$$

To this end, fix $x \in (-\pi, \pi)$ and consider the sequence in N of functions

$$S_N(x) := \frac{1}{2}a_0 + \sum_{n=1}^{N} \left[a_n \cos(nx) + b_n \sin(nx)\right]. \tag{11.38}$$

We plug in the values of the a_n and b_n to find

$$\begin{aligned} S_N(x) &= \frac{1}{2}a_0 + \sum_{n=1}^{N} \left[a_n \cos(nx) + b_n \sin(nx)\right] \\ &= \frac{1}{2\pi}\int_{-\pi}^{\pi} \phi(y)\,dy + \sum_{n=1}^{N}\left[\left(\frac{1}{\pi}\int_{-\pi}^{\pi} \phi(y)\cos(ny)\,dy\right)\cos(nx)\right. \\ &\qquad\qquad \left. + \left(\frac{1}{\pi}\int_{-\pi}^{\pi} \phi(y)\sin(ny)\,dy\right)\sin(nx)\right] \\ &= \frac{1}{2\pi}\int_{-\pi}^{\pi}\left[1 + 2\sum_{n=1}^{N}\left(\cos(ny)\cos(nx) + \sin(ny)\sin(nx)\right)\right]\phi(y)\,dy \\ &= \frac{1}{2\pi}\int_{-\pi}^{\pi}\left[1 + 2\sum_{n=1}^{N}\cos(n(x-y))\right]\phi(y)\,dy. \end{aligned}$$

To arrive at the second to last line, we simply rearranged terms, using the fact that the finite sum of the integrals is simply the integral of the finite sum. To pass to the last line, we employed the trigonometric double angle identity,

$$\cos(n(x-y)) = \cos(ny)\cos(nx) + \sin(ny)\sin(nx).$$

Now we are in a better position to ask the following:

For a fixed $x \in (-\pi, \pi)$, does $S_N(x)$ tend to $\phi(x)$ as $N \to \infty$?

The only possible way this could happen is if, as a function of y defined on the interval $(-\pi, \pi)$, we have

$$\frac{1}{2\pi}\left[1 + 2\sum_{n=1}^{N}\cos(n(x-y))\right] \xrightarrow{N\to\infty} \delta_x \text{ in the sense of distributions on } (-\pi,\pi)^7.$$

So somehow this combination of cosines (as a function of y) **concentrates (focuses)** at $y = x$. Let us now explore this combination further with the following important definition.

Definition of the Dirichlet Kernel

For each $N = 1, 2, \ldots$, we define the **Dirichlet kernel** to be the 2π periodic function

$$K_N(x) := 1 + 2\sum_{n=1}^{N}\cos(nx). \tag{11.39}$$

[7] By definition this means test functions ϕ are in $C_c^\infty((-\pi,\pi))$. However, as we have seen many times, this convergence to δ_x holds for a larger class of test functions ϕ (cf. Theorem 11.6).

We have shown that the partial sums of the Fourier series of ϕ are simply a convolution of K_N with ϕ; i.e.,

$$\boxed{S_N(x) = \frac{1}{2\pi}\int_{-\pi}^{\pi} K_N(x-y)\,\phi(y)\,dy.} \tag{11.40}$$

The Dirichlet kernel has the following properties:

(i) By definition, for each $N \in \mathbb{N}$, $K_N(0) = 1 + 2N$.

(ii) For each $N \in \mathbb{N}$, $K_N(x)$ is an even function.

(iii) For each $N \in \mathbb{N}$, we have

$$\frac{1}{2\pi}\int_{-\pi}^{\pi} K_N(x)\,dx = 1. \tag{11.41}$$

This is easy to check from its definition (all the cosine terms integrate to 0).

(iv) For each $N \in \mathbb{N}$, we have the equivalent form valid for all $x \neq 0$:

$$\boxed{K_N(x) = \frac{\sin((N+1/2)x)}{\sin(x/2)}.}$$

To see this, let us rewrite K_N as

$$K_N(x) = 1 + \sum_{n=1}^{N}\left[e^{inx} + e^{-inx}\right] = \sum_{n=-N}^{N} e^{inx}.$$

This mean that $K_N(x)$ is a finite geometric series which, for $x \neq 0$, can be summed as follows:

$$\begin{aligned} K_N(x) = \sum_{n=-N}^{N} e^{inx} &= \frac{e^{-iNx} - e^{i(N+1)x}}{1-e^{ix}} \\ &= \frac{e^{ix/2}}{e^{ix/2}}\,\frac{e^{-i(N+1/2)x} - e^{i(N+1/2)x}}{e^{-ix/2} - e^{ix/2}} \\ &= \frac{\sin((N+1/2)x)}{\sin(x/2)}. \end{aligned}$$

In Figure 11.6, we plot $K_N(x)$ over one period $x \in (-\pi, \pi)$ for different values of N. It certainly appears that for large N, $K_N(x)$ focuses its "mass" close to $x = 0$. Away from $x = 0$, it appears to oscillate rapidly between a and $-a$, for some small positive number a. Based on these observations, it seems reasonable to expect that $K_N(x) \to \delta_0$ in the sense of distributions on $(-\pi, \pi)$[8]. We will prove this shortly; in fact, we prove a stronger statement as our test functions need not be C^∞ but C^1. However, we will **not** be able to prove the convergence with test functions which are just continuous: this is in fact false.

[8]If one considered the periodic functions $K_N(x)$ on **all** of $\mathbb{R}$, then the distributional convergence on $\mathbb{R}$ as $N \to \infty$ would be to

$$\sum_{n=-\infty}^{\infty} \delta_{2\pi n}.$$

Recall Section 5.5.3 where we proved convergence in the sense of distributions (on $(-\infty, \infty)$) of a particular sequence of functions, f_N to δ_0. In that proof, we simply needed the test functions to be continuous; that is, for any continuous test function ϕ,

$$\int_{-\infty}^{\infty} f_N(x)\phi(x)\,dx \;\to\; \phi(0) \qquad \text{as } N \to \infty.$$

The following three properties of the sequence f_N made the proof easy:

(i) The functions f_N always integrated to 1.

(ii) The functions were always positive.

(iii) The concentration property — the part of the domain upon which f_N was not close to 0 concentrated at $x = 0$ as $N \to \infty$.

In the present situation note that the functions $K_N(x)$ are 2π periodic, and hence we focus on the distributional convergence on $x \in (-\pi, \pi)$ (one period interval). More importantly, the functions $K_N(x)$ are not always positive. The integral of $\frac{1}{2\pi}K_N(x)$ over its period is always 1 but there is cancellation. Judging from Figure 11.6, there is "a type" of concentration around 0; however, away from zero, the $K_N(x)$ oscillate about 0 with a roughly fixed amplitude but a frequency which increases with N. In other words, the tails of K_N do not tend to zero in a pointwise sense, but rather in the sense of distributions. In fact, they do so in a similar way to the distributional convergence of the sequence of functions $\sin nx$ to zero.

The fact that the K_N are not always positive will mean that we will not be able to prove that

$$\frac{1}{2\pi}\int_{-\pi}^{\pi} K_N(x)\phi(x)\,dx \;\to\; \phi(0) \qquad \text{as } N \to \infty,$$

for test functions ϕ which are only continuous. Indeed, we will need the differentiability (C^1) on the test functions. This means that we will not prove that the Fourier series of any continuous function converges pointwise. In fact, this is false and the cancellation effects in the Dirichlet kernel entail that more thorough treatments of the

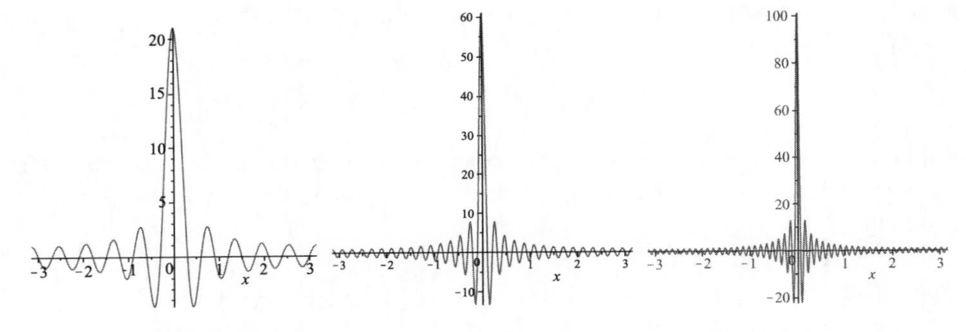

Figure 11.6. Plots of $K_N(x)$ on one period $(-\pi, \pi)$ for $N = 10$, 30, and 50, respectively. **Warning:** For these plots of the Dirichlet kernel, the software has chosen a vertical scale based upon the maximum value at $x = 0$. For this reason, it may look like the amplitude in the tails is tending to zero as $N \to \infty$. In fact, **they are not**: The oscillations are increasing but the amplitude is essentially fixed.

convergence of Fourier series are very difficult and delicate. Luckily for us, we can conveniently leave these difficulties and subtleties to certain mathematicians, referred to as *harmonic analysts*. In this text, the following convergence theorem will suffice. Since the theorem will apply to functions with jump discontinuities, the reader should first recall from (11.11) the notation for $\phi(x-)$ and $\phi(x+)$, the left and right limits.

Pointwise Convergence of the Full Fourier Series

Theorem 11.6. *Suppose that ϕ and ϕ' are piecewise continuous on $(-\pi, \pi)$ and that $\phi((-\pi)+)$, $\phi'((-\pi)+)$, $\phi(\pi-)$, and $\phi'(\pi-)$ exist. Extend ϕ periodically to all of $\mathbb{R}$. Then, for any $x \in \mathbb{R}$, the classical full Fourier series converges pointwise to*

$$\frac{\phi(x+) + \phi(x-)}{2};$$

that is, for all $x \in \mathbb{R}$,

$$S_N(x) \longrightarrow \frac{\phi(x+) + \phi(x-)}{2},$$

where the partial sums $S_N(x)$, defined by (11.38) *and* (11.37), *can be written in terms of the Dirichlet kernel as in* (11.40). *In particular, at any point x where the periodic extension of ϕ is continuous, $S_N(x)$ converges to $\phi(x)$.*

Note that the reason why we state the theorem for the periodic extension defined on all of $\mathbb{R}$ is that it facilitates a convenient and elegant way of also addressing the convergence of the Fourier series at the boundary points $x = \pm\pi$. We present the standard proof in the next subsection. Unfortunately, it is not a proof which directly addresses the concentration of the Dirichlet kernel into the delta function: Rather it is a *sneaky* proof wherein the convergence is proven using **Bessel's inequality**! It will also invoke the infamous L'Hôpital's Rule from first year calculus. While this proof is rather standard, we follow closely the treatment from Strauss [**6**].

11.5.3. • The Proof of Pointwise Convergence of the Full Fourier Series.

Proof of Theorem 11.6. Part 1: We first prove this in the case where ϕ and ϕ' are continuous on $(-\pi, \pi)$ and

$$\phi((-\pi)+) = \phi(\pi-), \qquad \phi'((-\pi)+) = \phi'(\pi-).$$

Note that the latter boundary conditions ensure that the periodic extension of ϕ is C^1 on all of $\mathbb{R}$. **Fix** an $x \in \mathbb{R}$. We change variables in the integral of (11.40) via $\theta = y - x$ to obtain

$$S_N(x) = \frac{1}{2\pi}\int_{-\pi}^{\pi} K_N(x-y)\,\phi(y)\,dy = \frac{1}{2\pi}\int_{-\pi}^{\pi} K_N(\theta)\,\phi(x+\theta)\,d\theta.$$

Above we used the fact that K_N was an even function. Also note that the limits of integration were kept fixed under the change of variables. This is perfectly fine since all functions in the integrand are periodic with period 2π. Hence all integrals over intervals of length 2π result in the same value. Using the property (11.41) of the Dirichlet kernel, we have

$$
\begin{aligned}
|S_N(x) - \phi(x)| &= \left| \frac{1}{2\pi} \int_{-\pi}^{\pi} K_N(\theta)[\phi(x+\theta) - \phi(x)]\, d\theta \right| \\
&= \left| \frac{1}{2\pi} \int_{-\pi}^{\pi} \psi(\theta) \sin((N+1/2)\theta)\, d\theta \right|, \qquad (11.42)
\end{aligned}
$$

with ψ defined by

$$\psi(\theta) := \frac{\phi(x+\theta) - \phi(x)}{\sin(\theta/2)}.$$

Thus far, the only thing we have done is rearrange terms and define ψ. Our goal remains to show that the right-hand side of (11.42) tends to 0 as $N \to \infty$. It turns out that our previous work with respect to the orthogonality of eigenfunctions associated with $\mathcal{A}$ and symmetric boundary conditions, and Bessel's inequality, is all we need. To this end, first note that the set of functions $\{X_N\}$ where

$$X_N(\theta) = \sin\left(\left(N + \frac{1}{2}\right)\theta\right), \qquad N = 0, 1, \ldots, \qquad (11.43)$$

constitutes an orthogonal set since they are simply the eigenfunctions of $\mathcal{A}$ on the interval $(0, \pi)$ corresponding to mixed boundary conditions (see Exercise **11.10**). Since they are all odd, they are also an **orthogonal set of functions on the entire interval** $(-\pi, \pi)$. A direct calculation (do it!) yields that for each $N = 0, 1, \ldots$, the X_N on $(-\pi, \pi)$ have squared length $\|X_N\|^2 = \pi$. Hence, the coefficients associated with Bessel's inequality are given by

$$A_N = \frac{1}{\pi} \int_{-\pi}^{\pi} \psi(\theta) \sin((N+1/2)\theta)\, d\theta.$$

If we can show $A_N \to 0$ as $N \to \infty$, we are done as this will mean that the right-hand side of (11.42) will also tend to 0. But this amounts to proving that ψ is in $L^2(-\pi, \pi)$; i.e.,

$$\int_{-\pi}^{\pi} |\psi(\theta)|^2\, d\theta < \infty. \qquad (11.44)$$

Indeed, if (11.44) holds true, then applying Bessel's inequality (Theorem 11.3) to the orthogonal family of functions $\{X_N\}$ gives

$$\sum_{N=1}^{\infty} A_N^2 \|X_N\|^2 \leq \int_{-\pi}^{\pi} |\psi(\theta)|^2\, d\theta \; < \infty.$$

Thus, $A_N \to 0$ as $N \to \infty$.

Hence it remains to prove (11.44). How can this integral blow up? Both the numerator and the denominator are continuous on $(-\pi, \pi)$ and, hence, the only problem could be when the denominator is 0, that is when $\theta = 0$. However, using **L'Hôpital's**

Rule one finds that

$$\lim_{\theta\to 0}\psi(\theta) = \lim_{\theta\to 0}\frac{\phi(x+\theta)-\phi(x)}{\sin(\theta/2)} = \lim_{\theta\to 0}\frac{\phi'(x+\theta)}{\frac{1}{2}\cos(\theta/2)} = 2\phi'(x) < \infty. \tag{11.45}$$

Hence, ψ remains bounded as θ approaches 0; therefore the integral on the left of (11.44) must be finite. This completes the proof for ϕ with continuous extension.

Part 2: We now prove pointwise convergence to $(\phi(x+)+\phi(x-))/2$ in the case where ϕ may either have jump discontinuities in $(-\pi,\pi)$ or have mismatched values at $x=\pm\pi$. All the work has actually been done! We note that the kernel $K_N(\theta)$ is symmetric about the vertical axis and hence,

$$\frac{1}{2\pi}\int_{-\pi}^{0} K_N(\theta)\,d\theta = \frac{1}{2\pi}\int_{0}^{\pi} K_N(\theta)\,d\theta = \frac{1}{2}.$$

Thus, instead of (11.42), we have

$$\begin{aligned}&\left|S_N(x)-\left(\frac{\phi(x+)+\phi(x-)}{2}\right)\right|\\ &= \left|\frac{1}{2\pi}\int_0^{\pi} K_N(\theta)\,[\phi(x+\theta)-\phi(x+)]\,d\theta + \frac{1}{2\pi}\int_{-\pi}^{0} K_N(\theta)\,[\phi(x+\theta)-\phi(x-)]\,d\theta\right|\\ &= \left|\frac{1}{2\pi}\int_0^{\pi} \psi_1(\theta)\sin((N+1/2)\theta)\,d\theta + \frac{1}{2\pi}\int_{-\pi}^{0} \psi_2(\theta)\sin((N+1/2)\theta)\,d\theta\right| \end{aligned}\tag{11.46}$$

where ψ_1 and ψ_2 are defined by

$$\psi_1(\theta) := \frac{\phi(x+\theta)-\phi(x+)}{\sin(\theta/2)} \quad\text{and}\quad \psi_2(\theta) := \frac{\phi(x+\theta)-\phi(x-)}{\sin(\theta/2)}.$$

We need to show that both integrals on the right-hand side of (11.46) tend to 0 as $N\to\infty$. However, as we previously mentioned, the X_N given by (11.43) are orthogonal on $(0,\pi)$. Since they are odd, they are also orthogonal on $(-\pi,0)$. Hence if we can show that ψ_1 and ψ_2 are in L^2 on $(0,\pi)$ and $(-\pi,0)$, respectively, then Bessel's inequality (Theorem 11.3) will hold on each interval and the result will follow. But this follows from exactly the same argument with (11.45) replaced by

$$\lim_{\theta\to 0^+}\psi_1(\theta) = 2\phi'(x+) < \infty \quad\text{and}\quad \lim_{\theta\to 0^-}\psi_2(\theta) = 2\phi'(x-) < \infty.$$

Note that since ϕ' is also piecewise continuous, both $\phi'(x+)$ and $\phi'(x-)$ exist. The proof is now complete. □

11.6. Term-by-Term Differentiation and Integration of Fourier Series

Given that it is easy to differentiate and integrate sines and cosines (indeed, they are eigenfunctions of the second derivative), it is very tempting to perform these operations on any function by operating **separately** on each term in its Fourier series. This way, each sine term becomes a constant times a cosine term and vice versa. However, this requires some care, as while it is true that the derivative of a **finite** sum is the finite sum of the derivatives, it is **not** in general true that the derivative of an **infinite** sum is the infinite sum of the derivatives.

The following intriguing example will help illustrate the dangers that one may encounter. Let us consider the function e^x on $(0, l)$. We think you will agree that this is a very nice function (C^∞). We consider the Fourier cosine series of e^x on an interval $(0, l)$ for some $l > 0$:

$$e^x = \frac{A_0}{2} + \sum_{n=1}^{\infty} A_n \cos\frac{n\pi x}{l}, \qquad \text{where} \quad A_n = \frac{2}{l}\int_0^l e^x \cos\left(\frac{n\pi x}{l}\right) dx. \tag{11.47}$$

Note that we are justified in writing the equality above as for any $x \in (0, l)$ this series converges to e^x.

Our aim now is to find the A_n in an indirect fashion using the fact that any derivative of e^x is also e^x. Indeed, differentiating both sides of the first equation in (11.47) with respect to x, one would expect that for any $x \in (0, l)$,

$$e^x = \sum_{n=1}^{\infty} -\frac{n\pi}{l} A_n \sin\frac{n\pi x}{l}.$$

We determined this by differentiating the Fourier cosine series term by term; i.e, we assumed the derivative of the infinite sum is the sum of the derivatives. Now suppose we differentiated one more time to find that for any $x \in (0, l)$

$$e^x = \sum_{n=1}^{\infty} -\frac{n^2\pi^2}{l^2} A_n \cos\frac{n\pi x}{l}. \tag{11.48}$$

But wait, we are back to a Fourier cosine series for e^x on $(0, l)$. Comparing the two Fourier cosine series in (11.47) and (11.48), we must have equality of the respective Fourier coefficients. This means that for all n, we must have $A_n = 0$!

So where is the mistake? The mistake is the term-by-term differentiation of the second sine series. Let us be more careful. Take ϕ to be a C^2 function on $(0, l)$ and consider its Fourier sine series:

$$\phi(x) = \sum_{n=1}^{\infty} a_n \sin\frac{n\pi x}{l} \qquad \text{where} \quad a_n = \frac{2}{l}\int_0^l \phi(x) \sin\frac{n\pi x}{l}\, dx.$$

Now, whatever $\phi'(x)$ is, we can consider its Fourier cosine series

$$\phi'(x) = \frac{b_0}{2} + \sum_{n=1}^{\infty} b_n \cos\frac{n\pi x}{l} \qquad \text{where} \quad b_n = \frac{2}{l}\int_0^l \phi'(x) \cos\frac{n\pi x}{l}\, dx.$$

What is the relationship between the a_n and b_n? Well,

$$b_0 = \frac{2}{l}\int_0^l \phi'(x)\, dx = \frac{2}{l}(\phi(l) - \phi(0)),$$

and via integration by parts,

$$\begin{aligned} b_n &= \frac{2}{l}\int_0^l \phi'(x)\cos\frac{n\pi x}{l}\,dx \\ &= \frac{2}{l}\left[\phi(x)\cos\frac{n\pi x}{l}\right]_0^l + \frac{n\pi}{l}\frac{2}{l}\int_0^l \phi(x)\sin\frac{n\pi x}{l}\,dx \\ &= \frac{2}{l}\left(\phi(l)(-1)^n - \phi(0)\right) + \frac{n\pi}{l}a_n. \end{aligned}$$

So, we do not have $b_n = \frac{n\pi}{l}a_n$ unless $\phi(l) = \phi(0) = 0$. On the other hand, for any n, we do have a relationship between a_n and b_n which allows us to determine A_n in (11.47) explicitly.

In this section, we will state and prove sufficient conditions on the function for which term-by-term differentiation and term-by-term integration are valid. In doing so, we will see an important and surprising relation between the decay of the Fourier coefficients and the smoothness of the function. Moreover, this fact will enable us to prove that under certain smoothness conditions, the pointwise convergence of the Fourier series is even stronger: It converges **uniformly**. For all these issues and results, we will focus on the full Fourier series on $(-\pi,\pi)$, that is, the classical Fourier series on $(-\pi,\pi)$ corresponding to periodic boundary conditions. These results immediately extend to the full Fourier series on interval $(-l,l)$ and, moreover, one can also generalize these results to other Fourier series.

11.6.1. Term-by-Term Differentiation. Consider, for example, the full Fourier series of a function $\phi(x)$ on $(-\pi,\pi)$,

$$\phi(x) = \frac{1}{2}a_0 + \sum_{n=1}^{\infty}\left[a_n\cos(nx) + b_n\sin(nx)\right]. \tag{11.49}$$

Suppose ϕ is differentiable on $(-\pi,\pi)$. By Theorem 11.6, we know the Fourier series will converge for any $x\in(-\pi,\pi)$. Some natural questions are: "What is the Fourier series of $\phi'(x)$, and is it simply obtained by differentiating term by term the Fourier series for ϕ?" In other words, do we have

$$\phi'(x) = \sum_{n=1}^{\infty}\left[n\,b_n\cos(nx) - n\,a_n\sin(nx)\right]? \tag{11.50}$$

As we saw with our opening example involving e^x, the answer is rather subtle. However, under certain additional hypotheses, we can simply use integration by parts to prove the following theorem.

Theorem 11.7 (Term-by-Term Differentiation). **(i)** *Let $\phi(x)$ be a continuous function on $[-\pi,\pi]$ with $\phi(-\pi)=\phi(\pi)$ and piecewise C^1 on $(-\pi,\pi)$. This means that the periodic extension of ϕ to all of $\mathbb{R}$ is a continuous and piecewise smooth function. Consider the full Fourier series of ϕ given by* (11.49) *and let A_n, $n=0,1,\ldots,$ and B_n, $n=1,2,\ldots,$ be the Fourier coefficients of $\phi'(x)$. Then*

$$A_0 = 0, \quad A_n = nb_n, \quad \textit{and} \quad B_n = -na_n, \quad n=1,2,\ldots.$$

(ii) *Suppose further that the extension of ϕ' is piecewise C^1 on any interval of $\mathbb{R}$ on which it is continuous. Then* (11.50) *holds true. If x is a point where ϕ' has a jump discontinuity, then the Fourier series on the right-hand side of* (11.50) *converges to*

$$\frac{\phi'(x+) + \phi'(x-)}{2}.$$

Proof. The proof of the first part is just integration by parts. Indeed, by definition,

$$A_n = \frac{1}{\pi}\int_{-\pi}^{\pi} \phi'(x) \cos nx \, dx.$$

Integrating by parts, we find for $n = 1, 2, \ldots$

$$\begin{aligned}\frac{1}{\pi}\int_{-\pi}^{\pi} \phi'(x) \cos nx \, dx &= \frac{1}{\pi}\phi(x) \cos nx \Big|_{x=-\pi}^{x=\pi} - \frac{1}{\pi}\int_{-\pi}^{\pi} \phi(x) (-n) \sin nx \, dx \\ &= n\frac{1}{\pi}\int_{-\pi}^{\pi} \phi(x) \sin nx \, dx \\ &= nb_n.\end{aligned}$$

Here we used the fact that $\phi(-\pi) = \phi(\pi)$ to dispense with the boundary terms. The case A_0 can be checked directly and the analogous argument follows for the B_n.

The proof of the second part now follows from Theorem 11.6 applied to $\phi'(x)$. □

Assuming enough smoothness of ϕ, one can extrapolate Theorem 11.7 to any number of derivatives of ϕ. If we combine this with the the Riemann-Lebesgue Lemma (Theorem 11.5), we find a rather surprising feature about the Fourier coefficients of very smooth functions: They must decay to 0 very fast. Indeed, if the periodic extension of ϕ is $C^\infty(\mathbb{R})$, then for any $k \in \mathbb{N}$, $n^k|a_n|$ and $n^k|b_n|$ must tend to zero as $n \to \infty$. In other words, $|a_n|$ and $|b_n|$ must tend to zero faster than $\frac{1}{n^k}$ as $n \to \infty$. Alternatively, if we know the rate of decay of the Fourier coefficient of some function, we know something about its smoothness (the number of derivatives which are continuous). This fact will enable us to prove the uniform convergence of certain Fourier series. The reader should note the analogue with Section 6.5 which discussed the relationship between smoothness and decay of the Fourier transform.

A good exercise (cf. Exercise **11.20**) is to write down and prove the analogous term-by-term differentiation theorems for the Fourier sine and cosine series.

Important Remark for Numerical Computation

Term-by-term differentiation of Fourier series is suggestive of what turns out to be a very powerful tool in scientific computation. Assuming we have a fast and efficient method for calculating Fourier coefficients, then numerical differentiation and approximately solving certain differential equations can be algebraically performed with a finite number of coefficients. This idea is at the root of a class of numerical methods known as **spectral methods**. They are widely used in science and engineering.

11.6.2. Term-by-Term Integration. Unlike differentiation, integration is a *smoothing process* and the requirements for term-by-term integration are far less restrictive. Indeed, we may relax the assumption on differentiability.

Theorem 11.8 (Term-by-Term Integration). *Let $\phi(x)$ be a piecewise continuous function on $[-\pi, \pi]$ and extend ϕ to $\mathbb{R}$ by periodicity (with period $= 2\pi$). Let $a_n, n = 0, 1, \ldots,$ and $b_n, n = 1, 2, \ldots,$ be the Fourier coefficients given by (11.37). Let Φ be the following primitive of ϕ:*

$$\Phi(x) := \int_0^x \phi(y)\, dy.$$

Then Φ is continuous, piecewise C^1, and at any x where ϕ is continuous, $\Phi'(x) = \phi(x)$. Moreover, the Fourier series for Φ is given by

$$\frac{1}{2}A_0 + \sum_{n=1}^{\infty}\left[\frac{-b_n}{n}\cos(nx) + \frac{a_n}{n}\sin(nx)\right] \quad \textit{where} \quad \frac{1}{2}A_0 = \frac{1}{2\pi}\int_{-\pi}^{\pi}\Phi(y)\, dy,$$

and it converges pointwise to Φ.

Proof. By what you know about the Fundamental Theorem of Calculus, the fact that $\Phi'(x) = \phi(x)$ should seem reasonable. Here, ϕ is only piecewise continuous, and there is some slightly more advanced mathematical analysis, which we omit, needed to prove that Φ is continuous and piecewise C^1. Accepting this, the remainder of the proof follows from Theorem 11.6 and the integration by parts argument in the proof of Theorem 11.7. These last short details are left to the reader. □

11.7. Convergence, III: Uniform Convergence

We have already addressed pointwise convergence of functions and the application to Fourier series. There is a stronger way functions can converge which is called **uniform convergence**.

11.7.1. Uniform Convergence of Functions. Pointwise convergence of functions f_N to f just says that for any input value x, the sequence of outputs $f_N(x)$ converge to the number $f(x)$. There is no information about the "**rates**" at which these values converge for **different** x. Roughly speaking, uniform convergence means they all converge at the "same rate". More specifically, a sequence of functions f_N **converges uniformly** to a function f on $[a, b]$ if and only if

$$\sup_{x\in[a,b]} |f_N(x) - f(x)| \xrightarrow{N\to\infty} 0.$$

Here we use sup to denote **the supremum**. However, in this section we will be concerned only with functions (the f_N and f) which are continuous and, in this case, $\sup_{x\in[a,b]}$ is simply the maximum value over $x \in [a, b]$. Alternatively, a sequence of functions f_N **converges uniformly** to a function f on $[a, b]$ if and only if for every $\epsilon > 0$ there exists $M > 0$ such that

$$\text{if } N \geq M, \quad \text{then} \quad |f_N(x) - f(x)| < \epsilon \quad \text{for all } x \in [a, b].$$

Uniform convergence is stronger than pointwise convergence as it is possible to have functions converge pointwise but at sufficiently different "rates" so that the convergence is not uniform. For example, consider the sequence of functions f_N defined on an interval $[a, b]$ by

$$f_N(x) := \begin{cases} 2Nx & \text{if } x \in \left[a, a + \frac{1}{2N}\right], \\ 2 - 2Nx & \text{if } x \in \left[a + \frac{1}{2N}, a + \frac{1}{N}\right], \\ 0 & \text{if } x \in \left[a + \frac{1}{N}, b\right]. \end{cases}$$

Do not be too bothered with this formula, but rather consider the graph of f_N depicted in Figure 11.7. For this sequence, $f_N \to 0$ pointwise on $[a, b]$ but not uniformly; indeed, for any n, we have

$$\max_{x \in [a,b]} |f_N(x) - 0| = 1.$$

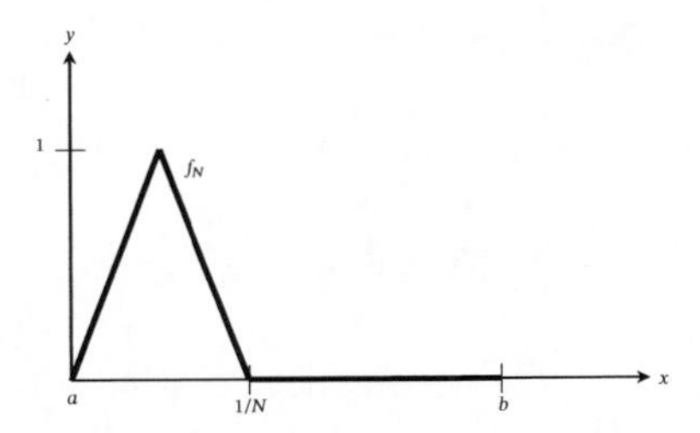

Figure 11.7. Depiction of the spike functions f_N on $[a, b]$.

Of course this example is rather special since for any $x \in [a, b]$, $f_N(x)$ is eventually exactly 0. However, as x gets closer to a, this "eventually" comes "later and later".

11.7.2. A Criterion for the Uniform Convergence of Fourier Series.

We want to consider uniform convergence of Fourier series which entails uniform convergence of the partial sums. To this end, let ϕ be a continuous function on $[-\pi, \pi]$ such that $\phi(-\pi) = \phi(\pi)$. Hence, the periodic extension is also continuous and bounded on all of $\mathbb{R}$. The Fourier series associated with ϕ will converge uniformly if and only if

$$\max_{x \in \mathbb{R}} |S_N(x) - \phi(x)| \xrightarrow{N \to \infty} 0.$$

To address uniform convergence, we must recall the notion of **absolute convergence** of an infinite series. Recall that an infinite series

$$\sum_{n=0}^{\infty} a_n \quad \textbf{converges absolutely} \text{ if } \sum_{n=0}^{\infty} |a_n| < \infty.$$

We can extrapolate to an infinite series of functions to say that the series $\sum_{n=0}^{\infty} g_n(x)$ converges absolutely if for all x (in the domain), we have $\sum_{n=0}^{\infty} |g_n(x)| < \infty$.

We now state a uniform convergence criterion.

Uniform Convergence of the Full Fourier Series

Theorem 11.9. *Suppose that ϕ is continuous and piecewise C^1 on $[-\pi, \pi]$ with $\phi(-\pi) = \phi(\pi)$. Then the classical full Fourier series converges absolutely and uniformly to ϕ on $[-\pi, \pi]$. In fact, the full Fourier series converges absolutely and uniformly to the periodic extension of ϕ on all of $\mathbb{R}$.*

For students with some background in mathematical analysis, we provide a proof of the theorem in the next subsection. If you choose to skip the proof, that is fine, but do **not** for a moment dismiss uniform convergence as something which is only of concern to overly pedantic mathematicians: The failure of uniform convergence is extremely important to anyone using Fourier series and is related to what is known as the Gibbs phenomenon, briefly discussed after the proof.

11.7.3. The Proof of Theorem 11.9.

Proof. The full Fourier series of ϕ is given by

$$\frac{1}{2}a_0 + \sum_{n=1}^{\infty} [a_n \cos(nx) + b_n \sin(nx)]$$

and by Theorem 11.6 we know that for any fixed $x \in \mathbb{R}$, this series converges to $\phi(x)$, where ϕ is understood to be the periodic extension to all of $\mathbb{R}$. To prove absolute convergence, note that for any $n = 1, 2, \ldots$ and $x \in \mathbb{R}$,

$$\begin{aligned} |a_n \cos(nx) + b_n \sin(nx)| &\leq |a_n \cos(nx)| + |b_n \sin(nx)| \\ &\leq |a_n| + |b_n|. \end{aligned}$$

Therefore, it suffices to prove that

$$\sum_{n=1}^{\infty} |a_n| + |b_n| < \infty. \tag{11.51}$$

Note that this does not *immediately* follow from Bessel's inequality (Theorem 11.3). Indeed, since ϕ is in $L^2((-\pi, \pi))$ (why?), Bessel's inequality implies that

$$\sum_{n=1}^{\infty} a_n^2 + b_n^2 < \infty.$$

This is not the same thing, nor does it directly imply (11.51): Recall that

$$\sum_{n=1}^{\infty} \frac{1}{n^2} < \infty \quad \text{but} \quad \sum_{n=1}^{\infty} \frac{1}{n} = \infty.$$

To show (11.51) we need to know that the a_n and b_n decay sufficiently fast. This decay rate will now be shown via the fact that we know something about the Fourier series of ϕ' from the term-by-term differentiation theorem, Theorem 11.7. Indeed, by hypothesis, ϕ' is in $L^2((-\pi, \pi))$, and so if A_n and B_n denote its Fourier coefficients, Bessel's inequality (Theorem 11.3) implies

$$\sum_{n=1}^{\infty} A_n^2 + B_n^2 < \infty. \tag{11.52}$$

By Theorem 11.7, $|A_n| = n|b_n|$ and $|B_n| = n|a_n|$. So

$$\sum_{n=1}^{\infty} |a_n| + |b_n| = \sum_{n=1}^{\infty} \frac{1}{n}(|A_n| + |B_n|).$$

What we require now is a very useful inequality (Lemma 11.7.1 below) which is called the Cauchy-Schwarz inequality for infinite series. We state and prove this lemma at the end of this proof. By Lemma 11.7.1, we find

$$\begin{aligned}\sum_{n=1}^{\infty} \frac{1}{n}(|A_n| + |B_n|) &\leq \left(\sum_{n=1}^{\infty} \frac{1}{n^2}\right)^{1/2} \left(\sum_{n=1}^{\infty} (|A_n| + |B_n|)^2\right)^{1/2} \\ &\leq \left(\sum_{n=1}^{\infty} \frac{1}{n^2}\right)^{1/2} \left(\sum_{n=1}^{\infty} 2(A_n^2 + B_n^2)\right)^{1/2}.\end{aligned}$$

To pass to the last line, we used the trivial inequality[9] $2|A_n||B_n| \leq A_n^2 + B_n^2$. But the right-hand side is finite by (11.52) and the fact that $\sum_{n=1}^{\infty} \frac{1}{n^2} < \infty$. Thus, (11.51) holds and the Fourier series converges absolutely.

With (11.51) in hand and the fact that Theorem 11.6 implies pointwise convergence, we can easily prove uniform convergence. To this end, since for any $x \in \mathbb{R}$,

$$\phi(x) = \frac{1}{2}a_0 + \sum_{n=1}^{\infty} [a_n \cos(nx) + b_n \sin(nx)],$$

we find that

$$\begin{aligned}\max_{x\in\mathbb{R}} |\phi(x) - S_N(x)| &= \max_{x\in\mathbb{R}} \left| \sum_{n=N+1}^{\infty} a_n \cos(nx) + b_n \sin(nx) \right| \\ &\leq \max_{x\in\mathbb{R}} \sum_{n=N+1}^{\infty} |a_n \cos(nx)| + |b_n \sin(nx)| \\ &\leq \sum_{n=N+1}^{\infty} |a_n| + |b_n|.\end{aligned}$$

But by (11.51) and the fact that the tails of a convergent series must tend to zero,

$$\sum_{n=N+1}^{\infty} (|a_n| + |b_n|) \longrightarrow 0 \quad \text{as} \quad N \to \infty.$$

Hence

$$\max_{x\in\mathbb{R}} |\phi(x) - S_N(x)| \longrightarrow 0 \quad \text{as} \quad N \to \infty.$$

□

[9] Trivial since it is a direct consequence of the fact that $(|A_n| - |B_n|)^2 \geq 0$.

Lemma 11.7.1 (Cauchy-Schwarz Inequality for Infinite Series). *For any sequences of real numbers* $\{a_n\}_{n=1}^\infty$, $\{b_n\}_{n=1}^\infty$ *such that* $\sum_{n=1}^\infty a_n^2 < \infty$ *and* $\sum_{n=1}^\infty b_n^2 < \infty$, *we have*

$$\sum_{n=1}^{\infty} a_n b_n \leq \left(\sum_{n=1}^{\infty} a_n^2\right)^{1/2} \left(\sum_{n=1}^{\infty} b_n^2\right)^{1/2}.$$

Proof. We first prove that for any natural number N,

$$\sum_{n=1}^{N} a_n b_n \leq \left(\sum_{n=1}^{N} a_n^2\right)^{1/2} \left(\sum_{n=1}^{N} b_n^2\right)^{1/2}. \tag{11.53}$$

There are many ways to prove (11.53). Here is a simple, purely algebraic proof. We have

$$\begin{aligned}
0 &\leq \sum_{n=1}^{N} \sum_{m=1}^{N} (a_n b_m - a_m b_n)^2 \\
&= \sum_{n=1}^{N} a_n^2 \sum_{m=1}^{N} b_m^2 + \sum_{n=1}^{N} b_n^2 \sum_{m=1}^{N} a_m^2 - 2 \sum_{n=1}^{N} a_n b_n \sum_{m=1}^{N} b_m a_m \\
&= 2\left(\sum_{n=1}^{N} a_n^2\right)\left(\sum_{n=1}^{N} b_n^2\right) - 2\left(\sum_{n=1}^{N} a_n b_n\right)^2.
\end{aligned}$$

This proves (11.53). Taking the limit as $N \to \infty$ on both sides of (11.53) completes the proof. □

11.7.4. The Gibbs Phenomenon.

We discuss here what is known as the **Gibbs**[10] **phenomenon**. Let us illustrate this with an example.

Consider the full Fourier series of the function ϕ defined on the interval $(-1, 1)$ by

$$\phi(x) := \begin{cases} 1 & \text{if } x \in (-1, 0], \\ -1 & \text{if } x \in (0, 1). \end{cases}$$

Note that this full Fourier series is exactly the same as the Fourier sine series of 1 on the interval $(0, 1)$ which we found in Example 11.1.1 (why?). We plot a partial sum of this full Fourier series in Figure 11.8. While any partial sum is infinitely smooth, its pointwise limit is the function

$$\begin{cases} 0 & \text{if } x = n,\ n \in \mathbb{Z}, \\ 1 & \text{if } x \in (2n, 2n+1),\ n \in \mathbb{Z}, \\ -1 & \text{if } x \in (2n-1, 2n),\ n \in \mathbb{Z}. \end{cases}$$

The convergence of this Fourier series is not uniform and we observe two remarkable features:

- For any noninteger values of x, there is pointwise convergence to $\phi(x)$ **but** it appears that the convergence is slow.

[10] **Josiah Willard Gibbs** (1839–1903) was an American physicist, chemist, and mathematician who made fundamental contributions to many fields including thermodynamics. While Gibbs's name is attached to many concepts in science, it is worth noting that the "Gibbs phenomenon" was actually discovered earlier by the English mathematician **Henry Wilbraham**.

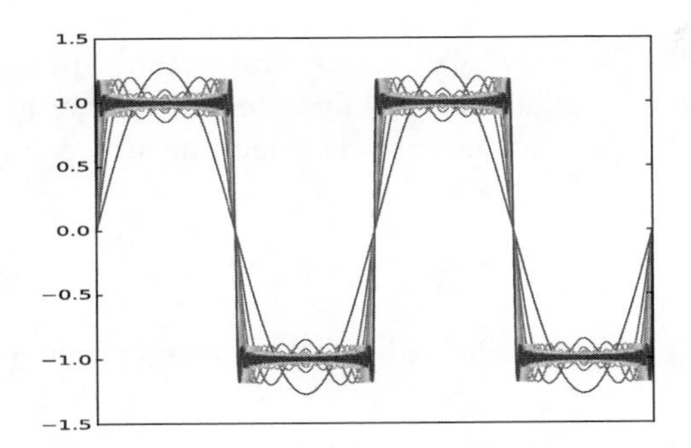

Figure 11.8. Plot of the partial sums $S_N(x)$ for the full Fourier sine series of $\phi(x) = 1$ on $(-2, 2)$.

- Near $x = 0$ (or $x = \pm 1$), there seems to be an **overshoot** of 1 or -1. Indeed, this overshoot always has a constant height above and below, approximately 9% of the jump discontinuity; further, the overshoot concentrates towards the jump discontinuity as $N \to \infty$. Of course, in the limit, the partial sums will "break", converging at integer values of x to exactly halfway between 1 and -1.

It turns out that these observations are **generic** for the full Fourier series of a function ϕ when **either** ϕ itself is discontinuous on the sampled domain $(-l, l)$ **or** ϕ is not periodic, i.e., $\phi(-l) \neq \phi(l)$, yielding a discontinuous periodic extension. These observations have profound consequences in the applications of Fourier series (for example, in signal processing) and are known as the **Gibbs phenomenon**.

We conclude with comments related to the first observation wherein we claimed that the convergence at noninteger values of x was slow. What exactly do we mean by slow and, more generally, by speed of convergence? If a function ϕ on $(-l, l)$ is smooth and periodic, then it can be shown that there exists an $\alpha > 0$ such that

$$\max_{x \in [-l,l]} |S_N(x) - \phi(x)| \leq e^{-\alpha N}.$$

This means the convergence is uniform and the partial sums converge **exponentially fast**. Check this out for yourself: You will observe that just an "average" number N of partial sums yields an excellent approximation. On the other hand, consider a piecewise smooth function whose periodic extension is discontinuous. Then at a point x away from these discontinuities, we would have pointwise convergence **but** one can only prove

$$|S_N(x) - \phi(x)| \leq \frac{C}{N},$$

for some constant C depending only on ϕ. This is what we would call slow convergence.

11.8. What Is the Relationship Between Fourier Series and the Fourier Transform?

We remind the reader that the present chapter was written to be independent from the previous Chapter 6 on the Fourier transform. Here we recall the Fourier transform and ask the natural question: What is the relationship?

Fourier series allow us to encapsulate/reconstruct the entire structure of a function defined on a finite interval with a countable sequence of numbers (the Fourier coefficients). On the other hand, recall the Fourier transform of an integrable function $f(x)$ on $\mathbb{R}$

$$\hat{f}(k) := \int_{-\infty}^{\infty} f(x)\, e^{-ikx}\, dx.$$

The Fourier inversion formula tells us how to *reconstruct* (*synthesize*) the function f from its Fourier transform $\hat{f}$:

$$f(x) = \frac{1}{2\pi}\int_{-\infty}^{\infty} \hat{f}(k)\, e^{ixk}\, dk. \tag{11.54}$$

In this case, we need a continuum of numbers ($\widehat{f}(k)$, $k \in \mathbb{R}$) to synthesize the function, with the function being synthesized by an integral rather than an infinite sum (as it was in a Fourier series). As we described in Section 6.12.2, formula (11.54) can also be written as an expression of f in terms of sines and cosines with frequency k. Thus, clearly there must be a more direct link between Fourier series and the Fourier transform. But how can we precisely describe/capture this link? We discuss two ways.

11.8.1. Sending $l \to \infty$ in the Full Fourier Series. Consider the full Fourier series for a function f on an interval, say, $(-l, l)$, and ask what happens as l gets large? That is, we consider the function (information) on larger and larger intervals. We will provide **an intuitive, yet convincing connection** which can be made rigorous but with much work. To this end, given that the Fourier transform is expressed as a complex-valued function, it would seem natural to write the Fourier series of f on $(-l, l)$ in the complex form as we did in (11.13):

$$f(x) = \sum_{n=-\infty}^{\infty} c_n e^{in\pi x/l} \quad \text{with} \quad c_n = \frac{1}{2l}\int_{-l}^{l} f(y)\, e^{-in\pi y/l}\, dy.$$

Note again that this is exactly the same as the full Fourier series of f on $(-l, l)$ and, even though the c_n and $e^{in\pi x/l}$ are complex valued, the resulting sum is real valued. We place the definition of the coefficients into the series to find

$$f(x) = \sum_{n=-\infty}^{\infty} \left(\frac{1}{2l}\int_{-l}^{l} f(y)\, e^{-in\pi y/l}\, dy\right) e^{in\pi x/l}.$$

We let $k = \frac{n\pi}{l}$; the choice of notation k is deliberate. With a bit of reordering, we have

$$f(x) = \frac{1}{2\pi}\sum_{n=-\infty}^{\infty} \left(\int_{-l}^{l} f(y)\, e^{-iky}\, dy\right) e^{ikx}\, \frac{\pi}{l}.$$

Now we are ready to ask the following: What happens as $l \to \infty$? Clearly the integral in the brackets will tend to the integral of the same function but from $-\infty$ to ∞. What about the sum? Intuitively, as $l \to \infty$, k tends to 0. At least this is true for a fixed n; however, note that in the sum n goes from $-\infty$ to ∞ making this all very subtle and difficult to parse. On the other hand, we can think of the above sum as a Riemann sum for the function of k,

$$\int_{-l}^{l} f(y)\, e^{-iky}\, dy.$$

In fact, the sum just samples this function at a discrete set of values for k and multiplies each by $\frac{\pi}{l}$ which is exactly Δk, the distance between successive values of this discrete set of k. Thus in the limit where $\Delta k \to 0$, this Riemann sum becomes an integral. That is,

$$\begin{aligned} f(x) &= \frac{1}{2\pi} \sum_{n=-\infty}^{\infty} \left(\int_{-l}^{l} f(y)\, e^{-iky}\, dy \right) e^{ikx}\, \frac{\pi}{l} \\ &\xrightarrow{l\to\infty} \frac{1}{2\pi} \int_{-\infty}^{\infty} \left(\int_{-\infty}^{\infty} f(y)\, e^{-iky}\, dy \right) e^{ikx}\, dk. \end{aligned} \tag{11.55}$$

So, in the limit of $l \to \infty$, the Fourier coefficients, initially indexed by n, become a continuum of coefficients, indexed by k, which is exactly the Fourier transform. The analogue of the Fourier series is exactly the Fourier inversion formula which tells us how to recover the original function as a certain integral involving the Fourier transform.

We reiterate that the above is only a rough, intuitive treatment of the limit in (11.55). In some sense, it is correct, or at least can be made correct under certain assumptions. Note that the necessary property of integrability for f over $\mathbb{R}$ (which includes decay to 0 as $x \to \pm\infty$) is hidden.

11.8.2. Taking the Distributional Fourier Transform of a Periodic Function. Another approach to the Fourier series - Fourier transform connection is via our distributional interpretation of the Fourier transform. To this end, consider a periodic function f on $[-l, l]$ which is C^1 on $(-l, l)$. Extend this function to all of $\mathbb{R}$ by periodicity — we also call the extension f. We have seen that all the information in f is contained in a countable sequence of numbers: its Fourier coefficients which, in the complex form, are given by c_n where

$$f(x) = \sum_{n=-\infty}^{\infty} c_n e^{in\pi x/l} \quad \text{with} \quad c_n = \frac{1}{2l} \int_{-l}^{l} f(x)\, e^{-in\pi x/l}\, dx. \tag{11.56}$$

Now what would happen if we took the Fourier transform of the extension f? Would we still need the continuum of information (through $\hat{f}(k)$) to reconstruct f? Well first off, f is neither integrable nor square integrable over $\mathbb{R}$ and hence the usual definition of the Fourier transform does not make any sense. However, f is locally integrable and we can hence take its Fourier transform in the sense of tempered distributions. To this end, let us use the Fourier series description of f given in (11.56). Then assuming that the distributional Fourier transform of the infinite sum is the sum of the Fourier transforms, we find

$$\mathcal{F}\{f(x)\} = \mathcal{F}\left\{ \sum_{n=-\infty}^{\infty} c_n\, e^{in\pi x/l} \right\} = \sum_{n=-\infty}^{\infty} c_n \mathcal{F}\left\{ e^{in\pi x/l} \right\}.$$

Now recall (6.47) which stated that in the sense of tempered distributions,

$$\mathcal{F}\left\{e^{in\pi x/l}\right\} = 2\pi\delta_{\frac{n\pi}{l}}.$$

Hence, in the sense of tempered distributions,

$$\boxed{\widehat{f} = \mathcal{F}\{f(x)\} = \sum_{n=-\infty}^{\infty} 2\pi\, c_n\, \delta_{\frac{n\pi}{l}},}$$

a weighted, infinite sum of delta functions with concentrations at a countable number of points uniformly distributed on $\mathbb{R}$ with spacing equal to $\frac{\pi}{l}$. As l gets larger, these concentration points get closer and closer, eventually (in the limit) "filling" out the entire real line. Thus for a periodic function f, its (distributional) Fourier transform is captured with only a countable amount of information, in contrast to a continuum of information (through $\widehat{f}(k)$) needed to reconstruct an integrable function on $\mathbb{R}$.

11.9. Chapter Summary

- There are **three classical Fourier series**: the Fourier sine series, the Fourier cosine series, and the full Fourier series. Each series is based upon sampling a function ϕ on a set interval *against* certain trigonometric functions. The choice of sampling interval $(0, l)$ for the Fourier sine and cosine series and $(-l, l)$ for the full series are for simplicity of expression. In other words, these choices yield eigenvalues of the simple form $n^2\pi^2/l^2$. In principle, one can find analogues of any of these Fourier series for a function ϕ on any interval (a, b). Note also that for obvious reasons, $l = \pi$ is a natural choice in which to illustrate results.

  Under certain smoothness conditions on ϕ we have pointwise convergence to ϕ on the interval upon which the coefficients are defined. At all points where the extension is continuous, the respective Fourier series will converge pointwise to the following: the odd periodic extension of ϕ, the even periodic extension of ϕ, and the period extension of ϕ. At points for which the periodic extension has a jump discontinuity, the Fourier series converges to the average of the limit from the left and from the right. All three extensions are always periodic functions of period $2l$.

- The trigonometric functions in all three Fourier series are examples of **eigenfunctions** associated to $\mathcal{A} = -\frac{d^2}{dx^2}$ with certain **symmetric boundary conditions**, that is, solutions to $X'' + \lambda X = 0$ on (a, b) with boundary conditions on $X(a)$ and $X(b)$ satisfying (11.23). Four common symmetric boundary conditions follow:

  Dirichlet boundary conditions on $[0, l]$: $X(0) = 0 = X(l)$ give eigenfunctions and eigenvalues

$$X_n(x) = \sin\frac{n\pi x}{l} \quad \text{with} \quad \lambda_n = \frac{n^2\pi^2}{l^2},\ n = 1, 2, \ldots,$$

  and correspond to the components of the **Fourier sine series** of ϕ on $(0, l)$.

Neumann boundary conditions on $[0, l]$: $X'(0) = 0 = X'(l)$ give eigenfunctions and eigenvalues

$$X_n(x) = \cos\frac{n\pi x}{l} \quad \text{with} \quad \lambda_n = \frac{n^2\pi^2}{l^2}, \; n = 0, 1, \ldots,$$

and correspond to the components of the **Fourier cosine series** of ϕ on $(0, l)$.

Periodic boundary conditions on $[-l, l]$: $X(-l) = X(l)$ and $X'(-l) = X'(l)$ give eigenfunctions and eigenvalues

$$X_n(x) = \cos\left(\frac{n\pi x}{l}\right), \; n = 0, 1, \ldots, \quad \text{and} \quad Y_n(x) = \sin\left(\frac{n\pi x}{l}\right), \; n = 1, 2, \ldots,$$

with

$$\lambda_n = \frac{n^2\pi^2}{l^2}, \; n = 0, 1, \ldots.$$

They correspond to the components of the **full Fourier series** of ϕ on $(-l, l)$.

Mixed boundary conditions on $[0, l]$: $X(0) = 0 = X'(l)$ give eigenfunctions and eigenvalues

$$X_n(x) = \sin\frac{(n+\frac{1}{2})\pi x}{l} \quad \text{with} \quad \lambda_n = \frac{(n+\frac{1}{2})^2\pi^2}{l^2}, \; n = 0, 1, 2, \ldots.$$

- The eigenfunctions corresponding to different eigenvalues of $\mathcal{A}$ with a given symmetric boundary condition have two amazing properties:
  (1) They are mutually orthogonal. This is easy to see from basic calculus (Green's Second Identity in 1D).
  (2) The set of all eigenfunctions span all "reasonable" functions in the sense that any "reasonable" function can be written as a possibly infinite linear combination of them. This is subtle.

  Hence, one can construct a **general Fourier series** of a "reasonable" function ϕ on (a, b) from the eigenfunctions associated with a specific symmetric boundary condition:

$$\sum_{n=1}^{\infty} A_n X_n(x) \quad \text{where} \quad A_n = \frac{(\phi, X_n)}{\|X_n\|^2} = \frac{\int_a^b \phi(x) X_n(x)\,dx}{\int_a^b X_n^2(x)\,dx}.$$

  This idea of the eigenfunctions "spanning" all functions (more precisely, the completeness of the eigenfunctions) entails a notion of convergence. We considered two types of convergence and in each case gave a criterion for a "reasonable" function.

- **L^2 theory:** We first note that if ϕ is an L^2 function on an interval (a, b), then any orthogonal family of functions $\{X_n\}$ on (a, b) satisfies **Bessel's inequality**

$$\sum_{n=1}^{\infty} A_n^2 \int_a^b |X_n(x)|^2\,dx \le \int_a^b |\phi(x)|^2\,dx, \qquad \text{where} \quad A_n = \frac{(\phi, X_n)}{\|X_n\|^2}.$$

  If we apply this to any general Fourier series, we see that the series will converge in L^2 (mean square convergence) to ϕ if we have equality in Bessel's inequality. This equality is called **Parseval's equality**. While a complete proof was not provided, Parseval's equality holds for every L^2 function ϕ: Hence, the general Fourier series of an L^2 function ϕ converges to ϕ in the L^2 sense.

- **Pointwise theory:** Under certain assumptions on the smoothness of the function ϕ, the classical Fourier series of ϕ will **converge pointwise** to $\phi(x)$ at a point of continuity. Otherwise, it converges to the average of the limit from the left and from the right. We proved this for the full Fourier series of a piecewise C^1 function. The proof was based upon the fact that the partial sums of the Fourier series entail a convolution of ϕ with a certain function, called the **Dirichlet kernel**, which converges in the sense of distributions on $[-\pi, \pi]$ to a **delta function**. In fact, our proof allows us to extend this convergence for the Dirichlet kernel for test functions which are only C^1. A key ingredient of the proof was the application of Bessel's inequality on a certain set of orthogonal functions.
- Under certain assumptions on the function ϕ, we can perform **term-by-term differentiation** and **term-by-term integration** of the classical Fourier series.
- If the periodic extension of ϕ is continuous on all of $\mathbb{R}$, then under certain additional conditions, any classical Fourier series will **converge uniformly** to ϕ. If, on the other hand, the periodic extension has jump discontinuities, then we cannot have uniform convergence. Moreover, around a jump discontinuity we observe the **Gibbs phenomenon**.

Exercises

11.1 Show by direct integration the orthogonality relations (11.2), (11.3), and (11.6). Also show by direct integration that $\int_{-l}^{l} \sin(m\pi x/l)\cos(n\pi x/l)\,dx = 0$, for any pair of integers m and n.

11.2 Show that Dirichlet, Neumann, and periodic boundary conditions on any interval $[a, b]$ are all symmetric. In other words, show that in each case, the right-hand side of (11.22) is 0.

11.3 Let $\phi(x)$ be defined on $(0, 4]$ as follows: $f(x) = 1$ if $0 < x \le 1$; $\phi(x) = 2$ if $1 < x \le 2$; $\phi(x) = 3$ if $2 < x \le 3$; and $\phi(x) = 4$ if $3 < x \le 4$. Extend $\phi(x)$ by periodicity to all $x \in (-\infty, \infty)$; i.e., let $\phi(x+4) = \phi(x)$ for all x. Consider the full Fourier series for ϕ:

$$\phi(x) \sim \frac{1}{2}a_0 + \sum_{n=1}^{\infty} a_n \cos\left(\frac{n\pi x}{2}\right) + b_n \sin\left(\frac{n\pi x}{2}\right).$$

(a) To what values will this Fourier series converge at $x = 0$, $x = 1$, $x = 4$, $x = 7.4$, and $x = 40$?

(b) Does the Fourier series converge uniformly to $\phi(x)$? Explain.

(c) Find a_0.

11.4 Consider the following functions: (i) $\phi(x) = x^2$; (ii) $\phi(x) = \sin \pi x$; (iii) $\phi(x) = \sin x$.

(a) For each case, find the Fourier cosine and Fourier sine series on the interval $(0, 2)$.

(b) For each case, find the Fourier cosine and Fourier sine series on the interval $(0, 1)$.

In all cases, plot partial sums of these Fourier series on the interval $(-4, 4)$.

To find the coefficients, you can use any mathematical software or online integration tables.

11.5 Find the full Fourier series for the following functions defined on the interval $(-1, 1)$: **(a)** $\phi(x) = \cos(26\pi x) - 4 - 3\sin(\pi x)$, **(b)** $\phi(x) = x$, **(c)** $\phi(x) = |x| + 1$, **(d)** $\phi(x) = x^3 + x^2$. Plot the partial sums corresponding to the first 3, 5, and then 30 terms. Use any software you like to superimpose your plots against that of the original function. To find the coefficients, you can use any mathematical software or online integration tables.

11.6 Write down the full Fourier series for the following functions defined on the interval $(-\pi, \pi)$, leaving the coefficients as integrals. Using any mathematical software or online integration tables, compute a few coefficients in order to plot the partial sums corresponding to the first 3, 5, and then 30 terms. Use any software you like to superimpose your plots against that of the original function.

$$\textbf{(a)}\ \phi(x) = \begin{cases} 0 & \text{if } -\pi < x < 0, \\ x^2 & \text{if } 0 \le x < \pi, \end{cases} \qquad \textbf{(b)}\ \phi(x) = \begin{cases} 0 & \text{if } -\pi < x < 0, \\ \sin x & \text{if } 0 \le x < \pi, \end{cases}$$

$$\textbf{(c)}\ \phi(x) = |\sin x|, \qquad \textbf{(d)}\ \phi(x) = x\cos x.$$

11.7 **(a)** Let

$$f_1(x) = \frac{a_0}{2} + \sum_{n=1}^{\infty} a_n \cos n\pi x, \qquad f_2(x) = \sum_{n=1}^{\infty} b_n \sin n\pi x,$$

and

$$f_3(x) = \frac{c_0}{2} + \sum_{n=1}^{\infty} (c_n \cos n\pi x + d_n \sin n\pi x)$$

where

$$a_n = 2\int_0^1 (x^3 + 1)\cos n\pi x\, dx, \qquad b_n = 2\int_0^1 (x^3 + 1)\sin n\pi x\, dx,$$

and

$$c_n = \int_{-1}^1 (x^3 + 1)\cos n\pi x\, dx, \qquad d_n = \int_{-1}^1 (x^3 + 1)\sin n\pi x\, dx.$$

Sketch the graphs of f_1, f_2, and f_3 for $x \in (-3, 3)$; i.e., sketch the pointwise limits of the respective Fourier series. Make sure your graph clearly indicates the values for $x = -2, -1, 0, 1, 2$.

(b) What is $\lim_{n\to\infty} d_n$? What is $c_0^2 + \sum_{n=1}^{\infty} c_n^2 + d_n^2$?

(c) Let

$$f_4(x) = \sum_{n=0}^{\infty} b_n \sin\left(n + \frac{1}{2}\right)\pi x \qquad \text{where } b_n = 2\int_0^1 x^3 \sin\left(n + \frac{1}{2}\right)\pi x\, dx.$$

Sketch the graphs of f_4 for $x \in (-5, 5)$; i.e., sketch the pointwise limits of the respective Fourier series.

11.8 As we have mentioned, our choice of interval $(0, l)$ for defining the Fourier sine and cosine series and $(-l, l)$ for the full Fourier series was for convenience: This choice afforded simple expressions for the eigenvalues and eigenfunctions. This exercise will help illustrate this point by asking you to find the three Fourier series for a function on some interval (a, b). To this end, let us consider the interval to be $(2, 3)$ and the sampled function to be $\phi(x) = x^2$.
(a) Find the Fourier sine series of this function on the interval $(2, 3)$; that is, find an infinite series involving only sine functions which converges pointwise to ϕ on the interval $(2, 3)$.
(b) Find the Fourier cosine series of this function on the interval $(2, 3)$; that is, find an infinite series involving only cosine functions which converges pointwise to ϕ on the interval $(2, 3)$.
(c) Find the full Fourier series of this function on the interval $(2, 3)$; that is, find an infinite series involving both sine and cosine functions which converges pointwise to ϕ on the interval $(2, 3)$.
In all cases, you can leave the coefficients as integrals but compute enough of them so that you can plot a few partial sums over the interval $(0, 5)$.

11.9 Find the complex version of the full Fourier series of $f(x) = e^x$ on $(-1, 1)$.

11.10 Find all eigenvalues and eigenfunctions to the following eigenvalue problems on $[0, l]$:
(a) Neumann boundary conditions $X'' + \lambda X = 0$, $X'(0) = 0 = X'(l)$.
(b) Mixed boundary conditions $X'' + \lambda X = 0$, $X(0) = 0$, $X'(l) = 0$.
(c) Periodic boundary conditions $X'' + \lambda X = 0$, $X(0) = X(l)$ and $X'(0) = X'(l)$.

11.11 Using Parseval's identity, show the following:

(a) $\displaystyle\sum_{n=0}^{\infty} \frac{1}{(2n+1)^2} = \frac{\pi^2}{8}$. Hint: Consider $f(x) = |x|$ on $(-\pi, \pi)$.

(b) $\displaystyle\sum_{n=1}^{\infty} \frac{1}{n^2} = \frac{\pi^2}{6}$ and $\displaystyle\sum_{n=1}^{\infty} \frac{(-1)^{n+1}}{n^2} = \frac{\pi^2}{12}$.
Hint: Consider $f(x) = x$ on $(-\pi, \pi)$.

11.12 Find the positive eigenvalues and corresponding eigenfunctions for the operator $\frac{d^4}{dx^4}$ on the interval $[0, l]$ with the boundary conditions $X(0) = X(l) = X''(0) = X''(l) = 0$.

11.13 Using Parseval's identity, find $\sum_{n=1}^{\infty} \frac{1}{n^4}$. **Hint:** Consider $f(x) = x^2$ on $(-\pi, \pi)$.

11.14 (**Convolution and Fourier Series**) First, recall the notion of convolution of integrable functions on $\mathbb{R}$ and how it related to the Fourier transform. These ideas all have analogues in Fourier series. Let f, g be integrable functions on $[-\pi, \pi)$. Extend f and g to all of $\mathbb{R}$ by periodicity. We define the convolution function $f * g$ to be

$$f * g\,(x) = \frac{1}{2\pi} \int_{-\pi}^{\pi} f(x-y)\, g(y)\, dy.$$

(a) Show that $f * g$ is a periodic function with period 2π.

(b) Show that $f * g = g * f$.
(c) Let a_n^i, b_n^i, $i = 1, 2, 3$, denote the full Fourier series coefficients of, respectively, f, g, and $f * g$. Show that

$$a_n^3 = a_n^1 a_n^2, \ n = 0, 1, \dots, \qquad \text{and} \qquad b_n^3 = b_n^1 b_n^2, \ n = 1, 2, \dots.$$

(d) (**For Students Interested in Mathematical Analysis**) Show that $f * g$ is integrable on $[-\pi, \pi)$ with

$$\int_{-\pi}^{\pi} |(f * g)(x)| \, dx \le \left(\int_{-\pi}^{\pi} |f(x)| \, dx \right) \left(\int_{-\pi}^{\pi} |g(x)| \, dx \right).$$

11.15 (**Cauchy-Schwarz Inequality in L^2**) Suppose $f, g \in L^2((a, b))$. Prove that

$$|(f, g)| \le \|f\| \|g\|.$$

Note that this is the generalization of the fact that if $\mathbf{a}, \mathbf{b} \in \mathbb{R}^N$, then $|\mathbf{a} \cdot \mathbf{b}| \le |\mathbf{a}||\mathbf{b}|$. **Hint:** Minimize the function of t, $G(t) := \|f + tg\|^2$, over all $t \in \mathbb{R}$.

11.16 (**Smoothness and Decay of the Coefficients**) This exercise presents the counterpart to Section 6.5 on the Fourier transform and addresses the duality between smoothness of the function and decay of its Fourier coefficients. Let ϕ be a 2π periodic function which is in $C^k(\mathbb{R})$. Since it is convenient to work with the complex version of the Fourier series for ϕ, i.e., (11.13) with $l = \pi$, we may as well assume that ϕ is complex valued. Prove that there exists a constant C depending only on ϕ (not on n) such that

$$|c_n| \le \frac{C}{|n|^k} \qquad \text{for all} \quad n \in \mathbb{Z}. \tag{11.57}$$

So for example, the Fourier coefficients of a C^1 function decay to 0 (as $|n| \to \infty$) like $1/|n|$, whereas the Fourier coefficients of a C^2 function decay faster to 0 (like $1/n^2$). **Hint:** (i) Prove that for any complex-valued 2π periodic smooth functions ϕ and ψ, we have $(\phi, \psi') = (\phi', \psi)$ where $(\cdot, \cdot)$ denotes the $L^2((-\pi, \pi))$ complex-valued inner product defined by (11.19). (ii) Now repeat (i) k times with $\psi = e^{inx}$ to prove that

$$c_n = \frac{1}{2\pi (in)^k} (\phi^{(k)}, e^{inx}),$$

where $\phi^{(k)}$ is the k-th derivative of ϕ. (iii) Use the Cauchy-Schwarz inequality to prove (11.57), explicitly finding the constant C in terms of $\phi^{(k)}$.

11.17 (**Fourier Series and Scaling**) We highly recommend the following exercise. Let ϕ be a continuous function on $\mathbb{R}$ which is periodic with period 1, and assume ϕ is C^1 on $(-\frac{1}{2}, \frac{1}{2})$. Let a_n ($n = 0, 1, \dots$) and b_n ($n = 1, 2, \dots$) denote the coefficients for the full Fourier series of ϕ on $(-\frac{1}{2}, \frac{1}{2})$. For any $k = 1, 2, \dots$, define

$$\phi^{[k]}(x) := \phi(kx).$$

(a) For any fixed k, describe in words the relationship between the functions $\phi^{[k]}(x)$ (for any k) and $\phi(x)$, making clear the notion of "change of scale".
(b) Let $a_n^{[k]}$ ($n = 0, 1, \dots$) and $b_n^{[k]}$ ($n = 1, 2, \dots$) denote the coefficients for the full Fourier series of $\phi^{[k]}(x)$. For any fixed k, what is the relationship between the respective Fourier coefficients $a_n^{[k]}, b_n^{[k]}$ and a_n, b_n?

11.18 **(Pointwise Convergence for the Fourier Sine Series)** State the analogous pointwise convergence result to Theorem 11.6 for the Fourier sine series of a function ϕ on $(0, \pi)$. Then present the proof as a simple corollary of Theorem 11.6.

11.19 **(A Proof of L^2 Convergence for Smooth Functions)** While we were not able to provide a complete proof for the L^2 convergence of the full Fourier series of any L^2 function, you can easily formulate a proof for a function ϕ which is continuous and piecewise C^1 on $[-\pi, \pi]$ with $\phi(-\pi) = \phi(\pi)$. Hint: Use Theorem 11.9.

11.20 State and prove the analogue of Theorem 11.7 (term-by-term differentiation of Fourier series) for the Fourier sine series and the Fourier cosine series.

11.21 This exercise involves many different features from past sections: convergence in the sense of distributions to the delta function, the Dirichlet kernel, term-by-term differentiation of Fourier series, and the Gibbs phenomenon. Consider the step function on $(-\pi, \pi)$

$$\phi(x) = \begin{cases} -1 & \text{if } -\pi < x < 0, \\ 1 & \text{if } 0 \leq x < \pi. \end{cases}$$

(a) Show that the full Fourier series of ϕ on $(-\pi, \pi)$ is

$$\frac{4}{\pi} \sum_{n \text{ odd}} \frac{\sin nx}{n} = \frac{4}{\pi}\left(\sin x + \frac{\sin 3x}{3} + \frac{\sin 5x}{5} + \cdots\right).$$

(b) Consider the partial sums $S_N(x)$, here, defined as the sum of the first N nonzero terms of the series:

$$S_N(x) := \frac{4}{\pi}\left(\sin x + \frac{\sin 3x}{3} + \cdots + \frac{\sin(2N-1)x}{2N-1}\right).$$

Plot $S_N(x)$ for $N = 3, 10, 20$ and qualitatively observe the Gibbs phenomenon.

(c) Recall that $\phi' = 2\delta_0$ in the sense of distributions on $(-\pi, \pi)$. Since ϕ does not satisfy the hypotheses of Theorem 11.7, we cannot differentiate the full Fourier series term by term. However, we can proceed by differentiating the partial sums and examining the limit of S_N' in the sense of distributions. To this end, first show that for any N,

$$S_N'(x) = \frac{2}{\pi} \frac{\sin 2Nx}{\sin x}.$$

Then using the results of Section 11.5, prove that S_N' converges to $2\delta_0$ in the sense of distributions on $(-\pi, \pi)$.

(d) Lastly, let us use the previous parts to **quantitatively** observe the Gibbs phenomenon, i.e., observe the mysterious 9%. Note that the *claimed* overshoot occurs at the first (local) maximum of $S_N(x)$ to the right of the discontinuity at $x = 0$. Show that for any N, this occurs at

$$x_* = \frac{\pi}{2N}.$$

Next, estimate $S_N(x_*)$ as follows. First show that

$$S_N(x_*) \longrightarrow \frac{2}{\pi} \int_0^{\pi} \frac{\sin x}{x}\, dx \qquad \text{as } N \to \infty.$$

Hint: S_N is a Riemann sum for the integral. Lastly, approximate this integral by expanding $\sin x$ in its Taylor series. Since the Taylor series for $\frac{\sin x}{x}$ converges very fast, you need only use six terms to obtain sufficient accuracy. Observe that the first two decimal places of $\frac{2}{\pi}\int_0^{\pi} \frac{\sin x}{x}\, dx$ are 1.18, that is, an overshoot past 1 by 9% of the jump 2.

11.22 (**The Fejér Kernel and Cesàro Means**) For each $J = 1, 2, \ldots$, the Fejér kernel is defined by

$$F_J(x) \ := \frac{1}{J} \sum_{N=1}^{J-1} K_N(x),$$

where $K_N(x)$ denotes the Dirichlet kernel defined by (11.39). Note that the Fejér kernel can be thought of as an average of the Dirichlet kernels.

(a) Show that for each $J = 1, 2, \ldots$, we have

$$F_J(x) \ = \ \frac{1}{J}\left(\frac{\sin\left(\frac{Jx}{2}\right)}{\sin\left(\frac{x}{2}\right)}\right)^2 .$$

In particular note that $F_J(x) \geq 0$ for all x. Note also that for any $J = 1, 2, \ldots$, $\int_{-\pi}^{\pi} F_J(x) = 2\pi$. The Fejér kernel shares the same property as the Dirichlet kernel in that

$$F_J(x) \longrightarrow \delta_0 \qquad \text{in the sense of distributions on } (-\pi, \pi).$$

However, the fact that it is always positive means that the convergence also holds true for functions which need not have any smoothness. Part (b) examines this further and is for students interested in mathematical analysis.

(b) Prove that for any ϕ which is continuous on $(-\pi, \pi)$, we have for each $x \in (-\pi, \pi)$

$$\int_{-\pi}^{\pi} F_J(x-y)\,\phi(y)\,dy \ \longrightarrow \ \phi(x).$$

In fact by researching online, the reader can see that the above holds true for functions ϕ which are simply integrable with the caveat that one must potentially remove certain values of x in a set of measure zero (a “bad” set).

Chapter 12

The Separation of Variables Algorithm for Boundary Value Problems

Even thirty years ago, an undergraduate course in PDEs would have most likely focused on one method: separation of variables. While analytical and applied studies of PDEs have substantially developed beyond this one technique, it nevertheless remains an important technique with a vast number of extensions. Moreover, its premise — **expanding in terms of the eigenfunctions of a symmetric differential operator** — is ubiquitous in mathematics and physics. When the differential operator is simply $-\frac{d}{dx^2}$ with a symmetric boundary condition on an interval, we descend directly into the realm of Fourier series (the previous chapter). Unfortunately, it is beyond the scope of this book to present this theory in any sort of generality (this is the realm of functional analysis and mathematical physics). What many undergraduate PDE texts do is present a vast exposition of examples related to the one- and higher-dimensional Laplacian in a variety of domains with a variety of boundary conditions. The purpose of this chapter is not to present all of these[1], but to simply address the basic algorithm for a selection of prototypical boundary value problems.

12.1. • The Basic Separation of Variables Algorithm

Given a **linear** PDE with boundary conditions and/or initial conditions, **the Separation of Variables Algorithm** is based upon the following steps. For simplicity, let us assume the PDE involves time and one spatial variable x, for example, a BVP/IVP on an interval for the diffusion or wave equation.

[1] See the excellent text of Pinsky [5] for a more exhaustive treatment.

(i) We look for **separated solutions** to the PDE **and** the boundary conditions **of the form** $u(x,t) = X(x)T(t)$. We note that finding these separated solutions reduces to solving **eigenvalue problems** for each of the components X and T with the **same** eigenvalue.

(ii) The boundary conditions carry over to the eigenvalue problem involving $X(x)$. We solve this boundary value / eigenvalue problem to find countably many eigenvalues λ_n for which there exist nontrivial solutions $X_n(x)$.

(iii) We solve the eigenvalue problem of $T(t)$ for each eigenvalue λ_n found in the previous step. We thus arrive at countably many separated solutions $u_n(x,t) = X_n(x)T_n(t)$ to the PDE and the boundary conditions.

(iv) We note that any finite **linear combination** u_n of these separated solutions will **also** be a solution to the PDE and the boundary conditions. We boldly consider an infinite linear combination of the form

$$\sum_{n=1}^{\infty} a_n X_n(x)T_n(t),$$

with coefficients a_n.

(v) We note that achieving the initial conditions amounts to choosing coefficients appropriately. When the eigenvalue problems are for $-\frac{d}{dx^2}$ on some interval with a symmetric boundary condition, we arrive at what we previously called **a general Fourier series** for the data. We find these coefficients by exploiting orthogonality and spanning properties of the eigenfunctions. This effectively means we find each coefficient via projection onto the respective eigenfunction.

In a nutshell, steps (i)–(iii) amount to finding the eigenvalues and eigenfunctions, and steps (iv)–(v) amount to finding the coefficients. The reader should not be worried if reading these steps for the first time is slightly overwhelming. The best way to absorb these steps is via the following simple examples. After going through a few of them, you will quickly master the algorithm.

12.1.1. • The Diffusion Equation with Homogeneous Dirichlet Boundary Conditions. We start with the following boundary value problem (BVP):

$$\begin{cases} u_t = \alpha u_{xx}, & 0 < x < l, t > 0, & \textbf{PDE}, \\ u(0,t) = u(l,t) = 0, & t > 0, & \textbf{boundary conditions}, \\ u(x,0) = \phi(x), & 0 < x < l, & \textbf{initial condition}. \end{cases} \tag{12.1}$$

Recall the physical interpretation of each line, i.e., u, x, t, the parameters l and k, the PDE, and boundary and initial conditions. Note here that the temperature at the ends is fixed and set to equal 0 at all times $t > 0$. These are called **homogeneous boundary conditions** and we now describe the method for such boundary conditions. We will then show how to modify the method for inhomogeneous boundary conditions. Since this is our first example, we enumerate the steps as in the previous general description.

Step (i): We look for **solutions of the PDE and the boundary conditions** of the form $X(x)T(t)$. These are referred to as separated solutions. For a separated product

$X(x)T(t)$ to solve the PDE, we require

$$X(x)T'(t) = \alpha X''(x)T(t) \qquad \text{or} \qquad -\frac{T'(t)}{\alpha T(t)} = -\frac{X''(x)}{X(x)}. \tag{12.2}$$

We now make a **crucial observation**: In the second equation of (12.2), the left-hand side is purely a function of t while the right-hand side is purely a function of x. Since equality holds for all x and t,

$$-\frac{T'(t)}{\alpha T(t)} = -\frac{X''(x)}{X(x)} \qquad \text{must be a constant (i.e., independent of } x \text{ and } t).$$

Call this constant λ. This leads us to **two eigenvalue problems** associated with the **same** eigenvalue:

$$X'' + \lambda X = 0 \qquad \text{and} \qquad T' + \lambda\alpha T = 0.$$

Step (ii): Next, we turn to the boundary conditions. The boundary conditions imply that $X(0)T(t) = X(l)T(t) = 0$. In order to get a nonzero solution (which would be the case if $T(t) = 0$ for all t), we must have $X(0) = X(l) = 0$. Thus, we have the eigenvalue problem

$$X'' + \lambda X = 0, \qquad X(0) = 0 = X(l).$$

Step (iii): We have already solved this problem; there are countably many eigenvalues and eigenfunctions:

$$\lambda_n = \left(\frac{n\pi}{l}\right)^2, \qquad X_n(x) = C\sin\left(\frac{n\pi x}{l}\right), \qquad n = 1, 2, \ldots. \tag{12.3}$$

For now, we take the value of the arbitrary constant C to be 1 (you will soon see why). We have now fixed the possible values for λ. With these values in hand, we turn our attention to the temporal part:

$$T' + \lambda_n \alpha T = 0 \qquad \text{or} \qquad T' + \left(\frac{n\pi}{l}\right)^2 \alpha T = 0.$$

This gives

$$T_n(t) = C\exp\left(-\frac{n^2\pi^2\alpha t}{l^2}\right).$$

Take $C = 1$ here.

We have found a **separated solution** for each $n = 1, 2, \ldots$ to the PDE (the heat equation) and the boundary conditions. Indeed, it is

$$u_n(x,t) = X_n(x)T_n(t) = \sin\left(\frac{n\pi x}{l}\right)\exp\left(-\frac{n^2\pi^2\alpha t}{l^2}\right).$$

Steps (iv) and (v): What about the initial conditions? How can we ensure they hold? Because the PDE is linear, we note that linear combinations of the separated solutions u_n are also solutions to the PDE. Due to the homogeneity in the boundary conditions (i.e., the 0's), a linear combination will also satisfy the boundary conditions. Now, we make the **bold** step of considering an **infinite** linear combination of these separated solutions. That is, consider

$$u(x,t) = \sum_{n=1}^{\infty} b_n \sin\left(\frac{n\pi x}{l}\right)\exp\left(-\frac{n^2\pi^2\alpha t}{l^2}\right), \tag{12.4}$$

for some choice of constants $b_n, n = 1, 2, \ldots$. The question is, **Can we choose these constants so that u satisfies the initial conditions?** In other words, can we find b_n such that

$$\phi(x) = u(x,0) = \sum_{n=1}^{\infty} b_n \sin\left(\frac{n\pi x}{l}\right)?$$

The answer is, miraculously, **yes** by considering the **Fourier sine series** of ϕ. Hence,

$$b_n = \frac{2}{l}\int_0^l \phi(x)\sin\left(\frac{n\pi x}{l}\right)dx. \tag{12.5}$$

Conclusion: The solution to the initial value/ boundary value problem (12.1) is (12.4) where the b_n are given by (12.5).

This is often the first solution formula for the diffusion equation that students encounter. At first glance, it looks very different from the "convolution"-based formula (7.8) of Chapter 7. It is not that different as the "convolution" flavor is in the coefficients. Indeed, by writing out the coefficients and interchanging the summation and integration at will, we find (cf. Exercise **12.1**) that (12.4), (12.5) can be written as

$$u(x,t) = \int_0^l \Phi(x,y,t)\,\phi(y)\,dy \tag{12.6}$$

where

$$\Phi(x,y,t) := \frac{2}{l}\sum_{n=1}^{\infty} \sin\left(\frac{n\pi x}{l}\right)\sin\left(\frac{n\pi y}{l}\right)\exp\left(-\frac{n^2\pi^2\alpha t}{l^2}\right). \tag{12.7}$$

12.1.2. • The Diffusion Equation with Homogenous Neumann Boundary Conditions. Here, our goal is to solve the boundary value problem

$$\begin{cases} u_t = \alpha u_{xx}, & 0 < x < l, t > 0, & \textbf{PDE}, \\ u_x(0,t) = 0 = u_x(l,t), & t > 0, & \textbf{boundary conditions}, \\ u(x,0) = \phi(x), & 0 < x < l, & \textbf{initial condition}. \end{cases} \tag{12.8}$$

The method of separation of variables immediately leads to solving the eigenvalue problem

$$X'' + \lambda X = 0, \qquad X'(0) = 0 = X'(l).$$

This means that the eigenvalues and spatial eigenfunctions in (12.3) are now

$$\lambda_n = \left(\frac{n\pi}{l}\right)^2, \qquad X_n(x) = \cos\left(\frac{n\pi x}{l}\right), \qquad n = 0, 1, \ldots.$$

Hence, we now have a term corresponding to $\lambda = 0$ whose associated temporal eigenfunction is a constant. Repeating the previous steps, we then arrive at

$$u(x,t) = \frac{1}{2}a_0 + \sum_{n=1}^{\infty} a_n \cos\left(\frac{n\pi x}{l}\right)\exp\left(-\left(\frac{n\pi}{l}\right)^2 \alpha t\right). \tag{12.9}$$

The initial condition $u(x,0) = \phi(x)$ requires us to choose a_n as the **Fourier cosine coefficients** of $\phi(x)$; that is,

$$a_n = \frac{2}{l}\int_0^l \phi(x)\cos\left(\frac{n\pi x}{l}\right)dx, \qquad n = 0, 1, \ldots. \tag{12.10}$$

Note that the long-time solution (steady-state solution) is simply

$$\lim_{t\to\infty} u(x,t) = \frac{a_0}{2} = \frac{1}{l}\int_0^l \phi(x)\,dx,$$

which is the **average** initial temperature. Thus, the solution to (12.8) is (12.9) where the coefficients a_n are given by (12.10).

12.2. • The Wave Equation

12.2.1. • The Wave Equation with Homogeneous Dirichlet Boundary Conditions. Recall from Section 3.10 the BVP/IVP for the wave equation with Dirichlet boundary conditions, modeling the vibrations of a **finite string with fixed ends**. While we argued that the D'Alembert/reflection approach was rather complicated, we can now readily solve this problem via separation of variables. To this end, consider

$$\begin{cases} u_{tt} = c^2 u_{xx}, & 0 < x < l,\ t > 0, \\ u(0,t) = u(l,t) = 0, & t \geq 0, \\ u(x,0) = \phi(x), \quad u_t(x,0) = \psi(x), & 0 < x < l. \end{cases} \tag{12.11}$$

In accordance with the algorithm, we first look for solutions of the PDE and the boundary conditions of the form $X(x)T(t)$. Solving the PDE requires

$$X(x)T''(t) = c^2 X''(x)T(t), \qquad \text{or} \qquad -\frac{T''(t)}{c^2 T(t)} = -\frac{X''(x)}{X(x)}. \tag{12.12}$$

Again, the crucial observation is as follows: In the second equation of (12.12), the left-hand side is purely a function of t while the right-hand side is purely a function of x. Since equality holds for all x and t,

$$-\frac{T''(t)}{c^2 T(t)} = -\frac{X''(x)}{X(x)} \qquad \text{must be a constant (independent of } x \text{ and } t).$$

Calling this constant λ leads us to two eigenvalue problems associated with the **same** eigenvalue:

$$X'' + \lambda X = 0, \qquad T'' + \lambda c^2 T = 0.$$

Next, we turn to the boundary conditions, which imply that

$$X(0)T(t) = X(l)T(t) = 0 \qquad \text{for all } t \geq 0.$$

In order to get a nonzero solution (which would be the case if $T(t) = 0$ for all t), we must have $X(0) = X(l) = 0$. Thus, we have the eigenvalue problem

$$X'' + \lambda X = 0, \qquad X(0) = 0 = X(l).$$

Again, we have already solved this problem. There are countably many eigenvalues and eigenfunctions:

$$\lambda_n = \left(\frac{n\pi}{l}\right)^2, \qquad X_n(x) = \sin\left(\frac{n\pi x}{l}\right), \qquad n \in \mathbb{N}.$$

With these values of λ in hand, we turn our attention to the temporal part:

$$T'' + \lambda_n c^2 T = 0 \qquad \text{or} \qquad T'' + \left(\frac{n\pi}{l}\right)^2 c^2 T = 0.$$

This gives

$$T_n(t) = c_n \cos\left(\frac{n\pi ct}{l}\right) + d_n \sin\left(\frac{n\pi ct}{l}\right),$$

where c_n and d_n are any constants. Thus,

$$u_n(x,t) = X_n(x)T_n(t) = \left[c_n \cos\left(\frac{n\pi ct}{l}\right) + d_n \sin\left(\frac{n\pi ct}{l}\right)\right] \sin\left(\frac{n\pi x}{l}\right).$$

We have now found, up to multiplication by constants, countably many **separated solutions** to the PDE (the wave equation) and the boundary conditions. What about the initial conditions? To this end, first note that by the principle of superposition, any finite combination (sum) of these separated solutions is also a solution. Now, we make the *bold* step of considering an infinite sum of these separated solutions. That is, consider

$$u(x,t) = \sum_{n=1}^{\infty} \left[c_n \cos\left(\frac{n\pi ct}{l}\right) + d_n \sin\left(\frac{n\pi ct}{l}\right)\right] \sin\left(\frac{n\pi x}{l}\right),$$

and ask if we can find values of c_n, d_n such that the initial conditions are satisfied. Since $u(x,0) = \phi(x)$, we would require

$$\phi(x) = \sum_{n=1}^{\infty} c_n \sin\left(\frac{n\pi x}{l}\right).$$

But this is precisely the Fourier sine expansion of $\phi(x)$, and hence

$$c_n = \frac{2}{l}\int_0^l \phi(x) \sin\left(\frac{n\pi x}{l}\right) dx.$$

For the condition $u_t(x,0) = \psi(x)$, we first differentiate the series term by term (assume we can do this) and then evaluate at $t = 0$ to find

$$\psi(x) = \sum_{n=1}^{\infty} \frac{n\pi c}{l} d_n \sin\left(\frac{n\pi x}{l}\right),$$

which is the Fourier sine expansion of $\psi(x)$ with coefficient $n\pi c d_n/l$. We can easily find the d_n by dividing the respective Fourier sine coefficients of ψ by $n\pi c/l$. That is,

$$d_n = \frac{2}{n\pi c}\int_0^l \psi(x) \sin\left(\frac{n\pi x}{l}\right) dx.$$

Example 12.2.1. Now recall the simple yet illustrative example we gave in Section 3.10: Let

$$\phi(x) = \sin(27\pi x/l) \quad \text{and} \quad \psi \equiv 0.$$

This implies that $d_n = 0 \;\forall\; n \in \mathbb{N}$ and that $c_{27} = 1$ and $c_n = 0$ for $n \neq 27$. Hence, the solution is simply

$$u(x,t) = \sin\left(\frac{27\pi x}{l}\right) \cos\left(\frac{27\pi ct}{l}\right).$$

In particular, this tells us something very important about the eigenfunctions and eigenvalues. Let us view the problem in terms of a finite vibrating string with fixed end points. If we input an eigenfunction as the initial displacement, then the solution at any time t is simply a constant times the initial displacement, i.e., some amplification of the initial displacement. In particular, the basic shape is preserved. We call such eigenfunctions **normal modes** of the string. On the other hand, if we fixed a position x on the string, we would see that the displacement oscillates with a frequency equal to $\frac{27\pi c}{l} = c\sqrt{\lambda_{27}}$. We call these frequencies the **natural frequencies (or harmonics)** of the vibrating string. There are countably many of these special solutions; indeed, for

$$\phi(x) = \sin(n\pi x/l) \quad \text{and} \quad \psi \equiv 0, \quad n = 1, 2, \ldots,$$

we have the solutions

$$u(x,t) = \sin\left(\frac{n\pi x}{l}\right) \cos\left(\frac{n\pi ct}{l}\right) \quad n = 1, 2, \ldots.$$

What is rather amazing is that **every** solution is a linear combination of these special ones.

12.2.2. The Wave Equation with Homogeneous Neumann Boundary Conditions. Suppose that we have Neumann boundary conditions. That is, we wish to solve

$$\begin{cases} u_{tt} = c^2 u_{xx}, & 0 < x < l,\ t > 0, \\ u_x(0,t) = u_x(l,t) = 0, & t > 0, \\ u(x,0) = \phi(x), \quad u_t(x,0) = \psi(x), & 0 < x < l. \end{cases} \tag{12.13}$$

The analogous steps give rise to eigenvalue problems

$$X'' + \lambda X = 0,\ X'(0) = X'(l) = 0, \qquad T'' + \lambda c^2 T = 0.$$

As before, we can show that we only have nonnegative eigenvalues λ and they are given by $\lambda_n = \left(\frac{n\pi}{l}\right)^2, n = 0, 1, \ldots.$ Note that the difference here is that 0 is an eigenvalue since the corresponding eigenfunction $X_0(x) = 1$ is nontrivial. The corresponding eigenfunctions are

$$X_n(x) = \cos\left(\frac{n\pi x}{l}\right), \quad n = 0, 1, \ldots.$$

For $\lambda = 0$, we find

$$X_0 = \textbf{constant}, \qquad T''(t) = 0 \Longrightarrow T(t) = c_0 + d_0 t.$$

Thus, the full solution can be written as

$$u(x,t) = \frac{1}{2}c_0 + \frac{1}{2}d_0 t + \sum_{n=1}^{\infty} \left(c_n \cos\left(\frac{n\pi ct}{l}\right) + d_n \sin\left(\frac{n\pi ct}{l}\right)\right) \cos\left(\frac{n\pi x}{l}\right). \tag{12.14}$$

We proceed, as before, to find the c_n and d_n corresponding to initial data ϕ and ψ. For example, we would choose the c_n such that

$$\phi(x) = \frac{1}{2}c_0 + \sum_{n=1}^{\infty} c_n \cos\left(\frac{n\pi x}{l}\right).$$

That is, we would find the Fourier cosine series for ϕ,

$$c_n = \frac{2}{l}\int_0^l \phi(x)\cos\left(\frac{n\pi x}{l}\right) dx, \quad n = 0, 1, 2, \ldots.$$

Taking a term-by-term derivative with respect to t in (12.14) and setting $t = 0$ would essentially entail finding the Fourier cosine series of ψ.

12.3. • Other Boundary Conditions

One can carry out the algorithm for many other homogeneous and inhomogeneous boundary conditions. Indeed, recalling Section 11.3, it would seem that the algorithm should work for any symmetric boundary conditions. For simplicity, we illustrate the steps for the diffusion equation with a few other boundary conditions.

12.3.1. • Inhomogeneous Dirichlet Boundary Conditions.

Consider

$$\begin{cases} u_t = \alpha u_{xx}, & 0 < x < l, t > 0, \\ u(0,t) = T_0, \quad u(l,t) = T_1, & t > 0, \\ u(x,0) = \phi(x), & 0 < x < l. \end{cases}$$

In this case, we look for a steady-state solution $u(x,t) = v(x)$ with $v'' = 0$ and $v(0) = T_0$, $v(l) = T_1$, but we notice that this is a simple boundary value problem for which the solution is given by $v(x) = Ax + B$ with $v(0) = B = T_0$ and $v(l) = Al + T_0 = T_1$. This leads us to

$$v(x) = T_0 + \frac{T_1 - T_0}{l} x.$$

Hence, the system wants to tend to a linear interpolation between the two temperatures. Now, let $w(x,t) = u(x,t) - v(x)$ so that it now satisfies

$$\begin{cases} w_t = \alpha w_{xx}, & 0 < x < l, t > 0, \\ w(0,t) = 0, \quad w(l,t) = 0, & t > 0, \\ w(x,0) = \phi(x) - v(x), & 0 < x < l. \end{cases}$$

We call w the **transient solution** which boasts homogeneous boundary conditions. Hence, we may proceed as before to solve for $w(x,t)$ and then simply add to it the steady-state solution $v(x)$ to arrive at the solution $u(x,t)$. Note that

$$\lim_{t\to\infty} w(x,t) = 0 \implies \lim_{t\to\infty} u(x,t) = v(x).$$

Thus, the steady-state solution v is the *eventual* or *long-time state* of the system and w measures how far we are away from the eventual steady-state solution.

12.3.2. • Mixed Homogeneous Boundary Conditions.

Consider

$$\begin{cases} u_t = \alpha u_{xx}, & 0 < x < l, t > 0, \\ u(0,t) = 0, \quad u_x(l,t) = 0, & t > 0, \\ u(x,0) = \phi(x), & 0 < x < l. \end{cases}$$

In this case, the eigenvalue problem for $X(x)$ is

$$X'' + \lambda X = 0, \qquad X(0) = 0 = X'(l),$$

which has eigenvalues and eigenfunctions

$$\lambda_n = \frac{\left(n + \frac{1}{2}\right)^2 \pi^2}{l^2}, \qquad X_n(x) = \sin \frac{\left(n + \frac{1}{2}\right)\pi x}{l}, \qquad n = 0, 1, 2, \ldots.$$

The corresponding temporal solutions are

$$T_n(t) = c_n \exp(-\lambda_n k t).$$

Hence, taking all linear combinations of the separated solutions, we arrive at

$$u(x,t) = \sum_{n=0}^{\infty} c_n \sin\left(\frac{\left(n+\frac{1}{2}\right)\pi x}{l}\right)\exp\left(-\frac{\left(n+\frac{1}{2}\right)^2\pi^2\alpha t}{l^2}\right).$$

Imposing the initial conditions requires

$$\phi(x) = \sum_{n=0}^{\infty} c_n \sin\left(\frac{\left(n+\frac{1}{2}\right)\pi x}{l}\right).$$

Remarkably, the orthogonal family of eigenfunctions, X_n, spans all reasonable functions. Further, note that $\|X_n\|^2 = \frac{l}{2}$, and hence, the coefficients are given by

$$c_n = \frac{2}{l}\int_0^l \phi(x)\sin\left(\frac{\left(n+\frac{1}{2}\right)\pi x}{l}\right)dx.$$

12.3.3. • Mixed Inhomogeneous Boundary Conditions.

Consider

$$\begin{cases} u_t = \alpha u_{xx}, & 0 < x < l, t > 0, \\ u_x(0,t) = 0, \quad u(l,t) = T_0, & t > 0, \\ u(x,0) = \phi(x), & 0 < x < l. \end{cases}$$

We **first** look for a steady-state solution $u(x,t) = v(x)$, i.e., a solution to $v'' = 0$ and $v'(0) = 0, v(l) = T_0$. The general solution is $v(x) = Ax + B$ with the additional boundary conditions, $v'(0) = A = 0$ and $v(l) = B = T_0$, yielding $v(x) = T_0$.

The transient solution $w(x,t) := u(x,t) - v(x)$ satisfies

$$\begin{cases} w_t = \alpha w_{xx}, & 0 < x < l, t > 0, \\ w_x(0,t) = 0, \quad w(l,t) = 0, & t > 0, \\ w(x,0) = \phi(x) - v(x), & 0 < x < l. \end{cases}$$

We can now solve for w with its homogenous boundary conditions and find u. Note that we now have a constant (T_0) eventual (long-time) temperature distribution.

12.3.4. • Inhomogeneous Neumann Boundary Conditions.

Consider

$$\begin{cases} u_t = \alpha u_{xx}, & 0 < x < l, t > 0, \\ u_x(0,t) = A, \quad u_x(l,t) = B, & t > 0, \\ u(x,0) = \phi(x), & 0 < x < l, \end{cases} \tag{12.15}$$

where A and B are constants. First off, what is the physical interpretation of the inhomogeneous Neumann boundary conditions? Here we are prescribing the heat flux at the ends which means we are pumping or removing heat from the bar at prescribed rates.

Suppose we proceed, as before, with finding a steady-state solution. Since the steady-state solution to the heat equation has the form $a + bx$, we would need $b = A = B$; otherwise, there would be no solution. However, even if $A = B$, we would not know what to take for a.

So what should we do? Here we first find a particular solution (not necessarily steady state) to the heat equation and the boundary conditions. In other words, a solution which depends on both x and t. How do we find such a solution? Let us look for a solution of the form

$$v(x,t) = ax^2 + bx + ct.$$

In order for this to solve the heat equation, we need $c = 2\alpha a$. Noting that $v_x(x,t) = 2ax + b$, we impose the boundary conditions to find that a and b must satisfy

$$b = A, \qquad 2al + A = B.$$

Solving these equations, we arrive at

$$b = A \qquad \text{and} \qquad a = \frac{B-A}{2l} \qquad \text{and} \qquad c = \frac{2\alpha(B-A)}{2l}.$$

Thus, we have

$$v(x,t) = \frac{B-A}{2l}x^2 + Ax + \frac{2\alpha(B-A)}{2l}t.$$

Now we define $w(x,t)$ such that

$$u(x,t) = w(x,t) + v(x,t).$$

Since we want u to solve (12.15), the new function $w(x,t)$ must solve

$$\begin{cases} w_t = \alpha w_{xx}, & 0 < x < l, t > 0, \\ w_x(0,t) = 0, \quad w_x(l,t) = 0, & t > 0, \\ w(x,0) = \phi(x) - v(x,0), & 0 < x < l. \end{cases}$$

Since we know how to solve for $w(x,t)$, we are done. The reader should pause to contemplate the physical interpretations of $v(x,t)$ and $w(x,t)$.

12.3.5. • The Robin Boundary Condition for the Diffusion Equation. Now things start to get a bit messy! Recall from the short Section 7.6.2, the Robin boundary condition and its interpretation in terms of heat transfer. The method of separation of variables will again work but there is a catch: The eigenvalues will need to be either calculated numerically or approximated with a simple formula. We provide one example with a simple Robin boundary condition on the right end. Let us work with $l = 1$ and consider

$$\begin{cases} u_t = \alpha u_{xx}, & 0 < x < 1, t > 0, \\ u(0,t) = 0, \quad u(1,t) + u_x(1,t) = 0, & t > 0, \\ u(x,0) = \phi(x), & 0 < x < 1. \end{cases} \tag{12.16}$$

Recall the interpretation via Newton's law of cooling and heat transfer at the right boundary $x = 1$. The method of separation of variables yields the following eigenvalue problem for $X(x)$:

$$X'' + \lambda X = 0, \qquad X(0) = 0, \qquad X(1) + X'(1) = 0.$$

We give an aside comment: We have yet to solve this eigenvalue problem so we will take a time out to address it. The reader can check that, as before, there are no negative eigenvalues. We look for positive eigenvalues $\lambda > 0$. Recall that the general solution to $X'' + \lambda X = 0$ with $\lambda > 0$ is

$$X(x) = A\sin\sqrt{\lambda}x + B\cos\sqrt{\lambda}x,$$

for any constants A, B. Since $X(0) = 0$, we must have $B = 0$. On the other hand, $X(1) + X'(1) = 0$ indicates that

$$\sin\sqrt{\lambda} + \sqrt{\lambda}\cos\sqrt{\lambda} = 0.$$

This means

$$\sqrt{\lambda} = -\tan\sqrt{\lambda}.$$

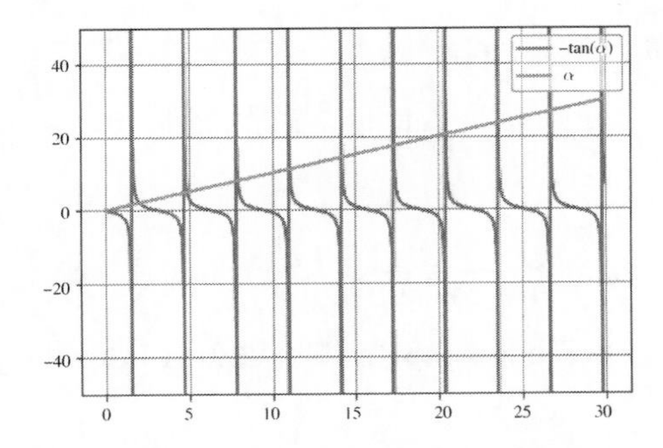

Figure 12.1. The graphs of $y = \alpha$ and $y = -\tan\sqrt{\lambda}$ with intersections at $\alpha_n, n = 1, 2, \ldots$.

Are there $\lambda > 0$ for which this is true? Certainly: Letting $\alpha = \sqrt{\lambda}$, we are looking for points u where

$$\alpha = -\tan\alpha.$$

There exists (cf. Figure 12.1) a sequence of such values $\alpha_n, n = 1, 2, \ldots$, hence a sequence of $\lambda_n = \alpha_n^2$. However, we have no simple closed form solution for these intersection points and can only approximate them numerically. For example, $\lambda_1 \approx 4.116, \lambda_2 \approx 24.14, \lambda_3 \approx 63.66$, after which, we have the approximate formula

$$\lambda_n \approx \frac{(2n-1)^2\pi^2}{4}, \qquad n = 4, 5, \ldots.$$

The corresponding eigenfunctions are

$$X_n(x) = \sin\sqrt{\lambda_n}\, x, \quad n = 1, 2, \ldots.$$

With these spatial eigenfunctions in hand, we can now complete the Separation of Variables Algorithm. The corresponding temporal terms are

$$T_n(t) = e^{-\lambda_n \alpha t}.$$

Thus putting it all together, the solution form is

$$u(x,t) = \sum_{n=1}^{\infty} b_n \sin\sqrt{\lambda_n}\, x\, e^{-\lambda_n \alpha t},$$

for constants b_n. The remaining issue is the initial data. Can we find b_n such that

$$\phi(x) = \sum_{n=1}^{\infty} b_n \sin\sqrt{\lambda_n}\, x\,?$$

The answer is miraculously yes! We have already shown (why?) that these eigenfunctions are orthogonal. As it turns out they also do span all reasonable functions. This is messy to prove directly but easily follows from the beautiful Spectral Theorem for compact operators in a Hilbert space. Hence,

$$b_n = \frac{1}{\|X_n(x)\|^2} \int_0^1 \phi(x) \sin\sqrt{\lambda_n}\, x\, dx.$$

A direct calculation (for $l = 1$) yields

$$\|X_n(x)\|^2 = \frac{1 + \cos^2 \lambda_n}{2}.$$

Thus,

$$b_n = \frac{2}{1 + \cos^2 \lambda_n} \int_0^1 \phi(x) \sin\sqrt{\lambda_n}\, x\, dx.$$

There are **many other combinations of Robin boundary conditions** of the general form:

$$c_1 u(0,t) + u_x(0,t) = c_2 \qquad \text{and/or} \qquad c_3 u(l,t) + u_x(l,t) = c_4.$$

While we will not address them here, the basic reasoning (i.e., the method) remains the same. There is even one case where you would find a negative eigenvalue! See Section 4.3 of Strauss [**6**] for a complete discussion.

12.4. Source Terms and Duhamel's Principle for the Diffusion and Wave Equations

In this section, we consider the addition of source terms to BVPs for the diffusion and wave equations. We illustrate the method via the diffusion equation with homogeneous Dirichlet boundary conditions. Our goal is to solve

$$\begin{cases} u_t = \alpha u_{xx} + f(x,t), & 0 < x < l, t > 0, \\ u(0,t) = 0 = u(l,t), & t > 0, \\ u(x,0) = \phi(x), & 0 \le x \le l. \end{cases} \tag{12.17}$$

By superposition it suffices to solve

$$\begin{cases} u_t = \alpha u_{xx} + f(x,t), & 0 < x < l,\ t > 0, \\ u(0,t) = 0 = u(l,t), & t > 0, \\ u(x,0) = 0, & 0 \le x \le l, \end{cases} \tag{12.18}$$

and then add it to the solution of

$$\begin{cases} u_t = \alpha u_{xx}, & 0 < x < l,\ t > 0, \\ u(0,t) = 0 = u(l,t), & t > 0, \\ u(x,0) = \phi(x), & 0 \le x \le l. \end{cases}$$

Duhamel's Principle states that the solution to (12.18) can be found as follows: For any $s > 0$, we consider the solution $u(x,t;s)$ to

$$\begin{cases} u_t(x,t;s) = \alpha u_{xx}(x,t;s), & 0 < x < l,\ t > s, \\ u(0,t;s) = 0 = u(l,t;s), & t > s, \\ u(x,s;s) = f(x,s), & 0 \le x \le l, \end{cases} \tag{12.19}$$

and then obtain the solution to (12.18) by integrating over s from 0 to t. That is, the solution to (12.18) is given by

$$u(x,t) = \int_0^t u(x,t;s)\, ds. \tag{12.20}$$

As with all the previous verifications of Duhamel's Principle, one can indeed check using rules of calculus that (12.20) solves (12.18). The reader is encouraged to verify the details in this context (cf. Exercise **12.8**).

To write down an explicit formula, let us solve (12.19) via separation of variables. The only minor issue here is the parameter s. For each $s > 0$, we obtain for $x \in \mathbb{R}$ and $t \ge s$

$$u(x,t;s) = \sum_{n=1}^{\infty} b_n(s) \sin\frac{n\pi x}{l}\, e^{-\alpha(\frac{n\pi}{l})^2(t-s)}$$

where

$$b_n(s) = \frac{2}{l}\int_0^l f(x,s)\sin\frac{n\pi x}{l}\, dx.$$

Assuming certain smooth assumptions on the source function $f(x,t)$, there will be sufficient decay in the coefficients $b_n(s)$ as $n \to \infty$ so that the infinite series will converge "nicely". Hence, from (12.20), the solution to (12.18) is given by

$$\begin{aligned} u(x,t) &= \int_0^t \sum_{n=1}^{\infty} b_n(s)\sin\frac{n\pi x}{l}\, e^{-\alpha(\frac{n\pi}{l})^2(t-s)}\, ds \\ &= \sum_{n=1}^{\infty} \sin\frac{n\pi x}{l} \int_0^t b_n(s)\, e^{-\alpha(\frac{n\pi}{l})^2(t-s)}\, ds. \end{aligned} \tag{12.21}$$

12.5. • Laplace's Equations in a Rectangle and a Disk

12.5.1. • Rectangle. Our goal is to find equilibrium (steady-state) heat distributions in a rectangle $R = [0, a] \times [0, b]$ where we specify the temperature (heat) on the boundary. This means that we are given four functions (of one variable) f_1, f_2, f_3, and f_4 and wish to solve

$$\begin{cases} u_{xx} + u_{yy} = 0, & & 0 < x < a, 0 < y < b, \\ u(x,0) = f_1(x), & u(x,b) = f_2(x), & 0 \le x \le a, \\ u(0,y) = f_3(y), & u(a,y) = f_4(y), & 0 \le y \le b. \end{cases} \tag{12.22}$$

The first thing we note is that it suffices to solve four separate BVPs; in each BVP three of the functions f_i are identically 0. For example, one of these BVPs would be

$$\begin{cases} u_{xx} + u_{yy} = 0, & & 0 < x < a, 0 < y < b, \\ u(x,0) = 0, & u(x,b) = 0, & 0 \le x \le a, \\ u(0,y) = 0, & u(a,y) = f_4(y), & 0 \le y \le b. \end{cases} \tag{12.23}$$

Why does it suffice? Because we can simply add the four separate solutions to find a solution to (12.22).

Let $f(y)$ be a function defined on $[0, b]$. We now solve

$$\begin{cases} u_{xx} + u_{yy} = 0, & & 0 < x < a, 0 < y < b, \\ u(x,0) = 0, & u(x,b) = 0, & 0 \le x \le a, \\ u(0,y) = 0, & u(a,y) = f(y), & 0 \le y \le b. \end{cases} \tag{12.24}$$

We start by looking for a separated solution of the form

$$u(x, y) = X(x)Y(y)$$

to **the PDE and the three homogeneous boundary conditions**. The PDE implies that

$$X''(x)Y(y) + X(x)Y''(y) = 0,$$

or

$$\frac{X''(x)}{X(x)} = -\frac{Y''(y)}{Y(y)},$$

for all $0 < x < a, 0 < y < b$. The only way this equality can hold for all $0 < x < a, 0 < y < b$ is if both ratios are the same constant. Call this constant λ. This gives us two eigenvalue problems:

$$X'' - \lambda X = 0 \qquad \text{and} \qquad Y'' + \lambda Y = 0.$$

The three **homogeneous** boundary conditions translate as

$$X(0) = 0, \qquad Y(0) = 0 = Y(b).$$

We first focus on the Y equation and recall that the only solutions to

$$Y'' + \lambda Y = 0, \qquad Y(0) = 0 = Y(b),$$

are when

$$\lambda_n = \frac{n^2\pi^2}{b^2},$$

in which case

$$Y_n(y) = \sin\left(\frac{n\pi y}{b}\right).$$

Now with these values of λ, we solve the X equation

$$X'' - \frac{n^2\pi^2}{b^2}X = 0.$$

We know what the solutions are *exponentials.*

We give an an aside comment on exponential solutions to $X'' = \lambda X$ for $\lambda > 0$.

We know the general solution is $C_1 e^{\sqrt{\lambda}x} + C_2 e^{-\sqrt{\lambda}x}$. However, for implementing certain boundary conditions, it is more convenient to write this general solution in terms of

$$\cosh\sqrt{\lambda}x = \frac{e^{\sqrt{\lambda}x} + e^{-\sqrt{\lambda}x}}{2} \quad \text{and} \quad \sinh\sqrt{\lambda}x = \frac{e^{\sqrt{\lambda}x} - e^{-\sqrt{\lambda}x}}{2}.$$

In other words, we can equivalently say that the general solution has the form

$$C_1 \cosh\sqrt{\lambda}x + C_2 \sinh\sqrt{\lambda}x.$$

The general solution to $X'' - \frac{n^2\pi^2}{b^2}X = 0$ is

$$C_1 \cosh\left(\frac{n\pi x}{b}\right) + C_2 \sinh\left(\frac{n\pi x}{b}\right).$$

The third homogeneous boundary condition gives $X(0) = 0$, which implies that $C_1 = 0$. We have found for each $n = 1, 2, \ldots$ separated solutions of the form

$$u_n(x, y) = \sinh\left(\frac{n\pi x}{b}\right) \sin\left(\frac{n\pi y}{b}\right)$$

to Laplace's equation and the three homogeneous boundary conditions. What about the fourth **inhomogeneous** boundary condition $u(a, y) = f(y)$? In general, we cannot achieve this with one of these separated solutions; however, we can take an infinite combination of the form

$$u(x, y) = \sum_{n=1}^{\infty} c_n \sinh\left(\frac{n\pi x}{b}\right) \sin\left(\frac{n\pi y}{b}\right)$$

and ask if we can find c_n, such that

$$f(y) = u(a, y) = \sum_{n=1}^{\infty} c_n \sinh\left(\frac{n\pi a}{b}\right) \sin\left(\frac{n\pi y}{b}\right).$$

The answer is yes by finding the Fourier sine series coefficient for $f(y)$,

$$\frac{2}{b}\int_0^b f(y) \sin\left(\frac{n\pi y}{b}\right) dy,$$

and then dividing these by $\sinh\left(\frac{n\pi a}{b}\right)$. In other words,

$$c_n = \frac{2}{b \sinh\left(\frac{n\pi a}{b}\right)} \int_0^b f(y) \sin\left(\frac{n\pi y}{b}\right) dy. \tag{12.25}$$

To conclude, the solution to (12.24) is

$$u(x,y) = \sum_{n=1}^{\infty} c_n \sinh\left(\frac{n\pi x}{b}\right) \sin\left(\frac{n\pi y}{b}\right) \quad \text{where the } c_n \text{ are given by (12.25).}$$

A good exercise is to consider the modifications when the inhomogeneous boundary condition is placed in one of the other three sides of the rectangle (cf. Exercise **12.14**).

12.5.2. • The Disk. Consider the BVP

$$\begin{cases} \Delta u = 0, & \mathbf{x} \in D = \{(x,y) : x^2 + y^2 < a^2\}, \\ u = h, & \mathbf{x} \in \partial D. \end{cases}$$

Let us solve this problem via separation of variables. Switching to polar coordinates, let $u = u(r,\theta)$ and note that $\mathbf{x} \in \partial D$ is the same as $r = a$, so that $u = h(\theta)$ on $r = a$. Rewriting the Laplacian operator in terms of r and θ yields

$$\Delta u = u_{rr} + \frac{1}{r} u_r + \frac{1}{r^2} u_{\theta\theta}.$$

Therefore, we are looking to solve

$$\begin{cases} u_{rr} + \frac{1}{r} u_r + \frac{1}{r^2} u_{\theta\theta} = 0, & r < a, \\ u = h(\theta), & r = a. \end{cases}$$

We look for separated solutions of the form $u(r,\theta) = R(r)\Theta(\theta)$, for which the PDE requires that

$$R''\Theta + \frac{1}{r} R'\Theta + \frac{1}{r^2} R\Theta'' = 0.$$

Dividing by $R\Theta$ and multiplying by r^2, we obtain

$$r^2 \frac{R''}{R} + r\frac{R'}{R} = -\frac{\Theta''}{\Theta} = \lambda.$$

Stopping here, we ask ourselves where are the boundary conditions associated with one of the eigenvalue problems? The sole boundary condition on ∂D for R will be used in place of the "initial conditions" via a Fourier series. In this instance, the boundary conditions associated with one of the eigenvalue problems are **implicit**. That is, because of the interpretation of θ, the Θ equation $\Theta'' + \lambda\Theta = 0$ naturally requires $\Theta(0) = \Theta(2\pi)$ and $\Theta'(0) = \Theta'(2\pi)$ (periodic boundary conditions). Thus we solve the Θ problem **first** to find eigenvalues and eigenfunctions

$$\lambda_n = n^2, \qquad \Theta_n(\theta) = A_n \cos(n\theta) + B_n \sin(n\theta), \qquad n = 1, 2, \ldots,$$

and for $n = 0$, we get that $\Theta_0(\theta) = A_0$.

The eigenvalue equation for R reads $r^2R'' + rR' - \lambda R = 0$ for $\lambda_n = n^2, n = 1, 2\dots$. Recall from your ODE class that one approaches such ODEs by looking for a solution of the form $R(r) = r^\alpha$. This gives

$$\alpha(\alpha-1)r^\alpha + \alpha r^\alpha - n^2 r^\alpha = 0 \iff \alpha^2 - n^2 = 0 \iff \alpha = \pm n$$

and so

$$R_n(r) = c_n r^n + d_n r^{-n}.$$

Next, we look at the description of the problem in order to remove some of these unwanted solutions. The solution should be defined and continuous at $r = 0$; hence, since $r^{-n} \to \infty$ as $r \to 0$, we must set $d_n = 0$ for all n. For $\lambda = 0$, we find $R_0(r) = c_0 + d_0 \ln(r)$; but again, the solution blows up at $r = 0$ so we must set $d_0 = 0$ to satisfy $u \in C^2([0, r])$.

Collecting all these admissible separated solutions and taking an infinite sum, we obtain

$$u(r,\theta) = \frac{1}{2}A_0 + \sum_{n=1}^{\infty} r^n \left(A_n \cos(n\theta) + B_n \sin(n\theta)\right).$$

At $r = a$, we must have $u(r,\theta) = h(\theta)$, and so

$$h(\theta) = \frac{1}{2}A_0 + \sum_{n=1}^{\infty} \left[(a^n A_n)\cos(n\theta) + (a^n B_n)\sin(n\theta)\right],$$

and thus,

$$A_n = \frac{1}{a^n\pi}\int_0^{2\pi} h(\phi)\cos(n\phi)\,d\phi, \qquad B_n = \frac{1}{a^n\pi}\int_0^{2\pi} h(\phi)\sin(n\phi)\,d\phi.$$

So, the solution is

$$\begin{aligned} u(r,\theta) &= \frac{1}{2\pi}\int_0^{2\pi} h(\phi)\,d\phi + \sum_{n=1}^{\infty}\frac{r^n}{\pi a^n}\int_0^{2\pi} h(\phi)[\cos(n\theta)\cos(n\phi) + \sin(n\theta)\sin(n\phi)]\,d\phi \\ &= \frac{1}{2\pi}\int_0^{2\pi}\left[1 + 2\sum_{n=1}^{\infty}\left(\frac{r}{a}\right)^n \cos(n(\theta-\phi))\right] h(\phi)\,d\phi. \end{aligned}$$

We have found the solution as an infinite series; however, in this special case, we can continue by noting that

$$1 + 2\sum_{n=1}^{\infty}\left(\frac{r}{a}\right)^n \cos(n(\theta-\phi))$$

is a geometric series which can readily be summed. To see this, let us write the expression in terms of complex exponentials. We have

$$\begin{aligned} 1 + 2\sum_{n=1}^{\infty}\left(\frac{r}{a}\right)^n \cos(n(\theta-\phi)) &= 1 + \sum_{n=1}^{\infty}\left(\frac{r}{a}\right)^n e^{in(\theta-\phi)} + \sum_{n=1}^{\infty}\left(\frac{r}{a}\right)^n e^{-in(\theta-\phi)} \\ &= 1 + \frac{re^{i(\theta-\phi)}}{a - re^{i(\theta-\phi)}} + \frac{re^{-i(\theta-\phi)}}{a - re^{-i(\theta-\phi)}} \\ &= \frac{a^2 - r^2}{a^2 - 2ar\cos(\theta-\phi) + r^2} \end{aligned}$$

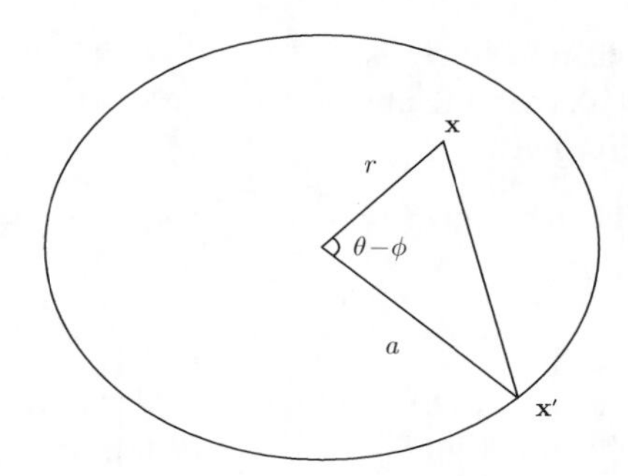

Figure 12.2. Depiction of $\mathbf{x}$ and $\mathbf{x}'$ on the disk of radius a.

so that

$$\boxed{u(r,\theta) = \frac{a^2 - r^2}{2\pi}\int_0^{2\pi} \frac{h(\phi)}{a^2 - 2ar\cos(\theta-\phi) + r^2}\, d\phi.}$$

If we consider two vectors $\mathbf{x}$ and $\mathbf{x}'$ where $\mathbf{x}'$ lies on the boundary of the disk and the angle between $\mathbf{x}$ and $\mathbf{x}'$ is $\theta - \phi$, then the law of cosines gives us

$$|\mathbf{x} - \mathbf{x}'|^2 = a^2 + r^2 - 2ar\cos(\theta - \phi).$$

Hence, noting that $ds_{\mathbf{x}} = a\, d\phi$, we have

$$\boxed{u(\mathbf{x}) = \frac{a^2 - |\mathbf{x}|^2}{2\pi a}\int_{\partial B(\mathbf{0},a)} \frac{h(\mathbf{x}')}{|\mathbf{x} - \mathbf{x}'|^2}\, ds_{\mathbf{x}'}.}$$

We notice that separation of variables yields the exact same formula, **Poisson's formula**, that we would obtain with Green's functions from Chapter 10!

12.6. • Extensions and Generalizations of the Separation of Variables Algorithm

This very short section is a prelude to the next four sections on extensions and generalizations of the Separation of Variables Algorithm. While we will not be exhaustive or comprehensive with all the possible extensions, we provide a few important ones. In most cases, we will explore the theory behind such extensions through particular examples of BVPs. However, most of what we present pertains to **eigenvalues and eigenfunctions of the Laplacian** which we briefly addressed in Section 8.7.

Recall what was stated without proof in Section 8.7. If $\Omega \subset \mathbb{R}^N$ was a bounded domain with piecewise smooth boundary, then there existed a sequence of positive eigenvalues of $-\Delta$ with Dirichlet boundary conditions which tended to infinity, **and** the associated eigenfunctions formed a basis for all reasonable functions (that is, they comprise a *complete set*).

One can use this powerful fact to apply the Separation of Variables Algorithm for problems involving the Laplacian in higher space dimensions using either rectangular, cylindrical, or spherical coordinates. While the basic approach remains the same, the associated eigenvalue problems are more complicated than those for sines and cosines, having nonconstant coefficients and lower-order terms. The solutions to these eigenvalue problems include some very important "**special functions**" which are solutions to important ODEs (alternatively, eigenvalue problems). Examples include the following:

- Bessel functions, previously introduced in Section 6.11.2.
- Legendre polynomials.

These special functions can be used to form an orthogonal "basis" for most functions; that is, they do indeed span reasonable functions and, as previously mentioned, are often referred to as a complete set of functions. Proving this directly for each class of functions is rather tedious, and hence we will not present any proofs. However, let me once more advertise that all of this theory falls under a beautiful and rigorous mathematical treatment via the **Spectral Theorem for a compact symmetric operator on a Hilbert space**.

We discuss Bessel functions in Section 12.8.1 in the particular setting of solving, via separation of variables, the wave equation on a 2D disk (the vibrating drum). In Section 12.9, we introduce Legendre polynomials in the particular setting of solving, via separation of variables, the Dirichlet problem in the 3D ball. A certain product involving Legendre polynomials (which we later define as **spherical harmonics**) will emerge and form an orthogonal "basis" for square-integrable functions defined on the sphere in $\mathbb{R}^3$. We also show how spherical harmonics can be used to solve the diffusion or wave equation on a 3D ball.

Lastly, we address a classical generalization of separation of variables which encompasses all the previous eigenvalue problems. In Section 12.10, we consider a very general form of eigenvalue problems known as **Sturm-Liouville problems**. These problems are also relevant for solving BVPs in inhomogeneous media.

12.7. • Extensions, I: Multidimensional Classical Fourier Series: Solving the Diffusion Equation on a Rectangle

In the previous two subsections, we considered BVPs in two space dimensions where our 1D theory for particular classes of 1D eigenfunctions (classical Fourier series) was sufficient. Now let us explore the idea of directly generalizing the classical Fourier sine, cosine, and full series to higher dimensions. Rather than doing this in any sort of generality, we focus on solving the 2D diffusion equation on a rectangle $R = [0, a] \times [0, b]$. Precisely, we wish to solve the following BVP:

$$\begin{cases} u_t = \alpha(u_{xx} + u_{yy}) & \text{for } (x, y) \in R,\ t > 0, \\ u(x, y, t) = 0 & \text{for } (x, y) \in \partial R,\ t > 0, \\ u(x, y, 0) = \phi(x, y). \end{cases}$$

Following the separation of variables idea, we look for separated solutions of the PDE and boundary conditions of the form

$$u(x, y, t) = X(x)Y(y)T(t).$$

Placing this form into the diffusion equation, we arrive at the conclusion that

$$-\frac{T'(t)}{\alpha T(t)} = -\frac{X''(x)}{X(x)} - \frac{Y''(y)}{Y(y)} \quad \text{must be a constant.} \tag{12.26}$$

We call this constant λ. But then, following the same logic we find

$$-\frac{X''(x)}{X(x)} = \frac{Y''(y)}{Y(y)} + \lambda \qquad \text{must be a constant.}$$

Call this constant β. This results in three eigenvalue problems:

$$X'' + \beta X = 0, \qquad Y'' + (\lambda - \beta)Y = 0, \qquad T' + k\lambda T = 0.$$

Focusing on the first two, we note that the boundary condition $u(x, y, t) = 0$ for $(x, y) \in \partial R$ and $t > 0$ implies that

$$X(0) = 0 = X(a) \qquad \text{and} \qquad Y(0) = 0 = Y(b).$$

We first solve

$$X'' + \beta X = 0, \qquad X(0) = 0 = X(a)$$

to find a sequence of solutions

$$\beta_n = \frac{n^2\pi^2}{a^2}, \qquad X_n(x) = \sin\left(\frac{n\pi}{a}x\right), \qquad n = 1, 2, \ldots.$$

Now for each $n = 1, 2, \ldots$, we must consider the eigenvalue problem for $Y(y)$,

$$Y'' + \left(\lambda - \frac{n^2\pi^2}{a^2}\right)Y = 0, \qquad Y(0) = 0 = Y(b).$$

For each n, we will have a sequence of nontrivial solutions, indexed by $m = 1, 2, \ldots$, as long as

$$\lambda - \frac{n^2\pi^2}{a^2} = \frac{m^2\pi^2}{b^2},$$

in which case the eigenfunction is

$$Y_m(y) = \sin\left(\frac{m\pi}{b}y\right).$$

This means that for each n and m, we have an eigenvalue of the form

$$\boxed{\lambda_{n,m} = \frac{n^2\pi^2}{a^2} + \frac{m^2\pi^2}{b^2}.}$$

With these values of $\lambda_{n,m}$, we have from (12.26) temporal solutions of the form

$$T_{n,m} = e^{-\alpha\left(\frac{n^2\pi^2}{a^2} + \frac{m^2\pi^2}{b^2}\right)t}.$$

Hence for any $n = 1, 2, \ldots$ and $m = 1, 2, \ldots$, we have a separated solution to the PDE and boundary conditions

$$u_{n,m}(x, y, t) = X_n(x)Y_m(y)T_{n,m}(t) = \sin\left(\frac{n\pi}{a}x\right)\sin\left(\frac{m\pi}{b}y\right)e^{-\alpha\left(\frac{n^2\pi^2}{a^2} + \frac{m^2\pi^2}{b^2}\right)t}.$$

We now make the bold move of taking an infinite linear combination of these,

$$\begin{aligned} u(x,y,t) &= \sum_{n=1}^{\infty}\sum_{m=1}^{\infty} b_{n,m} X_n(x) Y_m(y) T_{n,m}(t) \\ &= \sum_{n=1}^{\infty}\sum_{m=1}^{\infty} b_{n,m} \sin\Big(\frac{n\pi}{a}x\Big)\sin\Big(\frac{m\pi}{b}y\Big)\, e^{-\alpha\left(\frac{n^2\pi^2}{a^2}+\frac{m^2\pi^2}{b^2}\right)t}, \end{aligned}$$

and ask if we can find values for the coefficients $b_{n,m}$ such that

$$u(x,y,0) = \phi(x,y).$$

This amounts to asking whether we can find $b_{n,m}$ such that

$$\phi(x,y) = \sum_{n=1}^{\infty}\sum_{m=1}^{\infty} b_{n,m} \sin\Big(\frac{n\pi}{a}x\Big)\sin\Big(\frac{m\pi}{b}y\Big).$$

The answer is again (miraculously) yes, with the coefficients given by

$$b_{n,m} = \frac{4}{ab}\int_0^a\int_0^b \phi(x,y)\sin\Big(\frac{n\pi}{a}x\Big)\sin\Big(\frac{m\pi}{b}y\Big)\,dy\,dx.$$

While the fact that the set of functions

$$\Big\{\sin\Big(\frac{n\pi}{a}x\Big)\sin\Big(\frac{m\pi}{b}y\Big),\ \ n=1,2,\dots,\ \ m=1,2,\dots\Big\}$$

spans all reasonable functions $\phi(x,y)$ on R (i.e., completeness) is more subtle, we can easily address the fact that these functions are mutually orthogonal on R. In other words, for all distinct pairs (n_1,m_1) and (n_2,m_2), we have

$$\int_0^a\int_0^b \Big(\sin\Big(\frac{n_1\pi}{a}x\Big)\sin\Big(\frac{m_1\pi}{b}y\Big)\Big)\Big(\sin\Big(\frac{n_2\pi}{a}x\Big)\sin\Big(\frac{m_2\pi}{b}y\Big)\Big)\,dy\,dx = 0. \qquad (12.27)$$

This is immediate since

$$\begin{aligned} &\int_0^a\int_0^b \sin\Big(\frac{n_1\pi}{a}x\Big)\sin\Big(\frac{m_1\pi}{b}y\Big)\sin\Big(\frac{n_2\pi}{a}x\Big)\sin\Big(\frac{m_2\pi}{b}y\Big)\,dy\,dx \\ &= \Big(\int_0^a \sin\Big(\frac{n_1\pi}{a}x\Big)\sin\Big(\frac{n_2\pi}{a}x\Big)\,dx\Big)\Big(\int_0^b \sin\Big(\frac{m_1\pi}{b}y\Big)\sin\Big(\frac{m_2\pi}{b}y\Big)\,dy\Big). \end{aligned}$$

Then, since the pairs (n_1,m_1) and (n_2,m_2) are distinct, either $n_1 \neq n_2$ or $m_1 \neq m_2$ (or both). Hence, our previous orthogonality result implies at least one of the two factors is 0. Note that one could also arrive at the orthogonality property (12.27) by applying Green's Second Identity to two eigenfunctions of $-\Delta$ (corresponding to distinct eigenvalues) on R.

12.8. • Extensions, II: Polar and Cylindrical Coordinates and Bessel Functions

Solving the 2D and 3D wave and diffusion equations using polar or cylindrical coordinates takes us further into the realm of Bessel functions. We give one example of solving the 2D wave equation on a disk.

12.8.1. • Vibrations of a Drum and Bessel Functions. In this subsection, we solve a BVP which gives rise to an important class of eigenfunctions called **Bessel functions**. Consider the wave equation in two dimensions as a model for a vibrating drum. We wish to solve

$$\begin{cases} u_{tt} = c^2(u_{xx} + u_{yy}), & \mathbf{x} \in D = \{(x,y) : x^2 + y^2 < 1\}, \\ u = 0, & \mathbf{x} \in \partial D, \\ u = g, u_t = h, & t = 0. \end{cases} \tag{12.28}$$

Switching to polar coordinates, the 2D Laplacian becomes

$$\Delta u = u_{rr} + \frac{1}{r}u_r + \frac{1}{r^2}u_{\theta\theta}.$$

There are now three separate independent variables: r, θ, and t. We look for a separated solution to the PDE of the form

$$u(r,\theta,t) = T(t)R(r)\Theta(\theta).$$

Placing this form into the PDE gives (after a little algebra)

$$\frac{T''(t)}{c^2T(t)} = \frac{R''(r)}{R(r)} + \frac{R'(r)}{rR(r)} + \frac{\Theta''(\theta)}{r^2\Theta(\theta)}.$$

Hence, both the left- and right-hand sides must be a constant, which we denote by $-\lambda^2$. We shall soon see that this constant is always negative and will appreciate the notational reason for the square. Since

$$\frac{R''(r)}{R(r)} + \frac{R'(r)}{rR(r)} + \frac{\Theta''(\theta)}{r^2\Theta(\theta)} = -\lambda^2,$$

we see that $\Theta''(\theta)/\Theta(\theta)$ must also be a constant; it cannot vary with θ since the other terms on the left are all independent of θ. We call the constant Θ''/Θ, $-\gamma$.

These conclusions yield three eigenvalue problems:

$$T'' + c^2\lambda^2 T = 0, \qquad \Theta'' + \gamma\Theta = 0, \qquad R'' + \frac{1}{r}R' + \left(\lambda^2 - \frac{\gamma}{r^2}\right)R = 0. \tag{12.29}$$

We first solve $\Theta'' + \gamma\Theta = 0$ together with the implicit periodic boundary conditions; we find

$$\gamma = n^2, \qquad n = 0, 1, 2, \ldots,$$

and corresponding eigenfunctions

$$\Theta_n(\theta) = C\cos n\theta + D\sin n\theta, \qquad n = 1, 2, \ldots, \tag{12.30}$$

for any constants C and D. For the case where $n = 0$, i.e., $\gamma = n^2 = 0$, we find $\Theta_0(\theta)$ is a constant.

With these choices of γ, the R equation becomes

$$R'' + \frac{1}{r}R' + \left(\lambda^2 - \frac{n^2}{r^2}\right)R = 0, \qquad R(1) = 0, \qquad n = 0, 1, 2, \ldots. \tag{12.31}$$

As we shall see, we will also need the implicit boundary condition $R(0) < \infty$. We note that for $\lambda = 0$, the only solution to the R equation with $R(1) = 0$ is the trivial solution. Hence we need to only look for possible values of $\lambda > 0$ for both the T problem in (12.29) and the R problem in (12.31). These eigenvalues will come from the R equation.

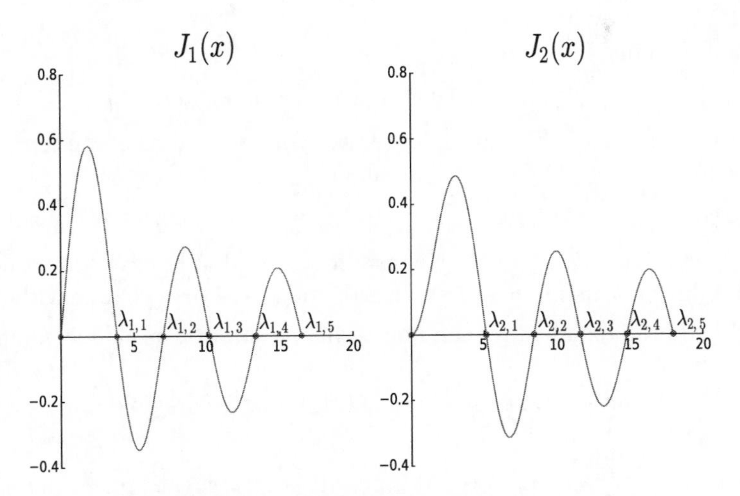

Figure 12.3. Bessel functions J_1, J_2 and their positive zeros.

Changing variables in (12.31) with $\rho = \lambda r$ gives

$$R_{\rho\rho} + \frac{1}{\rho}R_\rho + \left(1 - \frac{n^2}{\rho^2}\right)R = 0, \qquad n = 0, 1, \ldots. \tag{12.32}$$

Equation (12.32) is called **Bessel's equation** and, for each n, there exist two linearly independent solutions. However, only one of them will be bounded around $\rho = 0$. This smooth solution is called the **Bessel function** of order n and is denoted by $J_n(\rho)$. To find $J_n(\rho)$, you need to recall regular singular points and series solutions from your earlier class on ordinary differential equations. There, one finds that

$$J_n(\rho) = \sum_{i=0}^{\infty}(-1)^i \frac{(\rho/2)^{n+2i}}{i!\,(n+i)!}. \tag{12.33}$$

While this function is defined only by an infinite series, it is a C^∞ function which we are now naming! Sine and cosine are also functions which have given names but are precisely defined only via an infinite series.

In terms of these Bessel functions, we have eigenfunctions

$$R_n(r) = J_n(\lambda r), \qquad n = 0, 1, \ldots.$$

Our boundary condition tells us that $J_n(\lambda) = 0$. It turns out that for each $n = 0, 1, 2, \ldots$, $J_n(\rho)$ has an infinite number of positive roots (see Figure 12.3). Let us label these roots as

$$0 < \lambda_{n,1} < \lambda_{n,2} < \cdots.$$

In other words, $\lambda_{n,m}$ is the m-th root of the n-th order Bessel function $J_n(\rho)$.

Thus, for each $n = 0, 1, \ldots$ and $m = 1, 2, \ldots$, we have all the eigenvalues $\lambda_{n,m}$ with the corresponding R eigenfunctions

$$R_{n,m}(r) = J_n(\lambda_{n,m} r)$$

and T eigenfunctions

$$T_{n,m}(t) = A\cos\lambda_{n,m}ct + B\sin\lambda_{n,m}ct, \tag{12.34}$$

for any constants A and B. The full solution will consist of sums of the separated solutions of the form (indexed by $n = 0, 1, \ldots$ and $m = 1, 2, \ldots$)

$$\left(A_{n,m}\cos\lambda_{n,m}ct + B_{n,m}\sin\lambda_{n,m}ct\right) J_n(\lambda_{n,m}r)\left(C_{n,m}\cos n\theta + D_{n,m}\sin n\theta\right),$$

where $A_{n,m}, B_{n,m}, C_{n,m}, D_{n,m}$ are any constants. Note that here we have adjusted for the fact that the constants A, B, C, D in (12.30) and (12.34) can change with n and m.

Bringing this all together, we can now write the solution as a double sum:

$$\begin{aligned} u(r,\theta,t) = {} & \sum_{m=1}^{\infty} J_0(\lambda_{0,m}r)\left(A_{0,m}\cos\lambda_{0,m}ct + C_{0,m}\sin\lambda_{0,m}ct\right) \\ & + \sum_{n=1}^{\infty}\sum_{m=1}^{\infty} J_n(\lambda_{n,m}r)\Big[\left(A_{n,m}\cos n\theta + B_{n,m}\sin n\theta\right)\cos\lambda_{n,m}ct \\ & \qquad + \left(C_{n,m}\cos n\theta + D_{n,m}\sin n\theta\right)\sin\lambda_{n,m}ct\Big]. \end{aligned} \tag{12.35}$$

Finally, we introduce the initial conditions. For any $n = 0, 1, \ldots$ and $m = 1, 2, \ldots$, the product functions

$$u_{n,m}(r,\theta) = J_n(\lambda_{n,m}r)\cos n\theta \qquad \text{and} \qquad v_{n,m}(r,\theta) = J_n(\lambda_{n,m}r)\sin n\theta \tag{12.36}$$

are eigenfunctions of $-\Delta$ on D with Dirichlet boundary conditions. One can show that they are orthogonal on D and are complete; i.e., they span reasonable functions. Hence, by following the usual argument we can find the coefficients to satisfy the initial conditions. In Exercise **12.17** you are asked to write out these coefficients in terms of integrals involving the data functions $g(r,\theta)$ and $h(r,\theta)$ written in terms of polar coordinates.

We end with two comments.

1. **Bessel functions are not to be feared.** At first glance, the presence of the seemingly *uncanny* Bessel functions and nonexplicit values of the $\lambda_{n,m}$ might raise concerns as to the application of the solution formula (12.35). Bessel functions are not to be feared and with enough hands-on experience become almost as "user friendly" as sines and cosines. Almost all mathematical software, high-level computing languages like *Python* and *Matlab*, have commands to call on any particular Bessel function. Moreover, their zeros (the $\lambda_{n,m}$) are easily accessible to any required decimal place accuracy. Hence computation of the coefficients and approximation of the solution (12.35) can readily be performed with any mathematical software. In Exercise **12.19** you are asked to numerically approximate the solution at different times for the case where

$$g(r,\theta) = 1 - r^2, \qquad h \equiv 0.$$

2. **Modes, nodal lines, and natural frequencies of the drum.** Recall Example 12.2.1 where we discussed the normal modes and natural frequencies of a vibrating string (for example, a violin string). The same notions are there for the vibrating drum but the situation is far more complicated. The normal modes are given

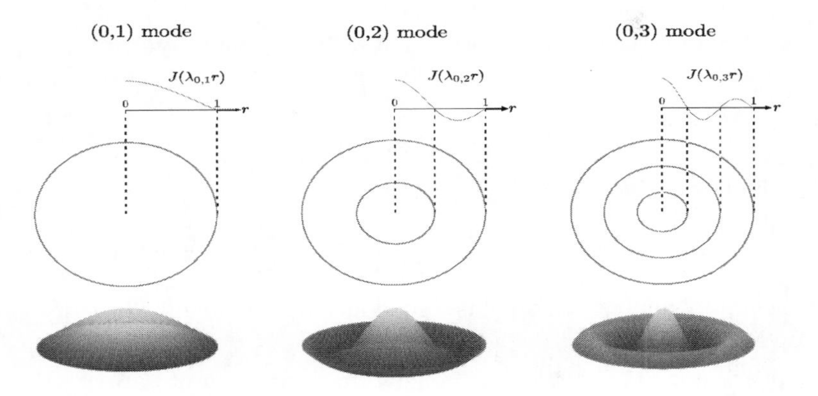

Figure 12.4. Three $n = 0$ modes illustrating the nodal lines which correspond to zeros of the function $J_0(\lambda_{0,m}r)$ for $r \in [0, 1]$.

by the spatial eigenfunctions (12.36), and the natural frequencies are $\lambda_{n,m}c$. One often uses the indices (n, m) to refer to a particular mode. These modes can be viewed as *natural shapes* for the drum whose *specific character* is determined by its size (here 1) and c (which involves the material parameters density and tension). For example, suppose for some n, m, we started with initial data

$$g(r, \theta) = J_n(\lambda_{n,m}r) \cos n\theta, \qquad h \equiv 0.$$

Then at any later time, the solution would simply be a constant times the initial shape $g(r, \theta)$; namely the solution would be

$$u(r, \theta, t) = J_n(\lambda_{n,m}r) \cos n\theta \sin \lambda_{n,m}ct.$$

The zeros of these modes are called **nodal lines** and represent places on the drum which remain fixed (no displacement) with time. Whereas for the vibrating string the nodal points were always at a fixed distance apart (a consequence of the uniformity for the zeros of sines and cosines), the nodal lines for a drum are not at fixed distances (a consequence of spacing for the zeros of the Bessel functions). In Figure 12.4 we focus on the modes for $n = 0$ which are purely radial and just involve the Bessel function J_0. For $n \geq 1$, the modes also depend on θ and Exercise **12.21** asks you to investigate a few of these.

12.9. Extensions, III: Spherical Coordinates, Legendre Polynomials, Spherical Harmonics, and Spherical Bessel Functions

We start by considering the 3D Laplace equation in spherical coordinates and we separate variables. In doing so, we will encounter a family of special functions $P_l^m(x)$ called the **associated Legendre**[2] **polynomials**. In addition, we will be introduced to a fundamental and ubiquitous family of functions $Y_l^m(\phi, \theta)$, which are known as the **spherical harmonics** and are defined over the unit sphere.

[2]**Adrien-Marie Legendre** (1752–1833) was a French mathematician who made many contributions but perhaps is best remembered for Legendre polynomials and the Legendre transform.

12.9.1. Separation of Variables for the 3D Laplace Equation in Spherical Coordinates. The 3D Laplacian Δ, which in rectangular coordinates (x, y, z) is given by

$$\Delta = \partial_{xx} + \partial_{yy} + \partial_{zz},$$

can be written in spherical coordinates (r, ϕ, θ) as

$$\Delta = \frac{1}{r^2}\partial_r\left(r^2\partial_r\right) + \frac{1}{r^2 \sin\phi}\partial_\phi\left[\sin\phi\,\partial_\phi\right] + \frac{1}{r^2\sin^2\phi}\partial_{\theta\theta}. \tag{12.37}$$

At some point in your education, it is a good idea for you to directly verify this using the chain rule.

In spherical coordinates, we will follow the Separation of Variables Algorithm, seeking nontrivial separated solutions to the 3D Laplace equation $\Delta u = 0$ of the form

$$u(r, \phi, \theta) = R(r)\Phi(\phi)\Theta(\theta).$$

We begin by first separating the radial variable from the two angular variables; namely, we seek harmonic functions of the form

$$u(r, \phi, \theta) = R(r)Y(\phi, \theta).$$

Using (12.37), we plug this separated ansatz into Laplace's equation to find

$$\Delta u = \frac{Y}{r^2}\frac{d}{dr}\left(r^2 R'\right) + \frac{R}{r^2\sin\phi}\partial_\phi\left[(\sin\phi)\partial_\phi Y\right] + \frac{R}{r^2\sin^2\phi}\partial_{\theta\theta}Y = 0.$$

As usual, we seek to extract ODEs in one variable from this equation which we hope will have the familiar form of eigenvalue problems. To this end, reorganizing the above equation gives

$$\frac{1}{R}\frac{d}{dr}\left[r^2R'\right] = -\frac{1}{Y\sin\phi}\partial_\phi\left[(\sin\phi)\partial_\phi Y\right] - \frac{1}{Y\sin^2\phi}\partial_{\theta\theta}Y \tag{12.38}$$

which needs to hold *for all* r, ϕ, and θ. Noting that the left-hand side is only a function of r and the right-hand side is only a function of (ϕ, θ), we conclude that there must exist a constant λ such that both sides are identically equal to λ. This provides us with our first ODE, an eigenvalue problem in r,

$$\frac{d}{dr}\left[r^2R'\right] = \lambda R, \tag{12.39}$$

which is known as a second-order **Cauchy-Euler equation**, having solutions of the form

$$R(r) = r^l, \quad \text{for } l > 0 \text{ with } \lambda = l(l+1).$$

Note that we could have "guessed" this by noting the scaling invariance of the ODE: letting $\tilde{r} = \alpha r$ yields an identical ODE. We have eliminated negative values of l by noting the requirement that $u = R\Phi\Theta$ must be twice differentiable everywhere, including at the origin.

For the right-hand side of (12.38) we obtain a PDE in ϕ, θ:

$$\frac{\sin\phi}{Y}\partial_\phi\left[(\sin\phi)\partial_\phi Y\right] + \frac{1}{Y}\partial_{\theta\theta}Y = -\lambda\sin^2\phi. \tag{12.40}$$

In order to solve (12.40) we will invoke separation of variables again and seek solutions of the form

$$Y(\phi, \theta) = \Phi(\phi)\Theta(\theta).$$

This turns (12.40) into

$$\frac{\sin\phi}{\Phi}\partial_\phi\Big[(\sin\phi)\Phi'\Big] + \frac{\Theta''}{\Theta} = -\lambda\sin^2\phi$$

which can be reorganized into

$$\frac{\sin\phi}{\Phi}\frac{d}{d\phi}\Big[(\sin\phi)\Phi'\Big] + \lambda\sin^2\phi = -\frac{\Theta''}{\Theta}.$$

With the usual reasoning we conclude that there must be a constant μ such that both the right- and left-hand sides are identically equal to μ, for all θ and all ϕ, respectively.

We supplement these two eigenvalue problems for Θ and Φ with the **implicit** boundary condition of periodicity: Both functions must be 2π periodic for the construction $u = R(r)\Phi(\phi)\Theta(\theta)$ to be well-defined. For the eigenvalue problem

$$\Theta'' = -\mu\Theta$$

this implies that μ must be positive and furthermore the square of a whole number; i.e.,

$$\mu = m^2 \qquad \text{for some positive integer } m.$$

Hence, we find a countable number of solutions to the eigenvalue problem for Θ:

$$\Theta(\theta) = e^{\pm im\theta} \qquad \text{for positive integers } m.$$

Note that we are using the complex form for the standard full Fourier series (with sines and cosines), even though the function u itself will be real (cf. Section 11.1.7).

Finally, we turn to the eigenvalue problem for Φ, with $\lambda = l(l+1)$ and $\mu = m^2$, which can be written as

$$\frac{d}{d\phi}\Big[(\sin\phi)\Phi'\Big] + \big(l(l+1)\sin\phi\big)\Phi = m^2\frac{\Phi}{\sin\phi}. \tag{12.41}$$

We make the important change of variable

$$x = \cos\phi \qquad \text{with } x \in [-1, 1]$$

noting that

$$\frac{d}{d\phi} = -\sin\phi\frac{d}{dx}.$$

The ODE for Φ now becomes

$$\boxed{\frac{d}{dx}\Big[(1-x^2)\frac{d}{dx}\Phi\Big] + \Big[l(l+1) - \frac{m^2}{1-x^2}\Big]\Phi = 0.} \tag{12.42}$$

This ODE is known as the **general Legendre equation**. Note that $x = \pm 1$, corresponding to $\cos\phi = \pm 1$, are singularities of the equation. By using the Frobenius method we have the existence of solutions (finite at $x = \pm 1$) for integer values of l with

$l \geq |m|$. In the next subsection, we discuss these solutions. At this stage, note that our characterization of all relevant eigenvalues is now complete:

- One set of eigenvalues is the squares of integers m^2.
- The other set is $\lambda = l(l+1)$ for any positive integer $l \geq |m|$.

The solutions of the general Legendre equation (12.42) (eigenvalue problem for Φ) are known, for any choice of positive integers l and m with $l \geq m$, as the **associated Legendre polynomials**, or the **associated Legendre functions**, since as we shall see they are polynomials **only** for certain values of m. These solutions are written as $P_l^m(x)$, and hence our eigenfunctions for Φ take the form

$$\Phi(\phi) = P_l^m(\cos\phi), \quad \text{for positive integers } l \text{ and integers } m \text{ with } 0 \leq m \leq l.$$

Putting this all together, we have obtained a countable number of separated solutions to original Laplace equation of the form

$$r^l \, e^{\pm im\theta} \, P_l^m(\cos\phi).$$

Now, the natural question arises as to the orthogonality and spanning properties of these eigenfunctions (i.e., completeness), in particular the associated Legendre polynomials. Miraculously, they hold! We will discuss orthogonality and appropriate normalizations in the next subsection. With these properties, one can now construct the series

$$\boxed{u(r,\phi,\theta) = \sum_{l=0}^{\infty} \sum_{m=-l}^{l} a_l^m \, r^l \, e^{im\theta} \, P_l^{|m|}(\cos\phi),}$$

for constants a_l^m appropriately chosen to accommodate data (e.g., boundary conditions), as the full solution to Laplace's equation. To repeat, the negative m are there because we are using the complex form of the full Fourier series in θ; if we used real eigenfunctions $\sin m\theta$ and $\cos m\theta$, we would only include nonnegative m.

12.9.2. Legendre Polynomials and Associated Legendre Polynomials. In this subsection, we discuss solutions of the general Legendre equation (12.42) and some of their properties. We recall that we are only interested in the case where m and l are both integers with $l \geq |m|$ and our domain of interest is $x \in [-1,1]$.

Case 1: $m = 0$.

First, we consider the special case where $m = 0$, which is known simply as the *Legendre equation*:

$$\frac{d}{dx}\left[(1-x^2)\frac{d}{dx}P\right] + n(n+1)P = 0 \tag{12.43}$$

for some positive integer n. The solutions to this ODE are, up to a normalizing constant, special functions known as **Legendre polynomials** $P_n(x)$. They are indeed polynomials in x and are characterized by an explicit formula, which we will accept without proof, known as the **Rodrigues formula**:

$$\boxed{P_n(x) = \frac{1}{2^n n!}\frac{d^n}{dx^n}(x^2-1)^n.}$$

This formula can be further expanded to give term-by-term coefficients of Legendre polynomials of any order.

The family of polynomials $\{P_n(x)\}_{n\in\mathbb{N}}$ are orthogonal in the sense that for $n' \neq n$ we have

$$\int_{-1}^{1} P_n(x)P_{n'}(x)dx = 0.$$

We obtain this by first multiplying the ODE for P_n by $P_{n'}$ and the ODE for $P_{n'}$ by P_n, then subtracting the two to obtain

$$\frac{d}{dx}\Big[(1-x^2)(P_n'P_{n'} - P_{n'}'P_n)\Big] = \Big[n(n+1) - n'(n'+1)\Big]P_nP_{n'}.$$

Now integrating both sides and noting that $(1-x^2)$ vanishes at both boundary points gives the desired result.

We also note without proof that Legendre polynomials form a complete basis for the space $L^2[-1,1]$. This means that for any function $f \in L^2[-1,1]$, there exists a sequence of real numbers a_n such that

$$f(x) = \sum_{n=0}^{\infty} a_n P_n(x)$$

where the series limit is taken in $L^2[-1,1]$. Furthermore, orthogonality of P_n implies that the coefficients a_n can be obtained by simply projecting f on the corresponding Legendre polynomial.

Case 2: $m \neq 0$.

We now return to the general Legendre equation (12.42). Its solutions will be denoted as P_l^m and are called **associated Legendre polynomials** or associated Legendre functions. They are only exact polynomials for even m. In fact, one can derive the following formula:

$$\boxed{P_l^m(x) = (-1)^m(1-x^2)^{\frac{m}{2}} \frac{d^m}{dx^m} P_l(x).} \tag{12.44}$$

It is convenient to define these functions for negative integers $-m$ as a certain constant times $P_l^m(x)$: Indeed, guided by the Rodrigues formula, one defines

$$P_l^{-m}(x) = (-1)^m \frac{(l-m)!}{(l+m)!} P_l^m(x).$$

Associated Legendre polynomials also enjoy an orthogonality property: The P_l^m are orthogonal for fixed m or fixed l, but *not if both parameters vary*. Specifically, we have

$$\int_{-1}^{1} P_l^m P_l^{m'} dx = 0$$

for any $m \neq m'$ and

$$\int_{-1}^{1} P_l^m P_{l'}^m dx = 0$$

for any $l \neq l'$. Finally, the normalization constant needed when projecting functions on associated Legendre polynomials is given by

$$\int_{-1}^{1} (P_l^m)^2 dx = \frac{2(l+m)!}{(2l+1)(l-m)!}.$$

It is worth noting that the $(-1)^m$ term in (12.44) is a source of notational confusion, since different scientific communities include or exclude it. If this term is excluded from the definition of P_l^m, it is then placed on the definition of spherical harmonics (see below). This term is commonly referred to as the **Condon-Shortley phase** with its historical roots in quantum mechanics. In order to mitigate this confusion, it is common to use the notation $P_{lm}(x)$ to refer to the associated Legendre polynomial *without* the Condon-Shortley phase term and reserve $P_l^m(x)$ for the definition above.

12.9.3. Spherical Harmonics. Recalling the solutions we obtained for $Y(\phi, \theta)$ in the Separation of Variables Algorithm for the Laplace equation, let us define

$$\boxed{Y_l^m(\phi, \theta) := c_l^m e^{im\theta} P_l^m(\cos\phi)}$$

where the c_l^m are normalization constants only depending on m and l and are chosen in such a way as to make the spherical harmonics an **orthonormal set**:

$$c_l^m = \sqrt{\frac{(2l+1)}{4\pi}\frac{(l-m)!}{(l+m)!}}.$$

The functions Y_l^m, defined for positive integers l and $m = -l, \ldots, 0, \ldots, l$, are called **spherical harmonics** and are ubiquitous in science and engineering. Here we briefly mention some of their fundamental properties.

First, spherical harmonics corresponding to distinct pairs (m, l) are orthogonal. Note the subtle difference between this and the weaker orthogonality of associated Legendre polynomials; spherical harmonics inherit this stronger orthogonality by virtue of including the extra $\sin(m\theta)$ and $\cos(m\theta)$ terms (the real and imaginary parts of $e^{im\theta}$) in addition to P_l^m. Furthermore, spherical harmonics Y_l^m form a complete basis for $L^2(S^2)$, the space of square-integrable functions on S^2, the two-dimensional unit sphere. This means that for any square-integrable function f on S^2, we can obtain a spherical harmonic expansion

$$f(\phi, \theta) = \sum_{l=0}^{\infty} \sum_{m=-l}^{l} f_l^m \, Y_l^m(\phi, \theta),$$

for suitable constants f_l^m and where the convergence is taken in the L^2 sense.

Similar to Fourier series expansion, the coefficients f_l^m can be obtained by projecting f onto the complex conjugate of the corresponding basis function; that is,

$$f_l^m = \int_{S^2} f(\phi, \theta) \, \overline{Y_l^m}(\phi, \theta) \, dS.$$

One can easily verify that for the spherical harmonic expansion series to yield a real number, the coefficients f_l^m must satisfy the relationship

$$f_l^m = (-1)^m \overline{f_l^{-m}}.$$

12.9.4. Solving the 3D Diffusion Equation on the Ball. We now apply the Separation of Variables Algorithm to solve

$$\begin{cases} u_t = \alpha\Delta u, & \mathbf{x} \in B(\mathbf{0},1),\ t>0, \\ u(\mathbf{x},t) = 0, & \mathbf{x} \in \partial B(\mathbf{0},1),\ t>0, \\ u(\mathbf{x},0) = g(\mathbf{x}), & \mathbf{x} \in B(\mathbf{0},1), \end{cases}$$

where $B(\mathbf{0},1)$ is the unit ball in $\mathbb{R}^3$. For simplicity of notation we set $\alpha = 1$.

We will again look for separated solutions in spherical coordinates of the form

$$u(r,\phi,\theta,t) \;=\; R(r)\,\Phi(\phi)\,\Theta(\theta)\,T(t).$$

We will find countably many separated solutions which satisfy the explicit boundary condition (and any implicit boundary conditions). Hopefully, we will have orthogonality and completeness of the eigenfunctions $\Phi(\phi)\,\Theta(\theta)$ which will allow us to construct appropriately weighted infinite sums which satisfy the initial conditions. This should all sound familiar. So what is new here? When we place the separated ansatz into the diffusion equation, we will obtain a familiar eigenvalue problem for T. However, the spatial terms will now be related to eigenvalues of the Laplacian in spherical coordinates. To this end, we begin by placing the ansatz

$$u(r,\phi,\theta,t) \;=\; U(r,\phi,\theta)\,T(t)$$

into the diffusion equation to obtain

$$\boxed{U(r,\phi,\theta)\,T'(t) \;=\; \Delta U(r,\phi,\theta)\,T(t) \qquad \text{or} \qquad \frac{T'(t)}{T(t)} = \frac{\Delta U(r,\phi,\theta)}{U(r,\phi,\theta)}.}$$

By the usual reasoning both sides of the above equation on the right must be a constant, say, $-\lambda$. This gives rise to

$$T' = -\lambda T \qquad \text{and} \qquad -\Delta U = \lambda U.$$

The T equation is easy to solve, while the second equation asks for eigenfunctions and eigenvalues of the Laplacian which satisfy the boundary condition $U = 0$ on $r = 1$. To find these eigenfunctions and eigenvalues, we again separate variables with the ansatz

$$U(r,\phi,\theta) \;=\; R(r)\,\Phi(\phi)\,\Theta(\theta).$$

This, in turn, will yield eigenvalue problems for R, Φ, and Θ, respectively. Those for Φ and Θ are the same as in the previous section, with the respective products being the spherical harmonics. The R equation will be different from the previous section: It will be similar to Bessel's equation, with solutions known as **spherical Bessel functions**.

Let us provide the details for the spherical eigenfunctions of the Laplacian:

$$-\Delta U \;=\; \lambda U, \qquad\qquad U(1,\phi,\theta) = 0.$$

As we did in Section 12.9.1, let us first separate the dependence on r from the angles; that is, we consider the separated ansatz

$$U(r,\phi,\theta) \;=\; R(r)\,Y(\phi,\theta)$$

into the PDE $-\Delta U = \lambda U$, using the formula (12.37) for the Laplacian in spherical coordinates. This yields

$$-\Delta U = -\frac{Y}{r^2}\frac{d}{dr}\left(r^2 R'\right) - \frac{R}{r^2 \sin\phi}\partial_\phi\left[(\sin\phi)\partial_\phi Y\right] - \frac{R}{r^2\sin^2\phi}\partial_{\theta\theta}Y = \lambda RY.$$

With a little algebra this yields

$$\frac{1}{R}\frac{d}{dr}\left(r^2 R'\right) + \lambda r^2 + \frac{1}{Y\sin\phi}\partial_\phi\left[(\sin\phi)\partial_\phi Y\right] + \frac{1}{Y\sin^2\phi}\partial_{\theta\theta}Y = 0.$$

The usual argument implies for a constant γ,

$$\frac{1}{R}\frac{d}{dr}\left(r^2 R'\right) + \lambda r^2 = \gamma \quad \text{and} \quad \frac{1}{Y\sin\phi}\partial_\phi\left[(\sin\phi)\partial_\phi Y\right] + \frac{1}{Y\sin^2\phi}\partial_{\theta\theta}Y = -\gamma.$$

As we saw in Section 12.9.1, the PDE for Y can be separated with

$$Y(\phi,\theta) = \Phi(\phi)\,\Theta(\theta)$$

to yield two eigenvalue problems:

$$\boxed{\Theta'' = -\mu\Theta \quad \text{and} \quad \frac{d}{d\phi}\left[(\sin\phi)\Phi'\right] + (\gamma\sin\phi)\Phi = \frac{\mu\Phi}{\sin\Phi}}$$

with yet another eigenvalue μ (this makes three λ, γ, μ). The fact that Θ must be periodic yields that $\mu = m^2$ for some integer m. As in Section 12.9.1, the eigenvalue problem for Φ can be transformed, with change of variable $x = \cos\phi$, into the Legendre equation

$$\frac{d}{dx}\left[\left(1-x^2\right)\frac{d}{dx}\Phi\right] + \left[\gamma - \frac{m^2}{1-x^2}\right]\Phi = 0. \tag{12.45}$$

We found this equation before in (12.42) with a set choice for γ. However, if we seek solutions to (12.45) which are bounded at ± 1, we immediately find that γ must be $l(l+1)$ for a nonnegative integer l such that $l \geq m$. The solutions for the Φ equation are

$$\Phi(\phi) = P_l^m(\cos\phi),$$

where the P_l^m are the Legendre polynomials (case $m = 0$) and associated Legendre polynomials ($m \neq 0$) previously discussed in Section 12.9.2.

Consider now the ODE for R, which we may write with $\gamma = l(l+1)$ as

$$\boxed{R'' + \frac{2}{r}R' + \left(\lambda - \frac{l(l+1)}{r^2}\right)R = 0.} \tag{12.46}$$

This equation, similar to Bessel's equation from (12.32), is often known as **the spherical Bessel equation.** Note the presence of the eigenvalue λ, which is as of yet undetermined. To solve (12.46), one first performs a change of variables in order to transform it into a regular Bessel equation, which one then solves via the method of series solutions. To this end, let

$$\overline{R}(r) := \sqrt{r}R(r).$$

In terms of $\overline{R}$, one can verify that (12.46) becomes

$$\overline{R}'' + \frac{1}{r}\overline{R}' + \left(\lambda - \frac{\left(l+\frac{1}{2}\right)^2}{r^2}\right)\overline{R} = 0.$$

This is Bessel's equation (12.32) but with fractional order $l + \frac{1}{2}$. The solution with the added implicit boundary condition that $\overline{R}(0)$ be finite is

$$\overline{R}(r) = J_{l+\frac{1}{2}}(\sqrt{\lambda} r),$$

where $J_{l+\frac{1}{2}}$ is the Bessel function of the first kind of order $l + \frac{1}{2}$ which can be found explicitly in terms of a convergent power series expansion. Reverting back to $R(r)$, we have

$$\boxed{R(r) = \frac{J_{l+\frac{1}{2}}(\sqrt{\lambda} r)}{\sqrt{r}}.}$$

Lastly, we find the possible eigenvalues λ by imposing the only explicit boundary condition $R(1) = 0$, i.e., solve, for λ, the equation

$$J_{l+\frac{1}{2}}(\sqrt{\lambda}) = 0.$$

As in Section 12.8.1, one has for any nonnegative integer l, the existence of countably many nonnegative solutions which we label as

$$\lambda_{l,1}, \lambda_{l,2}, \lambda_{l,3}, \ldots.$$

These values are not given explicitly but can be readily computed to machine precision.

So(!), what have we found so far? For any $l = 0, 1, \ldots, m = -l, \ldots, l$, and $j = 1, 2, \ldots$, we have found a separated solution to the diffusion equation and the boundary condition of the form

$$u_{lmj}(r, \phi, \theta, t) = e^{-\lambda_{l,j} t} \frac{J_{l+\frac{1}{2}}(\sqrt{\lambda_{l,j}}\, r)}{\sqrt{r}} P_l^{|m|}(\cos\phi)\, e^{im\theta}.$$

We now consider a weighted infinite sum of the u_{lmj}

$$u(r, \phi, \theta, t) = \sum_{l=0}^{\infty} \sum_{j=0}^{\infty} \sum_{m=-l}^{l} a_{lmj}\, e^{-\lambda_{l,j} t} \frac{J_{l+\frac{1}{2}}(\sqrt{\lambda_{l,j}}\, r)}{\sqrt{r}} P_l^{|m|}(\cos\phi)\, e^{im\theta} \tag{12.47}$$

as our candidate solution to the IVP. We will easily be able to choose the coefficients a_{lmj} to satisfy the initial condition $u(\mathbf{x}, 0) = g(\mathbf{x})$ provided that the set of eigenfunctions of the Laplacian

$$v_{lmj}(r, \phi, \theta) := \frac{J_{l+\frac{1}{2}}(\sqrt{\lambda_{l,j}}\, r)}{\sqrt{r}} P_l^{|m|}(\cos\phi)\, e^{im\theta}$$

are orthogonal and complete on the 3D ball $B(\mathbf{0}, 1)$. They are! Note that orthogonality means that if vectors (l, m, j) and (l', m', j') are different (i.e., at least one component differs), then

$$\int_0^{2\pi} \int_0^{\pi} \int_0^1 v_{lmj}(r, \phi, \theta)\, v_{l'm'j'}(r, \phi, \theta)\, r^2 \sin\phi\, dr\, d\phi\, d\theta = 0.$$

A few comments are in order:

- We have used the complex form of the full Fourier series in θ at the expense of introducing the negative values of m. Hence, v_{lmj} and the coefficients a_{lmj} will also be complex valued (even though in the end for real-valued g, u as defined by (12.47) will be real valued). Note that within the framework of the complex number, the usual inner product contains one complex conjugate; i.e., the inner product of two complex-valued functions f and g is $\iiint \overline{f(\mathbf{x})}\, g(\mathbf{x})\, d\mathbf{x}$. Thus in terms of Cartesian coordinates $\mathbf{x}$, we have (by orthogonality)

$$a_{lmj} = \frac{\iiint_{B(\mathbf{0},1)} \overline{v_{lmj}(\mathbf{x})}\, g(\mathbf{x})\, d\mathbf{x}}{\iiint_{B(\mathbf{0},1)} |v_{lmj}(\mathbf{x})|^2\, d\mathbf{x}},$$

or in terms of spherical coordinates,

$$a_{lmj} = \frac{\int_0^{2\pi}\int_0^{\pi}\int_0^1 \overline{v_{lmj}(r,\phi,\theta)}\, g(r,\phi,\theta)\, r^2 \sin\phi\, dr\, d\phi\, d\theta}{\int_0^{2\pi}\int_0^{\pi}\int_0^1 |v_{lmj}(r,\phi,\theta)|^2\, r^2 \sin\phi\, dr\, d\phi\, d\theta}.$$

- Each class of separate eigenfunctions for R, Φ, or Θ individually satisfies an orthogonality condition.
- One can rewrite (12.47) by using the spherical harmonics $Y_l^m(\phi,\theta)$.
- One can similarly solve the 3D wave equation on the ball in terms of spherical harmonics and Bessel functions (cf. Exercise **12.25**).
- On first exposure, the reader might find all of this to be *a big mess* and be dissatisfied with the challenging solution formula (12.47). Indeed, we do not have explicit values for the eigenvalues $\lambda_{l,j}$ which correspond to zeros of spherical Bessel functions, functions which are, in principle, "hard" to embrace (unlike, say, trigonometric functions). Moreover, the appearance of Legendre polynomials only adds to the seemingly "obtuse" nature of the solution formula. However, we emphasize that all these special functions have power series expansions and can also be computed to machine precision. Much is known about their qualitative properties and one can also use these series expansions to make certain approximate conclusions about the solution.

  Takeaway point: After some hands-on experience, the readers will be able to embrace such formulas in the same way as they have already embraced the solution for the 1D BVP as an infinite sum of familiar, and perhaps more user-friendly, exponentials and trigonometric functions.

12.10. Extensions, IV: General Sturm-Liouville Problems

Consider the following general boundary value/eigenvalue problem for a function $X(x)$ on $[0,l]$:

$$\boxed{(p(x)X'(x))' + q(x)X(x) + \lambda\sigma(x)X(x) = 0,} \tag{12.48}$$

or equivalently,

$$-(p(x)X'(x))' - q(x)X(x) = \lambda\sigma(x)X(x),$$

with boundary conditions of the form

$$\boxed{\alpha_1 X(0) + \alpha_2 X'(0) = 0, \qquad \alpha_3 X(l) + \alpha_4 X'(l) = 0.} \tag{12.49}$$

Here:

- $p(x)$, $q(x)$, and $\sigma(x) \geq 0$ are prescribed (fixed) functions on $[0, l]$ with $p(x) \in C^1(0, l)$.
- α_i, $i = 1, 2, 3, 4$, are prescribed (fixed) constants where at least one of α_1, α_2 and one of α_3, α_4 are nonzero.

The simplest eigenvalue problem resulting from the operator $\mathcal{A}$ of Section 11.2.3, which was the basis for Fourier series, would result from taking $p \equiv 1$ and $q \equiv 0$. This class of particular BVPs is commonly referred to as **Sturm-Liouville[3] problems** and comprise (with the exception of periodic boundary conditions) **all** the previous eigenvalue problems we have encountered thus far, in particular:

- The eigenvalue problems generated by separation of variables in one space dimension for the wave and diffusion equations. For these problems, the coefficient functions were constants.
- The eigenvalue problems generated by separation of variables for the diffusion and Laplace equations in higher dimensions. This includes those problems stemming from spherical coordinates. Note that Bessel's equation is an example where the coefficient functions are no longer constant.

There are other reasons for wanting to incorporate nonconstant coefficients, for example, the modeling of waves and diffusion in inhomogeneous media.

Notation: We have written the ODE in the general problem (12.48) with dependent variable X and independent variable x. Many other notations can be employed; for example, in Bessel's equation we used R in place of X and either r or ρ in place of x to emphasize their geometric interpretation. In other references, it is common to use ϕ or y in place of X.

Let us first discuss the notion of an eigenvalue problem in the context of (12.48). Recall that the classical Fourier series pertained to eigenvalues and eigenfunctions of the linear operator $\mathcal{A} = -\frac{d^2}{dx^2}$ on the interval $(0, l)$ for functions satisfying a symmetric boundary condition. In other words, for any admissible function $X(x)$, $\mathcal{A}X(x) = -X''(x)$. The operator was symmetric on the space of admissible functions in the sense that for all functions $X_1(x)$ and $X_2(x)$ on $x \in (0, l)$ which satisfy the boundary conditions, we have $(\mathcal{A}X_1, X_2) = (X_1, \mathcal{A}X_2)$, where the inner product of two functions f and g on $(0, l)$ is defined by $(f, g) = \int_0^l f(x)g(x)\,dx$. In the context of the more general problem (12.48), let us denote the linear operator by $\mathcal{L}$ where

$$\mathcal{L}\,X(x) := -(p(x)X'(x))' - q(x)X(x), \qquad \text{or simply } \mathcal{L}X := -(pX')' - qX.$$

We will show shortly that the operator $\mathcal{L}$ with the boundary conditions (12.49) is symmetric. We say $X(x)$ satisfying (12.49) is an **eigenfunction of $\mathcal{L}$ with weight $\sigma(x)$**

[3] **Jacques Charles Francois Sturm** (1803–1855) and **Joseph Liouville** (1809–1882) were French mathematicians particularly famous for their work on these types of eigenvalue problems.

and corresponding eigenvalue λ if

$$\mathcal{L}X(x) = \lambda\,\sigma(x)X(x), \quad \text{or simply} \quad \mathcal{L}X = \lambda\sigma X.$$

Note here the new presence of the weight function σ in the definition of the eigenvalue problem. Also note that this is completely equivalent to X solving (12.48).

We prove some simple facts about this class of general Sturm-Liouville problems. They are corollaries of a general calculus result which is the analogue for operator $\mathcal{L}$ of (11.22) for the operator $\mathcal{A}$. Recall that (11.22) was simply Green's Second Identity in 1D. In the context of the Sturm-Liouville operator $\mathcal{L}$, the identity is sometimes known as **Lagrange's identity**.

Theorem 12.1. *For any C^2 functions $X_1(x)$ and $X_2(x)$ on $[0, l]$, we have*

$$\int_0^l X_2(x)\,\mathcal{L}X_1(x) - X_1(x)\,\mathcal{L}X_2(x)\,dx = -p(x)X_1'(x)X_2(x)\big|_0^l + p(x)X_1(x)X_2'(x)\big|_0^l. \tag{12.50}$$

Proof. Unsurprisingly, the proof is just integration by parts. To this end,

$$\begin{aligned}
\int_0^l X_2(x)\,\mathcal{L}X_1(x)\,dx &= \int_0^l X_2(x)\Big(-(p(x)X_1'(x))' - q(x)X_1(x)\Big)\,dx \\
&= -p(x)X_1'(x)X_2(x)\big|_0^l + \int_0^l X_1'(x)p(x)X_2'(x)dx - \int_0^l X_1(x)q(x)X_2(x)\,dx \\
&= -p(x)X_1'(x)X_2(x)\big|_0^l + p(x)X_1(x)X_2'(x)\big|_0^l \\
&\qquad + \int_0^l X_1(x)\Big(-(p(x)X_2'(x))' - q(x)X_2(x)\Big)\,dx \\
&= -p(x)X_1'(x)X_2(x)\big|_0^l + p(x)X_1(x)X_2'(x)\big|_0^l + \int_0^l X_1(x)\,\mathcal{L}X_2(x)\,dx. \qquad \square
\end{aligned}$$

As a Corollary, we have the following:

Corollary 12.10.1. *The operator $\mathcal{L}$ with boundary conditions* (12.49) *is symmetric. That is, for all $X_1(x)$ and $X_2(x)$ which satisfy* (12.49) *we have*

$$(X_1, \mathcal{L}X_2) = (\mathcal{L}X_1, X_2);$$

that is,

$$\int_0^l X_1(x)\,\mathcal{L}X_2(x)\,dx = \int_0^l \mathcal{L}X_1(x)\,X_2(x)\,dx. \tag{12.51}$$

Proof. By Theorem 12.1, it suffices to show that

$$-p(x)X_1'(x)X_2(x)\big|_0^l + p(x)X_1(x)X_2'(x)\big|_0^l = 0. \tag{12.52}$$

To this end, first let us assume that both α_2 and α_4 are nonzero. Then (12.49) gives

$$X_1'(0) = -\frac{\alpha_1}{\alpha_2}X_1(0),\ X_1'(l) = -\frac{\alpha_3}{\alpha_4}X_1(l),\ X_2'(0) = -\frac{\alpha_1}{\alpha_2}X_2(0),\ X_2'(l) = -\frac{\alpha_3}{\alpha_4}X_2(l).$$

Hence

$$\begin{aligned}
-p(x)X_1'(x)X_2(x)\big|_0^l \; + \; p(x)X_1(x)X_2'(x)\big|_0^l
&= -p(l)X_1'(l)X_2(l) + p(0)X_1'(0)X_2(0) \\
&\quad + p(l)X_1(l)X_2'(l) - p(0)X_1(0)X_2'(0) \\
&= p(l)\frac{\alpha_3}{\alpha_4}X_1(l)X_2(l) - p(0)\frac{\alpha_1}{\alpha_2}X_1(0)X_2(0) \\
&\quad - p(l)\frac{\alpha_3}{\alpha_4}X_1(l)X_2(l) + p(0)\frac{\alpha_1}{\alpha_2}X_1(0)X_2(0) \\
&= 0.
\end{aligned}$$

Now if $\alpha_2 = 0$, we must have $X_1(0) = 0 = X_2(0)$, whereas if $\alpha_4 = 0$ we must have $X_1(l) = 0 = X_2(l)$. Thus (12.52) holds. □

An immediate consequence of Corollary 12.10.1 is the suitable orthogonality of eigenfunctions:

Corollary 12.10.2. *Let $X_1(x), X_2(x)$ be two eigenfunctions of $\mathcal{L}$ corresponding to distinct eigenvalues; i.e., X_1, X_2 with boundary conditions* (12.49) *satisfy*

$$\mathcal{L}X_1 = \lambda_1\,\sigma(x)X_1 \quad \textit{and} \quad \mathcal{L}X_2 = \lambda_2\,\sigma(x)X_2 \qquad \textit{with } \lambda_1 \neq \lambda_2.$$

Then

$$\int_0^l \sigma(x)X_1(x)\,X_2(x)\,dx = 0. \tag{12.53}$$

Proof. We have

$$\int_0^l X_1(x)\,\mathcal{L}X_2(x)\,dx = \int_0^l X_1(x)\,\lambda_2\,\sigma(x)\,X_2\,dx,$$

$$\int_0^l \mathcal{L}X_1(x)\,X_2(x)\,dx = \int_0^l \lambda_1\,\sigma(x)X_1(x)\,X_2\,dx.$$

By (12.51), this means

$$(\lambda_1 - \lambda_2)\int_0^l \sigma(x)X_1(x)\,X_2(x)\,dx = 0.$$

Since $\lambda_1 \neq \lambda_2$, we then have (12.53). □

There is an important dichotomy for Sturm-Liouville problems: regular and singular. Before discussing each type, let us just note that while (12.48) with periodic boundary conditions is not included in these classes of problems, similar results also hold true, for example, the orthogonality of the eigenfunctions.

12.10.1. Regular Sturm-Liouville Problems. For regular Sturm-Liouville problems, we require the functions p, q, and σ to be continuous on the closed interval $[0, l]$ and $p(x)$ and $\sigma(x)$ to be strictly positive on $[0, l]$. Except for periodic boundary conditions, all the eigenvalue problems in Sections 12.1 to 12.5.2 were regular Sturm-Liouville problems.

Regular Sturm-Liouville problems have a well-established theory, all of which mirrors previous results for Fourier series (i.e., for eigenfunctions of $\mathcal{A}$ with a symmetric boundary condition). Let us document some of the basic results:

1. All eigenvalues are real and there exists exactly a countable collection of them:
$$\lambda_1 < \lambda_2 < \cdots < \lambda_j < \cdots$$
with $\lambda_j \to \infty$ as $j \to \infty$.
2. For each eigenvalue λ_j, there exists a real eigenfunction $X_j(x)$ which is unique up to a scalar constant. In other words, the multiplicity of each eigenvalue is one.
3. Each eigenfunction X_j has exactly $j - 1$ zeroes on $(0, l)$.
4. Eigenfunctions corresponding to distinct eigenvalues are orthogonal with weight σ in the sense of Corollary 12.10.2.
5. The eigenfunctions are **complete** in the following senses. Let ϕ be a function on $(0, l)$ and suppose we define
$$S_N(x) := \sum_{j=1}^{N} c_j X_j(x), \qquad \text{where } c_j := \frac{\int_0^l \sigma(x)\,\phi(x)\,X_j(x)\,dx}{\int_0^l \sigma(x)\,X_j^2(x)\,dx}.$$
Then:
   (i) **L^2 (mean square) convergence.** If $\phi \in L^2(0, l)$, then $S_N(x)$ converges to ϕ in $L^2(0, l)$.
   (ii) **Pointwise convergence.** If ϕ is piecewise smooth on $(0, l)$, i.e., ϕ is continuous and differentiable with its derivative being piecewise continuous, then for all $x \in (0, l)$,
$$S_N(x) \longrightarrow \phi(x).$$

Properties 1, 3, and especially 5 are by no means easy to show. In the classical theory, their proofs can be found in books such as Coddington and Levinson[4] and Courant and Hilbert[5]. On the other hand, as we have already mentioned, they are all a consequence of fundamental results in functional analysis [**10**], [**19**]. These properties amount to saying that we can "do Fourier series" with the X_j; that is, we can write a suitably general function ϕ as

$$\phi = \sum_{j=1}^{\infty} c_j X_j(x), \qquad \text{where } c_j := \frac{\int_0^l \sigma(x)\,\phi(x)\,X_j(x)\,dx}{\int_0^l \sigma(x)\,X_j^2(x)\,dx}.$$

[4] **E. Coddington and N. Levinson**, *Theory of Ordinary Differential Equations*, McGraw-Hill, NY, 1966.
[5] **R. Courant and D. Hilbert**, *Methods of Mathematical Physics*, two volumes, Wiley-Interscience, NY, 1991.

12.10.2. Singular Sturm-Liouville Problems. Singular Sturm-Liouville problems[6] are defined to be those Sturm-Liouville problems in which either the coefficient functions are not continuous or they vanish somewhere in $[0, l]$. These include the classes of Bessel's equation and the Legendre equation. For example, the spherical Bessel equation of Section 12.9.4 is

$$R'' + \frac{2}{r}R' + \left(\lambda - \frac{l(l+1)}{r^2}\right)R = 0.$$

Here we use R and r for the dependent and independent variables, respectively. This equation can be readily put into form (12.48) as it is equivalent to

$$(r^2R')' - \gamma R + \lambda r^2 R = 0.$$

The Legendre equation of Section 12.9.4, written in notation $\Phi(\phi)$,

$$\frac{d}{d\phi}\Big[(\sin\phi)\Phi'\Big] + (\gamma\sin\phi)\Phi = \frac{\mu\Phi}{\sin\Phi}$$

for $\phi \in [0, \pi]$ is also a singular Sturm-Liouville problem.

Similar properties to 1–5 for regular Sturm-Liouville problems also hold for these and other singular Sturm-Liouville problems.

12.11. Separation of Variables for the Schrödinger Equation: Energy Levels of the Hydrogen Atom

> *"We would like now, however, to show you how one of the most remarkable consequences of Schrödinger's equation comes about — namely, the surprising fact that a differential equation involving only continuous functions of continuous variables in space can give rise to quantum effects such as the discrete energy levels in an atom."*
>
> **- Richard Feynman**[a]
>
> [a]From the third volume of *The Feynman Lectures on Physics*, Addison Wesley.

This section is a perfect way to end the chapter and also the new material of this text. It also serves to complete the previous short Section 7.11 which introduced the Schrödinger equation. We will apply separation of variables to the Schrödinger equation for the hydrogen atom, not on a bounded domain but rather on all of space with an additional integral constraint (a "boundedness condition"). In doing so we will achieve a rather remarkable feat: We will compute the quantized energy levels of the hydrogen atom, yielding numerical values which concur with experiments!

In the hydrogen atom there is one electron (of charge $-e$ and mass m_e) which is in motion about the nucleus (of mass m_n). The nucleus is comprised of exactly one proton of charge $+e$ and is assumed to be centered at the origin. One of the great scientific observations was that there exists a discrete (i.e., not continuous) number of states in which the hydrogen atom can exist. These states are associated with energy levels,

[6]Sturm-Liouville problems which are posed on an infinite interval are often also called singular.

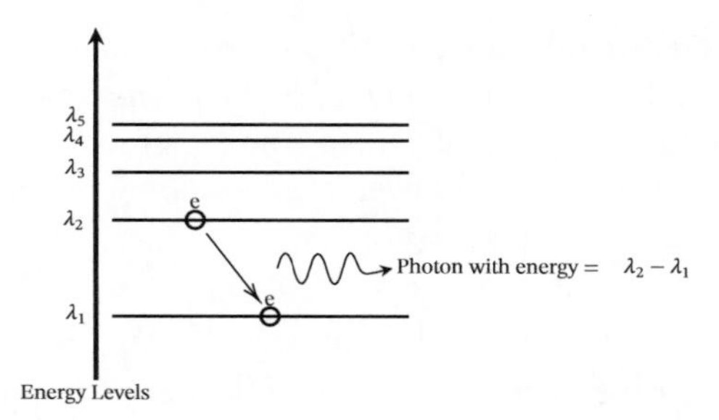

Figure 12.5. Emission spectral lines of the hydrogen atom whose frequencies are proportional (with proportionality constant $1/\hbar$) to the quantized energy levels λ_n. As an electron moves from energy level λ_2 to λ_1, it gives off energy. Precisely, it emits a photon with energy $\lambda_2 - \lambda_1 > 0$.

experimentally observed by emission spectral lines (see Figure 12.5 and its caption). In 1885, **Balmer**[7] used measurements of the hydrogen spectral lines by **Ångström**[8] and calculated a formula for their wavelengths. **Bohr**[9] provided the first theoretical agreement for these energy levels using what is known as the "old formulation" of quantum mechanics. Here we will show how the modern theory of quantum mechanics, specifically Schrödinger's equation, also correctly predicts these discrete energy levels.

By Coulomb's Law (cf. Sections 10.6.1 and 10.6.2), the potential resulting from the electron at position $\mathbf{x}$ and proton at the origin is given by

$$V(\mathbf{x}) = -k_e \frac{e^2}{r}, \qquad \text{where } r = |\mathbf{x}|,$$

where k_e is Coulomb's constant defined in (10.55). The associated Schrödinger equation for the motion of the electron is thus

$$\boxed{i\hbar u_t \;=\; -\frac{\hbar^2}{2m}\Delta u \;-\; k_e \frac{e^2}{r} u,} \tag{12.54}$$

where $\hbar$ is the reduced Planck constant, $r = |\mathbf{x}|$ denotes the distance of the electron from the origin (center of the nucleus), and m is the reduced mass[10] of the atom; that is, $m = \frac{m_e m_n}{m_e + m_n}$.

To begin, let us look for a solution separated in space and time of the form $u(\mathbf{x}, t) = T(t)U(\mathbf{x})$. Placing this form into (12.54), we find by the usual argument that

$$2i\hbar \frac{T'(t)}{T(t)} \;=\; -\frac{\hbar^2}{m}\frac{\Delta U(\mathbf{x})}{U(\mathbf{x})} \;-\; \frac{2k_e e^2}{r} \quad \text{must be a constant,}$$

which we denote by λ. Note that $\hbar$ has dimensions of energy $\times$ time and λ above has dimensions of energy. The resulting ODE for T is easy to solve yielding $T(t) = e^{-i\frac{\lambda}{2\hbar}t}$.

[7]**Johann Jakob Balmer** (1825–1898) was a Swiss mathematics school teacher known for his computation of the energy lines of the hydrogen atom, often known as *the Balmer series*.

[8]**Anders Jonas Ångström** (1814–1874) was a Swedish physicist and one of the founders of spectroscopy.

[9]**Niels Henrik David Bohr** (1885–1962) was a Danish physicist who made foundational contributions to our understanding of atomic structure and quantum theory. He received the Nobel Prize in Physics in 1922.

[10]Since m_e is much smaller than m_n, one often takes the reduced mass to be simply m_e.

Note that the term in the exponential $\frac{\lambda}{2\hbar}t$ is indeed dimensionless. The spatial part $U(\mathbf{x})$ solves

$$-\frac{\hbar^2}{m}\Delta U - \frac{2k_e e^2}{r}U = \lambda U. \tag{12.55}$$

Following Section 8.7.1, the possible eigenvalues λ are directly related to energy of the system. We define the values of λ which give "suitably finite" solutions (see the two additional conditions below) as the **energy levels of the bound states** of the hydrogen atom. By bound state we mean states which remain localized in a neighborhood of the origin (the nucleus); that is, the electron cannot escape the potential resulting from the nucleus.

What are the possible values of λ? All the previous problems of this chapter were for BVPs with boundary conditions which were directly responsible for determining the eigenvalues. Here we want to solve (12.54) on all of $\mathbb{R}^3$; however, what will replace the boundary conditions (and effectively impose a countable number of eigenvalues) are two additional "physical" conditions:

- Because of the interpretation of $u(\cdot, t)$ as the probability density for the position of the electron, we require that for all $t \geq 0$,

$$\iiint_{\mathbb{R}^3} |u(\mathbf{x}, t)|^2 \, d\mathbf{x} < \infty. \tag{12.56}$$

- The value at the origin $u(\mathbf{0}, t)$ is always finite.

We claim that the only values of λ which give nontrivial solutions to (12.55) satisfying these two additional conditions are

$$\lambda_n := -\frac{mk_e^2 e^4}{2\hbar^2 n^2}, \qquad n = 1, 2, 3, \ldots. \tag{12.57}$$

The reader should verify that λ_n has dimensions of energy. You may wonder as to the **negative sign** of the energy levels. The free electron (no interaction with the nucleus) is assigned energy level 0 (which corresponds to $\lim_{n\to\infty} \lambda_n$); hence to "relieve" the electron (in any of its discrete states) from the potential of the attractive nucleus requires (positive) energy. The formula (12.57) is in agreement with both Balmer's experimental observations and Bohr's derivation for the energy levels of the hydrogen atom.

While we will not supply full details for this claim (i.e., (12.57)), let us outline the steps with reference to an exercise which fills in the missing details. The basic premise is, of course, the Separation of Variables Algorithm in spherical coordinates where we seek solutions of (12.55) of the form $U(\mathbf{x}) = R(r)Y(\theta, \phi)$. For the purposes of finding the energy levels, it suffices to look for **spherically symmetric** solutions $U(\mathbf{x}) = R(r)$.

Before continuing, let us reflect on the form of the eigenvalues in (12.57). There is the dependence on the physical constants but, more importantly, the discrete nature and dependence on n. It will thus simplify our calculations to "eliminate" the constants, working in units which effectively allow us to take the values of the physical constants to be 1; that is,

$$\hbar = e = m = k_e = 1.$$

Of course this means that the exact occurrences of these physical constants in the energy levels (12.57) will be hidden; however, one can recover them by repeating the subsequent arguments with the original constants, at the expense of a messier calculation (see the work done in Chapter 4 of Griffiths[11]).

If we adopt these units, then solving (12.55) for $U(\mathbf{x}) = R(r)$ reduces to finding solutions $R(r)$ to

$$-\left(R_{rr} + \frac{2}{r}R_r\right) - \frac{2}{r}R = \lambda R, \qquad r \in [0, \infty). \tag{12.58}$$

Our two additional conditions reduce to requiring that

$$\int_0^\infty |R(r)|^2 r^2 \, dr < \infty \qquad \text{and} \qquad R(0) < \infty. \tag{12.59}$$

So it all boils down to showing that the only nontrivial solutions to the eigenvalue problem (12.58) satisfying (12.59) are when

$$\lambda = -\frac{1}{n^2}, \qquad n = 1, 2, 3, \ldots. \tag{12.60}$$

To this end, we first assume all eigenvalues are negative, an assumption which can be verified with some work. It is then useful and natural (see for example Section 9.5 of Strauss [**6**]) to make the following change of dependent variable:

$$\tilde{R}(r) := e^{\beta r} R(r) \qquad \text{where } \beta := \sqrt{-\lambda}. \tag{12.61}$$

This change of variables will transform (12.58) into a modified form of **Laguerre's equation**[12] which for certain parameter values has polynomial solutions known as **Laguerre polynomals**. Indeed, one finds (cf. Exercise **12.32**(a)) that in terms of $\tilde{R}$ the equation (12.58) becomes

$$\frac{1}{2} r \tilde{R}_{rr} - (\beta r - 1)\tilde{R}_r + (1 - \beta)\tilde{R} = 0. \tag{12.62}$$

The question becomes, For what values of β do we have solutions satisfying the boundedness conditions of (12.59)? You may recall from your ODE class the topic (perhaps not your favorite) of **regular singular points** and **series solutions**; this is what we need now. We write the solution in the form of an infinite power series

$$\tilde{R}(r) = \sum_{k=1}^{\infty} a_k r^k \tag{12.63}$$

and place the series into the ODE (12.62). In doing so, we find relationships between the coefficients. If $\beta = \frac{1}{n}$ for some $n = 1, 2, \ldots$, then the coefficients are eventually zero; that is, the solution is a polynomial. Otherwise, one finds an infinite series which does not satisfy the boundedness conditions. We conclude that the only solutions satisfying the boundedness conditions are when (12.60) holds true. The details are left in Exercise **12.32**(b).

[11]**David J. Griffiths**, Introduction to Quantum Mechanics, second edition, Cambridge, 2016.

[12]In neutral variables $y(x)$, Laguerre's equation is

$$xy'' + (1 - x)y' + \alpha y = 0.$$

We end with a few remarks:

- Associated with these energy levels are the eigenfunctions $\tilde{R}_n(r)$ (which are polynomials) and the resulting radially symmetric separated solutions to the Schrödinger equation (12.54), which in atomic units have the form

$$u_n(\mathbf{x}, t) = e^{-\frac{it}{2n^2}} e^{\frac{|\mathbf{x}|}{n}} \tilde{R}_n(|\mathbf{x}|).$$

  As in Exercise **12.32**(c) it is instructive to write down explicitly the formulas for $n = 1, 2, 3$ and reflect on their structure. Of particular interest is the **ground state** corresponding to $n = 1$ (the first eigenvalue).

- There are eigenvalues and radially symmetric solutions to (12.54) corresponding to **unbounded states** which do not satisfy the boundedness conditions. In fact, there is a continuum of eigenvalues referred to as the **continuous spectrum**.

- There are **nonradial** solutions which can be found by separating variables for the term $Y(\theta, \phi)$. Not surprisingly, these angular terms are related to spherical harmonics. The resulting solutions to Schrödinger's equation are associated with the **orbital angular momentum** of the electron.

12.12. Chapter Summary

Given a second-order linear PDE in a finite spatial domain with boundary conditions and (perhaps) initial conditions, the separation of variables approach is as follows:

- We look for separated solutions to the PDE and (some of) the boundary conditions. This will yield eigenvalue problems for a symmetric second-order differential operator which reduces to solving a linear second-order ODE within the class of functions satisfying the boundary conditions.

- For a large class of problems, there exists a countably infinite number of eigenvalues with mutually orthogonal corresponding eigenfunctions. Moreover, these eigenfunctions are complete, in the sense that they span all reasonable functions.

- These properties of the eigenfunctions allow us to achieve the initial condition (or additional boundary condition) by taking a weighted infinite sum of the separated solution. Thus we arrive at a "formula" for the solution in the form of an infinite series.

- For the 1D diffusion and wave equation on a spatial interval, the eigenfunctions are sines and/or cosines. For the Robin boundary condition, there is no explicit formula for the eigenvalues (frequencies of the trigonometric eigenfunctions); however, they can be approximated to great accuracy.

- For higher-D diffusion, wave, and Laplace equations on higher D rectangular (box) domains, the eigenfunctions are products of sines and/or cosines in each of the spatial variables.

- For these higher-D PDEs on radially symmetric domains (e.g., disks and balls), we may have additional eigenfunctions which are *special functions*; these include Bessel functions and Legendre polynomials, with the latter forming one of the factors in spherical harmonics.
- All these eigenvalue problems are part of a class called Sturm-Liouville problems.

Exercises

12.1 Verify that the solution formula (12.4), (12.5) can be written as (12.6), (12.7).

12.2 Use separation of variables to solve the BVP (12.1) with $\alpha = 1, l = 1$, and $\phi(x) = 4x$. Your answer will be an infinite series but you must compute all the coefficients. By truncating the series with the first 5 terms, plot the profiles $u(x, t)$ for $t = 0, t = 1, t = 2, t = 10$.

12.3 Use separation of variables to solve the BVP with mixed boundary conditions:

$$\begin{cases} u_t = u_{xx}, & x \in (0, l), t > 0, \\ u(0, t) = T_0 \text{ and } \frac{\partial u}{\partial x}(l, t) = 0, & t > 0, \\ u(x, 0) = T_1, & x \in (0, l), \end{cases}$$

where T_0 and T_1 are constants. Your answer will be an infinite series but you must compute all the coefficients. By truncating the series with the first 5 terms, plot the profiles $u(x, t)$ for $t = 0, t = 1, t = 2, t = 10$.

12.4 Use separation of variables to solve the BVP for the plucked string; that is, solve (12.11) with $\psi \equiv 0$ and $\phi(x) = \frac{2x}{l}$ if $0 < x < \frac{l}{2}$ and $\phi(x) = 2 - \frac{2x}{l}$ if $\frac{l}{2} < x < l$. Your answer will be an infinite series but compute all the coefficients. By truncating the series with the first 10 terms and setting $c = 1, l = 1$, plot the profiles $u(x, t)$ for times $t = 0, 1, \ldots, 10$. You can also make a short movie illustrating the dynamics.

12.5 Consider the BVP for the diffusion equation $u_t = u_{xx}$ on $0 < x < 1, t > 0$, with initial data $u(x, 0) = e^x$ for $0 < x < 1$ and boundary condition

$$u(0, t) - u_x(0, t) = 20, \qquad u(1, t) = 10, \quad t > 0.$$

(a) Give a physical interpretation of the boundary conditions (what do they mean physically?).
(b) Find the steady-state solution $v(x)$.
(c) Should $v(0)$ be equal to 20? Is it? Give an explanation of why this makes physical sense.

12.6 Consider the BVP (12.16) with $\phi(x) = x(1-x)$. Using computer software, compute λ_n for $n = 1, \ldots, 10$. Then plot the solution for times $t = 0, 0.5, 2, 4, 10$.

12.7 Consider the BVP for the heat equation $u_t = \alpha u_{xx}$ with boundary conditions

$$u_x(0,t) = A, \qquad -\alpha\frac{\partial u}{\partial x}(\pi,t) = h(u(\pi,t) - T_1), \qquad t > 0,$$

for constants $A, T_1 > 0$, and $h > 0$. Explain what the boundary conditions mean. Give a physical scenario, explaining the roles and physical dimensions of the constants.

12.8 Verify Duhamel's Principle by showing that (12.20) solves (12.18).

12.9 (**A more direct approach to equations with source terms**) Consider the inhomogeneous BVP/IVP:

$$\begin{cases} u_t = \alpha u_{xx} + f(x,t), & 0 < x < l, \\ u(0,t) = 0 = u(l,t), & t > 0, \\ u(x,0) = \phi(x), & 0 \le x \le l, \end{cases}$$

where $f(x,t)$ is a given function. Solve this equation using the following steps and arrive at the solution formula (12.21).

(i) Write the unknown solution as $u(x,t) = \sum_{n=1}^{\infty} T_n(t) \sin\frac{n\pi x}{l}$, where $T_n(t)$ are to be determined. For each $t \ge 0$, write the function $f(\cdot,t)$ in a Fourier series $f(x,t) = \sum_{n=1}^{\infty} f_n(t)\sin\frac{n\pi x}{l}$.

(ii) Compute $u_t - \alpha u_{xx}$ of the Fourier series by differentiating term by term.

(iii) Equate the result with the Fourier series for $f(x,t)$, noting that the respective coefficients must be equal. This will result in ODEs for $T_n(t)$.

(iv) Use the Fourier series of ϕ to determine the initial conditions of these ODEs and then solve them.

(v) Put everything together to arrive at the solution formula (12.21).

12.10 Let a be a constant and consider the BVP

$$\begin{cases} u_t = u_{xx} + au, & 0 < x < \pi, t > 0, \\ u(0,t) = 0 = u(\pi,t), & t > 0, \\ u(x,0) = \phi(x), & 0 \le x \le \pi. \end{cases}$$

(a) Solve this BVP as follows. Using our separation of variables theme, look for a solution of the form $u(x,t) = \sum_{n=1}^{\infty} T_n(t)\sin nx$, where the $T_n(t)$ are to be determined. By assuming term-by-term differentiation, put this form into the PDE and derive an ODE for $T_n(t)$. Solve it and then use the initial conditions to find the constants.

(b) What is the long-time behavior $\lim_{t\to\infty} u(x,t)$? The answer depends on a.

12.11 (**Time-dependent inhomogeneous Dirichlet boundary conditions**) Give a physical interpretation in terms of temperature in a bar for the following BVP/IVP:

$$\begin{cases} u_t = \alpha u_{xx}, & & 0 < x < l, t > 0, \\ u(0,t) = T_0(t), & u(l,t) = T_1(t), & t > 0, \\ u(x,0) = \phi(x), & & 0 < x < l, \end{cases}$$

where $T_0(t)$ and $T_1(t)$ are given functions. Solve this problem. **Hint:** Let $u(x,t) = w(x,t) - v(x,t)$ where $v(x,t) = T_0(t) + x\left(\frac{T_1(t)-T_0(t)}{l}\right)$. What BVP/IVP does $w(x,t)$ solve?

12.12 (Time-dependent inhomogeneous Neumann boundary conditions) Give a physical interpretation in terms of temperature in a bar for the following BVP/IVP:

$$\begin{cases} u_t = \alpha u_{xx}, & & 0 < x < l, t > 0, \\ u_x(0,t) = A(t), & u_x(l,t) = B(t), & t > 0, \\ u(x,0) = \phi(x), & & 0 < x < l, \end{cases}$$

where $A(t)$ and $B(t)$ are given functions. Solve this problem. **Hint:** Let $u(x,t) = w(x,t) - v(x,t)$ where $v(x,t) = A(t)x + (B(t) - A(t))\frac{x^2}{2l}$. What BVP/IVP does $w(x,t)$ solve?

12.13 (Time-dependent inhomogeneous Robin boundary conditions) Consider the BVP/IVP

$$\begin{cases} u_t = \alpha u_{xx}, & & 0 < x < l, t > 0, \\ au(0,t) + bu_x(0,t) = A(t), & u(l,t) = 0, & t > 0, \\ u(x,0) = \phi(x), & & 0 < x < l, \end{cases}$$

for some constants a, b and function $A(t)$. Give a physical interpretation and then solve this problem as follows. First look for a function of the form $v(x,t) = (c_1 + c_2 x + c_3 x^2)A(t)$ which satisfies only the boundary conditions. In other words, find the constants c_i. Then consider $w(x,t) = u(x,t) - v(x,t)$. What BVP does $w(x,t)$ solve? Hint: w will now solve the diffusion equation with an inhomogeneous right-hand side (cf. Exercise **12.9**).

12.14 (Laplace's Equation on a Rectangle with Dirichlet Boundary Conditions) Solve the BVP

$$\begin{cases} u_{xx} + u_{yy} = 0, & 0 < x < a,\ 0 < y < b, \\ u(x,0) = f(x),\ u(x,b) = 0, & 0 \le x \le a, \\ u(0,y) = 0,\ u(a,y) = 0, & 0 \le y \le b. \end{cases}$$

Give an example of a nonconstant function $f(x)$ for which the solution can be written with only one separated product (no infinite or finite sum).

12.15 (Laplace's Equation on a Rectangle with Neumann Boundary Conditions) Suppose $f(y)$ is a function such that $\int_0^b f(y)dy = 0$. Solve the BVP

$$\begin{cases} u_{xx} + u_{yy} = 0, & 0 < x < a,\ 0 < y < b, \\ u_y(x,0) = 0,\ u_y(x,b) = 0, & 0 \le x \le a, \\ u_x(0,y) = 0,\ u_x(a,y) = f(y), & 0 \le y \le b. \end{cases}$$

Interpret the BVP (i.e., the boundary conditions) as a steady-state temperature distribution in the rectangle. Based upon this interpretation, explain why there the BVP has no solution if $\int_0^b f(y)dy \neq 0$.

12.16 **(Laplace's Equation on a Rectangle with Mixed Boundary Conditions)** Solve the BVP

$$\begin{cases} u_{xx} + u_{yy} = 0, & 0 < x < a,\ 0 < y < b, \\ u(x,0) = 0,\ u(x,b) = f(x), & 0 \le x \le a, \\ u_x(0,y) = 0,\ u_x(a,y) = 0, & 0 \le y \le b. \end{cases}$$

12.17 Write down the orthogonality relations for the eigenfunctions (12.36). Then write expressions for all the coefficients in (12.35) in terms of these eigenfunctions.

12.18 **(a)** Solve the problem for the vibrating drum (12.28) in the case where $g \equiv 0$ and $h(r,\theta) = 1$ if $r \le 1/2$ and 0 otherwise. You need to write the coefficients in terms of Bessel functions. Use any software (or tables) to compute the first 10 coefficients and plot the solution for $t = 1, 4, 8$.
(b) Create a movie illustrating the solution.

12.19 **(a)** Solve the problem for the vibrating drum (12.28) in the case where $g(r,\theta) = 1 - r^2$ and $h \equiv 0$. You need to write the coefficients in terms of Bessel functions. Use any software (or tables) to compute the first 10 coefficients and plot the solution for $t = 1, 4, 8$.
(b) Create a short movie illustrating the solution.

12.20 **(a)** Solve the following IVP for 2D wave equation on a rectangle $R = [a,b] \times [c,d]$:

$$\begin{cases} u_{tt} = c^2(u_{xx} + u_{yy}), & (x,y) \in R,\ t > 0, \\ u(x,y,t) = 0, & (x,y) \in \partial R,\ t > 0, \\ u(x,y,0) = g,\ u_t(x,y,0) = h, & (x,y) \in R. \end{cases}$$

Your answer will of course have coefficients defined in terms of integrals involving g, h, and certain eigenfunctions.
(b) What are the fundamental modes? For a few of them, sketch the nodal lines.

12.21 For the vibrating drum of Section 12.8.1, use any software to sketch the nodal lines corresponding to the $(1,1)$, $(1,2)$, and $(2,2)$ modes. Make a short movie of the solution for initial data $g(r,\theta) = J_1(\lambda_{1,1} r)\cos\theta$ and $h \equiv 0$.

The next two exercises pertain to properties of Bessel functions $J_n(\rho)$ (cf. (12.33)).

12.22 The Bessel functions $J_n(\rho)$ have the following **integral representation**: For each n,

$$J_n(\rho) = \frac{1}{2\pi} \int_0^{2\pi} e^{i(\rho \sin\theta - n\theta)}\, d\theta.$$

Use this integral representation to prove the following recursive formula for Bessel functions: $\rho J_{n+1}(\rho) = nJ_n(\rho) - \rho J_n'(\rho)$. In particular, note that we can use this formula to compute any Bessel function of order $n \ge 1$ from J_0.

12.23 Recall the trigonometric formula for sine $\sin(\alpha+\beta) = \sin\alpha\cos\beta+\cos\alpha\sin\beta$. Use (12.33) to show the corresponding formula for Bessel functions:

$$J_n(\alpha+\beta) = \sum_{m=-\infty}^{\infty} J_m(\alpha)J_{n-m}(\beta),$$

where for $n = 0, 1, 2, \ldots$, we define $J_{-n} := (-1)^n J_n$.

12.24 (a) In an inhomogeneous medium with spatially varying conductivity $\sigma(\mathbf{x})$, explain why Maxwell's equations yield the PDE $\operatorname{div}(\sigma(\mathbf{x})\nabla u) = 0$ for the electric potential u.
(b) In two dimensions, consider a spatially varying conductivity defined on $B(\mathbf{0}, 1)$ by

$$\sigma(\mathbf{x}) = \begin{cases} 2 & \text{if } 0 \le |\mathbf{x}| < \frac{1}{2}, \\ 1 & \text{if } \frac{1}{2} \le |\mathbf{x}| < 1. \end{cases}$$

Use separation of variables to explicitly solve the BVP

$$\begin{cases} \operatorname{div}(\sigma(\mathbf{x})\nabla u) = 0 & \text{on } B(\mathbf{0}, 1) \\ u = g & \text{on } \partial B(\mathbf{0}, 1), \end{cases}$$

where $g(\mathbf{x}) = g(r, \theta) = \cos\theta$. Here one has the additional "implicit" boundary conditions at $r = \frac{1}{2}$: $u(\mathbf{x})$ and $\sigma\frac{\partial u}{\partial r}$ are continuous at $|\mathbf{x}| = r = \frac{1}{2}$. Explain why we enforce these "implicit" boundary conditions.

12.25 Repeat the arguments of Section 12.9.4 to solve the IVP/BVP for the 3D wave equation on the ball:

$$\begin{cases} u_{tt} = c^2\Delta u, & \mathbf{x} \in B(\mathbf{0}, 1),\ t > 0, \\ u(\mathbf{x}, t) = 0, & \mathbf{x} \in \partial B(\mathbf{0}, 1),\ t > 0, \\ u(\mathbf{x}, 0) = \phi(\mathbf{x}),\ u_t(\mathbf{x}, 0) = \psi(\mathbf{x}), & \mathbf{x} \in B(\mathbf{0}, 1). \end{cases}$$

12.26 Recall from Exercise **8.6** the notion of a homogeneous harmonic polynomial in 3D. Find a relationship between the spherical harmonics and these homogeneous harmonic polynomials.

12.27 This question assumes familiarity with the basic notions of electrostatics in Section 10.6. Suppose the electrostatic potential on the conducting surface of a sphere (centered at the origin with radius a) is given (in spherical coordinates) by $\Phi = 4\cos 2\phi$. Assume there are no charges inside or outside the sphere.
(a) Find the electrostatic potential inside and outside the sphere.
(b) Find the surface charge density on the sphere.

The next four exercises pertain to **eigenvalues and eigenfunctions of the Laplacian** $-\Delta$. Recall from Section 8.7 the key properties and notation for the eigenvalues and eigenfunctions of $-\Delta$ on a bounded domain Ω with Dirichlet boundary conditions on the piecewise smooth boundary.

12.28 (a) Prove that $-\Delta$ is a symmetric operator on the class of functions $C^2(\overline{\Omega})$ with zero boundary values; that is, prove that $\int\cdots\int_\Omega(-\Delta u)\,v\,d\mathbf{x} = \int\cdots\int_\Omega u\,(-\Delta v)\,d\mathbf{x}$.
(b) Assuming the properties of Section 8.7, write down the "general Fourier series" of an arbitrary ("reasonable") function $\phi(\mathbf{x})$ defined on Ω.

12.29 In each of the following domains, write down the eigenvalues and the corresponding orthogonal set of eigenfunctions.
(a) In dimension $N = 1$, the interval $[0, l]$.
(b) In dimension $N = 2$, the rectangle $[0, a] \times [0, b]$.
(c) In dimension $N = 2$, $B(\mathbf{0}, a)$, the disc of radius a.
(d) In dimension $N = 3$, the square box $[0, a] \times [0, b] \times [0, c]$.
(e) In dimension $N = 3$, $B(\mathbf{0}, a)$, the ball of radius a.
(f) In dimension $N = 3$, the inside of the cylinder $x^2 + y^2 \leq a^2, 0 \leq z \leq l$.

12.30 (**Weyl's Asymptotic Formula**) Let λ_n denote the sequence of eigenvalues of $-\Delta$ on $\Omega \subset \mathbb{R}^N$. Weyl's asymptotic formula states that

$$\lim_{n\to\infty} \frac{\lambda_n}{n^{2/N}} = \frac{4\pi^2}{(\omega_N |\Omega|)^{2/N}},$$

where ω_N denotes the N-dimensional "volume" of the unit ball in $\mathbb{R}^N$ and $|\Omega|$ denotes the N-dimensional "volume" of Ω. Verify Weyl's asymptotic formula for the interval $[0, l] \subset \mathbb{R}$ and for the rectangle $[0, a] \times [0, b] \subset \mathbb{R}^3$.

12.31 (**Neumann Boundary Conditions**) Now consider the eigenvalue problem:

$$\begin{cases} -\Delta u = \lambda u & \text{in } \Omega, \\ \frac{\partial u}{\partial \mathbf{n}} = 0 & \text{on } \partial\Omega. \end{cases} \tag{12.64}$$

As with the Dirichlet boundary conditions, there exists a countable number of nonnegative eigenvalues and associated eigenfunctions which constitute a complete set (i.e., they span all "reasonable" functions). But this time 0 is the first eigenvalue.
(a) Find the eigenfunction corresponding to $\lambda = 0$ and show that there is only one such linearly independent eigenfunction. Hence, repeating the eigenvalues with multiplicity we have $0 = \lambda_1 < \lambda_2 \leq \lambda_3 \leq \cdots \leq \lambda_n \leq \cdots$.
(b) Write down and prove the analogous statements of Theorems 8.4 and 8.5 of Section 8.7 on the relationship between the λ_i and minimizers of the Rayleigh quotient. Hint: First do Exercise **8.24** on the Neumann being the "natural" boundary conditions. You will then see that we need **not** put any constraints on the boundary values in the definition of the admissible class $\mathcal{A}$.

12.32 (a) Show that under the change of dependent variable (12.61), the ODE (12.58) reduces to **Laguerre's equation** (12.62).
(b) Solve (12.62) by the method of series solutions. This means consider power series solutions of the form (12.63), placing this power series into the ODE. By differentiating term by term, equate like powers of r^k. Recall that if $\sum_{k=1}^{\infty} b_k r^k \equiv 0$ on some interval, then $b_k = 0$ for all k. Use this fact to obtain a recursive equation for a_k and a_{k-1}. Show that for $\beta = \frac{1}{n}$ for some $n \in \mathbb{N}$, then the recursive relation implies that all but a finite number of the a_k are zero. In this case the solution is a polynomial (called a Laguerre polynomial) and satisfies the additional boundedness conditions. Show that if $\beta \neq \frac{1}{n}$, then the series does not terminate and the solution fails to satisfy the boundedness conditions.

(c) Write out the polynomial solutions for $n = 1, 2, 3$. Write down the associated radially symmetric solutions $u(r, t)$ for the Schrödinger equation (with atomic units), and plot $|u(r, t)|$ as a function of r for $t = 0, 1, 2$.

12.33 (**The Harmonic Oscillator**) Recall the vibrating spring, one of the simplest physical systems. If we attach a mass m to the end of the spring which at rest lies at the origin on the x-axis, then the position $x(t)$ of the mass at time t solves $mx''(t) = kx$, where k is the spring constant. The right-hand side is the restoring force of the spring which can also be written as $-\nabla V(x)$ where $V(x) = -\frac{1}{2}kx^2$ is the potential. This simple system is often referred to as a harmonic oscillator. In the quantum mechanical analogue, we do not know exactly the position of the mass but rather its complex-valued wave function $u(x, t)$ whose modulus gives the probability density for its position. The analogous equation is the 1D Schrödinger equation

$$i\hbar u_t = u_{xx} - \frac{1}{2}kx^2 u,$$

which we solve for $x \in \mathbb{R}$ and $t > 0$.

(a) By imposing the additional boundedness condition that $|u| \to 0$ as $|x| \to \pm\infty$, find separated solutions $u(x, t) = T(t)X(x)$. You may work in units wherein $\hbar = m = \frac{1}{2}k = 1$. For the X equation, you will obtain the eigenvalue problem $X'' + (\lambda - x^2)X = 0$. Note that $\lambda = 1$ is an eigenvalue for eigenfunction $X(x) = e^{-\frac{x^2}{2}}$. Now, make the change of dependent variable $Y(x) = e^{\frac{x^2}{2}}X(x)$ to arrive at **Hermite's equation**

$$Y'' - 2xY' + (\lambda - 1)Y = 0.$$

Apply the method of series solutions as outlined in the previous exercise to find that the only solutions which satisfy the boundedness (decay) condition are when λ is an odd integer. In these cases, show the associated solution $Y(x)$ are polynomials, which up to normalizing constants are known as Hermite polynomials.

(b) Show that up to a constant factor, the Hermite polynomials $H_k(x)$ satisfy the formula

$$H_k(x) = (-1)^k e^{x^2} \frac{d^k}{dx^k} e^{-x^2}.$$

(c) By researching online, discuss the orthogonality and completeness properties of the Hermite polynomials. Find a relationship between the Hermite polynomials and the Laguerre polynomials of the previous exercise.

12.34 One can write the general form of a **Sturm-Liouville eigenvalue problem in higher dimensions** on a bounded domain Ω as

$$\nabla \cdot (p(\mathbf{x})\nabla u) + q(\mathbf{x})u + \lambda\sigma(\mathbf{x})u = 0, \qquad \mathbf{x} \in \Omega,$$

with some boundary condition on $\partial\Omega$. Here, p, q, σ are functions defined on Ω with $\sigma \geq 0$. Consider the homogenous Dirichlet boundary condition.

(a) Show that in dimension three, eigenfunctions u_i, u_j associated with distinct eigenvalues λ_i and λ_j are orthogonal in the sense that

$$\iiint_\Omega u_i(\mathbf{x})u_j(\mathbf{x})\sigma(\mathbf{x})\, d\mathbf{x} = 0.$$

(b) By researching online (or in reference books), state (but do not prove) a result concerning the number and possible multiplicity of the eigenvalues, and the completeness of the eigenfunctions.
(c) Write down and prove the analogous statement of Theorems 8.4 for the **generalized Rayleigh quotient** in dimension three,

$$\mathcal{R}[w] := \frac{\iiint_\Omega \left(p(\mathbf{x})|\nabla w(\mathbf{x})|^2 - q(\mathbf{x})w^2(\mathbf{x})\right) d\mathbf{x}}{\iiint_\Omega \sigma w^2(\mathbf{x})\, d\mathbf{x}}.$$

12.35 (**Steklov Eigenvalues**) Let Ω be a bounded domain in $\mathbb{R}^N$ and consider for a real number σ the BVP

$$\begin{cases} \Delta u = 0, & \text{in } \Omega, \\ \frac{\partial u}{\partial \mathbf{n}} = \sigma u & \text{on } \partial\Omega. \end{cases} \tag{12.65}$$

If there exists a nontrivial solution to this BVP, we refer to σ as a Steklov[13] eigenvalue. It turns out that there is an infinite sequence of Steklov eigenvalues $0 = \sigma_0 \le \sigma_1 \le \sigma_2 \le \cdots$ tending to $+\infty$ and the associated eigenfunctions, solutions to (12.65), form an orthogonal set which is complete in $L^2(\partial\Omega)$; that is, any square-integrable function defined over the boundary $\partial\Omega$ has a generalized Fourier series in terms of these eigenfunctions which converges in L^2.
(a) First recall the Dirichlet-Neumann map defined in Exercise **10.16**. How are the Steklov eigenvalues related to the eigenvalues of the Dirichlet-Neumann map?
(b) Let $N = 1$ and $\Omega = (0, l)$. Find all the Steklov eigenvalues.
(c) Let $N = 2$ and $\Omega = B(\mathbf{0}, a)$, the ball of radius a centered at the origin. Find all the Steklov eigenvalues.

[13] **Vladimir Steklov** (1864–1926) was a Russian and Soviet mathematician and physicist.

Chapter 13

Uniting the Big Three Second-Order Linear Equations, and What's Next

The main purpose of the short and final chapter is to unify the big three second-order linear equations: the wave, diffusion, and Laplace equations. We first argue that there is essentially a trichotomy of classes for second-order linear PDEs and each of these equations is a canonical representative of their respective class. While both the techniques and behavior of the solutions were rather different for each of the big three, we show that the notion of a fundamental solution and Green's functions extends to all of them. Finally, we end by asking, "What's next?", and briefly addressing the approaches and material which we aim to present in a future volume planned for this subject.

13.1. Are There Other Important Linear Second-Order Partial Differential Equations? The Standard Classification

13.1.0.1. *Conic Sections.* We begin by recalling (perhaps from high school) the notion of a conic section. Conic sections have been studied since antiquity and are curves obtained as the intersection of a cone (more precisely the surface of the cone) and a plane. As illustrated in Figure 13.1, the possible curves are ellipses (including circles), parabolas, hyperbolas, and two lines (the degenerate cases).

In terms of Cartesian coordinates (x, y), we can write all such curves in the form

$$Ax^2 + Bxy + Cy^2 + Dx + Ey + F = 0,$$

where A to F are real numbers (not all 0). After excluding the degenerate cases (lines), we can classify the conic section via *the discriminant* $B^2 - 4AC$:

- If $B^2 - 4AC < 0$, we have an ellipse.
- If $B^2 - 4AC = 0$, we have a parabola.
- If $B^2 - 4AC > 0$, we have a hyperbola.

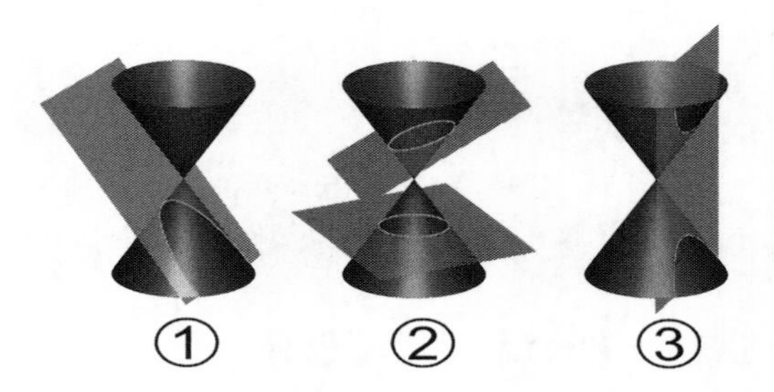

Figure 13.1. Cartoon illustrating a few conic sections: (1) parabola, (2) ellipse and circle, (3) hyperbola. Source: Wikimedia Commons, author Pbroks13, licensed under the Creative Commons Attribution 3.0 Unported (https://creativecommons.org/licenses/by/3.0/deed.en) license.

13.1.1. Classification of Linear Second-Order Partial Differential Equations.

We consider a general homogeneous linear second-order PDE in two variables:

$$a(x,y)u_{xx} + b(x,y)u_{xy} + c(x,y)u_{yy} + d(x,y)u_x + e(x,y)u_y + f(x,y)u = 0. \quad (13.1)$$

For simplicity, let us further assume that the coefficients are constants and focus on PDEs of the form

$$au_{xx} + bu_{xy} + cu_{yy} + du_x + eu_y + fu = 0, \quad (13.2)$$

where we assume at least one of the constants a, b, c is nonzero.

Given that the Fourier transform will effectively transform (13.2) into an algebraic equation in two independent variables, it is not surprising that the above trichotomy for conic sections might also prove pertinent. Indeed, we will make the following definitions:

- If $b^2 - 4ac < 0$, we call the PDE (13.2) **elliptic**.
- If $b^2 - 4ac = 0$, we call the PDE (13.2) **parabolic**.
- If $b^2 - 4ac > 0$, we call the PDE (13.2) **hyperbolic**.

Note that this definition is independent of the lower-order coefficients d, e, and f. The respective **canonical** examples are:

- **Laplace's equation** $u_{xx} + u_{yy} = 0$ $(a = 1, b = 0, c = 1)$ which is **elliptic**.
- The **diffusion equation** $u_y - u_{xx} = 0$ $(a = -1, b = 0, c = 0)$ which is **parabolic**.
- The **wave equation** $u_{yy} - u_{xx} = 0$ $(a = -1, b = 0, c = 1)$ which is **hyperbolic**.

As we have seen, the solutions to all three equations have rather different properties and behaviors with respect to boundary and initial values. It turns out that all elliptic equations have similar properties to Laplace's equation; all parabolic equations have similar properties to the diffusion equation; and all hyperbolic equations have similar properties to the wave equation. To provide some details, we must first set the following premise: If one adds lower-order terms (i.e., nonzero constants d, e, f) to either the Laplace, diffusion, or wave equation, one does not change the essential character of the PDE (for example, the way the solutions depend on data). Through a few examples, we certainly have some evidence to support this premise.

The following theorem states that for any PDE of the form (13.2), there exists a change of (the independent) variables to new coordinates (ζ, η) which transforms the PDE into either the Laplace, diffusion, or wave equation with lower-order terms. To state this precisely, we will as usual abuse notation by using the same u to denote $u(x, y)$ and $u(\zeta, \eta)$.

Theorem 13.1. *Consider a PDE of the form* (13.2) *where a, b, and c are not all* 0. *Then there exists a linear change of variables $\zeta(x, y), \eta(x, y)$ such that in the new coordinates (ζ, η), the PDE is transformed as follows:*

(1) *If* $b^2 - 4ac < 0,\quad u_{\zeta\zeta} + u_{\eta\eta} + F(u_\zeta, u_\eta, u) = 0.$

(2) *If* $b^2 - 4ac = 0,\quad u_{\eta\eta} + F(u_\zeta, u_\eta, u) = 0.$

(3) *If* $b^2 - 4ac > 0,\quad u_{\zeta\eta} + F(u_\zeta, u_\eta, u) = 0.$

In each case, F is some linear function of three variables.

The proof is a straightforward change of variables with guidance from conic sections (cf. almost any other book on PDEs). The PDEs for (1), (2), and (3) are often referred to as the **canonical form** of a second-order linear elliptic, parabolic, and hyperbolic PDE, respectively.

We conclude with two remarks.

(i) Variable coefficients. The identical classification can be made for PDEs (13.1) with variable coefficients with the caveat that the classification (elliptic, parabolic, or hyperbolic) can vary throughout the xy-plane. More precisely, Theorem 13.1 holds true in any domain where the discriminant $b^2 - 4ac$ remains either negative, positive, or zero. However, in each of the three cases the function F is now a function of five variables $F(u_\zeta, u_\eta, u, \zeta, \eta)$ which is linear in the first three arguments.

An important example is the **tricomi equation**

$$u_{xx} + x u_{yy} = 0$$

which is elliptic in the right half-plane $\{(x, y) \mid x > 0\}$ and hyperbolic in the left half-plane $\{(x, y) \mid x < 0\}$. The equation is useful in the study of *transonic flow* (flight at or near the speed of sound).

(ii) More than two independent variables. The same classification can be made for second-order linear PDEs in N variables. We can write such equations (homogeneous with constant coefficients) as

$$\sum_{i,j=1}^{N} a_{ij} u_{x_i x_j} + \sum_{i=1}^{N} a_i u_{x_i} + a_0 u = 0,$$

where $a_{ij} = a_{ji}$. The latter condition is natural as we assume the order of differentiation does not matter. The characterization can be phrased entirely in terms of the eigenvalues of the $N \times N$ matrix $\mathbf{A} = (a_{ij})$, which because of symmetry are all real valued.

For example, if all the eigenvalues of **A** are nonzero and have the same sign (either all positive or all negative), then the PDE is elliptic and, with a change of independent variables, can be transformed into

$$\Delta u + \cdots = 0,$$

where $\cdots$ indicates lower-order terms.

13.2. Reflection on Fundamental Solutions, Green's Functions, Duhamel's Principle, and the Role/Position of the Delta Function

This section presents a nice *coda* to this text and actually summarizes a good chunk of **linear** PDEs! Thus far, we have encountered and exploited the concept of a fundamental solution in the context of both Laplace's equation and the diffusion equation. For each equation, the fundamental solution was linked to **point sources** captured by the **delta function**. Green's functions were introduced in Chapter 10 for the Laplacian and were specific to a particular spatial domain with boundary conditions (we mostly focused on the Dirichlet boundary condition). In dimensions $N \geq 2$, one may view the fundamental solution for the Laplacian as the Green's function with Dirichlet "boundary conditions" on $\mathbb{R}^N$; here, "boundary conditions" are understood as vanishing at infinity; i.e., $\lim_{|\mathbf{x}|\to\infty} \Phi(\mathbf{x}) = 0$.

Fundamental solutions and Green's functions (with different boundary conditions) are actually common to **all** linear PDEs, not just the Laplacian. So, in a nutshell, what is a Green's function? One can think of the PDE as modeling some system, and the Green's function as capturing **the response of the system to a point source**.

In this section we will address, summarize, and connect features of fundamental solutions and Green's functions for *"the big three"*: Laplace, diffusion, and wave equations. In all cases, we will focus **only** on Green's functions with the Dirichlet boundary condition. We begin by recalling and connecting these notions for the Laplace and diffusion equations, noting the essentially equivalent placement (in the initial data vs. the inhomogeneous right-hand side) of the point source for the diffusion equation. We then explore these ideas in slightly more detail for the **wave equation**, the other second-order linear PDE which we extensively studied in Chapters 3 and 4, but **not** from the perspective of fundamental solutions or Green's functions. While diffusion in higher space dimensions is rather similar (with an analogous fundamental solution in each dimension), solutions to the wave equation have a completely different behavior in dimension 3 than in dimension 1. The propagation in 3D on sharp fronts (cf. the Huygens Principle) means that the fundamental solution and Green's functions can **not** be captured by functions; yet they are distributions!

Notational Warning: In each subsection, we use "Φ" to denote the fundamental solution; hence, Φ changes from subsection to subsection and with the space dimension.

13.2.1. Fundamental Solutions/Green's Functions for the Laplacian. We begin by recapping the fundamental solution for **Laplace's equation** (cf. Definition 8.5.1). It was a different function Φ in each space dimension; namely:

$$\boxed{\Phi(\mathbf{x}) \;:=\; \begin{cases} \frac{1}{2}|x| & \text{for } N = 1, \\ \frac{1}{2\pi}\log|\mathbf{x}| & \text{for } N = 2, \\ -\frac{1}{4\pi|\mathbf{x}|} & \text{for } N = 3, \\ -\frac{1}{N(N-2)\omega_N|\mathbf{x}|^{N-2}} & \text{for } N > 3. \end{cases}} \tag{13.3}$$

In all space dimensions it had a unifying property: It solved

$$\Delta\Phi(\mathbf{x}) \;=\; \delta_0 \qquad \text{in the sense of distributions on } \mathbb{R}^N,$$

or more generally, for any fixed $\mathbf{y} \in \mathbb{R}^N$,

$$\Delta_{\mathbf{x}}\,\Phi(\mathbf{x}-\mathbf{y}) \;=\; \delta_{\mathbf{y}} \qquad \text{in the sense of distributions on } \mathbb{R}^N. \tag{13.4}$$

The fundamental solution was the key object in solving Poisson's equation

$$\Delta u \;=\; f \quad \text{on } \mathbb{R}^N. \tag{13.5}$$

Indeed, the solution was simply the **convolution** of Φ with f (cf. Section 10.2.1 and Theorem 10.2); for example, in $N = 3$ dimensions

$$u(\mathbf{x}) \;=\; \iiint_{\mathbb{R}^3} \Phi(\mathbf{x}-\mathbf{y})\,f(\mathbf{y})\,d\mathbf{y}. \tag{13.6}$$

This solution formula can informally be explained from (13.4) as follows:

- Put a source (think of an electric charge) at a fixed position $\mathbf{y} \in \mathbb{R}^3$.
- Weight this source with factor $f(\mathbf{y})$. Thus the full charge is $f(\mathbf{y})\delta_{\mathbf{y}}$.
- From (13.4), the resulting electrostatic potential at position $\mathbf{x}$, i.e., the solution to

$$\Delta_{\mathbf{x}}\,u(\mathbf{x}) \;=\; f(\mathbf{y})\delta_{\mathbf{y}},$$

  is $u(\mathbf{x}) = f(\mathbf{y})\Phi(\mathbf{x}-\mathbf{y})$.
- To obtain the total potential at position $\mathbf{x}$, resulting from a continuum of such sources (charges), i.e., the solution to (13.5), we add the results, which effectively means integrate. This is exactly (13.6).

Key to this explanation is the linearity of the PDE and the resulting principle of superposition (cf. Section 1.4.3).

Using Φ and its derivatives, we were also able to write the values of a harmonic function u on a domain Ω at a point $\mathbf{x} \in \Omega$ in terms of boundary values of u and its normal derivative; precisely (cf. Theorem 10.3) if $\Delta u = 0$ in Ω and $\mathbf{x}_0 \in \Omega$, then

$$u(\mathbf{x}_0) \;=\; \iint_{\partial\Omega} \left[u(\mathbf{x})\,\frac{\partial\,\Phi(\mathbf{x}-\mathbf{x}_0)}{\partial\mathbf{n}} \;-\; \Phi(\mathbf{x}-\mathbf{x}_0)\,\frac{\partial u(\mathbf{x})}{\partial\mathbf{n}} \right] dS_{\mathbf{x}}.$$

The key point here is that Φ tells us exactly how the value of u inside Ω depends on values of u (and its normal derivative) on the boundary $\partial\Omega$. Note that, as in Section 10.3, we switch to using the notation $\mathbf{x}_0, \mathbf{x}$ in place of $\mathbf{x}, \mathbf{y}$. Green's functions for the Laplacian on a domain Ω with Dirichlet boundary conditions were modified fundamental solutions which vanished on the boundary $\partial\Omega$ (cf. Section 10.3). More precisely they were functions $G(\mathbf{x}, \mathbf{x}_0)$ such that $\Delta_{\mathbf{x}} G(\mathbf{x}, \mathbf{x}_0) = \delta_{\mathbf{x}_0}$ in the sense of distributions on Ω and $G(\mathbf{x}, \mathbf{x}_0) = 0$ for all $\mathbf{x} \in \partial\Omega$. With the Green's function G in hand, we were able to solve the Dirichlet problem for Laplace's equation

$$\begin{cases} \Delta u = 0 & \text{on } \Omega, \\ u = g & \text{on } \partial\Omega \end{cases}$$

in terms of a boundary integral involving g with the normal derivative of the Green's function on $\partial\Omega$ (cf. (10.15)): precisely for any $\mathbf{x}_0 \in \Omega$, we have

$$u(\mathbf{x}_0) = \iint_{\partial\Omega} g(\mathbf{x}) \frac{\partial}{\partial \mathbf{n}} G(\mathbf{x}, \mathbf{x}_0)\, dS_{\mathbf{x}}.$$

One could also solve the Dirichlet problem for Poisson's equation on a domain Ω,

$$\begin{cases} \Delta u = f & \text{on } \Omega, \\ u = g & \text{on } \partial\Omega, \end{cases}$$

with the inclusion of a bulk integral over Ω involving G and f (cf. (10.19)): precisely for any $\mathbf{x}_0 \in \Omega$, we have

$$u(\mathbf{x}_0) = \iiint_{\Omega} G(\mathbf{x}, \mathbf{x}_0)\, f(\mathbf{x})\, d\mathbf{x} + \iint_{\partial\Omega} g(\mathbf{x}) \frac{\partial}{\partial \mathbf{n}} G(\mathbf{x}, \mathbf{x}_0)\, dS_{\mathbf{x}}.$$

Conclusion: Regardless of whether we are looking for a function with vanishing Laplacian or a function with prescribed Laplacian, the key object is the fundamental solution Φ defined by (13.3), or, for a BVP, a modified version in the form of a Green's function. However, the strong caveat is that for BVPs, there were very few domains for which we could actually "find" the Green's function.

13.2.2. Fundamental Solutions/Green's Functions of the Diffusion Equation. Now let us recall the **diffusion equation** where, for simplicity of notation only, we consider $N = 1$ (one space dimension). We defined the fundamental solution $\Phi(x, t)$ in (7.7), for $x \in \mathbb{R}$ and $t > 0$, as

$$\boxed{\Phi(x, t) := \frac{1}{\sqrt{4\pi\alpha t}}\, e^{-\frac{x^2}{4\alpha t}}.}$$

It solved

$$\begin{cases} u_t - \alpha u_{xx} = 0, & x \in \mathbb{R}, t > 0, \\ u(x, 0) = \delta_0. \end{cases} \tag{13.7}$$

There are two ways we can interpret this IVP: (i) The smooth function $\Phi(x, t)$ is a classical solution to the diffusion equation for all $x \in \mathbb{R}$ and $t > 0$, and as $t \to 0^+$, $\Phi(x, t)$ converges in the sense of distributions to the one-dimensional delta function δ_0. (ii) Alternatively, we can follow the approach of Section 9.7.4 to define a distributional solution to the full IVP, i.e., incorporate the data (which is no longer a function) in the distributional statement.

We called $\Phi(x,t)$ the fundamental solution as it generated all solutions to the general IVP. Namely, we obtain the solution to

$$\begin{cases} u_t - \alpha u_{xx} = 0, & x \in \mathbb{R}, t > 0, \\ u(x,0) = g(x), \end{cases} \tag{13.8}$$

by the spatial convolving of the initial data g with $\Phi(x,t)$; i.e., the solution to (13.8) is

$$u(x,t) = \int_{-\infty}^{\infty} \Phi(x-y,t)\, g(y)\, dy. \tag{13.9}$$

Note that from (13.7) we have that for any fixed source y,

$$\Phi(x-y,t) \quad \text{solves} \quad \begin{cases} u_t - \alpha u_{xx} = 0, & x \in \mathbb{R}, t > 0, \\ u(x,0) = \delta_y. \end{cases}$$

Based upon these point sources, the solution formula (13.9) can be informally explained as follows:

- Put a source (think of temperature) at position y at time $t = 0$.
- Weight this source with factor $g(y)$. Thus the initial temperature is $g(y)\delta_y$.
- From (13.7) and its solution, the resulting temperature at position x and time t is $g(y)\Phi(x-y,t)$.
- To obtain the temperature, at position x and time t, resulting from a continuum of such sources, we add the results (which effectively means integrate). This is exactly (13.9).

We repeat that the above crucially makes use of **linearity** of the diffusion equation through the principle of superposition.

In comparing with the previous subsection for the Laplacian, the following related questions should (hopefully) pop into the readers mind:

(i) In terms of fundamental solutions, the delta function appears in the data for the diffusion equation, but as an inhomogeneous right-hand side for Laplace's equation. What's up with this?

(ii) Can't we also use the fundamental solution for the diffusion equation, $\Phi(x,t)$, to solve the inhomogeneous diffusion equation?

(iii) What if we put the delta function as a source term on the right-hand side of the diffusion equation?

(iv) Wait! Doesn't Duhamel's Principle "kind of" say that we can place the source term (the right-hand side of the PDE) in the initial data?

First off, let us directly address (iii) thereby addressing (i) as well. Suppose we extend the fundamental solution $\Phi(x,t)$ for all $t \in \mathbb{R}$ as follows:

$$\Phi(x,t) = \begin{cases} \frac{1}{\sqrt{4\pi\alpha t}}\, e^{-\frac{x^2}{4\alpha t}} & \text{if } t > 0, \\ 0 & \text{if } t \le 0. \end{cases} \tag{13.10}$$

Then one can show (cf. Exercise **9.23**) that the extended $\Phi(x,t)$ solves

$$u_t - \alpha u_{xx} = \delta_0 \qquad \text{in the sense of distributions on } \mathbb{R}^2. \tag{13.11}$$

Note that here δ_0 is the 2D delta function which we can informally write as "$\delta_0(x)\delta_0(t)$". Conclusion: For the diffusion equation, it does not matter where the delta function is!

Next, let us address (ii) and (iv) (connecting with (i) and (iii)). Recall **Duhamel's Principle** from Section 7.8 which states that the solution to the inhomogeneous IVP

$$\begin{cases} u_t - \alpha u_{xx} = f(x,t), & x \in \mathbb{R}, t > 0, \\ u(x,0) = 0 \end{cases}$$

is given by

$$u(x,t) = \int_0^t \int_{-\infty}^{\infty} f(y,s)\,\Phi(x-y,t-s)\,dy\,ds.$$

This is a spatio-temporal convolution of $f(x,t)$ with $\Phi(x,t)$: Note that by definition of the extension, we have

$$\int_0^{\infty} \int_{-\infty}^{\infty} f(y,s)\,\Phi(x-y,t-s)\,dy\,ds = \int_0^t \int_{-\infty}^{\infty} f(y,s)\,\Phi(x-y,t-s)\,dy\,ds. \tag{13.12}$$

So, yes, the fundamental solution can also be used to solve the inhomogeneous diffusion equation. Note that informally taking the inhomogeneous function to be "$f(x,t) = \delta_0(x)\delta_0(t)$" and placing it in the right-hand side of (13.12) yields the solution $u(x,t) = \Phi(x,t)$. This is in agreement with the fact that Φ solves (13.11). These observations explain why many authors would define the fundamental solution by its extension (13.10).

All these observations/statements directly carry over to higher space dimensions using the multi-D fundamental solution $\Phi(\mathbf{x},t)$ defined by (7.46), trivially extended by 0 for $t \le 0$.

Conclusion: The key (in fact only) object required to solve the **full** IVP,

$$\begin{cases} u_t - \alpha \Delta u = f(\mathbf{x},t), & \mathbf{x} \in \mathbb{R}^N,\ t > 0, \\ u(\mathbf{x},0) = g(\mathbf{x}), & \mathbf{x} \in \mathbb{R}^N, \end{cases}$$

is the fundamental solution $\Phi(\mathbf{x},t)$.

Just as with Laplace's equation, one can *adapt/modify* the fundamental solution of the diffusion equation to any spatial domain Ω and create **Green's function for the diffusion equation** on Ω. One can think of the fundamental solution as the Green's function on all of space. For example, one can consider the BVP

$$\begin{cases} u_t - \alpha u_{xx} = 0 & \text{for} \quad 0 \le x \le l,\ t > 0, \\ u(0,t) = 0 = u(l,t) & \text{for} \quad t > 0, \\ u(x,0) = g(x) & \text{for} \quad 0 \le x \le l. \end{cases} \tag{13.13}$$

For $0 \leq x \leq l, t > 0, 0 < y < l$, the Green's function for this problem is the function $G(x, t; y)$[1] which for each $y \in (0, l)$ satisfies

$$\begin{cases} G_t(x,t;y) - \alpha G_{xx}(x,t;y) = 0 & \text{for} \quad 0 \leq x \leq l, \ t > 0, \\ G(0,t;y) = 0 = G(l,t;y) & \text{for} \quad t > 0, \\ G(x,0;y) = \delta_y, \end{cases} \tag{13.14}$$

where this BVP (with the distributional initial values) is interpreted as in Section 13.2.1. You should think of $y \in (0, l)$ as a separate parameter; i.e., for each y we find the function $G(x, t; y)$ of x and t. With G in hand, the solution to (13.13) is given by

$$u(x,t) = \int_0^l G(x,t;y)\, g(y)\, dy.$$

Now, how do we find G? As with the Laplacian, we can use the method of images by placing ghost sources outside the interval $[0, l]$. Exercise **13.1** outlines the procedure and has the reader check that the resulting solution formula agrees with that obtained by separation of variables.

As with the Laplacian, the same caveat holds true concerning our ability to find Green's function **only** for a few very simple domains.

13.2.3. Fundamental Solutions/Green's Functions of the 1D Wave Equation.

We consider the 1D wave equation $u_{tt} - c^2 u_{xx} = 0$ and its IVP

$$\begin{cases} u_{tt} - c^2 u_{xx} = 0, & -\infty < x < \infty, \ t > 0, \\ u(x,0) = \phi(x), \quad u_t(x,0) = \psi(x), & -\infty < x < \infty. \end{cases}$$

Following the previous approach for the diffusion equation, we will first place the (1D) delta function in the data. But for the wave equation, there are two data functions. Let us start with the velocity and consider the following:

$$\begin{cases} u_{tt} - c^2 u_{xx} = 0, & -\infty < x < \infty, \ t > 0, \\ u(x,0) = 0, \quad -\infty < x < \infty, & \text{and} \quad u_t(x,0) = \delta_0. \end{cases} \tag{13.15}$$

One can easily derive a potential solution candidate by informally treating the delta function as a true function and simply "placing" it into D'Alembert's formula (cf. (3.6)). In doing so, one obtains

$$\text{“} \frac{1}{2c} \int_{x-ct}^{x+ct} \delta_0(y)\, dy \text{”}.$$

Intuitively, the above equals $1/2c$ if the region of integration contains $y = 0$, and 0 otherwise. This suggests the solution, denoted by $\Phi(x, t)$, should be[2]

$$\boxed{\Phi(x,t) := \frac{1}{2c} H(ct - |x|) = \begin{cases} \frac{1}{2c} & \text{if } |x| < ct, \ t > 0, \\ 0 & \text{if } |x| \geq ct, \ t > 0, \end{cases}} \tag{13.16}$$

where H is the Heaviside function. But how should we interpret this candidate solution? Unlike the fundamental solution for the diffusion equation, $\Phi(x, t)$ is not a

[1]Note the different notation from the Laplacian.

[2]One can also derive the same formula by taking the Fourier transform (in the sense of tempered distributions) with respect to x (cf. Exercise **13.2**).

smooth function. One can show that the function $\Phi(x,t)$ is a solution to the wave equation in the sense of distributions, and its time derivative converges to δ_0 in the sense of distributions. Alternatively, one can follow the approach of Section 9.7.4 to define a distributional solution to the full IVP (13.15). In Exercise **13.4** you are asked to define such a solution and verify that Φ satisfies the conditions.

We call $\Phi(x,t)$, defined by (13.16), **the fundamental solution of the 1D wave equation**. Why? For starters, it can immediately be used to solve the IVP

$$\begin{cases} u_{tt} - c^2 u_{xx} = 0, & -\infty < x < \infty,\ t > 0, \\ u(x,0) = 0, \quad u_t(x,0) = \psi(x), & -\infty < x < \infty, \end{cases} \tag{13.17}$$

for **any** $\psi(x)$. Indeed, just as in (13.9) for the diffusion equation, the solution to (13.17) is given by

$$u(x,t) = \int_{-\infty}^{\infty} \Phi(x-y,t)\,\psi(y)\,dy.$$

While this is hardly surprising given that Φ is a (distributional) solution to (13.15), we can also simply place (13.16) into the above to find

$$u(x,t) = \int_{-\infty}^{\infty} \Phi(x-y,t)\,\psi(y)\,dy = \frac{1}{2c}\int_{x-ct}^{x+ct} \psi(y)\,dy,$$

which (full circle) takes us back to D'Alembert's solution.

Now suppose we have nonzero data for ϕ (the initial displacement) and wish to solve

$$\begin{cases} u_{tt} - c^2 u_{xx} = 0, & -\infty < x < \infty,\ t > 0, \\ u(x,0) = \phi(x), \quad u_t(x,0) = 0, & -\infty < x < \infty. \end{cases} \tag{13.18}$$

The solution can also be obtained from Φ and is given by

$$u(x,t) = \frac{\partial}{\partial t}\int_{-\infty}^{\infty} \Phi(x-y,t)\,\phi(y)\,dy.$$

To this end, we find

$$\begin{aligned} \frac{\partial}{\partial t}\int_{-\infty}^{\infty} \Phi(x-y,t)\,\phi(y)\,dy &= \frac{\partial}{\partial t}\frac{1}{2c}\int_{x-ct}^{x+ct} \phi(y)\,dy \\ &= \frac{1}{2}\left[\phi(x+ct) + \phi(x-ct)\right], \end{aligned}$$

which, again, takes us back to D'Alembert's solution.

Lastly, suppose we put the delta function as an inhomogeneous term in the wave equation and consider

$$u_{tt} - c^2 u_{xx} = \delta_{\mathbf{0}} \qquad \text{in the sense of distributions on } \mathbb{R}^2, \tag{13.19}$$

where $\delta_{\mathbf{0}}$ is the 2D delta function, informally written as "$\delta_0(x)\delta_0(t)$". We discussed and solved this exact PDE in Section 9.7.3 and found the solution to be $\Phi(x,t)$, trivially extended to all $t \in \mathbb{R}$:

$$\Phi(x,t) = \begin{cases} \frac{1}{2c} & \text{if } |x| < ct, t > 0, \\ 0 & \text{if } |x| \geq ct, t > 0, \\ 0 & \text{if } t \leq 0. \end{cases} \tag{13.20}$$

With this in hand, we can solve the inhomogeneous IVP

$$\begin{cases} u_{tt} - c^2 u_{xx} = f(x,t), & -\infty < x < \infty,\ t > 0, \\ u(x,0) = 0, \quad u_t(x,0) = 0, & -\infty < x < \infty, \end{cases}$$

via the spatio-temporal convolution of Φ with f; i.e.,

$$u(x,t) = \int_0^\infty \int_{-\infty}^\infty \Phi(x-y, t-\tau)\, f(y,\tau)\, dy\, d\tau.$$

Indeed, using the explicit form (13.20) of Φ, we find

$$u(x,t) = \frac{1}{2c} \iint_D f(y,\tau)\, dy\, d\tau,$$

where D is precisely the domain of dependence associated with (x,t), i.e., the triangle in the xt-plane with top point (x,t) and base points $(x-ct, 0)$ and $(x+ct, 0)$. This is exactly what we found before via Duhamel's Principle in Section 3.6.

Conclusion: The key (in fact only) object required to solve the **full** IVP,

$$\begin{cases} u_{tt} - c^2 u_{xx} = f(x,t), & -\infty < x < \infty,\ t > 0, \\ u(x,0) = \phi, \quad u_t(x,0) = \psi, & -\infty < x < \infty, \end{cases}$$

is the fundamental solution Φ defined by (13.16).

Just as with the diffusion equations, for boundary value problems there exist modified fundamental solutions which are usually called **Green's functions**. For example, the BVP on domain $(0,l)$

$$\begin{cases} u_{tt} - c^2 u_{xx} = 0 & \text{for} \quad 0 \le x \le l,\ t > 0, \\ u(0,t) = 0 = u(l,t) & \text{for} \quad t > 0, \\ u(x,0) = \phi(x),\ u_t(x,0) = \psi(x) & \text{for} \quad 0 \le x \le l \end{cases} \tag{13.21}$$

has a Green's function (with Dirichlet boundary conidtions) $G(x,t;y)$ which satisfies, for any $y \in (0,l)$,

$$\begin{cases} G_{tt}(x,t;y) - c^2 G_{xx}(x,t;y) = 0 & \text{for} \quad 0 \le x \le l,\ t > 0, \\ G(0,t;y) = 0 = G(l,t;y) & \text{for} \quad t > 0, \\ G(x,0;y) = 0, \quad G_t(x,0;y) = \delta_y. \end{cases} \tag{13.22}$$

As before, the BVP (13.22) with its distributional initial values is interpreted as in Section 13.2.1. With G in hand, the solution to (13.21) is given by

$$u(x,t) = \int_0^l G(x,t;y)\, \psi(y)\, dy + \frac{\partial}{\partial t} \int_0^l G(x,t;y)\, \phi(y)\, dy.$$

Exercise **13.5** asks you to find the Green's function, first by the method of images and second using separation of variables.

13.2.4. Fundamental Solutions/Green's Functions of the 3D Wave Equation. We start again with the IVP

$$\begin{cases} u_{tt} - c^2 \Delta u = 0, & \mathbf{x} \in \mathbb{R}^3,\ t > 0, \\ u(\mathbf{x}, 0) = 0, \quad \mathbf{x} \in \mathbb{R}^3, & \text{and} \qquad u_t(\mathbf{x}, 0) = \delta_{\mathbf{0}}, \end{cases} \tag{13.23}$$

where here $\delta_{\mathbf{0}}$ is the 3D spatial delta function, informally written as

$$\text{“}\delta_0(x_1)\delta_0(x_2)\delta_0(x_3)\text{”}.$$

As before, one can derive a potential solution candidate by informally treating the delta function as a true function and simply "placing" it into Kirchhoff's formula (cf. (4.11)). In doing so, one obtains

$$\text{“}\frac{1}{4\pi c^2 t^2} \iint_{\partial B(\mathbf{x},ct)} t\delta_{\mathbf{0}}(\mathbf{y})\, dS_{\mathbf{y}} \;=\; \frac{1}{4\pi c^2 t} \iint_{\partial B(\mathbf{x},ct)} \delta_{\mathbf{0}}(\mathbf{y})\, dS_{\mathbf{y}}\text{”}.$$

But now what do we make of this? At first sight it seems that even this informal expression is problematic: We are integrating a 3D delta function over a 2D surface. To this end, let us argue (informally) as follows: For any function $\psi(\mathbf{y})$

$$\frac{1}{4\pi c^2 t} \iint_{\partial B(\mathbf{x},ct)} \psi(\mathbf{y})\, dS_{\mathbf{y}} \;=\; \text{“}\frac{1}{4\pi c^2 t} \iiint_{\mathbb{R}^3} \delta_0(ct - |\mathbf{y} - \mathbf{x}|)\psi(\mathbf{y})\, d\mathbf{y}\text{”}.$$

The idea here is that the delta function gives concentration on the sphere $\partial B(\mathbf{x}, ct)$ (a time slice of the light cone). Now if we informally replace $\psi(\mathbf{y})$ with the 3D delta function, we obtain

$$\text{“}\frac{1}{4\pi c^2 t} \iiint_{\mathbb{R}^3} \delta_0(ct - |\mathbf{y} - \mathbf{x}|)\delta_{\mathbf{0}}(\mathbf{y})\, d\mathbf{y} \;=\; \frac{1}{4\pi c^2 t}\delta_0(ct - |\mathbf{x}|)\text{”}.$$

For each time $t > 0$, the latter expression is a 3D distribution which concentrates on the sphere $\frac{|\mathbf{x}|}{c} = t$. For each $t > 0$, let us precisely **define this distribution**, labeling it as $\frac{1}{4\pi c^2 t}\, \delta_0(ct - |\mathbf{x}|)$: For any $\phi \in C_c^\infty(\mathbb{R}^3)$,

$$\left\langle \frac{1}{4\pi c^2 t}\, \delta_0(ct - |\mathbf{x}|),\, \phi(\mathbf{x}) \right\rangle \;:=\; \frac{1}{4\pi c^2 t} \iint_{\partial B(\mathbf{0},ct)} \phi(\mathbf{x})\, dS_{\mathbf{x}}. \tag{13.24}$$

Note that the labeling of this distribution has no functional meaning; however, it is suggestive as to what the distribution does. We now define the fundamental solution in 3D to be, at each $t > 0$, as a distribution

$$\boxed{\Phi(t) \;:=\; \frac{1}{4\pi c^2 t}\, \delta_0(ct - |\mathbf{x}|).} \tag{13.25}$$

So unlike in 1D, the fundamental solution cannot be captured by any true function of space; hence, we do **not** include a spatial argument for Φ. This is a direct consequence of Huygen's Principle which states that information propagates on sharp fronts. In fact, one can think of the fundamental solution as uniformly distributed delta functions on the light cone in space-time.

How do we use this fundamental solution (which is not a function) to solve the IVP

$$\begin{cases} u_{tt} - c^2 \Delta u = 0, & \mathbf{x} \in \mathbb{R}^3,\ t > 0, \\ u(\mathbf{x}, 0) = 0 \quad \text{and} \quad u_t(\mathbf{x}, 0) = \psi(\mathbf{x}), & \mathbf{x} \in \mathbb{R}^3\,? \end{cases} \tag{13.26}$$

The answer is again convolution. In Section 6.9.8 we defined the convolution of a one-dimensional (tempered) distribution with a smooth function. The analogous definition holds true in higher dimensions: Given a 3D distribution F and a smooth function $\psi : \mathbb{R}^3 \to \mathbb{R}$, we define the distribution $\psi * F$ by

$$\langle \psi * F\,,\, \phi \rangle \;:=\; \langle F\,,\, \psi^- * \phi \rangle \qquad \text{for any test function } \phi,$$

where $\psi^-(\mathbf{x}) := \psi(-\mathbf{x})$.

Now let ψ be the initial speed from the IVP (13.26). We claim that for each $t > 0$, the convolution $\psi * \Phi(t)$ is a distribution generated by a function $u(x, t)$ which solves (13.26). To this end, for any $\phi \in C_c(\mathbb{R}^3)$, we have

$$\begin{aligned}
\langle \psi * \Phi(t)\,,\, \phi \rangle \;&:=\; \langle \Phi(t)\,,\, \psi^- * \phi \rangle \\
&= \frac{1}{4\pi c^2 t} \iint_{\partial B(\mathbf{0}, ct)} (\psi^- * \phi)(\mathbf{y})\, dS_{\mathbf{y}} \\
&= \frac{1}{4\pi c^2 t} \iint_{\partial B(\mathbf{0}, ct)} \left(\iiint_{\mathbb{R}^3} \psi^-(\mathbf{y} - \mathbf{x}) \phi(\mathbf{x})\, d\mathbf{x} \right) dS_{\mathbf{y}} \\
&= \frac{1}{4\pi c^2 t} \iint_{\partial B(\mathbf{0}, ct)} \left(\iiint_{\mathbb{R}^3} \psi(\mathbf{x} - \mathbf{y}) \phi(\mathbf{x})\, d\mathbf{x} \right) dS_{\mathbf{y}} \\
&= \iiint_{\mathbb{R}^3} \left(\frac{1}{4\pi c^2 t} \iint_{\partial B(\mathbf{0}, ct)} \psi(\mathbf{x} - \mathbf{y})\, dS_{\mathbf{y}} \right) \phi(\mathbf{x})\, d\mathbf{x} \\
&= \iiint_{\mathbb{R}^3} \left(\frac{1}{4\pi c^2 t} \iint_{\partial B(\mathbf{x}, ct)} \psi(\mathbf{y})\, dS_{\mathbf{y}} \right) \phi(\mathbf{x})\, d\mathbf{x} \\
&= \iiint_{\mathbb{R}^3} u(\mathbf{x}, t)\, \phi(\mathbf{x})\, d\mathbf{x} \;=\; \langle u(\cdot, t)\,,\, \phi(\cdot) \rangle,
\end{aligned}$$

where

$$u(\mathbf{x}, t) \;=\; \frac{1}{4\pi c^2 t} \iint_{\partial B(\mathbf{x}, ct)} \psi(\mathbf{y})\, dS_{\mathbf{y}} \;=\; \frac{1}{4\pi c^2 t^2} \iint_{\partial B(\mathbf{x}, ct)} t\psi(\mathbf{y})\, dS_{\mathbf{y}}. \tag{13.27}$$

By Kirchhoff's formula, the function $u(\mathbf{x}, t)$ is indeed the solution to (13.26).

We can also use the fundamental solution to solve the IVP

$$\begin{cases} u_{tt} - c^2 \Delta u = 0, & \mathbf{x} \in \mathbb{R}^3,\ t > 0, \\ u(\mathbf{x}, 0) = \phi(\mathbf{x}) \quad \text{and} \quad u_t(\mathbf{x}, 0) = 0, & \mathbf{x} \in \mathbb{R}^3. \end{cases} \tag{13.28}$$

We obtain the solution by differentiating the convolution $\phi * \Phi(t)$ with respect to t. Indeed, since the convolution is actually the function

$$\frac{1}{4\pi c^2 t^2} \iint_{\partial B(\mathbf{x}, ct)} t\phi(\mathbf{y})\, dS_{\mathbf{y}},$$

we compute using the change of variables $\mathbf{y} = \mathbf{x} + ct\mathbf{z}$:

$$\begin{aligned}
\frac{\partial}{\partial t}\left(\frac{1}{4\pi c^2 t^2}\iint_{\partial B(\mathbf{x},ct)} t\phi(\mathbf{y})\, dS_{\mathbf{y}}\right) &= \frac{\partial}{\partial t}\left(\frac{1}{4\pi}\iint_{\partial B(\mathbf{0},1)} t\phi(\mathbf{x}+ct\mathbf{z})\, dS_{\mathbf{z}}\right)\\
&= \frac{1}{4\pi}\iint_{\partial B(\mathbf{0},1)} \left(\phi(\mathbf{x}+ct\mathbf{z}) \,+\, t\nabla\phi(\mathbf{x}+ct\mathbf{z})\cdot c\mathbf{z}\right) dS_{\mathbf{z}}\\
&= \frac{1}{4\pi c^2 t^2}\iint_{\partial B(\mathbf{x},ct)} \left(\phi(\mathbf{y}) \,+\, t\nabla\phi(\mathbf{y})\cdot\frac{\mathbf{y}-\mathbf{x}}{t}\right) dS_{\mathbf{y}}\\
&= \frac{1}{4\pi c^2 t^2}\iint_{\partial B(\mathbf{x},ct)} \left(\phi(\mathbf{y}) \,+\, \nabla\phi(\mathbf{y})\cdot(\mathbf{y}-\mathbf{x})\right) dS_{\mathbf{y}}.
\end{aligned}$$

This is Kirchhoff's solution to (13.28).

Lastly, suppose we place a 4D (space and time) delta function as an inhomogeneous term in the wave equation and consider the PDE

$$u_{tt} \,-\, c^2\Delta u \,=\, \delta_{\mathbf{0}} \qquad \text{in the sense of distributions on } \mathbb{R}^4, \tag{13.29}$$

where $\delta_{\mathbf{0}}$ is 4D delta function, informally written as "$\delta_0(x_1)\delta_0(x_2)\delta_0(x_3)\delta_0(t)$". Note that here we are looking for a solution for all $t \in \mathbb{R}$. As with the previous cases, the solution is simply the extended fundamental solution:

$$\begin{cases} \Phi(t) & \text{if } t > 0,\\ 0 & \text{if } t \leq 0.\end{cases}$$

Then for a function $f \in C_c(\mathbb{R}^4)$, one obtains the solution to

$$\begin{cases} u_{tt} - c^2\Delta u = f(\mathbf{x},t), & \mathbf{x}\in\mathbb{R}^3,\ t>0,\\ u(\mathbf{x},0) = 0, \quad u_t(\mathbf{x},0) = 0, & \mathbf{x}\in\mathbb{R}^3,\end{cases} \tag{13.30}$$

by the spatio-temporal convolution of $f(\mathbf{x},t)$ with the extended fundamental solution. What exactly do we mean by this given that the fundamental solution is, for each time, a distribution? This spatio-temporal convolution can be **informally written** as

$$\text{“}\int_0^t \iiint_{\mathbb{R}^3} \frac{1}{4\pi c^2(t-\tau)}\,\delta_0(c(t-\tau) - |\mathbf{x}-\mathbf{y}|)\, f(\mathbf{y},\tau)\, d\mathbf{y}\, d\tau\text{”}.$$

From the previous calculation, we know that the spatial convolution with a test function results in a function, namely the right-hand side of (13.27). Hence, for $t > 0$, this convolution is actually the function

$$u(\mathbf{x},t) \,:=\, \int_0^t \frac{1}{4\pi c^2(t-\tau)}\iint_{\partial B(\mathbf{x},c(t-\tau))} f(\mathbf{y},\tau)\, dS_{\mathbf{y}}\, d\tau. \tag{13.31}$$

In principle, note that the temporal convolution integrates over $(0,\infty)$; however the definition of the extended fundamental solution reduces the interval to $(0,t)$. The function u defined in (13.31) is indeed a solution to (13.30); in fact, applying Duhamel's Principle yields exactly the same solution. To see this, let us change variables (with

$s = c(t-\tau)$) to put (13.31) into a simpler form:

$$\begin{aligned} u(\mathbf{x},t) &= \frac{1}{4\pi c^2}\int_0^t \iint_{\partial B(\mathbf{x},c(t-\tau))} \frac{f(\mathbf{y},\tau)}{t-\tau}\, dS_{\mathbf{y}}\, d\tau \\ &= -\frac{1}{4\pi c^2}\int_{ct}^0 \iint_{\partial B(\mathbf{x},s)} \frac{f\left(\mathbf{y}, t-\frac{s}{c}\right)}{s}\, dS_{\mathbf{y}}\, ds \\ &= \frac{1}{4\pi c^2}\int_0^{ct} \iint_{\partial B(\mathbf{x},s)} \frac{f\left(\mathbf{y}, t-\frac{s}{c}\right)}{s}\, dS_{\mathbf{y}}\, ds \\ &= \frac{1}{4\pi c^2}\iiint_{B(\mathbf{x},ct)} \frac{f\left(\mathbf{y}, t-\frac{|\mathbf{x}-\mathbf{y}|}{c}\right)}{|\mathbf{x}-\mathbf{y}|}\, d\mathbf{y}. \end{aligned}$$

The last expression is often called a *retarded potential*.

Conclusion: The key (in fact only) object required to solve the **full** IVP,

$$\begin{cases} u_{tt} - c^2\Delta u = f(\mathbf{x},t), & \mathbf{x}\in\mathbb{R}^3,\ t>0, \\ u(\mathbf{x},0) = \phi(\mathbf{x}), \quad u_t(\mathbf{x},0) = \psi(\mathbf{x}), & \mathbf{x}\in\mathbb{R}^3, \end{cases}$$

is the distribution $\Phi(t)$ defined by (13.25), the fundamental solution to the 3D wave equation.

For boundary value problems there exist modified fundamental solutions (distributions) which are usually called Green's functions (yes, Green's functions as opposed to the correct description of Green's distributions!). The distribution $\Phi(t)$ can be thought of as the **Green's function for the wave equation on $\mathbb{R}^3$**. In Exercise **13.7** You are asked to find the Green's function for the wave equation on the unit ball with homogeneous Dirichlet boundary conditions.

Exercises

13.1 Consider the Green's function G for the 1D diffusion equation on the bar $[0,l]$ with homogenous Dirichlet boundary conditions, i.e., the solution to (13.14).
(a) Find G via the method of images.
(b) Find G via separation of variables and reconcile with the result of (a).

13.2 By taking the Fourier transform with respect to x in the sense of tempered distributions, solve (13.15) to obtain (13.16).

13.3 By taking the Fourier transform with respect to x in the sense of tempered distributions, find the fundamental solution for the 1D Klein-Gordon equation; that is, find the solution to (13.15) where the wave equation is replaced with the 1D Klein-Gordon equation

$$u_{tt} - c^2 u_{xx} + M^2 u = 0 \quad \text{where} \quad M = \frac{mc^2}{\hbar}.$$

Note that the inverse Fourier transform that you will need has a simple representation in terms of a Bessel function (look this up online).

13.4 Use the approach of Section 9.7.4 to define a distributional solution to the full IVP (13.15). Verify that Φ defined by (13.16) satisfies the conditions.

13.5 Repeat Exercise **13.1** for the Green's function of the 1D wave equation. Here, think of a finite string $[0, l]$ with fixed ends.

13.6 Show that $\Phi(t)$ defined by (13.25) is a solution in the sense of distributions to the 3D wave equation on $\mathbb{R}^3 \times (0, \infty)$. Note that $\Phi(t)$ is a four-dimensional distribution on space-time which only has a functional representation in time.

13.7 (a) Find the Green's function for the 3D wave equation on the unit ball $B(\mathbf{0}, 1)$ with homogeneous Dirichlet boundary conditions.
(b) Use it to explicitly solve the IVP/BVP

$$\begin{cases} u_{tt} - c^2 \Delta u = f(\mathbf{x}, t), & \mathbf{x} \in B(\mathbf{0}, 1),\ t > 0, \\ u(\mathbf{x}, t) = 0, & \mathbf{x} \in \partial B(\mathbf{0}, 1),\ t > 0, \\ u(\mathbf{x}, 0) = \phi(\mathbf{x}), \quad u_t(\mathbf{x}, 0) = \psi(\mathbf{x}), & \mathbf{x} \in B(\mathbf{0}, 1). \end{cases}$$

13.8 (**Fundamental Solution/Green's Functions for the 2D Wave Equation**) Repeat everything from Section 13.2.3 for the 2D wave equation. Note that, as expected (cf. wave propagation in 2D), the fundamental solution in 2D will be generated by a discontinuous function, as in the 1D case.

13.9 This is based upon Exercise **10.22**. Recall the informal approach to find the fundamental solution of

$$Lu := a_1^2 u_{x_1x_1} + a_2^2 u_{x_2x_2} + a_3^2 u_{x_3x_3} = 0.$$

There we assumed that the a_i were real and strictly positive constants. Suppose we choose $a_3 = 1$ but a_1, a_2 to be the complex number i. Then we would recover the 2D wave equation (where $x_3 = t$). When you plug these complex numbers into the fundamental solution found in Exercise **10.22**, do you obtain the fundamental solution of the 2D wave equation found in the previous exercise?

The next two exercises pertain to **Green's functions for second-order linear ODEs**. Given the simpler nature of an ODE, the reader might want to start with these two exercises. To do so, they will need to recall the method of *variation of parameters* and the notion of the *Wronskian* from their previous ODE class/text.

13.10 (**Green's Function for a Sturm-Liouville Problem**) To link with Section 12.10, let us consider a second-order linear ODE for $y(x)$ in the form of a regular Sturm-Liouville BVP:

$$(p(x)y'(x))' + q(x)y(x) = f(x), \quad x \in (0, l), \qquad y(0) = 0 = y(l), \tag{13.32}$$

for given functions $p(x) > 0$, q, and f. Let us write (13.32) in the form

$$\mathcal{L}_x y = f(x) \quad \text{where} \quad \mathcal{L}_x := \frac{d}{dx}\left(p \frac{d}{dx}\right) + q.$$

The Green's function for this problem is the function $G(x, z)$ defined on $[0, l] \times (0, l)$ such that $\mathcal{L}_x G(x, z) = \delta_z$ in the sense of distributions and $G(0, z) = 0 = G(l, z)$ for $z \in (0, l)$. In this exercise you will find G indirectly.

(a) First show that if such a G exists, then the solution to the BVP (13.32) is given by

$$y(x) = \int_0^l G(x,z)\, f(z)\, dz.$$

(b) Now find G by solving the BVP (13.32) by the method of variation of parameters and comparing with the form in part (a).
(c) Verify that this G is symmetric on $(0,l) \times (0,l)$.
(d) Find G for the specific BVP $y'' + c^2 y = f(x), x \in [0,1]$, and $y(0) = 0 = y(1)$.

13.11 Now consider the IVP

$$(p(x)y'(x))' + q(x)y(x) = f(x), \ x \in (0,l), \qquad y(0) = y_0, \ y'(0) = y_1.$$

Show that essentially the same Green's function of the previous exercise can be used to solve this IVP.

13.3. What's Next? Towards a Future Volume on This Subject

We plan for a future volume on this subject. There are several avenues which are natural to pursue. The first three pertain to nonlinear equations while the fourth returns to linear equations but from a much broader perspective.

(i) A vast class of PDEs (both linear and nonlinear) which have a particular structure in terms of minimization problems. We call such PDEs **variational** and the study of this class belongs to a subfield of mathematics called the **calculus of variations**.

(ii) Two classes of nonlinear first-order equations: **hyperbolic conservation laws** and **Hamilton-Jacobi equations**, their unique intricacies, connection, and structure.

(iii) A class of nonlinear PDEs known as **reaction-diffusion equations** (which are usually systems of PDEs) and their connection with **pattern formation** in biological and physical systems.

(iv) **General second-order elliptic linear equations** via techniques in **functional analysis**.

For (ii), these two classes of PDEs are connected and we saw versions of both in Chapter 2. For these equations, classical smooth solutions will either fail to exist or exist only for a fixed amount of time. Hence we will introduce appropriate notions of "weak" solutions. For conservation laws this will amount to considering a particular subset of solutions in the sense of distributions. This class of weak solutions can be discontinuous (**shock waves**). For the Hamilton-Jacobi equations, our "weak" solutions will all be continuous but their derivatives can be discontinuous.

For (iv), we will return to second-order linear equations but in far more generality via techniques from **functional analysis**. Both of the previous sections, Sections 13.1 and 13.2, link (in different ways) second-order linear PDEs, including the big three. However, from the perspective of finding a solution to BVPs, we are severely limited by only being able to find Green's functions for a few domains. Even for the vanilla

Dirichlet problem for the Laplacian on a general domain Ω, we do not as of yet know that a solution exists.

To approach such BVPs, we will introduce general methods which on one hand bypass Green's functions and sources (captured by the delta function), but on the other hand are directly linked to a weak notion of a solution in the spirit of the theory of distributions. In a certain sense we follow the **higher-level path** taken by von Neumann in order to present quantum mechanics with **no** mention of the delta function (see the footnote at the start of Section 5.2). We have already seen a little of this. Recall from Section 11.5.2 that the pointwise convergence of the full Fourier series was directly tied to the convergence of the Dirichlet kernel to a delta function. However, when it came to actually proving the result in Section 11.5.3, the key ingredient was Bessel's inequality (Theorem 11.3), an inequality derived in the space L^2 with simple least squares estimates. In other words, we presented a very indirect and *hands-off* approach for capturing the concentration effects of the delta function.

While these methods of functional analysis will require the notion of the Lebesgue integral (for the completeness of function spaces), it is remarkable that linear algebra in $\mathbb{R}^N$, and in particular the theory behind solving $\mathbf{Ax} = \mathbf{b}$, plays the central guiding force.

The curious reader can explore any of these topics now by venturing into the graduate texts cited in the bibliography.

Appendix

Objects and Tools of Advanced Calculus

This appendix consists primarily of necessary results in advanced calculus. The reader should consult the appropriate sections as needed. For more details and proofs of the results, we recommend the excellent text of Folland.[1]

A.1. Sets, Domains, and Boundaries in $\mathbb{R}^N$

In this text, the space dimension N will almost always be one, two, or three. We focus here on the notation and intuitive definitions for the following notions and properties: **points and vectors** and the distance between points; **balls and spheres** (the basic reference examples of sets); **open and closed sets** and **domains**; **bounded and unbounded sets**; **the boundary of a set and its closure**.

Points in $\mathbb{R}^N$: We denote points in $\mathbb{R}^N$ (also known as N-dimensional Euclidean space) via $\mathbf{x} = (x_1, \ldots, x_N)$. In the cases of $\mathbb{R}^2$ or $\mathbb{R}^3$, we also often use the notation (x, y) and (x, y, z), respectively. We denote the origin in $\mathbb{R}^N$ by $\mathbf{0} = (0, \ldots, 0)$. Given any point $\mathbf{x} \in \mathbb{R}^N$, let $|\mathbf{x}|$ denote the Euclidean norm; i.e., $|\mathbf{x}| := \left(\sum_{i=1}^N x_i^2\right)^{1/2}$. This norm induces a **distance**; namely the distance between two points $\mathbf{x}, \mathbf{y} \in \mathbb{R}^N$ is $|\mathbf{x} - \mathbf{y}|$.

A Side Comment on Points vs. Vectors: A natural confusion can exist in differentiating between points and vectors in $\mathbb{R}^N$. Let us illustrate with $N = 2$. We use the notation (x_0, y_0) for a point in $\mathbb{R}^2$ and $\langle x_0, y_0 \rangle$ for a two-dimensional vector. It is usually best to keep these two notions separate. For a point (x_0, y_0) in the plane, we simply specify the x_0 and y_0 coordinates of the point subject to a reference frame of two orthogonal axes. On the other hand, you should view the vector $\langle x_0, y_0 \rangle$ as a **set of directions**. Here what we mean by "directions" is in the sense of "directions to the closest subway station". Indeed, $\langle x_0, y_0 \rangle$ means you go x_0 units in the x-direction

[1]**Gerald B. Folland**, *Advanced Calculus*, Pearson, 2001.

and y_0 units in the y-direction. As such it makes complete sense to **add** two directions (vectors) $\mathbf{u}_0 = \langle x_0, y_0 \rangle$ and $\mathbf{u}_1 = \langle x_1, y_1 \rangle$, by simply adding the directions in each component; that is, $\mathbf{u}_0 + \mathbf{u}_1 = \langle x_0, y_0 \rangle + \langle x_1, y_1 \rangle = \langle x_0 + x_1, y_0 + y_1 \rangle$. Note that it really does not make sense to "add two points".

We can picture or illustrate a vector as a directed line segment, and one can view the addition of vectors geometrically by starting vector $\mathbf{u}_1$ at the end point of vector $\mathbf{u}_0$. Note that a vector has no inherent starting and ending point. Of course if you choose the starting point to be the origin $(0, 0)$ and then follow the vector $\langle x_0, y_0 \rangle$ (the directions) you end up at the point (x_0, y_0). Hence one **sometimes** views the vector $\langle x_0, y_0 \rangle$ as the **position vector of the point** (x_0, y_0). This is why, in certain instances, it is convenient, and deliberate, to not make the distinction. For example, we will often use the boldface variable $\mathbf{x}$ to denote both the associated point $(x_1, \ldots, x_N)$ and the vector $\langle x_1, \ldots, x_N \rangle$.

Balls and Spheres in $\mathbb{R}^N$: For any point $\mathbf{x} \in \mathbb{R}^N$ and $r > 0$, we define the **ball** centered at $\mathbf{x}$ with radius r to be the set

$$B(\mathbf{x}, r) := \left\{ \mathbf{y} \in \mathbb{R}^N \,\middle|\, |\mathbf{y} - \mathbf{x}| < r \right\}.$$

Strictly speaking, this is the definition of the **open ball** and represents the following: (i) an open solid "ball" in 3D, (ii) an open disc in 2D, and (iii) an open interval in 1D.

The sphere centered at $\mathbf{x}$ with radius r is the set

$$\left\{ \mathbf{y} \in \mathbb{R}^N \,\middle|\, |\mathbf{y} - \mathbf{x}| = r \right\}.$$

It represents the following: a "true" sphere in 3D, a circle in 2D, and two points in 1D.

General Open and Closed Sets in $\mathbb{R}^N$: A subset $\Omega \subset \mathbb{R}^N$ is **open** (or more precisely **open in** $\mathbb{R}^N$) if, roughly speaking, around every point $\mathbf{x} \in \Omega$ there is room to move in every direction while remaining in Ω. More precisely, the following property holds:

> For each point $\mathbf{x} \in \Omega$, there is some ball centered at $\mathbf{x}$ which lies inside of Ω. That is, for each point $\mathbf{x} \in \Omega$, there exists an $r > 0$ such that $B(\mathbf{x}, r) \subset \Omega$.

A set is called **closed** (or more precisely **closed in** $\mathbb{R}^N$) if and only if its complement is open. Closed sets have the property that if a sequence of point $\mathbf{x}_n \in \Omega$ converges in $\mathbb{R}^N$, then its limit must also be in Ω.

Boundaries of Sets in $\mathbb{R}^N$: Given any set $\Omega \subset \mathbb{R}^N$, we define **the boundary of** Ω (denoted by $\partial\Omega$) to be all points $\mathbf{x} \in \mathbb{R}^N$ with the following property:

> For every $r > 0$, the ball $B(\mathbf{x}, r) \cap \Omega \neq \emptyset$ **and** $B(\mathbf{x}, r) \cap \Omega^c \neq \emptyset$. Here Ω^c denotes the complement of Ω in $\mathbb{R}^N$.

In general, note that points on the boundary of Ω may or may not be in Ω. However, closed sets always contain their boundary.

Based on our reference examples, the ball $B(\mathbf{x}, r)$ is an open set. The **closed ball**

$$\left\{ \mathbf{y} \in \mathbb{R}^N \,\middle|\, |\mathbf{y} - \mathbf{x}| \leq r \right\}$$

is a closed set. The boundary of both the ball and the closed ball is the sphere; i.e.,

$$\partial B(\mathbf{x}, r) = \left\{ \mathbf{y} \in \mathbb{R}^N \,\middle|\, |\mathbf{y} - \mathbf{x}| = r \right\}.$$

We refer to the **closure of a set** Ω as the union of the set and its boundary. The closure of Ω is denoted by $\overline{\Omega}$ and by its very definition is a **closed set in** $\mathbb{R}^N$. For example,

$$\overline{B(\mathbf{x}, r)} = \left\{ \mathbf{y} \in \mathbb{R}^N \,\middle|\, |\mathbf{y} - \mathbf{x}| \leq r \right\}.$$

Bounded and Unbounded Sets: We say a domain Ω is **bounded** if there exists some $R > 0$ such that $\Omega \subset B(\mathbf{0}, R)$. In other words, the set is confined to lie is **some**, possibly quite big, ball (cf. Figure A.1, right).

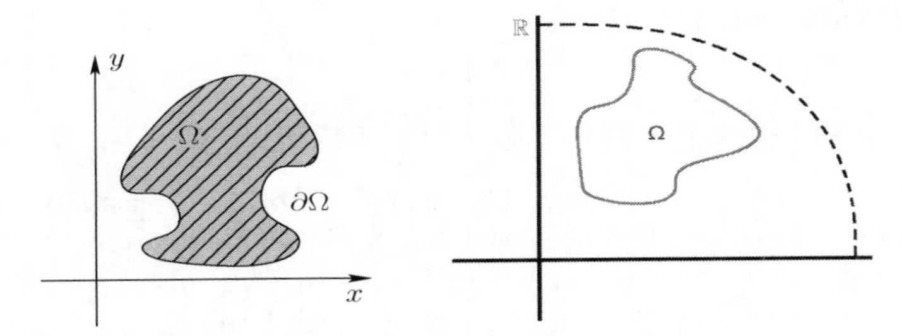

Figure A.1. Left: A domain Ω in $\mathbb{R}^2$ and its boundary $\partial\Omega$. In three dimensions, Ω would be a solid region in $\mathbb{R}^3$ and the boundary $\partial\Omega$ would be a surface. Right: A bounded domain fits inside a disc (ball) $B(\mathbf{0}, R)$ for some choice of R.

Note that we may also speak of the boundary of an unbounded set. For example, the boundary of the upper half-plane $\{\mathbf{x} = (x_1, x_2) \,|\, x_2 > 0\}$ is the x_1-axis, i.e, the set $\{\mathbf{x} = (x_1, x_2) \,|\, x_2 = 0\}$.

Connected Open Sets: While there are more general topological definitions, the following definition, usually referred to as **path connected**, will suffice. An open set Ω is (path) **connected** if for any two points $\mathbf{x}, \mathbf{y} \in \Omega$, there exists a continuous curve from $\mathbf{x}$ to $\mathbf{y}$ which lies entirely in Ω. Of course this definition also holds for sets which are not necessarily open.

An important concept in this book is a special subset of $\mathbb{R}^N$ which we call a domain.

Definition of a Domain

Definition A.1.1. A **domain** is an **open, connected subset** of $\mathbb{R}^N$ which has a **piecewise** C^1 **boundary**. Throughout this text, we will use Ω to denote a domain.

What do we mean by a piecewise C^1 boundary? In dimension $N = 2$ this means that the boundary is a piecewise smooth curve. More precisely, near any point on the boundary curve $\partial\Omega$, one can view the boundary as the graph of a C^1 function of one of the two variables. In dimension $N = 3$ this means that the boundary is a piecewise smooth surface. Precisely, near any point on the boundary surface $\partial\Omega$, one can view the boundary as the graph of a C^1 function of two of the three variables. Recall that piecewise means the boundary can have a finite number of singularities (corners or kinks) analogous to the boundary of a polygonal shape. For example, we could have a domain in the shape of a square or a cube; its boundaries would not be C^1 at the corner points (edges). Domains can be either bounded or unbounded. Even an unbounded domain can still have a boundary; for example the upper half-plane in $\mathbb{R}^2$ has the x-axis as its boundary. We will agree that if the domain is all of $\mathbb{R}^N$, then it has no boundary (more precisely its boundary is the empty set).

While these notions of boundaries and open and closed sets are hopefully intuitive, they are far more subtle than they might first appear. In this text you will only need the basic intuition associated with these definitions, guided by the canonical example of spheres and open and closed balls. If you are curious to find out more, see **[16]** or look up any of these terms online.

A.2. Functions: Smoothness and Localization

Functions form the basis of any study on PDEs. A function is an input-output machine. We will be mostly interested in functions where the input are points in $\mathbb{R}^N$, or in some domain $\Omega \subset \mathbb{R}^N$, and whose output are real numbers (i.e., in $\mathbb{R}$). So a function $f : \mathbb{R}^N \to \mathbb{R}$ is simply a rule which assigns each n-tuple of numbers, $\mathbf{x} \in \mathbb{R}^N$, to a real number $f(\mathbf{x})$. If $N = 1$, we refer to **a function of one variable**, and for $N > 1$, we refer to **a function of** N **variables**, or simply **a function of several variables**.

While there is a very well-developed modern theory of function spaces (classes of functions), here we will just present some simple notions. For more precise definitions the reader is encouraged to look them up online.

A.2.1. Function Classes Sorted by Smoothness. One of the most useful ways to classify functions pertains to their **smoothness**. Continuity and differentiability are the features by which one assesses smoothness. We make the assumption that the reader is very familiar and comfortable with the concepts of continuity and differentiability. The class of continuous functions on $\mathbb{R}^N$ is denoted either by $C(\mathbb{R}^N)$ or by $C^0(\mathbb{R}^N)$, and continuous functions which possess continuous first-order partial derivatives constitute the class $C^1(\mathbb{R}^N)$. More generally, $C^k(\mathbb{R}^N)$ is the class of continuous functions whose partial derivatives up to order k exist, and these partial derivatives are themselves continuous functions on $\mathbb{R}^N$. If **all** partial derivatives (of any order) of a function are continuous on $\mathbb{R}^N$, then we say the function belongs to the class $C^\infty(\mathbb{R}^N)$. Note that, by definition, we have

$$C^\infty(\mathbb{R}^N) \subset \cdots \subset C^2(\mathbb{R}^N) \subset C^1(\mathbb{R}^N) \subset C(\mathbb{R}^N),$$

and all the subset inclusions are proper, meaning that in any inclusion, there are functions in the right class which are not in the left class. In dimension $N = 1$, any polynomial, the sine and cosine functions, and the exponential function are all in $C^\infty(\mathbb{R})$. However, the continuous function $f(x) = |x| \in C(\mathbb{R})$ is not $C^1(\mathbb{R})$, since its derivative is not continuous; that is, $f'(x)$ has a jump discontinuity at $x = 0$.

The analogous definitions hold for a function defined on a domain Ω in $\mathbb{R}^N$, but with continuity and differentiability on $\mathbb{R}^N$ replaced with continuity and differentiability in Ω. These classes are denoted by $C(\Omega), C^k(\Omega), C^\infty(\Omega)$.

Throughout this text (and elsewhere) it is common to use the informal phrase **sufficiently smooth**. Roughly speaking, this amounts to saying that *enough* continuous derivatives exist for the problem at hand. For example, if we are dealing with a second-order PDE, then without any other information we would certainly like $u \in C^2(\Omega)$. In other words, we want the precise amount of smoothness (no more, no less) required so that all the derivatives we are working with are continuous functions. This will allow us, for example, to interchange the order of partial derivatives and, under further integrability conditions, differentiate under the integral sign (Section A.9). In certain cases, our analysis will require higher derivatives of the solution than the order of the PDE and, in these cases, we would assume the continuity of these higher derivatives as well.

A.2.2. Localization: Functions with Compact Support. An important subclass of $C^k(\mathbb{R}^N)$ $(k = 0, 1, \ldots, \infty)$ is those functions which are *localized* in the sense that they are identically zero except on a **confined** region of space. For any such function there exists a ball $B(\mathbf{0}, R)$, for some $R > 0$, such that $f \equiv 0$ on the complement of $B(\mathbf{0}, R)$. We denote the space of all such functions as $C_c^k(\mathbb{R}^N)$. The subscript c is used for the word *compact*. A more precise definition requires the notion of the support of a function and a compact subset of $\mathbb{R}^N$. To this end, we define the **support** of a function f on $\mathbb{R}^N$ to be

$$\text{the closure of the set } \{\mathbf{x} \in \mathbb{R}^N \mid f(\mathbf{x}) \neq 0\}. \tag{A.1}$$

We denote this closed set by supp f. A subset of $\mathbb{R}^N$ is called **compact** if it is closed and bounded. Hence, we say f has **compact support** if supp f is bounded; since by definition supp f is closed, this means it is a compact set. Figure A.2 shows the graph of three functions with compact support in $N = 1$ and $N = 2$. If a function is only defined on some domain Ω, one can define the notion of compact support in Ω and functions in $C_c^k(\Omega)$. Intuitively, this means that *their values become* 0 *before they hit* $\partial\Omega$. We will be more specific and discuss this further when needed (cf. Sections 5.7.1 and 9.6).

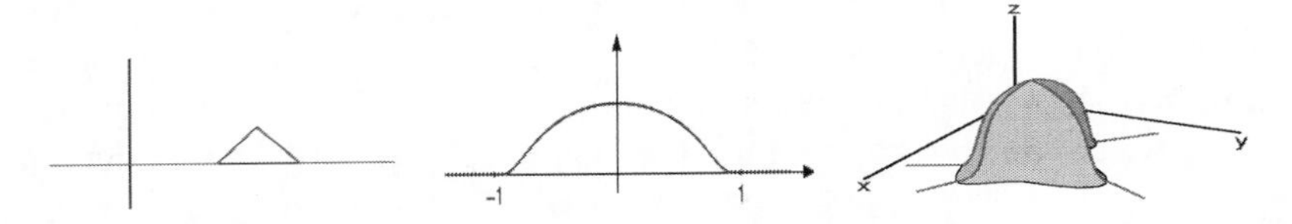

Figure A.2. Graph of three functions with compact support. The first depicts a function in $C_c(\mathbb{R})$ which is not C^1; the second depicts a function in $C_c^1(\mathbb{R})$; and the third depicts a function in $C_c(\mathbb{R}^2)$ (perhaps even smoother depending on your interpretation of the picture!).

A.2.3. Boundary Values for Functions Defined on a Domain. When encountering what is known as an initial value problem (IVP), we seek a function $u(\mathbf{x}, t)$ for all $t \geq 0$ by specifying both the values of $u(\mathbf{x}, t)$ at $t = 0$ and a PDE that u follows (solves) for all $t > 0$. In a boundary value problem (BVP) over a domain Ω, we seek a function $u(\mathbf{x})$ for $\mathbf{x} \in \overline{\Omega}$ by specifying both the values of u on the boundary of Ω and a PDE that u solves in Ω. Actually, one can think of the initial value problem as a boundary value problem for a function in $\mathbb{R}^{N+1}$ wherein $\Omega = \mathbb{R}^N \times (0, \infty)$ and $\partial\Omega = \mathbb{R}^N \times \{0\}$. For example if $N = 1$, Ω is the upper half-(x vs. t)-plane and $\partial\Omega$ is the x-axis.

But clearly any solution u to these boundary value problems must share a certain *integrity* with respect to its boundary values. What would you say if we just took any solution to the PDE inside Ω and then artificially extended it to $\overline{\Omega}$ by discontinuously defining it to be the desired boundary condition on $\partial\Omega$? We think you would say that we cheated. Hence, it is important to reflect on boundary values in the **context of continuity**.

If $u \in C(\Omega)$, then u may or may not be continuous up to the boundary. That is, for any point $\mathbf{x}_0 \in \partial\Omega$,

$$\lim_{\mathbf{x}\to\mathbf{x}_0} u(\mathbf{x}) \qquad \text{may or may not exist.} \tag{A.2}$$

If this limit exists for all $\mathbf{x}_0 \in \partial\Omega$, then it makes sense to extend u to the closure of Ω, i.e., to $\overline{\Omega}$, by defining the value of u at any point on the boundary to be the limit (A.2). If the limit does not exist, then it makes no sense to talk about boundary values of u. Indeed, there would be no choice of function g on $\partial\Omega$ for which the function

$$\overline{u}(\mathbf{x}) = \begin{cases} u(\mathbf{x}) & \text{if } \mathbf{x} \in \Omega, \\ g(\mathbf{x}) & \text{if } \mathbf{x} \in \partial\Omega \end{cases}$$

would be continuous on $\overline{\Omega}$. If these limits do exist for all $\mathbf{x}_0 \in \partial\Omega$, then we can extend u to all of $\overline{\Omega}$ by defining

$$u(\mathbf{x}_0) = \lim_{\mathbf{x}\to\mathbf{x}_0} u(\mathbf{x}) \qquad \text{for all } \mathbf{x}_0 \in \partial\Omega.$$

In this way, we say u is continuous on all of $\overline{\Omega}$ and write $u \in C(\overline{\Omega})$.

We can extend this notion to partial derivatives. For example, we say a function $u \in C^k(\overline{\Omega})$ if u and all its partial derivatives (up to and including order k) have boundary limits and the corresponding extended functions are continuous on $\overline{\Omega}$. One should keep in mind that very smooth functions on a domain Ω can have widely oscillatory behavior as we approach the boundary. For example, let $\Omega = (0, 1)$ (an open interval) and $u(x) = \sin\left(\frac{1}{x}\right)$. The function u is C^∞ in Ω. The boundary of Ω consists of two points 0 and 1. As $x \to 0^+$, it oscillates between ± 1 more and more often, and thus there is no limit to speak of. In higher space dimensions, oscillatory behavior is even more complicated.

So in conclusion, when we speak of the solution u to a boundary value problem (or initial value problem)

$$\begin{cases} \text{PDE for } u & \text{in } \Omega, \\ u = g & \text{on } \partial\Omega, \end{cases}$$

we mean a function u which satisfies the PDE in Ω and which can be extended continuously to the boundary to attain the boundary condition. That is, for any $\mathbf{x}_0 \in \partial\Omega$, we have

$$\lim_{\mathbf{x}\to\mathbf{x}_0} u(\mathbf{x}) \;=\; g(\mathbf{x}_0).$$

A.3. Gradient of a Function and Its Interpretations, Directional Derivatives, and the Normal Derivative

Given a function $f : \mathbb{R}^N \to \mathbb{R}$ (or defined on some domain in $\mathbb{R}^N$), the gradient at a point $\mathbf{x}$ is simply the vector in $\mathbb{R}^N$ whose components are the respective partial derivatives at $\mathbf{x}$.

Definition of the Gradient

Definition A.3.1. The **gradient** of a function $f(\mathbf{x})$ is defined as

$$\nabla f \;:=\; \left\langle \frac{\partial f}{\partial x_1}, \ldots, \frac{\partial f}{\partial x_N} \right\rangle.$$

Since the gradient is an N-dimensional vector which depends on the point $\mathbf{x}$, it may be viewed as a **vector field**, i.e., a map from the domain of the function f to the space of N-dimensional vectors.

So, for example, in two and three dimensions, with the independent variables written as $f(x, y)$ and $f(x, y, z)$, we have $\nabla f := \left\langle \frac{\partial f}{\partial x}, \frac{\partial f}{\partial y} \right\rangle$ and $\nabla f := \left\langle \frac{\partial f}{\partial x}, \frac{\partial f}{\partial y}, \frac{\partial f}{\partial z} \right\rangle$, respectively.

The gradient is a fundamental object in multivariable calculus, and we cannot overemphasize the importance of the next calculation and its implications. In particular, it implies that the gradient encapsulates **all the information** about first-order derivatives.

A.3.1. The Fundamental Relationship Between the Gradient and Directional Derivatives.

If ν is a unit vector, a natural question is to ask is, *What is the instantaneous rate of change of f with respect to distance in the ν direction?* We call this rate the **directional derivative of f in the direction** ν and it is denoted by $D_\nu f$. In formulas, this means

$$D_\nu f(\mathbf{x}) \;:=\; \frac{d}{dt} f(\mathbf{x} + t\nu)\Big|_{t=0}.$$

We can compute this derivative using **the chain rule**[2] to find

$$\begin{aligned} D_\nu f(\mathbf{x}) \;&=\; \frac{d}{dt} f(\mathbf{x} + t\nu)\Big|_{t=0} \\ &=\; \sum_{i=1}^{N} \frac{\partial}{\partial y^i} f(\mathbf{y} = \mathbf{x} + t\nu)\Big|_{t=0} \frac{dy^i}{dt} \\ &=\; \nabla f(\mathbf{x}) \cdot \nu. \end{aligned}$$

[2]If you have forgotten the various versions of the chain rule, review them now.

Hence we have the basic identity

$$D_{\nu} f = \nabla f \cdot \nu. \tag{A.3}$$

The implications of the simple identity (A.3) are far-reaching and are the basis for much of what is to follow in this text. Moreover, recall the equality that for any two vectors **a** and **b** in 2D or 3D (or N-D),

$$\mathbf{a} \cdot \mathbf{b} = |\mathbf{a}||\mathbf{b}| \cos\theta \qquad \text{where } \theta \text{ is the angle between the vectors.}$$

Combining this with (A.3) has two important consequences:

- The gradient points in the direction in which the directional derivative is largest.
- The value of this maximal directional derivative is $|\nabla f|$.

One can actually view these two properties as defining the gradient vector field. See Exercise **A.10** for an example of the utility of this more general definition.

The basic identity (A.3) also provides a geometric interpretation of the gradient with respect to an important class of subsets of the domain of f called **level sets.**

Definition A.3.2. We define a **level set** (or **level curve**) of a function f as the set of all points in the domain upon which f is identically constant. So, for example, in 2D, given any constant C, the level set (curve) corresponding to C is the set

$$\mathcal{C} := \left\{(x, y) \in \mathbb{R}^2 \,\middle|\, f(x, y) = C\right\}.$$

At any point on a level set the gradient must be **perpendicular** to the level set. Why? By our definition of the level set (a set upon which the values of f do **not** change), any directional derivative tangent to the level set (i.e., instantaneously in the direction of the level set) must be zero.

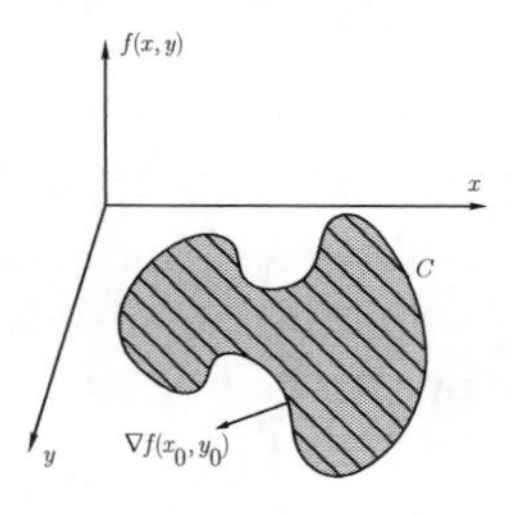

Figure A.3. Level curve $\mathcal{C}$ and gradient ∇f.

A.3.2. Lagrange Multipliers: An Illuminating Illustration of the Meaning of the Gradient. The previous geometric interpretations of the gradient are beautifully illustrated by considering the problem of **constrained optimization**. Suppose we wish to maximize or minimize a C^1 function $f(x, y)$, not on some domain in $\mathbb{R}^2$, but rather over some curve, defined as the level set of a C^1 function $g(x, y)$. In other words, the goal is to **maximize or minimize** the function $f(x, y)$ **subject to** the constraint $g(x, y) = 0$. How should one proceed? We look for points (x_0, y_0) which lie on the level curve $\mathcal{C}$, i.e., $g(x_0, y_0) = 0$, for which the instantaneous rate of change of f with respect to movement on the curve $\mathcal{C}$ is zero. These points are **critical points** with respect to movement on (or "life on") the curve $\mathcal{C}$. We make two observations:

- We seek points on the curve where the directional derivative is zero in the direction of the curve. This happens when $\nabla f(x_0, y_0)$ is perpendicular to $\mathcal{C}$ (see Figure A.3). Indeed, if $\nabla f(x_0, y_0)$ is not perpendicular to $\mathcal{C}$, then the component of the gradient tangential to the curve indicates a direction in which to proceed in order to increase the value of f.
- On the other hand, $\mathcal{C}$ is the level set of the function g, and hence $\nabla g(x_0, y_0)$ must also be perpendicular to the curve $\mathcal{C}$.

Combining these two observations, we look for points (x_0, y_0) such that $g(x_0, y_0) = 0$ where $\nabla f(x_0, y_0)$ and $\nabla g(x_0, y_0)$ are parallel; i.e.,

$$\nabla f(x_0, y_0) = \lambda \nabla g(x_0, y_0),$$

for some scalar λ called the **Lagrange multiplier**.

Alternatively, one can define a new function that absorbs the Lagrange multiplier

$$L(x, y; \lambda) := f(x, y) - \lambda g(x, y).$$

This function is sometimes called the **Lagrangian** and its optimization takes into account the constraint. In particular, we seek a solution $(x_0, y_0; \lambda)$ to the three equations

$$\begin{cases} \nabla L(x_0, y_0; \lambda) = 0, \\ g(x_0, y_0) = 0, \end{cases}$$

where ∇ denotes the two-dimensional spacial gradient.

A.3.3. An Important Directional Derivative: The Normal Derivative on an Orientable Surface. Given a surface $\mathcal{S}$ lying in the domain of a function u, there is an important directional derivative of u which measures the rate of change directly out of $\mathcal{S}$. For this to make sense the surface $\mathcal{S}$ must be **orientable**. You can think of this as saying the surface has a clearly defined outer and inner part. For example, a Mobius strip is not an orientable surface. For an orientable surface, there exists a well-defined unit normal $\mathbf{n}$ at every point on the surface, and this normal vector varies **continuously** as we traverse the surface. Given such an orientable surface, there are two unit normal vector fields which we could choose. The choice determines what we regard as the top of the surface. In many cases, the surface will be the boundary of some "reasonably regular" domain Ω. We call these closed surfaces. In these cases the boundary is clearly orientable with a choice of either the outer normal (which points out of Ω) or the inner normal (which points into Ω).

Now consider a function u defined on $\mathbb{R}^3$ and an orientable surface $\mathcal{S}$ in $\mathbb{R}^3$ with normal vector field $\mathbf{n}$. An important directional derivative at any point on the surface is the derivative in the normal direction $D_{\mathbf{n}}u$. We call this the **normal derivative** and denote it by

$$\boxed{\frac{\partial u}{\partial \mathbf{n}} = \nabla u \cdot \mathbf{n}.}$$

Note that this definition also applies in 2D wherein the boundary is now a curve.

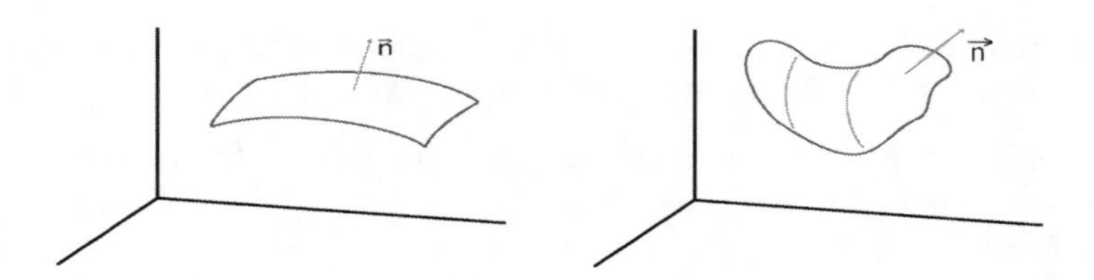

Figure A.4. Left: A unit normal to a surface. Right: The outer unit normal to the boundary of a domain.

As an example, let us compute the normal derivative of the radially symmetric function $u = |\mathbf{x}|^3$ on the surface $\partial B(\mathbf{0}, a) \subset \mathbb{R}^3$ (for some fixed $a > 0$) with the outer normal $\mathbf{n}$. Note that the function u is radially symmetric (only depends on $|\mathbf{x}|$) and the sphere, defined by the equation $|\mathbf{x}| = a$, itself is also radially symmetric. Hence, you might anticipate that the normal derivative will be constant at all points on the sphere. This will be the case; indeed, by Exercise **A.6**, we have

$$\nabla u = 3|\mathbf{x}|^2 \frac{\mathbf{x}}{|\mathbf{x}|} = 3|\mathbf{x}|\,\mathbf{x}.$$

On the other hand, the unit outer normal at any point $\mathbf{x} \in \partial B(\mathbf{0}, a)$ is simply

$$\mathbf{n} = \frac{\mathbf{x}}{|\mathbf{x}|}.$$

Thus,

$$\nabla u \cdot \mathbf{n} = 3\mathbf{x} \cdot \mathbf{x} = 3|\mathbf{x}|^2 = 3a^2,$$

where in the last equality we used the fact that we were evaluating the derivative at a point $\mathbf{x} \in \partial B(\mathbf{0}, a)$; i.e., $|\mathbf{x}| = a$.

One can check that one's intuition correctly confirms that on a spherical surface, the normal derivative of a radial function is simply the radial derivative evaluated at the radius of the sphere. That is, if $f(\mathbf{x}) = g(r)$ where $r = |\mathbf{x}|$, then at any point $\mathbf{x}$ on $\partial B(\mathbf{0}, r)$, we have

$$\frac{\partial f(\mathbf{x})}{\partial \mathbf{n}} = g'(r). \tag{A.4}$$

A.4. Integration

Integration naturally plays an important role in addressing PDEs. With the exception of some simple calculations in polar or spherical coordinates, it is **not** important that the readers know how to **evaluate** particular integrals using their repertoire of techniques. Rather, what will prove vital is knowledge of the following:

- What various types of integrals mean geometrically and/or physically.
- How to use and manipulate them in the analysis of physical and geometric quantities. As we shall see, "manipulation" will be mostly with respect to changes of variables and differentiation.

When you encounter any integral, you should first ground yourself by asking the following questions:

- What are you integrating (the integrand) and what does it represent?
- What is the region of integration (i.e., what are you integrating over)?
- What are we integrating with respect to — length, area, volume, time, area along a surface?
- What physical or geometric quantity does the integral yield?

When we speak of an integral, we mean the **Riemann integral** which you were exposed to in calculus. This classical notion of integration will suffice for all the functions we will consider in this text. While the richer, more complete, theory of integration, called the **Lebesgue theory** (see for example [**17**]), is indispensable for functional analysis, we can do an amazing amount of PDE theory without it. However, the reader should recall that the notion of a Riemann integral is based upon the existence of a **limit** of partial sums; it is easy to forget this fundamental fact by focusing solely on the computation of integrals via iterated integrals and the Fundamental Theorem of Calculus (looking for antiderivatives).

A.4.1. Bulk, Surface, and Line Integrals. We will frequently use two basic types of integrals (bulk and surface) and occasionally line integrals:

- **Bulk integrals** are integrals over (i) domains in $\mathbb{R}^3$ which have volume, (ii) domains in $\mathbb{R}^2$ which have area, and (iii) intervals in $\mathbb{R}$ which have length. One can, of course, consider bulk integrals over regions in $\mathbb{R}^N$, $N \geq 4$, as well, but this will be rare in this text. It is instructive to think of bulk integrals as **density integrals** in that the integrand will often denote a certain density over some region (domain) in $\mathbb{R}^N$.
- **Surface integrals** are integrals defined over surfaces in $\mathbb{R}^3$. They will often be associated with a notion of **flux**, discussed in the next subsection.
- **Line integrals** are defined over curves in either $\mathbb{R}^3$ or $\mathbb{R}^2$. In this text, we will only occasionally consider line integrals over curves in $\mathbb{R}^2$. You should view such a line integral in $\mathbb{R}^2$ as a "surface" integral in two dimensions, i.e., an integral over the boundary of some two-dimensional domain.

We adopt the following **notation** for these integrals. For bulk integrals in which we integrate a function $f(\mathbf{x})$ on some domain of $\Omega \subset \mathbb{R}^N$, we use

$$\int \cdots \int_{\Omega} f(\mathbf{x})\, d\mathbf{x},$$

where $d\mathbf{x} = dx_1, \ldots, dx_N$. The number of integral signs reflects the dimension N and a boldface $\mathbf{x}$ is used whenever the dimension of the independent variable in greater than

1. So our notation for bulk integrals in dimensions $N = 3, 2$, and 1 are, respectively,

$$\iiint_{\Omega} f(\mathbf{x})\, d\mathbf{x} = \iiint_{\Omega} f(\mathbf{x})\, dx_1 dx_2 dx_3, \qquad \iint_{\Omega} f(\mathbf{x})\, d\mathbf{x} = \iint_{\Omega} f(\mathbf{x})\, dx_1 dx_2,$$

$$\int_{\Omega} f(x)\, dx = \int_a^b f(x)\, dx.$$

When convenient, we will use x, y in place of x_1, x_2, or x, y, z in place of x_1, x_2, x_3.

If $\mathcal{S} \subset \Omega$ denotes a surface in $\mathbb{R}^3$ and $f(\mathbf{x})$ denotes a function defined on $\mathcal{S}$ (note that $\mathbf{x} \in \mathbb{R}^3$), then the surface integral of f over $\mathcal{S}$ is denoted by

$$\iint_{\mathcal{S}} f(\mathbf{x})\, dS.$$

Since $\mathcal{S}$ is a two-dimensional object, we use two integral signs. We use dS as the surface area increment along the surface $\mathcal{S}$. In particular, to compute such an integral requires a parametrization of the surface with two parameters α, β. We can then write the surface increment dS in terms of $d\alpha$ and $d\beta$, and the entire surface integral as a two-dimensional iterated integral. It will be quite common to consider surface integrals over a closed surface; i.e., $\mathcal{S} = \partial\Omega$, the boundary of a some domain $\Omega \subset \mathbb{R}^3$.

In $\mathbb{R}^2$, the analogue to a surface is a curve $\mathcal{C}$ and the surface integral becomes a line integral which we denote by

$$\int_{\mathcal{C}} f(\mathbf{x})\, ds.$$

We use ds as the increment of arc length along $\mathcal{C}$.

A.4.2. Flux Integrals. Let $\mathcal{S}$ be an orientable surface with its well-defined unit normal vector field $\mathbf{n}$ varying continuously across $\mathcal{S}$. We will interpret the direction of this normal as **out of the surface**. Now consider a vector field $\mathbf{F}$, defined on the surface $\mathcal{S}$, representing **flow per unit area**. The vector field may or may not be also defined at the points surrounding the surface. There are two pieces of information in the flow density $\mathbf{F}$: the direction of the flow (the direction of the vector $\mathbf{F}$) and the magnitude of the flow ($|\mathbf{F}|$). The dimensions of $|\mathbf{F}|$ are

$$\frac{\text{the dimensions for the amount of the quantity}}{\text{time} \times \text{length}^2};$$

that is, $|\mathbf{F}|$ is a measure of quantity per unit time per unit area. For example, $\mathbf{F}$ could measure the flow of a fluid across a surface, in which the dimensions for the quantity could be either mass or volume (length^3).

A natural calculation is to measure the **instantaneous flow out of the surface**. This is obtained by integration:

$$\iint_{\mathcal{S}} \mathbf{F} \cdot \mathbf{n}\, dS. \tag{A.5}$$

Note that the quantity $\mathbf{F} \cdot \mathbf{n}$ represents the component of the flux which points out of the surface. This surface integral gives the amount of the quantity per unit time and therefore has dimensions[3]

$$\frac{\text{the dimensions for the amount of the quantity}}{\text{time}}.$$

We call the vector field $\mathbf{F}$ the **flux** or the **flux field**, and the integral (A.5) is called a **flux integral** or simply the **total flux across $\mathcal{S}$.**

For closed surfaces $\mathcal{S}$ which are the boundary of a domain $\Omega \subset \mathbb{R}^3$, we will denote the surface by $\partial\Omega$. The unit normal is called the **outer normal** if it points **out** of Ω. In this case the flux integral $\iint_{\partial\Omega} \mathbf{F} \cdot \mathbf{n}\, dS$ gives the instantaneous rate at which the quantity leaves the domain Ω.

A.4.3. Improper Integrals, Singularities, and Integrability.

Recall from your calculus days the notion of an **improper integral**. This means an integral

$$\int \cdots \int_{\Omega} f(\mathbf{x})\, d\mathbf{x} \tag{A.6}$$

where **either** the domain $\Omega \subset \mathbb{R}^N$ is **unbounded and/or** the function f has a **singularity** (e.g., a blow-up) at some point in $\overline{\Omega}$, the closure of Ω.

Improper integrals will be pervasive in our study of PDEs. Not all the functions that we will encounter will be continuous everywhere. Indeed, singularities play an important role and are often the key element in finding nice smooth solutions to certain PDEs. For example, the integration of the function in dimension three with singularity at $\mathbf{x} = \mathbf{0}$,

$$f(\mathbf{x}) = \frac{1}{|\mathbf{x}|} = \frac{1}{\sqrt{x_1^2 + x_2^2 + x_3^2}},$$

plays a central role in finding infinitely smooth solutions to the Laplace and Poisson equations. On the other hand, regardless of the smoothness of f it is often necessary to compute its integral over the entire space $\mathbb{R}^N$.

Recall from calculus (i.e., Riemann integration) that we make sense of the improper integral as a **limit** of (proper) Riemann integrals. The particular limit we choose can be thought of as the way in which we "build up" or "add up" parts of the integral over pieces of the domain Ω. For example in dimension $N = 1$,

$$\int_0^1 \frac{1}{\sqrt{x}}\, dx = \lim_{a \to 0^+} \int_a^1 \frac{1}{\sqrt{x}}\, dx \qquad \text{and} \qquad \int_1^\infty \frac{1}{x^2}\, dx = \lim_{L \to \infty} \int_1^L \frac{1}{x^2}\, dx.$$

A natural question arises as to whether an improper integral is well-defined; that is, Does it matter the way in which we build up the integral via a limit? For example, given a function $f(x)$ on $\mathbb{R}$, will it be true that

$$\lim_{L\to\infty} \int_{-L}^{L} f(x)\, dx = \lim_{L\to\infty} \int_{-L}^{L^2} f(x)\, dx = \lim_{L\to\infty} \left[\int_{-L}^{-\frac{1}{L}} f(x)\, dx + \int_{\frac{1}{L^2}}^{\sqrt{L}} f(x)\, dx \right] ?$$

[3]Note that we are integrating over a surface with respect to surface area and hence the dimensions of the integrand get multiplied by the dimension of area.

For functions of, say, two or three variables, one can only imagine the infinite possibilities for building up an improper double or triple integral via limits. An important fact is that if the function f has only one sign, for example $f(\mathbf{x}) \geq 0$ for all $\mathbf{x}$, then there are only two possibilities for an improper integral: **Either** all possible limits tend to the same finite number **or** all possible limits tend to $+\infty$. For positive integrands there are no cancellation effects in the integration, and thus regardless of how we choose to build up the integral (i.e., which limit we take), we attain the same value. Thus for positive (or negative) functions, an improper integral is well-defined with an unambiguous value. This motivates the following definition of an important property pertaining to all functions, including those with singularities: **integrability**.

Definition of an Integrable Function

Definition A.4.1. A function f defined on a bounded or unbounded domain Ω in $\mathbb{R}^N$ is **integrable on** Ω if and only if the integral

$$\int \cdots \int_{\Omega} |f(\mathbf{x})|\, d\mathbf{x}$$

exists and is a finite number.

We often write the statement above as

$$\int \cdots \int_{\Omega} |f(\mathbf{x})|\, d\mathbf{x} < \infty. \tag{A.7}$$

If Ω is the entire space $\mathbb{R}^N$, then we just say the function is **integrable**.

Note the presence of the absolute value sign in (A.7); in general, a function can be both positive and negative. Improper integrals of functions which can be both positive and negative are far more subtle because of possible cancellation effects in the integration process. In certain cases, these cancellation effects can lead to ambiguities; different limits can lead to different answers. Here is a good example where $N = 1$, $\Omega = \mathbb{R}$, and $f(x) = \frac{1}{x}$. Would you say

$$\int_{-\infty}^{\infty} \frac{1}{x}\, dx < \infty \quad \text{or} \quad \int_{-\infty}^{\infty} \frac{1}{x}\, dx = \infty \quad \text{or} \quad \int_{-\infty}^{\infty} \frac{1}{x}\, dx = -\infty\ ?$$

On one hand, we should be able to capture this integral via the limit

$$\lim_{L\to\infty} \left(\int_{-L}^{-\frac{1}{L}} \frac{1}{x}\, dx + \int_{\frac{1}{L}}^{L} \frac{1}{x}\, dx \right).$$

Since $\frac{1}{x}$ is an odd function, for each $L > 1$ the integrals in the parentheses cancel out, and hence one would be tempted to conclude the limit is 0. On the other hand the same integral should be achieved by

$$\lim_{L\to\infty} \left(\int_{-L}^{-\frac{1}{L}} \frac{1}{x}\, dx + \int_{\frac{1}{L}}^{L^2} \frac{1}{x}\, dx \right). \tag{A.8}$$

In this case we find for any $L > 1$,

$$\begin{aligned}
\int_{-L}^{-\frac{1}{L}} \frac{1}{x}\,dx + \int_{\frac{1}{L}}^{L^2} \frac{1}{x}\,dx &= \int_{-L}^{-1} \frac{1}{x}\,dx + \int_{-1}^{-\frac{1}{L}} \frac{1}{x}\,dx + \int_{\frac{1}{L}}^{1} \frac{1}{x}\,dx + \int_{1}^{L^2} \frac{1}{x}\,dx \\
&= \int_{-L}^{-1} \frac{1}{x}\,dx + \int_{1}^{L^2} \frac{1}{x}\,dx \\
&= -\ln L + \ln L^2 \\
&= -\ln L + 2\ln L = \ln L,
\end{aligned}$$

and the limit in (A.8) would appear to be $+\infty$. It is even worse than this! For any number $\alpha \in [-\infty, \infty]$, one can find a limit (a way of calculating this integral) in such a way that it results in the conclusion that the integral is α. Yes, bizarre indeed, and a consequence of two facts:

- $\frac{1}{x}$ is **not** integrable; no matter which limit you take to build up $\int_{-\infty}^{\infty} \frac{1}{|x|}\,dx$, you get $+\infty$.
- The function is negative for $x < 0$ and positive for $x > 0$, and on each of these subdomains the area between the curve and the x-axis is $+\infty$. Hence an integral over the entire real line forces us to interpret the ambiguous "$\infty - \infty$".

This simple example is indeed worrisome but here is the punchline:

Given any function, it is exactly integrability which ensures that the improper integral (A.6) *is well-defined in the sense that all limits based upon building up the integral will converge to the same number. Integrals of nonintegrable functions are certainly important in harmonic analysis, PDEs, and, indeed, physics, but they need to be handled with great care. In the vast majority of this text, all of our functions will be either integrable or locally integrable (see below).*

One does need to turn to Lebesgue theory and the Lebesgue integral to truly appreciate and provide a simple proof of this statement. Indeed, given that any Riemann integral is itself a limit of partial sums, it may seem overkill to introduce another limit for the definition of an improper integral. The Lebesgue theory cleans all this up beautifully, and mathematically oriented students are encouraged to venture into it as soon as possible.

For our purposes, there are two ways a function f may **fail** to be integrable, i.e., two ways we can have

$$\int \cdots \int_{\Omega} |f(\mathbf{x})|\,d\mathbf{x} = +\infty.$$

Either (i) Ω is unbounded, and as $|\mathbf{x}| \to \infty$, $|f|$ does not decay "fast enough" to zero; **and/or** (ii) $|f|$ has a singularity at some $\mathbf{x}_0 \in \Omega$ wherein the function blows up "too slowly" as $\mathbf{x} \to \mathbf{x}_0$. To this end, it is instructive to recall integrability in 1D of the class of rational functions $f(x) = \frac{1}{|x|^p}$ for a fixed $p > 0$. Here, the larger the p, the faster the decay to 0 as $|x| \to \infty$ (issue (i)) but also the slower the blow-up to $+\infty$ as $x \to 0$ (issue (ii)). In particular, there is no value of $p > 0$ for which the function is integrable over all of $\mathbb{R}$. The reader should reflect (cf. Exercise **A.3**) on the analogous behavior for $f(\mathbf{x}) = \frac{1}{|\mathbf{x}|^p}, \mathbf{x} \in \mathbb{R}^N$.

It is often convenient to focus only on issue (ii) by considering the larger class of functions which are **locally integrable**.

Definition of a Locally Integrable Function

Definition A.4.2. A function $f : \mathbb{R}^N \to \mathbb{R}$ is **locally integrable** if and only if for every $K \subset \mathbb{R}^N$ which is closed and bounded, we have

$$\int \cdots \int_K |f(\mathbf{x})| \, d\mathbf{x} < \infty. \tag{A.9}$$

For example, $f(x)$ is locally integrable on $\mathbb{R}$ if and only if for any $R > 0$, $\int_{-R}^{R} |f(x)| \, dx < \infty$. Thus the function $f(x) \equiv 1$ is locally integrable but not integrable on all of $\mathbb{R}$.

A similar definition can be made for local integrability of a function on a bounded or unbounded domain Ω. Here we require (A.9) to hold true for all $K \subset \Omega$ which are closed (in Ω) and bounded.

A.5. Evaluation and Manipulation of Integrals: Exploiting Radial Symmetry

You may recall, from third- and fourth-semester calculus, the tedious evaluation of volume, surface, and line integrals. These often involved different techniques of integration, change of coordinates, complicated parametrization of regions of integration, etc. To this end, the **good news** is you will **not** need to do any such tedious calculations in this text. The **bad news** is that, despite having many such calculations under your belt, you may lack, as yet, a certain **grounding** in exactly what volume, surface, and line integrals signify and how to **work with them**. This will be the case if you have been guided by symbolic manipulation of objects and symbols rather than geometric (and physical) intuition and reasoning.

In this text, it is our hope that regardless of your past experience, you will eventually develop a certain grounding in integration, based upon geometric and physical reasoning. It is important to realize now that many of the integrals which we actually need to evaluate and/or analyze will pertain to situations which exhibit **radial symmetry**. In particular:

- The domain in question will be radially symmetric, e.g., a ball, sphere, or the entire space.
- In certain cases, the function $f(\mathbf{x})$ in question will also be radially symmetric and hence will only depend on $|\mathbf{x}|$.

Evaluation and manipulation of such integrals is straightforward, with the caveat that you have some underlying geometric and physical intuition behind integration. In fact, in many cases, the actual integration steps will degenerate into simple multiplication.

In this section, we begin by recalling spherical coordinates in $\mathbb{R}^3$ which consist of a radius r and two angles θ and ϕ. We then make an important observation; for most of our purposes **the explicit dependence** on the angles θ and ϕ is **irrelevant**. This will also often be the case when integrating functions which are **not** radially symmetric: These angular variables are of course implicitly there, but for our purposes, we will not need to explicitly deal with them. Hence, while we conveniently pose most of our results in one, two, or three space dimensions, this type of analysis will naturally generalize to all space dimensions. The following subsections should help illustrate these points.

A.5.1. Spherical (Polar) Coordinates in $\mathbb{R}^3$. Instead of describing points in space via their Cartesian coordinates x, y, and z or $\mathbf{x} = (x_1, x_2, x_3)$, we use the variables r, θ, and ϕ. The variable r (often also denoted by ρ) simply denotes the distance of the point from the origin $\mathbf{0}$. Thus, spherical coordinates are very origin focused[4]. The other two variables[5] are angles best explained with reference to the diagram in Figure A.5.

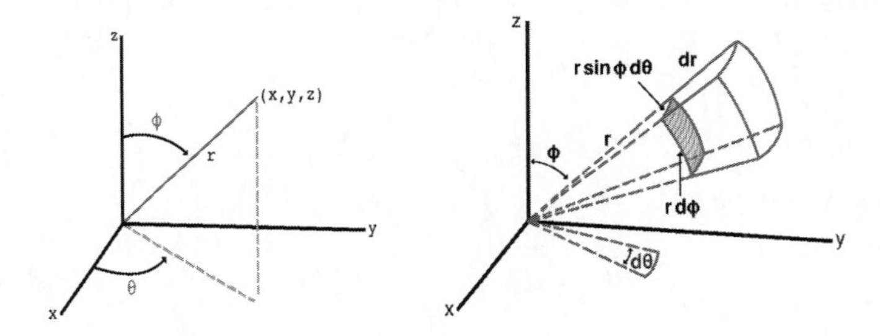

Figure A.5. Left: Spherical coordinates (r, θ, ϕ) and Cartesian coordinates (x, y, z). Right: The spherical rectangular volume spanned out by $dr, d\theta, d\phi$ at the point (r, θ, ϕ). Its volume is $dV = r^2 \sin\phi \, dr \, d\theta \, d\phi$. The spherical rectangular area spanned out by $d\theta, d\phi$ (fixing r) at the point (r, θ, ϕ) is $dS = r^2 \sin\phi \, d\theta \, d\phi$. This area is shaded.

The variable r tells us which sphere (centered at the origin) the point is on, and, analogous to latitude and longitude coordinates on earth, (θ, ϕ) tell us where the point on the sphere is. Note the ranges for r, θ, and ϕ are $0 \le r < \infty, 0 \le \theta \le 2\pi, 0 \le \phi \le \pi$. While r has dimensions of length, the other two coordinates θ and ϕ are dimensionless.

When transforming a triple integral (i.e., a bulk volume integral) over some solid region into spherical coordinates one must do the following:

(i) Describe (parametrize) the region in terms of r, θ, ϕ.

(ii) Write the integrand in terms of r, θ, ϕ.

(iii) Write the volume increment dV in terms of $dr \, d\theta \, d\phi$.

For step (iii), we note that in rectangular coordinates, dV is simply $dxdydz$. In spherical coordinates, it is not the case that dV is simply $drd\theta d\phi$. Note that dx, dy, and dz all represent small increments in the respective spatial variables, and hence $dxdydz$

[4]You can of course always make a translation to make any point the origin.

[5]**Warning:** In the physics literature the roles of θ and ϕ are reversed.

has dimensions of length cubed (volume). On the other hand, $d\theta$ and $d\phi$ represent increments in angles and therefore are dimensionless, giving the product $dr\,d\theta\,d\phi$ dimensions of length. By considering the volume of the spherical region spanned out by perturbing a point r, θ, ϕ by dr, $d\theta$, and $d\phi$ (cf. Figure A.5, right), one finds[6] that

$$dV = r^2 \sin\phi\, dr\, d\theta\, d\phi.$$

Note that the right-hand side does indeed have dimensions of volume. For a surface integral over a sphere of radius r (centered at the origin), the surface area increment swept out on the sphere of radius r by perturbing a point r, θ, ϕ by $d\theta$ and $d\phi$ is given by

$$dS = r^2 \sin\phi\, d\theta\, d\phi.$$

A.5.2. Integration of a Radially Symmetric Function. A radially symmetric function $f(\mathbf{x})$ is one which only depends on $|\mathbf{x}|$; thus, it can be written in spherical coordinates as $g(r)$ for some function g of one variable. Here, let us use $\mathbf{x}$ instead of (x, y, z).

Bulk Integrals: It is straightforward to integrate a radially symmetric function over a radially symmetric region. For example, let us compute the bulk integral over $\Omega = B(\mathbf{0}, a)$

$$\iiint_{B(\mathbf{0},a)} f(\mathbf{x})\, d\mathbf{x}.$$

In spherical coordinates this becomes

$$\int_0^{\pi}\int_0^{2\pi}\int_0^{a} g(r) r^2 \sin\phi\, dr\, d\theta\, d\phi = \int_0^a r^2 g(r)\, dr \left(\int_0^{\pi}\int_0^{2\pi} \sin\phi\, d\theta\, d\phi\right).$$

The integral in the parentheses on the right is simply the surface area of the unit sphere, which is 4π. Convince yourself of this, and you will, henceforth, be more comfortable working in spherical coordinates. Thus, we have

$$\boxed{\iiint_{B(\mathbf{0},a)} f(\mathbf{x})\, d\mathbf{x} = 4\pi \int_0^a r^2 g(r)\, dr.}$$

In other words, we have reduced the integration to a 1D integral.

For example, if we set $f(\mathbf{x}) = |\mathbf{x}|^p$ ($p > 0$), we have

$$\iiint_{B(\mathbf{0},a)} |\mathbf{x}|^p\, d\mathbf{x} = 4\pi \int_0^a r^2 r^p\, dr = 4\pi \int_0^a r^{p+2}\, dr = 4\pi \frac{a^{p+3}}{p+3}.$$

The beauty of this calculation is that it applies in any space dimension. For example, let $N > 3$ and let $B(\mathbf{0}, a)$ be the ball centered at the origin of radius a which lies in $\mathbb{R}^N$. Then, for a radial function $f(\mathbf{x}) = g(|\mathbf{x}|)$ defined on $\mathbb{R}^N$, we have

$$\int \cdots \int_{B(\mathbf{0},a)} f(\mathbf{x})\, d\mathbf{x} = \int_0^a r^{N-1} g(r)\, dr \cdot \left(\text{surface area of the unit sphere in } \mathbb{R}^N\right).$$

While the surface area of the unit sphere in dimensions $N = 1, 2, 3$ is clear, there is also a simple formula in N dimensions involving the **Gamma function** Γ.

[6] More precisely, this amounts to computing the Jacobian associated with the change of variables.

Surface Integrals: Suppose we wish to perform a surface integral whereby we integrate a radially symmetric function f over the sphere of radius a (the boundary of the ball $B(\mathbf{0}, a)$). Since the function will be constant on the sphere, we have

$$\iiint_{\partial B(\mathbf{0},a)} f(\mathbf{x})\, dS = g(a) \iiint_{\partial B(\mathbf{0},a)} 1\, dS = 4\pi a^2 g(a).$$

Note that there is no need to explicitly introduce all the spherical coordinates here though if we did, it would lead to the same answer:

$$\iiint_{\partial B(\mathbf{0},a)} f(\mathbf{x})\, dS = \int_0^{\pi} \int_0^{2\pi} g(a)\, a^2 \sin\phi\, d\theta\, d\phi = 4\pi a^2 g(a),$$

as the area element $dS = a^2 \sin\phi\, d\theta\, d\phi$ on the sphere of radius a.

We present two more examples.

Example A.5.1. We address the integrability of the function $\frac{1}{|\mathbf{x}|^p}$ for $p > 0$ on $\mathbb{R}^3$. Note that there are two issues to address: the blow-up discontinuity at the origin and its rate of decay as $|\mathbf{x}| \to \infty$. To separate these two effects, we break up $\mathbb{R}^3$ into the unit ball $B(\mathbf{0}, 1)$ in $\mathbb{R}^3$ and its complement

$$B^c(\mathbf{0}, 1) = \left\{ \mathbf{x} \in \mathbb{R}^3 \,\middle|\, |\mathbf{x}| \geq 1 \right\}.$$

We ask, For which values of $p > 0$ are the integrals

$$\iiint_{B(\mathbf{0},1)} \frac{1}{|\mathbf{x}|^p}\, d\mathbf{x} \qquad \text{and} \qquad \iiint_{B^c(\mathbf{0},1)} \frac{1}{|\mathbf{x}|^p}\, d\mathbf{x} \qquad \text{finite?}$$

To this end, we have for $p \neq 3$,

$$\iiint_{B(\mathbf{0},1)} \frac{1}{|\mathbf{x}|^p}\, d\mathbf{x} = 4\pi \int_0^1 \frac{r^2}{r^p}\, dr = 4\pi \int_0^1 r^{2-p}\, dr = 4\pi \left(\frac{r^{3-p}}{3-p} \right) \Big|_0^1,$$

which is finite if and only if $3 - p > 0$; in other words, $p < 3$. On the other hand,

$$\iiint_{B^c(\mathbf{0},1)} \frac{1}{|\mathbf{x}|^p}\, d\mathbf{x} = 4\pi \int_1^{\infty} \frac{r^2}{r^p}\, dr = 4\pi \int_1^{\infty} r^{2-p}\, dr = 4\pi \left(\frac{r^{3-p}}{3-p} \right) \Big|_1^{\infty},$$

which is finite if and only if $3 - p < 0$; in other words, $p > 3$. The critical case of $p = 3$ causes both integrals to blow up. It is a good exercise to repeat these calculations in dimension N.

Example A.5.2. We compute the flux integral

$$\iint_{\partial\Omega} \mathbf{F} \cdot \mathbf{n}\, dS \qquad \text{where} \qquad \mathbf{F}(\mathbf{x}) = \mathbf{x} \quad \text{and} \quad \Omega = B(\mathbf{0}, a),$$

with the outer normal. This is actually a very simple integral to evaluate. Indeed, since the outer normal at a point $\mathbf{x}$ on the sphere $\partial B(\mathbf{0}, a)$ is simply

$$\mathbf{n} = \frac{\mathbf{x}}{|\mathbf{x}|} = \frac{\mathbf{x}}{a},$$

we have

$$\mathbf{F} \cdot \mathbf{n} = \mathbf{x} \cdot \frac{\mathbf{x}}{|\mathbf{x}|} = \frac{|\mathbf{x}|^2}{a} = a.$$

Hence,

$$\iint_{\partial B(\mathbf{0},a)} \mathbf{F}\cdot\mathbf{n}\,dS \;=\; a\iint_{\partial B(\mathbf{0},a)} 1\,dS \;=\; a\,4\pi a^2 \;=\; 4\pi a^3.$$

A.5.3. Integration of General Functions over a Ball via Spherical Shells. When integrating over a radially symmetric region, like a ball of radius r centered at any point $\mathbf{x}_0 \in \mathbb{R}^3$, a very useful way of building up the integral is via spherical shells. Yes, it is true that this is directly related to using spherical coordinates (which we just discussed). However, even when the function is **not** radially symmetric, we will rarely be concerned with the details of the θ and ϕ integration, and hence it is convenient to simply write the shell (polar) decomposition as follows: If f is any integrable function, then one has

$$\iiint_{B(x_0,r)} f(\mathbf{x})\,d\mathbf{x} \;=\; \int_0^r \left(\iint_{\partial B(x_0,s)} f(\mathbf{x})\,dS\right) ds. \tag{A.10}$$

Here, the inner integral is a surface integral over the sphere $\partial B(x_0, s)$. If f is constant on the sphere, it would simply be that constant times $4\pi s^2$. If f was not constant along the sphere and one did in fact want to evaluate the integral, then one could parametrize the sphere with the two spherical coordinates θ and ϕ, here, translating the origin to the center x_0 and writing f in terms of s, θ, and ϕ. However, note one important consequence of (A.10) which is a sort of “Fundamental Theorem of Calculus for Integrals over a Ball”:

$$\boxed{\frac{d}{dr}\left(\iiint_{B(x_0,r)} f(\mathbf{x})\,d\mathbf{x}\right) \;=\; \iint_{\partial B(x_0,r)} f(\mathbf{x})\,dS.} \tag{A.11}$$

The statements (A.10) and (A.11) have direct analogues in any space dimension N.

A.5.4. Rescalings and Translations. For this subsection, you should recall from advanced calculus the basic change of variables formula and the role of the Jacobian. We will sometimes need to perform simple rescalings and translations (change of variables). For example, when one has an integral over a ball (or sphere) of radius a centered at some point $\mathbf{x}_0$, it is often convenient to shift and rescale to an integral over the unit ball (or sphere) centered at the origin. This would, for example, facilitate the use of spherical coordinates.

For example, for the triple integral

$$\iiint_{B(\mathbf{x}_0,a)} f(\mathbf{x})\,d\mathbf{x},$$

we can make the change of variable

$$\mathbf{y} = \frac{\mathbf{x}-\mathbf{x}_0}{a}.$$

In the $\mathbf{y}$ coordinates the region of integration becomes the unit ball centered at the origin and, since we are in 3D, the volume increment changes by[7]

$$d\mathbf{y} \;=\; \frac{1}{a^3}\,d\mathbf{x}.$$

[7]This may be a good time to review **the change of variables formula** involving the **Jacobian**, computing the Jacobian in this simple example.

Thus

$$\iiint_{B(\mathbf{x}_0,a)} f(\mathbf{x})\, d\mathbf{x} \;=\; \iiint_{B(\mathbf{0},1)} f(a\mathbf{y}+\mathbf{x}_0)\, a^3\, d\mathbf{y}.$$

On the other hand, suppose we were given the surface integral over the sphere of radius a centered at $\mathbf{x}_0$

$$\iint_{\partial B(\mathbf{x}_0,a)} f(\mathbf{x})\, dS.$$

Then we can make the same change of variables, transforming the region of integration to the unit sphere centered at the origin. However, now the area increment will change by (cf. Exercise **A.9**)

$$dS_{\mathbf{y}} \;=\; \frac{1}{a^2}\, dS_{\mathbf{x}}. \tag{A.12}$$

Here we use $dS_{\mathbf{y}}$ to denote the area increment over the unit sphere which is parametrized with variable $\mathbf{y}$, and $dS_{\mathbf{x}}$ for the sphere of radius a which is parametrized with variable $\mathbf{x}$. Thus

$$\iint_{\partial B(\mathbf{x}_0,a)} f(\mathbf{x})\, dS_{\mathbf{x}} \;=\; \iint_{\partial B(\mathbf{0},1)} f(a\mathbf{y}+\mathbf{x}_0)\, a^2\, dS_{\mathbf{y}}.$$

A.6. Fundamental Theorems of Calculus: The Divergence Theorem, Integration by Parts, and Green's First and Second Identities

Note: The results of this section all pertain to a **bounded** domain.

A.6.1. The Divergence Theorem. For simplicity of notation, let $N = 3$. Our basic object is a three-dimensional vector field defined on three-dimensional space

$$\mathbf{F}(x,y,z) = (F_1(x,y,z), F_2(x,y,z), F_3(x,y,z)).$$

We define the **divergence** of $\mathbf{F}$ to be[8]

$$\operatorname{div} \mathbf{F} \;=\; \frac{\partial F_1}{\partial x} + \frac{\partial F_2}{\partial y} + \frac{\partial F_3}{\partial z}.$$

In words, to compute the divergence, we differentiate each i-th component function with respect to the i-th coordinate and then add the results together. This may seem like a naive choice for a combination of derivatives; for example, why do we not include $\frac{\partial F_2}{\partial x}$ or $\frac{\partial F_1}{\partial y}$? Why this combination of derivatives? The answer is provided by the Divergence Theorem, which you can regard as **the Fundamental Theorem of Multivariable Calculus**.

[8]Physicists often adopt the following more suggestive notation for the divergence. Think of the gradient as a vector-valued operator called **del**,

$$\nabla \;=\; \left(\frac{\partial}{\partial x}, \frac{\partial}{\partial y}, \frac{\partial}{\partial z}\right).$$

By an operator we mean something which acts on functions and produces another function (here vector-valued). We may write the divergence as the dot product of this operator and the vector field $\mathbf{F}$:

$$\operatorname{div} \mathbf{F} \;=\; \nabla \cdot \mathbf{F}.$$

The Divergence Theorem

Theorem A.1. *Let* **F** *be a smooth vector field on a bounded domain* Ω *with outer normal* **n**. *Then*

$$\iiint_\Omega \operatorname{div} \mathbf{F}\, dx\, dy\, dz \;=\; \iint_{\partial\Omega} \mathbf{F}\cdot\mathbf{n}\, dS.$$

Recall that the surface integral above has a very important interpretation: It denotes the *flux* of the vector field **F** out of Ω. Hence, the Divergence Theorem can be restated by saying that at any point (x, y, z), div $\mathbf{F}(x, y, z)$ is the instantaneous amount per unit volume that the vector field **F** is "diverging" from (x, y, z).

If div $\mathbf{F}(x, y, z)$ is positive, then the vector field is instantaneously "diverging" **out of** the point (x, y, z), a **source** point. If it is negative, then the vector field is "diverging" (or rather "converging") **into** the point (x, y, z), a **sink** point. Recall that the dimensions of flux are

$$\frac{\text{the dimension for the amount of the quantity}}{\text{time} \times \text{length}^2}.$$

Hence, the dimensions of div **F** have an additional power of length in the denominator. In other words, div **F** measures the amount of change in the quantity per unit time **per unit volume**.

A.6.2. Two Consequences of the Divergence Theorem: Green's Theorem and a Componentwise Divergence Theorem. You will also recall **Green's Theorem** associated with a bounded domain Ω in the plane with boundary curve $\partial\Omega = \mathcal{C}$ oriented counterclockwise:

Theorem A.2 (Green's Theorem). *Let* $P(x, y)$ *and* $Q(x, y)$ *be smooth functions on a bounded 2D domain* Ω. *Then*

$$\iint_\Omega \frac{\partial Q(x,y)}{\partial x} - \frac{\partial P(x,y)}{\partial y}\, dx\, dy \;=\; \int_{\mathcal{C}} P(x,y)dx \,+\, Q(x,y)dy.$$

As an easy exercise (cf. Exercise **A.1**), you should derive Green's Theorem from the 2D Divergence Theorem by considering an appropriate vector field.

By applying the Divergence Theorem to a vector field which has only nonzero components in the i-th position, we readily obtain the following **Componentwise Divergence Theorem**:

Theorem A.3 (Componentwise Divergence Theorem). *Given a smooth function* $f = f(\mathbf{x})$ *on a bounded domain* $\Omega \subset \mathbb{R}^3$, *we have for any* $i \in \{1, 2, 3\}$,

$$\iiint_\Omega f_{x_i}(\mathbf{x})\, d\mathbf{x} \;=\; \iint_{\partial\Omega} f\, n_i\, dS, \tag{A.13}$$

where n_i *denotes the i-th component of the outer unit normal* **n**.

Theorem A.3 can easily be stated in any space dimension.

A.6.3. A Match Made in Heaven: The Divergence + the Gradient = the Laplacian. According to the Divergence Theorem, the divergence of a vector field plays a key role in computing a flux integral. On the other hand, given any function $u : \mathbb{R}^N \to \mathbb{R}$, the gradient of u also plays a key role. So let us couple the two! That is, let us compute the divergence of a gradient:

$$\operatorname{div} \nabla u = \operatorname{div}(u_{x_1}, \ldots, u_{x_N}) = \nabla \cdot \nabla u = \sum_{i=1}^{N} u_{x_i x_i}.$$

We call this combination of second-order partial derivatives the **Laplacian** of u. Mathematicians denote it by Δu whereas physicists tend to use the more suggestive notation $\nabla^2 u$, which really stands for $\nabla \cdot \nabla u$.

A.6.4. Integration by Parts and Green's First and Second Identities. Integration by parts is a core tool in the analysis of PDEs. Recall from your first-year calculus class that if f and g are smooth functions of one variable x, then

$$\boxed{\int_a^b f(x)\, g'(x)\, dx = \left[f(x)g(x)\right]\Big|_a^b - \int_a^b f'(x)\, g(x)\, dx.} \tag{A.14}$$

We refer to the application of such a formula as *integration by parts*. It is a direct consequence of the product rule for differentiation. Here, $[f(x)g(x)]_a^b$ stands for $f(b)g(b) - f(a)g(a)$ which one can think of as the evaluation of the product fg on the boundary of the interval (a, b). So (A.14) states that in terms of integration, we can place the derivative in the product $f(x)\,g'(x)$ on f instead of g at the expense of (i) a minus sign and (ii) an additional boundary term. Note that there are two components to deriving formula (A.14)[9]:

- the product rule in differentiation,
- the Fundamental Theorem of Calculus.

The (one independent variable) integration by parts simple formula (A.14) has many analogues for functions of several dimensions. The analogous components to all their derivations are the following:

- the product rule in differentiation applied to each variable,
- the Divergence Theorem.

We present three such formulas. The first involves a vector field. Consider a C^1 vector field $\mathbf{u} = (u^1, u^2, u^3)$ and a C^1 function v **defined on** some domain $\Omega \subset \mathbb{R}^3$, which are continuous on $\overline{\Omega}$. Here we use superscripts for the components of $\mathbf{u}$ to avoid confusion with the subscript notation for the partial derivatives. By the product rule, we have for each $i = 1, 2, 3$,

$$(u^i v)_{x_i} = u^i v_{x_i} + u^i_{x_i} v.$$

Summing over i gives

$$\sum_{i=1}^{3} (u^i v)_{x_i} = \sum_{i=1}^{3} u^i v_{x_i} + \sum_{i=1}^{3} u^i_{x_i} v$$

[9]The reader is strongly encouraged to perform this derivation now before reading on.

and hence

$$\sum_{i=1}^{3} u^i v_{x_i} = -\sum_{i=1}^{3} u^i_{x_i} v + \sum_{i=1}^{3} (u^i v)_{x_i}.$$

By definition of the divergence and gradient this means

$$\mathbf{u} \cdot \nabla v = -(\operatorname{div} \mathbf{u})\, v + \operatorname{div}(\mathbf{u}v).$$

We integrate over Ω to find

$$\iiint_{\Omega} \mathbf{u} \cdot \nabla v \, d\mathbf{x} = -\iiint_{\Omega} (\operatorname{div} \mathbf{u})\, v \, d\mathbf{x} + \iiint_{\Omega} \operatorname{div}(\mathbf{u}v)\, d\mathbf{x}.$$

Finally, we apply the Divergence Theorem to the last integral in order to obtain the necessary calculus equality/integration by parts formula:

Vector Field Integration by Parts Formula

$$\iiint_{\Omega} \mathbf{u} \cdot \nabla v \, d\mathbf{x} = -\iiint_{\Omega} (\operatorname{div} \mathbf{u})\, v \, d\mathbf{x} + \iint_{\partial\Omega} (v\mathbf{u}) \cdot \mathbf{n} \, dS. \qquad (A.15)$$

Here $\mathbf{n}$ denotes the outer unit normal to $\partial\Omega$.

Note that this is a generalization of integration by parts to $3D$: We can place the derivatives involved in ∇v onto the vector field $\mathbf{u}$ (which amounts to taking its divergence) at the expense of boundary terms. The analogous statement holds in any other space dimension. We stress that the sole ingredients involved in proving this equality were the product rule and the Divergence Theorem.

The next two integration by parts formulas involve only scalar functions and are known, respectively, as **Green's First and Second Identities**. They pertain to the integration of $(\Delta u)v = (\operatorname{div} \nabla u)\, v$ over a bounded domain Ω. As in (A.14), we would like to place the divergence derivatives on v at the expense of an additional boundary term. By writing the divergence in terms of the partial derivatives and applying the integration by parts (A.14) in each variable, the derivatives will be placed on v in the form of $-\nabla v \cdot \nabla u$ (check this!). However, what about the boundary terms? Now the boundary is a set $\partial\Omega$. The key here is to use the Divergence Theorem.

Let $N = 3$ (for simplicity of notation), $\mathbf{x} = (x_1, x_2, x_3)$, and $u, v \in C^2(\mathbb{R}^3)$. Then the product rule implies that for each $i = 1, 2, 3$

$$\left(v u_{x_i}\right)_{x_i} = v_{x_i} u_{x_i} + v u_{x_i x_i}.$$

If we now sum over $i = 1, 2, 3$, we find

$$\operatorname{div}(v\nabla u) = \nabla v \cdot \nabla u + v\Delta u.$$

Integrating both sides over a bounded domain Ω and using the Divergence Theorem on the left, we find

$$\iint_{\partial\Omega} v\nabla u \cdot \mathbf{n} \, dS = \iiint_{\Omega} \nabla u \cdot \nabla v \, d\mathbf{x} + \iiint_{\Omega} v\Delta u \, d\mathbf{x}.$$

Writing the left-hand side as

$$\iint_{\partial\Omega} v \frac{\partial u}{\partial \mathbf{n}} \, dS,$$

we arrive at the following:

Green's First Identity

$$\iiint_{\Omega} v\Delta u \, d\mathbf{x} = \iint_{\partial\Omega} v\frac{\partial u}{\partial \mathbf{n}}\, dS - \iiint_{\Omega} \nabla u \cdot \nabla v \, d\mathbf{x}. \tag{A.16}$$

One should think of this as an integration by parts formula for the $\Delta = \operatorname{div}\nabla$: The divergence on ∇u (on the left-hand side) can be "placed" on v in the form of a gradient (note the minus sign) together with some boundary term. The boundary term is $(v\nabla u)\cdot \mathbf{n}$ integrated over $\partial\Omega$.

Next, we apply (A.16) to u and v and then to v and u and subtract. In doing so, we find the following:

Green's Second Identity

$$\iiint_{\Omega} (v\Delta u - u\Delta v)\, d\mathbf{x} = \iint_{\partial\Omega} \left(v\frac{\partial u}{\partial \mathbf{n}} - u\frac{\partial v}{\partial \mathbf{n}}\right) dS. \tag{A.17}$$

In other words, Green's Second Identity states that we may interchange the Laplacian in the integrals

$$\iiint_{\Omega} u\,\Delta v\, d\mathbf{x} \qquad \text{or} \qquad \iiint_{\Omega} v\,\Delta u\, d\mathbf{x}$$

at the expense of certain boundary terms. Here, there are two derivatives transferred in "the integration by parts". Note that both identities hold true for any pair u, v which are defined in $\overline{\Omega}$ such that:

- $u \in C^2(\Omega)$ (so we can speak of their Laplacians).
- $u \in C^1(\overline{\Omega})$ (so we can speak of the boundary values of u, v and their derivatives).

A.7. Integral vs. Pointwise Results

We will often be faced with the situation of having information about integrals of a function from which we wish to infer pointwise information. To this end, we will benefit from the so-called IPW (integral to pointwise) theorems and the Averaging Lemma.

A.7.1. IPW (Integral to Pointwise) Theorems.

The following **IPW theorems** are useful in passing from equations involving integrals to equations involving functions which hold at each value of the independent variable. The proofs are left as exercises.

Theorem A.4 (One-Dimensional IPW Theorem (i)). *Suppose that f is a continuous function and that*

$$\int_a^b f(x)\, dx = 0$$

for every $a, b \in \mathbb{R}$ with $a < b$. Then we must have $f \equiv 0$; that is, $f(x) = 0$ for all $x \in \mathbb{R}$.

Alternatively, the result can be phrased in terms of integration against arbitrary functions (as opposed to integration over arbitrary domains):

Theorem A.5 (One-Dimensional IPW Theorem (ii)). *Suppose that f is a continuous function on $\mathbb{R}$ and that*

$$\int_{-\infty}^{\infty} f(x)\, g(x) dx = 0,$$

for every continuous function g on $\mathbb{R}$ with compact support. Then we must have $f \equiv 0$; that is, $f(x) = 0$ for all $x \in \mathbb{R}$.

The analogous theorems hold for any domain (possibly unbounded) in any space dimension.

Theorem A.6 (General IPW Theorem (i)). *Let $\Omega \subset \mathbb{R}^N$ be a domain and let $f(\mathbf{x}) \in C(\Omega)$ (i.e., f is a continuous function on Ω). Suppose that*

$$\int \cdots \int_{W} f(\mathbf{x})\, d\mathbf{x} = 0$$

for every bounded subdomain $W \subset \Omega$. Then we must have $f(\mathbf{x}) \equiv 0$ on Ω; i.e., $f(\mathbf{x}) = 0$ for all $\mathbf{x} \in \Omega$.

Theorem A.7 (General IPW Theorem (ii)). *Let $\Omega \subset \mathbb{R}^N$ be a domain and let $f(\mathbf{x}) \in C(\Omega)$ (i.e., f is a continuous function on Ω). Suppose that **for all** $g \in C(\Omega)$ with compact support, we have*

$$\int \cdots \int_{\Omega} f(\mathbf{x}) g(\mathbf{x})\, d\mathbf{x} = 0.$$

Then we must have $f(\mathbf{x}) \equiv 0$ on Ω; i.e., $f(\mathbf{x}) = 0$ for all $\mathbf{x} \in \Omega$.

For our purposes, the continuity assumptions on the functions in Theorems A.4–A.7 will suffice. They also make these theorems relatively straightforward to prove. However, with some measure theory and the notion of equality almost everywhere, one can relax these continuity assumptions. See **[17]** for more details.

A.7.2. The Averaging Lemma. The second result is based upon the notion of the average of a function. It also applies in any space dimension, but for convenience let us fix the dimension to be 3. Given a continuous function ϕ on $\mathbb{R}^3$ and a ball $B(\mathbf{0}, r)$ with spherical boundary $\partial B(\mathbf{0}, r)$, the average values of ϕ over the ball and the sphere are given by

$$\frac{3}{4\pi r^3} \iiint_{B(\mathbf{0},r)} \phi(\mathbf{x})\, d\mathbf{x} \qquad \text{and} \qquad \frac{1}{4\pi r^2} \iint_{\partial B(\mathbf{0},r)} \phi(\mathbf{x})\, dS, \quad \text{respectively.}$$

Note that in the former case we have divided the volume integral by $\frac{4}{3}\pi r^3$, the volume of the ball $B(\mathbf{0}, r)$, whereas in the latter we have divided the surface integral by $4\pi r^2$, the surface area of the sphere $\partial B(\mathbf{0}, r)$. What happens if we let r tend to 0, i.e., if we take averages over smaller and smaller sets?

Lemma A.7.1 (Averaging Lemma). *Suppose that ϕ is continuous on $\mathbb{R}^3$; then we have*

$$\phi(\mathbf{0}) = \lim_{r\to 0+} \frac{3}{4\pi r^3} \iiint_{B(\mathbf{0},r)} \phi(\mathbf{x})\, d\mathbf{x} = \lim_{r\to 0+} \frac{1}{4\pi r^2} \iint_{\partial B(\mathbf{0},r)} \phi(\mathbf{x})\, dS. \tag{A.18}$$

Note that the Averaging Lemma holds for any point $\mathbf{x}_0$ (not just the origin) by taking balls and spheres centered at $\mathbf{x}_0$. Let us provide an intuitive explanation for the second equality in (A.18) followed by a proof. The proof of the first equality is almost identical.

Intuitive Explanation: In either case we are considering an integral over a region (either a ball of radius r or a sphere of radius r) which is small and encompasses the origin. Since ϕ is a continuous function, on any such small region, its variation from $\phi(\mathbf{0})$ is small. In other words, for r small, $\phi(\mathbf{x})$ is "close to" $\phi(\mathbf{0})$ for $\mathbf{x}$ in either $B(\mathbf{0}, r)$ or $\partial B(\mathbf{0}, r)$. Hence, focusing on the sphere, we have some *very small* fixed r_0:

$$\begin{aligned}
\lim_{r\to 0+} \frac{1}{4\pi r^2} \iint_{\partial B(\mathbf{0},r)} \phi(\mathbf{x})\, dS \sim \frac{1}{4\pi r_0^2} \iint_{\partial B(\mathbf{0},r_0)} \phi(\mathbf{x})\, dS \quad &\sim \quad \frac{1}{4\pi r_0^2} \iint_{\partial B(\mathbf{0},r_0)} \phi(\mathbf{0})\, dS \\
&= \phi(\mathbf{0}) \frac{1}{4\pi r_0^2} \iint_{\partial B(\mathbf{0},r_0)} 1\, d\mathbf{x} \\
&= \phi(\mathbf{0}) \frac{1}{4\pi r_0^2} 4\pi r_0^2 = \phi(\mathbf{0}),
\end{aligned}$$

where we loosely use the symbol $\sim$ to denote "approximately equals".

Now let us turn this intuition into a proper proof by using the continuity of our function to effectively control the function's fluctuations from $\phi(\mathbf{0})$ on a very small sphere.

Proof. Let $\epsilon > 0$ and note that $\phi(\mathbf{0}) = \frac{1}{4\pi r^2} \iint_{\partial B(\mathbf{0},r)} \phi(\mathbf{0})\, dS$. We want to show that there exists $\delta > 0$ such that if $r < \delta$, then

$$\left| \frac{1}{4\pi r^2} \iint_{\partial B(\mathbf{0},r)} \phi(\mathbf{x})\, dS - \phi(\mathbf{0}) \right| = \left| \frac{1}{4\pi r^2} \iint_{\partial B(\mathbf{0},r)} (\phi(\mathbf{x}) - \phi(\mathbf{0}))\, dS \right| < \epsilon.$$

To this end, by the continuity of ϕ there exists a $\delta > 0$ such that if $|\mathbf{x}| < \delta$, then $|\phi(\mathbf{x}) - \phi(\mathbf{0})| < \epsilon$. Hence if $r < \delta$, we have

$$\begin{aligned}
\left| \frac{1}{4\pi r^2} \iint_{\partial B(\mathbf{0},r)} (\phi(\mathbf{x}) - \phi(\mathbf{0}))\, dS \right| &\le \frac{1}{4\pi r^2} \iint_{\partial B(\mathbf{0},r)} |\phi(\mathbf{x}) - \phi(\mathbf{0})|\, dS \\
&< \frac{1}{4\pi r^2} \iint_{\partial B(\mathbf{0},r)} \epsilon\, dS \\
&= \epsilon \frac{1}{4\pi r^2} \iint_{\partial B(\mathbf{0},r)} 1\, dS \\
&= \epsilon \frac{1}{4\pi r^2} 4\pi r^2 \\
&= \epsilon.
\end{aligned}$$

□

We make a few remarks.

- The Averaging Lemma applies in any space dimension. In each case the average entails division by the appropriate "size" of the (hyper)ball or (hyper)sphere.
- One might wonder if there is something special about balls and spheres. The answer is no. For example, the analogous result holds for averaging over smaller and smaller cubes.
- It is instructive to recall the Fundamental Theorem of Calculus and to think of the Averaging Lemma as a result about *differentiation*. Indeed, one can view these limits as "derivatives" of an integral.
- The Averaging Lemma is very robust in that there is a version which applies to "pretty much every" function (not only continuous functions). Students interested in mathematical analysis should look up online an amazing result: the *Lebesgue Differentiation Theorem.*

A.8. Convergence of Functions and Convergence of Integrals

Given a sequence $\{f_n(x)\}_{n=1}^{\infty}$ of functions defined on $\mathbb{R}$ and another function $f(x)$, we say the sequence of functions $f_n(x)$ **converges pointwise** to $f(x)$ if, for any fixed x, the sequence of numbers $f_n(x)$ converge to the number $f(x)$. So, for example,

$$f_n(x) = \begin{cases} 1 & \text{if } x \in [n, n+1], \\ 0 & \text{otherwise} \end{cases} \tag{A.19}$$

converges pointwise to $f(x) \equiv 0$. You should convince yourself of this; for any fixed x, say, x_0, there will always be a cutoff point for n, say, n_0, after which $f_n(x_0)$ is always 0. In other words, $f_n(x_0) = 0$ for $n \geq n_0$.

There are many other notions of convergence of functions, for example,

- uniform convergence,
- almost everywhere convergence,
- convergence in the sense of distributions,
- L^p convergence,
- convergence in some Sobolev space.

For the vast majority of this text we will only deal with pointwise convergence and convergence in the sense of distributions. Convergence in L^p for $p = 1, 2$ will also at times be relevant and will be introduced as needed. Many of these notions of convergence are associated with convergence of integrals. To this end, let us now document a useful result based upon a very natural question. Let $I \subset \mathbb{R}$ be a (possibly infinite) interval, and let $f_n(x)$ be a sequence of functions defined on I which converges pointwise on I to a function $f(x)$. When can we conclude that

$$\int_I f_n(x)\,dx \quad \text{converges to} \quad \int_I f(x)\,dx?$$

Note that, since the integral is a number, we are simply talking about convergence of numbers. Note further that this question can be rewritten as

$$\lim_{n\to\infty} \int_I f_n(x)\,dx = \int_I \lim_{n\to\infty} f_n(x)\,dx = \int_I f(x)\,dx.$$

In other words, we are asking whether or not we can **interchange the order of the limit and the integral**.

To give a really satisfactory answer to this question we will need to have a more robust notion of integration called Lebesgue integration. This theory results in two fundamental theorems, called the **Lebesgue Dominated Convergence Theorem** and the **Monotone Convergence Theorem** (see **[17]** for full details). For us, the following **watered-down** theorems will suffice. However, if the reader insists upon simple and elegant proofs, they are advised to learn some basic Lebesgue theory covered in any introductory course on measure theory. We begin with a version of the Lebesgue Dominated Convergence Theorem:

Theorem A.8. *Let $I \subset \mathbb{R}$ be a possibly infinite interval, let $\{f_n\}$ be a sequence of integrable functions on I, and let f be an integrable function on I. Suppose for all $x \in I$, $f_n(x)$ converges to $f(x)$ as $n \to \infty$. Suppose further that there exists an integrable function $g(x) \geq 0$ on I such that*

$$|f_n(x)| \leq g(x) \qquad \text{for all } n \text{ and for all } x \in I. \tag{A.20}$$

Then

$$\int_I f_n(x)\,dx \quad \longrightarrow \quad \int_I f(x)\,dx \qquad \text{as } n \to \infty.$$

The analogous theorem holds for functions of several variables.

We should at least give some motivation for the condition (A.20). Integrability alone of the f_n and f will not suffice. Consider the sequence of functions defined in (A.19). While these functions do converge pointwise to the constant function $f \equiv 0$, for each n,

$$\int_{-\infty}^{\infty} f_n(x)\,dx = 1, \qquad \text{which certainly does not converge to } 0 = \int_{-\infty}^{\infty} 0\,dx.$$

Thus somehow the mass under the curves for each f_n has *escaped* off to infinity and is no longer seen in the limit. Note that there is no integrable function g for which (A.20) holds. We could take $g \equiv 1$ but this would not be integrable. Hence the Dominated Convergence Theorem relies on the dominating function, $g(x)$, to trap all of the mass of the sequence of functions, allowing us to pass to the limit inside the integral.

Our version of the Monotone Convergence Theorem can be stated as follows:

Theorem A.9. *Let $I \subset \mathbb{R}$ be a possibly infinite interval, let f_n be a sequence of nonnegative integrable functions on I, and let f be an integrable function on I. Suppose for all $x \in I$, $f_n(x)$ converges to $f(x)$ as $n \to \infty$. Suppose further that for all $x \in I$, we have*

$$0 \leq f_1(x) \leq f_2(x) \leq f_3(x) \leq \cdots.$$

In other words, the functions are monotone increasing. Then we have

$$\int_I f_n(x)\,dx \quad \longrightarrow \quad \int_I f(x)\,dx \qquad \text{as } n \to \infty.$$

Both Theorems A.8 and A.9 have analogous statements in higher dimensions, i.e., for sequences of functions defined over a domain Ω in $\mathbb{R}^N$.

A.9. Differentiation under the Integral Sign

> I had learned to do integrals by various methods shown in a book that my high school physics teacher Mr. Bader had given me. It showed how to differentiate parameters under the integral sign — it's a certain operation. It turns out that's not taught very much in the universities; they don't emphasize it. But we caught on how to use that method, and we used that one damn tool again and again. If guys at MIT or Princeton had trouble doing a certain integral, then we come along and try differentiating under the integral sign, and often it worked. So we got a great reputation for doing integrals, only because my box of tools was different from everybody else's, and they had tried all their tools on it before giving the problem to me.
>
> - **Richard Feynman**[a]
>
> [a] *Surely You Are Joking, Mr. Feynman*, Bantam, New York, 1985. We found out about this quote from the nice online article by Keith Conrad entitled "Differentiation under the Integral Sign".

In this book we will also benefit greatly from the ability to differentiate under the integral sign, however, not so much to "do integrals" (in the sense of the above quote) but rather in the following:

- derivations of PDEs from physical principles,
- derivations of solution formulas for PDEs,
- determining properties of solutions to PDEs.

In all these situations, we want to differentiate an integral with respect to a variable, **different** from the dummy variable of integration. It is then natural to ask if we may simply bring this derivative inside the integral, i.e., interchange the order of the differentiation and integration. As long as the function is sufficiently "nice" with respect to **smoothness and integrability**, then we are fine. **However**, it is remarkable, in light of Feynman's quote, that many fundamental areas, e.g., the field of potential theory and PDEs involving the Laplacian, rest upon situations where we **cannot** differentiate under the integral sign (or at least not with differentiation in the pointwise sense to which you are accustomed).

A.9.1. General Conditions for Legality. We state theorems whose proofs can be found in either Chapter XIII (Section 3) of Serge Lang's book on undergraduate analysis[10] or in Section 5.12 of Wendell Fleming's book on advanced calculus[11]. We state these in two separate results, one in which the integral depends on a single parameter denoted by t and another where the integral depends on a vector parameter $\mathbf{y} \in \mathbb{R}^M$. These are far from the most general (or most elegant) theorems on the subject; however, they will suffice for our needs.

Theorem A.10. *Let $\Omega \subset \mathbb{R}^N$ be a domain and let $f(\mathbf{x}, t)$ be a continuous function defined for $\mathbf{x} \in \Omega \subset \mathbb{R}^N$ and $t \in (a, b)$, for some $-\infty \le a < b \le \infty$. Suppose the following:*

(i) $\frac{\partial f}{\partial t}(\mathbf{x}, t)$ *is also continuous on $\Omega \times (a, b)$.*

(ii) *There exist integrable functions $g(\mathbf{x})$ and $h(\mathbf{x})$ defined on Ω such that*

$$|f(\mathbf{x}, t)| \le g(\mathbf{x}) \quad \text{and} \quad \left|\frac{\partial f}{\partial t}(\mathbf{x}, t)\right| \le h(\mathbf{x}) \quad \text{for all } t \in (a, b).$$

Then the function of t defined by

$$\int \cdots \int_{\Omega} f(\mathbf{x}, t)\, d\mathbf{x}$$

is a continuous and differentiable function on $t \in (a, b)$ and

$$\frac{d}{dt}\left(\int \cdots \int_{\Omega} f(\mathbf{x}, t)\, d\mathbf{x}\right) = \int \cdots \int_{\Omega} \frac{\partial f(\mathbf{x}, t)}{\partial t}\, d\mathbf{x}.$$

Theorem A.11. *Let $f(\mathbf{x}, \mathbf{y})$ be a continuous function defined for $\mathbf{x} \in \Omega_1 \subset \mathbb{R}^N$ and $\mathbf{y} \in \Omega_2 \subset \mathbb{R}^M$ where Ω_1 and Ω_2 are domains in $\mathbb{R}^N$ and $\mathbb{R}^M$, respectively. Suppose that for $i \in \{1, 2, \ldots, M\}$ the following hold:*

(i) $\frac{\partial f}{\partial y_i}(\mathbf{x}, \mathbf{y})$ *is also continuous on $\Omega_1 \times \Omega_2$.*

(ii) *There exist integrable functions $g(\mathbf{x})$ and $h(\mathbf{x})$ defined on Ω_1 such that*

$$|f(\mathbf{x}, \mathbf{y})| \le g(\mathbf{x}) \quad \text{and} \quad \left|\frac{\partial f}{\partial y_i}(\mathbf{x}, \mathbf{y})\right| \le h(\mathbf{x}) \quad \text{for all } \mathbf{y} \in \Omega_2.$$

Then for all $\mathbf{y} \in \Omega_2$, we have

$$\frac{\partial}{\partial y_i}\left(\int \cdots \int_{\Omega_1} f(\mathbf{x}, \mathbf{y})\, d\mathbf{x}\right) = \int \cdots \int_{\Omega_1} \frac{\partial f(\mathbf{x}, \mathbf{y})}{\partial y_i}\, d\mathbf{x}.$$

With enough smoothness and integrability on the function and its derivatives, one can extend these results to any combination of partial derivatives in variables which we do not integrate over. Note the appearance of the dominating functions g and h in the hypotheses. Why? Recall that a partial derivative amounts to computing a limit of a difference quotient. Thus whether or not we can differentiate under the integral sign effectively boils down to whether or not we can interchange the limit and the

[10]**Serge Lang**, *Undergraduate Analysis*, Springer Verlag, second edition, 1997.
[11]**Wendell H. Fleming**, *Functions of Several Variables*, Springer, 1977.

integration. This was the issue dealt with in Theorem A.8, the watered-down version of the Lebesgue Dominated Convergence Theorem[12]. Having the dominating functions g and h is fundamental for the application of this result.

In deriving Kirchhoff's formula for the 3D wave equation and to prove the mean value property for harmonic functions, we will also need differentiation under the integral sign for **surface integrals**. Rather than state a general theorem, we simply document the result that we will use. Suppose $f(\mathbf{y}, \mathbf{x}, t)$ ($\mathbf{y}, \mathbf{x} \in \mathbb{R}^3$, $t \in \mathbb{R}$) is smooth (say, C^k) in $\mathbf{x}$ and t and is continuous in $\mathbf{y}$. Define

$$F(\mathbf{x}, t) = \int_{\partial B(\mathbf{0},1)} f(\mathbf{y}, \mathbf{x}, t)\, dS_{\mathbf{y}}.$$

Then F is smooth (C^k) in $\mathbf{x}$ and t and to compute these partial derivatives, we may differentiate under the integral; for example,

$$\frac{\partial F(\mathbf{x}, t)}{\partial t} = \int_{\partial B(\mathbf{0},1)} f_t(\mathbf{y}, \mathbf{x}, t)\, dS_{\mathbf{y}}.$$

Important Convention: In this text, the understanding is that whenever smoothness and integrability are of no issue, we will perform such differentiation under the integral sign without even mentioning that we are using one of these theorems. In other words, the hypotheses of these theorems will either hold true or are assumed to hold true.

A.9.2. Examples Where It Is Illegal.

You might be tempted to conclude that results like differentiating under the integral sign are intuitively correct and perfectly doable since counterexamples where they do not work occur only in strange situations that we rarely encounter in real life. If all the mathematically and physically relevant functions were smooth, this might be the case. However, central to Chapter 10 on solving PDEs involving the Laplacian is a case where the fundamental function has a singularity and differentiation under the integral sign fails. This case lies at the heart of potential theory.

To give some detail on this issue, we present a few examples where differentiating under the integral sign is illegal, i.e., will lead us to a wrong answer. According to our Theorems A.10 and A.11, there are two issues which need to hold true for the integrand: (i) smoothness and (ii) the uniform integrable bounds.

Example A.9.1. Consider the function defined by

$$F(t) := \int_{-2}^{2} |x - t|(x^2 + 3)\, dx. \tag{A.21}$$

Note that, while the integrand is continuous, its partial derivatives will have a singularity due to the absolute value. Now suppose we want to find $F''(0) = \frac{d^2F}{dt^2}(0)$. We

[12] It is true that in Theorem A.8 we considered a sequential limit ($n \to \infty$), whereas to compute a partial derivative we consider a continuum limit (e.g., $\Delta t \to 0$); however, one does have analogous results to Theorem A.8 for continuum limits.

cannot do this by claiming that

$$\frac{d^2F}{dt^2}(t) = \int_{-2}^{2} \frac{\partial^2 |x-t|}{\partial t^2}\,(x^2+3)\,dx,$$

carrying out the differentiation, and then evaluating at $t = 0$. Indeed, what is $\frac{\partial^2|x-t|}{\partial t^2}$? For fixed x and as a function of t, it is 0 except where $t = x$, at which it is undefined. The issue here is, of course, the "corner" in the absolute value function. On the other hand (cf. Exercise **A.5**), one can fix t and integrate with respect to x in the right-hand side of (A.21). In doing so, one finds an explicit formula only in terms of t. One can then differentiate with respect to t and finally evaluate at $t = 0$. In doing so, one finds $F''(0) = 6$. Note that Theorem A.10 does not apply because the second derivative of the integrand with respect to the parameter t is not continuous (condition (i) of Theorem A.10).

The next example shows that, even if the integrand is smooth on the interval in question, one still needs some form of the bounds associated with condition (ii) of Theorem A.10.

Example A.9.2. Let

$$F(t) = \int_0^\infty \frac{\sin ty}{y}\,dy.$$

Note that for all $y \in (0,\infty)$, the integrand is smooth (infinitely differentiable with respect to t). Now, first note that by the change of variables $u = ty$, we have $\int_0^\infty \frac{\sin ty}{y}\,dy = \int_0^\infty \frac{\sin u}{u}\,du = \frac{\pi}{2}$. Hence, $F(t)$ is a constant and its derivative with respect to t must be 0. On the other hand, if we differentiate under the integral sign, we find that $F'(t) = \int_0^\infty \cos ty\,dy$. This latter integral is not defined for any t! The issue here is that we are not able to find integrable functions g and h such that condition (ii) of Theorem A.10 holds true. Note that $\frac{1}{y}$ is not integrable on $(0,\infty)$.

A more fundamental example (at the heart of Chapter 10) is the following:

Example A.9.3. Let f be a continuous function with compact support and consider the function

$$F(\mathbf{y}) = \iiint_{\mathbb{R}^3} \frac{1}{|\mathbf{x}-\mathbf{y}|}\,f(\mathbf{x})\,d\mathbf{x}.$$

Note that the integrand is discontinuous at $\mathbf{x} = \mathbf{y}$ but it is integrable around its singularity. Hence the function F is defined at all $\mathbf{y}$. What is its Laplacian? Note that the Laplacian with respect to $\mathbf{y}$ of the integrand is 0 except at one point. But it is **not true that**

$$\Delta_{\mathbf{y}}\left(\iiint_{\mathbb{R}^3} \frac{1}{|\mathbf{x}-\mathbf{y}|}\,f(\mathbf{x})\,d\mathbf{x}\right) = \iiint_{\mathbb{R}^3} \Delta_{\mathbf{y}}\left(\frac{1}{|\mathbf{x}-\mathbf{y}|}\right) f(\mathbf{x})\,d\mathbf{x}.$$

In fact, as we shall see, the left-hand side turns out to be $f(\mathbf{y})$. So why do the hypotheses of Theorem A.11 fail here? From the point of view of the integral, we can ignore the singular point; that is,

$$\iiint_{\mathbb{R}^3} \frac{1}{|\mathbf{x}-\mathbf{y}|}\,f(\mathbf{x})\,d\mathbf{x} = \iiint_{\mathbb{R}^3\setminus\{\mathbf{y}\}} \frac{1}{|\mathbf{x}-\mathbf{y}|}\,f(\mathbf{x})\,d\mathbf{x}.$$

The function $\frac{1}{|\mathbf{x}-\mathbf{y}|}$ is smooth away from its singularity; however, the blow-up of the function is "felt" by the integral. The reason why Theorem A.11 fails to hold true is the lack of integrability (cf. Section A.4.3) of the second derivatives of $\frac{1}{|\mathbf{x}-\mathbf{y}|}$; that is, we cannot satisfy condition (ii) of Theorem A.11.

What is underlying this example (and Example A.9.1) is the Dirac delta function which we discuss at length in Chapters 5 and 9.

A.9.3. A Leibnitz Rule. The following result is often known as a Leibnitz rule and will be useful in verifying Duhamel's Principle for the wave and diffusion equations (Sections 3.6.1 and 7.8, respectively). Let $f(t,s)$ be a smooth function in both variables, and define

$$F(t) \;:=\; \int_0^t f(t,s)\,ds.$$

Then a Leibnitz rule tells us how to compute $F'(t)$.

Theorem A.12. *We have*

$$F'(t) \;=\; f(t,t) \;+\; \int_0^t f_t(t,s)\,ds.$$

There are two natural ways to prove this: One is directly from the definition of the derivative as a limit of difference quotients; the other is via a change of variable. We choose the latter method as it will also be used in differentiating spherical means for the 3D wave equation (Section 4.2.4) and the Laplace equation (Section 8.4.1).

Proof. By the change of variable $\tau = \frac{s}{t}$, we have

$$F(t) \;=\; \int_0^t f(t,s)\,ds \;=\; \int_0^1 t f(t,t\tau)\,d\tau.$$

Hence by differentiation under the integral sign

$$\begin{aligned} F'(t) \;&=\; \int_0^1 \frac{d}{dt}\,(t f(t,t\tau))\,d\tau \\ &=\; \int_0^1 f(t,t\tau) \,+\, t f_t(t,t\tau) \,+\, t f_s(t,t\tau)\tau\,d\tau. \end{aligned}$$

Changing variables back to s, we find

$$\int_0^1 t f_t(t,t\tau)\,d\tau \;=\; \int_0^t f_t(t,s)\,ds.$$

On the other hand, by the classic Fundamental Theorem of Calculus,

$$\int_0^1 f(t,t\tau) + tf_s(t,t\tau)\tau \, d\tau = \int_0^1 \frac{d}{d\tau}\left(\tau f(t,t\tau)\right) d\tau = f(t,t). \qquad \square$$

A.10. Change in the Order of Integration

A natural question when dealing with double integrals is whether or not we can simply reverse the order of integration, i.e., compute the iterated integrals with respect to the second variable first and vice versa. As with differentiation under the integral sign there are general rules for when this is legal. However, there are also important cases when such change of order is illegal. One such case surrounds the Fourier inversion formula (cf. Section 6.2.3), where one observes that even though the direct change in order of integration is **not** justified, it still proves useful for informal insight.

A.10.1. The Fubini-Tonelli Theorem. One of the primary tools for computing iterative integrals is due to Fubini[13] and Tonelli[14]. For simplicity of notation, let us work with double integrals (i.e., work in $\mathbb{R}^2$). Although they both have their own separate theorems, they are often combined into the following result:

Theorem A.13 (The Fubini-Tonelli Theorem). *Let $f(x,y)$ be a (possibly) complex-valued function on $\mathbb{R}^2$. Then,*

$$\int_{-\infty}^{\infty}\left(\int_{-\infty}^{\infty} |f(x,y)|\, dx\right) dy = \int_{-\infty}^{\infty}\left(\int_{-\infty}^{\infty} |f(x,y)|\, dy\right) dx. \tag{A.22}$$

Furthermore, if either of the above iterated integrals in (A.22) *is finite (yielding that f is integrable), then*

$$\int_{-\infty}^{\infty}\left(\int_{-\infty}^{\infty} f(x,y)\, dx\right) dy = \int_{-\infty}^{\infty}\left(\int_{-\infty}^{\infty} f(x,y)\, dy\right) dx.$$

Hence to reverse the order of integration, we just need f to be integrable. Note that the results hold true if any (or all) of the limits of integration are finite. We remark that our statement of the Fubini-Tonelli Theorem is a particular case; the theorem in its full glory requires treatment from a measure-theoretic standpoint. Indeed, we have omitted the requirement that f is a measurable function. The general theorem, along with its proof, can be found in [**17**].

Important Convention: As with differentiation under the integral sign, the understanding in this text is that whenever integrability and smoothness are of no issue, we will perform changes in the order of integration without even mentioning that we are using this theorem. In other words, the hypotheses of this theorem will either hold true or are assumed to hold true.

---

[13] **Guido Fubini** (1879–1943) was an Italian mathematician known for this result, as well as the Fubini-Study metric.

[14] **Leonida Tonelli** (1885–1946) was an Italian mathematician known for his variation of Fubini's theorem.

A.10.2. Examples Where It Is Illegal. Here we present two examples where the order of the iterated integrals **cannot** be interchanged.

Example A.10.1. Consider the function defined by $f(x, y) := \frac{x^2-y^2}{(x^2+y^2)^2}$ on $[0, 1]\times[0, 1]$. Let us compute the iterated integral:

$$\int_0^1 \left(\int_0^1 \frac{x^2 - y^2}{(x^2 + y^2)^2} dx \right) dy = \int_0^1 \left[\frac{-x}{x^2 + y^2} \right]_{x=0}^1 dy = \int_0^1 \frac{-1}{1 + y^2} dy$$
$$= \frac{-\pi}{4}.$$

On the other hand,

$$\int_0^1 \left(\int_0^1 \frac{x^2 - y^2}{(x^2 + y^2)^2} dy \right) dx = \int_0^1 \left[\frac{y}{x^2 + y^2} \right]_{y=0}^1 dx = \int_0^1 \frac{1}{x^2 + 1} dx$$
$$= \frac{\pi}{4}.$$

Again the order of integration here matters. In fact, we have that $\int_0^1 \int_0^1 |\frac{x^2-y^2}{(x^2+y^2)^2}| dydx = \infty$, and so the conditions of A.13 do not hold.

Example A.10.2. The second example is closer to what we will encounter when addressing the Fourier inversion formula (cf. Section 6.2.3). Here we will not show that different orders give different numerical values to the integral. Rather with one order of integration, we argue that the integral is well-defined and converges to some finite number, while with the other order, it fails to converge.

Fix a $\phi(y) \in C_c^\infty(\mathbb{R})$ and consider $f(x, y) = \phi(y) \sin(xy)$ on $\mathbb{R}^2$. The iterated integral

$$\int_{-\infty}^{\infty} \int_{-\infty}^{\infty} \phi(y) \sin(xy)\, dy\, dx$$

does converge (is finite). This may not be immediately apparent, but it can be seen in two steps. First, for each fixed x the integral

$$\int_{-\infty}^{\infty} \phi(y) \sin(xy)\, dy$$

converges since ϕ has compact support. On the other hand, Theorem 5.4 states that as a function of x this integral tends to 0 as $x \to \pm\infty$. Following the proof of Theorem 5.4, one can show further that it decays very fast to 0 and is hence integrable over $x \in \mathbb{R}$.

On the other hand, switching the order gives issues, since clearly for each fixed y the integral

$$\int_{-\infty}^{\infty} \sin(xy)\, dx$$

does not converge. The point is that in the first order we exploited the compact support (a rather strong form of decay at $\pm\infty$) of ϕ.

Note here that $f(x, y)$ is not integrable on $\mathbb{R}^2$, and so the conditions of Theorem A.13 do not hold.

A.11. Thinking Dimensionally: Physical Variables Have Dimensions with Physical Units

Even if you see yourself as a pure math major, it is always instructive to keep in mind that variables denoting length, time, mass, etc., have **dimensions**, or more precisely **physical dimensions**. Here, *dimensions* indicates that they represent a physical quantity for which a numerical value needs to be supplemented with a **physical unit**. For example, x, y, and z will denote positions in space and hence will have dimensions of length. To say $x = 1$ is meaningless unless we give the physical unit; do we mean 1 meter (m), 1 millimeter (mm), 1 nanometer (nm), 1 kilometer (km), or 1 light year? Alternatively, if we say $x = 10^{-5}$, can we say "x is small"? Well, we think we might agree that, yes, if x is measured in meters, but no, if x is measured in light years. In these cases, the unit provides us with a **reference** upon which to assess the numerical value. The same is true for a temporal variable which represents *time*, or a variable which represents mass. Other physical variables (e.g., those in electricity and magnetism) have different dimensions but the **big three (length, time, and mass)** will go a long way for the majority of this text. Note that these three are sufficient to describe

area, volume, velocity, momentum, force, energy,

which have respective dimensions

$$\text{length}^2, \quad \text{length}^3, \quad \frac{\text{length}}{\text{time}}, \quad \frac{\text{mass length}}{\text{time}}, \quad \frac{\text{mass length}}{\text{time}^2}, \quad \frac{\text{mass length}^2}{\text{time}^2}.$$

On the other hand, some variables are **dimensionless**. A good case in point is certain ratios. One of the most famous ratios of all time is π. It represents the ratio of two quantities (each of which has dimensions of length): the circumference of a circle and its diameter. At some point during your quantitative education, you should contemplate what exactly an **angle** is. Angles are ratios and dimensionless quantities. An angle is a measure of *rotation* and, as such, is a ratio of two lengths: the ratio of the length of a circular arc spanned out *to the* corresponding radius. Be careful, neither a *radian* nor a *degree* is a **physical** unit. Note that the independent variable involved in either an exponential or a trigonometric function must be dimensionless (think of the Taylor series definitions of these functions!).

One should also keep in mind dimensional accounting when dealing with **integration and differentiation** with respect to dimensional variables. For example, when integrating a function $f(x, y, z)$ with respect to space variables $dxdxdy$, keep in mind that the increments (the things with the d in front of them) represent small changes in the associated variables and hence carry their dimensions. Thus in the process of integration, the dimensions of f get multiplied by length^3 (three-dimensional volume).

If one takes a derivative (or partial derivative) of a function f with respect to a spatial variable, whatever the dimensions of the function are (i.e., the output values of the function), they get multiplied by $length^{-1}$. For example, if x represents length along the x-axis and $f(x)$ represents length along the y-axis, $f'(x)$ is dimensionless. Note that this makes perfect sense as slope is a ratio of two lengths and, assuming we use one system of units for length, it is unambiguous to say the slope of a curve is 1.

A general rule is that, given any equation involving physical variables, the dimensions of the left-hand side must equal the dimensions of the right-hand side. This simple law goes a long way and, it alone can be used to determine the exact nature of certain physical relationships (equations). This procedure is known as **dimensional analysis** and the reader unfamiliar with this is encouraged to look it up online.

Exercises

A.1 Prove that Green's Theorem follows from the Divergence Theorem.

A.2 Let $\mathbf{x} = (x_1, x_2, x_3)$ denote a point in $\mathbb{R}^3$ and consider the function $f(\mathbf{x}) = \frac{1}{|\mathbf{x}|}$. Show that for all $\mathbf{x}$ (except $\mathbf{x} = \mathbf{0}$), we have $\Delta f = 0$. Show that the analogous statement in $\mathbb{R}^2$ is false (it is also false in $\mathbb{R}^N, N \geq 4$).

A.3 **(a)** Let $B(\mathbf{0}, 1)$ be the unit ball in $\mathbb{R}^3$. Explicitly compute $\iiint_{B(\mathbf{0},1)} \frac{1}{|\mathbf{x}|}\, d\mathbf{x}$ and $\iint_{\partial B(\mathbf{0},1)} \frac{1}{|\mathbf{x}|}\, dS$. For (b) and (c), let $B(\mathbf{0}, 1)$ be the unit ball in $\mathbb{R}^N$ and let $B^c(\mathbf{0}, 1)$ be its complement; i.e., $B^c(\mathbf{0}, 1) = \{\mathbf{x} \in \mathbb{R}^N \mid |\mathbf{x}| \geq 1\}$.
(b) For what values of $p > 0$ is the integral $\int \cdots \int_{B(\mathbf{0},1)} \frac{1}{|\mathbf{x}|^p}\, d\mathbf{x}$ finite?
(c) For what values of $p > 0$ is the integral $\int \cdots \int_{B^c(\mathbf{0},1)} \frac{1}{|\mathbf{x}|^p}\, d\mathbf{x}$ finite?

A.4 Prove the IPW Theorems, Theorems A.4, A.6, and A.7. For the latter two, you can take the space dimension to be either $N = 2$ or $N = 3$.

A.5 Prove by direct calculation that if $F(t)$ is defined by (A.21), then $F''(0) = 6$. First fix t (which, if you like, you can assume without loss of generally is in $(-2, 2)$) and integrate with respect to x. Now differentiate with respect to t and then let $t = 0$.

A.6 Let $f(\mathbf{x}) = |\mathbf{x}|$, where $\mathbf{x} \in \mathbb{R}^N$. Show that $\nabla f = \frac{\mathbf{x}}{|\mathbf{x}|}$. More generally, if $f(\mathbf{x})$ is a radially symmetric function on $\mathbb{R}^N$, that is, $f(\mathbf{x}) = g(|\mathbf{x}|)$, for some function g on $\mathbb{R}$, write down an expression for ∇f in terms of the g'. Use this to find $\nabla(\sin |\mathbf{x}|)$.

A.7 Let $\mathbf{x} \in \mathbb{R}^3$ and let $u(\mathbf{x}) = \frac{1}{|\mathbf{x}|^p}$, $p > 1$. Calculate the normal derivative $\nabla u \cdot \mathbf{n}$ on $\partial B(\mathbf{0}, a)$ with the outer normal to $B(\mathbf{0}, a)$.

A.8 Verify the Divergence Theorem in $\mathbb{R}^3$ for $\Omega = B(\mathbf{0}, a)$ and vector field $\mathbf{x}$.

A.9 Using spherical coordinates and the change of variables formula, verify (A.12).

A.10 Suppose we define the gradient vector field of u in the following purely geometric way: The vector field points in the direction of fastest increase with a magnitude equal to the maximal rate of change. Using only this definition, show that the gradient of $|\mathbf{x}|$ is the radial unit vector field, $\mathbf{x}/|\mathbf{x}|$. (Do not take derivatives as you did in Exercise **A.6**!)

A.11 Verify that, indeed, equation (A.4) holds true.

A.12 Given a radially symmetric function $u(\mathbf{x}) = v(r)$ on $\mathbb{R}^N$ where $r = |\mathbf{x}|$, show that its Laplacian is given by $\Delta u = v'' + \frac{(N-1)}{r} v'$ where primes denote radial differentiation. Show that this can be written as $\Delta u = \frac{1}{r^{N-2}} \frac{d}{dr}\left(r^{N-2} v'(r)\right)$.

A.13 Prove the following "product rule" identities for the gradient, divergence, and Laplacian:

(a) $\nabla(fg) = f\nabla g + g\nabla f$, for any $f, g \in C^1(\mathbb{R}^3)$.

(b) $\operatorname{div}(f\mathbf{u}) = f \operatorname{div} \mathbf{u} + \mathbf{u} \cdot \nabla f$ for any function $f \in C^1(\mathbb{R}^3)$ and vector field $\mathbf{u} \in C^1(\mathbb{R}^3, \mathbb{R}^3)$.

(c) $\Delta(fg) = f\Delta g + 2\nabla f \cdot \nabla g + g\Delta f$ for any $f, g \in C^2(\mathbb{R}^3)$.

Bibliography

Books on Partial Differential Equations at the Undergraduate Level

[1] R. Haberman, *Elementary applied partial differential equations: With Fourier series and boundary value problems*, 2nd ed., Prentice Hall, Inc., Englewood Cliffs, NJ, 1987. MR913939

[2] J. Ockendon, S. Howison, A. Lacey, and A. Movchan, *Applied partial differential equations*, revised edition, Oxford University Press, Oxford, 2003. MR1995978

[3] P. J. Olver, *Introduction to partial differential equations*, Undergraduate Texts in Mathematics, Springer, Cham, 2014, DOI 10.1007/978-3-319-02099-0. MR3136142

[4] Y. Pinchover and J. Rubinstein, *An introduction to partial differential equations*, Cambridge University Press, Cambridge, 2005, DOI 10.1017/CBO9780511801228. MR2164768

[5] M. A. Pinsky, *Partial differential equations and boundary-value problems with applications*, reprint of the third (1998) edition, Pure and Applied Undergraduate Texts, vol. 15, American Mathematical Society, Providence, RI, 2011. MR2849590

[6] W. A. Strauss, *Partial differential equations: An introduction*, 2nd ed., John Wiley & Sons, Ltd., Chichester, 2008. MR2398759

[7] H. F. Weinberger, *A first course in partial differential equations with complex variables and transform methods*, Blaisdell Publishing Co. Ginn and Co. New York-Toronto-London, 1965. MR0180739

Books on Partial Differential Equations at the First Year Graduate Level

[8] C. Carathéodory, *Calculus of variations and partial differential equations of first order*, 3rd ed., American Mathematical Society, Providence, RI.

[9] W. Craig, *A course on partial differential equations*, Graduate Studies in Mathematics, vol. 197, American Mathematical Society, Providence, RI, 2018, DOI 10.1090/gsm/197. MR3839330

[10] L. C. Evans, *Partial differential equations*, Graduate Studies in Mathematics, vol. 19, American Mathematical Society, Providence, RI, 1998. MR1625845

[11] G. B. Folland, *Introduction to partial differential equations*, Preliminary informal notes of university courses and seminars in mathematics, Mathematical Notes, Princeton University Press, Princeton, N.J., 1976. MR0599578

[12] M. Renardy and R. C. Rogers, *An introduction to partial differential equations*, Texts in Applied Mathematics, vol. 13, Springer-Verlag, New York, 1993. MR1211418

Books on Fourier Series, the Fourier Transform, and Distributions at the Undergraduate Level

[13] B. G. Osgood, *Lectures on the Fourier transform and its applications*, Pure and Applied Undergraduate Texts, vol. 33, American Mathematical Society, Providence, RI, 2019. MR3887604

[14] G. B. Folland, *Fourier analysis and its applications*, Pure and Applied Undergraduate Texts, vol. 4, American Mathematical Society, Providence, RI, 1992.

[15] L. Schwartz, *Mathematics for the physical sciences*, Dover, 1966.

Books on Mathematical and Functional Analysis
Undergraduate Level

[16] W. Rudin, *Principles of mathematical analysis*, McGraw-Hill Book Company, Inc., New York-Toronto-London, 1953. MR0055409

Graduate Level

[17] G. B. Folland, *Real analysis: Modern techniques and their applications*, Pure and Applied Mathematics, A Wiley-Interscience Publication, John Wiley & Sons, Inc., New York, 1984. MR767633

[18] E. H. Lieb and M. Loss, *Analysis*, Graduate Studies in Mathematics, vol. 14, American Mathematical Society, Providence, RI, 1997, DOI 10.2307/3621022. MR1415616

[19] H. Brezis, *Functional analysis, Sobolev spaces and partial differential equations*, Universitext, Springer, New York, 2011. MR2759829

Index

Selected Published Titles in This Series

54 **Rustum Choksi,** Partial Differential Equations, 2022
53 **Louis-Pierre Arguin,** A First Course in Stochastic Calculus, 2022
52 **Michael E. Taylor,** Introduction to Differential Equations, Second Edition, 2022
51 **James R. King,** Geometry Transformed, 2021
50 **James P. Keener,** Biology in Time and Space, 2021
49 **Carl G. Wagner,** A First Course in Enumerative Combinatorics, 2020
48 **Róbert Freud and Edit Gyarmati,** Number Theory, 2020
47 **Michael E. Taylor,** Introduction to Analysis in One Variable, 2020
46 **Michael E. Taylor,** Introduction to Analysis in Several Variables, 2020
45 **Michael E. Taylor,** Linear Algebra, 2020
44 **Alejandro Uribe A. and Daniel A. Visscher,** Explorations in Analysis, Topology, and Dynamics, 2020
43 **Allan Bickle,** Fundamentals of Graph Theory, 2020
42 **Steven H. Weintraub,** Linear Algebra for the Young Mathematician, 2019
41 **William J. Terrell,** A Passage to Modern Analysis, 2019
40 **Heiko Knospe,** A Course in Cryptography, 2019
39 **Andrew D. Hwang,** Sets, Groups, and Mappings, 2019
38 **Mark Bridger,** Real Analysis, 2019
37 **Mike Mesterton-Gibbons,** An Introduction to Game-Theoretic Modelling, Third Edition, 2019
36 **Cesar E. Silva,** Invitation to Real Analysis, 2019
35 **Álvaro Lozano-Robledo,** Number Theory and Geometry, 2019
34 **C. Herbert Clemens,** Two-Dimensional Geometries, 2019
33 **Brad G. Osgood,** Lectures on the Fourier Transform and Its Applications, 2019
32 **John M. Erdman,** A Problems Based Course in Advanced Calculus, 2018
31 **Benjamin Hutz,** An Experimental Introduction to Number Theory, 2018
30 **Steven J. Miller,** Mathematics of Optimization: How to do Things Faster, 2017
29 **Tom L. Lindstrøm,** Spaces, 2017
28 **Randall Pruim,** Foundations and Applications of Statistics: An Introduction Using R, Second Edition, 2018
27 **Shahriar Shahriari,** Algebra in Action, 2017
26 **Tamara J. Lakins,** The Tools of Mathematical Reasoning, 2016
25 **Hossein Hosseini Giv,** Mathematical Analysis and Its Inherent Nature, 2016
24 **Helene Shapiro,** Linear Algebra and Matrices, 2015
23 **Sergei Ovchinnikov,** Number Systems, 2015
22 **Hugh L. Montgomery,** Early Fourier Analysis, 2014
21 **John M. Lee,** Axiomatic Geometry, 2013
20 **Paul J. Sally, Jr.,** Fundamentals of Mathematical Analysis, 2013
19 **R. Clark Robinson,** An Introduction to Dynamical Systems: Continuous and Discrete, Second Edition, 2012
18 **Joseph L. Taylor,** Foundations of Analysis, 2012
17 **Peter Duren,** Invitation to Classical Analysis, 2012
16 **Joseph L. Taylor,** Complex Variables, 2011
15 **Mark A. Pinsky,** Partial Differential Equations and Boundary-Value Problems with Applications, Third Edition, 1998

For a complete list of titles in this series, visit the
AMS Bookstore at **www.ams.org/bookstore/amstextseries/**.